Renewable Energy

Physics, Engineering, Environmental Impacts, Economics & Planning

Fourth Edition

Bent Sørensen

AMSTERDAM • BOSTON • HEIDELBERG • LONDON
NEW YORK • OXFORD • PARIS • SAN DIEGO
SAN FRANCISCO • SINGAPORE • SYDNEY • TOKYO
Academic Press is an imprint of Elsevier

Academic Press is an imprint of Elsevier
30 Corporate Drive, Suite 400, Burlington, MA 01803, USA
The Boulevard, Langford Lane, Kidlington, Oxford, OX5 1GB, UK

Notices

Knowledge and best practice in this field are constantly changing. As new research and experience broaden our understanding, changes in research methods, professional practices, or medical treatment may become necessary.

Practitioners and researchers must always rely on their own experience and knowledge in evaluating and using any information, methods, compounds, or experiments described herein. In using such information or methods they should be mindful of their own safety and the safety of others, including parties for whom they have a professional responsibility.

To the fullest extent of the law, neither the Publisher nor the authors, contributors, or editors, assume any liability for any injury and/or damage to persons or property as a matter of products liability, negligence or otherwise, or from any use or operation of any methods, products, instructions, or ideas contained in the material herein.

Library of Congress Cataloging-in-Publication Data
Sørensen, Bent.
  Renewable energy / Bent Sørensen – 4th ed.
    p. cm.
  Includes bibliographical references and index.
  ISBN 978-0-12-375025-9 (alk. paper)
  1. Renewable energy sources. I. Title.
  TJ163.2.S66 2011
  621.042–dc22

                                        2010022708

British Library Cataloguing-in-Publication Data
A catalogue record for this book is available from the British Library.

For information on all Academic Press publications,
visit our Web site at *www.elsevierdirect.com*

Typeset by: diacriTech, Chennai, India

Printed in the United States of America
10 11 12 13   6 5 4 3 2 1

Contents

Part I
Renewable Energy Resources

1. Perspectives on Energy Resources

2. Origin of Renewable Energy Flows

3. Individual Renewable Energy Sources

Part II
Renewable Energy Technologies
4. The Energy Conversion Processes

Part III
Renewable Energy Impacts: Planning for Sustainability and Climate Change Aversion

6. Energy System Planning

7. Socioeconomic Assessment

8. Integrated Approaches

Preface to Fourth Edition

The ongoing commercialization of the most viable renewable energy systems has led to maturation of these systems that may not have introduced many novel fundamental insights, but has streamlined the components in the interest of cost, durability, and trouble-free operation. For renewable technologies not competitive yet, the fact that technologies like wind power have already reached the marketplace seems to have spurred an enhanced effort to reach the same status. The evidence for this is a simultaneous focus on lowering cost and on offering smooth integration into existing or new energy systems. Rapidly developing countries that are attempting to create a 100-fold increase in average wealth within a few decades are a special market focus. It is evidently not possible to realize this growth for nearly half the world's population if energy continues to be derived from fossil resources. The interesting option is then to jump directly to the long-range solution offered by renewable energy systems, even if the mix of such systems includes some technologies that are not instantaneously characterized by economic viability but still may be so in a life-cycle view. The foresight and courage to follow such a path are the hallmarks of responsible politicians and other decision-makers.

These observations have guided the latest update of this book. Both new technology and advances of existing technology are covered in Chapters 4 and 5, and the system implications are added to Chapters 6 and 7. Chapters 2 and 3 have only been lightly updated, e.g., by the newest IPCC estimates of climate changes (which of course may have a direct influence on distribution of renewable energy flows). The market overview in Chapter 1 has been updated with available production data. Much of these data are still publicly available, although industry efforts to conceal cost data have been increasing along with the creation of real (i.e., not government subsidized) markets. Overall, some of the material in the book has been reorganized to enhance the use of the book as a textbook. Finally, the publisher has decided to alter page layout and change the language of this edition from International English to U.S. English. I apologize for any inconvenience ensuing.

The remarks made in the prefaces to the earlier editions still stand and may prove useful to guide the reader through the arrangement of the material.

Bent Sørensen, Gilleleje, March 2010

The present edition has been updated in a number of renewable energy technology areas (Chapters 4 and 5), where progress has been made over the recent years. New solar simulation studies have been added to Chapter 6, and market considerations have been included in the overview in Chapter 1 and in discussing industry liberalization in Chapter 7. The remarks on advanced subjects made in the preface to the second edition are still valid. A new growth area is fuel cells for stationary and mobile uses of hydrogen and other fuels. Only modest updates have been done in this area, as it is the subject of a new, companion book to be published about a year after this one (Sørensen: *Hydrogen and Fuel Cells*, Elsevier/Academic Press). Some older material has been omitted or tidied up, and maneuvering through the book has been eased, both for reference and for textbook uses. The following diagrams may assist in following the tracks of interest through the book:

Topic-driven Paths

Chapter	Wind	Solar Power	Solar Heat	Biofuels	Others
1	1.1	1.1	1.1	1.1	1.1
2	2.3.1 (end), 2.4.1, 2C	2.2.2, 2.4.1	2.2.2, 2.4.1	2.4.1	2.3.2, 2.4.1, 2B, 2D
3	3.2	3.1	3.1	3.6	3.3–3.5, 3.7
4	4.1.4, 4.3	4.1.5, 4.2.3	4.2.1–2, 4.6	4.8	4.1.3, 4.1.6, 4.4–5, 4.7, 4.8
5	5.1.2, 5.2.2	5.1.2, 5.2.2	5.1.1, 5.2.1		
6	6.2.5, 6.3.2, 6.4	6.2.4, 6.4	6.3.1, 6.4	6.2.7, 6.4	6.4
7	7.4.12–13, 7.5	7.4.12–13		7.4.12–13	
8	8	8	8	8	8

Course Material Uses

Chapter	Resource Studies	Energy Engineering	Energy Planning	Energy Economics	Energy & Environment
1	1		1	1.1	1.2
2	2				2.4.1
3	3	as needed			3.4.2 (end)
4		4	as needed	4.8 (start)	4.5
5		5	as needed		
6			6	as needed	as needed
7			7	7	7.4
8	8	8	8	8	8

Bent Sørensen, Gilleleje, October 2003

When the first edition of *Renewable Energy* appeared in 1979, it was the first textbook and research monograph since the 1920s to deal with the renewable energy sources and systems at a scholarly level. Indeed, it was instrumental in establishing the now universally used term "renewable energy" for a new area of science, which emerged under names such as "regenerative energy" in Germany and "solar energy" in the United States of America. In many countries, renewable energy appeared in planning documents as "supplementary energy," based on a conviction by administrators that this could never become a major source of energy. My suggestion in the journal *Science* (Sørensen, 1975b) that renewable energy could potentially become a 100% solution was regarded as absurd by many. Things have changed today, where official energy plans of some countries call for over 50% renewable energy coverage by year 2030 (Danish Department of Environment and Energy, 1996), where the best renewable energy technologies are already economically competitive relative to fossil options, and where increased concern over greenhouse warming effects may well alter the perceived indirect costs of different energy solutions.

The structure of the first edition was determined by the aim of placing renewable energy on the academic agenda. It was my goal to show young scientists, engineers, and future planners that renewable energy was at least as interesting and challenging as nuclear energy, and I tried to do this by showing the depth of problems to be solved using advanced techniques, shying no complication of quantum mechanics or nonlinear mathematics. This was seen as positive by reviewers and colleagues, but may have limited the sales figures for the book! Today, the requirements are quite different: now many universities and polytechnic institutes have renewable energy courses in their curriculum, and the task at hand is to provide good teaching materials for the relevant levels of courses. Therefore, I have thoroughly revised the content and presentation in the second edition. The main sections of each chapter are now suited for introductory level study, with only very general prerequisites. Any topic requiring more background is deferred to special sections marked as ADVANCED topics at the top corner of each page. They can be added individually at the choice of the teacher, or they can be left for further study by the user of the book. My reflections on whether to separate elementary and advanced topics in two volumes or keep them together are as follows. Needing to go back to a topic for more detailed study, it is very convenient to be able to find it in a book that you have already worked with. The style and assumptions are known to you, and first of all, the book is on your shelf and need not be retrieved from somewhere else. Against the single-volume solution speaks the book price for those who find it unlikely that they shall need more than the elementary sections. However, we are all surprised by the growth of our needs, and the price of this second edition is even below that of the first edition, thanks to modern preparation and printing methods.

Another issue is the arrangement of material, which I have basically kept as in the first edition: first describing the origin of renewable energy, then its disposition and availability at different geographical locations on Earth, then the techniques of energy conversion systems and systems suitable for each type of renewable energy, and finally the evaluation of the total system, in terms of economic and environmental impacts. The logic of this sequence is evident, but it means that someone wanting to know only about wind power will have to jump from chapter to chapter. This is made much easier in this edition by the addition, on each bottom left page, of references to previous and following sections dealing with the same form of renewable energy. As in the first edition, extensive references and an index are found at the end. The index also serves as a pointer to specialist words and concepts by giving the page where they are first explained. After the table of contents, a list of units and abbreviations is given.

The content has been revised in those areas where new advances have been made, notably in the sections on energy from biomass and on photovoltaic energy conversion, and in the economic chapter on life-cycle analysis. As in the first edition, emphasis is on basic principles. Fortunately, they do not wear much with time, and several sections needed only a light brush-up, sometimes with some tidying effort to keep the size down. However, new data available today have made it possible to improve many of the illustrations, notably in the area of global energy flows. At the end of each chapter, there are topics for discussion, including new ones. They are basically of two kinds: simple topics for classroom discussion and mini-project ideas that can serve as a basis for problem-oriented work extending from a few days to several months in duration. This is a reflection of the different styles of teaching at different institutions, where small projects are often offered to individuals or groups of students for credit, with the indicated range of time devoted to each problem (and a corresponding difference in depth of investigation).

The Danish Energy Agency supported part of the work upon which the second edition updates are based. The author welcomes comments and suggestions, which may be addressed as indicated below.

Bent Sørensen, Allerød, 1998

Renewable energy is the collective name for a number of energy resources available to man on Earth. Their conversion has always played an important role for the inhabitants of the planet, and apart from a period of negligible length—relative to evolutionary and historical time scales—the renewable energy sources have been the only ones accessible to mankind.

Yet the study of renewable energy resources, their origin and conversion, may at present be characterized as an emerging science. During the past fifty years of scientific and technological revolution, much more effort has been devoted to the extraction and utilization of nonrenewable energy resources (fuels), than to the renewable ones. Only very recently have funds been made available to re-establish renewable energy research and development, and it is still unclear whether the technologies based on renewable energy sources will become able to constitute the backbone of future energy supply systems.

The purpose of the present book is to provide an incentive as well as a basis of reference for those working within the field of renewable energy. The discontinuity between earlier and present work on renewable energy, and the broadness of disciplines required for assessing many questions related to the use of renewable energy, have created a need for a comprehensive reference book, covering methods and principles, rather than specific engineering prescriptions of passing interest in a rapidly developing field.

A survey of renewable energy has to draw upon a large number of individual scientific disciplines, ranging from astrophysics and upper atmospheric science over meteorology and geology to thermodynamics, fluid mechanics, solid-state physics, etc. Specialists in each discipline often use a vocabulary recognized only by insiders, and they rarely emphasize the aspects pertinent to renewable energy. I have attempted to use a common language throughout, and to restrict the prerequisites for understanding to a fairly elementary level (e.g., basic physics). However, this does not mean that I have avoided any degree of complication considered relevant, and the reader must be prepared to face a number of challenges.

I envisage my reader as a research worker or student working somewhere within the field of renewable energy. Such work is currently undertaken at universities, engineering schools, and various offices and laboratories in the public or private sectors. However, since a substantial part of the book deals with *energy systems* comprising renewable energy elements, and with the management and economy of such systems, including environmental and social aspects, then I sincerely hope to attract also readers in the energy planning and management sectors, whether their concern is the physical planning and operation of energy supply systems or the socioeconomic assessment of such systems.

When used as a textbook, particular chapters may be more relevant than others. Cross-references are included in most cases where definitions or basic concepts have to be taken from a different chapter. Courses in engineering may place the emphasis around Chapter 4 (e.g., including Chapters 3–6), courses in "energy physics" or on energy in general may place more emphasis on Chapters 2 and 3, while courses on energy planning, systems aspects, and technological or economic assessments may find it natural to shift the emphasis to Chapters 6 and 7.

It should be stressed that the main purpose of the book is to provide general tools for treating problems relating to renewable energy. This is evident from the approach to energy conversion in Chapter 4 (stressing principles rather than describing individual pieces of equipment in detail), and from the treatment of supply systems in Chapter 6 (which contains no exhaustive reviews of possible system combinations, but illustrates basic modeling and simulation techniques by use of just a few, selected system examples). Energy storage and transmission (Chapter 5) are described in a highly condensed form, with the sole purpose of introducing the components for use in energy systems like those discussed in Chapter 6.

I have been motivated to engage in work on renewable energy and to see the possibility of an increasingly important role played by the associated technologies by reflections that are largely summarized in Chapter 1, and that to some extent lie behind those amendments to conventional economic theory for application to long-term energy planning, proposed in Chapter 7. The subjective nature of a number of interpretations made in these two chapters is recognized, and an effort has been made to ban such interpretations from the remaining five chapters, so that readers disagreeing with my interpretations may still find the bulk of the book useful and stimulating.

I thank the following for reading and commenting on portions of the draft version of the manuscript: Niels Balling, Henning Frost Christensen, E. Eliasen, Frede Hvelplund, Johannes Jensen, Marshal Merriam, B. Maribo Petersen, and Ole Ulfbeck.

Bent Sørensen, Allerød, January 1979

Powers of 10

Prefix	Symbol	Value	Prefix	Symbol	Value
atto	a	10^{-18}	kilo	k	10^3
femto	f	10^{-15}	mega	M	10^6
pico	p	10^{-12}	giga	G	10^9
nano	n	10^{-9}	tera	T	10^{12}
micro	μ	10^{-6}	peta	P	10^{15}
milli	m	10^{-3}	exa	E	10^{18}

G, T, P, E are called milliard, billion, billiard, trillion in Europe, but billion, trillion, quadrillion, quintillion in the United States. M is universally used for million.

SI Units

Basic Unit	Name	Symbol	
length	meter	m	
mass	kilogram	kg	
time	second	s	
electric current	ampere	A	
temperature	Kelvin	K	
luminous intensity	candela	cd	
plane angle	radian	rad	
solid angle	steradian	sr	

Derived Unit	Name	Symbol	Definition
energy	joule	J	$kg\ m^2\ s^{-2}$
power	watt	W	$J\ s^{-1}$
force	newton	N	$J\ m^{-1}$
electric charge	coulomb	C	$A\ s$
potential difference	volt	V	$J\ A^{-1}\ s^{-1}$
pressure	pascal	Pa	$N\ m^{-2}$
electric resistance	ohm	Ω	$V\ A^{-1}$
electric capacitance	farad	F	$A\ s\ V^{-1}$
magnetic flux	weber	Wb	$V\ s$
inductance	henry	H	$V\ s\ A^{-1}$
magnetic flux density	tesla	T	$V\ s\ m^{-2}$
luminous flux	lumen	lm	$cd\ sr$
illumination	lux	lx	$cd\ sr\ m^{-2}$
frequency	hertz	Hz	$cycle\ s^{-1}$

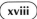

Conversion Factors

Type	Name	Symbol	Approximate Value
energy	electron volt	eV	1.6021×10^{-19} J
energy	erg	erg	10^{-7} J (exact)
energy	calorie (thermochemical)	cal	4.184 J
energy	British thermal unit	Btu	1055.06 J
energy	Q	Q	10^{18} Btu (exact)
energy	quad	q	10^{15} Btu (exact)
energy	tons oil equivalent	toe	4.19×10^{10} J
energy	barrels oil equivalent	bbl	5.74×10^{9} J
energy	tons coal equivalent	tce	2.93×10^{10} J
energy	m^3 of natural gas		3.4×10^{7} J
energy	liter of gasoline		3.2×10^{7} J
energy	kilowatthour	kWh	3.6×10^{6} J
power	horsepower	hp	745.7 W
power	kWh per year	kWh y	0.114 W
radioactivity	curie	Ci	3.7×10^{8} s^{-1}
radioactivity	becqerel	Bq	1 s^{-1}
radiation dose	rad	rad	10^{-2} J kg^{-1}
radiation dose	gray	Gy	J kg^{-1}
dose equivalent	rem	rem	10^{-2} J kg^{-1}
dose equivalent	sievert	Sv	J kg^{-1}
temperature	degree Celsius	°C	K −273.15
temperature	degree Fahrenheit	F	9/5 °C + 32
time	minute	m	60 s (exact)
time	hour	h	3600 s (exact)
time	year	y	8760 h
pressure	atmosphere	atm	1.013×10^{5} Pa
pressure	bar	bar	10^{5} Pa
mass	pound	lb	0.4536 kg
mass	ounce	oz	0.02835 kg
length	foot	ft	0.3048 m
length	mile (statute)	mi	1609 m
volume	liter	l	10^{-3} m^3
volume	gallon (US)		3.785×10^{-3} m^3

Renewable Energy Resources

Chapter 1

Perspectives on Energy Resources

1.1 CURRENT RENEWABLE ENERGY MARKET

The penetration of renewable energy into the energy system of human settlements on Earth is, from one point of view, nearly 100%. The energy system experienced by the inhabitants of the Earth is dominated by the environmental heat associated with the greenhouse effect, which captures solar energy and stores it within a surface-near sheet of topsoil and the atmosphere around the Earth. Only 0.02% of this energy system is currently managed by human society, as illustrated in Fig. 1.1. Within this economically managed part of the energy sector, renewable energy sources currently provide about 25% of the energy supplied. As the figure indicates, a large part of this renewable energy is in the form of biomass energy, either in food crops or in managed forestry providing wood for industrial purposes or for incineration (firewood used for heat and cooking in poor countries or for mood-setting fireplaces in affluent countries, residue and waste burning in combined power and heat plants or incinerators). The other sources of renewable energy exploited include hydro, wind, and solar. Hydropower is a substantial source, but its use is no longer growing due to environmental limits identified in many locations with potential hydro resources. Passive solar heating is a key feature of building design throughout the world, but active solar heat or power panels are still at a very minute level of penetration. Also, wind has both a passive and an active role. Passive use of wind energy for ventilation of buildings plays a significant role, and active power production by wind turbines is today a rapidly growing energy technology in many parts of the world. The highest penetration, reaching nearly 20% of total electricity provided, is found in Denmark, the country pioneering modern wind technology. Additional renewable energy technologies, so far with fairly small global penetration, include biofuels, such as biogas and geothermal power and heat. As indicated in Fig. 1.1, the dominant direct energy sources are still fossil fuels, despite the fact that they are depletable and a cause of frequent national conflicts, due to the mismatch between their particular geographical availability and demand patterns.

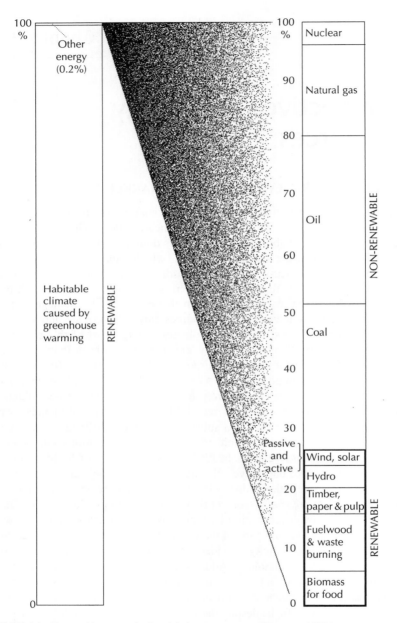

FIGURE 1.1   Renewable energy in the global energy system (Sørensen, 1992c).

From a business point of view, the total renewable energy flows, including free environmental heat, are, of course, not quite as interesting as the energy that can be traded in a market. Current renewable energy markets comprise both consumer markets and markets driven by government demonstration programs and policies aimed to diminish market distortions. The occasional initial subsidies are part of industrial policy aimed at helping new industry through market stimulation. Compensation for market distortions addresses the fact that conventional energy industries are not fully paying for the negative environmental effects of their products. This is a complex issue, partly because of the difficulty in exact determination of external costs and partly because most countries already levy taxes on energy products that may in part be contributing toward paying for any environmental damage, but often are just a government revenue not specifically used to offset the negative effects associated with using fossil or nuclear fuels (read more about these issues in Chapter 7).

The current penetration of active use of renewable energy in national energy systems is growing, and Figs. 1.3–1.15 show the values for the year 2000, which may serve as a reference year for assessing newer data. In the wind case, where the growth since 2000 is significant, a more recent figure is added; if available, the growth in the annual production is mentioned in the figure caption, showing the national distribution of markets. Because all renewable energy production levels are given in W/cap per country, the 2007 population per country is given for reference in Fig. 1.2 (cf. Fig. 6.1, showing population figures on an area basis). The growth in installed technology and energy production is significant for wind energy and for some biofuels. Renewable energy technologies still confined to niche or subsidized markets also show fairly high growth rates (e.g., solar electricity). The 2000 renewable energy production data, given explicitly in Table 1.1, are suitable for the growth comparisons mentioned.

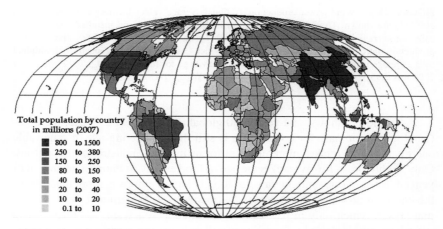

FIGURE 1.2 World population, per country, in 2007. (Based on WRI, 2008, UN 2010; for a few non-reporting countries, earlier data plus growth rates have been used.)

(a)

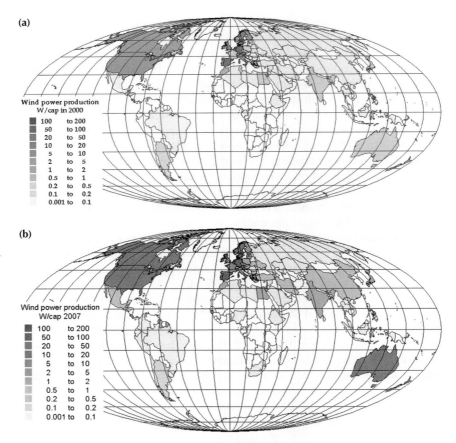

(b)

FIGURE 1.3a,b   Wind power production. National average values for the years 2000 (*a*) and 2007 (*b*) are shown. The 2000 data are from BTM (2001), based on installed capacity and using an average capacity factor of 0.3. The 2007 data are from WEC (2009), giving actual production data, except when not available, in which case the installed capacity is multiplied by an average capacity factor of 0.22. The world average for the year 2000 is 0.92 W/cap, for 2007 it is 2.95 W/cap; for 2008, it has risen to 4.12 W/cap (BTM, 2009).

As Table 1.1 shows, at a year-2000 global average of 222 W/cap, the traditional use of biomass for combustion is still the dominating use of renewable energy, although it takes more efficient forms in many industrialized countries. Only slightly less (146 W/cap) is the use of food energy in biomass of animal or vegetable origin (the nutrient value of food being, in any case, more than that of the energy it provides). Next comes hydropower (50 W/cap) and then geothermal power, which only in part can be classified as renewable (as many steam reservoirs are exploited at a rate that will exhaust the reservoir over periods of decades). At the level of 1 W/cap, that is, two orders of magnitude less than the energy in food intake, one finds biomass waste (used for power or heat), biogas, liquid biofuels (used in the transportation sector), wind power, and geothermal heat (used for district heating). Last are solar heat, tidal power, and solar power, with the latter below 0.01 W/cap. However, the fastest growing markets are wind and solar power, with both currently adding over 35% of installed power each year.

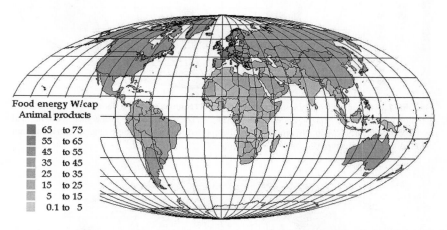

FIGURE 1.4 Biomass energy contained in human time-averaged food intake of animal products. National average values for the year 2000 are shown (FAO, 2003). The world average animal-based food intake for the year 2000 is 22.2 W/cap.

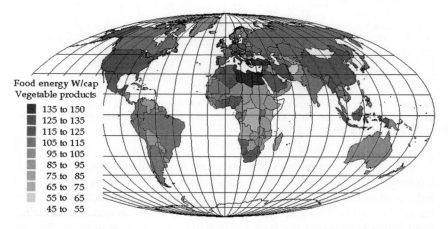

FIGURE 1.5 Biomass energy contained in human time-averaged food intake of vegetable products. National average values for the year 2000 are shown (FAO, 2003). The world average vegetable food intake for the year 2000 is 113.7 W/cap.

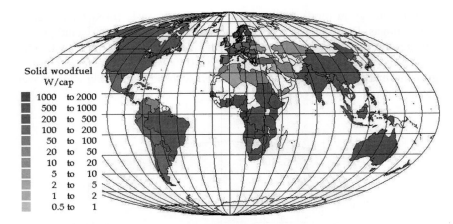

FIGURE 1.6 Biomass energy contained in woodfuel. National average values for the year 2000 are shown (OECD/IEA, 2002a, 2002b). The world average woodfuel use in the year 2000, implied by the data, is 221.9 W/cap. No woodfuel use is reported for Russia. Some other countries are not covered by the data source.

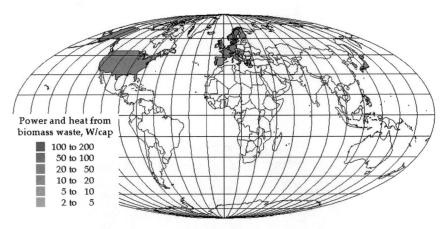

FIGURE 1.7 Energy in biomass waste (refuse) utilized for power or heat production. National average values for the year 2000 are shown, based upon OECD/IEA (2002a, 2002b). The world average for the year 2000 is 3.7 W/cap.

The market characteristics of the various renewable energy forms exhibit differences linked to the nature of each source. For food energy, the price is influenced by variations in production due to climatic variations, by the choices made in regard to area use, by livestock holdings, by fish quotas, and by the competitive behavior of the food processing and marketing industry. Yet the bulk prices of different commodities seem remarkably consistent with their energy content, varying only between some 70 US or euro-cents per kWh (heat value) and 200 c/kWh. Translating OECD data (OECD, 2002) to energy units, the current wholesale price of cereals like rice or wheat is around 70 c/kWh, while the wholesale price of typical meat and dairy products is about 100 c/kWh. Only specialized gourmet products obtain higher prices in the marketplace. Consumer retail prices are typically some five times higher than the bulk prices

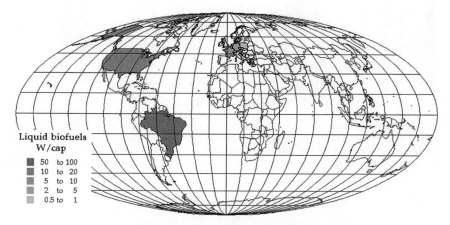

FIGURE 1.8   Energy in liquid biofuels (presently ethanol and biodiesel). National average values for the year 2000 are shown, based upon EC-ATLAS (2003) and OECD/IEA (2002a, 2002b). The world average for the year 2000 is 2.3 W/cap, increased to 5.0 W/cap by 2007 (UNEP, 2009).

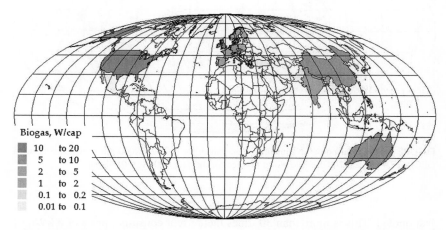

FIGURE 1.9   Energy in biogas. National average values for the year 2000 are shown, based upon converting the number of units (FAO-Asia, 2003), assuming on average for each family unit a biogas production of half that of an optimally working unit fed manure from the equivalent of 2.5 cows and producing 1736 W of biogas. Additional data are from OECD/IEA (2002a, 2002b). The world average for the year 2000 is 2.8 W/cap.

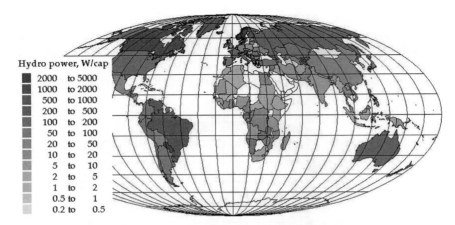

FIGURE 1.10   Hydropower. National average values for the year 2000 are shown, based upon OECD/IEA (2002a, 2002b) and installed power data from Aqua-Media (1997) for countries not covered by IEA, using estimated capacity factors of 0.2 to 0.4. The world average for the year 2000 is 50 W/cap, and in 2007 it is 52 W/cap (WEC, 2009), the new capacity being mainly in China.

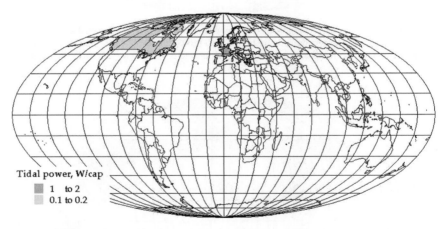

FIGURE 1.11   Tidal power. National average values for the year 2000 are shown, based upon OECD /IEA (2002a, 2002b). The world average for the year 2000 is 0.01 W/cap.

just quoted. This is more than 30 times the current consumer price of a kWh of electricity produced from fossil fuels, suggesting that it would be economically unacceptable to use food crops for energy. However, this is not always true, because food prices vary all the time according to local supply and demand (cf. the discussion in Chapter 7.1).

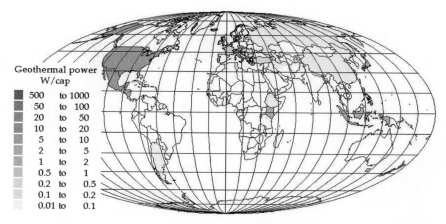

FIGURE 1.12   Geothermal power. National average values for the year 2000 are shown, based upon either 10% of heat input (from OECD/IEA, 2002a, 2002b) or 60% of installed power (from Barbier, 1999). The world average for the year 2000 is 9.3 W/cap. By 2007 it has risen to 9.7 W/cap (WEC, 2009).

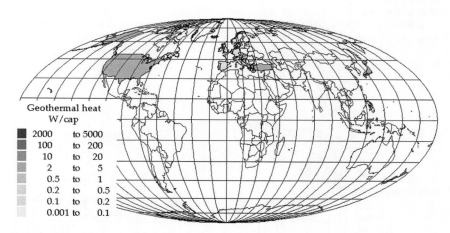

FIGURE 1.13   Geothermal heat (mainly district heating). National average values for the year 2000 are shown, based upon OECD/IEA (2002a, 2002b). The world average for the year 2000 is 0.5 W/cap.

Wholesale market prices for biomass waste and fuelwood range from about 1 c/kWh (of "burning value," i.e., energy of combustion) in India (FAO-Asia, 2003) to 2 c/kWh in industrialized countries (e.g., straw and wood chips 1.6 c/kWh and wood pellets 1.9 c/kWh; Danish Energy Agency, 1996; Alakangas *et al.*, 2002). For comparison, the cost of coal before considering externalities is 0.5 c/kWh (Danish Energy Agency, 2002).

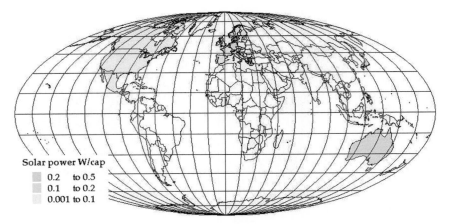

FIGURE 1.14 Solar power. National average values for the year 2000 are shown, based upon OECD/IEA (2002b) or an average energy production equal to 1.5% of the installed capacity given in IEA-PVPS (2001). This estimate takes the annual average incident radiation as 15% of the maximum (full sunlight along the normal to the collector plane), and multiplies it by a 10% average collector efficiency. The world average for the year 2000 is 0.007 W/cap. The growth rate from 2000 to 2001 was 35% (IEA-PVPS, 2002), and by 2007 the estimated production had grown to 0.01 W/cap (IEA-PVPS, 2008). Unfortunately, there is no collection of global production data, which would require metering of each installation, including those in private homes.

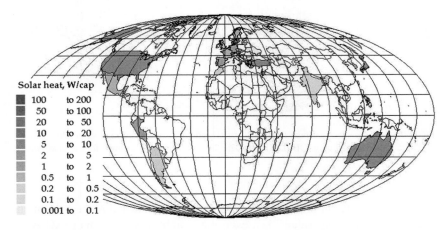

FIGURE 1.15 Solar heat. National average values for the year 2000 are shown, based upon IEA (2002). Both building-integrated and central district heating thermal systems are included. The world average for the year 2000 is 0.11 W/cap. Production estimates and metering at the end-user are even more difficult than for solar power, because the energy yield depends on the instantaneous state of the entire heating system (cf. the solar heating system modeling studies in sections 4.4 and 6.5).

Country W/cap or number	2000 Pop/1000	Hydro	Geoth. Power	Geoth. Heat	PV	Solar Heat	Tidal	Wind	Bio-Res	Bio Solid	Bio Liq	Bio-Gas	Anim. Food	Veg. Food
Afghanistan	17270	3.6	0	0	0	0	0	0	0	0	0	0	11.67	62.86
Albania	3400	164.13	0	0	0	0	0	0	0	23.45	0	0	37.68	101.02
Algeria	30400	1.8	0	0	0	0	0	0.04	0	3.5	0	0	14.77	127.85
Andorra	70	0	0	0	0	0	0	0	0	0	0	0	44.41	121.07
Angola	13100	8.11	0	0	0	0	0	0	0	572.03	0	0	7.65	84.5
Anguilla	10	0	0	0	0	0	0	0	0	0	0	0	12	65
Antarctica	0	0	0	0	0	0	0	0	0	0	0	0	0	0
Antigua Barb.	70	0	0	0	0	0	0	0	0	0	0	0	38.69	77.34
Argentina	37000	89.05	0.01	0	0	0.22	0	0.13	0	105.93	0	0	48.43	105.62
Armenia	3800	38.46	0	0	0	0	0	0	0	0	0	0	15.01	79.08
Aruba	70	0	0	0	0	0	0	0	0	0	0	0	12	96.85
Australia	19160	100.06	0	0	0.15	6.8	0	0.47	0	349.36	0	9.57	50.8	102.95
Austria	8110	591.74	0.15	0.82	0.05	7.7	0	2.6	9.34	453.31	2.13	4.75	59.23	122.71
Azerbaijan	8000	21.59	0	0	0	0	0	0.05	0	0	0	0	17.53	101.99
Azores Port.	260	0	0	0	0	0	0	0	0	0	0	0	48.43	121.07
Bahamas	320	0	0	0	0	0	0	0	0	0	0	0	34.43	83.87
Bahrain	700	0	0	0	0	0	0	0	0	0	0	0	38.74	121.07
Bangladesh	131100	0.81	0	0	0	0	0	0	0	77.02	0	0	3.24	98.55
Barbados	260	0	0	0	0	0	0	0	0	0	0	0	33.95	112.4
Belarus	10000	0.2	0	0	0	0	0	0.04	0	131.54	0	0	38.79	101.74
Belgium	10250	5.06	0.01	0.13	0	0.26	0	0.6	15.94	33.96	0	3.89	54.29	124.94
Belize	250	0	0	0	0	0	0	0	0	0	0	0	29.06	110.8
Benin	6300	0	0	0	0	0	0	0	0	375.39	0	0	4.84	119.03
Bermuda	70	0	0	0	0	0	0	0	0	0	0	0	37	110.07
Bhutan	1840	64	0	0	0	0	0	0	0	0	0	0	4.84	96.85
Bolivia	8300	27.21	0	0	0	0	0	0	0	115.26	0	0	17.19	90.27
Bosnia Herzeg	4000	146.15	0.33	0	0	0	0	0	0	59.79	0	0	17.53	111.33
Botswana	1480	0	0	0	0	0	0	0	0	0	0	0	18.11	91.09
Brazil	170400	204.29	0	0	0	0	0	0.04	0	324.52	52	0	29.78	114.77
Br. Virgin Isl.	30	0	0	0	0	0	0	0	0	0	0	0	24.21	96.85
Brunei Daruss.	300	0	0	0	0	0	0	0	0	88.58	0	0	24.65	112.49
Bulgaria	8200	37.27	0	0	0	0	0	0	0	93.98	0	0	33.41	86
Burkina Faso	11100	0.7	0	0	0	0	0	0	0	0	0	0	5.62	105.42
Burundi	6030	0	0	0	0	0	0	0	0	0	0	0	1.74	75.98
Cambodia	6820	0.03	0	0	0	0	0	0	0	0	0	0	8.81	91.43
Cameroon	14900	26.75	0	0	0	0	0	0	0	444.07	0	0	6.44	102.76
Canada	30750	1331.4	0	0	0.02	0	0.13	1.36	0	484.27	0	0	45.42	108.28
Cape Verde	440	0	0	0	0	0	0	0.1	0	0	0	0	22.37	136.37
Cayman Isl.	30	0	0	0	0	0	0	0	0	0	0	0	24.21	96.85
Cen. African R	2790	1.7	0	0	0	0	0	0	0	0	0	0	9.49	84.7
Chad	6670	0	0	0	0	0	0	0	0	0	0	0	6.92	92.15
Chile	15200	143.35	0	0	0	0	0	0	0	369.75	0	0	30.22	109.35
China	1262500	20.13	0.02	0	0	0	0	0.08	0	225.66	0	5	28.23	118.45
Colombia	42300	86.69	0	0	0	0	0	0	0	165.22	0	0	20.63	105.18
Comoros	3800	171.33	29.37	0	0	5.59	0	0	0	87.41	0	0	4.46	80.44
Congo	3000	13.29	0	0	0	0	0	0	0	261.3	0	0	6.39	101.26
Cook Islands	20	0	0	0	0	0	0	0	0	0	0	0	24.21	96.85
Costa Rica	2500	74	0	0	0	0	0	6.1	0	0	0	0	25.18	109.64
Croatia	4400	154	0.3	0	0	0	0	0	0	111.73	0	0	24.94	95.3
Cuba	11200	1.19	0	0	0	0	0	0	0	347.58	0	0	17	107.17
Cyprus	800	0.17	0	0	0	59.79	0	0	0	16.61	0	0	46.15	111.62
Czech R	10270	19.54	0	0	0	0	0	0.05	0	41.14	5.95	4.66	40.53	109.78
Denmark	5340	0.5	0.02	0	0.03	1.99	0	131.5	145.55	220.94	8.46	17.17	63.24	101.21
Djibouti	70	0	0	0	0	0	0	0	0	0	0	0	12.83	86.44
Dominica	80	21	0	0	0	0	0	0	0	0	0	0	33.41	111.53
Dominican R	8400	11.07	0	0	0	0	0	0	0	215.11	0	0	16.46	96.17
Congo/Zaire	50900	12.27	0	0	0	0	0	0	0	355.26	0	0	2.28	71.04
Ecuador	12600	68.54	0	0	0	0	0	0	0	73.81	0	0	20.97	109.44
Egypt	64000	25.33	0	0	0	0	0	0.32	0	27.61	0	0	12.4	149.64
El Salvador	6300	21.09	14.34	0	0	0	0	0	0	291.04	0	0	14.67	106.54
Equat. Guinea	430	0	0	0	0	0	0	0	0	0	0	0	4.84	96.85
Eritrea	4100	0	0	0	0	0	0	0	0	165.27	0	0	4.99	75.64
Estonia	1400	0	0	0	0	0	0	0	0	474.51	0	0	42.52	120.97
Ethiopia	64300	2.89	0.04	0	0	0	0	0.03	0	359.95	0	0	5.08	92.88
Falkland Isl.	0	0	0	0	0	0	0	0	0	0	0	0	24.21	96.85
Fiji	730	23	0	0	0	0	0	0	0	0	0	0	26.97	111.57
Finland	5180	323.44	0	0	0.05	0	0	2.3	11.54	1659.77	0	4.62	55.79	100.48
France	60430	126.58	0.27	0	0.02	0.55	1.08	0.3	39.82	202.03	5.78	3.83	65.18	108.72
Fr. Guiana	130	0	0	0	0	0	0	0	0	0	0	0	24.21	96.85
Fr. Polynesia	210	0	0	0	0	0	0	0	0	0	0	0	40	98.16
Gabon	1200	66.43	0	0	0	0	0	0	0	1018.62	0	0	17.14	107.02
Gambia	690	0	0	0	0	0	0	0	0	0	0	0	5.67	114.14
Georgia	5000	132.86	0	0	0	0	0	0	0	18.6	0	0	19.23	97.63

TABLE 1.1 Year 2000 average renewable energy use (W/cap; cf. Figs. 1.3–1.15).

Country W/cap or number	2000 Pop/1000	Hydro	Geoth. Power	Geoth. Heat	PV	Solar Heat	Tidal	Wind	Bio-Res	Bio Solid	Bio Liq	Bio-Gas	Anim. Food	Veg. Food
Germany	82170	30.22	0	0.16	0.14	1.47	0	22.3	28.47	79.49	2.64	9.01	50.12	116.95
Ghana	19300	39.24	0	0	0	0	0	0	0	366.24	0	0	5.81	124.84
Gibraltar	30	0	0	0	0	0	0	0	0	0	0	0	4.84	121.07
Greece	10560	40.01	0	0.25	0	12.46	0	7.8	0	118.77	0	0.13	41.02	138.4
Greenland	50	200	0	0	0	0	0	0	0	0	0	0	48.43	96.85
Grenada	90	0	0	0	0	0	0	0	0	0	0	0	32.64	101.21
Guadeloupe	440	0	0	0	0	0	0	0	0	0	0	0	24.21	96.85
Guam	120	0	0	0	0	0	0	0	0	0	0	0	14.53	96.85
Guatemala	11400	23.31	0	0	0	0	0	0	0	454.54	0	0	10.07	95.06
Guinea	4050	6	0	0	0	0	0	0	0	0	0	0	3.87	110.07
Guinea Bissau	920	0	0	0	0	0	0	0	0	0	0	0	7.85	105.18
Guyana	870	0	0	0	0	0	0	0	0	0	0	0	20.29	104.75
Haiti	8000	3.32	0	0	0	0	0	0	0	252.44	0	0	6.44	93.12
Honduras	6400	39.44	0	0	0	0	0	0	0	276.11	0	0	16.66	99.32
Hong Kong	6800	0	0	0	0	0	0	0	0	9.77	0	0	38.74	96.85
Hungary	10020	1.99	0	0.66	0	0	0	0	2.65	47.21	0	0	53.9	113.56
Iceland	280	2595.6	623.51	2984.7	0	0	0	0	4.75	0	0	0	67.65	94.19
India	1015900	8.37	0	0	0	0.17	0	0.4	0	263.64	0	2.3	9.39	108.18
Indonesia	210400	4.93	1.44	0	0	0	0	0.02	0	299.95	0	0	5.67	134.87
Iran IR	63700	6.67	0	0	0	0	0	0.06	0	16.48	0	0	13.03	128.04
Iraq	23300	2.85	0	0	0	0	0	0.08	0	1.71	0	0	4.26	102.13
Ireland	3790	25.59	0	0	0	0	0	9.7	0	47.68	0	9.82	54.38	120.58
Israel	6200	0.2	0	0	0.007	127.72	0	0.08	0	0	0	0	31.96	140.53
Italy	57730	87.5	6.7	0	0.03	0.25	0	2.2	7.69	36.71	0	2.97	45.28	132.01
Ivory Coast	16000	12.46	0	0	0	0	0	0	0	350.43	0	0	4.75	120.68
Jamaica	2600	5.11	0	0	0	0	0	0	0	245.29	0	0	18.98	111.43
Japan	126920	78.55	3.01	2.33	0.25	8.41	0	0.34	10.56	47.64	0	0	27.55	106.2
Jordan	4900	0.3	0	0	0	17.62	0	0.08	0	0	0	0	15.79	117.34
Kazakhstan	14900	57.96	0	0	0	0	0	0.04	0	6.24	0	0	31.23	113.61
Kenya	30100	4.86	1.63	0	0	0	0	0	0	519.98	0	0	11.33	83.78
Kiribati	60	0	0	0	0	0	0	0	0	0	0	0	18.84	124.36
Korea	47280	9.7	0	0	0.008	1.18	0	0.06	41.84	4.83	0	1.1	22.47	127.31
Korea DPR	22300	109.03	0	0	0	0	0	0	0	59.58	0	0	5.96	99.81
Kuwait	2000	0	0	0	0	0	0	0	0	0	0	0	34.53	117.14
Kyrgyzstan	4900	319.96	0	0	0	0	0	0.02	0	0	0	0	26.73	112.3
Laos	2960	13	0	0	0	0	0	0	0	0	0	0	7.7	102.08
Latvia	2400	132.86	0	0	0	0	0	0	0	548.06	0	0	33.32	104.94
Lebanon	4300	12.36	0	0	0	2.16	0	0	0	40.17	0	0	19.56	133.22
Lesotho	1980	0.4	0	0	0	0	0	0	0	0	0	0	4.84	106.54
Liberia	2510	7	0	0	0	0	0	0	0	0	0	0	3.24	97.29
Libya ArabJam	5300	0	0	0	0	0	0	0.04	0	35.1	0	0	17.82	142.23
Liechtenstein	30	0	0	0	0	0	0	0	0	0	0	0	48.43	106.59
Lithuania	3700	10.77	0	0	0	3.59	0	0	0	226.23	0	0	45.28	113.12
Luxembourg	440	30.2	0	0	0	0	0	4.09	81.53	48.31	0	3.02	54.29	124.94
Macedonia FY	2000	66.43	1.99	0	0	0	0	0	0	139.51	0	0	24.21	121.31
Madagascar	8460	5	0	0	0	0	0	0	0	0	0	0	9.59	87.65
Malawi	9860	0	0	0	0	0	0	0	0	0	0	0	2.37	103.2
Malaysia	23300	34.21	0	0	0	0	0	0	0	144.27	0	0	27.41	113.95
Maldives	270	0	0	0	0	0	0	0	0	0	0	0	31.82	93.7
Mali	8070	1.1	0	0	0	0	0	0	0	0	0	0	4.84	106.25
Malta	400	0	0	0	0	0	0	0	0	0	0	0	44.21	127.36
Marshall Isl.	50	0	0	0	0	0	0	0	0	0	0	0	24.21	96.85
Martinique	450	0	0	0	0	0	0	0	0	0	0	0	24.21	96.85
Mauritania	2250	5	0	0	0	0	0	0	0	0	0	0	15.84	107.02
Mauritius	1220	11	0	0	0	0	0	0	0	0	0	0	20.68	123.87
Mexico	97220	38.94	6.94	0	0.014	0.59	0	0.02	0	109.81	0	0.08	28.23	125.04
Micronesia	0	0	0	0	0	0	0	0	0	0	0	0	24.21	96.85
Moldova Rep	4300	3.09	0	0	0	0	0	0	0	18.54	0	0	18.98	114.87
Monaco	30	0	0	0	0	0	0	0	0	0	0	0	45.28	108.72
Mongolia	2310	0.08	0	0	0	0	0	0	0	0	0	0	45.67	50.27
Morocco	28700	2.78	0	0	0	0.28	0	0.56	0	20.37	0	0	10.36	133.17
Mozambique	17700	45.04	0	0	0	0	0	0	0	496.18	0	0	2.32	90.99
Myanmar	47700	4.46	0	0	0	0	0	0	0	255.7	0	0.03	6	131.62
Namibia	1800	88.58	0	0	0	0	0	0	0	125.48	0	0	12.88	115.4
Nauru	10	0	0	0	0	0	0	0	0	0	0	0	19.37	96.85
Nepal	23000	8.09	0	0	0	0	0	0	0	388.77	0	3.7	7.75	110.22
Netherlands	15920	1	0	0	0.08	1.34	0	8.9	41.81	28.38	0	11.02	57.38	102.18
New Caledonia	170	94	0	0	0	0	0	0	0	0	0	0	31.72	101.02
New Zealand	3830	734.39	82.56	164.43	0	0	0	2.74	0	286.89	0	10.41	52.59	104.89
Nicaragua	5100	5.21	3.13	0	0	0	0	0	0	369.94	0	0	8.77	99.08
Niger	8650	0	0	0	0	0	0	0	0	0	0	0	5.42	95.74
Nigeria	126900	5.23	0	0	0	0	0	0	0	757.29	0	0	4.21	133.8
Niue	0	0	0	0	0	0	0	0	0	0	0	0	14.53	96.85

TABLE 1.1   (Continued)

Country W/cap or number	2000 Pop/1000	Hydro	Geoth. Power	Geoth. Heat	PV	Solar Heat	Tidal	Wind	Bio-Res	Bio Solid	Bio Liq	Bio-Gas	Anim. Food	Veg. Food
N Mariana Isl.	30	0	0	0	0	0	0	0	0	0	0	0	14.53	96.85
Norway	4490	3603.3	0	0	0.13	0	0	0.87	36.69	354.8	0	1.48	56.13	109.15
Oman	2400	0	0	0	0	0	0	0	0	0	0	0	33.9	96.85
Pakistan	138100	14.24	0	0	0	0	0	0	0	231.09	0	0.01	20.77	97.97
Palau Islands	20	0	0	0	0	0	0	0	0	0	0	0	19.37	96.85
Panama	2900	123.7	0	0	0	0	0	0	0	210.75	0	0	28.09	92.4
Papua N Guin.	4420	8	0	0	0	0	0	0	0	0	0	0	10.12	95.25
Paraguay	5500	1111.2	0	0	0	0	0	0	0	553.2	0	0	28.09	94.58
Peru	25700	71.86	0	0	0	2.74	0	0	0	115.29	0	0	16.8	110.31
Philippines	75600	11.78	17.57	0	0	0	0	0.02	0	167.66	0	0.01	17.14	98.06
Poland	38650	6.22	0	0	0	0	0	0.1	0	123.31	0.7	1.03	43.1	120.39
Portugal	10010	129.28	0.92	0.13	0.009	2.39	0	3.3	23.1	249.14	0	0.13	51.19	128.72
Puerto Rico	4940	4.6	0	0	0	0	0	0	0	0	0	0	48.43	121.07
Qatar	600	0	0	0	0	0	0	0	0	0	0	0	38.74	96.85
Reunion	650	38	0	0	0	0	0	0	0	0	0	0	29.06	96.85
Romania	22400	75.33	0.12	0	0	0	0	0	0	169.05	0	0	32.74	125.81
Russia Fed	185500	108	0.04	0	0	0	0	0.02	0	0	0	0	31.38	109.88
Rwanda	145600	128.76	0.05	0	0	0	0	0	0	63.97	0	0	2.47	98.11
Saint Lucia	150	0	0	0	0	0	0	0	0	0	0	0	32.11	105.33
San Marino	30	0	0	0	0	0	0	0	0	0	0	0	45.04	132.01
S Tome&Princ.	150	0	0	0	0	0	0	0	0	0	0	0	4.41	111.33
Saudi Arabia	20700	0	0	0	0	0	0	0.04	0	0	0	0	21.6	117.63
Senegal	9500	0	0	0	0	0	0	0	0	240.55	0	0	9.64	99.66
Seychelles	80	0	0	0	0	0	0	0	0	0	0	0	23.97	93.8
Sierra Leone	3480	0.2	0	0	0	0	0	0	0	0	0	0	3.15	87.12
Singapore	4000	0	0	0	0	21.26	0	0	0	0	0	0	29.06	96.85
Slovak R	5400	99.89	0	0	0	0	0	0.04	0	19.68	0	0	38.16	113.56
Slovenia	2000	219.23	0.66	0	0	0	0	0	0	305.59	0	0	45.08	108.33
Solomon IsL.	350	0	0	0	0	0	0	0	0	0	0	0	10.27	100
Somalia	10510	0	0	0	0	0	0	0	0	0	0	0	29.88	48.96
South Africa	42800	3.73	0	0	0	0	0	0	0	393.32	0	0	17.92	121.84
Spain	39930	81.19	0	0.2	0.02	1.03	0	21.3	7.75	130.47	0	3.79	44.41	117.92
Sri Lanka	19400	18.49	0	0	0	0	0	0	0	291.75	0	0.1	7.51	108.91
St Kitts&Nevis	50	0	0	0	0	0	0	0	0	0	0	0	36.17	93.85
St Vincent&Gr.	120	0	0	0	0	0	0	0	0	0	0	0	22.23	102.66
Sudan	31100	4.27	0	0	0	0	0	0	0	602.37	0	0	22.08	91.62
Suriname	420	166	0	0	0	0	0	0	0	0	0	0	17.68	110.75
Swaziland	700	12	0	0	0	0	0	0	0	0	0	0	18.89	107.99
Sweden	8870	1017.1	0	0	0.03	0.75	0	8.97	60.22	1155.48	0	4.19	49.83	100.78
Switzerland	7190	585.42	0	16.82	0.21	4.25	0	0.12	140.26	92.76	0	11.64	52.88	106.59
Syria Arab Rep	16200	65.61	0	0	0	0	0	0.08	0	0	0	0	19.85	127.26
Taiwan Teipei	22200	45.49	0	0	0	0	0	0.02	0	0.6	0	0	29.06	96.85
Tajikistan	6200	255.01	0	0	0	0	0	0	0	0	0	0	7.17	76.13
Tanzania UR	33700	7.49	0	0	0	0	0	0	0	567.73	0	0	5.91	86.39
Thailand	60700	11.38	0	0	0	0	0	0.04	0	312.13	0	0.15	13.75	107.6
Togo	4500	7	0	0	0	0	0	0	0	307.06	0	0	3.87	108.91
Tonga	100	0	0	0	0	0	0	0	0	0	0	0	29.06	96.85
Trinidad&Tob.	1300	0	0	0	0	0	0	0	0	30.66	0	0	21.16	113.32
Tunisia	9600	1.38	0	0	0	0.28	0	0.34	0	171.62	0	0	16.8	142.95
Turkey	66840	52.8	0.13	3.56	0	5.21	0	0.09	0	128.31	0	0.1	18.21	147.22
Turkmenistan	5200	0.4	0	0	0	0	0	0	0	0	0	0	21.94	107.6
Turks&Caicos I	10	0	0	0	0	0	0	0	0	0	0	0	29.06	96.85
Tuvalu	10	0	0	0	0	0	0	0	0	0	0	0	29.06	96.85
Uganda	16670	0	0	0	0	0	0	0	0	0	0	0	6.83	107.41
Ukraine	48500	26.85	0	0	0	0.03	0	0.04	0	7.12	0	0	28.33	110.7
U. Arab Emir.	2900	0	0	0	0	0	0	0	0	0	0	0	38.16	116.47
U. Kingdom	59760	9.76	0	0.02	0.003	0.24	0	2.13	6.23	18.65	0	17.88	48.52	112.93
United States	275420	103.04	6.09	2.5	0.05	7.32	0	2.84	34.29	290.69	15.55	15.91	50.51	132.16
Uruguay	3300	245.5	0	0	0	0	0	0	0	169.1	0	0	46.88	92.49
US Virgin Isl.	100	0	0	0	0	0	0	0	0	0	0	0	43.58	121.07
Uzbekistan	24800	27.32	0	0	0	0	0	0	0	0	0	0	20.97	93.85
Vanuatu	150	0	0	0	0	0	0	0	0	0	0	0	23.15	102.13
Vatican City	0	0	0	0	0	0	0	0	0	0	0	0	45.04	132.01
Venezuela	24200	296.47	0	0	0	0	0	0	0	29.65	0	0	17.19	92.06
Vietnam	78500	21.16	0	0	0	0	0	0.02	0	383.02	0	0.02	13.17	111.91
W Sahara	90	0	0	0	0	0	0	0	0	0	0	0	15.84	96.85
W Samoa	190	0	0	0	0	0	0	0	0	0	0	0	29.06	96.85
Yemen	17500	0	0	0	0	0	0	0	0	0	0	0	6.59	92.11
Serbia Monten	10600	130.36	0	0	0	0	0	0	0	31.34	0	0	48.23	76.22
Zambia	10100	88.14	0	0	0	0	0	0	0	674.84	0	0	4.5	88.09
Zimbabwe	12600	29.53	0	0	0	0	0	0	0	589.45	0	0	6.83	95.64

TABLE 1.1   (Continued)

The production cost of biogas is 3.6–7.0 c/kWh (Danish Energy Agency, 1992), while that of wind power is 3–7 c/kWh (depending on wind conditions) and that of photovoltaic solar power is 40–130 c/kWh (IEA-PVPS, 2002, 2008). The photovoltaic market enjoys substantial public start-up subsidies (often in the form of subsidized customer investments or attractive buy-back rates for excess solar power). This is the case in countries like Germany and Japan, while in Switzerland the market has largely been created by industries' buying photovoltaic panels for reasons of aesthetics or image greening.

Hydropower costs 1–5 c/kWh, while coal- and gas-based power costs about 5 c/kWh to produce (Danish Energy Agency, 2002). Added to this are costs of distribution from centralized production units to the customers and, in many countries, taxes and environmental externality payments, leading to customer prices in excess of 14 c/kWh. As a result, in many countries wind power, being exempt from pollution and CO_2 fees, and biomass-based power are sold at prices very similar to those of fossil-based power. Also, geothermal power is usually competitive with other forms of electricity, while the viability of geothermal heat depends on local costs of district heating distribution.

Current oil production costs vary from less than 1 c/kWh at some Middle East wells to nearly 2 c/kWh from offshore facilities in the North Sea. The bulk sales price (in February 2003 around 2 c/kWh) is not strongly coupled to production prices, but is determined by market and political considerations. Some countries are willing to wage war against oil-producing countries in order to control prices. Refined products like gasoline are currently sold at prices around 4 c/kWh, with diesel fuel slightly lower, plus taxes and environmental fees where they apply (Danish Energy Agency, 2002; IEA 2002). Liquid biofuels have production costs of 3–7 c/kWh (ethanol from sugar cane the lowest, ethanol from sugar beet the highest, methanol from woody biomass at 4–5 c/kWh). Hydrogen from woody biomass could become about 3 c/kWh (Turkenburg *et al.*, 2000). Natural gas market prices are currently 10% higher than those of oil (IEA, 2002).

Because of the cost of the advanced equipment often required, it is clear that only in particular cases can the prices of renewable energy match those of fossil fuels at around 50% resource exhaustion, even with expected progress in renewable energy technology and production facilities. The case for increasing the role of renewable energy sources is therefore linked to uncertainty of future fossil fuel prices (for political and resource depletion reasons) and increased awareness of the indirect costs of pollution caused by fossil and nuclear fuels, including, in the fossil fuel case, emissions of substances contributing to excess greenhouse gases and global warming (cf. Chapter 7).

The 2000 source data for Figs. 1.3–1.15 are given in Table 1.1. It should be kept in mind that several of the numbers involve estimates and modeling, as direct energy production is not always monitored.

1.2 PAST AND PRESENT ENERGY RESOURCES

From a scientific point of view, an issue more essential than the place of renewable energy in the marketplace is its place within the physical universe. This view is developed in the following paragraphs, as a prerequisite for estimating the amounts of energy that can be extracted for use by human society at a rate that qualifies the process as renewable. Other views are philosophical and economic, the latter of which is taken up in Chapter 7.

The speed of the Earth in its orbit around the Sun is about 3×10^4 m s^{-1}, corresponding to a kinetic energy of some 2.7×10^{33} J. The Earth further rotates around its axis with an angular velocity of about 7.3×10^{-5} rad s^{-1}, furnishing an additional kinetic energy of some 2.2×10^{29} J. The work required to pull the Earth infinitely far away from the Sun, against gravitational attraction, is about 5.3×10^{33} J, and the corresponding work required to separate the Earth from its moon is of the order of 8×10^{28} J.

These are some of the external conditions for our planet, spelled out in energy units. It is a little more difficult to obtain reliable estimates for the amount of energy residing within the Earth itself. The kinetic energy of molecular motion, heat energy, is of the order of 5×10^{30} J. This estimate represents the total heat energy, relative to the absolute zero temperature. It is extrapolated from the value 4×10^{30} J, given in section 3.4.2, for the heat energy in the interior of the Earth relative to the average surface temperature of 287 K.

The materials forming the Earth carry further energy, in addition to the heat energy corresponding to their temperature. About 10^{21} J is, on average, present as kinetic energy in the atmospheric and oceanic circulation (cf. section 2.3), and the potential energy of the continental height-relief, relative to sea level, is about 2×10^{25} J, taking into account density variations in the crust (Goguel, 1976). Much larger amounts of energy are involved in the chemical and nuclear bindings, which determine the state and structure of matter. The carbon compounds of biological material provide an example of chemical energy. During earlier periods of the Earth's history, fossilization of biological material created the deposits of coal, oil, and natural gas, of which at least 10^{23} J is presently believed to be recoverable in a form suitable for fuel uses (see sections 2.3 and 4.6). Current standing crops of biomass correspond to an average of 1.5×10^{22} J (cf. sections 2.3 and 3.5).

Nuclear energy may be released in large quantities from nuclear reactions, such as fission of heavy nuclei or fusion of light nuclei. Except for spontaneously fissioning nuclear isotopes in the Earth's crust, which release about 4×10^{20} J y^{-1}, an initial amount of energy must be provided in order to get the energy-releasing fission or fusion processes going. Set-ups for explosive release of nuclear energy involving both types of processes are used for military purposes. Only the fission process has as yet been demonstrated as a basis for controlled energy supply systems, and, with necessary additional improvements in the technology of fast breeder reactors, recoverable resources of nuclear fuels are estimated to be of the order of 10^{24} J (see Sørensen, 1979 and 2005). If fusion of deuterium nuclei to form helium nuclei could be made viable on the basis of deuterium present in seawater, this resource alone would amount to more than 10^{31} J.

Energy conversion processes depleting certain materials of the Earth may be said to constitute irreversible processes. This is often true from a practical point of view, even if the reverse process may be theoretically possible.

The terms *energy use*, *spending energy*, etc., which are commonly used in energy literature as well as in everyday language, are of course imprecise expressions describing energy conversion processes. Such processes are in most cases associated with an increase in entropy. Entropy (S) is a property of a system that quantifies the "quality" of the energy contained in the system. The system may be, for example, an amount of fuel, a mass of air in motion, or the entire Earth–atmosphere system.

The entropy change for a process (e.g., an energy conversion process) that brings the system from a state 1 to a state 2 is defined as

$$\Delta S = \int_{T_1}^{T_2} T^{-1} dQ, \qquad (1.1)$$

where the integral is over successive infinitesimal and reversible process steps (not necessarily related to the real process, which may not be reversible), during which an amount of heat dQ is transferred from a reservoir of temperature T to the system. The imagined reservoirs may not exist in the real process, but the initial and final states of the system must have well-defined temperatures T_1 and T_2 in order for (1.1) to be applicable.

Conversion of certain forms of energy, such as electrical or mechanical energy, among themselves may, in principle, not change entropy, but in practice some fraction of the energy always gets converted into heat. The energy processes characteristic of human activities on Earth involve a series of successive conversion processes, usually ending with all the converted energy in the form of heat, radiated to space or released to the atmosphere, from which radiation of heat energy into space also takes place. The temperatures involved (the T_2) are typically 200–300 K. The processes involved are discussed in more detail in Chapter 2.

Stored energy of any form that may be converted to heat that is ultimately lost to space could then be called a *non-renewable energy resource*. The term *renewable energy resource* is used for energy flows that are replenished at the same rate as they are "used." The prime renewable energy resource is thus solar radiation intercepted by the Earth, because the Earth (i.e., the Earth–atmosphere system) re-radiates to space an amount of heat equal to the amount of solar radiation received (see Chapter 2). To utilize solar energy thus means converting it in a way convenient for man, but the net result is the same as if man had not interfered: that is, ultimately to convert solar radiation into heat radiated to space. Such usage may involve a delay in returning the heat, either as a part of man's conversion scheme or by a natural process. For this reason, energy stores, which are part of the natural process of converting solar energy into heat re-radiation, are also considered *renewable energy resources.*

Renewable energy is not narrowly defined here, and it may be taken to include the usage of any energy storage reservoir that is being "refilled" at rates comparable to that of extraction.

The amount of solar energy intercepted by the Earth and hence the amount of energy flowing in the *solar energy cycle* (from incident radiation flux via reflection, absorption, and re-radiation to heat flux away from the Earth) is about 5.4×10^{24} J per year.

Energy fluxes of other than solar origin that occur naturally at the surface of the Earth are numerically much smaller. For example, the heat flux from the interior of the Earth through the surface is about 9.5×10^{20} J y^{-1} (cf. section 3.4), and the energy dissipated in connection with the slowing down of the Earth's rotation (due to tidal attraction by other masses in the solar system) is of the order of 10^{20} J y^{-1} (cf. section 2.5).

1.2.1 Energy History

The minimum human energy requirement may be taken as the amount of "exchangeable" chemical energy that can be associated with the amount of food necessary to maintain life processes for someone performing a minimum of work and not losing weight. This minimum depends on the temperature of the surroundings, but for an adult human is generally considered to lie in the region of 60–90 W on average for extended periods, corresponding to (6 to 8) $\times 10^{6}$ J day^{-1}. The total intake requirements for sustaining human life are, of course, more than energy, comprising adequate supplies of water, nutrients, etc.

In order to perform any (muscle) work not purely vegetative, additional energy must be supplied in the form of food, or energy stored in the body will become depleted. The efficiency in converting stored energy into work typically ranges from 5% to 50%, with the lower efficiencies being associated with activities involving large fractions of static conversion (e.g., carrying a weight, which requires the conversion of body energy even if the weight is not being moved). The percentage complementary to the efficiency is released as various forms of heat energy.

The maximum average rate of food energy intake that a human being can continue for extended periods is about 330 W, and the maximum average rate at which work can be delivered for extended periods is of the order of 100 W (Spitzer, 1954). During work periods, the "man-power" output level may be 300–400 W, and the maximum power that can be delivered by an adult male for a period of about a minute is roughly 2,000 W.

Although it is not certain that the rates of energy conversion by the human body have remained constant during human evolution, it may be reasonable to assume that the average amount of "muscle power" used by the earliest members of the genus *Homo,* which evidence suggests lived some 4×10^6 years ago in Africa (Leakey, 1975), was of the order of 25 W.

The total energy flux received by an individual man in a food-gathering or hunting society is then the sum of the energy in the food, averaging say 125 W, and the absorbed flux of radiation and heat from the surroundings, which may reach considerably larger values, but is highly dependent on clothing, climate, and the nature of the surroundings (cf. Budyko, 1974). The outgoing energy flux again consists of heat and radiation fluxes, as well as turnover of organic material, plus the amount of energy converted into work. For growing individuals, the net flux is positive and the mass of biological material increases, but also for adult individuals with zero net energy flux, new biomass continues to be produced to replace "respiration losses."

Humans have successively developed new activities that have allowed them to gain access to larger amounts of energy. Solar energy may have been used for drying purposes, and as soon as fires became available, a number of activities based on firewood energy may have started, including heating, food preparation, and process heat for tool making. The earliest evidence for fires used in connection with dwellings is from Hungary; it dates from 350 000–400 000 years ago (H. Becker, 1977, personal communication).

A good fire in open air, using some 10–50 kg of firewood per hour, may convert energy at a rate of 10^4–10^5 W, whereas indoor fires are likely to have been limited to about 10^3 W. Several persons would presumably share a fire, and it would probably not burn continuously at such a power level, but rather would be relit when required (e.g., from glowing embers). It is thus difficult to estimate the average fire energy per person, but it would hardly exceed 100 W in primitive societies. The efficiency of delivering energy for the desired task is quite low, in particular for open-air fires.

A recent estimate of the energy used by Neanderthals during the warm interglacial Eem period some 125 000 years ago is an average of 135 W/cap for a location in Northern Europe with an average temperature of about 8 °C (Sørensen, 2009). Table 1.2 gives an estimate of activities requiring energy use beyond the basic metabolic rate (which for Neanderthal males averaged 92 W, for females 77 W), based on a group with 10 adult members and 15 children. Activities include hunting, wood provision, and tool making; fires were used for cooking and heating the cave or hut used for dwelling; but without woollen covers and some clothes and footwear, survival at Northern latitudes would not be possible.

Example of Time Use and Corresponding Rate of Monthly Average Energy Use (W/cap)	4 Males	1 Male	1 Female	4 Females
Hunt: tracking down prey (8 h, 1 day in month)	2.04	2.04	1.71	
Hunt: prey killing (1 h, 1 day in month)	0.45	0.45	0.37	
Hunt: parting mammoth, drying, (3 h, 1 day in month)	0.96	0.96	0.80	
Hunt: eat, sleep, rest at hunt site (12 h, 1 day in month)	1.69	1.69	1.41	
Hunt: eat, watch, cut, scrape, sleep (24 h, 12 days in month)		44.16	36.96	
Hunt: carrying meat back (10 h, 7 or 1 days in month)	26.83	3.83	3.21	
Hunt: sleep, rest at home (14 h, 7 or 1 days in month)	13.77	1.97	1.65	
Hunt: returning to hunt site from home (8 h, 6 days in month)	12.27			
Hunt: eat, sleep, rest at hunt site (16 h, 6 days in month)	14.72			
Home: wood cutting (8 h, 5 days in month)	15.33			
Home: stone flaking, tools construction, clothes making (8 h, 11, 16, 30 d)	22.49	32.71	27.38	61.33
Home: fire attention, child rearing, food prep., leisure, eat (8 h, 16, 30 d)	22.90	22.90	19.16	42.93
Home: sleep (8 h, 16 or 30 days in month)	16.36	16.36	13.69	25.67
Monthly average energy expenditure, adult humans (W/cap)*	149.81	127.06	106.35	129.93

Summary:	Male	Female
Average adult minimum energy requirement (W/cap)	145.26	125.22
Total average adult energy requirement, W for whole group of 10 adults	1352.37	
Children's average energy requirement (W for whole group of 15 children)	400.00	
Equivalent meat intake (loss 30%), W and GJ per month for whole group of 25 people	2503 W or 6.5 GJ/month	
Equivalent meat intake (loss 30%), in kg per month for whole group of 25 people	813 kg/month	
Fires: 5 cooking fires 8 h, 30 d (346 kg dry wood), average over month	1667 W	
Fires: large outdoor fire 8 h, 30 d (622 kg dry wood)	3000 W	
Fires: Possible fire at hunt site 12 h, 14 days in month (168 kg)	810 W	

TABLE 1.2   Estimate of activities requiring metabolic food conversion, muscle energy, or heat from fires in a Northern Eem Neanderthal group (Sørensen, 2009).

The next jump in energy utilization is generally considered to have been associated with the taming of wild animals to serve as livestock and the introduction of agriculture. These revolutions have been dated to about 10^4 years ago for the Near East region (cf. DuRy, 1969), but may have developed at about the same time in other regions, such as in Thailand and Peru (Pringle, 1998). The introduction of livestock would have promoted the tendency to settle at a given place (or vice versa), increasing in turn the requirement for food beyond the capacity of a hunting society. Agriculture was based at first on wild varieties of wheat, for example, and it is believed that artificial irrigation was necessary at many of the sites where evidence of agriculture (various tools) has been found. The power for water transport and, later, pumping would then be derived from suitable draught animals in the livestock pool, as a substitute for man's own muscle power. The transition from a hunting to an agricultural society, often called the *Neolithic* or *new Stone Age,* occurred several thousand years later in the temperate zones of northern America and Europe.

The creation of cultures of growing size and increasing levels of sophistication, leading to the formation of large cities (for example, at the Euphrates, Tigris, and Nile rivers), from about 7000 years ago, witnessed a growing use of energy for plowing, irrigation, grinding, and transport (of food supplies and of materials, for example, in connection with buildings and monuments), as well as the harvest of solar energy through agricultural crops. It is not known exactly how much of the physical work was performed by men and how much by animals, but it is likely that another 100–200 W was added to the average energy usage per capita in the most developed regions.

Figure 1.16 gives an example of a recent reconstruction of energy use from 125 000 to 1100 years ago, for Denmark (Sørensen, 2011a). The energy use is taken at the end-user and delivered energy would be higher, due to losses (e.g., from open fires). There are no specific energy data for the period, but energy is derived from indirect data, such as number of farms and taxes paid in Medieval times, and from burial sites in the Stone Age. An attempt to maintain consistency is made by use of a demographic model, where parameters are changed only when identifiable events suggest so. The resulting variations agree well with the "free-hand" estimate for the early period in Fig. 1.18, which was made for the first (1979) edition of this book.

It is important to bear in mind that there must have been large differences in energy use between different societies and, at least in later times, between individuals within a given society. Throughout man's history (the term *history* not being restricted to the presence of written records) there have been individuals whose access to energy was largely limited to that converted by their own bodies. In large regions of Asia and Africa today, the average energy spent per person is only a few hundred watts above the muscle-power level (with firewood as an important energy source). This means that parts of the population today use no more energy than the average person during the Neolithic period.

Thus, neolithic energy sources were direct solar radiation, environmental heat, and animal biomass, as well as primary (plant) biomass in the form of food and then as firewood, plus mechanical work from the muscle power of animals.

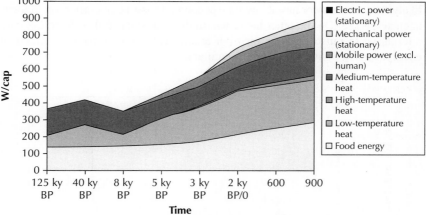

FIGURE 1.16   Estimated end-use energy in Northern Europe (after nations formed in the region some 1.3 ky BP just Denmark) 125 to 1.1 ky BP, distributed in energy-use categories (*a*) or in energy qualities (*b*). The increase in food energy is associated with harder labor (in the fields, in wars) (Sørensen, 2011a).

In the Near East, oil was used for lighting, and bitumen had non-energy uses. Boat travel in the open sea (the Mediterranean) is believed to have started over 9000 years ago (Jacobsen, 1973), and there is evidence of wind energy utilization by means of sails in Egypt about 4500 years ago (Digby, 1954). Per person, wind energy may not at this time have contributed a significant proportion of the total energy use in the Mediterranean region, but later, when trade became more developed (about 4000 years ago), the total amount of energy spent on transportation on land and at sea constituted a less negligible share (maybe a few percent) of the total amount of energy spent in the "developed regions" of the world at the time.

The building of houses in many cases implied the creation of a required indoor climate with utilization of solar energy. In low-latitude regions, structures with high heat capacities were employed in order to smooth out day-to-night temperature variations, and often houses were built partly underground and the evaporation of soil moisture was utilized to create cool environments for living (during hot periods) and food storage (Bahadori, 1977). In regions with a colder climate, insulating[*] building materials (e.g., straw for roofs) were employed to reduce heat losses, and heat production not involving fires was increased by keeping livestock within the living area of the houses, so as to benefit from their respirational heat release.

Water mills and windmills (e.g., the vertical axis panemone type, probably derived from waterwheels, or the sail-wing type, presumably copied from ships' sails) also played a role after a certain stage in development. The earliest mention of windmills in actual use is from India about 2400 years ago (Wulff, 1966). Considering these windmills' low efficiency and overall size, it is unlikely that wind power has at any time accounted for a large proportion of average energy use. On the other hand, windmills and water mills offered the only alternative to muscle power for high-quality (i.e., low-entropy) mechanical energy, until the invention of the steam engine.

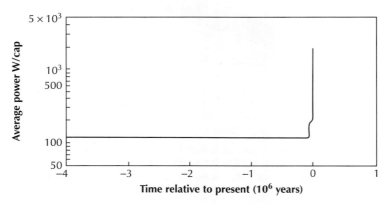

FIGURE 1.17 Trends in average rate of energy conversion per capita, not including fluxes associated with the local thermal environment.

---

[*]The term *insulating* is taken to include suppression of convective heat transfer.

The Industrial Revolution 200–300 years ago placed at man's disposal amounts of power capable of producing work far beyond his own muscle power. However, at that time, firewood was barely a renewable resource in the developed regions of the world, despite quite extensive programs to plant new forests to compensate for usage. The increase in energy usage made possible by growing industrialization did not really accelerate, therefore, before large amounts of coal became available as fuel. In the 20th century, the large growth in energy consumption was made possible by the availability of inexpensive fossil fuels: coal, natural gas, and oil.

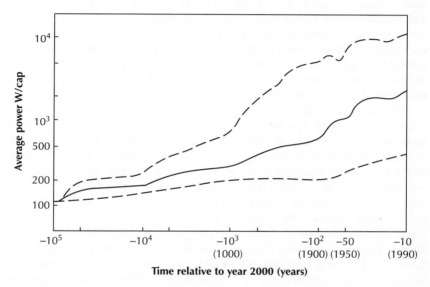

FIGURE 1.18   Trends in average rate of energy conversion per capita (*solid line*), not including fluxes associated with the local thermal environment (same as Fig. 1.1, but on a logarithmic time scale). Corresponding trends (*dashed lines*) for the societies that at a given time have the highest and lowest average energy usage. For the more recent period, data from Darmstadter *et al.* (1971) and European Commission (1997) have been used, in a smoothed form.

An outline of the possible development in energy usage up to the present is presented in Figs. 1.17–1.19. Only over the past century or two have reliable worldwide data on energy usage been recorded, and even for this period the data comprise mainly direct use of commercial fuels, supplemented with incomplete information on biomass and other renewables. One reason for this is that it is more difficult to specify the remaining energy use, because, for example, solar collectors are often not individually monitored, local biomass use is not quantified in energy units, environmental heat gains vary from day to day, and so on. In Figs. 1.17–1.19, which are anyway only indicative, fuels are included in terms of their gross energy value, independently of end-use efficiency. The use of renewable energy flows, on the other hand, is given as an estimated net energy at the primary conversion stage, that is, the energy in food intake rather than the total amount of energy absorbed by the plants or the total biomass of plants and animals. The environmental energy contribution to maintaining man's body temperature as well as the regulation of indoor climate by the choice of materials and building systems ("passive energy systems") are excluded.

Figure 1.17 shows the trend in average rate of energy conversion per capita, on a linear time scale, and Fig. 1.18 shows the same trend on a logarithmic time scale, extending backward from the year 2000. Figure 1.18 also indicates the estimated spread in energy usage, with the upper curve representing the societies with highest energy use at a given time, and the lower curve representing the societies with the lowest energy use. These curves, which do not reflect any great degree of accuracy, do not represent rigorous limits, and values outside the interval may certainly be appropriate for individuals of a given society— the very rich or the very poor.

The energy conversion rate for food alone has been taken as 125 W throughout the time interval. The increase in energy usage from about -10^5 y is associated with access to fire. The amount of energy derived from fires depends on whether fires were used only for cooking and for heating as well. The choice of the average curve also rests on the assumption that between -7×10^4 and -10^4 y (i.e., during the latest ice age; cf. section 2.4) about half of the world's population used fires for heating purposes.

In the time interval -10^4 to -10^3 y, human settlements developed into a variety of societies, some of which had a very high degree of organization and urbanization. The increase in energy usage was mainly associated with more systematic heating and cooking practices, with tool production (e.g., weapons) and with transportation (e.g., by riding or by draught animals). With increasing population density, materials that previously had been available in the immediate natural surroundings had to be transported from far away, or substitutes had to be manufactured; either way, additional energy had to be spent. In several of the societies in question, mechanical work was performed not only by animals but also by human slaves, so that the average per capita energy usage was less affected. The trends of the curves also reflect the differences in development characterizing different geographical regions. Simultaneously with the culmination of the civilizations in Mesopotamia and Egypt, northern Europe and northern America entered the Neolithic period, with warm climatic conditions quite different from those of the preceding several thousand years.

During the last 1000 years, the increasing energy usage is, in part, due to the shift in population distribution toward higher latitudes, and to overall increased requirements for space heating in such regions (the *little Ice Age*, cf. section 2.4). It should also be mentioned that the efficiency of converting the energy of firewood (supplemented by animal dung and later by peat) into useful heat for cooking, craft work, hot water, and space heating was quite low, for example, in 16th-century Europe, but gradually improved nearing the 20th century (Bjørnholm, 1976). During the period 1500–1900, the curves are a result of this feature (in particular, the early high maximum value attained for the most affluent societies) combined with increased energy demand (e.g., larger proportions of the population acquiring energy-demanding habits or lifestyles, such as taking hot baths, drinking hot beverages, washing clothes in hot water, etc.). The development in the last century is dominated by the energy consumption of the industrialized countries (industrial process heat, transportation, increased room temperature, refrigeration, lighting, etc.). During this period, the top curve in Fig. 1.18 represents the energy use of an affluent American, while the lowest curve represents the average energy use in the poor regions of Africa or India, including non-commercial fuels, such as cow dung and stray wood (which used to be absent from official statistics, as first noted by Makhijani, 1977).

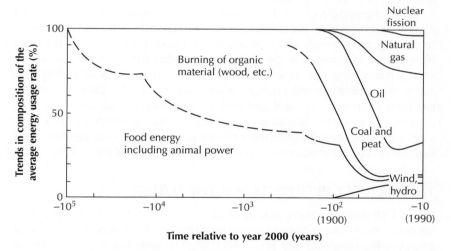

FIGURE 1.19 Trends in the distribution of the average rate of energy use on different types of energy resources. The most recent period is based on smoothed data from Darmstadter *et al.* (1971) and European Commission (1997), and the basis for the estimates pertaining to earlier periods is explained in the text. Needless to say, such estimates should be regarded as very tentative, and the definition of average use is itself uncertain, particularly for the early periods (e.g., the 20% contribution from fires 50 000 years ago depends sensitively on the fraction of the world population living in regions where space heating was desirable).

In Fig. 1.19, the distribution of the energy consumption on different sources of energy is sketched. Again, only for the past century or two have actual data been used. The shape of the curve describing the diminishing share of food energy starting about 10^5 years ago is again dependent on emerging cultures and geographical distribution of the population, outlined above. It is clear, however, that the energy basis for human societies has been renewable energy sources until quite recently. Whether all the wood usage should be counted as renewable is debatable. Early agricultural practice (e.g., in northern Europe) involved burning forest areas for farming purposes and repeating the process in a new area after a few years, as the crop yield diminished owing to nutrient deficiency in the soil. Most forests not being converted into permanent agricultural land survived this exploitation, owing to the low population density and the stability of the soils originating from glacier deposits. Similar overuse, or overgrazing by livestock, would be (and was in fact) disastrous in low-latitude regions with a very shallow soil layer, which would simply be eroded away if the vegetation cover were removed (cf. section 2.4). Replanting forests has been common in northern Europe during the last few centuries, but the strongly increasing demand for wood over the last century (not just for fuel purposes), as well as construction work associated with urbanization, has led to an actual decrease in forest area in most parts of the world.

From the middle of the 19th century, non-renewable fossil fuels have rapidly increased their share of total energy usage, to the present 80%–90%. In the beginning, fossil fuels replaced wood, but they soon became the basis for exponential growth in energy use, associated with a number of novel energy-demanding activities. During the same period, usage of hydropower has increased, and recently nuclear fission power passed the 1% level. Growth has been interrupted by wars and periods of economic recession. The high dependence on non-renewable energy sources has developed over a very short period of time. The briefness of this era compared with the history of man on Earth stands out clearly on the linear scale used in Fig. 1.17.

1.3 RESOURCE PROSPECTS FOR THE FUTURE

Figure 1.18 shows the very large difference between the energy use of the affluent countries or affluent persons within countries and the energy use of the least energy-using inhabitants of poor countries. The situation may currently be changing, as the level of global interaction increases and every world citizen becomes aware of the kind of lifestyle that is "possible." However, current development does not seem to indicate a diminishing gap between the energy use of those using the most and those using the least energy. This is also true of other commodities related to living standard.

Energy use and resource depletion do not, of course, constitute the primary goals of any society or any individual within a society. For example, the average European or Japanese uses about half as much energy as the average North American, but the former have a living standard no lower than that of the average North American. This fact demonstrates that, while living standard and welfare depend on having primary needs (food, shelter, relations) met, there are also secondary standards based on individual preference and implementation that can have different implications for energy use.

The relationship between economic activities and social welfare has been debated for a considerable period of time, as has the possibility of physical limits to growth in material exploitation of the resources of a finite planet. Conventional economists argue that the inventiveness of man will lead to substitution of new materials for those threatened by exhaustion, in an ever-ongoing process. Recognizing the finiteness of fossil and nuclear energy sources, this argument leads to the general prediction that renewable energy sources must take over at some stage, and the only debate is on how soon it will happen.

Most geologists believe that oil and natural gas production will peak sometime within the next one or two decades. After that, prices are bound to rise, thereby easing the introduction of alternative energy sources. The predicted higher price of energy also suggests that energy should be used more efficiently, in order to prevent higher energy cost's slowing down improvement of human welfare. Measures for improving energy efficiency are already available that will raise efficiency by a factor of 3 to 5 without any substantial increase in cost (Sørensen, 1991, 2008b).

Carbon capture could extend the use of coal, which is the fossil commodity that seems most abundant (see section 4.6). However, if coal is used for combustion, there might still be pollution beyond greenhouse gas emissions to worry about. Of course, the same is true for biofuels, unless the conversions are not to hydrocarbons but to pure hydrogen (producing only water by combustion and possibly not even combusted but rather used in fuel cells).

Nuclear fuel resources are no more abundant than oil or gas, when considered for use in conventional light-water reactors. Breeder technologies are currently no longer actively developed, since they proved to present problems with safety, capacity factors, and cost (Sørensen, 2005), and nothing suggests that fusion reactors, should they become workable, would have fewer radioactivity problems than fission reactors. The hope entertained by some fusion researchers, that it may some day be possible to find materials for the confinement structure that do not present a radioactivity danger after exposure to the operating conditions for nuclear fusion, seems to be inconsistent with the physics of fusion.

1.4 GLOBAL TEMPERATURE IMPACTS AND OTHER CLIMATE IMPACTS

Currently, the interest in renewable energy is closely tied to the discussion of global warming, which is caused by the increased injection of greenhouse gases into the atmosphere (cf. section 2.4). However, it should not be forgotten that the prime reason for choosing renewable solutions is renewability, a property very closely linked to the desire to establish sustainable solutions that do not just replace old problems with new ones. Sustainability means exploiting flows, rather than static resources.

The greenhouse effect itself is a completely understood physical phenomenon, to which we owe the habitability of the Earth, as explained in Chapter 2 (see also Fig. 1.1). However, human interference is altering the magnitude of the greenhouse effect, and it is the climate change induced by this that should worry us. Human interference includes land-use change, emission into the atmosphere of light-absorbing gases, emission of particulate matter with different size distributions, and injection of chemical pollutants into the atmosphere at various levels. All of these cause changes in temperature and solar radiation budget, in a pattern that varies with latitude and longitude and can be positive or negative. The climate models used to predict the changes all include the basic greenhouse effect, but differ in how many of the other possible climate impacts they model. Furthermore, the models' accuracy is limited by two approximations: One is use of an integration mesh that is much more coarse than some known variations in atmospheric variables (the model grid size has diminished from 250 to 23 km over the last 20 years, but the accuracy of the model predictions is up to five times poorer than the grid dimensions, e.g., presently no better than around 100 km). The other approximation is that the climate variables used in models are averages of the true variables, such as wind velocity at a given location in the atmosphere. The true variables are the sum of a mean value and a deviation from the mean value, but the latter is neglected. In consequence, so are also the terms in the circulation equations that couple average motion and small-scale motion (i.e., the deviations from averages, cf. section 2.3). Finally, only a subset of the effects involving more than just air velocity, moisture, and temperature are included. Particles, chemical pollutants, and land use are quantities that can be treated theoretically, whereas the effects of localized volcanic eruptions or sunspot flares are more difficult to deal with. So people who do not "believe" in greenhouse warming can plainly be ignored, but those concerned about the accuracy of climate models are legitimate discussants. Even so, over the last 40+ years, each addition to the climate models has yielded results indicating a man-made, global warming (global average temperature increase for a given increase of greenhouse gases in the atmosphere), although justified question marks are attached to specific regional model predictions or statements on the frequency of extreme atmospheric events.

It is important to realize that the Earth–atmosphere system is not in its most stable state. The present state is meta-stable, and a more stable situation could occur if snow and ice covered the entire Earth's surface (thus causing maximum reflection of sunlight, see section 2.4). This means that nudging the present system could trigger a change disproportionate to the amount of initial action. However, it is known that the present state possesses a level of stability, because neither seasonal variations nor annual climatic fluctuations nor long-term dispersal of particulates from major volcanic eruption has been able to induce a transition to another quasi-stable or fundamental state of the Earth–atmosphere system. On the other hand, the claim (IPCC, 2007a) that extreme events are becoming more frequent, and that the amplitude of excursions is increasing, suggests a finite (although likely small) probability of a climate change far beyond the causal implication of greenhouse warming.

The bottom line is clearly "Don't play with the climate"! This dictum over-rules suggestions like spraying soot on the Arctic/Antarctic ice or injecting radiation-absorbing dust into the atmosphere, for a "least-cost" suppression of global warming, without having to change any fossil energy habits. Although the suggested actions appear to have an effect opposite to that of greenhouse gas emissions, the differing geographical and height dependences of the intended effect make it difficult to ensure (and beyond present model-building capabilities to predict) that the result would not be different than the one aimed for, such as inducing a new glacial period. Past ice ages have not been equidistant in time and although the main causes (in terms of orbital parameters, cf. section 2.4) are known, the timing and strength of glaciation onsets have varied in ways not presently understood.

1.5 ROLE OF ENVIRONMENTAL AND SOCIAL ISSUES

The development in energy use is linked to another factor that may serve to accelerate the energy transition, namely, increased awareness of the negative environmental impacts of energy production and use. Early man was capable of causing environmental disturbance only on a very local scale. However, extensive burning of forests, for example, to provide land for agriculture, which would later be abandoned when overexploitation diminished the crop or grazing yields, may have been instrumental in creating the desert and semi-desert regions presently found at low latitudes (Bryson, 1971). This is an early example of a possibly man-made climatic change. Recently, man has reached a technological level enabling him to convert energy at rates that can be maintained over extended areas and that are no longer small compared to the energy fluxes of solar origin that are responsible for the climate.

The average heat flux of anthropogenic origin (i.e., from fossil fuels) in an industrial and urban area like the Los Angeles Basin (about 10^{10} m^2) was estimated in 1970 to be 7 Wm^{-2} (Lees, 1970). The global average value in 1970 was 0.015 Wm^{-2}, and the average solar flux absorbed by the Earth–atmosphere system is 240 Wm^{-2} (see section 2.4). In comparison, a forest fire, burning down an area of fertile, tropical forests in one week, would release a heat flux of about 1000 Wm^{-2}. Yet the average heat flux from forest fires in all continental regions, the average being over several years, is less than the average anthropogenic heat flux. The nuclear weapons arsenal built up during the last 50 years is somewhere in the range of 10^4–10^5 megatons (Feld, 1976), with the higher figure corresponding to about 4.4×10^{20} J. If these weapons were detonated within a 24-hour interval, the average energy flux would be 5×10^{15} W, and if the target area were 10^{12} m^2, the average heat flux would be 5000 Wm^{-2}. The destructive effects would not be confined to those of the immediate energy release. Radioactive contamination of the environment would cause additional death and decay and would establish other mechanisms for climatic disturbance (e.g., destruction of the stratospheric ozone shield), in addition to threatening human survival as the dominant species on the planet.

In the recent century, energy habits have been formed largely by those who sell fossil energy and energy-using equipment. They are responsible for consumer behavior biased toward buying an additional kWh of energy instead of saving one by efficiency measures, even if they are cheaper. This is linked to the "growth paradigm" introduced by the simplistic market economy in the 18th century. Even today, all financial sector accounting is done in terms of growth rates and all economic figures are calculated as a percentage of last year's, a procedure that makes sense only during periods of exponential growth. Such periods must necessarily end and be replaced by more quiet variations, or they may even lead to declines. Today's liberal paradigm also counts on "globalization" to achieve the lowest cost of welfare. The idea is that labor is done where it is cheapest, implying that a typical European or North American consumer product has been shipped back and forth to the Far East a number of times, in order to have each part produced, or each process performed, as cheaply as possible (so far, keeping the artistic designs at home). Similarly, food comes from areas where it appears cheapest to produce, and fruits are shipped from the opposite side of the Earth, at least during winter at the location of the consumer. This behavior rests on a fundamental assumption: Transportation around the globe, and the energy required for it, is a negligible part of the price of the goods in question. So far, fossil fuels have supported this assumption, because they cost little to extract (especially in the Middle East) and the negative environmental and climate impacts have not been reflected in intercontinental transportation costs. European nations have tried to levy taxes on fossil fuels consumed within their borders (to at least partially compensate for the negative impacts), but transport by ship or plane has continually escaped taxation (as the result of the United States' insistence and its domination of the World Trade Organization). These factors make it clear why the majority of financial and political decision-makers still regard fossil fuels as the backbone of the energy supply for decades to come. As long as they are cheap, they cannot be scarce! Of course, such arguments are invalid in malfunctioning economic markets like the one prevailing today: In the 18th century, liberalism made it clear that the theory behind a market economy was valid only if markets consisted of many small actors and only if they all had full access to the knowledge needed to make the right decisions. Neither of these conditions is even approximately fulfilled in the present world economic system. The sad conclusion is that the proper energy solutions (as well as policies compatible with climate stabilization) will not prevail until the economic paradigm has been changed in a positive direction.

Some would say that certain energy supply systems and economic organizations go better together than other ones. However, it is probably an exaggeration to imagine that the introduction of one kind of energy technology rather than another will determine or solve such institutional problems. What may be true, though, is that certain types of technology allow a meaningful discussion of how societies could be organized differently from their present organization. At least some of the renewable energy technologies fit well with the needs of sophisticated, decentralized societies based upon information technology and with the needs of the underprivileged parts of the world.

1.6 THE SUSTAINABILITY TEST

Science and technology literature contains a range of suggestions for handling future energy demands. In the past, some of the technologies brought forward as "technically feasible" have actually been developed to commercial viability, and others not, for a variety of reasons. Over the last few decades, renewable energy has passed from the level of technical feasibility to a level of cautious introduction into the marketplace and not least into long-term government planning. One reason for its slow penetration is that some influential funding institutions, including the European Commission, have continued to use a large fraction of their R&D funds, as well as loan and aid money, on coal, fission, and fusion, ignoring the risk of pollution and long-range radioactive waste problems and hoping to obtain short-term industry advantages in export of outdated technology to former Eastern bloc and developing nations. If funds were wholeheartedly aimed at a rapid transition from the fossil to the renewable era, progress could be made much faster. This has been demonstrated by a number of recent scenario studies, some of which are described in Chapter 6. The general question of who controls technology development has been discussed by Elliott and Elliott (1976) and by Sørensen (1983, 2001a). In recent decades, members of a number of "grassroot" movements have advocated use of renewable energy, and ideally their preferences will prevail as a new generation makes their way into decision-making positions. The reality of such hopes is probably bleak, as indicated in section 1.5.

Renewable energy sources are typically characterized by a theoretical maximum rate at which energy may be extracted in a "renewable" mode—that is, the rate at which new energy is arriving or flowing into the reservoirs associated with many of the renewable energy flows. In some cases, the additional loop on a given renewable energy cycle, caused by man's utilization of the source, will by itself modify the rate at which new energy is arriving. For instance, utilization of temperature differences in the oceans may alter surface evaporation rates and the velocities of ocean currents, which in both cases means the mechanisms for establishing the temperature differences may be altered (cf. section 3.3). The geothermal energy flux from the interior of the Earth is not a renewable resource, since the main part of the flux is associated with cooling of the interior (section 3.4). On the other hand, it is a very small fraction of the heat that is lost annually (2.4×10^{-10}), so for practical purposes geothermal energy behaves as a renewable resource. Only in case of overexploitation, which has characterized some geothermal steam projects, is renewability not ensured.

Sustainability is discussed further in Chapter 8. The general structure of the chapters in this book is:

In Chapter 2, the nature and origin of renewable energy sources are discussed in what may resemble an odyssey through the sciences of astrophysics, atmospheric physics and chemistry, oceanography, and geophysics. The importance of connecting all the pieces into an interlocking, overall picture becomes evident when the possible environmental impact of extended use of the renewable energy sources in the service of mankind is investigated in Chapter 7.

Chapter 3 provides, for each renewable energy source, an estimate of the size of the resource, defined as the maximum rate of energy extraction that on an annual average basis can be renewed, independently of whether it is possible to extract such energy by known devices. Issues of power density and variability are also discussed in this chapter.

Chapter 4 opens with some general features of energy conversion devices and then describes a number of examples of energy conversion equipment suitable for specific renewable energy sources.

Chapter 5 gives an overview of various methods of energy transport and storage, which, together with the energy conversion devices, form the components of the total energy supply systems discussed in Chapter 6.

Chapter 6 discusses modeling the performance of individual renewable energy devices as well as whole systems and, finally, scenarios for the global use of renewable energy, with consideration of both spatial and temporal constraints in matching demand and supply.

In Chapter 7, renewable energy resources are first placed in the framework of current economic thinking, as a preliminary to quantifying some of the considerations that should go into constructing a viable energy supply system. Next, the chapter presents a survey of indirect economic factors to be considered, which leads to the description of the methodology of life-cycle analysis, which, together with the scenario technique, constitutes the means for an up-to-date economic analysis. Finally, concrete examples of systems assessment are given.

Chapter 8 concludes with a general discussion of sustainability, climate, and the position of renewable energy technologies in a world dominated by outdated environmental and economic behavior.

Chapter 2

Origin of Renewable Energy Flows

In this chapter, renewable energy is followed from the sources where it is created—notably the Sun—to the Earth, where it is converted into different forms, e.g., solar radiation to wind or wave energy, and distributed over the Earth–atmosphere system through a number of complex processes. Essential for these processes are the mechanisms for general circulation in the atmosphere and the oceans. The same mechanisms play a role in distributing pollutants released to the environment, whether from energy-related activities like burning fossil fuels or from other human activities. Because the assessment of environmental impacts plays an essential part in motivating societies to introduce renewable energy, the human interference with climate is also dealt with in this chapter, where it fits naturally.

2.1 SOLAR RADIATION

At present the Sun radiates energy at the rate of 3.9×10^{26} W. At the top of the Earth's atmosphere an average power of 1353 W m^{-2} is passing through a plane perpendicular to the direction of the Sun. As shown in Fig. 2.1, regular

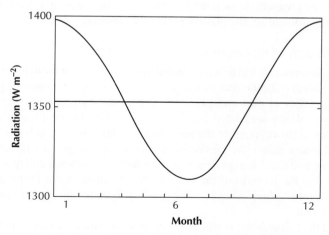

FIGURE 2.1 Yearly variations in the "solar constant" (the head-on solar radiation flux at the Earth's distance; NASA, 1971).

Renewable Energy, Fourth Edition

oscillations around this figure are produced by the changes in the Earth–Sun distance, as the Earth progresses in its elliptical orbit around the Sun. The average distance is 1.5×10^{11} m and the variation is $\pm 1.7\%$. Further variation in the amount of solar radiation received at the top of the atmosphere is caused by slight irregularities in the solar surface, in combination with the Sun's rotation (about one revolution per month), and by possible time variations in the surface luminosity of the Sun.

2.1.1 Energy Production in Main-sequence Stars Like the Sun

The energy produced by nuclear reactions in the interior of the Sun must equal the amount of energy radiated from the surface, since otherwise the Sun could not have been structurally stable over long periods of time. Evidence for the stability of the Sun comes from several sources. Stability over a period of nearly 3×10^9 years is implied by the relative stability of the temperature at the Earth's surface (oxidized sediments and fossil remains indicate that water in its fluid phase has been present throughout such periods). Stability over an even longer time is implicit in our understanding of the evolution of the Sun and other similar stars. As an indication of this stability, Fig. 2.2 shows the variations in the radius of the Sun believed to have taken place since its assumed formation from clouds of dust and gas.

The conversion of energy contained in the atomic constituents of main-sequence stars like the Sun, from heat of nuclear reactions (which transform hydrogen into helium) to radiation escaping from the surface, is largely understood. The basis for regarding such radiation as a renewable source is that it may continue essentially unaltered for billions of years. Yet there is also a possibility of tiny variations in solar energy production that may have profound implications for life on the planets encircling the Sun.

The Birth and Main-sequence Stage of a Star

Stars are believed to have been created from particulate matter present in a forming galaxy. Contraction of matter is achieved by gravitational processes involving clouds of dust and gas, but counteracted by the increasing pressure of the gases as they are heated during contraction (cf. the equilibrium equations given below). However, once the temperature is high enough to break up molecules and ionize atoms, the gravitational contraction can go on without increase of pressure, and the star experiences a "free fall" gravitational collapse. If this is considered to be the birth of the star, the development in the following period consists of establishing a temperature gradient between the center and the surface of the star.

While the temperature is still relatively low, transport of energy between the interior and the surface presumably takes place mainly through convection of

FIGURE 2.2 Variation in the solar radius as a function of time (*bottom*) together with selected milestones in the development on Earth, associated with the build-up of oxygen in the Earth's atmosphere (*top*). The rapid development of phytoplankton in the upper layers of the oceans at a relative oxygen concentration of 10^{-2} is associated with the formation of an ozone shield in the atmosphere, cutting down the ultraviolet part of the solar spectrum. When the oxygen level has reached 10^{-1}, the atmospheric ultraviolet absorption is strong enough to allow life on land. (Based on Herbig, 1967; Berkner and Marshall, 1970; Cloud and Gibor, 1970.)

hot gases (Hayashi, 1966). This leads to a fairly constant surface temperature, while the temperature in the interior rises as the star further contracts, until a temperature of about 10^7 K is reached. At this point the nuclear fusion processes start, based on a thermonuclear reaction in which the net result is an energy generation of about 25 MeV (1 MeV = 1.6×10^{-13} J) for each four hydrogen nuclei (protons) undergoing the chain of transformations resulting in the formation of a helium nucleus, two positrons and two neutrinos,

$$4\,^1_1\mathrm{H} \rightarrow {}^4_2\mathrm{He} + 2e^+ + 2\nu + 26.2\,\mathrm{MeV}.$$

After a while, the temperature and pressure gradients have built up in such a way that the gravitational forces are balanced at every point and the gravitational contraction has stopped. The time needed to reach equilibrium depends on the total mass. For the Sun it is believed to have been about 5×10^7 years (Herbig, 1967). Since then, the equilibrium has been practically stable, exhibiting smooth changes due to hydrogen burning, such as the one shown in Fig. 2.2 for the radius.

Nuclear Reactions in the Sun

Most of the nuclear reactions responsible for energy production in the Sun involve two colliding particles and two, possibly three, "ejectiles" (outgoing particles). For example, one may write

$$A + a \rightarrow B + b + Q, \qquad \text{or} \quad A(a, b)B, Q,$$
$$A + a \rightarrow B + b + c + Q, \qquad \text{or} \quad A(a, bc)B, Q,$$

where Q is the energy release measured in the center-of-mass coordinate system (called the "Q-value" of the reaction). Energy carried by neutrinos will not be included in Q. The main processes in the solar energy cycle are

$$^1_1H\,(^1_1H, e^+\nu)\,^2_1H, \qquad Q = 1.0 \text{ to } 1.4\,\text{MeV}, \qquad Q_{av} = 1.2\,\text{MeV},$$

$$^2_1H\,(^1_1H, \gamma)\,^3_2He, \qquad Q = 5.5\,\text{MeV}.$$

It is highly probable (estimated at 91% for the present Sun) that the next step in the hydrogen-burning process is
    Branch 1:

$$^3_2He\,(^3_2He, 2\,^1_1H)\,^4_2He, \quad Q = 12.9\,\text{MeV}.$$

This process requires the two initial ones given above to have taken place twice. The average total energy liberated is 26.2 MeV. In an estimated 9% of cases (Bahcall, 1969), the last step is replaced by processes involving ^4He already present or formed by the processes of branch 1,

$$^3_2He\,(^4_2He, \gamma)\,^7_4Be, \quad Q = 1.6\,\text{MeV}.$$

The following steps may be either
    Branch 2:

$$^7_4Be\,(e^-, \nu)\,^7_3Li, \qquad Q = 0 \text{ or } 0.5\,\text{MeV}, \qquad Q_{av} = 0.1\,\text{MeV}$$

$$^7_3Li\,(^1_1H, ^4_2He)\,^4_2He, \qquad Q = 17.4\,\text{MeV},$$

or, with an estimated frequency among all the hydrogen-burning processes of 0.1%,
    Branch 3:

$$^7_4Be\,(^1_1H, \gamma)\,^8_5B, \quad Q = 0.1\,\text{MeV},$$

followed by the decay processes

$$^8_5B \rightarrow \,^8_4Be + e^+ + \nu, \quad Q = 4 \text{ to } 18 \text{ MeV}, \quad Q_{av} = 10.8 \text{ MeV},$$
$$^8_4Be \rightarrow 2\,^4_2He \quad\quad Q = 0.1 \text{ MeV}.$$

The average Q-values for processes involving neutrino emission are quoted from Reeves (1965). The initial pp-reaction (^1_1H = hydrogen nucleus = p = proton) receives competition from the three-body reaction

$$^1_1H + e^- + \,^1_1H \rightarrow \,^2_1H + \nu.$$

However, under solar conditions, only 1 in 400 reactions of the pp-type proceeds in this way (Bahcall, 1969).

In order to calculate the rates of energy production in the Sun, the cross-sections of each of the reactions involved must be known. The cross-section is the probability for the reaction to occur under given initial conditions, such as a specification of the velocities or momenta.

Among the above-mentioned reactions, those that do not involve electrons or positrons are governed by the characteristics of collisions between atomic nuclei. These characteristics are the action of a Coulomb repulsion of long range, as the nuclei approach each other, and a more complicated nuclear interaction of short range, if the nuclei succeed in penetrating their mutual Coulomb barrier. The effective height of the Coulomb barrier increases if the angular momentum of the relative motion is not zero, according to the quantum mechanical description. The height of the Coulomb barrier for zero relative (orbital) angular momentum may be estimated as

$$V_{coul} \approx Z_1 Z_2 e^2 / R(Z_2,\, A_2) \gtrsim 1 \text{ MeV},$$

where the radius of one of the nuclei has been inserted in the denominator (A_i is the mass of the nucleus in units of the nucleon mass; Z_i is the nuclear charge in units of the elementary charge e, and $e^2/4\pi\varepsilon_0 = 1.44 \times 10^{-15}$ MeV m). Thus, the thermal energy even in the center of the Sun, $T(0) = 1.5 \times 10^7$ K,

$$\sigma(E) = S(E) \exp(-2\pi\eta(E))\,/\,E,$$

is far below the height of the Coulomb barrier (1 MeV = 1.6×10^{-13} J), and the nuclear reaction rate will depend strongly on the barrier penetration, i.e., the quantum tunneling through the barrier. If the energy of the approaching particle does not correspond to a nuclear resonance (see Fig. 2.3), then the part of the cross-section due to barrier penetration (and this will contain most of the energy dependence) can be estimated by a simple quantum approximation, called the WKB method (Merzbacker, 1970). In the absence of relative angular momentum, the nuclear reaction cross-section may then be written, as a function of energy,

$$\sigma(E) = E^{-1}S(E) \exp(-2\pi\eta(E)),$$

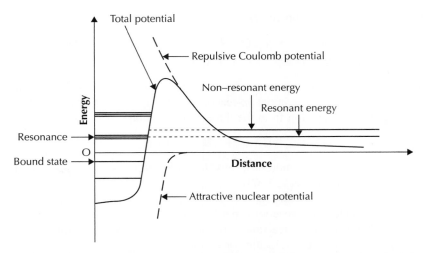

FIGURE 2.3 The nuclear potential seen by an impinging particle, as a function of relative distance. A few nuclear states are indicated to the left of this highly schematic picture, in order to illustrate the concepts of resonant and nonresonant energies. The incident particle kinetic energy at solar temperatures is much lower than the height of the Coulomb barrier, so only quantum tunneling can lead to a nuclear reaction.

where $S(E)$ is a slowly varying function of E, depending on the nuclear interactions, while the exponential function expresses the barrier penetration factor in terms of the Coulomb parameter

$$\eta(E) = 2\pi Z_1 Z_2 e^2 h^{-1} (\mu/2E)^{1/2},$$

with the reduced mass being

$$\mu = A_1 A_2 M_p / (A_1 + A_2).$$

The reactions involving electrons are governed not by nuclear (so-called "strong") interactions, but by the "weak" interaction type, which is available to all kinds of particles. The smallness of such interactions permits the use of a perturbational treatment, for which the cross-section may be written (Merzbacker, 1970)

$$v\sigma = (2\pi)^2 \, h^{-1} \, |<H>|^2 \rho(\text{final}).$$

Here $v\sigma$ is the transition probability, i.e., the probability that a reaction will occur, per unit time and volume. If the formation of the positron–neutrino, electron–antineutrino or generally lepton–antilepton pair is not the result of a decay but of a collision process, then the transition probability may be split into a reaction cross-section σ times the relative velocity, v, of the colliding particles [for example, the two protons in the $^1H(^1H, e^+v)\,^2H$ reaction]. ρ (final) is the density of final states, defined as the integral over the distribution of momenta of the outgoing particles. When there are three particles in the final state, this will constitute a function of an undetermined energy, e.g., the neutrino energy. Finally, $<H>$ is a matrix element, i.e., an integral over the quantum wave functions of all the particles in the initial and final states, with the interaction operator in the middle and usually approximated by a constant for weak interaction processes.

The reaction rate for a reaction involving two particles in the initial state is the product of the numbers n_1 and n_2 of particles 1 and 2, respectively, each per unit volume, times the probability that two such particles will interact to form the specified final products. This last factor is just the product of the relative velocity and the reaction cross-section. Since the relative velocity of particles somewhere in the Sun is given by the statistical distribution corresponding to the temperature, the local reaction probability may be obtained by averaging over the velocity distribution. For a non-relativistic gas, the relative kinetic energy is related to the velocity by $E = 1/2\mu v^2$, and the velocity distribution may be replaced by the Maxwell–Bolzmann distribution law for kinetic energies (which further assumes that the gas is nondegenerate, i.e., that kT is not small compared with the thermodynamic chemical potential, a condition that is fulfilled for the electrons and nuclei present in the Sun),

$$<\sigma v> = (8/\pi\mu)^{1/2}(kT)^{-3/2} \int \sigma(E)E \exp(-E/kT)\mathrm{d}E.$$

The reaction rate (number of reactions per unit time and unit volume) becomes

$$P = n_1 n_2 <\sigma v>(1 + \delta_{12})^{-1},$$

where δ_{12} (= 1 if the suffixes 1 and 2 are identical, otherwise 0) prevents double counting in case projectile and "target" particles are identical.

The number densities n (number per unit volume) may be re-expressed in terms of the local density $\rho(r)$ and the mass fraction of the ith component, X (e.g., the hydrogen fraction X considered earlier),

$$n_i = X_i \rho(r)/(A_i M_p).$$

The central part of the reaction rate, $<\sigma v>$, is shown in Fig. 2.4 for the branch 1 reactions of the pp-cycle, as well as for the ^{3}He + ^{4}He reaction opening the alternative branches. The basic cross-sections $S(0)$ used in the calculations are quoted in the work of Barnes (1971). The barrier penetration effects are calculated in a modified form, taking account of the partial screening of the nuclear Coulomb field by the presence of electrons. Such penetration factors are included for the weak interaction processes as well as for the reactions caused by nuclear forces.

One notes the smallness of the initial, weak interaction reaction rate, the temperature dependence of which determines the overall reaction rate above

$T = 3 \times 10^6$ K. The four to five orders of magnitude difference in the rates of the ^{4}He + ^{3}He and the ^{3}He + ^{3}He processes account for the observed ratio of branch 2 (and 3) to branch 1, when the oppositely acting ratio of the ^{4}He to ^{3}He number densities is also included.

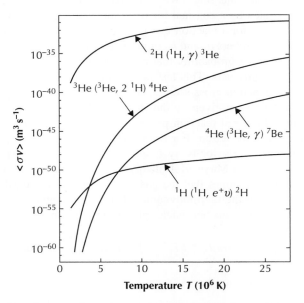

FIGURE 2.4   Reaction rates for various nuclear reactions of the *pp*-chain, as functions of temperature, and assuming that there is just one of each of the reacting particles per unit volume (denoted $<\sigma v>$). (Based on Barnes, 1971.)

The rate of energy production to be entered in equation (2.7) below for thermal equilibrium is obtained from the reaction rate by multiplying with the energy release, i.e., the Q-value. Since the energy production rate appearing in (2.7) is per unit mass and not volume, a further division by ρ is necessary,

$$\varepsilon(\rho, T, X_i) = \sum_j \varepsilon_j(\rho, T, X_i) = \sum_j n_1 n_2 <\sigma v> \frac{Q(j)}{\rho(1 + \delta_{12})}. \qquad (2.1)$$

The sum over j extends over all the different branches; $Q(j)$ is the energy released in branch j; and the indices 1 and 2, for which n_1 and n_2 are taken and $<\sigma v>$ evaluated, may be any isolated reaction in the chain, since the equilibrium assumption implies that the number densities have adjusted themselves to the differences in reaction cross-section. For decay processes, the reaction rate has to be replaced by the decay probability, in the manner described above.

FIGURE 2.5 Rate of energy production by the *pp*- and CNO-chains, as a function of temperature, for a choice of composition (relative abundance of the constituents) typical for main-sequence stars like the Sun. (Based on Iben, 1972.)

Figure 2.5 shows the temperature dependence of the energy production rate for the *pp*-chain described above, as well as for the competing CNO-chain, which may come into play at temperatures slightly higher than those prevailing in the Sun's interior. The curves should be regarded as typical for stars of composition similar to that of the Sun, but as (2.1) indicates, the energy production depends on the density as well as on the abundance of the reacting substances. Hence a change in composition or in central density will change the rate of energy production. For the *pp*-chain, Fig. 2.5 assumes a hydrogen fraction $X = 0.5$ and a density of 10^5 kg m^{-3}, believed to be reasonable for the center of the Sun. For the CNO-chain, additional assumptions have been made that the carbon, nitrogen, and oxygen abundances are $X_C = 0.003$, $X_N = 0.001$, and $X_O = 0.012$ (Iben, 1972). These figures apply to stellar populations similar to, but more advanced than, the Sun, whereas much smaller CNO-abundances and production rates are found for the so-called population-II stars. It should also be mentioned that, owing to the decrease in density and change in composition as one goes away from the center of the star, the average energy production for the energy-producing region will be lower than the value deduced from the conditions near the center.

Equilibrium Processes in the Sun

The present near-equilibrium state of the Sun may be described by a number of physical conditions. The first one is the condition of hydrostatic equilibrium, which states that the pressure gradient at every place must exactly balance the gravitational force. Assuming the Sun to be spherically symmetric, the condition may be written

$$dP/dr = -G M(r) \rho(r)/r^2, \tag{2.2}$$

where $M(r)$ is the integrated mass within a sphere of radius r, $\rho(r)$ is the local density, and $G = 6.67 \times 10^{-11}$ m^3 kg^{-1} s^{-2} is the constant of gravitation.

The pressure $P(r)$ may be related to the local temperature $T(r)$ by an equation of state. The conditions everywhere in the Sun are close to those of an ideal gas, implying

$$P(r) = n(r) k T(r),$$

where $n(r)$ is the number of free particles per unit volume and $k = 1.38 \times 10^{-23}$ J K^{-1} is the Boltzmann constant. Alternatively, one may express n in terms of the gas constant, $\mathcal{R} = 8.315$ J K^{-1} mol^{-1}, the density, and μ, the mean molecular weight (i.e., the mass per mole of free particles),

$$n(r) = \mathcal{R}\rho(r)/(k\mu).$$

For stars other than the Sun the ideal gas law may not be valid, as a result of electrostatic interactions among the charges (depending on the degree of ionization, i.e., the number of free electrons and other "unsaturated" charges) and radiation pressure.

Cox and Giuli (1968) estimate the maximum pressure corrections to be 1.5% from the electrostatic interactions and 0.3% from the radiation pressure, calculated from the blackbody radiation in the Sun's center. The mean molecular weight may be obtained from assuming the Sun to consist of hydrogen and helium in mass fractions X and $(1 - X)$. If all hydrogen and helium is fully ionized, the number of free particles is $2X + 3(1 - X)/4$ and the mass per mole of free particles is

$$\mu = 4N_A M_p/(5X + 3),$$

where $N_A = 6 \times 10^{23}$ is Avogadro's number (the number of particles per mole) and 1.67×10^{-27} kg is the proton mass. The Sun's hydrogen fraction X is about 0.7.

The potential energy of a unit volume at the distance r from the Sun's center, under influence of the gravitational forces, is

$$\varepsilon_{pot}(r) = -G M(r) \rho(r)/r,$$

while the kinetic energy of the motion in such a unit volume, according to statistical mechanics, is

$$\varepsilon_{kin}(r) = 1.5 n(r) k T(r).$$

Since ε_{pot} is a homologous function of order -1, the virial theorem of classical mechanics (see, for example, Landau and Lifshitz, 1960) states that for the system as a whole,

$$2\,\overline{\varepsilon}_{kin} = -\overline{\varepsilon}_{pot}.$$

The bars refer to time averages, and it is assumed that the motion takes place in a finite region of space. Inserting the expressions found above, using $R = k\,N_A$ and integrating over the entire Sun and approximating $\int(M(r)/r)\mathrm{d}M(r)$ by M/R (M and R being the total mass and radius), one finds an expression for the average temperature,

$$T = 4GMM_p/(3(5X+3)kR).$$

An estimate for the average temperature of the Sun may be obtained by inserting the solar radius $R = 7 \times 10^8$ m and the solar mass $M = 2 \times 10^{30}$ kg, leading to

$$T = 5 \times 10^6 \text{ K}.$$

The rigorousness of the hydrostatic equilibrium may be demonstrated by assuming a 1% deviation from equilibrium, implying an inward acceleration

$$\mathrm{d}^2r/\mathrm{d}t^2 = \rho(r)^{-1}(GM(r)\rho(r)/r^2 - \mathrm{d}P/\mathrm{d}r) = -0.01\,GM(r)/r^2.$$

The time needed for a 10% contraction $(\Delta r/R = -0.1)$ under the influence of this acceleration is

$$\Delta t = 7200\,\text{s} = 2\,\text{h},$$

at the solar surface $R = 7 \times 10^8$ m. The observed stability of the solar radius (see Fig. 2.2) demands a very strict fulfilment of the hydrostatic equilibrium condition throughout the Sun.

The Energy Transport Equations

In order to derive the radial variation of the density and the temperature, energy transport processes have to be considered. Energy is generated by the nuclear processes in the interior. The part of the energy from nuclear processes that is carried by neutrinos would escape from a star of the Sun's assumed composition. One may express this in terms of the mean free path of neutrinos in solar matter (the mean free path being large compared with the Sun's radius) or in terms of the mass absorption coefficient,

$$\kappa = \frac{-\mathrm{d}I}{\rho\,I\mathrm{d}s} = \frac{1}{\rho}\sum_i \sigma_i,$$

with this quantity being very small for neutrinos in the Sun. I denotes the energy flux per unit solid angle in the direction $\mathrm{d}s$, and s denotes the path length in the matter. The alternative way of writing the mass absorption coefficient involves the cross-sections σ_i per volume of matter of all the absorption or other processes that may attenuate the energy flux.

The energy not produced in association with neutrinos is in the form of electromagnetic radiation (γ-rays) or kinetic energy of the particles involved in the nuclear reactions. Most of the kinetic energy is carried by electrons or positrons. The positrons annihilate with electrons and release the energy as electromagnetic radiation, and also part of the electron kinetic energy is gradually transformed into radiation by a number of processes. However, the electromagnetic radiation also reacts with the matter, notably the electrons, and thereby converts part of the energy to kinetic energy again. If the distribution of particle velocities locally corresponds to that predicted by statistical mechanics (e.g., the Maxwell–Boltzmann distribution), one says that the system is in local thermodynamic or statistical equilibrium.

Assuming that thermodynamic equilibrium exists locally, at least when averaged over suitable time intervals, the temperature is well defined, and the radiation field must satisfy the Planck law,

$$\frac{dF}{d\nu} = \frac{2h\nu^3}{c^2}\frac{1}{e^{h\nu/kT}-1}. \tag{2.3}$$

Here F is the power radiated per unit area and into a unit of solid angle, ν is the frequency of radiation, $h = 6.6 \times 10^{-34}$ J s is Planck's constant, and $c = 3 \times 10^8$ m s^{-1} is the velocity of electromagnetic radiation (light) in vacuum.

The appropriateness of the assumption of local thermodynamic equilibrium throughout the Sun's interior is ensured by the shortness of the mean free path of photons present in the Sun, as compared to the Sun's radius (numerically the ratio is about 10^{-13}, somewhat depending on the frequency of the photons). The mean free path, l, is related to the mass absorption coefficient by

$$l\,(\text{photons}) = 1/\rho\kappa.$$

This relation as well as the relation between l and the cross-sections of individual processes may be taken to apply to a specific frequency ν. κ is sometimes referred to as the opacity of the solar matter.

The radiative transfer process may now be described by an equation of transfer, stating that the change in energy flux along an infinitesimal path in the direction of the radiation equals the difference between emission and absorption of radiation along that path,

$$dI(\nu)/(\rho\,ds) = e(\nu) - \kappa(\nu)\,I(\nu).$$

Integrated over frequencies and solid angles, this difference equals, in conditions of energy balance, the net rate of local energy production per unit mass. In complete thermodynamic equilibrium one would have Kirchhoff's law, $e(\nu) = \kappa(\nu)\,I(\nu)$ (see, for example, Cox and Giuli, 1968), but under local thermodynamic equilibrium, $e(\nu)$ contains both true emission and also radiation scattered into the region considered. According to Cox and Giuli (1968), neglecting refraction in the medium,

$$e(v) = \kappa_t(v)dF/dv + \kappa_s(v)I_{av}(v), \quad \kappa_t(v) + \kappa_s(v) = \kappa(v).$$

Here κ_t and κ_s are the true absorption and scattering parts of the mass absorption coefficient, and I_{av} is I averaged over solid angles. Cox and Giuli solve the equation of transfer by expressing $I(v)$ as a Taylor series expansion of $e(v)$ around a given optical depth (a parameter related to κ and to the distance r from the Sun's center or a suitable path length s). Retaining only the first terms, one obtains, in regions of the Sun where all energy transport is radiative,

$$I(v) = \frac{dF(v)}{dv} - \frac{1}{\kappa(v)\rho} \frac{d}{ds} \frac{dF(v)}{dv} \cos\theta.$$

θ is the angle between the direction of the energy flux $I(v)$ and ds. The first term dominates in the central region of the Sun, but gives no net energy transport through a unit area. If the path s is taken as the radial distance from the center, the net outward flux obtained by integrating $I(v)\cos\theta$ over all solid angles becomes

$$\int I(v)\cos\theta \, 4\pi \, d\cos\theta = -\frac{4\pi}{3\kappa(v)\rho} \frac{d}{dr} \frac{dF(v)}{dv}.$$

The luminosity $L(r)$ of a shell of radius r is defined as the net rate of energy flow outward through a sphere of radius r, i.e., $4\pi r^2$ times the net outward flux. Introducing at the same time the temperature gradient, one obtains

$$L_v(r) = -\frac{(4\pi r)^2}{3\kappa(v)\rho} \left(\frac{\partial}{\partial T} \frac{dF(v)}{dv} \right) \frac{dT(r)}{dr}$$

for the spectral luminosity and

$$L(r) = -\frac{(4\pi r)^2 \, 4\sigma T^3}{3 \, \pi\kappa\rho} \frac{dT(r)}{dr} \tag{2.4}$$

for the total luminosity.

Here $\sigma = 5.7 \times 10^{-8}$ W m^{-2} K^{-4} is Stefan's constant in Stefan–Boltzmann's law for the total power radiated into the half-sphere, per unit area,

$$\int\int \frac{dF(v)}{dv} \cos\theta \, 2\pi \, d\cos\theta \, dv = \sigma T^4. \tag{2.5}$$

From (2.4) it may be expected that dT/dr will be large in the central regions of the Sun, as well as near the surface, where the opacity κ is large. If dT/dr becomes sufficiently large, the energy cannot be transported away by radiation alone, and convective processes can be expected to play a role. A proper description of turbulent convection will require following all the eddies formed according to the general hydrodynamic equations. In a simple, approximate treatment, convective transport is described in a form analogous to the first-order [in terms of $l(\text{photons}) = (\kappa\rho)^{-1}$] solution (2.4) to the equation of radiative transfer. The role of the mean free path for photons is played by a mixing length $l(\text{mix})$ (the mean distance that a "hot bubble" travels before it has given off its surplus heat and has come into thermal equilibrium with its local surroundings). The notion of "hot bubbles" is thus introduced as the mechanism for convective energy (heat) transfer, and the "excess heat" is to be understood as the amount of heat that cannot be transferred by adiabatic processes. The excess energy per unit volume that is given away by a "hot bubble" traversing one mixing length is

$$\rho\, c_P \left(\frac{\partial T}{\partial s} - \left(\frac{\partial T}{\partial s}\right)_{adiabatic} \right) l(\text{mix}),$$

where c_P is the specific heat at constant pressure. The energy flux is obtained by multiplying by an average bubble speed v_{conv} (see, for example, Iben, 1972):

$$F_{conv} \approx \rho\, c_P \left| \frac{\partial T}{\partial s} - \left(\frac{\partial T}{\partial s}\right)_{adiabatic} \right| l(\text{mix})\, v_{conv}.$$

In this expression also the derivatives of the temperature are to be understood as average values. The similarity with expression (2.4) for the radiative energy flux $F_{rad} = L(r)/(4\pi r^2)$ is borne out if (2.4) is rewritten slightly,

$$F_{rad} = - \frac{\partial P_{rad}}{\partial T} \frac{\partial T}{\partial r}\, l(\text{photons})\, c,$$

where the radiation pressure P_{rad} equals one-third the energy density of black-body radiation,

$$P_{rad} = \frac{1}{3} a T^4,$$

and where use has been made of the relation $4\sigma = ac$ between Stefan's constant σ and the radiation constant a to introduce the photon velocity c (velocity of light) as an analogy to the average speed of convective motion, v_{conv}.

The radial bubble speed v_{conv} may be estimated from the kinetic energy associated with the buoyant motion of the bubble. The average density deficiency is of the order

$$\Delta\rho = \left(\frac{\partial\rho}{\partial r} - \left(\frac{\partial\rho}{\partial r}\right)_{adiabatic} \right) l(\text{mix}),$$

and the average kinetic energy is of the order

$$\tfrac{1}{2}\, \rho\, (v_{conv})^2 = \frac{1}{2} \frac{GM(r)}{r^2} \Delta\rho\, l(\text{mix})$$

(taken as half the maximum force, since the force decreases from its maximum value to zero, times the distance traveled). When p is inserted from the equation of state, one obtains the estimate for the bubble speed,

$$v_{conv} = \frac{\rho\, c_P}{r} \left(\frac{GM(r)}{T} \left| \frac{\partial T}{\partial r} - \left(\frac{\partial T}{\partial r}\right)_{adiabatic} \right| \right)^{1/2}.$$

The convective energy flux can now be written

$$F_{conv} \approx \frac{\rho\, c_P}{r} \left(\frac{GM(r)}{T}\right)^{1/2} \left|\frac{\partial T}{\partial r} - \left(\frac{\partial T}{\partial r}\right)_{adiabatic}\right|^{3/2} (l(\mathrm{mix}))^2. \qquad (2.6)$$

For an ideal gas, the adiabatic temperature gradient is

$$\left(\frac{\partial T}{\partial r}\right)_{adiabatic} = \frac{c_P - c_V}{c_P} \frac{T}{P} \frac{dP}{dr},$$

where c_V is the specific heat at fixed volume. For a monatomic gas (complete ionization), $(c_P - c_V)/c_P = 2/5$.

The final equilibrium condition to be considered is that of energy balance, at thermal equilibrium,

$$dL(r)/dr = 4\pi r^2 \rho(r)\varepsilon(\rho, T, X, ...), \qquad (2.7)$$

stating that the radiative energy loss must be made up for by an energy source term, ε, which describes the energy production as function of density, temperature, composition (e.g., hydrogen abundance X), etc. Since the energy production processes in the Sun are associated with nuclear reactions, notably the thermonuclear fusion of hydrogen nuclei into helium, then the evaluation of ε requires a detailed knowledge of cross-sections for the relevant nuclear reactions.

Before touching on the nuclear reaction rates, it should be mentioned that the condition of thermal equilibrium is not nearly as critical as that of hydrostatic equilibrium. Should the nuclear processes suddenly cease in the Sun's interior, the luminosity would, for some time, remain unchanged at the value implied by the temperature gradient (2.4). The stores of thermal and gravitational energy would ensure an apparent stability during part of the time (called the "Kelvin time") needed for gravitational collapse. This time may be estimated from the virial theorem, since the energy available for radiation (when there is no new energy generation) is the difference between the total gravitational energy and the kinetic energy bound in the thermal motion (i.e., the temperature increase associated with contraction),

$$t_{Kelvin} = (\varepsilon_{gravit} - \bar{\varepsilon}_{kin})/L(R) = -(\varepsilon_{pot} + \bar{\varepsilon}_{kin})/L(R) = \bar{\varepsilon}_{kin}/L(R) = 3 \times 10^7 \text{ y}.$$

On the other hand, this time span is short enough to prove that energy production takes place in the Sun's interior.

A Model of the Solar Processes

In order to determine the radial distribution of energy production, mass, and temperature, the equilibrium equations must be integrated. Using the equation of state to eliminate $\rho(r)$ [expressing it as a function of $P(r)$ and $T(r)$], the basic equations may be taken as (2.2), (2.4) in regions of radiative transport, (2.6) in regions of convective transport, and (2.7), plus the equation defining the mass inside a shell,

$$dM(r)/dr = 4\pi r^2 \rho(r) = 4\pi r^2 \mu\, P(r)/(\mathscr{R}T(r)). \qquad (2.8)$$

In addition to the equation of state, two other auxiliary equations are needed, namely, (2.1) in order to specify the energy source term and a corresponding equation expressing the opacity as a function of the state and composition of the gas. The solution to the coupled set of differential equations is usually obtained by smooth matching of two solutions, one obtained by integrating outward with the boundary conditions $M(0) = L(0) = 0$ and the other one obtained by inward integration and adopting a suitable boundary condition at $r = R$. At each step of the integration, the temperature gradient must be compared to the adiabatic gradient in order to decide which of the equations, (2.6) or (2.7), to use.

Figure 2.6 shows the results of a classical calculation of this type (Schwarzschild, 1958). The top curve shows the prediction at a convective

FIGURE 2.6 Radial model of the Sun. The *solid curves* represent (from top to bottom) temperature gradient relative to adiabatic, temperature, density, hydrogen abundance, and relative luminosity. The *dashed curves* represent opacity, pressure, and relative mass fraction inside a given radius. The opacity has not been calculated in the outer convective region. (Based on Schwarzschild, 1958.)

region $0.85R < r < R$. The temperature gradient also rises near the center, but not sufficiently to make the energy transport in the core convective. This may happen for slight changes in parameters, such as the initial abundances, without impinging on the observable quantities, such as luminosity and apparent surface temperature. It is believed that the Sun does have a small, convective region near the center (Iben, 1972). The bottom curves in Fig. 2.6 show that the energy production takes place in the core region, out to a radius of about $0.25R$, and that roughly half the mass is confined in the energy-producing region. Outside this region the luminosity is constant, indicating that energy is only being transported, not generated.

One significant feature of the time-dependent models of solar evolution is the presence of a substantial amount of helium (in the neighborhood of 25%) before the onset of the hydrogen-burning process. At present, about half the mass near the center is in the form of helium (cf. Fig. 2.6). The initial abundance of helium, about 25%, is in agreement with observations for other stars believed to have just entered the hydrogen-burning stage. There is reasonable evidence to suggest that in other galaxies, too, in fact everywhere in the Universe, the helium content is above 25%, with the exception of a few very old stars, which perhaps expelled matter before reaching their peculiar composition.

2.1.2 Spectral Composition of Solar Radiation

Practically all of the radiation from the Sun received at the Earth originates in the photosphere, a thin layer surrounding the convective mantle of high opacity. The depth to which a terrestrial observer can see the Sun lies in the photosphere. Owing to the longer path-length in the absorptive region, the apparent brightness of the Sun decreases toward the edges. The photosphere consists of atoms of varying degree of ionization, plus free electrons. A large number of scattering processes take place, leading to a spectrum similar to the Planck radiation (2.3) for a blackbody in equilibrium with a temperature $T \approx 6000$ K. However, this is not quite so, partly because of sharp absorption lines corresponding to the transitions between different electron configurations in the atoms present (absorption lines of over 60 elements have been identified in the solar spectrum), and partly because of the temperature variation through the photosphere, from around 8000 K near the convective zone to a minimum of 4300 K at the transition to the chromosphere (see Fig. 2.7). Yet the overall picture, shown in Fig. 2.8, is in fair agreement with the Planck law for an assumed effective temperature $T_{eff} \approx 5762$ K, disregarding in this figure the narrow absorption lines in the spectrum.

The Structure of the Solar Surface

The turbulent motion of the convective layer underneath the Sun's surface manifests itself in the solar disc as a granular structure interpreted as columns of hot, vertical updrafts and cooler, downward motions between the grain-like structures (supported by observations of Doppler shifts). Other irregularities of solar luminosity include bright flares of short duration, as well as sunspots, regions near the bottom of the photosphere with lower temperature, appearing and disappearing in a matter of days or weeks in an irregular manner, but statistically periodic (with an 11-year period). Sunspots first appear at latitudes at or slightly above 30°, reach maximum activity near 15° latitude, and end the cycle near 8° latitude. The "spot" is characterized by churning motion and a strong magnetic flux density (0.01–0.4 Wb m^{-2}), suggesting its origin from vorticity waves traveling within the convective layer, and the observation of reversed magnetic polarity for each subsequent 11-year period suggests a true period of 22 years.

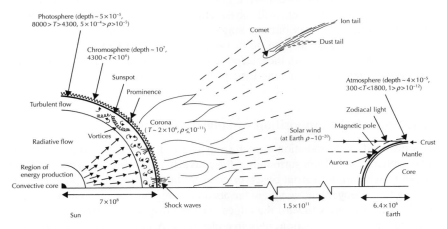

FIGURE 2.7   Schematic picture of solar layers, starting from the center of the Sun at the left. The solar radius is defined by the bottom of the visible Sun. All distances are in meters, all temperatures T are in K, and all densities ρ are in kg m^{-3}. The solar corona continues into an accelerating stream of particles, the solar wind. At the Earth's distance from the Sun (center to the right, note the two changes of scale), the solar wind gives rise to magnetic induction and aurorae, and the extension of the corona is seen near the horizon as zodiacal light, in the absence of direct or scattered light from the solar photosphere. The solar wind is also responsible for comet tails' being directed away from the Sun (whereas radiation pressure can move only the lightest material in the tail). The tails of comets usually have an ion part and a dust part, with the latter moving more slowly and being deflected as a result of the Sun's rotation (a period of around 25 days). The inner part of the Earth is only sketched. The mantle is believed to consist of an outer part (silicates of Mg and Fe) and an inner part (oxides of Mg and Fe). Similarly, the core has an outer part (probably liquid FeS) and an inner part (a solid iron–nickel alloy).

Above the photosphere is a less dense gas with temperature increasing outward from the minimum of about 4300 K. During eclipses, the glow of this chromosphere is visible as red light. This is because of the intense H_α line in the chromospheric system, consisting primarily of emission lines.

After the chromosphere comes the corona, still less dense (on the order of 10^{-11} kg m^{-3} even close to the Sun), but of very high temperature (2×10^6 K, cf. Fig. 2.7). The mechanism of heat transfer to the chromosphere and to the corona is believed to be shock waves originating in the turbulent layer (Pasachoff, 1973). The composition of the corona (and chromosphere) is believed to be similar to that of the photosphere, but owing to the high temperature in the corona, the degree of ionization is much higher, and, for example, the emission line of Fe^{13+} is among the strongest observed from the corona (during eclipses). A continuous spectrum (K-corona) and Fraunhofer absorption lines (F-corona) are also associated with the corona, although the total intensity is only 10^{-6} of that of the photosphere, even close to the Sun (thus the corona cannot be seen from the Earth's surface except during eclipses, owing to the atmospheric scattering of photospheric light). Because of its low density, no continuous radiation is produced in the corona itself, and the K-corona spectrum is due to scattered light from the photosphere, where the absorption lines have been washed out by random Doppler shifts (as a result of the high kinetic energy).

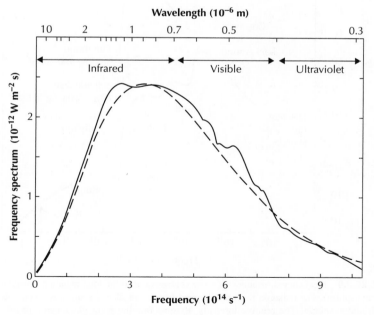

FIGURE 2.8 Frequency spectrum of solar radiation received on a unit area at the Earth's mean distance from the Sun, the area facing the Sun. *Solid line*: measured values (smoothed out over absorption lines; based on NASA, 1971). *Dashed line*: Planck law corresponding to an effective temperature of 5762 K, normalized to the experimental curve.

The corona extends into a dilute, expanding flow of protons (ionized hydrogen atoms) and electrons known as *solar wind*. The increasing radial speed at increasing distances is a consequence of the hydrodynamic equations for the systems (the gravitational forces are unable to balance the pressure gradient; Parker, 1964). Solar wind continues for as long as the momentum flow is large enough not to be deflected appreciably by the magnetic fields of interstellar material. Presumably, solar wind penetrates the entire solar system.

Radiation Received at the Earth

At the top of the Earth's atmosphere, solar wind has a density of about 10^{-20} kg m^{-3}, corresponding to roughly 10^7 hydrogen atoms per m^3. The

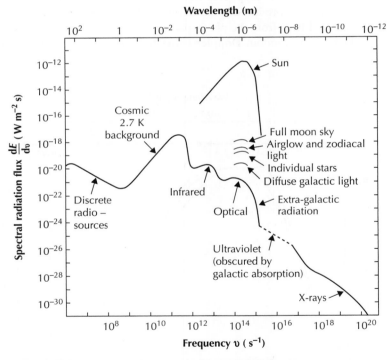

FIGURE 2.9 Frequency distribution of some contributions to the total radiation (integrated over the hemisphere). The radiation from the Sun and from extragalactic sources is given outside the Earth's atmosphere. The relative intensity of some familiar light phenomena in the visible frequency region is indicated (some of these sources, e.g., the solar corona, also emit radiation in the ultraviolet and X-ray regions). In some regions of the spectrum, the information on the extragalactic radiation is incomplete and the curve represents an estimate. (Based on NASA, 1971; Peebles, 1971; Allen, 1973; Longair and Sunyaev, 1969.)

ions are sucked into the Earth's magnetic field at the poles, giving rise to such phenomena as the aurorae borealis and to magnetic storms. Variations in solar activity affect the solar wind, which in turn affects the flux of cosmic rays reaching the Earth (many solar wind hydrogen ions imply larger absorption of cosmic rays).

Cosmic ray particles in the energy range 10^3–10^{12} MeV cross interstellar space in all directions. They are mainly protons, but produce showers containing a wide range of elementary particles when they hit an atmosphere.

Figure 2.9 summarizes a number of radiation sources contributing to the conditions at the Earth. Clearly, radiation from the Sun dominates the spectral distribution as well as the integrated flux. The next contributions, some six orders of magnitude down in the visible region, even integrated over the hemisphere, are also of solar origin, such as moonlight, airglow, and zodiacal light (originating in the Sun's corona, being particularly visible at the horizon just before sunrise and just after sunset). Further down, in the visible regions of the spectrum, are starlight, light from our own galaxy, and, finally, extragalactic light.

The peak in the spectral distribution of the extragalactic radiation is in the microwave region. This is the universal background radiation approximately following a Planck shape for 2.7 K, that is, the radiation predicted by, for example, the Big Bang theory of the expanding Universe.

The main part of solar radiation can be regarded as unpolarized at the top of the Earth's atmosphere, but certain minor radiation sources, such as the light scattered by electrons in the solar corona, do have a substantial degree of polarization.

2.2 DISPOSITION OF RADIATION ON THE EARTH

Solar radiation, which approximately corresponds to the radiation from a black-body of temperature 6000 K, meets the Earth–atmosphere system and interacts with it, producing temperatures at the Earth's surface that typically vary in the range of 220 to 320 K. The average (over time and geographical location) temperature of the Earth's surface is presently 288 K.

As a first approach to understanding the processes involved, one may look at the radiation flux passing through unit horizontal areas placed either at the top of the atmosphere or at the Earth's surface. The net flux is the sum (with proper signs) of the fluxes passing the area from above and from below. The flux direction toward the center of the Earth will be taken as positive, consistent with reckoning the fluxes at the Sun as positive, if they transport energy away from the solar center.

Since the spectral distributions of blackbody radiation (2.3) at 6000 and 300 K, respectively, do not substantially overlap, most of the radiation fluxes involved can be adequately discussed in terms of two broad categories, called short-wavelength (*sw*) and long-wavelength (*lw*) or thermal radiation.

2.2.1 Radiation at the Top of the Atmosphere

The flux of solar radiation incident on a surface placed at the top of the atmosphere depends on time (t) and geographical location (latitude ϕ and longitude λ) and on the orientation of the surface,

$$E_{0+}^{sw}(t, \phi, \lambda) = S(t) \cos \theta(t, \phi, \lambda).$$

Here $S(t)$ is the "solar constant" at the distance of the Earth (being a function of time, due to changes in the Sun–Earth distance and due to changes in solar luminosity) and θ is the angle between the incident solar flux and the normal to the surface considered. The subscript "0" on the short-wavelength flux E^{sw} through the surface indicates that the surface is situated at the top of the atmosphere, and "+" indicates that only the positive flux (in the "inward" direction) is considered. For a horizontal surface, θ is the zenith angle z, obtained by

$$\cos z = \sin \delta \, \sin \phi + \cos \delta \, \cos \phi \, \cos \omega,$$

where δ is the declination of the Sun and ω is the hour angle of the Sun (see Fig. 2.10). The declination is given approximately by (Cooper, 1969)

$$\delta = 0.4093 \sin \left(2\pi \frac{284 + day}{365} \right), \tag{2.9}$$

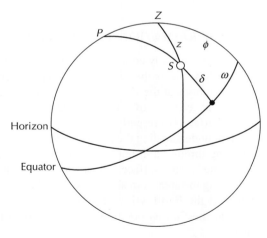

FIGURE 2.10   Coordinates in a geocentric picture. P represents the North Pole, Z is the zenith, and ϕ is the geographical latitude. The unit radius great circle shown through P and Z is the median of the observer. S represents the direction to the Sun, and the solar coordinates are the declination δ and the hour angle ω. Also shown is the zenith angle z (equal to $\pi/2$ minus the height of the Sun over the horizon).

where *day* is the actual day's number in the year. The hour angle (taken as positive in the mornings) is related to the local time t_{zone} (in hours) by

$$\omega = 2\pi(12 - t_{zone})/24 - (\lambda - \lambda_{zone}) - TEQ. \tag{2.10}$$

Here λ_{zone} is the longitude of the meridian defining the local time zone (longitudes are taken positive toward east; latitudes and declinations are positive toward north; all angles are in radians). The correction *TEQ* (*equation of time*) accounts for the variations in solar time caused by changes in the rotational and orbital motion of the Earth, e.g., due to speed variations in the elliptical orbit and to the finite angle (obliquity) between the axis of rotation and the normal to the plane or orbital motion. The main part of *TEQ* remains unaltered at the same time in subsequent years and may be expressed as a function of the day of the year (see Duffie and Beckman, 1974),

$$
\begin{aligned}
TEQ &= -0.0113 - 0.0019 \times day, \text{ if } day \leq 20,\\
&= -0.0227 - 0.0393 \cos(0.0357(day - 43)), \text{ if } 20 < day \leq 135,\\
&= -0.0061 + 0.0218 \cos(0.0449(day - 135)), \text{ if } 135 < day \leq 240,\\
&= 0.0275 + 0.0436 \cos(0.0360(day - 306)), \text{ if } 240 < day \leq 335,\\
&= -0.0020 \times (day - 359), \text{ if } day > 335.
\end{aligned}
$$

The resulting daily average flux on a horizontal surface,

$$E_{0+}^{sw} = \frac{1}{24} \int_{day} E_{0+}^{sw}(t, \phi, \lambda) \, dt,$$

which is independent of λ, is shown in Fig. 2.11. The maximum daily flux is about 40% of the flux on a surface fixed perpendicular to the direction of the Sun. It is obtained near the poles during midsummer, with light close to 24 hours of each day but a zenith angle of $\pi/2 - \max(|\delta|) \approx 67°$ (cf. Fig. 2.10, $|\delta|$ being the absolute value of δ). Daily fluxes close to 30% of the solar constant are found throughout the year at the Equator. Here the minimum zenith angle comes close to zero, but day and night are about equally long.

The Disposition of Incoming Radiation

A fraction of the incoming solar radiation is reflected back into space. This fraction is called the *albedo* a_0 of the Earth–atmosphere system. The year and latitude–longitude average of a_0 is about 0.35. This is composed of about 0.2 from reflection on clouds, 0.1 from reflection on cloudless atmosphere (particles, gases), and 0.05 from reflection on the Earth's surface. Figure 2.12 gives the a_0 distribution for two months of 1997, derived from ongoing satellite measurements (cf. Wilson and Matthews, 1971; Raschke *et al.*, 1973; NCEP-NCAR, 1998).

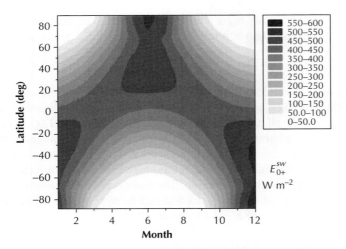

FIGURE 2.11 Daily average radiation, E_{0+}^{sw} (W m^{-2}), incident at the top of the Earth's atmosphere, as a function of time of the year and latitude (independent of longitude). Latitude is (in this and the following figures) taken as positive toward north. (Based on NCEP-NCAR, 1998.)

The radiation absorbed by the Earth–atmosphere system is

$$A_0 = E_{0+}^{sw}(1 - a_0).$$

Since no gross change in the Earth's temperature is taking place from year to year, it is expected that the Earth–atmosphere system is in radiation equilibrium, i.e., that it can be ascribed an effective temperature T_0, such that the blackbody radiation at this temperature is equal to $-A_0$,

$$T_0 = \left(\frac{\frac{1}{4}S(1 - a_0)}{\sigma} \right)^{1/4} \approx 253 \text{ K}.$$

Here S is the solar constant and $\sigma = 5.7 \times 10^{-8}$ W m^{-2} K^{-4} is Stefan's constant [cf. (2.4)]. The factor 1/4 is needed because the incoming radiation on the side of the Earth facing the Sun is the integral of $S \cos \theta$ over the hemisphere, while the outgoing radiation is uniform over the entire sphere (by definition when it is used to define an average effective temperature). The net flux of radiation at the top of the atmosphere can be written

$$E_0 = E_{0+}^{sw}(1 - a_0) + E_0^{lw}, \tag{2.11}$$

(a)

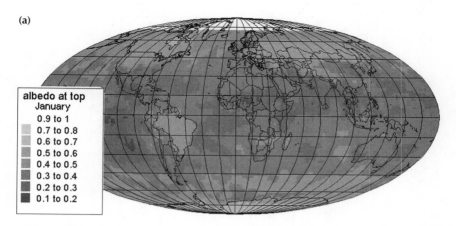

albedo at top
  January
  0.9 to 1
  0.7 to 0.8
  0.6 to 0.7
  0.5 to 0.6
  0.4 to 0.5
  0.3 to 0.4
  0.2 to 0.3
  0.1 to 0.2

FIGURE 2.12a,b   Top of the atmosphere albedo for the Earth–atmosphere system, derived from satellite data for January (*a*) and July (*b*) 1997 (NCEP-NCAR, 1998).[*]

(b)

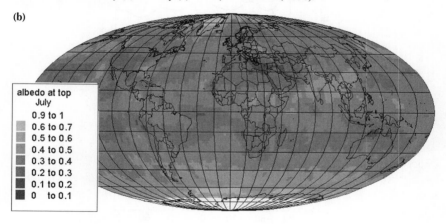

albedo at top
  July
  0.9 to 1
  0.6 to 0.7
  0.5 to 0.6
  0.4 to 0.5
  0.3 to 0.4
  0.2 to 0.3
  0.1 to 0.2
  0   to 0.1

where $E_0^{lw} = E_{0-}^{lw}$ on average equals $-\sigma T_0^{\,4}$ (recall that fluxes away from the Earth are taken as negative) and E_0 on average (over the year and the geographical position) equals zero.

The fact that T_0 is 34 K lower than the actual average temperature at the Earth's surface indicates that heat is accumulated near the Earth's surface, due to re-radiation from clouds and atmosphere. This is called the *greenhouse effect*, and it is further illustrated below, in the discussion of the radiation fluxes at the ground.

---

[*]These and most of the geographical figures in the following use the equal-area Mollweide projection, making it easy to compare quantities at different locations.

FIGURE 2.13   Latitude dependence (averaged over longitudes and over the year) of incident (E_{0+}^{sw}) and outgoing long-wavelength $(-E_{0-}^{lw})$ radiation at the top of the atmosphere, the amount of radiation absorbed (A_0) and reflected (R_0) by the Earth–atmosphere system, and the net radiation flux E_0 at the top of the atmosphere, in W m^{-2}. (Based on Sellers, 1965; Budyko, 1974.)

The net radiation flux at the top of the atmosphere, E_0, is a function of time as well as of geographical position. The longitudinal variations are small. Figure 2.13 gives the latitude distribution of $\overline{E_0}, \overline{A_0}$, and $\overline{R_0}$, where the reflected flux is

$$R_0 = E_{0+}^{sw} a_0 \quad (= -E_{0-}^{sw}),$$

as well as $\overline{E_0^{lw}}$.

Although $\overline{R_0}$ is fairly independent of ϕ, the albedo $\overline{a_0}$ increases when going toward the poles. Most radiation is absorbed at low latitudes, and the net flux exhibits a surplus at latitudes below 40° and a deficit at higher absolute latitudes. Since no progressive increase in equatorial average temperatures, nor any similar decrease in polar temperatures, is taking place, there must be a transport of energy from low latitudes in the directions of both poles. The mechanisms for producing this flow are ocean currents and winds. The necessary amount of poleward energy transport, which is immediately given by the latitude variation of the net radiation flux (since radiation is the only mode of energy transfer at the top of the atmosphere), is given in Fig. 2.10, which also indicates the measured share of ocean transport. The total amount of energy transported is largest around $|\phi| = 40°$, but since the total circumference of the latitude circles diminishes as latitude increases, the maximum energy transport across a unit length of latitude circle is found at $|\phi| \approx 50°$.

FIGURE 2.14 The poleward energy transfer required by the latitude variation of the net energy flux shown in Fig. 2.13. The *dashed curve* represents the same quantity divided by the length of the latitude circle, i.e., the energy transport required across a unit length of the latitude circle. These figures are annual and longitudinal averages. (Based on Sellers, 1965.) The ocean transport contribution is indicated as a dark blurred region, based on measurements with estimated uncertainties (Trenberth and Solomon, 1994).

Figures 2.15 and 2.16 give the time and latitude dependence at longitude zero of the long-wavelength re-radiation, E_{0-}^{lw} or E_0^{lw} (since the flux is only in the negative direction), and net flux, E_0, of the Earth–atmosphere system, as obtained from satellite measurements within the year 1997 (Kalnay *et al.*,

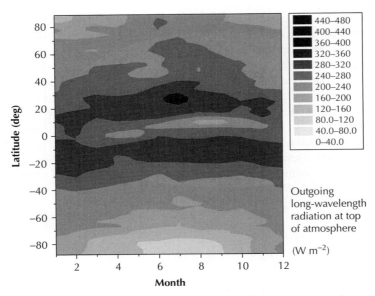

FIGURE 2.15   Long-wavelength radiation from the Earth–atmosphere system $(-E^{lw}_{0-})$.

FIGURE 2.16   Net radiation flux E_0 at the top of the atmosphere (NCEP-NCAR, 1998).

1996; NCEP-NCAR, 1998). A_0 may be obtained as the difference between E_{0+}^{sw} and E_{0-}^{sw}. The latter equals the albedo (Fig. 2.12) times the former, given in Fig. 2.11.

2.2.2 Radiation at the Earth's Surface

The radiation received at the Earth's surface consists of direct and scattered (plus reflected) short-wavelength radiation plus long-wavelength radiation from sky and clouds, originating as thermal emission or by reflection of thermal radiation from the ground.

Direct and Scattered Radiation

Direct radiation is defined as radiation that has not experienced scattering in the atmosphere, so that it is directionally fixed, coming from the disc of the Sun.

Scattered radiation is, then, the radiation having experienced scattering processes in the atmosphere. In practice, it is often convenient to treat radiation that has experienced only forward scattering processes together with unscattered radiation, and thus define direct and scattered radiation as radiation coming from, or not coming from, the direction of the Sun. The area of the disc around the Sun used to define direct radiation is often selected in a way depending on the application in mind (e.g., the solid angle of acceptance for an optical device constructed for direct radiation measurement). In any case, "direct radiation" defined in this way will contain scattered radiation with sufficiently small angles of deflection, due to the finite solid angle of the Sun's disc.

The individual scattering and absorption processes are discussed further below, but in order to introduce the modifications in the spectral distribution due to passage through the atmosphere, Fig. 2.17 shows the amount of radiation surviving at sea level for a clear day with the Sun in zenith. A large number of absorption lines and absorption bands can be seen in the low-frequency part of the spectrum (corresponding to wavelengths above 0.7×10^{-6} m). They are due to H_2O, CO_2, O_2, N_2O, CH_4, and other, minor constituents of the atmosphere. At higher frequencies, the continuous absorption bands dominate, in particular those of O_3. Around 0.5×10^{-6} m, partial absorption by O_3 produces a dip in the spectrum, and below a wavelength of 0.3×10^{-6} m (the ultraviolet part of the spectrum) or a frequency above 9.8×10^{14} s^{-1}, the absorption by ozone is practically complete.

Figure 2.17 also shows the scattered part of the radiation on a clear day (dashed line). The scattered radiation is mainly composed of the most energetic (high-frequency) radiation, corresponding to the blue part of the visible spectrum. The sky on a clear day is therefore blue. The scattered light is also polarized. The radiation from clouds, i.e., from a completely overcast sky, is also shown in Fig. 2.17. It has a broad frequency distribution in the visible spectrum, so that the cloud light appears white.

FIGURE 2.17 Frequency spectrum of solar radiation at the top of the atmosphere and at the surface (sea level and minimum air mass, corresponding to the Sun in zenith on a clear day). Also shown are typical spectra of scattered radiation in the extremes of pure sky radiation (cloudless day) and pure cloud radiation (completely overcast). (Based on NASA, 1971; Gates, 1966.)

Disposition of Radiation at the Earth's Surface

On average, roughly half of the short-wavelength radiation that reaches the Earth's surface has been scattered. Denoting the direct and scattered radiation at the Earth's surface as D and d, respectively, and using the subscript s to distinguish the surface level (as the subscript "0" was used for the top of the atmosphere), the total incoming short-wavelength radiation at the Earth's surface may be written

$$E_{s+}^{sw} = D + d,$$

for a horizontal plane. The amount of radiation reflected at the surface is

$$R_s = E_{s+}^{sw} a_s \quad (= -E_{s-}^{sw}),$$

where the surface albedo, a_s, consists of a part describing specular reflection (i.e., reflection preserving the angle between beam and normal to the rejecting surface) and a part describing diffuse reflection. Here diffuse reflection is taken to mean any reflection into angles different from the one characterizing beam reflection. An extreme case of diffuse reflection (sometimes denoted "completely diffuse" as distinguished from "partially diffuse") is one in which the angular distribution of reflected radiation is independent from the incident angle.

The total net radiation flux at the Earth's surface is

$$E_s = E_{s+}^{sw}(1-a_s) + E_{s-}^{lw},\qquad(2.12)$$

where the long-wavelength net radiation flux is

$$E_s^{lw} = E_{s+}^{lw} + E_s^{lw},$$

in terms of the inward and (negative) outward long-wavelength fluxes.

FIGURE 2.18 Incoming radiation fluxes on a horizontal plane at the Earth's surface. Averages over the year and over longitudes are shown for direct (D), scattered (d) and total radiation (E_{s+}^{sw}), as functions of latitude. (Adapted from Sellers, 1965.)

Figure 2.18 gives the latitude distribution of the incoming short-wavelength flux E_{s+}^{sw} and its division into a direct, D, and a scattered part, d. The fluxes are year and longitude averages (adapted from Sellers, 1965). Similarly, Fig. 2.19 gives the total annual and longitude average of the net radiation flux at the Earth's surface, and its components, as a function of latitude. The components are, from (2.12), the absorbed short-wavelength radiation,

$$A_s^{sw} = E_{s+}^{sw}(1-a_s),$$

and the net long-wavelength thermal flux E_s^{lw} (Fig. 2.19 shows $-E_s^{lw}$). The figures show that the direct and scattered radiation are of comparable magnitude for most latitudes and that the total net radiation flux at the Earth's surface is positive except near the poles. The variations in the long-wavelength difference between incoming (from sky and clouds) and outgoing radiation indicate that

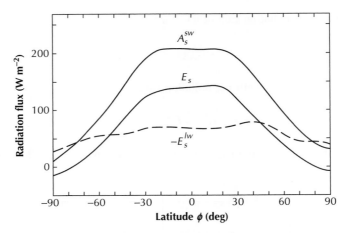

FIGURE 2.19   The latitude dependence of annual and longitudinal averages of net radiation flux on a horizontal plane at the Earth's surface (E_s) and its components, the amount of short-wavelength radiation absorbed (A_s^{sw}) and the net emission of long-wavelength radiation ($-E_s^{lw}$). (Adapted from Sellers, 1965.)

the temperature difference between low and high latitudes is not nearly large enough to compensate for the strong variation in absorbed short-wavelength radiation.

The conclusion is the same as the one drawn from the investigation of the Earth–atmosphere net radiation flux. A poleward transport of heat is required, either through the Earth's surface (e.g., ocean currents) or through the atmosphere (winds in combination with energy transfer processes between surface and atmosphere by means other than radiation).

Fig. 2.20 summarizes in a crude form the transfer processes believed to take place in the disposition of the incoming solar radiation. In the stratosphere, the absorption of short-wavelength radiation (predominantly by ozone) approximately balances the emission of long-wavelength radiation (from CO_2, H_2O, etc.). In the lower part of the atmosphere, the presence of clouds largely governs the reflection of short-wavelength radiation. On the other hand, particles and gaseous constituents of the cloudless atmosphere are most effective in absorbing solar radiation. All figures are averaged over time and geographical location.

Slightly over half the short-wavelength radiation that reaches the Earth's surface has been scattered on the way. Of the 27% of the solar radiation at the top of the atmosphere that is indicated as scattered, about 16% has been scattered by clouds and 11% by the cloudless part of the sky.

The albedo at the Earth's surface is shown in Fig. 2.21 for two months of 1997, based on a re-analysis of measured data, using climate modeling to

FIGURE 2.20 Schematic summary of the disposition of incoming radiation (E_{0+}^{sw}, put equal to 100 units) in the Earth–atmosphere system, averaged over time and geographical location. Symbols and details of the diagram are discussed in the text. (The data are adapted from Sellers, 1965; Budyko, 1974; Schneider and Dennett, 1975; Raschke *et al.*, 1973.)

ensure consistency. The model has not been able to achieve full consistency of fluxes, probably because of too high albedo values over ocean surfaces (Kalnay *et al.*, 1996). In Fig. 2.20 the average albedo is indicated as the smaller value $a_s \approx 0.12$. The albedo of a particular surface is usually a strongly varying function of the incident angle of the radiation. For seawater, the average albedo is about $a_s = 0.07$, rising to about 0.3 in case of ice cover. For various land surfaces, a_s varies within the range 0.05–0.45, with typical mean values 0.15–0.20. In case of snow cover, a_s lies in the range 0.4–0.95, with the smaller values for snow that has had time to accumulate dirt on its surface.

As an illustration of the difficulties in estimating mean albedos, Fig. 2.22 shows the albedo of a water surface for direct sunlight, as a function of the zenith angle of the Sun. For a plane water surface and perpendicular incidence, the albedo implied by the air-to-water refraction index is 0.024. However, for ocean water, the surface cannot be regarded as a plane, and typical mean albedos (over the year) are of the order of 0.06, except at high latitudes, where the value may be greater. This estimate includes direct as well as scattered incident light, with the latter "seeing" a much more constant albedo of the water, of about 0.09 (Budyko, 1974).

(a)

ground albedo
January

0.9 to 1
0.8 to 0.9
0.7 to 0.8
0.6 to 0.7
0.5 to 0.6
0.4 to 0.5
0.3 to 0.4
0.2 to 0.3
0.1 to 0.2

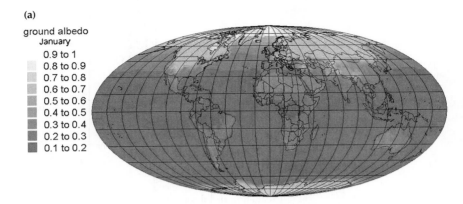

FIGURE 2.21a,b   Albedo at the Earth's surface, based on the analysis of ground-based and satellite measurements, for January (a) and July (b) of 1997. White areas: missing data, notably near poles (NCEP-NCAR, 1998).

(b)

surface albedo
July

0.9 to 1
0.8 to 0.9
0.7 to 0.8
0.6 to 0.7
0.5 to 0.6
0.4 to 0.5
0.3 to 0.4
0.2 to 0.3
0.1 to 0.2
0   to 0.1

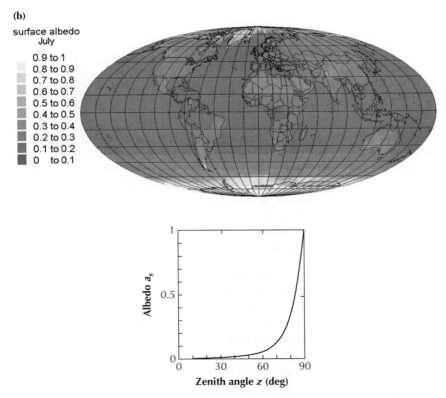

FIGURE 2.22   Albedo of water surface for direct radiation, as a function of zenith angle (i.e., angle between normal and the direction of radiation). (Based on Budyko, 1974.)

The total amount of direct and scattered light, which in Fig. 2.20 is given as reflected from the ground, is a net value representing the net reflected part after both the initial reflection by the ground and the subsequent back-reflection (some 20%) from the clouds above, and so on.

Disposition of Radiation in the Atmosphere

On average, only a very small fraction of the long-wavelength radiation emitted by the Earth passes unhindered to the top of the atmosphere. Most is absorbed in the atmosphere or by clouds, from which the re-radiation back toward the surface is nearly as large as the amount received from the surface.

The net radiation flux of the atmosphere,

$$E_a = E_a^{sw} + E_a^{lw} = A_a^{sw} + A_a^{lw} + S_a^{lw},$$

equals on average -29% of E_{0+}^{sw}. The short-wavelength radiation absorbed is $A_a^{sw} = E_s^{sw} = 21\%$, whereas the long-wavelength radiation absorbed and emitted in the lower atmosphere is $A_a^{lw} = 109\%$ and $S_a^{lw} = -159\%$, respectively. Thus, the net long-wavelength radiation is $E_a^{lw} = -50\%$.

The net radiation flux of the atmosphere is on average equal to, but (due to the sign convention adopted) of opposite sign to, the net radiation flux at the Earth's surface. This means that energy must be transferred from the surface to the atmosphere by means other than radiation, such as by conduction and convection of heat, by evaporation and precipitation of water and snow, etc. These processes produce energy flows in both directions, between surface and atmosphere, with a complex dynamic pattern laid out by river run-off, ocean currents, and air motion of the general circulation. The average deviation from zero of the net radiation flux at the surface level gives a measure of the net amount of energy displaced vertically in this manner, namely, 29% of E_{0+}^{sw} or 98 W m^{-2}. As indicated in Fig. 2.18, the horizontal energy transport across a 1-m-wide line (times the height of the atmosphere) may be a million times larger. The physical processes involved in the transport of energy are discussed further in section 2.3.

Annual, Seasonal, and Diurnal Variations in the Radiation Fluxes at the Earth's Surface

In discussing the various terms in the radiation fluxes, it has tacitly been assumed that averages over the year or over longitudes or latitudes have a well-defined meaning. However, as is well known, the climate is not completely periodic with a 1-year cycle, even if solar energy input is to a very large degree. The short-term fluctuations in climate seem to be very important, and the development over a time scale of a year depends to a large extent on the detailed initial conditions. Since, owing to the irregular short-term fluctuations, initial conditions at the beginning of subsequent years are not exactly the same, the development of the general circulation and of the climate as a whole does not repeat itself with a 1-year period. Although the gross seasonal characteristics are preserved as a result of the outside forcing of the system (the solar radiation), the year-average values of such quantities as mean temperature or mean cloudiness for a given region or for the Earth as a whole are not the same from one year to the next.

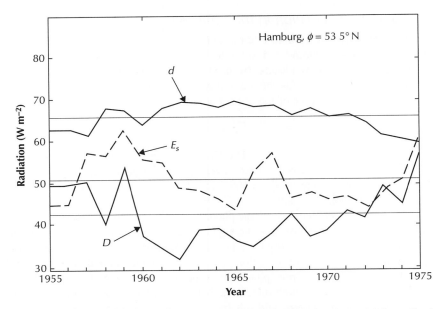

FIGURE 2.23   Variation of annual averages of direct and scattered radiation (*solid, heavy lines*) around their 21-year average value (*thin, horizontal lines*) for Hamburg. Also shown is the variation in net radiation flux (*dashed line*) around its 21-year average value. (Based on Kasten, 1977.)

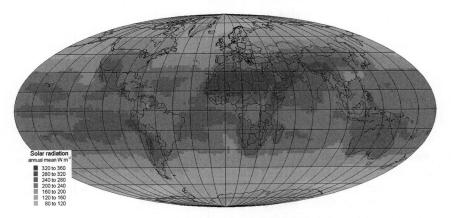

FIGURE 2.24   Annual average incoming short-wavelength radiation (E_{s+}^{sw}) on a horizontal surface at the Earth's surface, as a function of geographical position (W m^{-2}). (Based on NCEP-NCAR, 1998.) Seasonal variations are shown in Chapter 3, Figs. 3.1a–d.

This implies that the components of the net radiation flux do not have the same annual average values from year to year. In particular, the disposition of incoming radiation as rejected, scattered, direct, and absorbed radiation is very sensitive to variations in cloud cover, and, therefore, exhibits substantial changes from year to year. Figure 2.23 shows the variations of yearly means of direct and scattered radiation measured at Hamburg (Kasten, 1977) over a 20-year period. Also shown is the variation of the net radiation flux at the ground. The largest deviation in direct radiation, from the 20-year average, is over 30%, whereas the largest deviation in scattered radiation is 10%. The maximum deviation of the net radiation flux exceeds 20%.

(a)

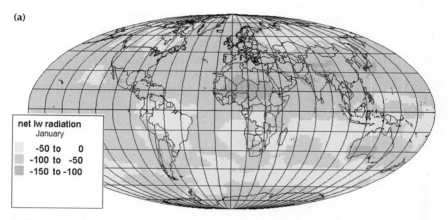

FIGURE 2.25a,b Average net (i.e., downward minus upward) long-wavelength radiation flux (E_s^{lw}) in January (*a*) and July (*b*), at the Earth's surface, as a function of geographical position (W m^{-2}). (Based on NCEP-NCAR, 1998.)

(b)

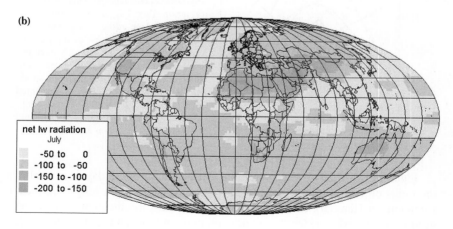

Figures 2.24 and 2.25 give, for different seasons in 1997, the total short-wavelength radiation $(D + d)$ and the long-wavelength radiation fluxes (E_s^{lw}) at the Earth's surface, as a function of latitude and longitude.

Figure 2.26 gives the results of measurements at a specific location, Hamburg, averaged over 19 years of observation (Kasten, 1977). Ground-based observations at meteorological stations usually include the separation of short-wavelength radiation into direct and scattered parts, quantities that are difficult to deduce from satellite-based data. Owing to their importance for solar energy utilization, they are discussed further in Chapter 3. The ratio of scattered to direct radiation for Hamburg is slightly higher than the average for the latitude of 53.5° (see Fig. 2.18), and the total amount of incident short-wavelength radiation is smaller than average. This may be due to the climate modifications caused by a high degree of urbanization and industrialization (increased particle emission, increased cloud cover). On the other hand, the amount of short-wavelength radiation reflected at the surface, R_s, exceeds the average for the latitude.

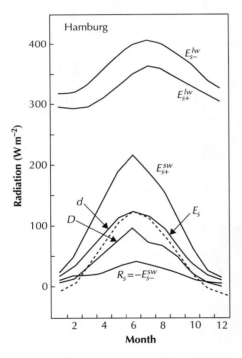

FIGURE 2.26 Components of the net radiation flux (E_s, *dashed line*) on a horizontal plane at the Earth's surface in Hamburg. Based on data for 19 consecutive years, the monthly averages have been calculated for direct (D), scattered (d), and total incident short-wavelength radiation (E_{s+}^{sw}), the rejected part ($R_s = -E_{s-}^{sw}$), and the flux of long-wavelength radiation received (E_{s+}^{lw}) and emitted (E_{s-}^{lw}). (Based on Kasten, 1977.)

Figures 2.27–2.29 give the variations with the hour of the day of several of the radiation contributions considered above, again based on long-term measurements at Hamburg (Kasten, 1977). In Fig. 2.27, the incident radiation at ground level is shown for the months of June and December, together with the direct and reflected radiation. The mean albedo is about 0.18 at noon in summer and about 0.3 in winter. In December, the amount of reflected radiation exceeds the amount of direct radiation incident on the plane. The long-wavelength components, shown in Fig. 2.28, are mainly governed by the temperature of the ground. The net outward radiation rises sharply after sunrise in June and in a weakened form in December. In June, the separate outward and inward fluxes have a broad maximum around 14 h. The delay from noon is due to the finite heat capacity of the ground, implying a time lag between the maximum incident solar radiation and the maximum temperature of the soil. The downward flux changes somewhat more rapidly than the upward flux, because of the smaller heat capacity of the air. In December, the 24-h soil temperature is almost constant, and so is the outgoing long-wavelength radiation, but the re-radiation from the atmosphere exhibits a dip around noon. Kasten (1977) suggests that this dip is due to the dissolution of rising inversion layers, which during the time interval 16–08 h contributes significantly to the downward radiation. The hourly variation of the net total radiation flux is shown in Fig. 2.29 for the same two months. During daytime this quantity roughly follows the variation of incident radiation, but at night the net flux is negative, and more so during summer when the soil temperature is higher.

FIGURE 2.27 For the months of June and December, the total short-wavelength radiation on a horizontal plane at the Earth's surface in Hamburg is shown, as well as the direct and reflected radiation, for a given hour of the day. The fluxes are hourly averages based on 10 years of data, shown as functions of the hour of the day in true solar time. (Based on Kasten, 1977.)

Penetration of Solar Radiation

In discussing the solar radiation absorbed at the Earth's surface, it is important to specify how deep into the surface material, such as water, soil, or ice, the radiation is able to penetrate. Furthermore, the reflection of sunlight may take place at a depth below the physical surface.

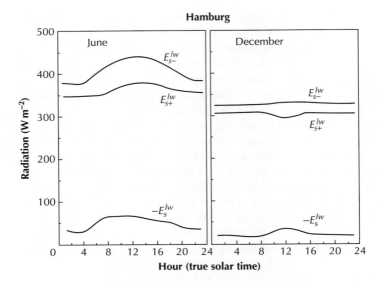

FIGURE 2.28   For June and December, the net long-wavelength radiation on a horizontal plane at the Earth's surface in Hamburg is shown, as well as the separate incoming and outgoing fluxes, for a given hour of the day. The fluxes are hourly averages based on 10 years of data, shown as functions of the hour of the day in true solar time. (Based on Kasten, 1977.)

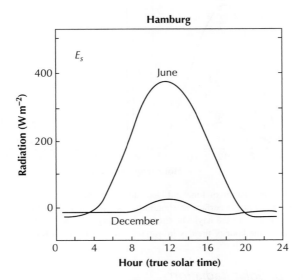

FIGURE 2.29   For the months of June and December, the total net radiation flux on a horizontal plane at the Earth's surface in Hamburg is shown for a given hour of the day. The fluxes are hourly average values based on 10 years of measurements, shown as functions of the hour of the day in true solar time. (Based on Kasten, 1977.)

The penetration into soil, vegetation, sand, and roads or building surfaces is usually limited to at most 1 or 2 mm. For sand of large grain size, the penetration may reach 1–2 cm, in particular for the wavelengths above 0.5×10^{-6} m. For pure water, shorter wavelengths are transmitted more readily than longer ones. Figure 2.30 shows calculated intensity variations as a function of depth below the water's surface and wavelength. The availability of solar radiation (although only about 5% of that incident along the normal to the water surface) at ocean depths of nearly 100 m is, of course, decisive for the photosynthetic production of biomass by phytoplankton or algae. For coastal and inland waters, the penetration of sunlight may be substantially reduced owing to dispersed sediments, as well as to eutrophication. A reduction in radiation intensity by a factor two or more for each meter penetrated is not uncommon for inland ponds (Geiger, 1961). The penetration of radiation into snow or glacier ice varies from a few tenths of a meter (for snow with high water content) to over a meter (ice).

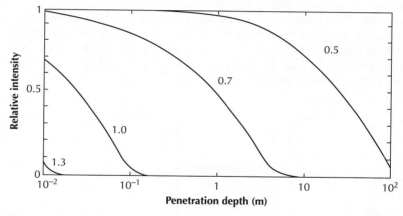

FIGURE 2.30 Penetration depth of monochromatic radiation through water as a function of wavelength of radiation (indicated on curves, in 10^{-6} m). The curves should only be considered rough indications of the behavior. (Based on Sellers, 1965.)

2.3 PROCESSES NEAR THE SURFACE OF THE EARTH

A very large number of physical and chemical processes take place near the Earth's surface. No exhaustive description is attempted here, but emphasis is placed on those processes directly associated with, or important for, the possibility of utilizing the renewable energy flows in connection with human activities.

In particular, this leads to a closer look at some of the processes by which solar energy is distributed in the soil–atmosphere–ocean system, e.g., the formation of winds and currents. Understanding these processes is also required in order to estimate the possible climatic disturbance that may follow certain types of interference with the "natural" flow pattern.

Although the processes in the different compartments are certainly coupled, the following subsections first list a number of physical processes specific for the atmosphere, for the oceans, and, finally, for the continents and then indicates how they can be combined to account for the climate of the Earth.

2.3.1 The Atmosphere

The major constituents of the present atmosphere are nitrogen, oxygen, and water. For dry air at sea level, nitrogen constitutes 78% by volume, oxygen 21%, and minor constituents 1%, the major part of which is argon (0.93%) and carbon dioxide (0.03%). The water content by volume ranges from close to zero at the poles to about 4% in tropical climates.

Typical variations of density, pressure, and temperature, as functions of altitude, are shown in Fig. 2.31. Common names for the different layers are indicated. They are generally defined by turning points in the temperature profile ("pauses") or by the lower border of constant temperature regions if no sharp change in the sign of the temperature gradient takes place. Actual temperature profiles are not independent of latitude and season and do not exhibit sharp "turning points," as indicated in Fig. 2.32.

FIGURE 2.31   Altitude dependence of temperature, pressure, and density for the U.S. standard atmosphere (U.S. Government, 1962).

Estimates of the variation with altitude of the average abundance of some gaseous constituents of the atmosphere are shown in Fig. 2.33. The increasing abundance of ozone (O_3) between 15 and 25 km altitude is instrumental in keeping the level of ultraviolet radiation down, thereby allowing life to develop on the Earth's surface (cf. Fig. 2.2). Above a altitude of about 80 km, many of the gases are ionized, and for this reason the part of the atmosphere above this level is sometimes referred to as the ionosphere. Charged particles occurring at the altitude of 60 000 km still largely follow the motion of the Earth, with their motion being mainly determined by the magnetic field of the Earth. This layer is referred to as the magnetosphere.

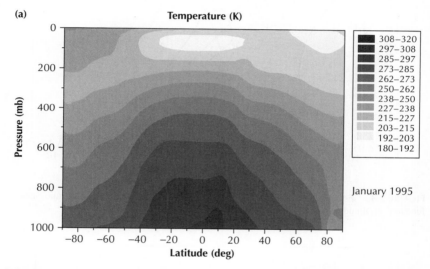

FIGURE 2.32a,b   Average temperature in the atmosphere as a function of altitude and latitude for January (*a*) and July (*b*) 1995, both at longitude zero. (Based on NCEP-NCAR, 1998.)

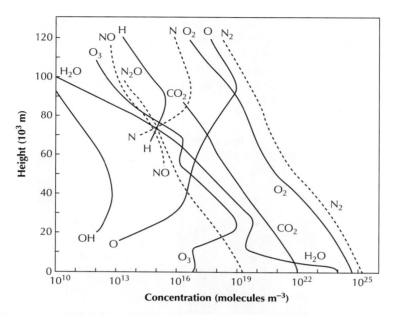

FIGURE 2.33   Estimated mean concentrations of some gaseous constituents of the atmosphere, as functions of altitude. (Based on Almquist, 1974.)

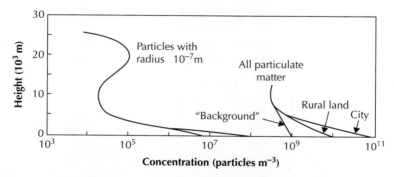

FIGURE 2.34   Trends of particle concentrations as functions of altitude, either for large particles only (*left*) or for all particles. The range of surface concentrations over different surfaces is indicated. "Background" refers to an extrapolated unpolluted and clean surface, such as the Arctic ice covers. (Based on Craig, 1965; Wilson and Matthews, 1971.)

Particles in the Atmosphere

In addition to gaseous constituents, water, and ice, the atmosphere contains particulate matter in varying quantity and composition. A number of mechanisms have been suggested for the production of particles varying in size from 10^{-9} to 10^{-4} m. One is the ejection of particles from the surface of the oceans, in the form of salt sprays. Others involve the transformation of gases into particles. This can occur as a result of photochemical reactions or as a result of adsorption to water droplets (clouds, haze), chemical reactions depending on the presence of water and finally evaporation of the water, leaving behind a suspended particle rather than a gas.

Dust particles from deserts or eroded soil also contribute to the particle concentration in the atmosphere, as does volcanic action. Man-made contributions are in the form of injected particles, from land being uncovered (deforestation, agricultural practice), industry, and fuel burning, and in the form of gaseous emissions (SO_2, H_2S, NH_3, etc.), which may be subsequently transformed into particles by the above-mentioned processes or by catalytic transformations in the presence of heavy metal ions, for example.

Wilson and Matthews (1971) estimate that yearly emission into, or formation within, the atmosphere of particles with a radius smaller than 2×10^{-5} m is in the range $(0.9-2.6) \times 10^{12}$ kg. Of this, $(0.4-1.1) \times 10^{12}$ kg is due to erosion, forest fires, sea salt sprays, and volcanic debris, while $(0.3-1.1) \times 10^{12}$ kg is due to transformation of non-anthropogenic gaseous emissions into particles. Anthropogenic emissions are presently in the range $(0.2-0.4) \times 10^{12}$ kg, most of which is transformed from gaseous emissions.

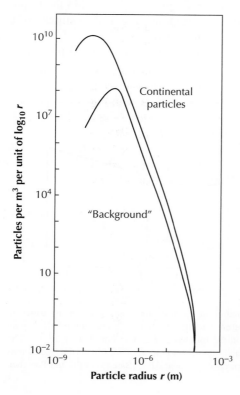

FIGURE 2.35 Size distribution of particles in the atmosphere. The estimated distribution over land is shown separately. The distribution over oceans resembles that of the clean "background" distribution, except in the radius interval 10^{-6}–10^{-5} m, where it approaches that for continental particles. (Based on Wilson and Matthews, 1971.)

Figure 2.34 gives a very rough sketch of the particle density as a function of altitude. The distribution of large particles (radii in the range $10^{-7} - 2 \times 10^{-5}$ m) is based on measurements at a given time and place (based on Craig, 1965), while the total particle counts are based on extrapolations of the size distribution to very small particles (see Fig. 2.35). It is clear that the total number of particles depends sensitively on the density of very small particles, which on the other hand do not contribute significantly to the total mass of particulate matter. The concentrations of rural and city particles in the right-hand part of Fig. 2.34 correspond to an extrapolation toward smaller radii of the size distributions in Fig. 2.35, such that the number of particles per m^3 per metric decade of radius does not decrease rapidly toward zero.

Returning to mass units, the total mass of particles in the atmosphere can be estimated from the emission and formation rates, if the mean residence time is known. The removal processes active in the troposphere include dry deposition by sedimentation (important only for particles with a radius above 10^{-6} m), diffusion, and impact on the Earth's surface, vegetation, etc., and wet deposition by precipitation. The physical processes involved in wet deposition are condensation of water vapor on particles, in-cloud adsorption of particles to water droplets (called rainout), or scavenging by falling rain (or ice) from higher-lying clouds (called washout). Typically, residence times for particles in the troposphere are in the range 70–500 h (Junge, 1963), leading to the estimate of a total particle content in the atmosphere of the order of 10^{14} kg.

Particles may also be removed by chemical or photochemical processes, transforming them into gases.

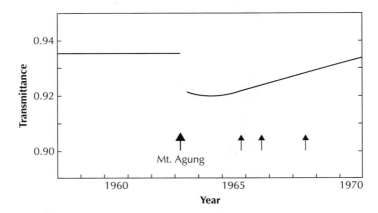

FIGURE 2.36   Transmittance of solar radiation in the direction of incidence, as measured in Hawaii near sunrise, during the periods before and after the volcanic eruption of Mt. Agung in Bali. Other, smaller volcanic eruptions in the following years are indicated by arrows. The curves have been smoothed over seasonal and other short-term variations, with the fluctuations being generally less than ±0.05 in the transmittance on clear days. (Based on Ellis and Pueschel, 1971.)

Particle residence times in the stratosphere are much longer than in the troposphere. Sulfates (e.g., from volcanic debris) dominate, and the particles are spread close to homogeneously over each hemisphere, no matter where they are being injected into the stratosphere. Residence times are of the order of years, and removal seems to be by diffusion into the troposphere at latitudes around 55°. This picture is based on evidence from major volcanic eruptions and from detonation of nuclear weapons of megaton size. Both of these processes inject particulate matter into the stratosphere. The particles in the stratosphere can be detected from the ground, because they modify the transmission of solar radiation, particularly close to sunrise or sunset, where the path-length through the atmosphere is at its maximum. Figure 2.36 shows atmospheric transmittance at sunrise, measured in Hawaii before and after the major volcanic eruption of Mt. Agung on Bali in 1963. The arrows indicate other eruptions that have taken place near the Equator during the following years and that may have delayed the return of the transmittance toward its value from before 1963.

Figure 2.37 shows the ground deposition of radioactive debris (fallout), at latitude 56 °N, after the enforcement of the test ban in 1963, which greatly reduced, but did not fully end, nuclear tests in the atmosphere. It can be concluded that the amount of radioactivity residing in the stratosphere is reduced by half each year without new injection. Observations of increased scattering (resulting in a shift toward longer wavelengths) following the very large volcanic eruption of Krakatoa in 1883 have been reported (sky reddening at sunrise and sunset). It is believed that the influence of such eruptions on the net radiation flux has had significant, although in the case of Krakatoa apparently transient, effects on the climate of the Earth.

FIGURE 2.37   Yearly fallout of ^{90}Sr following the atmospheric nuclear test ban in 1963, as measured in Denmark. (Based on Aarkrog, 1971.)

Absorption and Scattering in the Atmosphere

Absorption and scattering of radiation take place throughout the atmosphere. The radiation may cause electrons in atoms to go from one orbital to another, or it may cause transition between vibrational or rotational levels in molecules. Generally, the fundamental energy differences associated with rotational excitations are smaller than those associated with vibrational excitations, which again are smaller than those associated with low-lying electron states. There is only a fairly small probability that solar radiation is capable of ionizing atoms (expelling electrons) or dissociating molecules.

Since the wavelengths in the solar spectrum are much longer than the dimensions of atoms and molecules found in the atmosphere, the time-dependent fields of the atomic or molecular system can be approximated by the first few terms in a multipole expansion. In the dipole approximation, the probability of a transition between two levels i and f, accompanied by the emission or absorption of radiation with frequency equal to (or very close to)

$$v_{if} = |E_f - E_i|/h,$$

where $E_{i(f)}$ is the energy of level $i(f)$, is proportional to the square of the dipole transition moment d_{if} (an integral over the initial and final state quantal wavefunctions times the position vector; see, for example, Merzbacker, 1970). The absorption probability (or "rate") is further proportional to the spectral intensity at the required frequency.

$$\text{"Rate of absorption"} \propto |d_{if}|^2 I(v_{if}).$$

If the absorption and re-emission processes were sufficiently frequent, a state of thermal equilibrium would be reached, in the sense that the relative population of any two atomic or molecular levels would be given by the Boltzmann factor $\exp(-hv_{if}/(kT))$. If the radiation has a frequency distribution corresponding to the Planck law, the common temperature T will be that appearing in the Planck distribution for the radiation. In this way, a single atom or molecule can be in equilibrium with radiation and can be ascribed the same temperature as that of the radiation.

On the other hand, the thermodynamic temperature is defined from the distribution of kinetic energy in the "external" motion of the atoms or molecules. This distribution is maintained by collisions between the atoms or molecules. If the frequency of collisions is comparable to that of interactions with the radiation, the kinetic and radiation temperatures will become similar (e.g., the radiative transfer section of the Sun).

Where collisions are much more frequent than interactions with the radiation field, the radiative energy absorbed by a given atomic or molecular transition will become shared among all the atoms or molecules, and if the corresponding emission spectrum can be described by the blackbody law, the temperature entering should be the kinetic temperature, which need not be equal to the temperature of the radiation field (in fact, the kinetic temperature must be lower than the radiation temperature, if there are no other energy sources). Re-emission at frequencies corresponding to the same definite atomic or molecular transitions that caused absorption of radiation can be regarded as scattering of the incoming radiation.

In the Earth's atmosphere, the number of molecular collisions is generally large compared with the number of interactions with solar radiation. Thus, the radiation that is not re-emitted as scattered light will be redistributed over a large number of thermal degrees of freedom, i.e., the absorbed light is transformed into heat. The population of various molecular levels will not be in equilibrium with the temperature of the solar radiation, and the spectrum of scattered light will not be of Planck form (cf. Fig. 2.17).

The scattering process may be viewed as the creation of a dipole field in the atom or molecule, as a result of the radiation field, and a subsequent emission of radiation, as is well known from an oscillating dipole. The angular distribution of scattered light is proportional to $(1 + \cos^2\psi)$, where ψ is the angle between incident and scattered light.

Absorption Processes in Different Frequency Regions

Even in the thermosphere, most ultraviolet radiation with wavelength below 1.8×10^{-7} m is already being absorbed by N_2O and O_2. Owing to its low density, the thermosphere exhibits a rapid temperature response to variations in solar intensity, such as those caused by sunspots. However, for the same reason, the temperature changes have little or no effect on the lower-lying layers of the atmosphere. Figure 2.38 shows the penetration depth of ultraviolet radiation, confirming that the shortest wavelengths are being stopped at an altitude of 100 km or more.

FIGURE 2.38 The altitude, as a function of wavelength, at which the intensity of solar radiation is reduced to $1/e$. The main absorbing agents are indicated for different regions of the ultraviolet part of the spectrum. (Based on Ratcliffe, 1960.)

This figure also shows that most of the radiation in the wavelength interval 1.8×10^{-7} to 3×10^{-7}m is absorbed in the mesosphere or the stratosphere. The agents of absorption are O_2 and, in particular, O_3. The rate of absorption by ozone peaks around 2.5×10^{-7}m wavelength, and the amount of energy converted to heat is sufficient to explain the peak in the temperature profile at about 50 km (see Fig. 2.32).

The ozone concentration is not uniform as a function of geographical position, as indicated by Fig. 2.39. Ozone is formed in the stratosphere by photodissociation of molecular oxygen,

$$O_2 + \gamma \rightarrow 2O$$

$$O_2 + O \rightarrow O_3$$

(a)

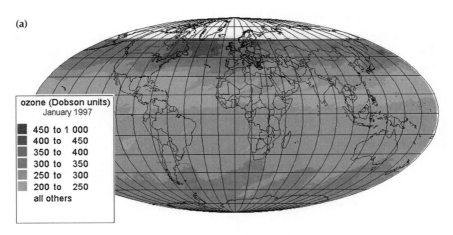

FIGURE 2.39a,b   Daily satellite measurements of ozone distributions for 1 January (*a*) and 1 April (*b*) 1997. The figures indicated are the volumes (in Dobson units, 10^{-5} m^3 m^{-2}) that the ozone would occupy at sea-level standard temperature and pressure, if all the ozone in a vertical column were put together. Areas obscured from the satellite are shown as white (NASA, 1998).

(b)

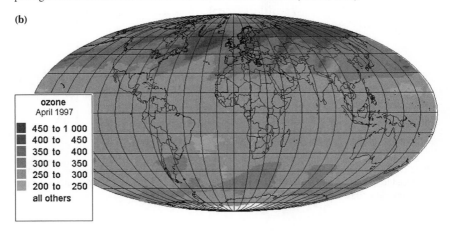

(Chapman, 1943). Natural removal processes include

$$O_3 + \gamma \rightarrow O_2 + O$$

$$O_3 + O \rightarrow 2O_2$$

$$O_3 + NO \rightarrow NO_2 + O_2$$

(Hesstvedt, 1973).

FIGURE 2.40 Spectral absorption efficiency of selected gaseous constituents of the atmosphere, and for the atmosphere as a whole (*bottom*). (From Fleagle and Businger, *An Introduction to Atmospheric Physics*, Academic Press, 1963.)

The concentrations of nitrogen oxides in the stratosphere may be increased by a number of human activities. The ozone concentrations vary between years (Almquist, 1974), but anthropogenic emissions of fluorocarbons are believed to dominate the picture, as they have been demonstrated to cause increased ozone depletion near the poles (cf. Fig. 2.39).

While the ultraviolet part of the solar spectrum is capable of exciting electronic states and molecular vibrational–rotational levels, little visible light is absorbed in the atmosphere. The energy is insufficient for most electronic excitations, and few molecular bands lie in this frequency region. Of course, harmonics of lower fundamental bands can be excited, but the cross-section for such absorption processes is very low.

In the infrared region, at wavelengths above 7×10^{-7} m, the fundamental vibrational and rotational bands of several molecules are situated. They give rise to clearly identifiable signatures of a large number of molecules, including H_2O, CO_2, N_2O, CH_4, CO, SO_2, H_2S, NO, NO_2, and NH_3. A few of these absorption spectra are shown in Fig. 2.40. Owing to the variations in water content, as well as the seasonal changes in the concentration of the molecules formed in biological cycles, the combined absorption spectrum is far from invariant.

Models for the Description of Scattered Radiation

The cross-section for scattering processes in the atmosphere is large enough to make multiple scattering important. The flux reaching a given scattering center is thus composed of a unidirectional part from the direction of the Sun, plus a distribution of intensity from other directions. On the basis of scattering of radiation on atoms and molecules much smaller than the wavelength (the so-called Rayleigh scattering, described above), the intensity distribution over the sky may be calculated, including the direct (unscattered) part as well as the simply and multiply scattered parts. Figure 2.41 gives the result of such a calculation.

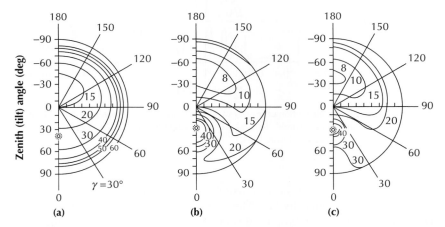

FIGURE 2.41 Luminance distribution over the sky (as a function of the zenith angle and azimuth relative to that of the Sun, for the direction of observation): (*a*) for pure Rayleigh atmosphere (multiple Rayleigh scattering on gases only), (*b*) for Mie atmosphere (including multiple Mie scattering on particles typical of continental Europe for a cloudless sky) and (*c*) measured on a clear day in the Swiss Alps. The three luminance distributions have been arbitrarily normalized to 15 at the zenith. The Sun's position is indicated by a circle. (Based on Möller, 1957.)

Assuming the solar radiation at the top of the atmosphere to be unpolarized, the Rayleigh scattered light will be linearly polarized, and the distribution of light reaching the Earth's surface will possess a component with a finite degree of polarization, generally increasing as one looks away from the direction of the Sun, until a maximum is reached 90° away from the Sun (Fig. 2.42).

The Rayleigh distribution of luminance (intensity) over the sky does not correspond to observations, even on seemingly clear days without clouds or visible haze (see Fig. 2.41). The reason for this is the neglect of scattering on particulate matter. It follows from the particle size distribution given in Fig. 2.35 that particles of dimensions similar to the wavelength of light in the solar spectrum are abundant, so that a different approach to the problem must be taken.

The theory of scattering of radiation on particles, taking into account reflection and refraction at the surface of the particle, as well as diffraction due to the lattice structure of the atoms forming a particle, was developed by Mie (1908). The cross-section for scattering on a spherical particle of radius r is expressed as

$$\sigma = \pi r^2 K(y, n),$$

where the correction factor K multiplying the geometrical cross-section depends on the ratio of the particle radius and the wavelength of the radiation through the parameter $y = 2\pi r / \lambda$ and on the refraction index (relative to air) of the material forming the particle, n.

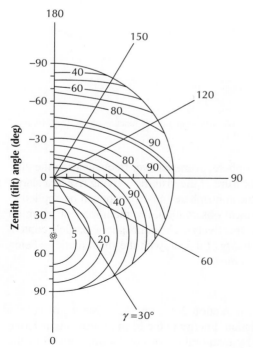

FIGURE 2.42   Distribution of polarization over the sky (in %) for a typical clear-sky atmosphere. (Based on Sekera, 1957.)

Numerical calculations of the function K (Penndorf, 1959) are shown in Fig. 2.43 for fixed refraction index. The behavior is very oscillatory, and the absorption may exceed the geometrical cross-section by as much as a factor of four.

Angular distributions of Mie scattered light are shown in Fig. 2.44 for water particles (droplets) of two different size distributions representative for clouds and haze. The Rayleigh scattering distribution is shown for comparison. The larger the particles, the more forward peaked is the angular distribution, and the higher the forward to backward ratio.

In Fig. 2.41b, the results of a Mie calculation of the luminance distribution over the sky are presented. A particle size distribution proportional to r^{-4} has been assumed, and the mean refraction index of the atmosphere has been fixed at $n = 1.33$ (Möller, 1957). Multiple scattering has been included, as in the case of pure Rayleigh scattering, shown in Fig. 2.41a. For comparison, Fig. 2.41 also shows the results of measurements of luminance performed under the "clean air" conditions of a high-altitude Alpine site. The calculation using the Mie model is in qualitative agreement with the measured distribution, while the Rayleigh calculation is in obvious disagreement with the measurements.

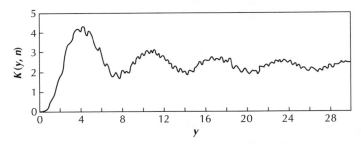

FIGURE 2.43   Cross-section function $K(y, n)$ for Mie scattering, as function of $y = 2\pi r/\lambda$ for fixed $n = 1.50$ (cf. text). (Based on Penndorf, 1959.)

The mean free path in the atmosphere, averaged over frequencies in the visible part of the spectrum, gives a measure of the visibility, giving rise to a clarity classification of the condition of the atmosphere (this concept can be applied to traffic control as well as astronomical observation, the main interest being on horizontal and vertical visibility, respectively). The atmospheric particle content is sometimes referred to as the *turbidity* of the air, with the small particles being called *aerosols* and the larger ones *dust*.

The Types of Stored Energy

The net radiation fluxes discussed in section 2.2 show the existence of energy transport processes other than radiation. Energy in the form of heat is transferred from the land or ocean surface by evaporation or conduction and from the atmosphere to the surface by precipitation, by friction, and again by conduction in small amounts. These processes exchange sensible[*] and latent heat between the atmosphere and the oceans and continents. The exchange processes in the atmosphere include condensation, evaporation, and a small amount of conduction. In addition, energy is added or removed by transport processes, such as convection and advection.

FIGURE 2.44 Angular distributions of radiation for simple (not multiple) scattering on water particles with different size distributions. Mie scattering theory was used in connection with the size distributions of clouds (comprising very large particles or droplets) and haze (smaller particle sizes but still comparable to, *or* larger than, the wavelength of solar radiation). For comparison, the angular distribution for Rayleigh scattering (particle sizes small compared to wavelength) is also shown. The intensity is averaged over different polarization directions. (Based on Hansen, 1975.)

The turbulent motion of the convection processes is often described in terms of overlaying eddies of various characteristic sizes. The advective motion is the result of more or less laminar flow. All motion in the atmosphere is associated with friction (viscosity), so that kinetic energy is constantly transformed into heat. Thus, the general circulation has to be sustained by renewed input of energy. The same processes are involved in the circulation in the oceans, but the quantitative relations between the different processes are different, owing to the different physical structure of air and water (density, viscosity, etc.). As mentioned earlier, the source of energy for the transport processes is the latitude variation of the net radiation flux. Additional transport processes may take place on the continents, including river and surface run-off as well as ground water flow. Direct heat transport in dry soil is very small, because of the smallness of the heat conductivity.

The amount of energy stored in a given volume of air, water, or soil may be written

$$W^{stored} = W^{pot} + W^{kin} + W^{sens} + W^{lat}, \tag{2.13}$$

---

[*]That is, due to heat capacity.

where W^{pot} is the geopotential energy, W^{kin} is the kinetic energy of external motion (flow), W^{sens} is the amount of sensible heat stored (internal kinetic motion), and W^{lat} is the amount of latent heat, such as the energies involved in phase changes (solid to liquid, fluid to gaseous, or other chemical rearrangement).[*] The zero point of stored energy is, of course, arbitrary. It may be taken as the energy of a state where all atoms have been separated and moved to infinity, but in practice it is customary to use some convenient average state to define the zero of energy.

The geopotential energy may be written

$$W^{pot} = \rho(r)g(r)r, \tag{2.14}$$

where ρ is the density (of air, water, or soil) and g is the gravitational constant at the distance r from the Earth's center (eventually r is replaced by $z = r - r_s$ measured from the surface r_s). Both ρ and g further have a weak dependence on the geographical location (ϕ, λ). The kinetic energy W^{kin} receives contributions both from the mean flow velocity and from turbulent (eddy) velocities.

The sensible heat is

$$W^{sens} = \rho c_P T, \tag{2.15}$$

where the heat capacity c_P (at constant pressure) depends on the composition, which again may be a function of altitude and position. All quantities may also depend on time. The latent heat of a given constituent such as water may be written

$$W^{lat} = L_m(m_w + m_v) + L_v m_v, \tag{2.16}$$

where L_v (2.27 J kg^{-1} for water) is the latent heat of vaporization and L_m (0.33 J kg^{-1} for ice) is the latent heat of melting. m_v is the mixing ratio (fractional content by volume) of water vapor and m_w is the mixing ratio of liquid water (applicable in oceans and soil as well as in the atmosphere). For other constituents it may be necessary to specify the chemical compound in which they enter in order to express the content of latent energy.

Total Energy Fluxes

As in the case of radiation, all relevant energy fluxes must now be specified, and the sum of radiation and non-radiation fluxes (E^{nonr}) gives the net total energy flux,

$$E^{total} = E^{rad} + E^{nonr} = E^{sw} + E^{lw} + E^{nonr}. \tag{2.17}$$

The sign convention introduced for radiative fluxes should still be applied.

---

[*]The sum of sensible and latent energy constitutes the total thermodynamic *internal energy* (cf. section 4.1).

Since many energy conversion processes (e.g., melting of snow and ice, run-off) take place at the surface of the Earth, the fluxes through strictly two-dimensional planes may vary rapidly for small vertical displacements of the defining plane and hence may be inconvenient quantities to work with. For this reason, the net total energy flux through such a plane will be, in general, different from zero, except possibly for year averages. For the upward boundary surface between Earth (continent or ocean) and atmosphere, the total net energy flux may be written

$$E_s^{total} = E_s^{rad} + E_s^{lat} + E_s^{sens} + E_s^{pot}, \qquad (2.18)$$

(a)

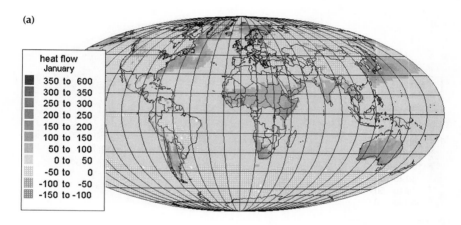

FIGURE 2.45a,b   Flux of sensible heat from the Earth's surface to the atmosphere $(-E_s^{sens})$ for January (*a*) and July (*b*) 1997, in W m^{-2}. (Based on NCEP-NCAR, 1998.)

(b)

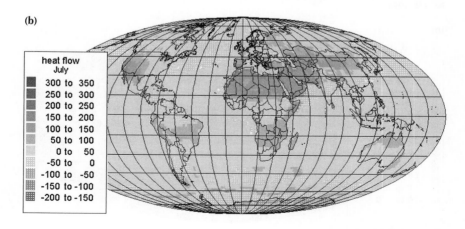

The kinetic energy flow has been neglected, since the vertical velocity of matter at the Earth's surface is usually close to zero (an exception is volcanic matter being expelled into the atmosphere). Yet transport of matter does take place, usually by means of convective eddy flow, with average vertical velocities that increase from zero as the height increases and by falling matter (including precipitation). The changes in geopotential energy involved in these flows are contained in the term E_s^{pot}. The radiation flux E_s^{rad} is given by (2.4). The latent energy flow, E_s^{lat}, is the rate of evaporation times the latent heat of vaporization, L_v. Finally, the net flux of sensible heat, E_s^{sens}, includes all transfer of heat in both directions, such as transfer by convection, conduction, and exchange of matter (e.g., precipitation or volcanic debris) with temperatures different from that of the surroundings.

(a)

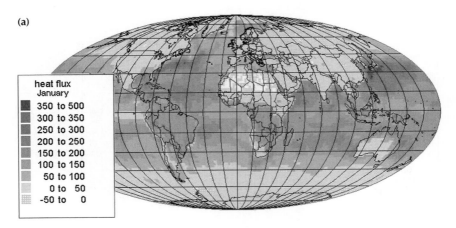

FIGURE 2.46a,b   Flux of latent heat from the Earth's surface to the atmosphere $(-E_s^{lat})$, i.e., L_v times the rate of evaporation, for January (*a*) and July (*b*) 1997, in W m^{-2}. (Based on NCEP-NCAR, 1998.)

(b)

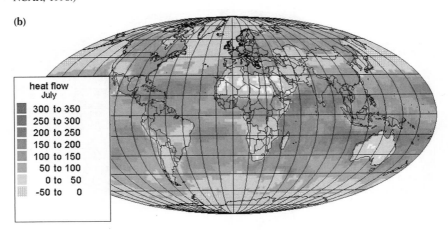

Figures 2.45 and 2.46 show the sensible and latent flux distributions for January and July 1997, and Fig. 2.47 shows the sum of the two fluxes, based on recent model evaluations within the re-analysis project of Kalnay *et al.* (1996). The analysis ensures balance between short- and long-wavelength radiation and nonradiative fluxes at the surface, and it is seen that there are significant areas where the latent and sensible heat fluxes are directed from the atmosphere toward the oceans or land masses. The sum of nonradiative fluxes behaves more smoothly than its components, indicating that the total energy surplus to be moved is split between evaporation and convective processes in ways depending strongly on local conditions (such as relative humidity).

(a)

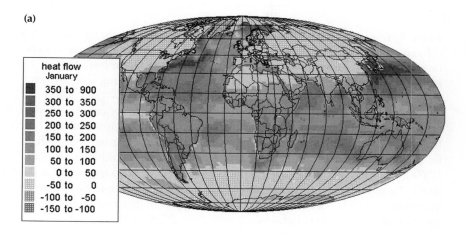

FIGURE 2.47a,b Sum of nonradiative fluxes (i.e., sensible and latent heat flows in Figs. 2.45 and 2.46) from the Earth's surface to the atmosphere for January (*a*) and July (*b*) 1997, in W m^{-2}. (Based on NCEP-NCAR, 1998.)

(b)

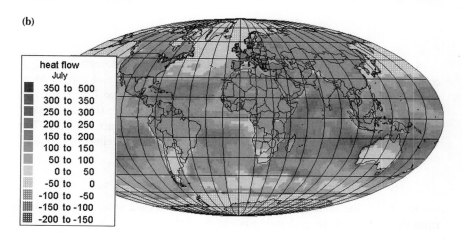

Energy Transfer Processes

Turning now to the behavior of the amount of energy contained in a vertical column, it is convenient to consider first the atmospheric and the continent–ocean parts of the column separately. The depth of the continent–ocean column is taken as one at which no energy is transferred in either of the two vertical directions. This is an idealization that neglects outward transfer of heat produced in the interior of the Earth, which is discussed in section 2.4.

Denoting the continent–ocean column by inferior *d*, the change in stored energy may be expressed in the form

$$\frac{\mathrm{d}W_d^{stored}}{\mathrm{d}t} = E_s^{total} + F_d, \tag{2.19}$$

where F_d represents the horizontal net inflow of energy through the sides of the column considered. For a continental column, F_d would comprise the net surface run-off and ground water loss multiplied by the latent heat content L_w and the possible surplus of sensible heat over the surrounding soil, all taken with opposite sign because of the sign convention in (2.19).

Typically, F_d is much less important for continents than for oceans, where it comprises currents and other horizontal advection. The currents transport heat from the equatorial regions toward the poles, and in particular the North Atlantic (Gulf Stream) and Northern Pacific (Kuro Shio). According to Budyko (1974), the continental F_d is negligible in comparison. Also neglected in the above discussion of F_d is the addition or removal of potential energy by change of density (e.g., when a continental column loses part of its ground water during summer, or in an ocean column as a result of the different densities of cold and warm water). According to (2.13)–(2.16), the change (2.18) in stored energy may involve a change in temperature, formation or melting of ice, etc.

Considering similarly a column in the atmosphere, bounded by the top of the atmosphere and the Earth's surface, the change in stored energy may be written

$$\frac{\mathrm{d}W_a^{stored}}{\mathrm{d}t} = E_0^{total} - E_s^{total} + F_a, \tag{2.20}$$

Since the flux of matter at the top of the atmosphere is negligible, E_0^{total} equals E_0^{rad}, which again equals zero if averaged over time and geographical position, but it depends on latitude and time, as indicated in Fig. 2.16. The quantity F_a represents the energy gain of the column in question, as a result of horizontal transport by the general circulation of the atmosphere. The winds carry both sensible and latent heat, with the latter mainly in the form of water vapor. F_a may thus be written

$$F_a = F_a^{sens} + F_a^{lat} \approx F_a^{sens} + F_s^{lat} + L_v r,$$

where $-E_s^{lat}$ is the latent heat added to the atmosphere as a result of evaporation from the surface of the underlying ocean or continent, and *r* is the amount of precipitation, so that $L_v r$ represents the net transformation of latent heat (bound in water vapor) into sensible heat by condensation processes in the atmosphere. In this way $E_s^{lat} + L_v r$ represents the net vertical loss of latent heat for the atmospheric column. This may be equated to the gain by horizontal inflow, F_a^{lat}, provided the changes in stored, latent heat can be neglected, i.e., that $dW_a^{lat}/dt \approx 0$. Such an approximation is likely to be reasonable if annual averages are taken.

The net latitudinal transport in the atmosphere consists of the longitude-averaged value of F_a and the net transport by ocean currents, F_d, as shown schematically in Fig. 2.48. The quantities $-F_a$ and $-F_d$ express the transport of energy away from a given latitude region, in latitudinal direction. Because they are annual average values, the sum of $-F_a$ and $-F_d$ should equal the net energy flux at the top of the atmosphere, E_0^{rad}. In order to see this, (2.19) and (2.20) are added to give the change in stored energy for an entire vertical column of the Earth–atmosphere system, as a function of time,

$$\frac{d(W_a^{stored} + W_d^{stored})}{dt} = E_0^{rad} + F_a + F_d.$$

The annual average of the left-hand side must be approximately zero, since otherwise the mean temperature in a given latitude zone would continue to increase

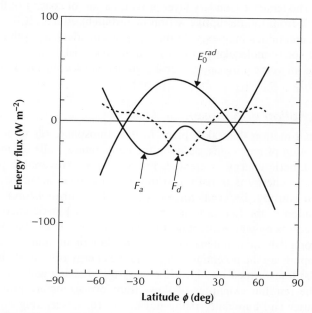

FIGURE 2.48 Schematic illustration of net energy flux components of the Earth–atmosphere system (longitude and time average). The net energy gains by a column at a given latitude, by transport of latent and sensible heat by atmospheric motion and ocean currents, respectively, are denoted F_a and F_d. (Based on Budyko, 1974.)

or decrease. However, there is a transport of energy from lower to higher latitudes, mediated by ocean currents and meridional wind transport, including processes like evaporation, transport of water vapor, and precipitation at another location. Details of the transport patterns can be deduced from data like those presented in Figs. 2.45–2.47, in conjunction with the wind data presented in the following. Variations with time are substantial, including interannual differences. At latitudes higher than 40°–60°, there is a net inflow of latent heat (implying that precipitation exceeds evaporation) as well as a sensible energy gain.

Vertical Transport in the Earth–Atmosphere Boundary Layer

Returning to vertical transport, the important components of the net energy flux E_s^{total} at the Earth's surface (2.18) are the net radiation flux, E_s^{rad}, the latent heat flux, E_s^{lat}, and the net sensible heat flux, E_s^{sens}. These are illustrated in Figs. 2.25 and 2.47.

The vertical transport of heat is predominantly achieved by turbulent motion. However, approaching the physical surface of the Earth, the turbulent air velocity must go to zero and, in a thin air layer close to the ground, no macroscopic motion takes place. This region, called the *laminar boundary layer* (typically only a fraction of a millimeter thick), allows vertical transport of heat, water vapor, and horizontal momentum only by molecular diffusion processes. The laminar boundary layer provides a sort of coating of the surface, protecting it from the disruptive action of turbulent eddies. If the surface is rough, discontinuities may exist in the laminar boundary layer thickness, but a region of solely molecular processes will always exist.

A model of vertical transport near the ground is described and discussed in section 2.5.1.

The Description of Atmospheric Motion

The external motion that takes place in the atmosphere (as opposed to the internal motion of molecules, defining the temperature) calls for mechanisms by which kinetic energy is created. Friction is an irreversible process by which kinetic energy is transformed into sensible heat, accompanied by an increase in entropy. External forces, such as the Coriolis force impressed by the rotation of the Earth, do not create or destroy kinetic energy. Neither do the processes by which latent heat is gained or lost. Thus, the only sources left to accomplish the maintenance of external kinetic motion despite the frictional damping are the potential energy and the energy of sensible heat, which can both be transformed into kinetic energy by adiabatic and reversible processes. The familiar examples of these transformations are gravitationally falling matter and buoyantly rising hot matter (or descending cool matter). In order to estimate the rate of creation and destruction of kinetic energy in the atmosphere, it is convenient first to collect the equations governing the general motion of the atmosphere.

These are: the equation of motion,

$$\rho \frac{d\boldsymbol{v}}{dt} = -\text{grad}\, P + \boldsymbol{F}_{viscous} + \boldsymbol{F}_{ext},$$

where the frictional force $\boldsymbol{F}_{viscous}$ depends on the internal properties of the fluid (the air), and the external force $\boldsymbol{F}_{ext}$ comprises the Coriolis force and the gravitational force, provided a coordinate system is employed that follows the rotation of the Earth; the equation of continuity,

$$\text{div}(\rho \boldsymbol{v}) = -\frac{\partial \rho}{\partial t};$$

and the law of energy conservation (first law of thermodynamics),

$$\frac{dW^{total}}{dt} = work\ plus\ heat\ added\ during\ time\ dt.$$

The work performed on the system is that of the forces appearing on the right-hand side of the equation of motion, and the heat added is partly due to radiation, and partly to molecular conduction processes. W^{stored} contains kinetic, potential, and sensible heat energy. The latent heat may be omitted in W^{stored}, provided any condensation or other process releasing latent heat is included in the heat added on the right-hand side, at the time when the latent heat is transformed into sensible heat.

The above equations would have to be complemented by the equation of state (e.g., the ideal gas law) and the source function specifying the external heat sources. These quantities are not independent of the constitution of the air, and in general the above equations are coupled to a large number of equations specifying the time development of the atmospheric composition (water vapor, ozone, etc.).

In practice, a simplified system of equations is used, based on an analysis of the relative importance of different terms, in particular of the forces, and depending on the scale of motion, which is considered interesting. The dominance of horizontal wind fields in the large-scale atmospheric motion makes it possible to simplify the vertical (z) component of the equation of motion, neglecting also the corresponding component of the viscous and Coriolis forces. This approximation leads to the hydrostatic equation,

$$\frac{\partial P}{\partial z} = -\rho g,$$

where g is the average acceleration of the Earth's gravity (neglecting anisotropy in the mass distribution inside the Earth and the latitude dependence associated with Coriolis corrections), $g \approx GM(r)/r^2 \approx 9.8\ \text{m s}^{-2}$. This does not imply that terms in the equation of motion, such as dw/dt, where w is the z-component of $\boldsymbol{v}$, may be neglected in other circumstances, but only that they are small in comparison with the two terms kept in the equation expressing the hydrostatic approximation. This equation states that the pressure gradient at any infinitesimal horizontal area must balance the weight of the entire air column above it, just as if there were no motion in the atmosphere.

Time Averaging

In order to describe the general circulation in the atmosphere, a time-averaging procedure is performed, over a time interval Δt that is small compared with the interesting large-scale circulation, but large compared with the bulk of turbulent eddy periods,

$$\overline{A} = \frac{1}{\Delta t} \int_{t_1}^{t_1 + \Delta t} A(t_2, \ldots) \mathrm{d}t_2,$$

for any quantity A. The deviation of A from its time-averaged value is denoted $\widetilde{A}$,

$$A = \overline{A} + \widetilde{A}. \tag{2.21}$$

The equations describing the system may be simplified by further introducing the density-weighted averages,

$$A^* = \overline{\rho A} / \rho, \tag{2.22}$$

and the corresponding splitting of A,

$$A = A^* + A'. \tag{2.23}$$

One should note that $\overline{\overline{A}} = 0$, but $\widetilde{A}^* \neq 0$ and that $\overline{\rho A'} = \overline{\rho(A - A^*)} = 0$.

The basic equations for motion of the atmosphere in terms of time-averaged variables are derived in section 2.5.3, which also discusses the experimental evidence for expecting the splitting of atmospheric motion into large-scale average motion and small-scale turbulence to work under certain conditions. It is well known from the quality of meteorological forecasts that the period of predictability is about 2–5 days. This means that the terms coupling the small- and large-scale motion are small enough to be neglected for periods of a few days, but eventually grow (at speeds depending on the actual weather pattern) and make any forecast unreliable. This fundamental limitation on meteorological forecasting does not imply that climatic forecasting is impossible. The weather system is forced by the seasonally periodic incoming solar radiation, which makes the long-term behavior (climate being defined as about 30-year averages of variables) stable and predictable, as a condition for maintaining an atmosphere at all. The boundary conditions of the system make overall stability coexist with unpredictability on short-time or short-distance scales, a feature of quasi-chaotic systems discussed in more precise terms in section 2.3.3.

Features of the Observed Atmospheric Circulation

Figure 2.49 shows the longitude-averaged component of the large-scale wind velocity along the x-axis, i.e., parallel to the latitude circles, for two seasons. This component is called *zonal wind*. At low latitudes, the zonal wind is directed toward the west, at mid-latitudes toward the east (reaching high speeds around 12 km altitude), and again toward the west near the poles.

In Fig. 2.50, the meridional circulation is exhibited, again for two different seasons and averaged over longitudes. The figure gives the streamfunction, related to the mass flux $\rho\,v^*$ by

$$\Psi(\phi,z) = \int \overline{\rho}\,v^* \cdot n r_s \cos\phi\;\mathrm{d}\lambda\;\mathrm{d}s = 2\pi r_s \cos\phi \int_0^{(\phi,\,z)} \overline{\rho}\,[v^*]\cdot n\;\mathrm{d}s,$$

(a)

FIGURE 2.49a,b   Zonal winds (at longitude zero) for January and April of 1997, as functions of latitude and pressure (height). (Based on NCEP-NCAR, 1998.)

(b)

(c)

FIGURE 2.49c,d   Zonal winds (at longitude zero) for July and October of 1997, as functions of latitude and pressure (height). (Based on NCEP-NCAR, 1998.)

where **n** is normal to the area element $d\lambda \, ds$, and the integration path s in the (y, z) plane connects the surface $\Psi = 0$ with the point (ϕ, z) considered. "[ ]" denotes longitude average. The sign of Ψ along a closed-loop streamline is positive if the direction of flow is toward the south at the ground and toward the north aloft. The loops are called meridional cell motion, and one notes a cell of large mass transport near the Equator. Its direction changes with season. On an annual average basis, there are two additional cells on each side of the

Equator. They are called Hadley cells (Hadley, 1735), with the northern one being negative and the southern one being positive, according to the above sign convention. Near the poles there is an even weaker set of meridional cells without seasonal sign changes. Data for 1997 indicate that all the cells extend up to about 200 mb, in contrast to Fig. 2.50 (NASA, 1998). The cell structure is not entirely stable and features transient cells of shorter duration.

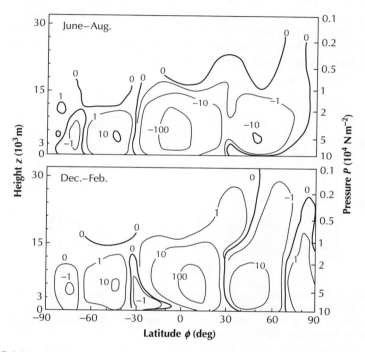

FIGURE 2.50   Summer (*top*) and winter (*bottom*) streamfunctions for the meridional mass transport (longitude average), given in units of 10^9 kg s^{-2}. Cells with a positive streamfunction have northward transport at top. (Based on Newell *et al.*, 1969.)

Combining the standing cell picture with the mean zonal winds, one observes the following regularities near the Earth's surface. In a region near the Equator, the westward zonal winds combine with meridional transport toward the Equator (from both sides) to form the *trade winds* (blowing toward the southwest in the northern Hemisphere and toward the northwest in the Southern Hemisphere). At mid-latitudes the eastward zonal winds combine with meridional cell motion away from the Equator to produce the *westerlies* (blowing toward the northeast in the Northern Hemisphere and toward the southeast in the Southern Hemisphere). Further away from the Equator, in the polar regions, the wind directions are as for the trade winds.

The deviation of the wind field from the longitude average is not unimportant. A part of it is quite regular, in the form of standing horizontal cells, which are usually denoted (large-scale) eddies. They play an important role in transporting angular momentum toward the poles. Figure 2.51 shows the annual and longitude average of this transport.

Angular momentum is added to the atmosphere due to friction and drag forces between the atmosphere and the Earth's surface. Figure 2.52 gives a sketch of the torque acting on the atmosphere as a result of these forces. The main part of the torque is due to friction, but the contribution from drag forces is not negligible in mountain regions. The figure is based on a rough estimate by Priestley (1951), but it is consistent with the angular momentum transport in the atmosphere, as mentioned by Lorenz (1967).

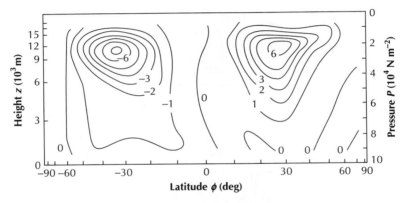

FIGURE 2.51   Northward transport of angular momentum by eddies, averaged over year and longitudes and expressed as 10^{18} kg m^2 s^{-1} per unit of pressure layer, i.e., per 10^4 N m^{-2}. Expressed per unit of height z, the magnitude of transport would increase less rapidly upward toward the two maxima. (Based on Buch, 1954; Obasi, 1963; also using Lorenz, 1967.)

The angular momentum added to the atmosphere at low latitudes is transported to a height of about 10 km by the meridional cell motion. Here it is taken over by the large-scale horizontal eddy motion, which brings it to midlatitudes. Practically no transport across latitudes is performed by the meridional cell motion itself, as can be seen from Fig. 2.53. The absolute angular momentum (i.e., measured in a coordinate system not following the Earth's rotation) of a certain volume of the atmosphere is proportional to the tangential velocity and to the distance to the Earth's axis of rotation, $r \cos\phi$. If the same volume of air is moved poleward, the value of $r \cos\phi$ decreases, and the volume of air must either increase its tangential velocity or get rid of some of its angular momentum. Both happen in the mid-latitude atmosphere. The regions of increased tangential velocity are easily recognized at a height of about 12 km and latitudes of 30–50° (pressure about 200 mb, cf. Fig. 2.49). Some angular momentum is lost by internal friction in the atmosphere, and some is transported back down to Earth level, where it compensates for the angular momentum losses occurring as a result of friction with the surface. The fact that the rotation of the Earth is not accelerating or decelerating shows that, on average, friction forces in opposite directions cancel each other out at low and mid-latitudes. At latitudes above roughly 60°, the transport of angular momentum reverses sign, becoming a net transport toward the Equator.

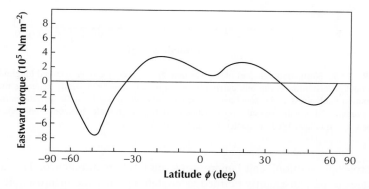

FIGURE 2.52 Estimate of the annual and longitude-averaged eastward torque acting on the atmosphere as a result of friction and momentum impact at the Earth's surface. (Based on Priestley, 1951, as quoted by Lorenz, 1967.)

It has been pointed out by Lorenz (1967) that the transport of angular momentum by large eddies cannot be regarded as a simple diffusion process, since the corresponding "eddy diffusion coefficient" k would in that case have to be negative over a large part of the atmosphere. Indeed, the transport of angular momentum has to be seen as an integral part of the general circulation, which cannot be considered to consist of just zonal and meridional motion. This is also inherent in the equation of motion written in terms of density-averaged quantities (see section 2.5.2), which contains the horizontal eddy motion as a combination of the velocity vector components V_x^* and V_y^*.

Energy Conversion Processes and the Separation of Scales of Motion

In order to include such essential properties as those mentioned above in a model of the circulation, it is important that the time interval Δt used in defining the average properties (2.21) and (2.22) be properly defined. This again poses the question of whether or not a rigorous division of atmospheric motion into large-scale and small-scale (turbulent eddy) motion is at all possible. This issue is dealt with in section 2.5.2, where it is seen that the wind velocity distribution exhibits two distinct categories, which can be denoted

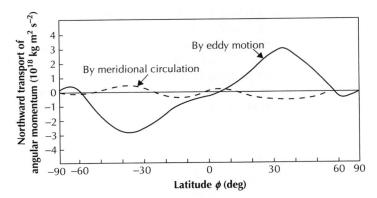

FIGURE 2.53   Northward transport of angular momentum averaged over year and longitude. The meridional contribution is the one carried by meridionally averaged winds, whereas the eddy contribution is obtained as the zonal average of the transport arising from the deviations from the zonal average of the wind velocity vector (thus including large-scale, horizontal eddies). (Based on Buch, 1954; Obasi, 1963, as quoted by Lorenz, 1967.)

"macroscopic motion" and "microscopic motion" or turbulence. This allows averages to be consistently defined as containing only the macroscopic components of motion.

In order to describe all the necessary transport of heat, any accurate model of the circulation of the atmosphere must include couplings to the heat transport within the ocean–continent system (ocean currents, rivers, run-off along the surface, and, to a lesser extent, as far as heat transport is concerned, ground water motion). Such coupled models are considered below in sections 2.3.2 and 2.5.2.

The kinetic energy of the atmospheric circulation is diminished by friction, leading to an irreversible transformation of kinetic energy into internal energy (heat). In order to make up for such frictional losses, new kinetic energy must be created in the atmosphere. This can be achieved essentially by two processes, both of which are reversible and may proceed adiabatically (details of the formulation are given in section 2.5.2). One is the conversion of potential energy into kinetic energy (by gravitational fall), and the other is the transformation of internal energy into kinetic energy by motion across a pressure gradient.

Creation and Destruction of Kinetic Energy

Direct measurements of heat and temperature distributions (cf. Fig. 2.32 and section 2.5.2) allow an estimate of the generation of internal energy available for possible conversion into kinetic energy. With the use of wind data (e.g., Figs. 2.49–2.51), it is possible to obtain an estimate of the conversion of available energy and large-scale horizontal eddy motion into zonal motion. Also, the available energy may be split into a zonal part corresponding to the use of the zonal mean temperatures and a (large-scale) eddy part derived from the deviations of the temperature from longitudinal average (Lorenz, 1967; Newell *et al.*, 1969). Separate estimates exist for the frictional losses from zonal and eddy motion as a function of height. An attempt to survey the main conversion processes (deriving poorly known quantities, such as the conversion of available eddy energy into kinetic eddy energy, from balance requirement) is given in Fig. 2.54, based on Lorenz (1967) and considering comments by Newell *et al.* (1969), who discuss the seasonal variations and uncertainties involved in the procedures used.

On an annual and global average basis, the creation of kinetic energy in the form of large-scale motion (V^*, w^*) amounts to 2.3 W m^{-2} or 0.7% of the solar radiation at the top of the atmosphere. For consistency, the frictional losses must be of equal magnitude, which is not quite consistent with direct estimates (4–10 W m^{-2}). Newell *et al.* argue that the value of about 2.3 W m^{-2} given in Fig. 2.54 is most likely to be correct (see also Sørensen, 2008a).

FIGURE 2.54 Estimated summary of the average energy (*boxes*) and main conversion processes in the atmosphere, for a vertical column, averaged over place and time. Boxes contain kinetic energy (W^{kin}) and the amounts of internal and potential energy available for conversion into kinetic energy (W^{avail}, "available" refers to energy above the reference state of the atmosphere). The compartments are further divided into the energy of zonally averaged motion and the average deviations from these, denoted eddy energy. (Based on Oort, 1964; Lorenz, 1967.)

The process of creating kinetic energy may be intuitively interpreted in terms of air parcels of a temperature different from the ambient temperature. Such air parcels will rise (if they are hotter than the surroundings) or sink (if they are colder) and give rise to a redistribution of the surrounding air (i.e., to kinetic motion) for as long as temperature differences exist at the same altitude. More precisely, the change in kinetic energy is due to movement of the air parcels across horizontal or vertical pressure gradients, as directly implied by the equations given in section 2.5.2.

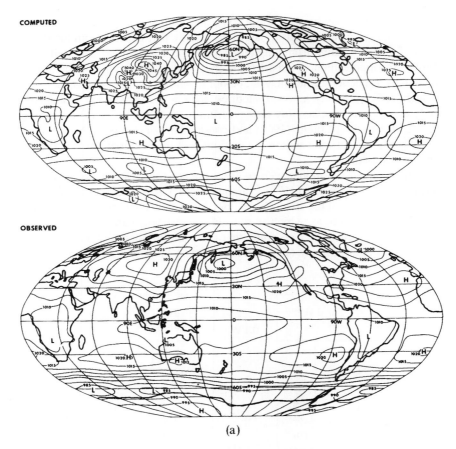

COMPUTED

OBSERVED

(a)

FIGURE 2.55a   Calculated (*top of each pair*) and observed (*bottom of each pair*) sea-level pressure in mb (1 mb being 10^2 N m^{-2}), (*a*) averaged over the period December–February (for *b*, see next page).

Models of General Atmospheric Circulation

Before going into the influence of the oceans and land masses on the atmospheric circulation, examples of early numerical simulation of the atmospheric circulation without including couplings to oceanic circulation are presented. (Based on the work of Manabe *et al.*, 1974; Manabe and Holloway, 1975; Hahn and Manabe, 1975.) The early simulations used a global, horizontal grid of about 250 km latitude and longitude steps and around ten vertical layers extending to a height of about 30 km, including equations of motion for wind, temperature, and water vapor, and using the hydrostatic approximation and the ideal gas equation of state. Continental topography is modeled, whereas the oceans are prescribed seasonal temperatures as boundary conditions for the atmospheric integration. The radiation source term is calculated as a function of the seasonal variation in extraterrestrial radiation, the state variables plus ozone and carbon dioxide data. The hydrological cycle includes a model for evaporation, condensation, and cloud distribution, as well as the processes involving ice and snow formation and accumulation as terrestrial snow cover or sea ice. Turbulent and sub-grid-scale convection and transport of vapor have been modeled.

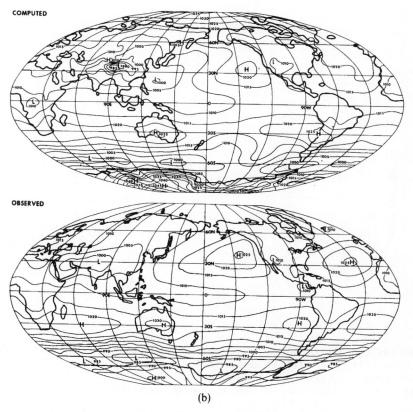

(b)

FIGURE 2.55*b*   Calculated (*top of each pair*) and observed (*bottom of each pair*) sea-level pressure in mb (1 mb being 10^2 N m^{-2}), (*b*) averaged over the period June–August (cf. *a* on previous page). (From S. Manabe and J. Holloway (1975), *J. Geophys. Res.* **80**, 1617–1649, copyright American Geophysical Union.)

Figures 2.55*a* and *b* compare the average ground-level pressures during the periods December to February and June to August with observed values. Figure 2.56 gives the mean zonal wind for July as a function of latitude and height. The longitudinal averaging extends only over the interval 80 °E to 95 °E, whereas the observed distribution is for 100 °E. Within the latitudes included in the figure, the agreement is fair, but in general the model is rather poor in predicting the longitude average of zonal winds in all height–latitude regions, in particular the magnitude and position of the strong mid-latitude jets observed at a height of about 12 km (an indication of the sensitivity of these features on the seasonal variation of radiation and on the model grid size may be inferred from a comparison with earlier work, e.g., Holloway and Manabe, 1971).

FIGURE 2.56 Observed (*top*) and calculated (*bottom*) zonal wind for July, in m s^{-1} (positive eastward). The lower silhouette represents the mountain topography, an idealized one being assumed in the calculation. The observed wind field is for 100 °E, whereas the calculated one has been averaged over the interval 80–95°E. (Based on Manabe *et al.*, 1974.)

Figure 2.57 gives the streamfunction corresponding to the calculated meridional circulation for January and July. Corresponding measured values, averaged over somewhat longer periods (three months), are shown in Fig. 2.50. In this case, the agreement is quite convincing, both in regard to the strength of the Hadley cells near the Equator and to the distribution of the weaker cells at higher latitudes. Figure 2.58 gives the northward transport of absolute angular momentum by eddies for July and January, averaged over longitudes. The corresponding measured values are found in Fig. 2.51, but averaged over the year.

Figure 2.58 shows that the calculated transport is strongest in winter, and the average of January and July generally concurs with the value derived from measurements.

In Fig. 2.59, the annual average precipitation rates derived from the model calculation are compared with measured values. Apparently, agreement between the two is very good, over both oceans and continents. A comparison with earlier efforts by Manabe's group (e.g., those neglecting the seasonal variation in the extraterrestrial radiation) shows that the precipitation rates are very sensitive indicators of the appropriateness of the assumptions regarding the hydrological cycle. The more accurate presentation of measured data in Fig. 2.61 indicates detailed features not captured by the limited early atmospheric model results.

As suggested by Fig. 2.54, the formation of kinetic energy mainly involves the eddy motion (i.e., deviations from zonal mean winds). Figure 2.60 shows the calculated latitude variations of the main conversion branches: the transformation of available eddy energy into eddy kinetic energy and of zonal kinetic energy into eddy kinetic energy (which is mostly negative, i.e., the process goes the opposite way) and the dissipation of eddy kinetic energy through friction (transformation of large-scale eddies into small-scale eddies, cf. section 2.5.2, and the small-scale eddies eventually into heat).

FIGURE 2.57   July (*top*) and January (*bottom*) streamfunctions of longitude-averaged meridional mass transport in units of 10^9 kg s^{-1}. (Based on Manabe and Holloway, 1975.)

FIGURE 2.58 Northward transport of angular momentum by eddies, averaged over longitude, for July (*top*) and January (*bottom*), in 10^{18} kg m^2 s^{-2} per unit of pressure, 10^4 N m^{-2}. (Based on Manabe *et al.*, 1974.)

Calculated height distributions indicate that the frictional dissipation is not confined to regions near the ground (Manabe *et al.*, 1974). Comparing the conversion rates with Fig. 2.54, it is evident that the order of magnitude is correct. The formation of eddy kinetic energy from eddy available energy and the frictional dissipation are both about 50% larger than the value deduced from observations, and the zonal to eddy kinetic energy is about three times too small. However, it should be remembered that there are no directly observed values for these conversions.

2.3.2 The Oceans and Continents

The Water Cycle

The processes by which water is transferred between different regions of the ocean–soil–atmosphere system may be described in much the same way as that used for the transfer of energy. Considering first a vertical column extending from the ocean–continent surface down to a level where no exchange of water takes place, the change in the amount of water present within the column, A_{wd}, can be written similarly to (2.19),

$$\frac{\mathrm{d}A_{wd}}{\mathrm{d}t} = E_{ws}^{total} + F_{wd}, \qquad (2.24)$$

where the total water flux across the atmosphere-to-ocean/continent boundary is

$$E_{ws}^{total} = r + d - e, \tag{2.25}$$

where r is the rate of precipitation, d is the formation of dew (usually negligible in comparison with the former), and e is the rate of evaporation. F_{wd} is the net inflow of water into the column due to advection, such as river or surface run-off and ground water motion on continents and ocean floors. In the oceans, the net inflow is usually zero, except for density changes.

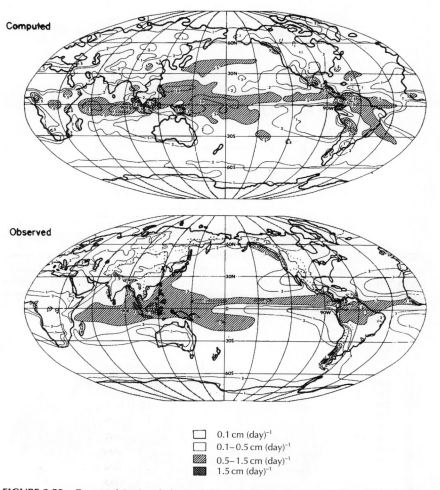

FIGURE 2.59 Computed (*top*) and observed (*bottom*) annual average rate of precipitation. The key to shading of different intervals is at the bottom, 1 cm/day being about 1.16×10^{-7} m^3 water equivalent per m^2 and per second. (From S. Manabe and J. Holloway, 1975: *J. Geophys. Res.* **80**, 1617–1649, copyright American Geophysical Union.)

Figures 2.45 and 2.46 show the net upward flux of sensible and latent heat. The dominating upward part of the latter is evaporation, whereas the precipitation r shown in Fig. 2.61 is only a part of the net sensible energy exchange. Assuming that dA_{wd}/dt is zero when averaged over the year, Sellers (1965) has used (2.24) to calculate the average net inflow of water into a column, F_{wd}, shown in Fig. 2.62 along with longitudinal averages of r and e.

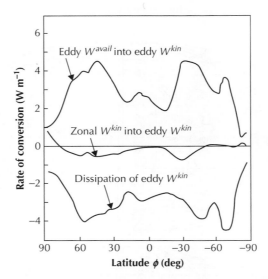

FIGURE 2.60 Calculated rates of eddy kinetic energy formation and destruction are shown, integrated over height and averaged over longitude and time. The middle curve is only approximately equal to the rate of conversion from zonal to eddy kinetic energy, since a contribution from the divergence of the eddy energy flux has been subtracted. (Based on Manabe *et al.*, 1974.)

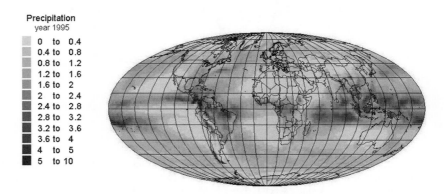

FIGURE 2.61 Annual precipitation 1995, m/y. (Based on NCEP-NCAR, 1998.)

For a vertical column extending from the surface upward into the atmosphere, the change in the amount of water present can be written

$$\frac{dA_{wa}}{dt} = -E_{ws}^{total} + F_{wa},$$ (2.26)

assuming that no water is leaving the column at the top of the atmosphere. If again dA_{wa}/dt is zero on an annual average basis, the transport of water horizontally through the atmosphere is given by

$$F_{wa} = -F_{wd}.$$

From the net influx into a vertical column one may calculate the required northward transport of water, which is very similar to the value estimated from measurements in the atmosphere, shown in Fig. 2.63. In order to construct an average water cycle, it is also necessary to split the ocean–continental transport term, F_{wd}, into its main components, i.e., the river and surface run-off (which

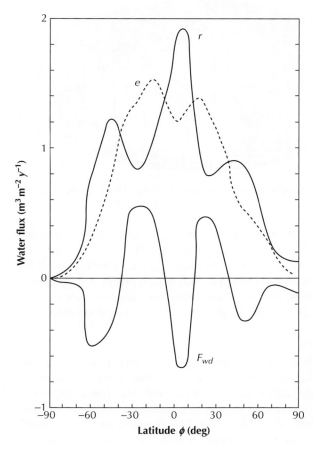

FIGURE 2.62 Latitude dependence (averaged over year and longitude) of water fluxes into a vertical column extending from the Earth's surface downward: precipitation r, evaporation e, and net inflow into the column, F_{wd}. (Based on Sellers, 1965.)

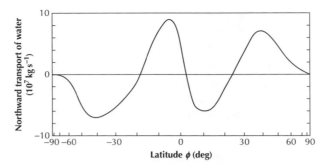

FIGURE 2.63   Northward transport of water, averaged over the year and summed over longitudes (computed from fluxes similar to those of Fig. 2.58, assuming the transport to be zero at the poles). (Based on Peixoto and Crisi, 1965.)

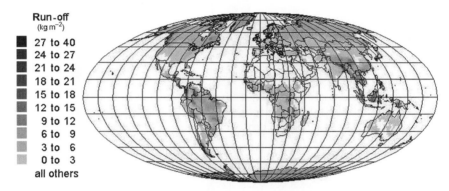

FIGURE 2.64   Accumulated run-off from land areas during 1997 (kg of water per m^2) (NCEP-NCAR, 1998).

can be estimated from observations, as shown in Fig. 2.64), and the ground water flow (which may be calculated as a residue). This has been done in Fig. 2.65, based on Kalinin and Bykow (1969) and Lvovitch (1977). Also indicated are the total amounts of water stored in each compartment. Dividing these by the corresponding fluxes, one obtains the turnover time for water in each compartment, a quantity that is very large for ice caps and deep oceans.

Evaporation Processes

The evaporation e (2.25) from a given surface depends on several factors, such as the temperature and moisture of the surface; the temperature, humidity, and wind velocity of the air above the surface; and the vertical gradients of these quantities. For soil evaporation, the moisture of the soil is considered to be of importance only if it is below a "critical" level. The physical parameter behind this "critical" value is likely to be the moisture tension, which plays the same role for evaporation as does the surface tension for the evaporation from the surface of a fluid (e.g., a water surface). The evaporation from vegetation is further complicated by the plant's ability to regulate the transportation of vapor from its interior, by opening and closing its pores (stomata). The release of water from plant-covered soil is called *evapotranspiration* (cf. Geiger, 1961; Sellers, 1965).

The vertical transport of moisture through the lower atmosphere is further described in section 2.5.1. The potential evaporation from land surfaces, inferred from measured data, is shown in Fig. 2.66 (Kalnay *et al.*, 1996). This is the evaporation that would take place if water transport from the surface were adequate. Evaporation from ocean surfaces has a dependence on wave and wind conditions, including the mechanism of whitecaps, from which water droplets evaporate before falling back into the sea. This evaporation process leaves salt particles that may serve as condensation nuclei for water vapor not carried away (up or to the side) quickly enough.

The State Variables of the Oceans

The state of the oceans, like that of the atmosphere, is given by the temperature T and the density $\rho = \rho_w$, but other variables are of great significance, primarily

FIGURE 2.65 Schematic summary of the water cycle, including free water to a depth of about 5 km below the Earth's surface. The transfer rates, given at the arrows in 10^{12} m^3 y^{-1}, are averaged over time. For each compartment (box), the average water volume is indicated. (Based on Budyko, 1974; Kalinin and Bykow, 1969; Lvovitch, 1977.)

the salt concentration and, in regard to biomass production, the amounts of dissolved oxygen and nutrients. The salinity S (salt fraction by mass) is necessary in order to obtain a relation defining the density, to replace the ideal gas law used in the atmosphere.

Figures 2.67–2.69 show measured temperature and salinity distributions along sections through the three major oceans, at latitude circles at longitudes 30 °W (Atlantic Ocean), 90 °E (Indian Ocean), and 150 °W (Pacific Ocean) (Levitus and Boyer, 1994).

Notable features of all three oceans are the large temperature gradients over the first 1000 m below the sea surface, observed at low latitudes. At latitudes of about 45°, the average temperature gradients become zero, and at higher latitudes they are generally small and irregular. The figures shown are annual average values, and the sea surface temperatures at the higher latitudes exhibit considerable seasonal variation, apart from regions covered by ice for most or all of the year. The regions in the upper water layers near the Equator generally have a high salinity, which can be understood in terms of the extensive evaporation, since most of the salt content is left behind by evaporation processes.

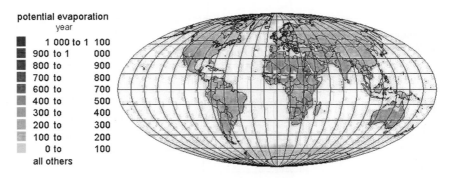

potential evaporation
year

■	1 000 to 1 100
■	900 to 1 000
■	800 to 900
■	700 to 800
■	600 to 700
■	400 to 500
■	300 to 400
■	200 to 300
■	100 to 200
□	0 to 100
	all others

FIGURE 2.66 Potential evaporation from land surfaces, expressed in energy units (by multiplication with L_v) and averaged over the year (NCEP-NCAR, 1998).

The asymmetry between the salinity distributions in the two hemispheres has to do with the geographical specificity of vertical mixing processes in the oceans. Convective overturn between the surface waters and deep-sea waters (2000 m or more) takes place only in the Atlantic Ocean just south of Greenland and possibly in the Norwegian Sea (Reid, 1971). These regions are characterized by a small density increase as a function of depth. When cool and dense water is formed in these regions near the surface, such water can sink or mix downward with little hydrostatic resistance. Also, in the Antarctic region, cool and dense water is formed in winter, but owing to substantial precipitation, the salinity is low and, as seen from Figs. 2.67–2.69, the temperature and salinity in open sea remain fairly stratified. Only along the continental (Antarctic) shelf does the cool water sink down, in an essentially non-turbulent manner (primarily in the Weddell Sea). This accounts for the difference between the conditions (in particular the salinity) in the Southern Hemisphere and the Northern Hemisphere, which contains the only regions of deep convective mixing and thus the possibility of adding to the bottom some water that has recently been in strongly interactive contact with the atmosphere. In the Norwegian Sea, large masses of water have a composition revealing equilibrium between atmosphere and water. This water does not immediately sink to the bottom of the Atlantic Ocean, owing to the shallow sill separating the Norwegian Sea from the main part of the ocean (see Fig. 2.67). Instead, water from the Norwegian Sea is pressed through the narrow passage in a highly turbulent mixing process, thus losing density and becoming unable to penetrate to the very bottom. It is therefore believed that the water near the bottom of the Atlantic Ocean is of Antarctic origin.

FIGURE 2.67a,b
Annual average temperature T (a) and salinity S (b) along a south–north section (30 °W) through the Atlantic Ocean. (Based on Levitus and Boyer, 1994; data source: University of Columbia, 1998.)

(a)

(b)

FIGURE 2.68a,b
Annual average temperature T (a) and salinity S (b) along a south–north section (90 °E) through the Indian Ocean. (Based on Levitus and Boyer, 1994; data source: University of Columbia, 1998.)

Scales of Oceanic Motion

The time scale of the oceanic circulation is very different from that of the atmosphere. A frequency spectrum analogous to the atmospheric one (cf. section 2.5.2) will reveal cycle times for the average circulation of a year or more, increasing with depth by about a factor of 10 and being another factor of 10 larger for motion in a vertical plane than for horizontal motion. Temperature adjustments in the upper ocean require times of the order of weeks.

As in the case of the atmosphere, it is thus possible to separate the general circulation from the short-term movements, which in the oceans, however, comprise not only quasi-random, turbulent motions (eddies), but also organized motion in the form of waves. As mentioned above, wave motion may play an important role in the process of transferring mechanical energy between the atmosphere and the ocean through the wind stress, the magnitude of which may itself depend on the waves created.

FIGURE 2.69a,b
Annual average temperature T (a) and salinity S (b) along a south–north section (150°W) through the Pacific Ocean. (Based on Levitus and Boyer, 1994; data source: University of Columbia, 1998.)

To provide a rough picture of the classes of wave motion, Fig. 2.70 indicates the frequency spectrum of wave motion in the oceans (Roll, 1957). The harmonic waves in the spectral decomposition may be characterized by their wavelength, λ_w, the phase velocity, U_w, and the amplitude at the water surface, a. Figure 2.66 shows a time-averaged estimate of the distribution of amplitudes, a, on cycle times, $T_w = \lambda_w U_w^{-1}$. For waters of depth h much larger than the wavelength λ_w, harmonic waves governed by the gravity and surface tension forces are characterized by the relation

$$U_w = \left(\frac{g\lambda_w}{2\pi} + \frac{2\pi\gamma_w}{\lambda_w\rho_w} \right)^{1/2}, \tag{2.27}$$

between wave velocity and wavelength, provided that the viscous forces in the water are neglected. The gravitational acceleration at the Earth's surface is $g = 9.81$ m s^{-1}, and the surface tension of water against air, γ_w, has a value of about 0.073 N m^{-1}. It follows from (2.27), which is shown graphically in Fig. 2.71,

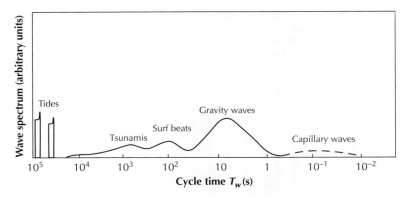

FIGURE 2.70   Sketch of the main features of the spectral decomposition of ocean wave amplitudes, averaged over time and position. (Based on Roll, 1957.)

that U_w must exceed a minimum value of about 0.23 s^{-1} and that two values of λ_w are associated with each U_w above the minimum value. The branch of wavelengths smaller than about 0.017 m (corresponding to the minimum U_w) is called *capillary waves*, since they are mainly governed by the surface tension. The branch of wavelengths above $\lambda_w = 0.017$ m are the *gravity waves*, for which the term involving g in (2.27) dominates.

At cycle times around 2 min, pulsations in the amplitude of gravity waves formed earlier by interaction with the wind field may give rise to the surf beat phenomenon observed at shores (Sverdrup and Munk, 1947; Munk, 1980). The

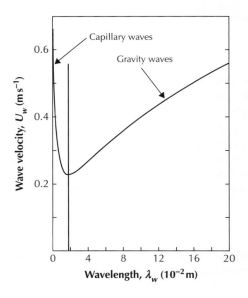

FIGURE 2.71   Theoretical relationship between wave (phase) velocity, U_w, and wavelength, λ_w, under the influence of gravity and surface tension, but neglecting viscosity. (Based on Roll, 1957.)

broad peak in the wave spectrum in Fig. 2.70, centered on $T_w = 20$ min, includes the flood-producing *tsunamis* occurring quite frequently, e.g., in the Pacific Ocean, as a result of sudden movements of the ocean floor caused by earthquakes (Neumann and Pierson, 1966).

The peaks at 12 and 24 h represent the tidal waves created by the time-dependence of the gravitational fields of various celestial bodies. The most important tides are caused by the Moon and the Sun in connection with the variations in distance from a given spot on the Earth's surface, resulting from the rotation of the Earth. Much longer tidal periods may be associated with orbital motion (lunar month, year, etc.). The theoretical description of waves and tides is discussed in section 2.5.3.

Joint Models of General Circulation in Oceans and Atmosphere

Early models coupling atmospheric and ocean circulation models did not include any detailed topography, had fixed amounts of cloud cover, and prescribed a radiation field corresponding to constant carbon dioxide levels, constant water vapor, and constant ozone. Oceanic equations included temperature, salinity, and ice, and the hydrological model was fairly detailed on precipitation, evaporation, soil moisture, snow cover, sea ice, melting processes, and river run-off (e.g., assuming that when soil moisture exceeded a certain value, the excess would run off to the nearest ocean). The spatial grids were typically around 500 km by 500 km at the Equator, with around 10 vertical layers, and the time steps were of the order of a month for oceans, but of the order of 6 h for the atmosphere (Bryan, 1969; Manabe, 1971; Wetherald and Manabe, 1972).

As illustrated by, for example, the ocean salinity calculation (Fig. 2.72), the general results were quite realistic (compare with an average of the salinity data shown in Figs. 2.67–2.69). The models also allowed a first orientation into the structure of major current systems in the oceans (Fig. 2.73).

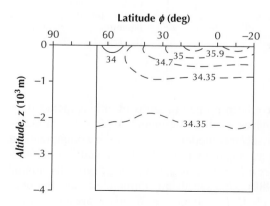

FIGURE 2.72   Average salinity ($g\ kg^{-1}$) calculated with an early joint ocean–atmosphere model without seasonal variations in radiation. (Based on Bryan, 1969.)

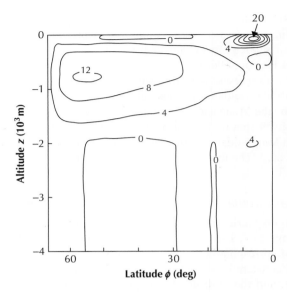

FIGURE 2.73 Streamfunction of meridional transport of water, averaged over year and longitude, calculated by a joint ocean–atmosphere model (unit: 10^9 kg of water per second). The sign convention is such that a meridional cell with positive streamfunction has northward transport at the top and southward below. (Based on Wetherald and Manabe, 1972.)

FIGURE 2.74 Coupled ocean–atmosphere model results for ocean salinity, longitudinally averaged (England *et al.*, 1993).

Over the years, models have been refined in terms of effects included, as well as in terms of the mesh sizes used. Some models replace the latitude–longitude compartments by a spectral model for the Fourier components of the variables [i.e., expanding all functions in terms of sine and cosine functions of a base frequency times n ($n = 1, ... n_{max}$), with n_{max} typically around 30] (Manabe and Stouffer, 1988). This speeds up computing times, but once realistic topography is included, the advantage of this representation is diminished.

(a)

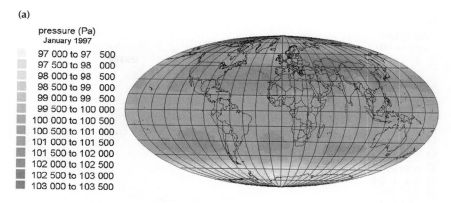

pressure (Pa)
January 1997

- 97 000 to 97 500
- 97 500 to 98 000
- 98 000 to 98 500
- 98 500 to 99 000
- 99 000 to 99 500
- 99 500 to 100 000
- 100 000 to 100 500
- 100 500 to 101 000
- 101 000 to 101 500
- 101 500 to 102 000
- 102 000 to 102 500
- 102 500 to 103 000
- 103 000 to 103 500

FIGURE 2.75a,b   Coupled ocean–atmosphere model results for mean sea-level pressure (Pa) in January 1997 (*a*) and July 1997 (*b*). HADCM2-SUL, the model used, which includes sulfate aerosols, has been integrated from pre-industrial times (Mitchell and Johns, 1997).

(b)

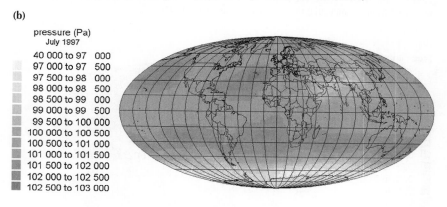

pressure (Pa)
July 1997

- 40 000 to 97 000
- 97 000 to 97 500
- 97 500 to 98 000
- 98 000 to 98 500
- 98 500 to 99 000
- 99 000 to 99 500
- 99 500 to 100 000
- 100 000 to 100 500
- 100 500 to 101 000
- 101 000 to 101 500
- 101 500 to 102 000
- 102 000 to 102 500
- 102 500 to 103 000

Current models include topography, a more complete water cycle with a cloud model, and a number of minor components in the atmosphere contributing to the radiation balance. The transfer of energy and momentum between ocean surfaces and the atmosphere has been a difficult process to model, and only since about 1995 has it been possible to formulate coupled ocean–atmosphere models not requiring some artificial correction for the mismatch between oceans and atmosphere to be introduced. This need was not only due to lack of knowledge of the processes involved, but also came from the different numerical treatment of ocean circulation and atmospheric circulation, notably the difference in time steps, and from having to start up the model by having it run for a period of time in order to gain stability (the finding that one could not model ahead from some known state of the atmosphere rests on the fact that input data were never fully consistent with the model structure and restrictions, so that an "artificial" initial state had to be created by running the model for several years, cf. the discussion in Gates *et al.*, 1996). Current models use spatial grids of 30–100 km spacing and around 30 atmospheric levels plus a similar number of ocean depth levels,

(a)

precipitation
January 1997

    0   to 0.04
  0.04 to 0.08
  0.08 to 0.12
  0.12 to 0.16
  0.16 to 0.2
  0.2 to 0.24
  0.24 to 0.28
  0.28 to 0.32
  0.32 to 0.36
  0.36 to 0.4
  0.4 to 0.5
  0.5 to 1
  ☐ all others

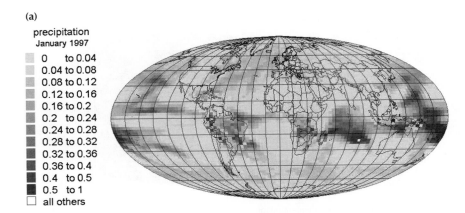

FIGURE 2.76a,b Coupled ocean–atmosphere model results for total precipitation (m/y) in January 1997 (*a*) and July 1997 (*b*). The HADCM2-SUL model, which includes sulfate aerosols, has been integrated from pre-industrial times (Mitchell and Johns, 1997).

(b)

precipitation (m)
July 1997

    0   to 0.04
  0.04 to 0.08
  0.08 to 0.12
  0.12 to 0.16
  0.16 to 0.2
  0.2 to 0.24
  0.24 to 0.28
  0.28 to 0.32
  0.32 to 0.36
  0.36 to 0.4
  0.4 to 0.5
  0.5 to 1
  ☐ all others

with temporal grids of around half an hour for the atmosphere and an hour or more for the ocean part (see, for example, UK Meteorological Office, 1997).

Models used near the year 2000 include the greenhouse effects of a number of gases as well as scattering and reflection effects of aerosols, such as those derived from sulfur dioxide emissions from human activity. Earlier models lumped these together as an effective change in solar radiation fluxes used in the models. Also, the interactions between the biosphere and the atmosphere and oceans are important for both heat and moisture fluxes, albedo, and surface roughness experienced by winds. This means that the seasonal variation in standing crops as well as agricultural and forestry practices become important for climate modeling, along with direct anthropogenic interference through emissions of polluting substances to the atmosphere, in addition to natural

(a)

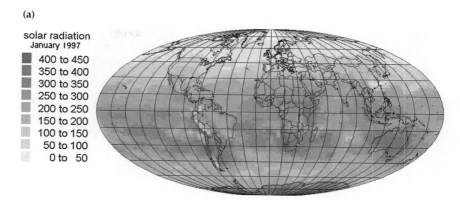

FIGURE 2.77a,b   Coupled ocean–atmosphere model results for surface solar radiation (W m^{-2}) on a horizontal surface in January 1997 (*a*) and April 1997 (*b*). The HADCM2-SUL model includes sulfate aerosols (Mitchell and Johns, 1997).

(b)

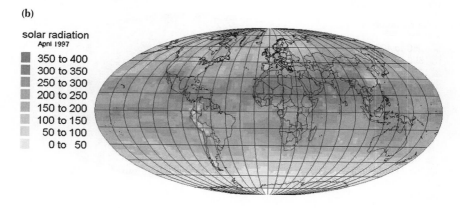

processes, such as volcanic eruptions. The human interference is further discussed below.

Figure 2.74 shows a newer model calculation of ocean salinity, considerably more accurate than the early model shown in Fig. 2.72. Figures 2.75–2.79 show mean sea-level pressure (i.e., pressure adjusted to zero height), precipitation, surface solar radiation, wind speed, and temperature from a coupled ocean–atmosphere model including sulfate aerosol effects and greenhouse gases corresponding to 1997 levels. Where seasonal variation is important, the calculation is shown for two or four seasons. Measured data for comparison are contained in Fig. 2.80 for sea-level pressure, with Fig. 2.81 giving the pressure levels at the actual surface for comparison, and Fig. 2.82 giving those for precipitation, while solar radiation on a horizontal plane is shown in Fig. 2.24 and seasonally in Fig. 3.1.

Figure 2.85 shows observed surface temperatures and Figs. 2.83 and 2.84 show observed wind speeds (*w*) and directions (*d*) at the surface (10 m).

(c)

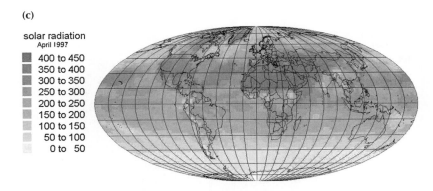

FIGURE 2.77c,d Coupled ocean–atmosphere model results for surface solar radiation on a horizontal surface in July 1997 (c) and October 1997 (d). The model includes sulfate aerosols (Mitchell and Johns, 1997).

(d)

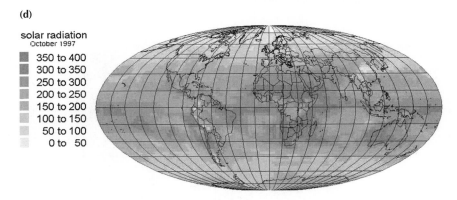

These two quantities are derived from the zonal (u) and meridional (v) components of the wind velocity,

$$w = (u^2 + v^2)^{1/2}, \quad d = A \tan (v/u) \text{ or } \pm 90° (\text{if } u = 0 \text{ and } v > 0 \text{ or } v < 0).$$

The sea-level pressure model results are generally consistent with the data, but detailed agreement is better on the Southern Hemisphere. In the Northern Hemisphere, the variations from North America over the North Atlantic Ocean to Europe are out of phase with data for January, as are the Europe to Asia variations in July. It should be kept in mind that the model does not aim at reproducing data for individual years, but the features of the Northern hemisphere measured pressures are fairly general, so a slight imbalance in the model treatment of oceans and land may be indicated.

The model results for precipitation rates are generally in better agreement with measured values, for both January and July of 1997. Only for an area northeast of Australia does the model predict much heavier rainfall than indicated by the observations.

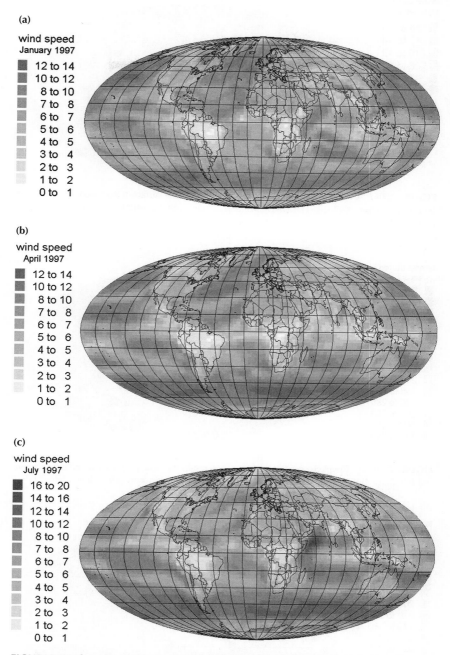

FIGURE 2.78a,b,c Coupled ocean–atmosphere model results for surface wind speed (m s^{-1}, height 10 m), in January (*a*), April (*b*), and July 1997 (*c*). The HADCM2-SUL model includes sulfate aerosols (Mitchell and Johns, 1997).

wind speed
October 1997

14 to 16
12 to 14
10 to 12
 8 to 10
 7 to 8
 6 to 7
 5 to 6
 4 to 5
 3 to 4
 2 to 3
 1 to 2
 0 to 1

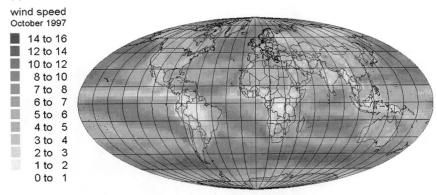

FIGURE 2.78*d* Coupled ocean–atmosphere model results for surface wind speed (m s^{-1}, height 10 m) in October 1997 (*d*). The HADCM2-SUL model includes sulfate aerosols (Mitchell and Johns, 1997).

(a)

temperature (K)
January 1997

200 to 230
230 to 240
240 to 250
250 to 260
260 to 265
265 to 270
270 to 275
275 to 280
280 to 285
285 to 290
290 to 295
295 to 300
300 to 305
310 to 310
310 to 340

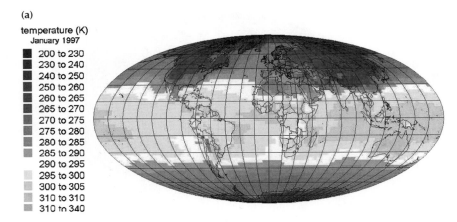

FIGURE 2.79*a,b* Coupled ocean–atmosphere model results for surface temperature (K) in January (*a*) and July (*b*) of 1997. The HADCM2-SUL model includes sulfate aerosols (Mitchell and Johns, 1997).

(b)

temperature (K)
July 1997

180 to 230
230 to 240
240 to 250
250 to 260
260 to 265
265 to 270
270 to 275
275 to 280
280 to 285
285 to 290
290 to 295
295 to 300
300 to 305
305 to 310
310 to 360

(a)

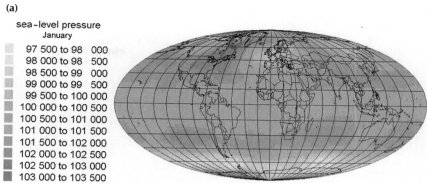

FIGURE 2.80a,b Observed mean pressure, reduced to sea level, for January (*a*) and July (*b*) of 1997 (NCEP-NCAR, 1998).

(b)

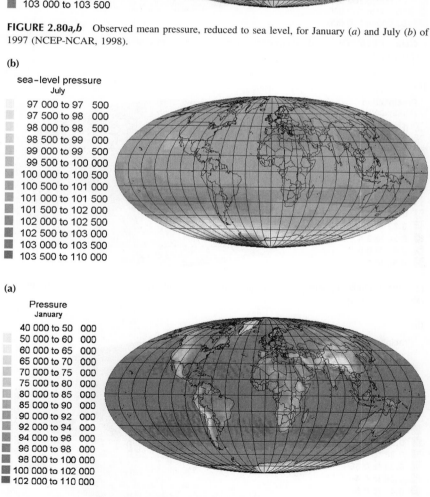

(a)

FIGURE 2.81a Observed mean pressure (Pa) at actual surface height for January of 1995 (NCEP-NCAR, 1998).

(b)

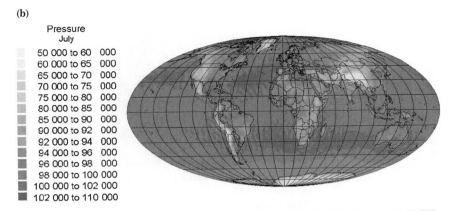

Pressure
July

50 000 to 60 000
60 000 to 65 000
65 000 to 70 000
70 000 to 75 000
75 000 to 80 000
80 000 to 85 000
85 000 to 90 000
90 000 to 92 000
92 000 to 94 000
94 000 to 96 000
96 000 to 98 000
98 000 to 100 000
100 000 to 102 000
102 000 to 110 000

FIGURE 2.81b Observed mean pressure (Pa) at actual surface height for July of 1995 (NCEP-NCAR, 1998).

(a)

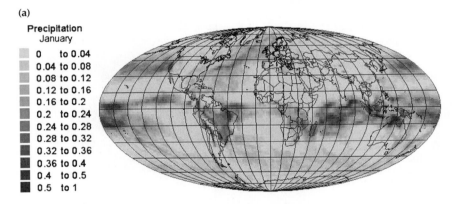

Precipitation
January

0   to 0.04
0.04 to 0.08
0.08 to 0.12
0.12 to 0.16
0.16 to 0.2
0.2 to 0.24
0.24 to 0.28
0.28 to 0.32
0.32 to 0.36
0.36 to 0.4
0.4 to 0.5
0.5 to 1

FIGURE 2.82a,b Observed total precipitation (m per month) for January (a) and July (b) of 1995 (NCEP-NCAR, 1998).

(b)

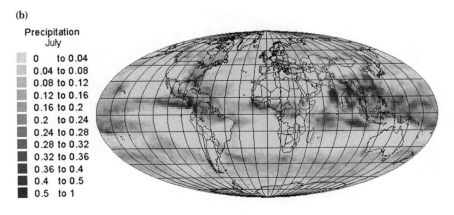

Precipitation
July

0   to 0.04
0.04 to 0.08
0.08 to 0.12
0.12 to 0.16
0.16 to 0.2
0.2 to 0.24
0.24 to 0.28
0.28 to 0.32
0.32 to 0.36
0.36 to 0.4
0.4 to 0.5
0.5 to 1

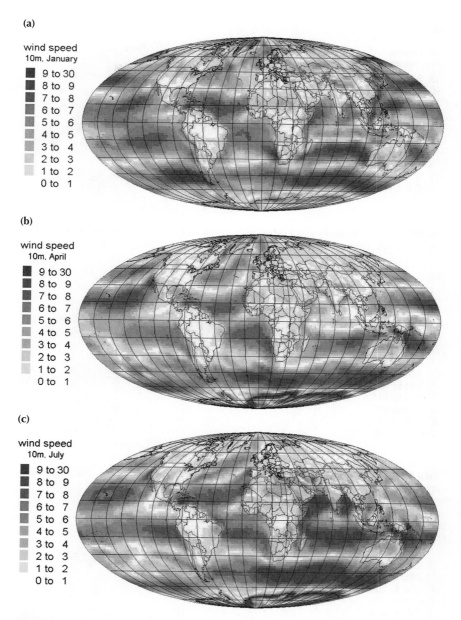

(a)

wind speed
10m, January

- 9 to 30
- 8 to 9
- 7 to 8
- 6 to 7
- 5 to 6
- 4 to 5
- 3 to 4
- 2 to 3
- 1 to 2
- 0 to 1

(b)

wind speed
10m, April

- 9 to 30
- 8 to 9
- 7 to 8
- 6 to 7
- 5 to 6
- 4 to 5
- 3 to 4
- 2 to 3
- 1 to 2
- 0 to 1

(c)

wind speed
10m, July

- 9 to 30
- 8 to 9
- 7 to 8
- 6 to 7
- 5 to 6
- 4 to 5
- 3 to 4
- 2 to 3
- 1 to 2
- 0 to 1

FIGURE 2.83a,b,c Observed mean wind speed (m s^{-1}) at surface height (10 m) for January (*a*), April (*b*), and July (*c*) of 1997 (NCEP-NCAR, 1998).

For wind speeds, the model calculations generally reproduce observed patterns, but the overall magnitude is on the low side, particularly for land locations in the Northern Hemisphere. This may be a problem with the data

(d)

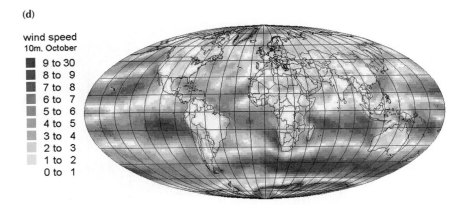

wind speed
10m, October

9 to 30
8 to 9
7 to 8
6 to 7
5 to 6
4 to 5
3 to 4
2 to 3
1 to 2
0 to 1

FIGURE 2.83d Observed mean wind speed (m s^{-1}) at surface height (10 m) for October of 1997 (NCEP-NCAR, 1998).

(a)

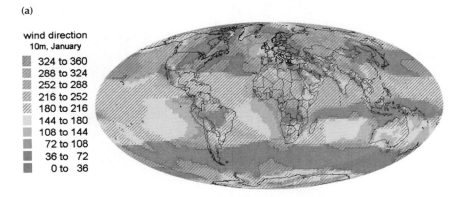

wind direction
10m, January

324 to 360
288 to 324
252 to 288
216 to 252
180 to 216
144 to 180
108 to 144
72 to 108
36 to 72
0 to 36

FIGURE 2.84a,b Observed mean wind direction at surface (height 10 m, deg. counter-clockwise, 0° is west to east) for January (a) and April (b) of 1997 (NCEP-NCAR, 1998).

(b)

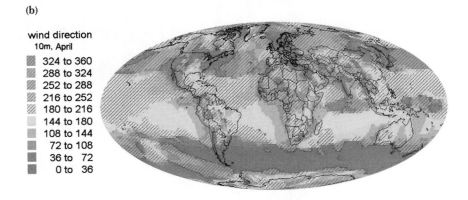

wind direction
10m, April

324 to 360
288 to 324
252 to 288
216 to 252
180 to 216
144 to 180
108 to 144
72 to 108
36 to 72
0 to 36

rather than with the calculation, because data are strongly influenced by land-based measuring stations (which are particularly abundant for Northern Hemisphere continents) and, as many stations are placed in sheltered locations, they may not represent the true average wind speeds. This has implications for the use of meteorological data to forecast production from wind turbines, as is further discussed in Chapter 3.

Figures 2.86 and 2.87 summarize calculated latitude distributions of the components of the net total energy flux (2.18), and the rates of poleward energy transport for a seasonal model, but averaged over seasons. The fluxes compare well with those based on observations (Figs. 2.25, 2.45–2.47), while the sum of ocean and atmospheric transport should match that required by the net radiation flux at the top of the atmosphere (Figs. 2.11, 2.12).

(c)

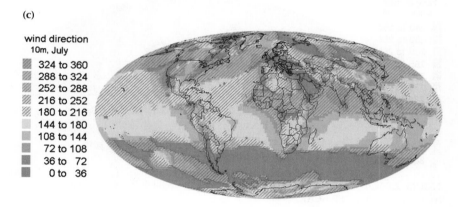

FIGURE 2.84c,d   Observed mean wind direction at surface (height 10 m, deg. counter-clockwise, 0° is west to east) for July (c) and October (d) of 1997 (NCEP-NCAR, 1998).

(d)

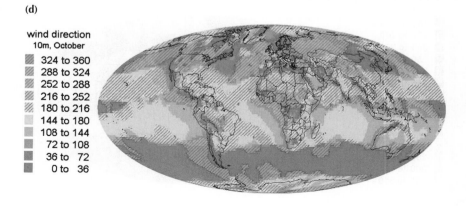

2.3.3 The Climate

In the preceding subsections, the state of the atmosphere and the ocean–continent system is described in terms of a finite set of state variables, such as temperature, pressure, salinity, moisture content of soil or air, ozone content, CO_2 content, etc. In addition, the motion of matter is described in terms of a velocity field, and the Earth–atmosphere system interacts with its surroundings through the exchange of radiation. The incoming radiation may at first be considered to constitute a fixed boundary condition, whereas the outgoing radiation may adjust itself as a result of the equations governing the system, i.e., the equations of motion and the rate equations for chemical processes, as well as the equations describing physical processes other than particle motion (e.g., the absorption and scattering of light).

(a)

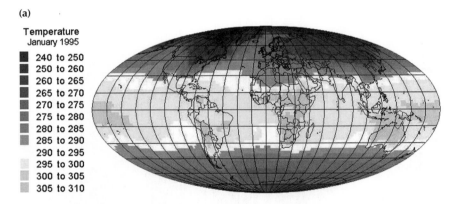

Temperature
January 1995

240 to 250
250 to 260
260 to 265
265 to 270
270 to 275
275 to 280
280 to 285
285 to 290
290 to 295
295 to 300
300 to 305
305 to 310

FIGURE 2.85a,b Observed surface temperatures (in K at height 1000 mb) for January (a) and April (b) of 1995 (NCEP-NCAR, 1998).

(b)

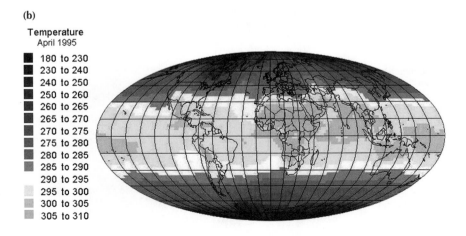

Temperature
April 1995

180 to 230
230 to 240
240 to 250
250 to 260
260 to 265
265 to 270
270 to 275
275 to 280
280 to 285
285 to 290
290 to 295
295 to 300
300 to 305
305 to 310

(c)

Temperature
July 1995

230 to 240
240 to 250
250 to 260
260 to 265
265 to 270
270 to 275
275 to 280
280 to 285
285 to 290
290 to 295
295 to 300
300 to 305
305 to 310
310 to 360

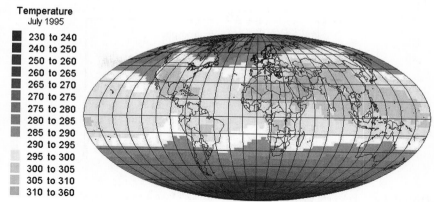

FIGURE 2.85c,d Observed surface temperatures (in K at height 1000 mb) for July (c) and October (d) of 1995 (NCEP-NCAR, 1998).

(d)

Temperature
October 1995

230 to 240
240 to 250
250 to 260
260 to 265
265 to 270
270 to 275
275 to 280
280 to 285
285 to 290
290 to 295
295 to 300
300 to 305
305 to 310
310 to 360

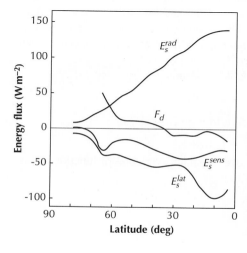

FIGURE 2.86 Annual and longitude average of calculated energy fluxes (in W m^{-2}) through a vertical column extending from the surface of the Earth downward: net radiation, E_s^{rad}, net sensible energy flux, E_s^{sens}, net latent energy flux, E_s^{lat}, and net flux into the column, F_d, taken as the upward flux into the top ocean level of the joint ocean–atmosphere model, from deeper-lying ocean levels. (Based on Wetherald and Manabe, 1972.)

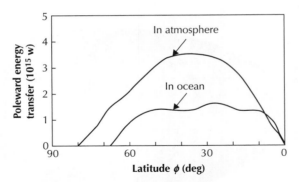

FIGURE 2.87   Poleward energy transfer (net) calculated from a joint ocean–atmosphere model. (Based on Wetherald and Manabe, 1972.)

It has been indicated that a state of overall equilibrium prevails between the Earth–atmosphere system and its surroundings. This equilibrium is characterized by the absence of a net energy flux between the system and its surroundings, on a suitable average basis (involving at least annual and latitude–longitude averaging). The state of equilibrium may vaguely be described as "the climate of the Earth."

Climate Variables

To arrive at a more accurate definition of climate, it is important to distinguish between micro- and macro-variables and to define the kind of averaging performed. From a physical point of view, the micro-variables may be thought of as variables describing the atomic or molecular state and motion, while macro-variables would be those associated with averages over statistical ensembles, such as pressure, temperature, and wind velocity. However, this scale may still be considered microscopic from a meteorological point of view. Higher levels of macroscopic scale would thus be defined by further averaging (e.g., introducing a distinction between eddy motion and synoptic scale motion, as discussed in section 2.5.2), and by making a distinction between zonal average circulation and deviations from it.

If the complete set of micro-variables is denoted

$$\{x_j \,|\, j = 1, ..., n\},$$

then the macro-variables corresponding to a given scale, s,

$$\{\bar{x}_j^s \,|\, j = 1, ..., n_s\},$$

are obtained by a suitable averaging over the micro-variables or functions of the micro-variables, e.g., time averages like

$$\overline{F}_j^s = \frac{1}{\Delta t_s} \int_{t_1}^{t_1 + \Delta t_s} F_j^s(\{x_i(t_2) \,|\, i = 1, ..., n\}) dt_2,$$

or by ensemble averages over N, identical systems,

$$<F_j^s> = \frac{1}{N_s} \sum_{m=1}^{N_s} F_j^s(\{x_i \,|\, i = 1, ..., n\})_m.$$

The number of macro-variables of scale s, n_s, may be smaller than the number of micro-variables, n, because not all variables survive the averaging procedure or because only a limited number are judged to be important (indeed, if the introduction of a macro-level description did not diminish the dimension, there would be little motivation for performing the averaging procedure at all). In general, it may be possible to define a very limited number of functions, K_j^s (to be denoted "climate functions"), which are *uniquely* determined from $\{x_j\}$ and depend only on the macro-variables of scale s,

$$K_k^s = G_k^s(\{x_j^s \,|\, j = 1, ..., n_s\}),$$

or a similar expression involving ensemble averages, for $k = 1, ..., m_s$.

The climate of scale s may then be defined as the set of functions $\{K_k^s \,|\, k = 1, ..., m_s\}$. If, alternatively, a definition of climate is sought that does not depend on whether unique definitions of the climate functions are possible, one might follow Lorenz (1968) in using the averages of the micro-variables directly as a "mathematical" climate, choosing $F_j^s(\{x_i\}) = x_j$,

$$\overline{x}_j^s = \frac{1}{\Delta t_s} \int_{t_1}^{t_1 + t_s} x_j(t_2) dt_2.$$

If $\{x_j\}$ are sufficiently well-behaved functions of time, these averages exist and may be called the mathematical climate of scale s. Letting Δt_s go toward infinity, one obtains what Lorenz calls "the climate of the variables x_j."

The equations from which the time dependence of x_j may be determined (the micro-level equations of state, of motion, etc.) are often expressed as differential equations of the first order, and the time development in general depends on the initial conditions. This means that the averages $\overline{x}_j^s$ depend on the time t_1, in which case the equations are called *intransitive*, and the climate functions are not unique.

If, on the other hand, the averages $\overline{x}_j^s$ are independent of t_1, the equations are called *transitive*, and the mathematical climate is uniquely defined. A somewhat weaker condition would be obtained if only the climate functions K_k^s have to be uniquely defined, i.e., independent of initial conditions. As mentioned, the K_k^s then approximately (to the extent that these climate functions represent the solutions to the original micro-level equations) describe the actual climate of scale s.

It is to be expected that the climate functions may gradually approach independence from initial conditions, as the time-averaging period increases. At $\Delta t_s = 1$ year, a certain stability is already obvious (due to the periodicity of radiative boundary conditions), but meteorologists normally define climate by requiring $\Delta t_s \approx 30$ years.

If time intervals of 30 years or longer are considered, it may no longer be possible to assume that the external boundary conditions remain constant. In that case, one would talk about climate change, a subject that is further dealt with in section 2.4.

Stability of Climate

The uniqueness of climate functions of a given scale s implies that the state of the system is stationary. By stationary, one should not mean that the time development has to be exactly periodic, but rather that the state of the system is periodic or constant, after fluctuations of frequency much higher than $(\Delta t_s)^{-1}$ have been filtered out (by the averaging procedure defining the scale s).

The climate may also be characterized by its response to a small perturbation of the state variables imposed at time t_1. If there is a finite probability that the climate functions will remain unaltered, the climate may be said to be stable, otherwise it is unstable. This follows directly from the independence (or dependence) on initial conditions at time t_1. A similar argument may be made for sequences of random perturbations, of closer resemblance to the actual "noise" introduced by fluctuations of the micro-variables around their average value (or more generally, fluctuations of time scales smaller than Δt_s).

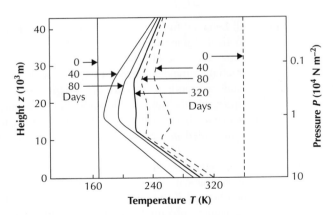

FIGURE 2.88 Approach toward equilibrium distribution of temperature for a one-dimensional model of the atmosphere with fixed external conditions. The initial condition is a uniform low (*solid curves*) or uniform high (*dashed curves*) temperature, and the different curves correspond to different periods of integration. (Based on Manabe and Strickler, 1964.)

A transitive system only possesses one stable climate. If a system has more than one stable climate, it is by definition intransitive, and the interesting parameter is the magnitude of perturbation that would cause the system to make a transition from one stable climate to another. However, as pointed out by Lorenz (1968), it is not always necessary to induce a perturbation in order to make the system perform the transition. Some systems may remain in one stable state for a substantial time and then change into a different "stable" state, where they would again remain for a long time, and then eventually make the transition back, and so on, without requiring any external perturbation. Such systems (examples of which have been discussed, for example, by Hide, 1958) may be called *near-intransitive*. Lorenz introduces this concept for systems that are strictly transitive in the limit $\Delta t_s \rightarrow \infty$, but for which the time averages over different finite time intervals may differ substantially. He has constructed numerical experiments with vacillating behavior (periodic transition between two patterns of circulation), and suggests that the atmosphere may be such a quasi-intransitive system. This might explain some of the climatic changes in the past, without recourse to external causes.

FIGURE 2.89 Time-averaged records of temperature from Godthaab, Greenland. The averaging period Δt_s is indicated for each curve. (Based on Dansgaard *et al.*, 1975.)

   The equations of section 2.5.2, used in sections 2.3.1 and 2.3.2 to describe
the general circulation in the atmosphere, are, in equilibrium calculations,
assumed to possess one stable solution, independent of initial conditions (at
least over a substantial range). The way in which the equations were solved
in early calculations (Manabe and Strickler, 1964) used this directly. Starting
with simple and completely unrealistic conditions, such as a constant temperature throughout the atmosphere, the equations are first integrated over a long
period of time, with fixed boundary conditions (such as radiation and ocean surface temperature). Then, after equilibrium has been reached, seasonal or other
variations in the boundary conditions, as well as the ocean–atmosphere couplings, are introduced. After a while, the situation is again stationary. Figure
2.88 shows, for a simple, one-dimensional model, how the stationary temperature profile is reached, starting from initial conditions of uniform high, or uniform low, temperature (Manabe and Strickler, 1964).

   In more general cases, for example, with time-dependent forces like those
associated with the solar radiation, one cannot expect the solutions to be independent of the initial conditions. Contemporary calculations use actual data as
initial conditions, and provided these are of sufficient quality, the transient
behavior of the solutions should faithfully describe the effects of variations in
external forcing of the climate system (IPCC, 1996b). This is important for calculations including greenhouse gas emissions into the atmosphere. Early calculations just aimed at deriving the equilibrium situation for a doubling of CO_2,
whereas current models are capable of deriving the dynamic behavior of climate
during a period of time with changing emissions and changing response of the
reservoirs included in the models.

   Figure 2.89 indicates the variation in climate (represented by the single variable T = temperature) actually observed at a given location, using different
averaging times Δt_s. It can be seen that the standard meteorological climate
scale, 30 years, still leaves some structure in the climate function T (Dansgaard
et al., 1975).

The Ergodic Hypothesis

It is expected that, for a stationary climate, time averaging and averaging over a
statistical ensemble of identical "climate experiments" will yield the same
value, when Δt_s and N_s become sufficiently large (in this case, the summation
over the N_s experiments may be replaced by an integral over each variable times
the probability density for that $\bar{a}$ variable). This identity is known as the *ergodic
hypothesis* (see, for example, Hinze, 1975),

$$\lim_{\Delta t_s = \infty} \bar{x}_j^s = \lim_{N_s = \infty} <x_j^s>.$$

Averaging over position vectors has been considered (e.g., in constructing
zonal means), but these would equal the time or statistical averages at a fixed

position only if the system were homogeneous, which the atmosphere is not. It should be clear that the macro-variables describing the climate need not be global averages, but may represent definite positions or zonal means. Also, the use of a large time-averaging interval, Δt_s, does not imply that the climate functions cannot comprise indicators of seasonal or diurnal changes, such as 30-year averages of May 18-humidity or of noon-pressure.

2.4 THE ENERGY CYCLE OF THE EARTH

In this section, the energy conversion processes taking place in the Earth–atmosphere system are summarized for the more or less stationary state constituted by the present climate. In doing so, the energy flow is taken as a "common denominator" for the variety of physical and chemical processes by which matter is transformed or moved. There are certainly contexts in which it is necessary to follow the flow of matter directly, but only examples of this are given.

The second question examined is the possible interference with this "natural" climate or energy flow caused (or possibly caused) by anthropogenic activity. The possibility of climatic changes resulting from the injection of CO_2 or particulate matter into the atmosphere has been widely discussed, and only a brief review is given. However, the climatic modeling techniques utilized in the discussion of such effects may also prove valuable in discussing the impact of an increasing use of the renewable energy flows for human purposes, since the models deal with precisely those flows that would be redirected or displaced by renewable energy conversions.

2.4.1 Flows of Energy and Matter

The Undisturbed Energy Cycle

Based on the material presented in sections 2.2 and 2.3, and in particular on the net radiation summary, Fig. 2.20, it is now possible to construct a schematic flow chart for the energy conversion and transport taking place in the Earth–atmosphere system (averaged over the year and geographical position). The result is shown in Fig. 2.90.

Different energy forms have been placed in different vertical columns, whereas the horizontal rows represent the different components of the Earth–atmosphere system. Compartments in which energy may become stored are represented by boxes, and the energy flows are indicated by lines with arrows, with the corresponding rates being given in units of terawatts (TW). The usefulness of diagrams of this type has been pointed out by King Hubbert (1971).

The first column contains the incoming short-wavelength solar radiation and its disposition into one part reflected into space and another part absorbed in various layers of the Earth–atmosphere system. The second column represents energy stored as (geo-)potential energy, latent energy of phase changes, and other chemical (molecular and atomic) energy, plus nuclear energy.

FIGURE 2.90 Schematic energy cycle without anthropogenic interference. The energy flows are in TW (10^{12} W). Numbers in parentheses are uncertain or rounded off. (Constructed with use of Odum, 1972; Budyko, 1974; Gregg, 1973; Lorenz, 1967; Manabe, 1969.)

Most of these energy quantities are bound within matter and do not participate in the undisturbed climatic processes. The third column represents storage of sensible heat, and the fourth column represents storage of kinetic energy in the general circulation of the atmosphere and the oceans. Certain compartments considered of little importance have been left out (e.g., kinetic energy of continents moving, tides of the solid Earth, and movements of ground water and of animals, as well as sensible energy stored in the biosphere). The final column contains the long-wavelength re-radiation into space. According to the assumption of a stationary state, the net energy exchange between the Earth–atmosphere system and its surroundings is zero, which is the basis for talking about an energy *cycle*: the outgoing flow can formally be joined to the incoming flow to form a closed cycle.

The top row contains the small contribution from gravitational attraction to other celestial bodies ("tidal energy"), in addition to the radiative fluxes.

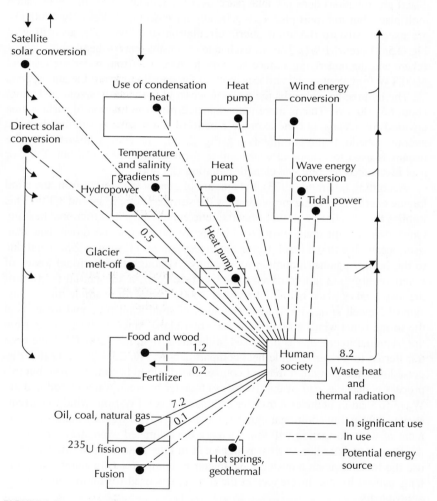

FIGURE 2.91   Past (1970s) and possible future human modification and points of interference with the energy cycle. Circles indicate possible sources of energy conversion for use in human society, while a heavy arrow indicates return flows. The unit is TW, and the background box layout corresponds to the compartments in Fig. 2.90. (Numerical entries based on Darmstadter *et al.*, 1971; European Commission, 1997.)

In the atmosphere, 41 400 TW, or about a quarter, of the incoming radiation is absorbed. Based on this and a similar amount of energy from condensation of water vapor (36 000 TW of which is latent energy of vapor evaporated from the oceans, 5400 TW being from continents), the atmosphere maintains a strong interaction with the Earth's surface, involving an intense exchange of long-wavelength radiation (see Fig. 2.20). The surface re-radiates a total of about 91 400 TW as long-wavelength radiation into space. The condensation of water vapor (and associated precipitation) does not take place over the regions where the evaporation took place, but transport processes generally have occurred. Yet only 1200 TW are used to sustain the atmospheric circulation (the "winds"), according to Fig. 2.54 (Lorenz, 1967). The net exchange of sensible energy between the atmosphere and the underlying surface amounts to 4000 TW from water surfaces and 4600 TW from continental surfaces, both in the direction toward the atmosphere.

The hydrosphere absorbs the major fraction of the incoming solar radiation, about 65 400 TW. This is more than expected from the fraction of surface area covered by oceans, but the average albedo of the oceans is smaller than the average albedo of the continents (cf. Fig. 2.21). The energy absorbed by the oceans is used to evaporate water and to form sensible heat (36 000 TW), or it is re-radiated as long-wavelength radiation (25 400 TW).

According to Gregg (1973), the amount of energy transferred from the wind kinetic energy to oceanic circulation, by the surface stress, is around 3 TW. This implies that practically all the loss of kinetic energy goes into frictional heating. One would assume that the frictional losses are more pronounced over land than over ocean, due to the much larger roughness length (cf. section 2.5.1) and due to the impact against mountain topography. Although the frictional losses of kinetic energy are not confined to the surface layer, a substantial fraction of the losses takes place below a height of 1000 m (see, for example, the discussion of Newell *et al.*, 1969; another region of significant frictional losses is at the strong zonal winds around an altitude of 12 km). Since the heat arising from the frictional interaction is dispersed in the air at first, the 1200 TW have been transferred to the air sensible heat compartment in Fig. 2.90. If the heat is dispersed near the ground, it may, of course, contribute to heating the soil, but this process is listed as part of the (two-way) transfer between water or soil and air. Wave formation includes a transfer estimated at 0.3 TW from wind to currents (Bryan, 1969), but additional transfer may be directly to waves. A realistic treatment of wave formation will thus increase the estimated wind stress and hence the energy transfer above what is indicated in Fig. 2.90. Climate models often use the transfer as an adjustable parameter to compensate for numerical problems caused by the differences in the time scales used for the oceans and the atmosphere.

The lithosphere has been divided into an upper and a lower part. The upper part is characterized by having an active heat exchange with the atmosphere, whereas little heat transport takes place in the lower part, only the geothermal fluxes and a little transport by way of ground water flow. Of the 15 600 TW estimated to be absorbed at the continental surfaces, 5400 TW go to evaporation and sublimation, and a net amount of 4600 TW are transferred to the atmosphere as sensible heat. Transfer from continents to oceans may take place by run-off if the water temperature differs from that of its environment. Such transfers are assumed to be negligible.

The polar ice is considered a part of the continents, and it is estimated that melting consumes 21 TW of the absorbed energy. The latent energy of melting is only formally transferred to the oceans, since it will normally not become available until the water has been evaporated into the atmosphere and condenses into snow.

The biosphere is presented as a separate compartment in Fig. 2.90, because of its importance in the following. Biomass production utilizes 133 TW or 0.08% of the solar radiation at the top of the atmosphere, according to Odum (1972). Photosynthesis by phytoplankton and green algae in the top 100 m of the oceans (cf. Fig. 2.30) accounts for about a third of biomass production. Almost as large a contribution comes from a much smaller area, namely, the wet tropical and subtropical forests (such as the region around the Amazon River). Grassland and pastures are responsible for roughly 10% of biomass production.

The amounts of energy contained in each of the compartments considered in Fig. 2.90 are generally much harder to estimate reliably than are the conversion rates. If both are known, the turnover time for energy in a given compartment can be obtained by dividing the amount of stored energy by the conversion rate.

The amount of kinetic energy contained in the atmospheric circulation has been studied in some detail. From Fig. 2.54, the *available potential energy* (cf. section 2.3.1) in the atmosphere is 2.8×10^{21} J, and the kinetic energy of the circulation is 7.5×10^{20} J, corresponding to turnover times of 27.5 days and 7.4 days, respectively. Available potential energy is defined as the potential plus internal energy relative to a reference state of the atmosphere, and thus represents a measure of the nonradiative energy produced in a given region by absorbing radiation, by moving sensible heat into the region, and by condensation of water vapor. The stored energy is denoted "available" if it can be converted into kinetic energy, and the reference state is one that cannot form any new kinetic energy (Lorenz, 1967; Newell *et al.*, 1969). If the stored energy in the atmosphere is not taken to be the available energy, very different values could be obtained. For instance, the latent heat of vaporization contained in all the water vapor present in the atmosphere in itself represents some 1.5×10^{22} J.

Much less is known about the magnitude of the energy stored in the oceans. An indication of the amount of energy stored in wave motion may be derived from the observed frequency distributions of wave height. In the North Atlantic ($\phi = 59\,°N$, $\lambda = 19\,°W$), the annual average wave amplitude is $\langle a \rangle = 1.23$ m, and the average of the squared amplitude is $\langle a^2 \rangle = 2.13$ m (Salter, 1974). From the discussion in section 2.5.3, rigorously valid only for sinusoidal waves, one then derives the average energy stored per m^2 ocean surface, $\overline{W}^{total} \approx 10^4\,\mathrm{J\,m^{-2}}$. (In Fig. 2.90, the wave compartment has been placed under "kinetic energy," although on average half the energy is in the form of potential energy, cf. section 2.5.3). If this amount of stored energy were found uniformly over the world's oceans, the total energy stored would be of the order of 10^{18} J. From similar order-of-magnitude arguments, based on measured velocities and estimated mass transports of surface currents (Sverdrup et al., 1942), the kinetic energy of horizontal currents would be of the order of 10^{17} J, and the kinetic energy of the thermohaline (meridional) motion, based on the calculated circulation shown in Fig. 2.73, would be of the order of 10^{16} J. These rough estimates, which should be viewed with considerable reservation, would imply turnover times of about a few days for the surface wave motion, increasing to several years for the motion in the meridional cells. In the deep sea, which was not adequately described in the calculation referred to (Bryan, 1969), the turnover time may be several hundreds of years.

The amount of energy stored in the biosphere may be taken as the total biomass in energy units. Based on Odum (1972), the average biomass may be estimated to be about 1.5×10^{22} J, giving an average turnover time of 3.5 years. This average conceals a considerable spread, with values of a few weeks for phytoplankton and several hundreds of years for certain trees being included.

Man's Interference with the Energy Cycle

Figure 2.91 summarizes the modifications of the energy cycle caused by human society today and identifies a number of points on the flow diagram of Fig. 2.90 that could contribute to serving man's purpose. Some of these conversion methods are well known but are only in modest use today. The compartments and main layout of Fig. 2.90 are repeated without text in Fig. 2.91, as a background for the lines connecting certain points along the energy flows or at the energy stores, with the box representing human society.

The main source of present energy conversion in human society is fossil fuels, such as oil, coal, and natural gas. This flow amounted to about 7.2 TW in the early 1970s, and about 12 TW in the late 1990s (European Commission, 1997), whereas the rate of forming new fuel of this type (peat, etc.) is negligible in comparison. Conversion of nuclear energy by induced fission of ^{235}U (plus other fissile material produced in the reactors) in a sustained neutron-rich environment presently accounts for about 0.7 TW (electric output), while hydropower converts about 0.5 TW of potential energy from the hydrological cycle (regulated through the formation of high-lying reservoirs).

The energy flow of organic matter from the biosphere to human society is about 1.2 TW, of which half is food. Not all the 0.6 TW transferred to human society as food actually serves as food intake, since part of it is lost in processing or is transferred to refuse. The other 0.6 TW of organic matter is mostly wood, but only a small fraction is burned directly. The rest goes into wood products or is lost in the production process. The amount of energy stored in wooden products is probably growing, although losses of stock take place when wooden (including paper) products are discarded and eventually burned.

Other conversion processes (indicated by dashed lines in Fig. 2.91) contribute less than 0.05 TW. They are geothermal energy (hot springs, etc.), tidal power, wind turbines, heat pumps, and solar collectors (thermal) or solar cells. The possibility of an increasing role played by conversion schemes based on renewable energy flows is a major theme in the following chapters.

A minor theme is the study of those areas of the energy cycle that have not yet been utilized by man but for which conversion methods have been proposed. In several of these areas, research is under way, although the practical viability of most of the proposed schemes remains to be demonstrated. The dot-and-dash lines in Fig. 2.91 indicate some such conversion possibilities, including wave energy, temperature and salinity gradients in the oceans, direct solar energy conversion outside the atmosphere, and hydropower stations in connection with glacier melt-off. Fusion of light elements to release nuclear energy is also included as an unproven process (that is, controlled), although its position might have been taken in the hydrosphere (deuteron–deuteron or deuteron–tritium processes) rather than in the lithosphere (bearing in mind that lithium enters the reaction chains primarily being studied today).

Energy conversion within human society produced about 8.2 TW of waste heat (in the early 1970s) in the form of sensible and latent energy emissions and long-wavelength radiation. The sensible energy emissions go into the hydrosphere (cooling water) or atmosphere, and the latent energy emissions (water vapor from cooling towers, etc.) go mostly into the atmosphere, where they add to the natural heat source function (see section 2.5.2). The long-wavelength radiation alters the net radiation flux at the level in question (most conversion presently being at ground level), which leads to a readjustment of the fluxes in both directions (upward and downward). Such changes have been observed locally ("heat islands" over metropolitan or heavily industrialized areas), whereas the evidence for global changes is at present inconclusive. The 8.2 TW constitutes 4.7×10^{-5} of the solar radiation at the top of the atmosphere. For comparison, the maximum relative change in the solar radiation at the top of the atmosphere due to changes in the eccentricity of the Earth's orbit around the Sun during the past 500 000 years has been 10^{-3} (Hays et al., 1976).

Energy is stored in a number of products within human society, and some are later exported, such as the 0.2 TW estimated to go back into the biosphere (agriculture) with fertilizers.

There is a net inflow of energy into human society, indicating a build-up of stored energy. The exact magnitude of the amount of energy stored is difficult to assess, since some of it is likely to be found in discarded products and other wastes, some of which may be on their way back into the biosphere or lithosphere, while other wastes cannot be expected to release stored energy in the near future (e.g., some manufactured products containing stored chemical energy). On the other hand, the stock of such energy (e.g., chemical) residing in the structures and materials forming our cities, transportation systems, etc., has certainly increased with time.

The amounts of nonrenewable fuels expected to be recoverable with technology not too different from today's (proven plus estimated recoverable reserves) are of the order of 10^{22} J for oil and gas, 10^{23} J for coal, and 10^{22} J for ^{235}U (Ion, 1975; Sørensen, 1999b). Ion states that not all known coal reserves may be mineable, and, of the uranium estimate given here, only about 10% has been verified so far.

Matter Cycles

Since the exchange of matter between the Earth–atmosphere system and its surroundings is small (cosmic particles are received and some atmospheric matter is probably lost to space), matter cycles are essentially irrelevant when conserved quantities like nonradioactive elements are considered. Normally, each element participates in several different chemical configurations during its cycle. Only in a few cases, such as the water cycle considered in Fig. 2.65, is it relevant to describe the flow of a chemical compound. Here the reason is that the gross flow of water in its different phases is little affected by chemical reactions between water and other elements, of which a number do take place in the atmosphere as well as in the lithosphere.

In most cases, a cycle would then be drawn for a single element, such as oxygen, carbon, or nitrogen. However, the flow of matter (the element) in itself does not give any indication of the processes involved, because it is often the chemical reactions that determine the flow rates in the system.

Of particular interest in connection with renewable energy flows are the flows of those materials that participate in the biological processes depending on photosynthesis. Examples of such elements are carbon, nitrogen, phosphorus, potassium, oxygen, and hydrogen, but a long list of other elements should be added to account for the structure of organic material, including the enzymes and hormones important for the control of biological processes.

The Carbon Cycle

Figure 2.92 shows the construction of a gross carbon cycle, based on a small number of highly aggregated compartments in the marine and terrestrial areas. The matter flow is expressed in units of 10^{12} kg of carbon per year, and the amounts of carbon stored in the compartments is expressed in units of 10^{12} kg of carbon.

The rates and stored quantities have been assumed to correspond to a stationary situation, before interference by human society. Over a longer period of historical development, as shown in Fig. 2.2 for oxygen, photosynthetic conversion can no

FIGURE 2.92   Schematic carbon cycle. The rates of transfer are given in 10^{12} kg of carbon per year, and the quantities stored are given in 10^{12} kg of carbon. (Based on Odum, 1972; Bolin, 1970; Wilson and Matthews, 1971; Gold and Soter, 1980.)

longer be assumed to be at a constant level. On this time scale, the most important processes adding or removing CO_2 from the atmosphere, aside from photosynthesis, are the weathering of limestone and the formation of such rock, by the combination of Ca^{2+} ions with CO_3^{2-} ions, and the exchange of CO_2 between the oceans and the atmosphere. The maximum CO_2 content of the atmosphere, which may have prevailed when there was no photosynthesis, is limited to about 1% (about 30 times the present level) by the $CaCO_3$ rate of formation (Rubey, 1951). Assuming the oceans to be present, the maximum amount of atmospheric oxygen would, in the absence of photosynthesis, be given by the rate of photodissociation of water in competition with the screening of this process by the oxygen already formed. The corresponding upper limit for atmospheric oxygen would be of the order of 0.02% (1000 times smaller than the present level) according to Berkner and Marshall (1970).

It follows that practically all the oxygen now present in the atmosphere has been formed by photosynthesis. The same seems to be true for the oxygen present in the lithosphere, and the ratio of carbon and oxygen in the atmosphere plus lithosphere (but excluding the hydrosphere, which contains oxygen in water) is very close to 0.5, indicating that all of it may have its origin in photosynthetic dissociation of CO_2 (Cloud and Gibor, 1970). A positive amount of CO_2 would have been transferred from oceans to atmosphere during most of the period in which photosynthesis has taken place outside the oceans (about 5×10^8 years).

Possible nonfossil carbon, in the form of methane, may be present in the Earth's crust (Gold and Soter, 1980), and it has been proposed that the sudden release of large amounts of methane could have been responsible for the climate change occurring 55 million years ago (Norris and Röhl, 1999).

Figure 2.92 indicates a surplus of CO_2 going into the atmosphere as a result of the burning of fossil fuels, the clearing of natural vegetation, and the introduction of agricultural practices, which expose larger volumes of soil to contact with the air (deep plowing), whereby they release more CO_2. Part of this surplus is absorbed by the oceans, but the process seems unable to keep pace with the rapid growth of CO_2 in the atmosphere, so the net result is that about half the extra CO_2 injected into the atmosphere is accumulating there. Also, the mixed upper layer of the oceans has a turnover time that is longer than the time scale of human interference with the carbon cycle, so the additional CO_2 absorbed by the oceans largely accumulates in the mixed layer, rather than reaching the long-term sink provided by the deep ocean.

The transfer rate of CO_2 across the atmosphere–ocean boundary, which in a stationary state would equal the ocean–atmosphere transfer rate, has been estimated on the basis of the depletion rate of ^{14}C injected into the atmosphere by nuclear weapons tests. Telegadas (1971) found that the effective half-life of the excess ^{14}C in the atmosphere was at an average value of 5.6 years during the period 1963–1966 and 8.2 years during the period 1966–1969. He interpreted the rising half-life as an approach to an equilibrium, in which re-emitted ^{14}C from biosphere and oceans must be subtracted from the fallout rate in order to give the effective half-life. The value $t_{1/2}^{eff} = 5.6$ y is thus an upper limit of the one-way fallout rate, which may become diminished if mechanisms of sufficiently short-time re-entry into the atmosphere are present. One such mechanism would be associated with the part of the biosphere respiration (cf. Fig. 2.92) derived from living plants or animals. The atmosphere–ocean transfer rate derived from these arguments is about 6×10^{13} kg C y^{-1}, but different interpretations of essentially the same data give transfer rates in the range $(3–14) \times 10^{13}$ kg C y^{-1} (Machta, 1971; Bolin, 1970; Fairhall, 1973). An average value of 9×10^{13} kg y^{-1} is used in Fig. 2.92.

The rates of new fossilization and formation of new rock (limestone) containing $CaCO_3$ have been estimated from the average history of deposits present today. Thus, the actual formation may have taken place over a shorter period, yielding a varying rate that is larger than that indicated under certain circumstances, but zero during other periods.

The Nitrogen Cycle

Another biologically important element is nitrogen, the tentative cycle of which is depicted in Fig. 2.93. Nitrogen gas (N_2) is a major constituent of the atmosphere, but in order for it to become available for building biological material, the nitrogen must first be "fixed." Nitrogen *fixation* is defined as a process by which gaseous nitrogen (N_2) is transformed into a constituent of a compound, such as ammonia (NH_3):

Nitrogen fixation

$$N_2 + \text{energy} \ (0.67 \times 10^6 \ \text{J mol}^{-1}) \to 2N,$$

$$2N + 3H_2 \to 2NH_3.$$

The fixation may be accomplished by plants, but not all plants are suited to perform the process. Leafy vegetables are particularly suitable, and the increased cultivation of such vegetables by human activity is estimated to be responsible for a nearly 50% increase in the total fixation of nitrogen in the terrestrial biosphere (14×10^9 kg y^{-1}). The actual set of chemical reactions involved in the biological fixation is much more complicated than the summary equations given above, since they are combined with metabolic processes by which the solar energy input is transformed into useful energy. Thus, the energy needed in the first step is derived from photosynthesis, while the second step, which is slightly exoergic, takes a different form, due to the absence of free hydrogen in practically all biological materials.

In the industrial nitrogen fixation process (the "Haber process"), the energy input may be from natural gas (methane) or another fossil fuel, which also provides the hydrogen for the second step. If renewable (other than biological) or nuclear energy is the starting point, hydrogen may be obtained by electrolysis. The rate of industrial fixation is presently around 30×10^9 kg y^{-1} (Delwicke, 1970). Most of the ammonia produced in this way is used to manufacture fertilizers, e.g., ammonium nitrate (NH_4NO_3) or urea ($CO(NH_2)_2$):

Fertilizer production

$$NH_3 + 2O_2 \to HNO_3 + H_2O,$$

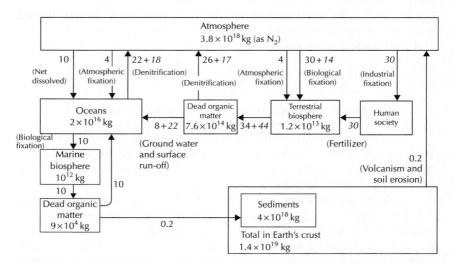

FIGURE 2.93 Schematic nitrogen cycle. The rates of transfer are given in 10^9 kg of nitrogen per year. Man-made contributions are shown in italics. (Constructed with use of Delwicke, 1970.)

$$HNO_3 + NH_3 \rightarrow NH_4NO_3,$$

or

$$2NH_3 + CO_2 \rightarrow CO(NH_2)_2 + H_2O.$$

The growth of plants that rely on nitrogen fixed by other plants can be greatly enhanced by administration of fertilizer. Nitrogen in fixed form appears to be a dominant limiting factor for such plants, but also the yield of plants that are themselves able to fix nitrogen can be enhanced by the addition of fertilizer. Since fertilizer can replace biological fixation, it saves the biosphere from spending the amount of energy that would otherwise be used to fix the corresponding amount of nitrogen. This is the basis for associating the use of fertilizer with an energy transfer from human society to the biosphere, as shown in Fig. 2.91. To the extent indicated by this figure, one may say that plants are grown on fossil fuels rather than on solar energy.

A third source of fixed nitrogen, both for continents and oceans, is shown in Fig. 2.93 under the heading "atmospheric fixation." It is due to fixation within the atmosphere by ionization of N_2 caused by radiation or lightning, and the nitrogen in fixed form is transferred to ocean or land surfaces by precipitation.

In a natural ecosystem that has reached a stationary state, no net removal of nitrogen from the atmosphere takes place. This is because the nitrogen fixed by natural processes is later returned to the atmosphere by denitrification processes. The dead organic matter, in which the nitrogen may be bound in very complex compounds, is decomposed by bacteria that release the nitrogen in the form of ammonia (*ammonifying* bacteria). Another set of bacteria, called *nitrifying* bacteria, oxidize ammonia to nitrite (NO_2^-) or nitrate (NO_3^-). Finally, the return of gaseous nitrogen, N_2, or N_2O to the atmosphere is achieved by *denitrifying* bacteria through highly exo-energetic processes based on sulfur or glucose, for example:

Denitrification processes

$$C_6H_{12}O_6 + 6KNO_3 \rightarrow 6CO_2 + 3H_2O + 6KOH + 3N_2O,$$

$$5C_6H_{12}O_6 + 24KNO_3 \rightarrow 30CO_2 + 18H_2O + 24KOH + 12N_2.$$

At the present level, denitrification processes do not balance the sum of natural and man-made fixation processes. The reason is not so much the time lag between the growing man-made nitrogen fixation and the formation of dead organic material, as the fact that only a fraction of the additional nitrogen in fixed forms adheres to the biological cycle all the way. A large quantity (in Fig. 2.93 estimated as 22×10^9 kg y^{-1}) leaves the terrestrial biosphere with ground water or surface run-off or is accumulated in soil (a build-up of 5×10^9 kg y^{-1} is indicated for the dead organic matter compartment on land). Some of the fertilizer administered to agricultural soils runs off with surface water even before it has been absorbed by plants, and some is carried away by ground water flows. Also, the conditions for a matching growth in denitrification do not appear to be fulfilled, for physical (the way in which the dead organic matter is deposited) or biological (the limiting factors governing growth of the population of denitrifying bacteria) reasons. The result is an accumulation of nitrogen in fixed forms in local water systems and in the oceans. According to the estimates made in Fig. 2.93, 4×10^9 kg is accumulated per year in the hydrosphere, and 9×10^9 kg y^{-1} is being removed from the atmosphere.

The above examples of matter cycles illustrate some of the problems introduced by the rapidly increasing amount of interference, on the part of human society, with stationary cycles that have taken much longer to establish than the time scales of the changes now induced. It has also been demonstrated that the construction of a reasonable model of such cycles requires an intimate knowledge of all the chemical and physical processes of importance for the element considered.

2.4.2 Climate Changes

The year-by-year fluctuations in climate parameters may appear as random fluctuations around an average climate when only a small number of years are considered. Over longer periods, evidence for systematic changes has been found, and over the entire geological history, it is certain that major changes in climate have taken place.

One important question is the extent to which human activities inadvertently cause climatic changes that would not otherwise have occurred or speed up changes that would have happened anyway. In the longer term, the use of planned anthropogenic intervention may be considered, in order to avert natural climate changes that would limit or reduce the possibility of sustaining a human society of the type desired. At present, however, the consequences of human interference with climate, planned or inadvert, cannot be predicted with a sufficient degree of certainty, so it would seem prudent to keep the amount of interference at the lowest possible level while more reliable information is being sought.

Energy use is the focus when addressing questions of anthropogenic interference with climate, as well as in the discussion of environmental pollution. The utilization of renewable energy flows is often considered a nonpolluting alternative (dams for hydroelectric installations perhaps excepted), which may be true in terms of the absence of gaseous and particulate emissions. However, since these flows are integral parts of the climate, it is by no means certain that any utilization—be it only a time delay in the return of long-wavelength radiation to space—could not have an effect on climate. In particular, agriculture is a way of utilizing solar energy that involves changes of large land areas and almost certainly has had climatic impacts during the history of cultivation (cf. the discussion in Chapter 1).

Studies applying general circulation models have been used to address the question of the possible climatic impact of fuel burning with CO_2 emissions, of activities that may reduce the stratospheric ozone content, and of general changes in net radiation, e.g., due to emissions altering cloud cover. Below, first an outline of climatic changes that have occurred in the past, inasmuch as they have been identified, is given.

Climatic History

There is no detailed account of the early climate of the Earth before the Palaeozoic period starting about 6×10^8 years BP (before the present). Palaeontological

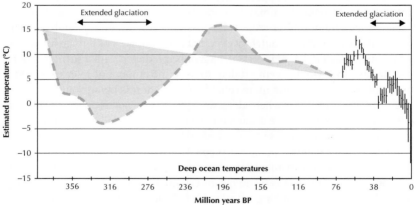

FIGURE 2.94   Trends in the surface temperature of the Earth. The lower part gives gross estimates of temperature based on paleontological evidence and, from 7×10^7 years BP, an estimate from drilled deep-sea cores with benthic data, assuming ice-free ocean surface. Glaciations have taken place in the period between 3.5×10^8 and 2.7×10^8 years BP, and again 3.8×10^7 years ago at the South Pole and 4×10^6 years ago at the North Pole. (Based on Shakleton and Opdyke, 1973; Crowley, 1998; Zachos *et al.*, 2001; Jansen *et al.*, 2007.) The top part shows, on an expanded scale, the trends in temperature variations during the last 5×10^6 years, also from benthic data (^{18}O to ^{16}O ratio and fossilized organisms). (Based upon Lisieki and Raymo, 2005.)

evidence suggests that even the poles must have been ice-free more than 90% of the time since then. However, about 3×10^8 years BP, a major glaciation took place at both poles, which at that time were covered by two vast continents, Laurasia in the north and Gondwana in the south. This period of glaciation, possibly with interruptions, lasted about 5×10^7 years and was followed by a warm period extending throughout the Mesozoic era (see Fig. 2.94). It is tempting to associate the cessation of this ice age with continental drift, allowing warm ocean water to be circulated into the Arctic regions.

From the Tertiary period (beginning about 6×10^7 years BP), a cooling finally led to new periods of glaciation within the Quaternary period (from about 10^6 years BP). During the last half of the Quaternary period, oscillatory behavior of the climate has been well established by studying layer sequences in ocean cores (Hays et al., 1976), as well as evidence derived from estimated extents of the ice cover. The glaciation period started about 5×10^6 years BP in the Antarctic region, but 2–3 million years later in the Arctic region, presumably due to the distribution of continents at the time.

The fluctuating extent of the ice cover from the beginning of the Quaternary period to the present has been subjected to various kinds of smoothing and spectral analysis. Based on ocean core data, Hays et al. (1976) found an overall frequency of the glaciations of about 10^{-5} y^{-1}. The shape of the temperature function is asymmetric, with a slow cooling from the interglacial maximum, taking about 9×10^4 y to reach the minimum (peak glaciation), followed by a rapid (10^4 years) rise of temperature back to the maximum value. The difference between maximum and minimum average temperature is believed to be about $10\,°C$. Superimposed on the glaciation cycle are oscillations corresponding to a cycle time of about 4×10^4 years and another set, possibly with more than one component, of cycle times in the range $(1.9–2.4) \times 10^4$ years.

Figure 2.95 shows estimated variations in the average temperature during the last 140 ky, based on ocean core data and (for the last 75 000 years) ice cores from Greenland (Dansgaard, 1977). The ice cores contain a more detailed record of climate, where seasonal variations can be followed. Each year corresponds to a depth interval of several centimeters, except for the very oldest (deepest) parts of the core. For the ocean cores one has to rely on estimated sedimentation rates. The temperature estimate is primarily based on the isotopic abundance of ^{18}O, which depends on the temperature prevailing in the atmosphere when the precipitation that formed the ice was falling. The reason is that $H_2 {}^{18}O$ has a lower vapor pressure than $H_2 {}^{16}O$, and thus the abundance of ^{18}O will be slightly higher in the precipitation than in the atmospheric vapor, implying a depletion of ^{18}O from the atmosphere. The lower the atmospheric temperature, the more ^{18}O has been removed from an air parcel having traveled for a while under the influence of condensation processes, so that a low ^{18}O content in the snow in Greenland indicates a low temperature throughout the "fetch region." It is clear that this type of temperature determination contains a number of uncertainties and that it is not possible to deduce the global average temperature from such data. However, the time sequence of warming and cooling trends should be very reliable, and comparisons with direct temperature measurements in recent periods indicate that these trends from the Greenland ice cores correlate well with

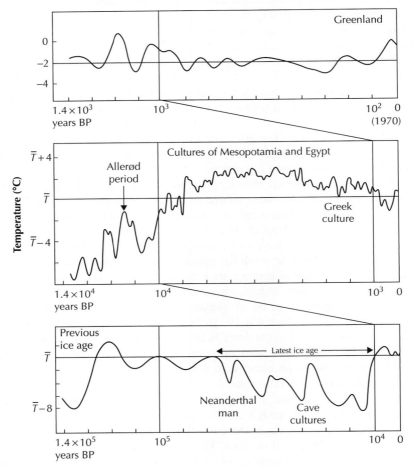

FIGURE 2.95 Trends in average surface temperature during the last 1.4×10^5 years. The upper panel contains the most recent 1.4×10^4 and 1.4×10^3 years on expanded scales, and different smoothing times Δt_s, which "filters out" the short-term variations in temperature. The last 75 000 years are based on ice cores from drilling in Greenland, while the older periods are based on a number of determinations, including deep-sea cores and pollen data. A tentative temperature scale relative to the average temperature of the most recent 10^3 years is indicated in the two lower figures. The top figure has a temperature scale derived from present-day measurements (Fig. 2.89), representative for Greenland. The amplitudes of the temperature variations during the most recent couple of hundred years have been smaller at mid-latitudes than in Greenland. (Based on Bolin, 1974; Dansgaard *et al.*, 1975; Dansgaard, 1977.)

the trends observed at northern mid-latitudes (Dansgaard *et al.*, 1975). Temperature scales similar to those indicated in Fig. 2.95 are suggested on the basis of other kinds of data, e.g., tree ring data, pollen records, and fossil plankton data (see Bolin, 1974).

The latest ice age built up over a period of at least 60 000 years and then disappeared over something like 4000 years, with a number of fluctuations of appreciable amplitude superimposed on the main trend. During this glaciation period, man developed societies of substantial sophistication, including medical use of plants and the art of painting. Major urbanizations are found throughout the warm period starting about 10^4 years ago ("the climatic optimum"). During the most recent 10^3 years, the greatest change in climate has occurred in the present century, exhibiting a warming trend until about 1940, then a slight cooling trend until 1980, and finally more warming (IPCC, 1996a).

The 1880–1940 temperature rise from the Greenland data (Fig. 2.95 *top*) is somewhat larger than that observed at mid-latitudes, with the latter being 0.4 °C according to IPCC (1996b).

Causes of Climatic Change

In the early history of the Earth, climate changes were integral parts of the evolution of continents and of the formation of the atmosphere and the oceans and their subsequent evolution. Major alterations in surface structure, the formation of mountain chains and volcanic activity, influenced climatic variables, and varying cloudiness and ozone content in the atmosphere modified the net radiation received at the surface, which itself, through changes in albedo, modified the radiation being reflected. Therefore, there are plenty of causes to choose from and to combine in order to account for the large-scale evolution of climate on Earth. As briefly mentioned above, the formation and dissolution of large continents near the poles may explain the late Palaeozoic glaciation and its subsequent disappearance. Superimposed on the large-scale trend of climatic behavior shown in Fig. 2.94, short-term temperature excursions are likely to have taken place. Budyko (1974) argues that if the size and frequency of volcanic eruptions were stochastically distributed, then there would, during a time period as long as the Mesozoic era, be several occurrences of global dust-triggered temperature drops of perhaps 5 or 10 °C, strong enough to last for several decades. The speculation is, then, that such episodes may have been responsible for the extinction of certain animal species (in particular species of dinosaurs, which appear to have had no thermoregulatory mechanism). Of course, the durations of these temperature excursions were too short to leave any geological traces. Alternative speculations on the extinction of dinosaurs involve meteorite hits.

Considering the Quaternary period, the physical state of the Earth (distribution of continents and oceans, composition of atmosphere) may be regarded as sufficiently stationary to exclude it as a likely cause of climatic changes. It is thus tempting to associate the periodic occurrence of ice ages and temperature oscillations (Fig. 2.94 *top*) with periodic variations in the external conditions serving as boundary conditions for the climate equations. Such variations would primarily be variations in incident radiation, which could be caused, for example, by the periodic changes in the Earth's orbital parameters, due to many-body interactions in the planetary system. Other suggestions invoke periodic passage through interplanetary dust (the difficulties of such theories have been pointed out by Dennison and Mansfield, 1976) or periodic changes in the Sun's radiation output (see Hammond, 1976), similar, but with different periods, to the well-known changes in magnetic activity of the Sun, e.g., the 11-year sunspot cycle. However, measured variations in solar output are much smaller than what seems needed, at least at present (IPCC, 1996b). Further, the oscillations in temperature during the Quaternary period might be interpreted in terms of Lorenz's model of near-intransitive, vacillating systems (see section 2.3.3), requiring no external causes.

An explanation in terms of orbital changes was first proposed by Milankovitch (1941). The frequency spectrum of such variations has a peak at 105×10^3 y (changes in the eccentricity of the Earth's orbit), 41 000 years (changes in obliquity, i.e., the angle between the equatorial and ecliptic planes), and a continuous shape with two peaks at about 23 000 and 19 000 years (changes in the longitude of the perihelion due to precession). The variations in obliquity and precession do not alter the total solar radiation received by the Earth, but lead to periodic variations at fixed geographical locations. The variation in eccentricity does alter the total amount of radiation, but by less than 0.1%. If this is considered negligible, the amount of radiation is unaffected, even locally, by the eccentricity variations.

In recent studies of ocean core samples, evidence for the presence of these spectral peaks, in the ^{18}O to ^{16}O ratio as well as in the sea surface temperature [estimated from the abundance of radiolarians (unicellular organisms) in the samples], has been strengthened (Hays *et al.*, 1976). The samples cover a time interval of 450 000 y back from the present (cf. Fig. 2.94). Whereas the periods around 4×10^4 years and around 2×10^3 years may then be interpreted in terms of variations in solar radiation, the coincidence of the main glaciation period and the period of eccentricity variations does not immediately suggest the nature of a possible causal relation. It is possible, however, that the 10^5 year period of climatic variations can be related to the amplitude of the precession components, since these amplitudes are directly proportional to the eccentricity. The coupling involved would have to be nonlinear.

The White Earth State

It is important to realize that, under present conditions, and probably throughout the Quaternary period, the most stable state of the Earth is one of complete glaciation (i.e., all oceans frozen, all continents snow-covered). The explanation is of course that the albedo of such a "white Earth" would exceed 0.7 and thus reflect more than twice as much of the incident radiation as at present. The estimated global average surface temperature in this state is about −70 °C (Budyko, 1974). Model calculations indicate that the average temperature for the entire atmosphere would be about 20 °C below the average surface temperature, but that stratospheric temperatures would be almost unchanged (Wetherald and Manabe, 1975).

FIGURE 2.96 The calculated variation in globally averaged surface temperature, as a function of changes in the solar constant, relative to the present. The full curve is based on the one-dimensional model of Budyko (1974), while the dashed curve is based on a three-dimensional model of Wetherald and Manabe (1975).

Various attempts have been made to estimate the size of perturbation that would trigger a transition from the present climate to a "white Earth." According to Budyko's (1974) simple, one-dimensional model, which parametrizes the components of the net radiation flux and neglects seasonal variations and variations in cloudiness, a decrease as small as 1.6% in the solar constant would cause the ice caps to exceed a critical size, beyond which they would continue to grow until the entire Earth was covered. This process comes about as a result of a feedback loop: temperature decrease → increased glaciation → increased albedo → more temperature decrease. Budyko's model contains latitudinal transport of energy, in a highly idealized way, with parameters adjusted to reproduce present latitude dependence of temperature. In Fig. 2.96, the mean surface temperature derived from Budyko's model is compared with the results of Wetherald and Manabe, using the three-dimensional atmospheric circulation model with idealized topography, described in section 2.3.2, but without the oceanic circulation and without seasonal variations or detailed cloud modeling. The hydrological model is included, but, as the authors emphasize, the results concerning the relationship between solar radiation and temperature (or glaciation) may not be more reliable than those of Budyko, because his phenomenological parameters are fitted to (present) data, while Wetherald and Manabe approximately solve the true equations for a system that resembles, but is not, the actual Earth. As seen in Fig. 2.96, their average temperature for the Earth at present is about 6 °C too high, and they predict a much slower dependence of the temperature on solar radiation, with no "white Earth" instability in the interval of solar constants considered, from 2% above the present actual value to 4% below. They warn that inclusion of the meridional transport by ocean currents may increase the sensitivity to the value of the solar constant. The extent of the maximum ice caps during the Quaternary period exceeded 75% of the way to Budyko's critical latitude of roughly 50°, which would have led to a white Earth.

From the above discussion, the long-term evolution of the Earth's climate may be seen in a new light when one realizes that an explanation is required to account for the presence of non-frozen water for more than 3×10^9 years, i.e., that the Earth could not have been in a stable state of complete glaciation during that time (or, if it was for shorter periods, an explanation is required to account for the large amount of energy necessary to melt the ice). The apparent contradiction is derived from the near certainty that the solar constant must have been increasing regularly from a value about 25% lower than the present value during the 4.5×10^9 years since the Sun became a main–sequence star. The gradual increase in solar luminosity is a result of the hydrogen depletion and radius increase, cf. Fig. 2.9. The magnitude of the relative increase in the solar constant, being thus about 0.05 per 10^9 years, is not strongly dependent on the particular solar model employed (Newman and Rood, 1977).

It is thus tempting to conclude that the change in solar constant must have been counteracted by a change in the Earth–atmosphere system, such that the average surface temperature has never been below the critical value leading to the white Earth, which again implies either that the Earth–atmosphere albedo was lower before or that the greenhouse effect was more pronounced. Even for an entirely ocean-covered Earth, the albedo decrease might not be sufficient, because increased cloudiness would counteract such a decrease. The greenhouse effect would be increased if the long-wavelength re-radiation from the atmosphere toward the Earth's surface increased, as a result of an atmosphere of a composition different from the present one. The larger CO_2 content possible in the periods with little or no photosynthesis would act in this direction, but according to Sagan and Mullen (1972), even the maximum possible CO_2

content (in equilibrium with the CO_2 in oceans and the carbonate-containing rocks) would be insufficient. Sagan and Mullen suggest that a small amount of NH_3 (mixing ratio 10^{-5}) was present, which would effectively absorb radiation of wavelengths of around 10^{-5}m, leading to a surface temperature increase of as much as 70 °C. At the same time, the atmospheric availability of NH_3 would enhance the chances of forming life, bypassing the step of nitrogen fixation. It may also be suggested that NH_3 was formed as a result of atmospheric fixation in periods of intense electrical activity in the atmosphere. If no new changes in the atmospheric composition take place, Sagan and Mullen estimate that the continued rise in the magnitude of the solar constant will lead to a 50 °C increase in average temperature in another 4.5×10^9 years (this in itself may change the atmosphere into one like that of Venus today). A similar calculation for the Martian atmosphere suggests that in 4.5×10^9 years it will have a temperature much like the Earth's today, which may serve as a consolation.

Man's Interference with Climate

The first significant anthropogenic influence on climate is probably the one associated with extensive destruction of natural vegetation, primarily aimed at transforming it into arable land or pastures. Presently, about 10% of the continental area is cultivated and about 30% is occupied by grassland or pastures (Odum, 1972). The transformation of steppe or forest into arable land has a number of consequences. The components of the net energy flux (2.18) are altered. The net radiation flux is diminished owing to a (modest) increase in surface albedo. The albedos for steppe and coniferous forest (in non-Arctic regions) are about 0.13, for deciduous forest and tropical forest during the dry season, 0.18, for tropical forest during wet season, 0.24, for savannahs and semi-deserts, 0.14 during the dry season and 0.18 during the wet season, and for deserts, 0.28 (Budyko, 1974); the mean albedo for cultivated land and pastures is 0.20. Considering the types of transformation taking place, a reduction in net radiation ranging from 3% to 20% has been reported for the establishment of agricultural land in the temperate zones (Wilson and Matthews, 1971). Also, land reclaimed from the sea has a higher average albedo. The evapotranspiration and hence the latent heat flux away from the surface is smaller for grassland and slightly smaller for dry arable land than for forest, while it is slightly larger for wet arable land (rice paddies, etc.). The upward sensible heat flux is very small for wet arable land, medium for dry arable land and deciduous forest, and quite high for grassland and coniferous forest. Thus, the transformation of forest to arable land is likely to reduce considerably the Bowen ratio, defined as the ratio of sensible and latent heat fluxes,

$$\text{Bowen ratio} = E_s^{sens}/E_s^{lat}.$$

It is probable that man's quest for agricultural land has had much larger unintended effects than the above-mentioned changes. It is believed that the savannah grasslands in the tropical regions are entirely man-made, since the natural vegetation of these semi-humid regions is dry, deciduous forest. The activities of man have greatly augmented the number of forest fires.

In the semi-arid regions, dry steppe is being transformed into semi-desert by overgrazing, which exposes the soil to wind erosion as well as to water erosion during the (short) rainy season. Calculations indicate that the albedo change due to overgrazing may induce severe drafts in such regions (Charney and Stone, 1975). Also, tropical forests have been, and still are being, destroyed by human activity. For instance, archaeological evidence suggests that the Rajputana desert in India was as fertile as a tropical forest only 1000 years ago, which implies that its Bowen ratio, which is now in the range 2–6, has decreased more than tenfold (Bryson, 1971).

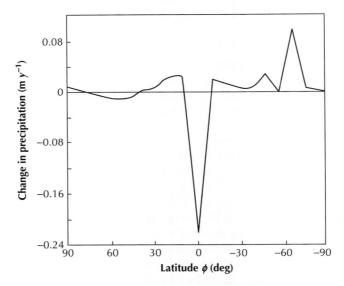

FIGURE 2.97   Calculated change in the rate of precipitation (averaged over longitudes), resulting from removing all tropical jungles. (Based on Potter *et al.*, 1975.)

When a tropical forest is being removed, the soil is no longer able to retain its humidity. Then precipitation runs off along the surface, and the soil becomes still drier. The humus layer is washed away and a solid crust of clay is developed. Weathering of this dry soil may greatly increase the particle content of the atmosphere, and by reflecting incoming radiation this dust may reduce the net radiation at the surface. On the other hand, cloud cover is likely to become reduced. According to a computer simulation (Potter *et al.*, 1975), the surface temperature may drop by about 0.4 °C, while the temperature at a height of 12–13 km may drop by 1 °C (year average). Figure 2.97 shows the calculated change in zonal precipitation resulting from the assumed complete removal of the tropical rain forests. It is worth noting that the induced climate changes are global. The model does not include the influence of the dust particles from the barren soil on the rate of precipitation.

Urbanization is another factor in climate modification due to anthropogenic land use that may be significant, although the area involved is smaller than that of agricultural utilization. Urbanization usually entails an increase in temperature, a decrease in wind speed (through increased roughness length for the wind profile described in section 2.5.1), and, at least for societies relying on burning fossil fuels and on polluting industries, an increase in atmospheric content of particles and trace gases and an increase in cloudiness and precipitation, but less incoming radiation at the surface and less relative humidity of the atmosphere over the urban area.

FIGURE 2.98   Schematic picture of human interference with river run-off. The rates are given in units of 10^9 m^3 of water per year. The "stable" part of the run-off is the part maintained by ground water flow plus the part associated with regulated reservoirs. (Based on Lvovitch, 1977.)

Man's interference with the hydrological cycle (Fig. 2.65) has become quite significant. Lvovitch (1977) divides the total river run-off of 39×10^{12} m^3 water per year into a "stable" part, being maintained by ground water flows (12×10^{12} m^3 y^{-1}) or being regulated by reservoirs (2×10^{12} m^3 y^{-1}), plus a variable or "unstable" part (25×10^{12} m^3 y^{-1}). Anthropogenic interference with these flows is summarized in Fig. 2.98. It appears that the largest usage is for irrigation, then industry, and then a small amount for household and livestock watering, but the significant measure for interference is probably the amount of water necessary to dilute wastewaters (which for some industrial enterprises may be very large), plus of course the amount of water really "used," i.e., left in an unretrievable form. As much as 45% of the stable run-off is being diverted through human society, plus around 5% of the unstable run-off. Human extraction of ground water enhances the hydrological cycle, and significant changes in the net water flow at the surface are induced by the creation of artificial lakes or reservoirs and (working in the opposite direction) by drainage of swamps and other humid soils.

The amounts of latent energy moved by irrigation and subsequent evaporation and condensation are several times larger than the waste heat presently associated with direct anthropogenic energy conversion. But rather than changing the energy content of the atmosphere (as in urban "heat islands"), the effect of irrigation is to speed up the hydrological cycle. This could have climatic effects beyond the hydrological cycle, which, as demonstrated, for example, by the model calculations described in section 2.3.1, is strongly coupled to the general circulation and temperature distribution.

The general circulation of the oceans has also been affected by human activities. In this context, the regulation of rivers by building dams (in connection with hydropower) or canals (usually for transportation) may play a role by causing an "unnatural" timing of the peak run-off into the oceans or by diverting the run-off to other parts of the oceans than those originally receiving the river water. Few such planned engineering enterprises have been undertaken in open oceans, but a number of proposals have been put forward that could invoke large-scale climatic changes. An example is the proposal for building a dam across the Bering Strait and pumping water from the Arctic Ocean into the Pacific Ocean. Thereby, new warm water would be pulled into the Arctic region from the Atlantic Ocean, where it might melt the sea ice (the same might be achieved by spraying soot on the ice to reduce its albedo). Another example is the artificial upwelling of cold and nutrient-rich water into surface regions, with the purpose of increasing biomass production (e.g., in the tropical ocean surfaces, where the net radiation is abundant and the supply of nutrients is the main limiting factor) or energy production (thermal or salinity gradient conversion, see Chapter 4). Primary climatic effects might be increased fogginess in summer and increased heat flux in winter (Wilson and Matthews, 1971).

Concern has been expressed over the effect of oil spills (from tankers, offshore drilling, etc.) on the ocean–atmosphere interface (Wilson and

Matthews, 1970). At large oil concentrations, the heat fluxes may become impeded, but even at small concentrations wave motion is damped (particularly the short-wavelength wave components near the crests), and thus the transfer of kinetic energy from the atmosphere to the oceans may be diminished, with consequences for the circulation in the oceans as well as in the atmosphere.

Direct interference with the energy cycle takes place in connection with rain making (seeding clouds with condensation or freezing nuclei, such as silver iodide crystals) or rain prevention (hail suppression, hurricane control). The possible climatic effects have not been adequately studied, and they presumably depend strongly on the circumstances of a given operation, e.g., whether the provoked precipitation is subsequently transferred by evapotranspiration or by surface run-off.

Particle emissions (e.g., from industry) in a certain size range may provide an unintended source of cloud condensation nuclei. It is estimated (Hobbs *et al.*, 1974) that a relatively small number of suitable particles may affect the hydrological cycle in a globally significant way. All activities that may affect cloud structure, cloud cover, and cloud height have the potential of inducing climatic changes brought about by the influence on the radiation climate (short- and long-wavelength) and on the hydrological cycle.

Aerosols

Only particles with suitable properties, related to crystal structure for freezing nuclei and solubility for water condensation nuclei, can serve as cloud nuclei. The concentration of such particles is estimated as $(4–10) \times 10^{-5}$ m^{-3} over oceans and 4×10^{-4} m^{-3} over continents, their production rate being estimated as 4×10^{28} y^{-1}. Their average mass is 10^{-20} kg (Twomey and Wojciechowski, 1969). This is only a small fraction of the total particle content in the atmosphere (cf. Fig. 2.34), and the production rate is only about 0.1% of total anthropogenic emissions (cf. section 2.3.1). The remaining particles influence climate primarily by scattering and absorption of radiation.

A layer of particles implies quite complex processes, in which radiation is (multiply) scattered, absorbed, or transmitted, with each subprocess depending on the wavelength of radiation as well as on the properties of the particle (aerosol) on which the radiation is scattered. The net result is a modified albedo of the Earth–atmosphere system containing an aerosol layer, as well as a modification in the amount of upward radiation from the Earth's surface into space. The modified albedo is expected to be larger when the aerosol layer is present, leading to less radiation incident on the Earth's surface, which again constitutes a cooling effect. On the other hand, the presence of the aerosol layer is expected to decrease the amount of radiation from the Earth's surface back into space. Therefore, the temperature at the surface increases, implying a heating effect. It has been estimated (Reck *et al.*, 1974) that the cooling effect dominates if the albedo a_s of the Earth's surface is below 0.6 and that the heating effect dominates if a_s is above 0.6 (snow and ice), fairly independently of other parameters (note that albedos used here all refer to short-wavelength radiation).

Globally averaged, there is thus little doubt that an increase in atmospheric aerosol content will lead to a decrease in temperature, but in the polar region there will be a delicate balance between cooling and heating effects. The average albedo at high latitudes is about 0.6 for the combined Earth–atmosphere system (Raschke *et al.*, 1973), but higher for continental ice sheets (Greenland and Antarctic). At low latitudes, the surface albedo a_s is considerably smaller than the Earth–atmosphere albedo a_0, due to the influence of clouds, but in polar regions, a_s and a_0 are more similar. It may thus be expected that the addition of particles in Arctic regions may have a slight warming (ice melting) effect, while the effect at lower latitudes is definitely cooling.

FIGURE 2.99 Absorption of short-wavelength radiation by the Earth–atmosphere system with an aerosol layer, as a function of the optical thickness L^{sw} of the aerosol layer in visible light. The absorbed flux A_0 is related to incoming radiation and albedo of the Earth–atmosphere system by $A_0 = E_{0+}^{sw}(1-a_0)$. The figure also shows the long-wavelength re-radiation into space, E_0^{lw}, as a function of the optical thickness L^{lw} of the aerosol layer in infrared light (*top scale*), for a range of surface temperatures T_s (*right-hand scale*). (Based on Rasool and Schneider, 1971.)

The magnitude of the cooling effect caused by increased aerosol load in the atmosphere was first studied by Rasool and Schneider (1971) and others at the same time. Figure 2.99 shows the amount of incoming (short-wavelength) radiation absorbed by the Earth–atmosphere system (including an aerosol layer), as a function of the dimensionless optical thickness L of the aerosol layer (a measure of the amount of aerosols present),

$$L = \int \sigma_{abs} \, ds = \int l^{-1} ds = \int dI/I,$$

where σ_{abs} is the absorption cross-section per unit volume, l is the mean free path of light due to aerosol absorption, and I is the intensity of the radiation. In all cases, averages over wavelengths should be performed, but often the calculation simply uses one "characteristic" wavelength, in the present case

$\lambda = 5.5 \times 10^{-7}$ m, corresponding to visible light. The corresponding optical thickness will be denoted L^{sw}. The interval of uncertainty indicated represents the lack of knowledge of the absorptive properties of the individual aerosols, the nature of which has not been specified. The calculation of Rasool and Schneider assumes that the aerosol layer lies between the surface and an altitude of 1000 m (cf. Fig. 2.34). It further assumes a 50% cloud cover and an aerosol size distribution proportional to r^{-4} for the (approximate) calculation of multiple Mie scattering (cf. section 2.3.1).

For long-wavelength radiation, the Mie scattering cross-section is much smaller than for wavelengths in the visible range, as seen from Fig. 2.43. Rasool and Schneider assume that the ratio of optical thicknesses for infrared (taken as $\lambda = 10^{-5}$ m) and visible radiation is

$$L^{lw}/L^{sw} = 0.108.$$

Thus, infrared radiation to space is little affected by the presence of an aerosol layer, relative to the influence on the transmission of visible light. In Fig. 2.99, an L^{lw} scale is added at the top, and curves representing the long-wavelength radiation to space are drawn (thin lines) for different values of the surface temperature, T_s, i.e., the Planck radiation temperature of the upward radiation spectrum [cf. (2.3)]. The optical aerosol thickness at present is believed to be around 0.1. The figure indicates that in order to decrease the average surface temperature by one degree, the aerosol content would have to be increased by a factor of 1.5–2.0. For the average residence time around 15 days mentioned for atmospheric particles in section 2.3.1, the particle emission, which increases the atmospheric load by a factor of two, can be taken as twice the present level. Thus, if 20% of the emissions (including emission of gases that later form particles) at present are of anthropogenic origin, then a doubling of the total amount would require a sixfold increase of the anthropogenic emissions.

The newest climate model calculations mentioned below under "carbon dioxide" all include a model for aerosol content in the atmosphere.

Carbon Dioxide

Gaseous emissions into the atmosphere may entail environmental side-effects and may affect human health. Such pollution has been demonstrated, for example, in connection with sulfur oxides from fossil fuel combustion and with photochemical smog. However, trace gases may also interfere directly with climate, by altering components of the radiation fluxes. As can be seen from Fig. 2.40, H_2O and CO_2 are very important absorbers in the infrared wavelength region, particularly for wavelengths above 10^{-5} m, where the peak of the long-wavelength radiation to space is situated. It is thus to be expected that an increase in the atmospheric content of these absorbers will lead to a higher temperature in the troposphere as well as on the Earth's surface. The present content of CO_2 is already big enough to imply a nearly total absorption at $\lambda = 1.5 \times 10^{-5}$ m.

It is therefore unlikely that even a dramatic increase in CO_2 could have consequences similar to the addition of a small amount of NH_3 (temperature changes globally averaging something of the order of 10 °C or more), discussed earlier in this section. However, a temperature change of just one or two degrees globally may have serious effects on climate, particularly if the local temperature change is amplified in sensitive regions, such as the polar regions.

As shown in Fig. 2.92, a little over half of the anthropogenic CO_2 emissions are accumulating in the atmosphere, but it is difficult to determine precisely the magnitude of future emissions that would lead to a specific temperature increase, owing to the nonlinear absorption response at the specific radiation wavelengths characterizing each gas contributing to greenhouse warming (cf. Fig. 2.40).

Numerous calculations estimating the climatic effects of doubling the atmospheric CO_2 content have been performed over the years. An early model by Manabe and Wetherald (1967) used a global average one-dimensional (height) temperature distribution, taking into account both the changes in radiation components and the change in the turbulent (convective) fluxes from the surface, as a function of CO_2 content. Assuming a fixed relative humidity and cloudiness, a doubling of the CO_2 concentration in this calculation implied a temperature increase at the Earth's surface amounting to 2.4 K. A later three-dimensional equilibrium model used by Manabe and Wetherald (1975) employed the techniques described in section 2.3.1, but with the idealized topography discussed in section 2.3.2 in connection with the joint ocean–atmosphere model of the same authors. No seasonal variations of radiation were considered in that calculation, with no oceanic heat transport, and the cloud cover was fixed at 0.5. Test computations starting with substantially different initial conditions gave equilibrium temperature distributions that concurred with each other to better than about 1 K. Figure 2.100 shows the hemispherical longitude averages of temperature for the calculation with CO_2 content doubled, minus the corresponding temperatures for a reference calculation (CO_2 mixing ratio 4.56×10^{-7} by mass). In the stratosphere, the doubled CO_2 content leads to a net cooling, while the troposphere is heated. The temperature at mid-heights is increased by about 3 K, while the surface temperature is increased by 2–3 K, except in the polar region, where the temperature increases by up to 10 K. The sensitivity of the polar region is associated with the possibility of large albedo changes, primarily connected with ice melting. The model predicts a poleward movement of the border of the ice cap by about 10°, plus a reduction in the thickness of snow cover. The latter is due to reduced snowfall (in the polar region), while the total amount of precipitation increases at all latitudes, as a result of doubling CO_2. For this reason, the latent heat upward flux (evaporation) increases significantly, and the warming of the troposphere by a greenhouse effect is due to the combined absorption of radiation by water vapor and CO_2. The effective temperature T_0 of the entire Earth–atmosphere system is increased by roughly 0.6 K, but the authors warn that the strong interconnection between the CO_2

cycle and the hydrological cycle that the model has indicated may necessitate the inclusion of a predictive equation for the cloud cover in their model.

Finally, Fig. 2.101 gives the results of a recent modeling effort using a grid with spatial resolution $2.5° \times 3.75°$ combined with 19 atmospheric height levels and 20 oceanic depth levels (Mitchell *et al.*, 1995; Mitchell and Johns, 1997). The model is transient, starting with realistic data from about the year 1860 and following the development under conditions of changing CO_2 emissions as well as taking into account the sulfate aerosol negative forcing. Model output for the present era substantially agrees with observed data, and Fig. 2.101 shows the geographical distribution of calculated change between surface temperatures around 2055 and those characterizing the 1990s for each season, averaged over period of 10 years.

FIGURE 2.100 Calculated change in longitude averaged temperature distribution (K) in the atmosphere over the Northern Hemisphere, as a result of doubling the concentration of carbon dioxide. (Based on Manabe and Wetherald, 1975.)

Compared to the earlier calculation, the only significant change is the cooling effect of particles in the atmosphere, which is strong at present and in the near future (giving a very complex temperature modification pattern), but diminishes in importance toward the end of the period because of assumed reductions in sulfate aerosol emissions from power plants and other activities, as environmental regulation introduced in many countries today is making itself felt globally. The average warming from the doubled CO_2 is about 2.5 °C.

There is a continued effort to include more and more realistic effects in the climate simulation models. The effort includes differentiating between different residence times of different greenhouse gases (rather that lumping them together as an "effective CO_2 emission"); improving modeling of cloud and water vapor in general, of oceanic transport, and of transfer of heat and particles between oceans and atmosphere; and, of course, refining the grid resolution and accuracy of topography.

Support for the correctness of the calculated persistence in the atmosphere of greenhouse gases has recently come from the analysis of a climatic change that occurred 55 million years ago and that is interpreted as having been caused by sudden release of methane from sedimentary hydrides. The residence time found by the sediment core study is 120 000 years (Norris and Röhl, 1999).

It is not the intention here to give an extensive list of possibilities for human interference with climate. In brief, it may be stated that life, and in particular human life, is sensitive to even very small changes in certain critical elements. Among these, a well-established case is that of ultraviolet radiation penetration to the surface, which, as discussed in section 2.3.1, depends on the presence of definite trace gases in the stratosphere (see Fig. 2.38). In particular, it has been suggested that the ozone concentration may be very sensitive to various human activities, such as aircraft combustion products (see, for example, Alyea *et al.*, 1975), chlorofluoromethanes (see, for example, Wofsy *et al.*, 1975), and nitrous oxide, N_2O, from increased use of fertilizers in agriculture (Johnston, 1977) (cf. section 2.4.1). In the two latter cases, the gases in question are very inert in the

(a)

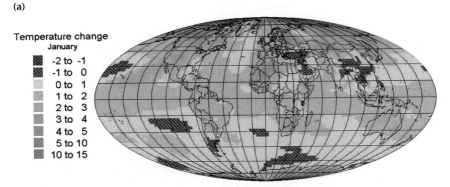

FIGURE 2.101*a,b*   January and April surface temperature changes for the mid-21st century, from a transient climatic model including both CO_2 increase (roughly doubling) and sulfate aerosol forcing (average 2050–2059 minus 1990–1999, in °C). (Based on HADCM2-SUL; Mitchell and Johns, 1997; data source IPCC DDC, 1998.)

(b)

(c)

Temperature change
July

▨	-10 to -5
▨	-5 to -4
▨	-4 to -3
▨	-3 to -2
▨	-2 to -1
▨	-1 to 0
	0 to 1
	1 to 2
	2 to 3
	3 to 4
	4 to 5
	5 to 10

FIGURE 2.101c,d July and October surface temperature changes for the mid-21st century, from a transient climatic model including both CO_2 increase (roughly doubling) and sulfate aerosol forcing (average 2050–2059 minus 1990–1999, in °C). (Based on HADCM2-SUL; Mitchell and Johns, 1997; data source IPCC DDC, 1998.)

(d)

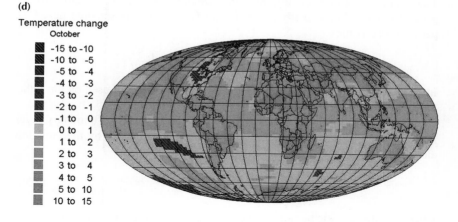

Temperature change
October

▨	-15 to -10
▨	-10 to -5
▨	-5 to -4
▨	-4 to -3
▨	-3 to -2
▨	-2 to -1
▨	-1 to 0
	0 to 1
	1 to 2
	2 to 3
	3 to 4
	4 to 5
	5 to 10
	10 to 15

troposphere, but may become catalysts for chemical processes in the stratosphere by which ozone is depleted. The modeling of these changes, and their induced modifications of the radiative spectrum (Cutchis, 1974), was one of the first climate calculations undertaken. It involves a detailed knowledge of the chemistry and photochemistry of the atmosphere, together with the general atmospheric models describing the circulation and other physical processes, in particular the mixing and diffusion processes between the stratosphere and the lower atmosphere.

As a result of these calculations, international agreements on the reduction of emissions affecting the ozone content in the atmosphere have been reached, and similar agreements may theoretically be forthcoming regarding the curbing of greenhouse gas emissions. The very substantial economic implications involved are discussed in Chapters 7 and 8.

2.5 INCLUSION OF NONRADIATIVE ENERGY FLOWS

2.5.1 Vertical Transport in the Atmosphere

If the transport of sensible or latent energy is governed by the gradients of temperature and humidity, it may be described by the diffusion equation (see, for example, Sellers, 1965)

$$\frac{dT}{dt} = \frac{\partial}{\partial z}\left(k\frac{\partial T}{\partial z}\right), \tag{2.28}$$

where k is the diffusion coefficient and the z-axis is in the vertical direction. Considering a quasi-stationary situation (the partial derivatives of T with respect to time t as well as with respect to horizontal directions x and y being considered negligible), (2.28) may be integrated to give

$$k\frac{\partial T}{\partial z} = \text{constant} = \frac{E_s^{sens}}{\rho(0)c_P}. \tag{2.29}$$

The change in stored sensible heat in a small volume $dx\,dy\,dz$ close to the ground is $dW^{sens}/dt = \rho(0)\,dx\,dy\,dz\,c_P\,(dT/dt)$, and the sensible heat flux per unit area is thus

$$dE_s^{sens} = \frac{1}{dx\,dy}\frac{dW^{sens}}{dt} = \rho(0)c_P dz\frac{dT}{dt}.$$

Insertion of this into (2.28) explains the form of the constant used on the right-hand side of (2.29).

The vertical transport of water vapor through the laminar boundary layer may also be taken as a diffusion-type process and is given by the equation analogous to (2.29),

$$k\frac{\partial q}{\partial z} = \frac{E_s^{lat}}{\rho(0)L_v}, \tag{2.30}$$

where q is the specific humidity. The molecular diffusion coefficient in air is of the order of $k = 2 \times 10^{-5}$ m^2 s^{-1}.

Considering the vertical change in the horizontal velocity u of the air, the expression analogous to (2.29) or (2.30),

$$\rho k\frac{\partial u}{\partial z} = \tau, \tag{2.31}$$

represents the shearing stress acting on the ground (or more generally on a horizontal unit area). τ has the dimension of force per unit area or pressure.

Above the laminar boundary layer, transport usually takes place by turbulent convection. The mass motion may be considered as consisting of overlaying eddies of varying size, eventually superimposed on a laminar flow. The fact that laminar flow is, in general, horizontal or nearly horizontal near the ground implies that all vertical transport is related to eddy motion.

It is now assumed (Rossby, 1932) that the expressions (2.29)–(2.31), based on molecular diffusion theory, can also be applied in the region of turbulent motion, provided that the coefficient k is given an effective value. This *eddy diffusion parameter* is no longer constant. It will depend on the quantity being transported and on the velocity field of the atmosphere, i.e., k will be a time-dependent function of the height z. The picture of turbulent transport given here is equivalent to the mixing length formulation used in the discussion in section 2.1.1 of convective transport in stellar atmospheres. The eddy diffusion parameter k is related to the mixing length l by

$$k = l^2 \frac{\partial u}{\partial z}.$$

If the temperature profile is approximately adiabatic,

$$\frac{dT}{dz} \approx \left(\frac{dT}{dz} \right)_{adiabatic} = \frac{c_P - c_V}{c_P} \frac{T}{P} \frac{dP}{dz} \tag{2.32}$$

(P is pressure, c_V is the heat capacity at constant volume), then it has been suggested that the mixing length be taken as proportional to or at least a linear function of z,

$$l = \kappa z$$

or

$$l = \kappa(z + z_0), \tag{2.33}$$

where κ close to 0.4 is von Kármáns constant (Prandtl, 1932). Using (2.33) together with the assumption that the shearing stress τ (2.31) is independent of z within the stable atmospheric layer, where the temperature gradient has been assumed to correspond to adiabaticity, then the horizontal velocity varies with height according to

$$u = \int_{z_0} \left(\frac{\tau}{\rho} \right)^{1/2} \frac{dz}{l} = \frac{1}{\kappa} \left(\frac{\tau}{\rho} \right)^{1/2} \log \left(\frac{l}{\kappa z_0} \right). \tag{2.34}$$

The lower limit of integration, z_0, is the height of the laminar boundary layer, for which the velocity u becomes zero if the first expression (2.33) for l is used, whereas for the second expression the velocity u becomes zero at the ground, $z = 0$. None of the expressions can be rigorously correct, since the second one, which gives the correct boundary condition $u(z = 0) = 0$, also predicts eddies that extend to the ground, $l(z = 0) \neq 0$, in contradiction to the basic assumptions regarding the laminar boundary layer. The first expression correctly predicts $l(0) = 0$, but cannot be used to estimate u if $z \lesssim z_0$. The coefficient $(\tau/\rho)^{1/2}$ has the dimension of velocity and is often referred to as the *friction velocity*.

If the eddy diffusion parameter k is left as a function of z [to be determined empirically rather than using (2.33)], and the heat flux E^{sens} is regarded as independent of z, then (2.29) may be integrated,

$$E_s^{sens} = \int_0^z \frac{dz}{k(z)} = \rho\, c_P (T(z) - T(0)).$$

According to Budyko (1974), the integral appearing on the left-hand side depends little on z, in contrast to k, so that the expression offers an empirical method of determining E_s^{sens} from temperature measurements. A similar expression for E_s^{lat} may be found from (2.30), again replacing the molecular diffusion coefficient k by the eddy diffusion parameter as a function of z.

Water Transport

The transport of water vapor through the lower part of the atmosphere may be described by equation (2.30), which may be read as an expression of the potential evaporation from a "wet surface" (i.e., a surface of moisture exceeding the above-mentioned "critical" value), depending on the distribution of humidity q in the air above the surface, and on the diffusion coefficient $k(z)$, which again may depend on the wind velocity, etc.,

$$e = -\frac{E_s^{lat}}{L_v} = -\rho k \frac{\partial q}{\partial z}. \tag{2.35}$$

Within the laminar boundary layer, k may be considered constant, but in the turbulent layer, k varies. If the integral

$$\int_0^z k^{-1} dz$$

can be considered as constant (Budyko, 1974), an integration of (2.35) shows that e varies linearly with the specific humidity q at some height z above the surface. On the other hand, if k is identified with the diffusion coefficient for horizontal wind velocity u (2.31), one obtains, similarly to (2.34),

$$\int_0^z k^{-1}\, dz = \kappa^{-1} \left(\frac{\tau}{\rho}\right)^{-1/2} \log\left(\frac{z}{z_0}\right) = \kappa^{-2} u^{-1} \left(\log\left(\frac{z}{z_0}\right)\right)^2$$

and

$$e \approx \rho\, (q(0)) - q(z))\, \kappa^2\, u \left(\log\left(\frac{z}{z_0}\right)\right)^{-2}. \tag{2.36}$$

Several expressions of this structure or similar ones have been compared with measurements, where they are found to agree overall but not in specific detail (see, for example, Sellers, 1965; Sørensen, 1975a).

The evaporation from ocean surfaces is not independent of the occurrence of waves. At high wind speeds, surface tension is no longer able to hold the water surface together, and spray droplets are ejected into the atmosphere (whitecaps). Some of these droplets evaporate before they fall back into the sea, thereby providing a source of vapor that does not follow the diffusion formulae given above.

Condensation of water vapor in the atmosphere may take place whenever the vapor pressure of water exceeds that of saturation at the prevailing conditions. However, if no condensation nuclei are present, the vapor may become supersaturated. One source of condensation nuclei that is of particular importance over oceans is the salt particles left over from the spray droplets discussed above, after the evaporation of the water. Typically, the mass of these salt particles is so small that gravitation can be disregarded, and the salt particles may remain suspended in the atmosphere for a considerable length of time. Other particles of suitable size and structure serve equally well as condensation nuclei, and globally the sea salt spray particles do not dominate the condensation processes (Junge, 1963).

The Disposition of Water and Heat in Soils

The penetration of water into soil and the movement of ground water depend strongly on local conditions, such as the sequence of soil types. Underground aquifers may transport water at moderate velocities, but the average turnover time even in the regions of active exchange is very slow, 200–300 years, as estimated from Fig. 2.65.

The ability of soils to store heat absorbed from solar radiation depends on the heat capacity of the soil, which may be written

$$C_s = C_s^{dry} m_s + C_w m_w, \tag{2.37}$$

in terms of the heat capacity of dry soil (C_s^{dry} typically in the range $2.0–2.5 \times 10^6 \, \text{J m}^{-3} \, \text{K}^{-1}$) and of water [$C_w \approx 4.2 \times 10^6 \, \text{J m}^{-3} \, \text{K}^{-1}$ or roughly half of this value for ice (frozen soil)]. The mixing ratios (by volume) m_s and m_w are much more variable. For soils with different air content, m_s may be $0.2–0.6$ (low for peat, medium for clay, and high for sand). The moisture content m_w spans the range from zero to the volume fraction otherwise occupied by air, m_a (equal to about 0.4 for clay and sand, but 0.9 for peat) (Sellers, 1965). In temperate climates an average value of m_w for the upper soil layer is around 0.16 (Geiger, 1961).

Since the absorption of radiation takes place at the soil surface, the storage of heat depends on the competition between downward transport and long-wavelength re-radiation plus evapotranspiration and convective transfer. The thermal conductivity of the soil, $\lambda(z)$, is defined as the ratio between the rate of heat flow and the temperature gradient, at a given depth z,

$$E_s = \lambda(z) \frac{\partial T}{\partial z}. \tag{2.38}$$

(Note that by the sign convention previously adopted, z is negative below the soil surface.) Figure 2.102 gives a rough idea of the dependence of λ on the moisture content for different soil types.

If the downward heat transport can be treated as a diffusion process, equation (2.28) is valid. Taking the partial derivative of (2.38) with respect to depth $(-z)$, and assuming that k and λ are independent of z (homogeneous soil),

$$\frac{\partial E_z}{\partial z} = \frac{\lambda}{k}\frac{dT}{dt}.$$

Together with the heat transport equation

$$\frac{\partial E_z}{\partial z} = C_s \frac{dT}{dt}$$

(cf. section 2.5.2; in the absence of heat sources, heat sinks, and a fluid velocity in the soil), this implies the relation

$$k = \lambda/C_s. \tag{2.39}$$

Figure 2.103 shows an example of the variation in the monthly average temperature as a function of depth in the soil. The time variation diminishes with increasing depth, and the occurrence of the maximum and the minimum is increasingly delayed. These features are direct consequences of the diffusion approach, as one can easily see, for example, by assuming a periodic temperature variation at the surface (Carslaw and Jaeger, 1959),

$$T(z = 0, t) = T_0 + T_1 \sin \omega t. \tag{2.40}$$

With this boundary condition, (2.28) can be satisfied by

$$T(z, t) = T_0 + T_1 \exp\left(z\left(\frac{\omega C_s}{2\lambda}\right)^{1/2}\right) \sin\left(\omega t + z\left(\frac{\omega C_s}{2\lambda}\right)^{1/2}\right), \tag{2.41}$$

and the corresponding heat flux is from (2.38)

FIGURE 2.102   Heat conductivity, λ, for various soil types, as a function of water content by volume, m_w. (Based on Sellers, 1965.)

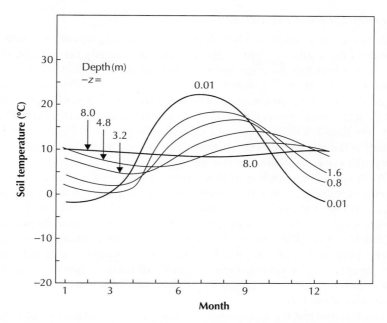

FIGURE 2.103 Seasonal variations of soil temperature at various depths measured at St. Paul, Minnesota. In winter, the average air temperature falls appreciably below the topsoil temperature ($z = -0.01$ m), due to the soil being insulated by snow cover. (Based on Bligh, 1976.)

$$E_z(t) = -T_1(\omega C_s \lambda)^{1/2} \exp\left(z\left(\frac{\omega C_s}{2\lambda}\right)^{1/2}\right) \sin\left(\omega t + z\left(\frac{\omega C_s}{2\lambda}\right)^{1/2} + \frac{\pi}{4}\right). \quad (2.42)$$

The amplitude of the varying part of the temperature is seen from (2.41) to drop exponentially with depth ($-z$), as the heat flux (2.42) drops to zero. For a fixed depth, z, the maximum and minimum occur at a time that is delayed proportional to $|z|$, with the proportionality factor $(\omega C_s/2\lambda)^{1/2}$. The maximum heat flux is further delayed by one-eighth of the cycle period (also at the surface).

The approximation (2.40) can be used to describe the daily as well as the seasonal variations in T and E_z. For the daily cycle, the surface temperature is maximum some time between 1200 and 1400 h. This temperature peak starts to move downward, but the highest flux away from the surface is not reached until 3 h later, according to (2.42). While the temperature peak is still moving down, the heat flux at the surface changes sign some 9 h after the peak, i.e., in the late evening. Next morning, 3 h before the peak temperature, the heat flux again changes sign and becomes a downward flux. Of course, the strict sine variation may only be expected at equinoxes.

Geothermal Heat Fluxes

In the above discussion, the heat flux of nonsolar origin, coming from the interior of the Earth, has been neglected. This is generally permissible, since the average heat flux from the interior is only of the order of 3×10^{12} W, or about 8×10^{-2} W m^{-2}. Locally, in regions of volcanoes, hot springs, etc., the geothermal heat fluxes may be much larger. However, it has been estimated that the heat transfer through the Earth's surface by volcanic activity contributes only less than 1% of the average flux (Gutenberg, 1959). Equally small or smaller is the amount of heat transmitted by seismic waves, and most of this energy does not contribute any flux through the surface.

Although the distribution of heat generation within the solid Earth is not directly measurable, it may be estimated from assumptions on the composition of various layers of the Earth's interior (cf. Lewis, 1974, and Fig. 2.11). Thus, one finds that most of the heat generation takes place within the rocks present in the Earth's crust. The source of heat is the decay of radioactive elements, chiefly potassium, uranium, and thorium. The estimated rate of generation (see, for example, Gutenberg, 1959) is roughly of the same order of magnitude as the outward heat flux through the surface, although there is considerable uncertainty about this.

It is believed that the temperature gradient is positive inward (but of varying magnitude) all the way to the center, so that the direction of heat transport is predominantly outward, with possible exceptions in regions of convection or other mass movement.

A more detailed discussion of the nature of the geothermal energy flux is deferred to section 3.4.2.

Temperature gradients also exist in the oceans, but it will become clear that they are primarily maintained not by the heat flux from the Earth's interior, but by extraterrestrial solar radiation coupled with conditions in the atmosphere and in the surface layers of the continents.

Momentum Exchange Processes Between Atmosphere and Oceans

The mixing and sinking processes sketched above constitute one of the two sources of oceanic circulation. The other one is the momentum exchange between the atmospheric circulation (winds) and the oceans. The interaction between wind and water is complex, because a large fraction of the kinetic energy is transformed into wave motion rather than directly into currents. Although the roughness length over water [z_0 of (2.33), (2.34)] is very small, a strong wind is capable of raising waves to a height of about 10 m.

The wind stress acting on the ocean surface may be taken from (2.33),

$$\tau = \rho_a \kappa^2 V(z_1)^2 \left(\log\left(\frac{z_1}{z_0}\right) \right)^{-2}, \tag{2.43}$$

where ρ_a is the density of air, κ is von Karman's constant, and z_1 is a reference height. When the height variation of the horizontal wind velocity V is logarithmic as in (2.34), τ is independent of height. This is true for the lower part of the turbulent layer (the Prandtl layer), within which z_1 must be taken, and the value of τ found in this way will remain constant through the laminar boundary layer, i.e., the force exerted on the ocean surface can be determined from measurements in a height z_1 of, say, 10 m.

However, the approach may be complicated by the motion of the water surface. If waves are propagating along the water surface in the direction of the wind, one may argue that the stress should be proportional to the square of the relative velocity $(V - U_w)^2$, where U_w is the wave velocity, rather than to the square of V alone. Further, the roughness parameter z_0 may no longer be a constant, but may depend on the wave velocity as well as on the wave amplitude. Given that the waves are formed as a result of the wind stress τ, this implies a coupled phenomenon, in which τ initiates wave motion, which again modifies τ as a result of a changed surface friction, etc. There is experimental evidence to suggest that z_0 is nearly constant except for the lowest wind velocities (and equal to about 6×10^{-3} m; Sverdrup, 1957).

The mechanism by which the surface stress creates waves is not known in detail, nor is the distribution on waves and currents of the energy received by the ocean from the winds, or any subsequent transfer (Pond, 1971). Waves may be defined as motion in which the average position of matter (the water "particles") is unchanged, whereas currents do transport matter. This may also be viewed as a difference in scale. Waves play a minor role in the atmosphere, but are important in the hydrosphere, particularly in the upper water layers, but also as internal waves. The reason is, of course, the higher density of water, which allows substantial potential energy changes to be caused by modest lifting heights.

2.5.2 Circulation Modeling

The Basic Equations in Terms of Time-averaged Variables

As outlined in section 2.3, the averaging procedure leading to (2.23) may now be applied to the general Eulerian equation of transport, for any relevant quantity A,

$$\frac{\partial}{\partial t}(\rho A) + \mathrm{div}(v \rho A) + \mathrm{div}(s_A) = S_A. \tag{2.44}$$

This equation states that the change in A within a given volume, per unit time, is due to gains from external sources, S_A, minus the macroscopic and molecular outflow (divergence terms). The macroscopic outflow is due to the velocity field v, and the microscopic transport is described by a vector s_A, the components of which give the transport by molecular processes in the directions of the co-ordinate axes.

Applying the definitions (2.21)–(2.23), the time-averaged version of (2.44) becomes

$$\frac{\partial}{\partial t}(\overline{\rho} A^*) + \text{div}(\overline{\rho} v^* A^* + \overline{\rho v'A'}) + \text{div}(\overline{s}_A) = S_A. \tag{2.45}$$

If A is taken as unity, (2.44) becomes the equation of continuity (of mass),

$$\frac{\partial}{\partial t}\overline{\rho} + \text{div}(\overline{\rho} v^*) = 0. \tag{2.46}$$

The density-averaging method is essential for obtaining the same form as for the instantaneous values. If A is taken as a component of the velocity vector, (2.45) becomes [with utilization of (2.46)]

$$\overline{\rho}\left(\frac{\partial v_i^*}{\partial t} + (v^* \cdot \text{grad})v_i^*\right) + \overline{((\text{div}v')\rho v_i')} = \sum_j \frac{\partial}{\partial x_j}\tau_{ij} + F_{ext,i}, \tag{2.47}$$

The sources of velocity are the external forces F_{ext} (gravitational and Coriolis forces), and the molecular transfer of velocity (or momentum) is given in terms of the derivatives of the stress tensor τ_{ij},

$$-(s_{v_j})_i = \tau_{ij} = \left(-P + \left(\eta' - \frac{2}{3}\eta\right)\rho\,\text{div}(v)\right)\delta_{ij} + \eta\rho\left(\frac{\partial v_i}{\partial x_j} + \frac{\partial v_j}{\partial x_i}\right),$$

where P is the thermodynamic pressure, η and η' are the kinematic and volume viscosities,* and δ_{ij} is the Kronecker delta (Hinze, 1975). In the last term on the left-hand side of (2.47), the divergence operator $\text{div} = \sum_j \partial/\partial x_j$ is supposed to act on all three factors $v_j' \rho v_i'$ following.

It is clear from the equation of motion (2.47) that the large-scale motion of the atmosphere, v^*, cannot generally be described by an equation that depends on only averaged quantities (for a discussion of early attempts to achieve such a description, see Lorenz, 1967). The divergence term on the left-hand side describes the influence of small-scale turbulent motion on the large-scale motion. It is usually referred to as the *eddy transport* term.

Apart from the well-known pressure gradient term, the contributions from the stress tensor describe the molecular friction forces. These may be important for calculations of energy transformations, but are often left out in calculations of general circulation (Wilson and Matthews, 1971). According to Boussinesq (1877), it may be possible to approximate the eddy transport term by an expression of the same form as the dynamic viscosity term [see (2.60) below], introducing an effective "eddy viscosity" parameter similar to the parameter k considered above in connection with the turbulent transport of heat from the Earth's surface to the boundary layer [the *Prandtl layer,* extending about 50 m above the laminar boundary layer and defined by an approximately constant shear stress τ in expressions like (2.34)].

---

*The viscosities are sometimes taken to include the factor ρ.

The difference between the scales of vertical and horizontal motion makes it convenient to define separate velocities w and V,

$$v = (w \cdot e_z)e_z + (v - (v \cdot e_z)e_z) = we_z + V,$$

and split the equation of motion (2.47) into a horizontal part,

$$\bar{\rho}\left(\frac{\partial V^*}{\partial t} + (V^* \cdot \text{grad})V^* + w^* \frac{\partial V^*}{\partial z}\right) + \frac{\partial}{\partial z}(\overline{\rho w' V'}) = -\text{grad}\,P - \rho f\,e_z \times V^*, \quad (2.48)$$

and a vertical part, which, as mentioned, can be approximated by the hydrostatic equation,

$$\frac{\partial P}{\partial z} = -g\rho. \tag{2.49}$$

In (2.48), the molecular friction forces have been left out, and only the vertical derivative is kept in the eddy term. This term thus describes a turbulent friction, by which eddy motion reduces the horizontal average wind velocity V^*. The last term in (2.48) is due to the Coriolis force, and

$$f = 2\Omega \sin\phi,$$

where Ω is the angular velocity of the Earth's rotation and ϕ is the latitude.

As a lowest-order approximation to the solution of (2.48), all the terms on the left-hand side may be neglected. The resulting horizontal average wind is called the *geostrophic wind*, and it depends on only the pressure gradient at a given height and a given geographical position. The smallness of the left-hand terms in (2.48), relative to the terms on the right-hand side, is, of course, an empirical relationship found specifically in the Earth's atmosphere. It may not be valid for other flow systems.

The large-scale horizontal wind approaches the geostrophic approximation when neither height nor latitude is too small. At lower heights the turbulent friction term becomes more important (up to about 1000 m), and at low latitudes the Coriolis term becomes smaller, so that it no longer dominates over the terms involving local time and space derivatives.

In analogy to (2.47), the averaging procedure may now be applied to the general transport equation for some scalar quantity, such as the mixing ratio for some minor constituent of the atmosphere. Equation (2.47) then becomes

$$\bar{\rho}\left(\frac{\partial A^*}{\partial t} + (V^* \cdot \text{grad})A^* + w^* \frac{\partial A^*}{\partial z}\right) + \frac{\partial}{\partial z}(\overline{\rho w' A'}) = S_A. \tag{2.50}$$

The source term on the right-hand side indicates whether the substance in question is created or lost, for example, by phase changes or chemical reactions. S_A will, in general, depend on all the state variables of the atmosphere, the air velocity, and the mixing ratios for other constituents. In an approximate manner, the source term may be taken to depend only on the average values of its parameters (Wilson and Matthews, 1971),

$$S_A = S_A(\rho, P, V^*, w^*, A_j^*).$$

The ideal gas approximation to the equation of state for the air reads, in averaged form,

$$P = \mathscr{R}\rho T^*/\mu, \qquad (2.51)$$

where $\mathscr{R}/\mu \cong 280$ m^2 s^{-2} K^{-1} up to a height z of about 90 km, above which level the molecular weight drops quickly. Eventually, a correction depending on humidity may be incorporated into (2.51).

Temperature is often replaced by a variable called the *potential temperature*, defined by

$$\theta = T\left(\frac{P_0}{P}\right)^{(c_P - c_V)/c_P} \quad \text{or} \quad \theta^* = T^*\left(\frac{P_0}{P}\right)^{(c_P - c_V)/c_P}, \qquad (2.52)$$

where the constant P_0 is usually taken as the standard sea-level pressure (10^5 N m^{-2}). The significance of the potential temperature is that it is constant during processes that do not change entropy. Figure 2.104 shows an example of latitude variations of θ^* and T^* based on the same data.

The first law of thermodynamics states that the change in internal energy per unit mass is given by

$$c_V\, dT = \rho^{-1} dW^{int} = -P\, d(\rho^{-1}) + Q\, dt, \qquad (2.53)$$

where Q is the amount of heat added to the system (an air parcel of unit mass) per second. As in (2.48), the molecular friction (which also produces heat) will be left out. From the definition of specific entropy s, one obtains

$$T\, ds = c_P\, dT - \rho^{-1} dP,$$

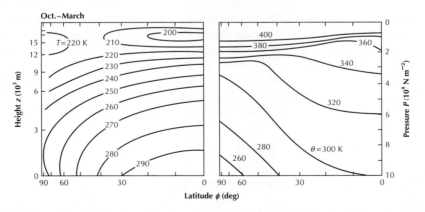

FIGURE 2.104 Distribution of seasonal (October–March) and longitudinal average of temperature T and potential temperature θ for the Northern Hemisphere. (Based on Lorenz, 1967.)

and by introducing the potential temperature (2.52),

$$\mathrm{d}s = c_P\,\theta^{-1}\mathrm{d}\theta.$$

Inserting this into (2.53), the time derivative of θ is obtained,

$$\frac{\mathrm{d}\theta}{\mathrm{d}t} = \frac{Q}{c_P}\left(\frac{P_0}{P}\right)^{(c_P-c_V)/c_P}.$$

This is the source term to be entered on the right-hand side of the expression (2.50) for the averaged potential temperature,

$$\bar{\rho}\left(\frac{\partial\theta^*}{\partial t} + (V^* \cdot \mathrm{grad})\theta^* + w^*\frac{\partial\theta^*}{\partial z}\right)\frac{\partial}{\partial z}\left(\overline{\rho w'\theta'}\right) = \frac{\overline{\rho Q}}{c_P}\left(\frac{P_0}{P}\right)^{(c_P-c_V)/c_P}. \qquad (2.54)$$

Again the last term on the left-hand side describes an amount of heat lost from a given volume by small-scale vertical eddy motion. The external heat sources are radiation and, according to the implicit way of treating latent heat, the heat of condensation of water vapor (plus possibly other phase change contributions),

$$\rho Q = R + C,$$

where R is the amount of heat added to a unit volume by radiation and C is the contribution from condensation processes.

For the approximations considered, a complete set of equations for the atmosphere is constituted by (2.46), (2.48), (2.49), (2.54) and a number of equations (2.50), the most important of which may be that of water vapor. Auxiliary relations include the equation of state, (2.51), and the equations defining the potential temperature (2.52), the heat sources, and the source functions for the other constituents that have been included in (2.50). The water vapor equation is important for calculating the condensation contribution to the heat source function Q. The calculation of the radiative contribution to Q involves, in addition to knowing the external radiation input from space and from continents and oceans as a function of time, a fairly detailed modeling of those constituents of the atmosphere that influence the absorption of radiation, e.g., ozone, water in all forms (distribution of clouds, etc.), and particulate matter.

The Atmospheric Heat Source Function

Alternatively, but with a loss of predictive power, one may take Q as an empirical function, in order to obtain a closed system of equations without (2.50), which—if the turbulent eddy term is also left out—allows a determination of the averaged quantities $V^*, w^*, \bar{\rho}, \bar{P}$ and T^* or θ^*. Any calculation including the dissipation by turbulent eddies must supplement the equations listed above with either the equations of motion for the fluctuations V' and w', or (as is more common) by some parametrization of, or empirical approximation to, the time averages $\overline{\rho w'V'}$, $\overline{\rho w'\theta'}$ and eventually $\overline{\rho w'A'}$.

FIGURE 2.105   Height and latitude distribution of total diabatic heating ($\overline{\rho Q}/c_P \overline{\rho}$), averaged over longitude and season (*left*: June–August, *right*: December–February). The unit is K (day)$^{-1}$. (Based on Newell *et al.*, 1969.)

Figure 2.105 shows an empirical heat source function for two different seasons, but averaged over longitudes. This function could be inserted on the right-hand side of (2.54) to replace $\overline{\rho Q}/(c_P \overline{\rho})$ (the proper SI unit would be K s^{-1} and not K day^{-1} as used in the figure). The different contributions to the heat source function are shown in Figs. 2.106 and 2.107. Figure 2.106

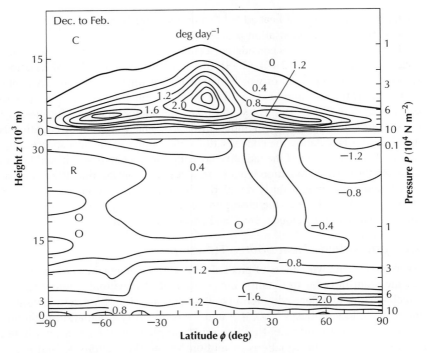

FIGURE 2.106   Height and latitude distribution of heat added to the atmosphere as a result of condensation of water vapor, C (*top*) and radiation, R (*bottom*). Both are in units of (c_P deg day^{-1}) and averaged over longitude and the period December–February. (Based on Newell *et al.*, 1969.)

shows the net radiation distribution, R, over the atmosphere (bottom) and the net condensation (condensation minus evaporation) of water vapor in the atmosphere, C (both in units of c_P). These two contributions constitute the truly external heat sources Q, but the total diabatic heating of the atmosphere illustrated in Fig. 2.105 also includes the addition of sensible heat by the turbulent eddy motion described by the last term on the left-hand side of (2.47). Figure 2.107 gives the magnitude of this "heating source" by the quantity

$$\frac{\partial}{\partial z}(\overline{\rho \, w' \, T'})/\rho.$$

The diabatic heating contributions integrated over a vertical column through the atmosphere are indicated in Fig. 2.108. The difference between radiation absorbed and re-emitted from the atmosphere, E_a^{rad}, is negative and fairly independent of latitude. The heat of condensation, which is approximated by the precipitation rate r_a times the latent heat of vaporization L_v, varies substantially with latitude, depending primarily on the distribution of land and ocean surfaces. The heat added to the atmosphere by turbulent convection is equal to the net sensible heat flow at the Earth's surface, with opposite sign, i.e., $-E_s^{sens}$. The sum of the three contributions shown in Fig. 2.108 is not zero, but has to be supplemented by the transport of sensible heat in the atmosphere, between the different latitude areas. However, this is exactly what the equations of motion are supposed to describe.

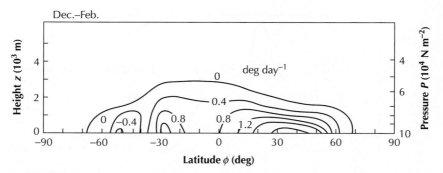

FIGURE 2.107   Height and latitude distribution of heat added to the atmosphere by turbulent transport of sensible heat from the Earth's surface, in units of (c_P deg day^{-1}) and averaged over longitude and the period December–February. (Based on Newell *et al.*, 1969.)

Separation of Scales of Motion

The averaging procedure discussed in conjunction with (2.23) presumes that a "large" and a "small" scale of motion are defined. This may be properly done by performing a Fourier analysis of the components of the wind velocity vector, or of the wind speed $|v|$. It is convenient to consider the variance of the wind speed (because it is proportional to a kinetic energy), defined by $\langle \widetilde{v} \cdot \widetilde{v} \rangle$.

The brackets <> denote average over a statistical ensemble, corresponding to a series of actual measurements. Writing the spectral decomposition in the form

$$\tfrac{1}{2}\langle \tilde{\mathbf{v}} \cdot \tilde{\mathbf{v}} \rangle = \int_0^\infty S(\omega)\,d\omega = \int_0^\infty \omega\,S(\omega)\,d(\log\omega),$$

one obtains a form-invariant spectral function $S(\omega)$, which is shown in Fig. 2.109, based on a modern version of a pioneering measurement effort made by van der Hoven (1957). The peaks exhibited by the spectrum vary in magnitude with the height of measurement (cf. Fig. 3.37).

A striking feature of the spectrum in Fig. 2.109 (and the analogues for other heights) is the broad gap between $\omega \approx 0.5\ \mathrm{h}^{-1}$ and $\omega \approx 20\ \mathrm{h}^{-1}$. A large number of measurements have confirmed that the existence of such a gap is an almost universal feature of the wind speed spectrum. Its significance is to provide a clear distinction between the region of large-scale motion ($\omega \lesssim 0.5\ \mathrm{h}^{-1}$) and the region of small-scale (eddy) motion ($\omega \gtrsim 5\ \mathrm{h}^{-1}$). The existence of the gap makes the time-averaging procedure in (2.21) easy and makes the exact choice of Δt insignificant over a reasonably large interval, so that the resulting large-scale motion is not critically dependent on the prescription for time averaging. One should be warned, however, not to conclude that the gap in Fig. 2.109 indicates that large-scale motion and small-scale motion are not coupled. It is not necessary for such couplings to involve all intermediate frequencies; on the contrary, distinct couplings may involve very different scales of motion in a

FIGURE 2.108 Components of the net energy flux passing the atmosphere, averaged over longitude and the year. E_a^{rad} is the net radiation flux, $L_v r_a$ is the heat gained by condensation of water vapor, and $-E_s^{sens}$ is the turbulent flux of sensible heat from the Earth's surface to the atmosphere (i.e., the same quantities appearing in Figs. 2.106 and 2.107, but integrated over height). The deviation of the sum of the three quantities from zero should equal the heat gain from horizontal transport of sensible heat (F_a^{sens}). (Based on Sellers, 1965.)

straightforward way. Examples of this are evident in the equations of motion, such as (2.48), which allows kinetic energy to be taken out of the large-scale motion (V^*) and be put into small-scale motion ($\overline{\rho w' V'}$) or vice versa.

Thus small- and large-scale motions in the atmosphere are well-defined concepts, and no accurate model of the atmospheric circulation can be obtained without treating both simultaneously. Current modeling, both for short-term behavior of the atmospheric circulation ("weather forecasts") and for long-term behavior ("climate change"), uses averaged values of terms coupling the scales and thus effectively only includes large-scale variables. This is the reason for the poor validity of weather forecasts: they remain valid only as long as the coupling terms involving turbulent (chaotic) motion do not change. For climate modeling, the constraints offered by the system boundaries make the model results valid in an average sense, but not necessarily the detailed geographical distribution of the magnitude of climate variables. This also means that the stability of the atmosphere, which as mentioned in section 2.4 is not in its lowest energy state, cannot be guaranteed by this type of calculation and that the atmosphere could experience transitions from one relatively stability state to another (as it has during its history, as evidenced by the transitions to ice ages and back).

FIGURE 2.109 Wind speed variance spectrum, based on three different measurements (the main part from Petersen, 1974, cf. Sørensen, 1995a).

Furthermore, since the motion within the atmosphere is not capable of carrying out all the necessary transport of heat (see Fig. 2.48), an accurate model of the circulation of the atmosphere is not possible without including the heat transport within the ocean–continent system (ocean currents, rivers, run-off along the surface, and, to a lesser extent, as far as heat transport is concerned, ground water motion). Such coupled models are considered below, after introducing the energy quantities of relevance for the discussion.

Energy Conversion Processes in the Atmosphere

The kinetic energy of the atmospheric circulation can be diminished by friction, leading to an irreversible transformation of kinetic energy into internal energy (heat). In order to make up for frictional losses, new kinetic energy must be created in the atmosphere. This can be achieved by one of two processes, both of which are reversible and may proceed adiabatically. One is the conversion of potential energy into kinetic energy (by gravitational fall), and the other is the transformation of internal energy into kinetic energy by motion across a pressure gradient. In terms of the averaged quantities (i.e., neglecting terms like $\overline{\rho V' \cdot V'}$), the kinetic, potential, and internal energies may be written

$$W^{kin} = \tfrac{1}{2}\overline{\rho}\left(V^* \cdot V^* + w^* w^*\right),$$

$$W^{pot} = \overline{\rho} g z, \tag{2.55}$$

$$W^{int} = \overline{\rho} c_V T^*,$$

and the corresponding changes in time

$$\frac{dW^{kin}}{dt} = V^* \cdot \text{grad}\, P - w^* \frac{\partial P}{\partial z} - w^* \overline{\rho} g - V^* \cdot \frac{\partial}{\partial z}\left(\overline{\rho w' V'}\right), \tag{2.56}$$

obtained from (2.46) by scalar multiplication with $v*$,

$$\frac{dW^{pot}}{dt} = \overline{\rho} g w^* \tag{2.57}$$

[from the definition in (2.55)] and

$$\frac{dW^{int}}{dt} = -P\text{div}(v^*) + \overline{\rho Q} + V^* \cdot \frac{\partial}{\partial z}\left(\overline{\rho w' V'}\right), \tag{2.58}$$

which follows from (2.53) by using

$$\rho \frac{d}{dt}\left(\frac{1}{\rho}\right) = \text{div}(v)$$

and adding the heat gained by turbulent convection on the same footing as the external heat sources ρQ (radiation and gain of latent heat by condensation). As noted by Lorenz (1967), not all potential and internal energy is available for conversion into kinetic energy. The sum of potential and internal energy must be measured relative to a reference state of the atmosphere. Lorenz defines the reference state as a state in which the pressure equals the average value on the isentropic surface passing through the point considered [hence the convenience of introducing the potential temperature (2.52)].

Modeling the Oceans

As mentioned in section 2.3, the state of the oceans is given by the temperature T, the density $\rho = \rho_w$ (or alternatively the pressure P), the salinity S (salt fraction by mass), and possibly other variables, such as oxygen and organic materials present in the water. S is necessary in order to obtain a relation defining the density that can replace the ideal gas law used in the atmosphere. An example of such an equation of state, derived on an empirical basis by Eckart (1958), is

$$P = \frac{x_1(T, S)\rho}{1 - x_0\rho} - x_2(T, S), \tag{2.59}$$

where x_0 is a constant and x_1 and x_2 are polynomials of second order in T and first order in S^*.

The measured temperature and salinity distributions are shown in Figs. 2.67–2.69. In some studies, the temperature is instead referred to surface pressure, thus being the potential temperature θ defined according to (2.52). By applying the first law of thermodynamics, (2.53), to an adiabatic process ($Q = 0$), the system is carried from temperature T and pressure P to the surface pressure P_0 and the corresponding temperature $\theta = T(P_0)$. The more complex equation of state (2.59) must be used to express $d(1/\rho)$ in (2.53) in terms of dT and dP,

$$d\left(\frac{1}{\rho}\right) = \frac{\partial X}{\partial P}dP + \frac{\partial X}{\partial T}dT,$$

where

$$X = \frac{x_1 + x_0(P + x_2)}{P + x_2} = \frac{1}{\rho}.$$

This is then inserted into (2.53) with $Q = 0$, and (2.53) is integrated from (T, P) to (θ, P_0). Because of the minimal compressibility of water (as compared to air, for example), the difference between T and θ is small.

Basic Equations Governing the Oceanic Circulation

Considering first a situation where the formation of waves can be neglected, the wind stress may be used as a boundary condition for the equations of motion analogous to (2.47). The "Reynold stress" eddy transport term may, as mentioned in connection with (2.47), be parametrized as suggested by Boussinesq, so that it gets the same form as the molecular viscosity term and—more importantly—can be expressed in terms of the averaged variables,

$$\overline{(\text{div}\,v')\rho v'} \approx -k_z\bar{\rho}\frac{\partial}{\partial z}\frac{\partial}{\partial z}V^* - k_y\bar{\rho}\frac{\partial}{\partial y}\left(\text{grad}\,V_y^* + \frac{\partial}{\partial y}V^*\right) - k_x\bar{\rho}\frac{\partial}{\partial x}\left(\text{grad}\,V_x^* + \frac{\partial}{\partial x}V^*\right). \tag{2.60}$$

$^*x_0 = 0.698 \times 10^{-3}\,\text{m}^3\text{kg}^{-1}$,
$x_1 = (177950 + l\,125\,T - 7.45\,T^2 - (380 + T)1000\,S)\,\text{m}^2\text{s}^{-2}$,
$x_2 = (5890 + 38\,T - 0.375\,T^2 + 3000\,S)10^5\,\text{N}\,\text{m}^{-2}$,
where T should be inserted in °C (Bryan, 1969).

Here use has been made of the anticipated insignificance of the vertical velocity, w^*, relative to the horizontal velocity, V^*. A discussion of the validity of the Boussinesq assumption may be found in Hinze (1975). It can be seen that (2.60) represents diffusion processes that result from introducing a stress tensor of the form (2.31) into the equations of motion (2.47). In (2.60), the diffusion coefficients have not been assumed to be isotropic, but in practice the horizontal diffusivities k_x and k_y are often taken to be equal (and denoted K; Bryan, 1969; Bryan and Cox, 1967). Denoting the vertical and horizontal velocity components in the ocean waters w_w and V_w, the averaged equations of motion, corresponding to (2.48) and (2.49) for the atmosphere, become

$$\frac{\partial \overline{P}_w}{\partial z} = -g\overline{\rho}_w, \tag{2.61}$$

$$\frac{\partial V_w^*}{\partial t} + (V_w^* \cdot \mathrm{grad}) V_w^* + w_w^* \frac{\partial V_w^*}{\partial z} - k_z \frac{\partial^2}{\partial z^2} V_w^* - K \left(\frac{\partial^2}{\partial x^2} + \frac{\partial^2}{\partial y^2} \right) V_w^*$$

$$- K \left(\frac{\partial}{\partial x} \mathrm{grad}\, V_{wx}^* + \frac{\partial}{\partial y} \mathrm{grad}\, V_{wy}^* \right) = \frac{1}{\overline{\rho}} \mathrm{grad}\, \overline{P}_w - f \mathbf{e}_z \times V_w^*, \tag{2.62}$$

where the Coriolis parameter is

$$f = 2\Omega \sin\phi + \frac{V_{wx}^*}{r_s} \tan\phi,$$

in terms of the angular velocity of the Earth's rotation, Ω, plus the longitudinal angular velocity of the mean circulation, $V_{wx}^*(r_s \cos\phi)^{-1}$, with r_s being the radius of the Earth.

The boundary conditions at the ocean's surface are

$$w_w^*(z = 0) = 0,$$

$$\overline{\rho}_w k_z \frac{\partial V_w^*(z = 0)}{\partial z} = \overline{\tau}, \tag{2.63}$$

where τ is the wind stress (2.43). The boundary condition (2.63) [cf. (2.31)] reflects the intuitive notion that, for a small downward diffusion coefficient, k_z, a non-zero wind stress will lead to a steep velocity gradient at the ocean surface. Knowledge of k_z is limited, and in oceanic calculations k_z is often taken as a constant, the value of which is around 10^{-4} m^2 s^{-1} (Bryan, 1969).

If the density of water, ρ_w, is also regarded as a constant over time intervals much larger than that used in the averaging procedure, then the equation of continuity (2.46) simply becomes

$$\mathrm{div}\, v_w^* = \frac{\partial V_{wx}^*}{\partial x} + \frac{\partial V_{wy}^*}{\partial y} + \frac{\partial w_w^*}{\partial z} = 0. \tag{2.64}$$

Since the pressure, P_w in (2.61), is related to temperature and salinity through the equation of state, (2.59), a closed set of equations for the oceanic circulation must encompass transport equations analogous to (2.50) for the average values of temperature and salinity,

$$\frac{\partial A^*}{\partial t} + (V_w^* \cdot \mathrm{grad})A^* + w_w^* \frac{\partial A^*}{\partial z} - k_z \frac{\partial^2}{\partial z^2} A^* - K\left(\frac{\partial^2}{\partial x^2} + \frac{\partial^2}{\partial y^2}\right)A^* = \frac{\overline{S_A}}{\rho_w}. \quad (2.65)$$

The turbulent diffusion coefficients, k_z and K, being phenomenological quantities, could eventually be chosen separately for temperature, $A^* = T_w^*$, and salinity, $A^* = S_w^*$, and not necessarily be identical to those used in (2.62) for momentum. If it is assumed that there are no sources or sinks of heat or salt within the ocean, the right-hand side of (2.65) may be replaced by zero. The exchange of heat and salt between the ocean and the atmosphere can then be incorporated into the boundary conditions, under which the two equations (2.65) are solved,

$$\overline{\rho_w} k_z \frac{\partial T_w^*(z=0)}{\partial z} = -\frac{E_s^{total}}{C_w}, \quad (2.66)$$

$$\overline{\rho_w} k_z \frac{\partial S_w^*(z=0)}{\partial z} = S_w^*(z=0)(e-r). \quad (2.67)$$

Among the chemical reaction terms that have been suggested for inclusion as source terms, S_A in (2.50), are the sulfate-forming reactions found in sea-salt spray particles (Laskin et al., 2003).

The surface net energy flux E_s^{total} [cf. (2.18)], the evaporation e, and precipitation r are variables that link the ocean with the atmosphere, as well as the wind stress appearing in (2.63). In (2.66), C_w is the specific heat of water.

The above expressions do not include the contributions from ice-covered water. Bryan (1969) has constructed boundary conditions allowing for formation and melting of sea ice. Pack ice extending to a depth of less than 3 m is able to move, so Bryan adds a transport equation for such sea ice in the general form of (2.65) and with a suitable source term. More recent models include a specific ice cover model and take into account both melting and sublimation processes (see, for example, NCAR, 1997).

Recent Model Applications Surveyed in the 4th IPCC Assessment

Climate models continuously improve, both with respect to resolution (currently down to about 25 km for the integration grids used, leading to some 125 km accuracy of the final calculation results) and with respect to effects included. Recent models treat sea ice formation and movements, as well as details of cloud cover and its influence on both water cycle and air properties. Still not included in the otherwise most advanced models are the coupling between glaciation, depression and upheaval of the crust, and the associated changed in sea shorelines (Peltier, 1994; 2004). These are important for understanding past climates but insignificant for modeling a few hundred years. Still, studies of past climates could be very relevant for the basic understanding of climate change, whether anthropologically induced or not (Kaspar and Cubasch, 2007; Jansen et al., 2007; NOAA, 2010; Sørensen, 2011).

Figure 2.110 shows the temperature results from a Japanese model included in the assessment of the fourth IPCC report (IPCC, 2007). The model has been run for the pre-industrial situation prevailing around 1860, and for various emission scenarios for future emissions of greenhouse gases. Figure 2.110 picks the A1B scenario, which is an update of the "business as usual" scenario of previous IPCC assessments. That scenario was an unlimited growth scenario with exponential increase in emissions from fossil resources. The A1B scenario still assumes the current 18th-century economic paradigm prevails, but recognizes that in some parts of the world there is a discussion of limited resources and carrying capability of the environment. The emission scenario assumes that a mixture of fossil, nuclear, and renewable energy resources will be employed, as dictated by market prices during the unrestrained economic growth. Miraculously (i.e., without identifying causes but presumably expecting the decline in fertility characterizing presently rich countries to spread, a feature that some would associate with the somewhat more equitable distribution of wealth obtained during the 20th-century period with socialist governments in many rich countries), world population stabilizes and the CO_2 concentration in the atmosphere stabilizes at 720 ppm toward the end of the 21st century.

Compared to the earlier model results shown in Fig. 2.97, for a doubling of CO_2 but relative to about 1995, the new 2055 numbers, expected to represent a comparable situation, show increased warming, and in contrast to the 2020 results and the old 2055 values of Fig. 2.97, there are now no regions with a lowering of temperature. Considering the different year of departure, the results can be considered consistent. These remarks are valid for both the January and the July results.

By 2090, warming is very strong, up to 25 °C at the poles, and over 5 °C when averaged over land areas. This is the expected outcome of the lack of political action characterizing the A1B emission scenario of the IPCC (see description at IPCC, 2007; and, in short form, IPCC, 2010). The warming is predicted to affect high-latitude Northern areas already in 2020, by 2055 most areas in the northern United States, Siberia, and China. Mountain areas like the Himalayas experience strong warming, and both they and the arctic areas are likely to lose large amounts of ice. By 2090, all inhabited parts of the world are strongly affected, including South America, Africa, and Australia, and in both January and July.

The IPCC Data Distribution Centre (2010) holds a considerable number of alternative model results, generally agreeing on the main conclusions.

(a)

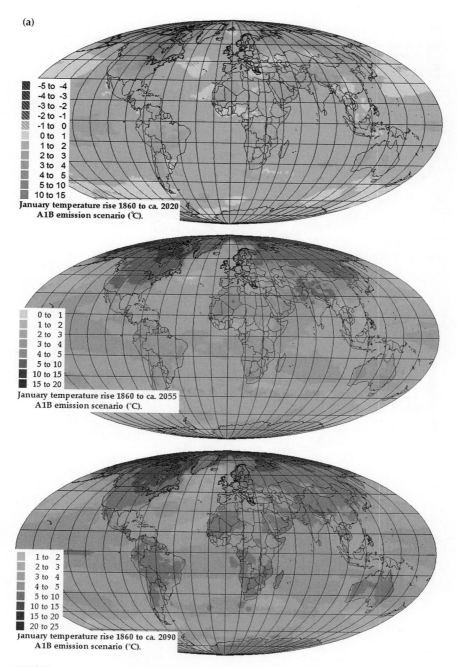

FIGURE 2.110a January temperature differences from pre-industrial level for 20-y periods centered at 2020, 2055, and 2090. A1B emission scenario, NIES-MIROC3.2-HI climate model (Hasumi and Emori, 2004; IPCC Data Distribution Centre, 2010).

(b)

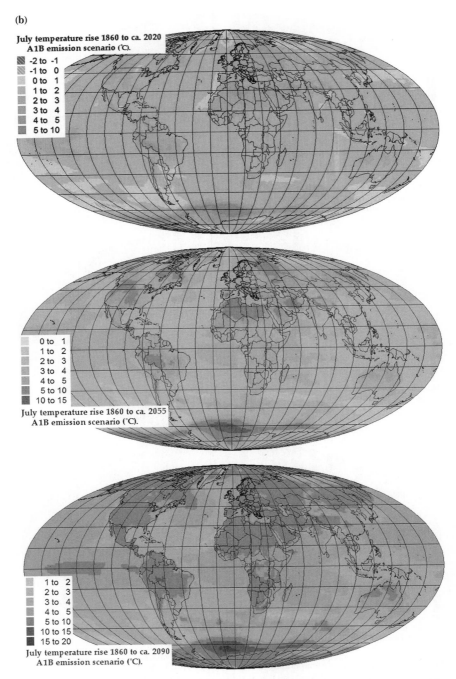

FIGURE 2.110*b* Same as Fig. 2.110a but for July temperature differences (*top*: 2020, *middle*: 2055, *bottom*: 2090).

(a)

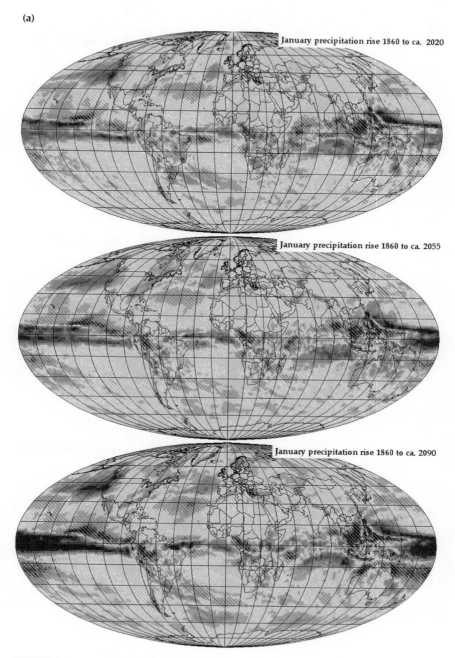

FIGURE 2.111a   January precipitation differences from pre-industrial level for 20-y periods centered at 2020, 2055, and 2090. A1B emission scenario, NIES-MIROC3.2-HI climate model (Hasumi and Emori, 2004; IPCC Data Distribution Centre, 2010).

(b)

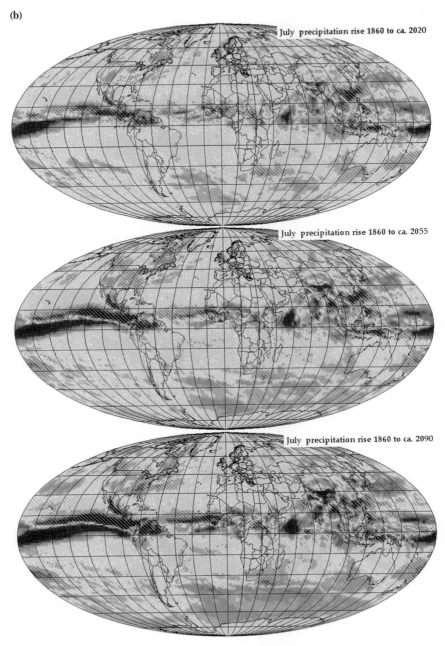

FIGURE 2.111*b*   Same as Fig. 2.111*a* but for July precipitation differences (*top*: 2020, *middle*: 2055, *bottom*: 2090). The scale used in Fig. 2.111*a* and *b* is shown at the top of the following page.

A1B emission scenario (kg m^{-2} s^{-1})

▨	-0.0001	to -0.00005
▨	-0.00005	to -0.00004
▨	-0.00004	to -0.00003
▨	-0.00003	to -0.00002
▨	-0.00002	to -0.00001
▨	-0.00001	to -0.000005
▨	-0.000005	to 0
▨	0	to 0.000005
▨	0.000005	to 0.00001
▨	0.00001	to 0.00002
▨	0.00002	to 0.00003
▨	0.00003	to 0.00004
▨	0.00004	to 0.00005
▨	0.00005	to 0.0001

FIGURES 2.111a and b   Scale used in figures on the two preceding pages.

The calculated effect of climate change on precipitation, shown in Fig. 2.111, shows a much stronger change than in temperature from pre-industrial times to the 2011–2030 period, and somewhat less during the rest of the 21st century, except for a pronounced reduction in the Eastern Equatorial Pacific and an even stronger increase in the Western Equatorial Pacific. A similar but more divided effect is seen in the Indian and Atlantic Oceans. This behavior is even stronger in July than in January. Generally, the dry areas in the northern half of Africa and in the Middle East become even dryer, and the high-latitude areas more wet. The U.S. Midwest is becoming dryer during July but is neutral in January, which could have implications for agriculture there, which is already considerably dependent on artificial irrigation. The opposite is the case for India, getting more summer rain in the southeast and considerably less in the northwest, including the already dry Rajastan desert.

There is no conservation of the water cycle through the atmosphere: The amount of water evaporated from land and sea surfaces and later falling as precipitation increases, as one would expect due to the general warming.

The studies assessed by the (presently four) IPCC Working Group I reports have had a considerable effect on making decision-makers aware of the climate change problem associated with anthropogenic emissions. The success is largely due to the scientific credibility of the assessments, which was ensured when IPCC was formed, by following its first chairman, Bert Bolin, in deviating from the usual United Nations procedure of having governments designate members of committees (leading to the nomination of family members of African leaders to committees that they had no expertise to be on, or to the nomination of energy–industry lobbyists). This was accomplished by not having IPCC subsumed under one U.N. agency, but sitting between two. When Bolin resigned after the second assessment report, things reverted to traditional U.N. procedures, and the role of IPCC has therefore been questioned (Hulme *et al.*, 2010). The main message, however, remains unchanged and attacks from various lobby groups have so far not shaken its substance. After all, the greenhouse effect is an established physical fact and the many additional effects that add to the behavior of climate are important but only as additions to the main effects. Those (e.g., some sunspot scientists) who think that finding new effects will automatically invalidate classical physics are clearly at variance with scientific rationality.

2.5.3 Tides and Waves

Consider a celestial body of mass M and distance R from the center of the Earth. The gravitational force acting on a particle of mass m placed at distance r from the Earth's center may, in a co-ordinate system moving with the Earth, be written

$$F^{tidal} = m\frac{d^2r}{dt^2} = m\frac{d^2}{dt^2}(R'-R) = -mMG\left(\frac{R'}{R'^3} - \frac{R}{R^3}\right),$$

where R' is the position vector of the mass m relative to M. The force may be decomposed into a radial and a tangential part (along unit vectors e_r and e_t), introducing the angle θ between $-R$ and r and assuming that $|r|$ is much smaller than $|R|$ and $|R'|$ in order to retain only the first terms in an expansion of R'^{-2} around R^{-2},

$$F^{tidal} \approx \frac{2mMGr}{R^3}(\cos^2\theta e_r + \sin\theta\cos\theta e_t). \tag{2.68}$$

If the particle m is at rest within or near to the Earth, it is further influenced by the gravitational force of the mass of the sphere inside its radius, with the center coinciding with that of the Earth. If the particle is following the rotation of the Earth, it is further subjected to the centrifugal force and, if it is moving relative to the Earth, to the Coriolis force. The particle may be ascribed a tidal potential energy, taken to be zero if the particle is situated at the center of the Earth (Bartels, 1957),

$$W^{tidal} = \int_0^r F^{tidal} \cdot e_r dr = \frac{mMGr^2}{R^3}\cos^2\theta.$$

If the particle is suspended at the Earth's surface in such a way that it may adjust its height, so that the change in its gravitational potential, mgz, can balance the tidal potential, then the change in height, $z = \Delta r$, becomes

$$z = \frac{MGr^2}{gR^3}\cos^2\theta.$$

If one inserts lunar mass and mean distance, this height becomes $z = 0.36$ m for $\cos^2\theta = 0$, i.e., at the points on the Earth facing the Moon or opposite (locations where the Moon is in zenith or in nadir). Inserting instead data for the Sun, the maximum z becomes 0.16 m. These figures are equilibrium tides that might be attained if the Earth's surface were covered with ocean and if other forces, such as those associated with the Earth's rotation, were neglected. However, since water would have to flow into the regions near $\theta = 0$ and $\theta = \pi$, and since water is hardly compressible, volume conservation would require lowering of the water level at $\theta = \pi/2$.

The estimate of "equilibrium tides" only serves the purpose of a zero-order orientation of the magnitude of tidal effects originating from different celestial bodies. The actual formation of tidal waves is a dynamic process that depends not only on the above-mentioned forces acting on the water particles, but also particularly on the topography of the ocean boundaries. The periods involved can be estimated by inserting into (2.68) the expression for the zenith angle θ of the tide-producing celestial body, in terms of declination δ and hour angle ω of the body,

$$\cos\theta = \sin\delta \sin\phi + \cos\delta \cos\phi \cos\omega$$

[cf. the discussion in connection with (2.9) and (2.10)]. The tidal force will have components involving $\cos\omega$ and $\cos^2\omega = (\cos(2\omega) + 1)/2$, or the corresponding sine functions, which proves the existence of periods for lunar tides equal to one lunar day (24 h and 50 min) and to half a lunar day. Similarly, the solar tides are periodic with components of periods of a half and one solar day. If higher-order terms had been retained in (2.68), periods of a third the basic ones, etc., would also be found, but with amplitudes only a small fraction of those considered here.

Owing to the inclination of the lunar orbital plane relative to that of the Earth–Sun system, the amplitudes of the main solar and lunar tidal forces (2.68) will add in a fully coherent way only every 1600 years (latest in 1433; Tomaschek, 1957). However, in case of incomplete coherence, one can still distinguish between situations where the solar tidal force enhances that of the Moon ("spring tide") and situations where the solar tidal forces diminish the lunar tides ("neap tide").

Tidal waves lose energy due to friction against the continents, particularly at inlets with enhanced tidal changes in water level. It is estimated that this energy dissipation is consistent with being considered the main cause for the observed deceleration in the Earth's rotation (the relative decrease being 1.16×10^{-10} per year). Taking this figure to represent the energy input into tidal motion, as well as the frictional loss, one obtains a measure of the energy flux involved in the tides as equal to about 3×10^{12} W (King Hubbert, 1971).

Gravity Waves in the Oceans

The surface of a wavy ocean may be described by a function $\sigma = \sigma(x, y, t)$, giving the deviation of the vertical coordinate z from zero. Performing a spectral decomposition of a wave propagating along the x-axis yields

$$\sigma(x, t) = \int_{-\infty}^{\infty} \exp(-i(kx - \omega(k)t)) S(k) \, dk, \tag{2.69}$$

where $S(k)$ represents the spectrum of σ. If ω is constant and positive, (2.69) describes a harmonic wave moving toward the right; if ω is constant and negative, the wave moves toward the left. By superimposing waves with positive and negative ω, but with the same spectral amplitude $S(k)$, standing waves may be constructed. If $S(k)$ in (2.69) is different from zero only in a narrow region of k-space, σ will describe the propagation of a disturbance (wave packet) along the ocean surface.

The wave motion should be found as a solution to the general equations of motion (2.61)–(2.64), removing the time averages and interpreting the eddy diffusion terms as molecular diffusion (viscous) terms, with the corresponding changes in the numerical values of $k_x = K$ into η_w (the kinematic viscosity of water, being around 1.8×10^{-6} m^2 s^{-1}). Owing to the incompressibility assumption (2.64), the equation of motion in the presence of gravity as the only external force may be written

$$\frac{\partial v_w}{\partial t} + (v_w \cdot \mathrm{grad})v_w = -ge_z - \rho_w^{-1}\mathrm{grad}\, P_w + \eta_w \,\mathrm{div\,grad}\, v_w. \tag{2.70}$$

For irrotational flow, rot $v = 0$, the velocity vector may be expressed in terms of a velocity potential, $\phi\,(x, y, z, t)$,

$$v_w = \mathrm{grad}\,\phi, \tag{2.71}$$

and the following integral to the equation of motion exists,

$$\frac{\partial \phi}{\partial t} + \frac{1}{2}v_w^2 + gz + \frac{P_w}{\rho_w} = 0, \tag{2.72}$$

where the constant on the right-hand side is only zero if proper use has been made of the freedom in choosing among the possible solutions ϕ to (2.71). In the case of viscous flow, rot v_w is no longer zero, and (2.71) should be replaced by

$$v_w = \mathrm{grad}\,\phi + v_w^{rot}. \tag{2.73}$$

A lowest-order solution to these equations may be found using the infinitesimal wave approximation (Wehausen and Laitone, 1960) obtained by linearizing the equations [i.e., disregarding terms like $\frac{1}{2}v_w^2$ in (2.72), etc.]. If the surface tension, which would otherwise appear in the boundary conditions at $z = 0$, is neglected, one obtains the first-order, harmonic solution for gravity waves, propagating in the x-direction,

$$\phi = a\,U_w e^{kz}\sin\!\big(k(x - U_w t)\big)\,\exp(-2k^2\,\eta_w\,t), \tag{2.74}$$

$$\sigma = a\cos\!\big(k(x - U_w t)\big)\,\exp(-2k^2\,\eta_w\,t), \tag{2.75}$$

where the connection between wave velocity (phase velocity) U_w, wave number k, and wavelength λ_w is

$$U_w = \left(\frac{g}{k}\right)^{1/2}; \quad k = \frac{2\pi}{\lambda_w} \tag{2.76}$$

[cf. the relation in the presence of surface tension (2.27)]. No boundary condition has been imposed at the bottom, corresponding to a deep ocean assumption, $\lambda_w \ll h$, with h being the depth of the ocean. If the viscous forces are disregarded, the lowest-order solutions are similar to those of (2.74) and (2.75), but with the last exponential factor omitted. Improved solutions can be obtained by including higher-order terms in the equations of motion, by perturbative iteration starting from the first-order solutions given here. The second-order correction to ϕ is zero, but the wave surface (2.75) becomes (for $\eta_w \approx 0$)

$$\sigma^{(2)} = a \cos(k(x - U_w t)) + 0.5\, a^2 k \cos(2k(x - U_w t)).$$

The correction makes the wave profile flatter along the bottom (trough) and steeper near the top (crest). Exact solutions to the wave equations, including surface tension but neglecting viscous forces, have been constructed (see, for example, Havelock, 1919). An important feature of these solutions is a maximum value that the ratio between wave amplitude a and wavelength λ_w can attain, the numerical value of which is determined as

$$ak = a\frac{2\pi}{\lambda_w} \leq 0.89. \tag{2.77}$$

FIGURE 2.112 Wave profile for pure gravity wave of maximum amplitude-to-wavelength ratio in a coordinate system following the wave. (Based on Wehausen and Laitone, 1960.)

As already suggested by Stokes (1880), the "corner" occurring at the crest of such a wave would have an angle of 120°. Figure 2.112 indicates the form of the gravity wave of maximum amplitude-to-wavelength ratio. For capillary waves, the profile has opposite characteristics: the crest is flattened, and a corner tends to develop in the trough.

The Formation and Dissipation of Energy in Wave Motion

The energy of wave motion is the sum of potential, kinetic, and surface contributions (if surface tension is considered). For a vertical column of unit area, one has

$$W^{pot} = \frac{1}{2}\rho_w g \sigma^2, \tag{2.78}$$

$$W^{kin} = \frac{1}{2}\rho_w \int_{-\infty}^{0} \left(\left(\frac{\partial \phi}{\partial x}\right)^2 + \left(\frac{\partial \phi}{\partial z}\right)^2 \right) dz, \tag{2.79}$$

$$W^{surf} = \frac{1}{2}\gamma_w \left(\frac{\partial \sigma}{\partial x}\right)^2. \tag{2.80}$$

Thus, for the harmonic wave (2.74) and (2.75), the total energy, averaged over a period in position x, but for a fixed time t, and neglecting W^{surf}, is

$$W^{total} = \frac{1}{2}\rho_w \, ga^2 \exp(-4k^2\eta_w t). \tag{2.81}$$

This shows that a wave that does not receive renewed energy input will dissipate energy by molecular friction, with the rate of dissipation

$$D = -\frac{d\overline{W}^{total}}{dt}\bigg|_{t=0} = 2\rho_w \, ga^2 \, k^2 \, \eta_w. \tag{2.82}$$

Of course, this is not the only mechanism by which waves lose energy. Energy is also lost by the creation of turbulence on a scale above the molecular level. This may involve interaction with the air, possibly enhanced by the breaking of wave crests, or oceanic interactions due to the Reynold stresses (2.60). Also, at the shore, surf formation and sand put into motion play a role in energy dissipation from wave motion.

Once the wind has created a wave field, it may continue to exist for a while, even if the wind ceases. If only frictional dissipation of the type (2.82) is active, a wave of wavelength $\lambda_w = 10$ m will take 70 h to be reduced to half the original amplitude, while the time is 100 times smaller for $\lambda_w = 1$ m.

The detailed mechanisms by which a wave field is created by the wind field, and subsequently transfers its energy to other degrees of freedom, are not understood in detail. According to Pond (1971), about 80% of the momentum transfer from the wind may be going initially into wave formation, implying that only 20% goes directly into forming currents. Eventually, some of the energy in wave motion is transferred to the currents.

The Energy Flux Associated With Wave Motion

Now reconsider a single, spectral component of the wave field. The energy flux through a plane perpendicular to the direction of wave propagation, i.e., the power carried by the wave motion, is given in the linearized approximation by (Wehausen and Laitone, 1960)

$$P = -\int_{-\infty}^{0} \rho_w \frac{\partial \phi}{\partial t} \frac{\partial \phi}{\partial x} dz - \gamma_w \frac{\partial \sigma}{\partial t} \frac{\partial \sigma}{\partial x}. \tag{2.83}$$

For the harmonic wave (2.74) and (2.75), neglecting surface tension, this gives

$$P = \frac{1}{4} \, \rho_w \, g \left(\frac{g}{k}\right)^{1/2} a^2,$$
(2.84)

when averaged over a period. Since $(g/k)^{1/2} = U_w$, this together with (2.81) for $\eta_w = 0$ gives the relationship

$$P = \frac{1}{2} \, U_w \, W^{total}.$$
(2.85)

Thus the energy is not transported by the phase velocity U_w, but by $1/2 \, U_w$. For a spectral distribution, the power takes the form

$$P = \frac{1}{2} \, \rho_w \, g \int S(k)^2 \, U_g(k) \, dk,$$
(2.86)

where $U_g(k)$ is the group velocity $d(kU_w)/dk$ (equal to $1/2 \, U_w$ for the ocean gravity waves considered above). It is no surprise that energy is transported by the group velocity, but (2.85) and (2.84) show that this is the case even if there is no group (or wave packet) at all, i.e., for a single, sinusoidal wave.

Wind-driven Oceanic Circulation

The generation of currents by a part of the wind stress, or maybe as an alternative dissipation process for waves (Stewart, 1969), can be described by the equation of motion (2.62), with the boundary condition (2.63) eventually reducing $\bar{\tau}$ by the amount spent in the formation of waves. Considering the Coriolis and vertical friction terms as the most important ones, Ekman (1902) writes the solution in the form

$$V^*_{wx} = V^*_0 e^{bz} \cos(bz + \alpha - \pi/4),$$

$$V^*_{wy} = V^*_0 e^{bz} \sin(bz + \alpha - \pi/4),$$

where z is the directional angle of the surface stress vector (measured from the direction of the x-axis) and

$$V^*_0 = |\bar{\tau}| \, \rho_w^{-1} (fk_z)^{-1/2}; \quad b = f^{1/2} (2k_z)^{-1/2}$$

(f is the Coriolis parameter). The surface velocity vector is directed 45° to the right of the stress vector, and it spirals down with decreasing length. The average mass transport is perpendicular to the wind stress direction, 90° to the right. The observed angle at $z \approx 0$ is about 10° rather than 45° (Stewart, 1969), supporting the view that more complicated processes take place in the surface region. The gross mass transport, however, is well established, for example, in the North Atlantic Ocean, where water flows into the Sargasso Sea both from the south (driven by the prevailing eastward winds at the Equator) and from the north (driven by the mid-latitude "westerlies").

In the Sargasso Sea, the assembled water masses are pressed downward and escape to the sides at a depth. All the water masses are, on average, following the rotation of the Earth, but the lateral squeezing process causes a change in rotational speed of the water involved, in particular a reduction in the vertical component of the angular speed vector. If the rotational speed of some mass of water does not fit with the rotation of the solid Earth at the latitude in question, the water mass will try to move to another latitude that is consistent with its angular speed (unless continents prevent it from doing so). For this reason, the water masses squeezed out below the Sargasso Sea move toward the Equator. Later, they have to return northward (otherwise, the polar part of the Atlantic Ocean would be emptied). Also, in this case, the rotation has to be adjusted to conserve angular momentum. This change in rotation is achieved by frictional interaction with the western continents. Owing to the sense of the Earth's rotation, these return flows (e.g., the Gulf Stream) have to follow western shores, whether the return flow is northward or southward.

A description is provided above of how the wind drives the Ekman flow in the Atlantic Ocean, which induces its major currents. Similar mechanisms are present in the other oceans. Models of these flows, as well as the flows generated by sinking cold water (i.e., not by the winds), have been constructed and integrated along with the atmospheric models considered in section 2.3.1 (Manabe, 1969; Bryan, 1969; Wetherald and Manabe, 1972).

Chapter 3

Individual Renewable Energy Sources

3.1 DIRECT SOLAR ENERGY

Assessment of the "magnitude" of solar radiation as an energy source will depend on the geographical location, including local conditions, such as cloudiness, turbidity, etc. In section 2.2.2 a number of features of the radiation flux at a horizontal plane are described, such as spectral distribution, direct and scattered parts, geographical variations, and dependence on time, from annual to diurnal variations at a given location. The seasonal variation in solar radiation on a horizontal plane is shown in Fig. 3.1, corresponding to the annual average of Fig. 2.24.

For actual applications, it is often necessary to estimate the amount of radiation received by tilted or complexly shaped devices, and it is useful to look at relations that allow relevant information to be extracted from some basic measured quantities. For instance, radiation data often exist only for a horizontal plane, and a relation is therefore needed to predict the radiation flux on an arbitrarily inclined surface. In regions at high latitudes, directing solar devices toward the Equator at fairly high tilt angles actually gives an increase in incident energy relative to horizontally placed collectors.

Only the incoming radiation is discussed in detail in this section, since the outgoing flux may be modified by the specific type of energy conversion device considered. In fact, such a modification is usually the very idea of the device. A description of some individual solar conversion devices is taken up in Chapter 4, and their potential yield in Chapter 6, in terms of combined demand and supply scenarios.

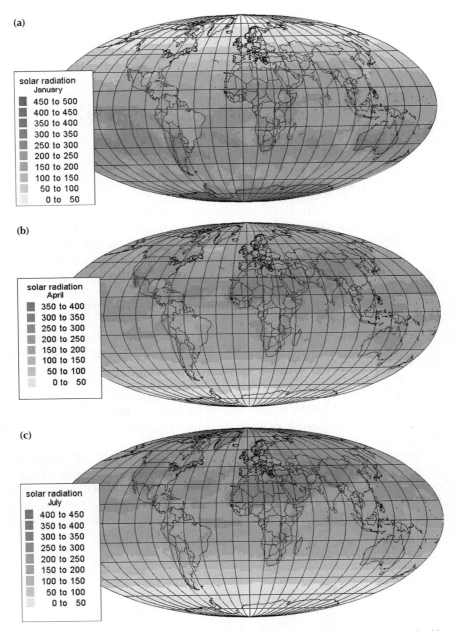

FIGURE 3.1a,b,c   Average short-wavelength solar radiation on a horizontal plane at the Earth's surface (W m^{-2}), for the months of January (*a*), April (*b*) and July (*c*) of 1997 (NCEP-NCAR, 1998).

(d)

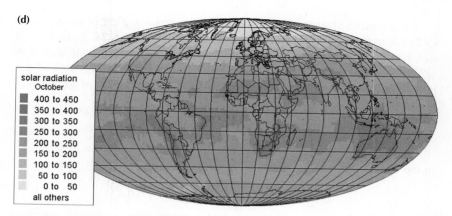

FIGURE 3.1d Average short-wavelength solar radiation on a horizontal plane at the Earth's surface (W m^{-2}), for the month of October of 1997 (NCEP-NCAR, 1998).

3.1.1 Direct Radiation

The inclination of a surface, e.g., a plane of unit area, may be described by two angles. The tilt angle, s, is the angle between vertical (zenith) and the normal to the surface, and the azimuth angle, γ, is the angle between the southward direction and the direction of the projection of the normal to the surface onto the horizontal plane; γ is considered positive toward east [in analogy to the hour angle (2.10)]. In analogy to the expression at the top of the atmosphere, given in section 2.2.1, the amount of direct radiation reaching the inclined surface characterized by (s, γ) may be written

$$D_{s,\gamma} = S_N \cos\theta, \tag{3.1}$$

where S_N is the "normal radiation," i.e., the solar radiation from the direction to the Sun. The normal radiation is entirely *direct* radiation, according to the definition of direct and scattered radiation given in section 2.2.2. The angle θ is the angle between the direction to the Sun and the normal to the surface specified by s and γ. The geometrical relation between θ and the time-dependent co-ordinates of the Sun, declination δ (2.9) and hour angle ω (2.10), is

$$\cos\theta = (SC - CSC)\sin\delta + (SS\sin\omega + (CC + SSC)\cos\omega)\cos\delta, \tag{3.2}$$

where the time-independent constants are given in terms of the latitude ϕ and the parameters (s, γ) specifying the inclination of the surface,

$$
\begin{aligned}
SC &= \sin\phi\cos s, \\
CSC &= \cos\phi\sin s\cos\gamma, \\
SS &= \sin s\sin\gamma. \\
CC &= \cos\phi\cos s, \\
SSC &= \sin\phi\sin s\cos\gamma.
\end{aligned}
\tag{3.3}
$$

For a horizontal surface ($s = 0$), θ equals the zenith angle z of the Sun, and (3.2) reduces to the expression for cos z given in section 2.2.1. Instead of describing the direction to the Sun by δ and ω, the solar altitude $h = 1/2\,\pi - z$ and azimuth Az (conventionally measured as positive toward the west, in contrast to the hour angle) may be introduced. The two sets of co-ordinates are related by

$$\sin h = \sin \delta \sin \phi + \cos \delta \cos \phi \cos \omega,$$
$$\sin Az \cos h = -\cos \delta \sin \omega. \tag{3.4}$$

In the height–azimuth co-ordinate system, (3.2) may be written

$$\cos \theta = \sin h \cos s + \cos h \sin s \cos (Az + \gamma). \tag{3.5}$$

Again, the sign conventions for Az and γ are opposite, so that the argument of the last cosine factor in (3.5) is really the difference between the azimuth of the Sun and the projection of the normal to the surface considered. If cos θ found from (3.2) or (3.5) is negative, it means that the Sun is shining on the "rear" side of the surface considered. Usually, the surface is to be regarded as "one-sided," and cos θ may then be replaced by zero, whenever it assumes a negative value, in order that the calculated radiation flux (3.1) becomes properly zero.

If data giving the direct radiation flux D on a horizontal plane are available, and the direct flux impinging on an inclined surface is wanted, the above relations imply that

$$D_{s,\gamma} = D \cos \theta / \cos z.$$

It is evident that care should be taken in applying this relation when the Sun is near the horizon.

More reliable radiation fluxes may be obtained if the normal incidence radiation flux S_N is measured (as function of time) and (3.1) is used directly. S_N is itself a function of zenith distance z, as well as a function of the state of the atmosphere, including ozone mixing ratio, water vapor mixing ratio, aerosol and dust content, and cloud cover. The dependence on zenith angle is primarily a question of the path that the radiation has taken through the atmosphere. This path-length is shortest when the Sun is in zenith (often denoted *air mass one* for a clear sky) and increases with z, being quite large when the Sun is near the horizon and the path is curved due to diffraction. The extinction in the atmosphere is normally reduced at elevated locations (or low-pressure regions, notably mountain areas), in which case the effective air mass may become less than one. Figure 3.2 gives some typical variations of S_N with zenith angle for zero, small, and heavy particle load (*turbidity*). Underlying assumptions are: cloudless sky, water vapor content throughout a vertical column equal to 0.02 $m^3\ m^{-2}$, standard sea-level pressure, and mean distance to the Sun (Robinson, 1966).

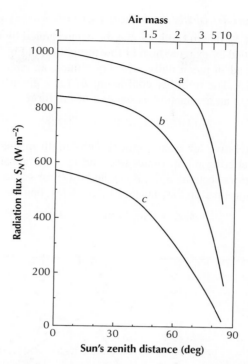

FIGURE 3.2   Normal incidence radiation as a function of zenith angle for a cloudless sky with different particle contents: (*a*), hypothetical atmosphere with zero turbidity [*B* = 0, no aerosols but still water vapor and molecular (Rayleigh) scattering]; (*b*), clear sky (*B* = 0.01); (*c*), atmosphere with heavy aerosol pollution (*B* = 0.4). *B* is the turbidity coefficient B_λ (defined in the text) averaged over wavelengths. Sea-level pressure (10^5 N m^{-2}), a water content of 0.02 m^3 m^{-2}, an ozone content of 0.0034 m^3 m^{-2} (both referred to standard temperature and pressure) and mean Earth–Sun distance have been assumed. At the top, an approximate air mass scale is provided, relative to the one for a vertical path from sea level. (Based on Robinson, 1966.)

Since a complete knowledge of the state of the atmosphere is needed in order to calculate S_N, such a calculation would have to be coupled to the equations of state and motion discussed in section 2.3.1. Only some average behavior may be described without doing this, and if, for example, hourly values of S_N are required in order to predict the performance of a particular solar energy conversion device, it would be better to use measured values of S_N (which are becoming available for selected locations throughout the world, cf. Turner, 1974) or values deduced from measurements for horizontal surfaces. Measurement techniques are discussed by Coulson (1975), among others.

Attempts to parametrize the solar radiation received at the ground are usually made for global radiation (direct plus scattered, and, for tilted planes, radiation reflected onto the surface), rather than separately for normal incidence and scattered radiation [cf. (2.20)].

Dependence on Turbidity and Cloud Cover

The variability of S_N due to turbidity may be demonstrated by noting the range of S_N values implied by extreme high or low turbidities, in Fig. 3.2, and by considering the spread in mean daily turbidity values, an example of which is shown in Fig. 3.3. The turbidity coefficient B_λ for a given wavelength may be defined through an attenuation expression of the form

$$S_N(\lambda)/E_{0+}^{sw}(\lambda) = 10^{-m_r(\tau_\lambda^s + \tau_\lambda^a + B_\lambda)}.$$

Here m_r is the relative air mass (optical path-length in the atmosphere), τ_λ^s describes the scattering on air molecules, and τ_λ^a is the absorption by ozone. In terms of the cross-sections $\sigma_s(\lambda)$, $\sigma_a(\lambda)$, and $\sigma_p(\lambda)$ for scattering on air molecules, ozone absorption, and attenuation by particles, one has

$$-m_r\tau_\lambda^s = \log_{10}\{\exp(\int\sigma_s(\lambda)ds)\},$$

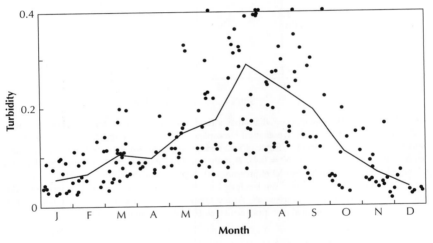

FIGURE 3.3　Turbidity coefficient B_λ at the wavelength $\lambda = 5 \times 10^{-7}$ m, as measured during 1972 at Raleigh (North Carolina) by Bilton *et al.* (1974). Daily means are indicated by dots, monthly means by the curve. Only days with a cloudless sky in the direction of the Sun have been included. Data points above $B_\lambda = 0.4$ are indicated at 0.4 (cf. the high turbidity curve of Fig. 3.2).

and similarly for τ_λ^a and B_λ. Figure 3.3 gives both daily and monthly means for the year 1972 and for a wavelength of 5×10^{-7} m (Bilton *et al.*, 1974). The data suggest that it is unlikely to be possible to find simple, analytical expressions for the detailed variation of the turbidity or for the solar fluxes, which depend on turbidity.

Another major factor determining the amount and distribution of different types of fluxes is the cloud cover. Both cloud distribution and cloud type are important. For direct radiation flux, the important questions are whether the path-line is obscured, and, if it is, how much attenuation is caused by the particular type of cloud. Figure 3.4 shows the total and scattered flux on a horizontal plane (and, by subtraction, the direct flux), for a clear sky and three different types of clouds, as a function of the zenith angle of the Sun. The cloud classification represents only a general indication of category. It is evident in this example that altocumulus and stratus clouds are almost entirely impermeable to direct radiation, whereas cirrus clouds allow the penetration of roughly half the direct radiation flux. Meteorological observations containing records of sunshine (indicating whether the direction to the Sun is obscured) and of cloud cover (as a percentage and identifying the cloud types and their estimated height) may allow a fairly reliable estimate of the fraction of direct radiation reaching a plane of given orientation.

FIGURE 3.4   The influence of clouds on the radiation flux reaching a horizontal plane (total short-wavelength flux, $D + d$, and scattered flux alone, d). (Based on measurements by J. Millard and J. Arvesen, as quoted in Turner, 1974.)

3.1.2 Scattered Radiation

The scattered radiation for a clear sky may be estimated from knowledge of the composition of the atmosphere, as described in section 2.3.1. The addition of radiation scattered from clouds requires knowledge of the distribution and type of clouds, and the accuracy with which a calculation of the scattered flux on a given plane can be made is rather limited. Even with a clear sky, the agreement between calculated and observed fluxes is not absolute, as seen, for example, in Fig. 2.41.

Assuming that the distribution of intensity, $S_{h,Az}^{scatt.}$, for scattered radiation as a function of the directional co-ordinates (h, Az) [or (δ, ω)] is known, then the total flux of scattered radiation reaching a plane tilted at an angle s and directed azimuthally, at an angle γ away from south, may be written

$$d_{s,\lambda} = \int S_{h,Az}^{scatt.} \cos\theta\, d\Omega = \int_0^{\pi/2} d(Az) \int_{h_{\min}(Az)}^{\pi/2} dh S_{h,Az}^{scatt.} \cos\theta(h, Az)\cos h. \quad (3.6)$$

Here $h_{min}(Az)$ is the smallest height angle, for a given azimuth, for which the direction defined by (h, Az) is on the "front" side of the inclined plane. The unit solid angle is $d\Omega = \sin z \, dz \, d(Az) = -\cos h \, dh \, d(Az)$. For a horizontal plane $\theta\,(h, Az) = z = 1/2\,\pi - h$.

If the scattered radiation is isotropic,

$$S_{h,Az}^{scatt.} = \text{constant} = S^{scatt.},$$

then the scattered radiation flux on a horizontal plane becomes

$$d = \pi S^{scatt.}, \qquad (3.7)$$

and the scattered radiation flux on an arbitrarily inclined surface may be written

$$d_{s,\gamma} = d \cos^2(s/2), \qquad (3.8)$$

an expression that can also be derived by considering the fraction of the sky "seen" by the tilted surface.

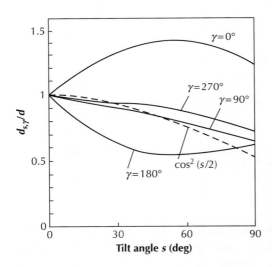

FIGURE 3.5 Ratio between scattered radiation flux on an inclined surface (tilt angle s, azimuth angle γ) and scattered radiation flux on a horizontal surface for a cloudless sky with solar height 15°. (Based on Kondratyev and Fedorova, 1976.)

A realistic distribution of scattered radiation intensity is not isotropic, and, as mentioned earlier, it is not even constant for a cloudless sky and fixed position of the Sun, but depends on the momentary state of the atmosphere. Figure 3.5 shows the result of measurements for a cloudless sky (Kondratyev and Fedorova, 1976), indicating that the assumption of isotropy would be particularly poor for planes inclined directly toward or away from the Sun (Az equal to 0 or π relative to the solar azimuth). Robinson (1966) notes, from observations such as the one shown in Fig. 2.41, that the main differences between observed distributions of scattered radiation and an isotropic one are: (a) increased intensity for directions close to that of the Sun, and (b) increased intensity near the horizon. In particular, the increased intensity in directions near the

Sun is very pronounced, as is also evident from Fig. 3.5, and Robinson suggests that about 25% of d, the scattered radiation on a horizontal plane, should be subtracted from d and added to the direct radiation, before the calculation of scattered radiation on an inclined surface is performed using the isotropic model. However, such a prescription would not be generally valid, because the increased intensity in directions near the Sun is a function of the turbidity of the atmosphere as well as of cloud cover.

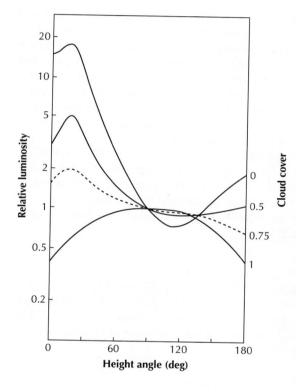

FIGURE 3.6   Luminance distribution along a great circle containing zenith as well as the direction toward the Sun, as a function of cloud cover. The distributions have been normalized at $h = 90°$. The data are based on several measurements with solar height 20° and mean turbidity, performed by Tonne and Normann (1960).

It is evident from Fig. 3.4 that the effect of clouds generally is to diminish direct radiation and increase scattered radiation, although not in the same proportion. Cirrus clouds and, in particular, altocumulus clouds substantially increase the scattered flux on a horizontal plane.

The radiation scattered by clouds is not isotropically distributed. Figure 3.6 gives the luminance distribution, i.e., the intensity of total radiation, along the great circle containing zenith as well as the direction to the Sun, the height angle of which was 20° at the time of measurement. For complete cloud cover, the luminance distribution is entirely due to scattered radiation from the clouds, and it is seen to be maximum at zenith and falling to 0.4 times the zenith value at the horizon.

3.1.3 Total Short-wavelength Radiation

For inclined surfaces or surfaces surrounded by elevated structures, the total short-wavelength (wavelengths below, say, 3×10^{-6} m) radiation flux comprises not only direct and scattered radiation, but also radiation reflected from the ground or from the surroundings onto the surface considered.

Reflected Radiation

The reflected radiation from a given area of the surroundings may be described in terms of an intensity distribution, which depends on the physical nature of the area in question, as well as on the incoming radiation on that area. If the area is specularly reflecting, and the incoming radiation is from a single direction, then there is also a single direction of outgoing, reflected radiation, a direction that may be calculated from the law of specular reflection (polar angle unchanged, azimuth angle changed by π). Whether the reflection is specular may depend not only on the fixed properties of the particular area, but also on the wavelength and polarization of the incoming radiation.

The extreme opposite of specular reflection is completely diffuse reflection, for which by definition the reflected intensity is isotropic over the hemisphere bordered by the plane tangential to the surface at the point considered, no matter what the distribution of incoming intensity. The total (hemispherical) amount of reflected radiation is in this case equal to the total incoming radiation flux times the albedo a of the area in question,

$$R = aE_+,$$

where, for horizontal surfaces, $E_+ = D + d$ (for short-wavelength radiation).

In general, the reflection is neither completely specular nor completely diffuse. In this case, the reflected intensity in a given direction, e.g., specified by height angle and azimuth, $S_{\Omega_r}^{refl.}$, depends on the distribution of incident radiation intensities $S_{\Omega_r}^{inc.}$,[*]

$$S_{\Omega_r}^{refl.} = \int_{hemisphere} \rho_2(\Omega_r, \Omega_i) S_{\Omega_i}^{inc.} \cos \theta(\Omega_i) d\Omega_i. \qquad (3.9)$$

---

[*]The intensity S_Ω is here defined as the energy flux passing through an infinitesimal area into an infinitesimal solid angle in the direction specified by Ω (e.g., h and Az). The infinitesimal area is perpendicular to the direction Ω, and the dimension of S_Ω is energy per unit time, unit area, and unit solid angle. The energy flux passing through a unit area is found by multiplying S_Ω by $\cos \theta$, with θ being the angle between the direction Ω and the normal to the plane, and by $d\Omega$, and then integrating over the hemisphere above or below the unit area considered (giving directional fluxes E_- or E_+). These definitions are consistent with those used in sections 2.1 and 2.2, remembering that the incident solar radiation represents a limiting case of infinitesimal solid angle for the definition of intensity, because the rays from the Sun are treated as parallel, and recalling that several of the energy fluxes considered in section 2.2 are global averages.

Here $\rho_2(\Omega_y, \Omega_i)$ is called the bi-angular reflectance (Duffie and Beckman, 1974; these authors include an extra factor π in the definition of ρ_2). Dividing by the average incoming intensity,

$$\int S_{\Omega_i}^{inc.} \cos \theta(\Omega_i) d\Omega_i \bigg/ \int \cos \theta(\Omega_i) d\Omega_i = E_+/\pi,$$

a reflectance depending only on one set of angles (of reflected radiation) is defined,

$$\rho_1(\Omega r) = \pi S_{\Omega_r}^{refl.}/E_+. \tag{3.10}$$

In general, ρ_1 is not a property of the reflecting surface, since it depends on incoming radiation, but if the incoming radiation is isotropic (diffuse or blackbody radiation), the incoming intensity can be divided out in (3.10).

The total hemispherical flux of reflected radiation may be found by integration of (3.9),

$$R = \int_{hemisphere} S_{\Omega_r}^{refl.} \cos \theta(\Omega_r) d\Omega_r, \tag{3.11}$$

and the corresponding hemispherical reflectance,

$$\rho = R/E_+ = a, \tag{3.12}$$

is equal to the albedo defined above.

All of the above relations have been written without reference to wavelength, but they are valid for each wavelength λ, as well as for appropriately integrated quantities.

In considering the amount of reflected radiation reaching an inclined surface, the surrounding surfaces capable of reflecting radiation are often approximated by an infinite, horizontal plane. If the reflected radiation is further assumed to be isotropic of intensity $S^{refl.}$, then the reflected radiation flux received by the inclined surface is

$$R_{s,\gamma} = \pi S^{refl.} \sin^2(s/2). \tag{3.13}$$

The factor $\sin^2(s/2)$ represents the fraction of the hemisphere above the inclined surface from which reflected radiation is received. This fraction is evidently complementary to the fraction of the hemisphere from which scattered radiation is received and, therefore, equal to $1 - \cos^2(s/2)$ from (3.8). The albedo (3.12) may be introduced, noting from (3.11) that $R = \pi S^{refl.}$ for isotropic reflection,

$$R_{s,\gamma} = aE_+ \sin^2(s/2). \tag{3.14}$$

For short-wavelength radiation, the flux E_+ on a horizontal plane equals $D + d$, the sum of direct and scattered short-wavelength fluxes.

If the geometry of the reflecting surroundings is more complicated (than a plane), or if the reflected intensity is not isotropic, the calculation of the amount of reflected radiation reaching a given inclined surface involves integration over the hemisphere seen by the inclined surface. Thus, for each direction of light reflected onto the inclined plane, a contribution to $S_{\Omega_r}^{refl.}$ is included from the first unobscured point on the line of sight capable of reflecting radiation. If semitransparent objects (such as a water basin) are present in the surroundings, the integration becomes three-dimensional, and the refraction and transmission properties of the partly opaque objects must be considered.

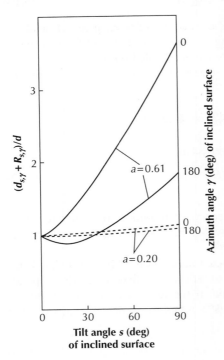

FIGURE 3.7 The ratio of scattered plus reflected radiation flux on inclined surfaces to that on a horizontal surface (scattered flux d only) for a clear sky with the Sun at height 34°. Measurements corresponding to two different ground albedos are depicted, based on summer and winter (snow-covered ground) conditions at a location in the Soviet Union (Kondratyev and Fedorova, 1976.)

Figure 3.7 shows an example of the enhancement of indirect radiation that may result from increasing albedo of the surroundings—in this case, due to snow cover in winter (Kondratyev and Fedorova, 1976). The rejected radiation on the inclined surface reaches a maximum value for the tilt angle $s = 90°$ (vertical).

Average Behavior of Total Short-wavelength Radiation

The sum of direct, scattered, and reflected radiation fluxes constitutes the total short-wavelength (*sw*) radiation flux. The total short-wavelength flux on a horizontal surface is sometimes referred to as the *global radiation*, i.e., $D + d$

(if there are no elevated structures to reflect radiation onto the horizontal surface). For an inclined surface of tilt angle s and azimuth γ, the total short-wavelength flux may be written

$$E_{s,\gamma}^{sw} = D_{s,\gamma} + d_{s,\gamma} + R_{s,\gamma}, \qquad (3.15)$$

with the components given by (3.1), (3.6) or (3.8), and (3.14) or a generalization of it. The subscript "+" on E^{sw} has been left out, since the direction of the flux is clear from the values of s and γ (the E_- flux would generally require $s > 90°$).

Examples of the influence of clouds, and of solar zenith angle, on global radiation $E_{s=0,\gamma}^{sw}$ are given in Fig. 3.4. Figure 3.8 illustrates the influence of cloud cover on the daily sum of total short-wavelength radiation received by

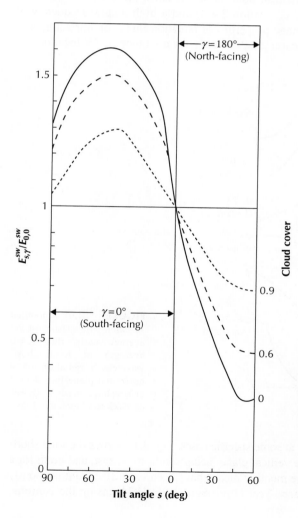

FIGURE 3.8 Daily sums of total short-wavelength radiation on an inclined plane, relative to that of a horizontal plane, for different cloud cover and as a function of tilt angle for north- and south-facing surfaces. (Based on measurements in the Soviet Union, Kondratyev and Fedorova, 1976.)

inclined surfaces, relative to that received by a horizontal surface. For south-facing slopes, the radiation decreases with increasing cloud cover, but for north-facing slopes the opposite takes place.

Monthly averages of total short-wavelength radiation for different geographical locations are shown in Figs 3.9 and 3.10 for a horizontal surface. In Fig. 3.10, two sets of data are compared, each representing a pair of locations with the same latitude. The two at $\phi \approx 43\,°N$ correspond to a coastal and a continental site, but the radiation, averaged for each month, is nearly the same. The other pair of locations are at $35°$–$36\,°N$. Albuquerque has a desert climate, while Tokyo is near the ocean. Here the radiation patterns are vastly different during summer, by as much as a factor of two. The summer solar radiation in Tokyo is also smaller than in both the $43\,°N$ sites, presumably due to the influence of urbanization (cf. section 2.4.2) and a high natural frequency of cloud coverage. Other average properties of radiation fluxes on horizontal surfaces are considered in section 2.2.2, notably for a latitude $53\,°N$ location.

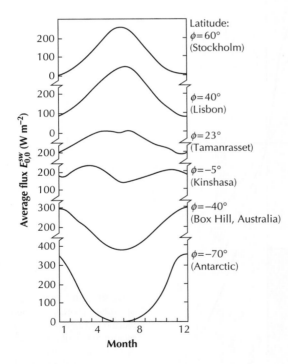

FIGURE 3.9 Smoothed variation with seasons of the average daily flux (24-h average) of total short-wavelength radiation on a horizontal plane for selected geographical locations. (Based on Flack and Morikofer, 1964.)

Looking more closely at some specific cases, Fig. 3.11 shows the total short-wavelength radiation on a vertical plane facing south, west, east, and north for a location at $\phi = 56\,°N$. The monthly means have been calculated from the hourly data of the Danish reference year (Andersen *et al.*, 1974), using the isotropic approximation (3.8) and (3.14) with an assumed albedo $a = 0.2$. The Danish reference year consists of selected meteorological data exhibiting "typical" fluctuations. This is achieved by selecting monthly sequences of actual data, with monthly averages of each of the meteorological variables in the vicinity of the 30-year mean. The different sequences that make up the reference year have been taken from different years. The variables pertaining to solar radiation are global radiation $(D + d)$, normal-incidence radiation (S_N), and scattered radiation on a horizontal plane (d). Only global radiation has been measured over extended periods of time, but the two other variables have been constructed so that the three variables, of which only two are independent, become reasonably consistent. Several other countries are in the process of constructing similar reference years, which will allow for easy comparison of, for example, performance calculations of solar energy devices or building insulation prescriptions. Monthly averages of the basic data of the Danish reference year, $D + d$ and D, are shown in Fig. 3.12. Reference years have subsequently been constructed for a number of other locations (European Commission, 1985).

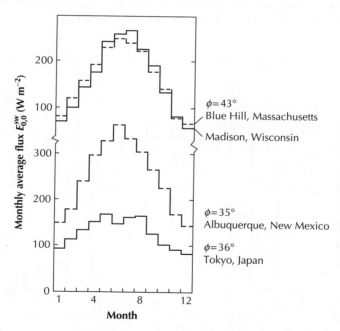

FIGURE 3.10 Monthly average flux on a horizontal plane for pairs of locations with similar latitude but different climate. [Based on data from NOAA (U.S. National Oceanic and Atmospheric Administration), quoted from Duffie and Beckman, 1974.]

Figure 3.13 shows the composition of the total short-wavelength flux on a vertical, south-facing surface, in terms of direct, scattered, and reflected radiation. It has been assumed that the albedo changes from 0.2 to 0.9 when the ground is covered by snow (a piece of information also furnished by the reference year). Snow cover is therefore responsible for the relative maximum in reflected flux for February. The proportion of direct radiation is substantially higher during winter for this vertical surface than for the horizontal surface (Fig. 3.12).

Inclined Surfaces

Since the devices used to capture solar radiation for heat or power production are often flat-plate collectors integrated into building facades or roofs, it is important to be able to estimate energy production for such inclined mountings as function of time and location.

Available satellite solar data comprise only down-going fluxes of total radiation on a horizontal plane at the ground surface level. They are calculated from solar fluxes at the top of the atmosphere and reflected radiation reaching the satellite, with use of measured cloud and turbidity data (ECMWF, 2008). At low latitudes, these may be appropriate for solar devices, but at higher latitudes, radiation on inclined surfaces has to be derived indirectly.

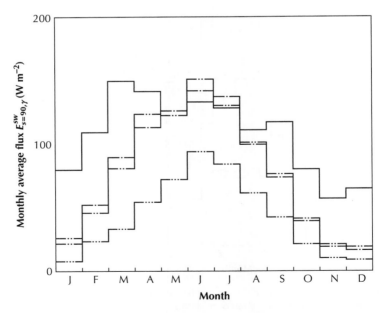

FIGURE 3.11 Monthly average flux on a vertical surface with different azimuthal orientations, based on the Danish reference year (latitude 56°) and an assumed ground albedo of 0.2 (0.9 with snow cover). Key: —— $\gamma = 0°$ (south); —·—·— $\gamma = -90°$ (west); —··—··— $\gamma = 90°$ (east); —···—···— $\gamma = 180°$ (north). This and the following calculations assume that the scattered radiation is isotropic.

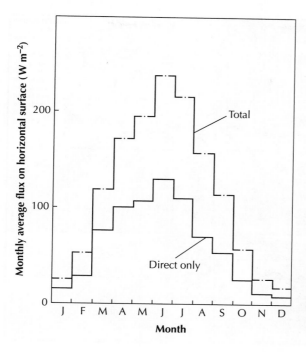

FIGURE 3.12 Monthly average short-wavelength flux on horizontal surface, and the direct part, D, based on the Danish reference year, $\phi = 56\,°N$ (Andersen *et al.*, 1974).

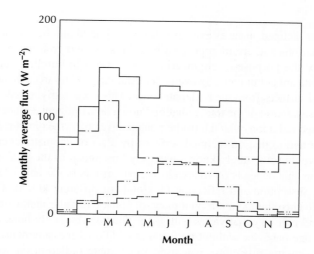

FIGURE 3.13 Components of monthly average fluxes on vertical, south-facing surface. Key: —— total short-wavelength radiation; —·— direct radiation $D_{s,\gamma}$; —··— scattered radiation $d_{s,\gamma}$; —···— reflected radiation $R_{s,\gamma}$ (in all cases $\gamma = 0$). The calculation is based on the Danish reference year, $\phi = 56\,°N$, and an assumed albedo of 0.2 (0.9 with snow cover).

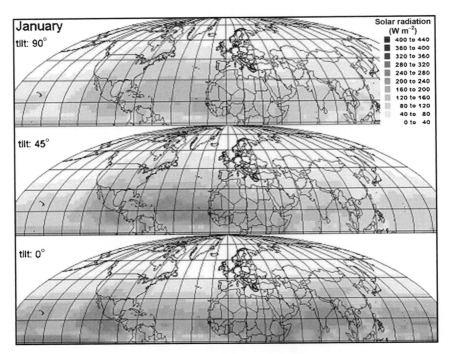

FIGURE 3.14   Estimated January solar radiation in W m^{-2} on differently tilted surfaces.

Data for inclined surfaces exist directly for a limited number of local meteorological stations and are incorporated into the so-called "reference data" constructed for the purpose of engineering applications in building construction. As these data are not universally available, I shall instead try to estimate radiation on inclined surfaces from the horizontal ones. This is not a trivial task, because it requires a separation of the total radiation known on a horizontal plane into direct and scattered radiation, which must be treated differently: the direct radiation can be transformed to that on inclined surfaces by a set of geometrical calculations given above, but it is rarely known precisely how much of the total radiation is scattered, as this depends on many elusive parameters (solar height, attenuation in the air, absorption and reflection on clouds of different kinds). These same factors influence the radiation on a plane of some particular inclination, but differently and in a way that varies with time. Thus one can only hope to describe some average behavior without performing additional measurements.

The following procedure was used to estimate radiation on south-facing planes tilted at 45 °S and 90 °S: 62.5% of the horizontal radiation, representing direct as well as nearly forward-scattered radiation (i.e., radiation scattered by small angles, known to be a large component of the indirect radiation) is treated by the straightforward angular transformations to normal-incidence radiation,

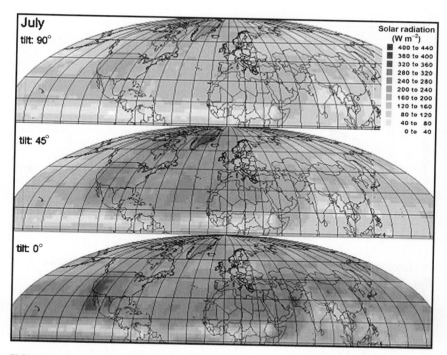

FIGURE 3.15   Estimated July solar radiation in W m^{-2} on surfaces of various tilt angles.

and then to radiation on the tilted surface. Strictly direct radiation is roughly 50% (see above), so 25% of the scattered radiation is assumed to arrive from directions close to that of the Sun. The remaining 37.5% of the horizontal radiation is distributed on the inclined surfaces by multiplication with a factor taken as 1.5 for the $s = 45°$ tilted surface and 1.25 for the $s = 90°$ tilted surface, based on measurements by Kondratyev and Fedorova (1976). In other words, the scattered radiation is far from isotropic, as this would give a simple $\cos^2(s/2)$-factor due to the reduced fraction of the sky seen by the surfaces at a tilt angle s.

The resulting monthly average distribution of radiation for January and July are given in Figs. 3.14 and 3.15 for the Northern hemisphere, using 6-hour data for the year 2000. The average values serve to provide a general overview and the full 6-hour data set is necessary for actual simulation studies. On the Southern hemisphere, appropriate tilt angles for solar collectors would be north-facing rather than south-facing. For those locations where measurements have been performed, the quality of the model can be assessed. Based on only a cursory comparison, the model seems to agree with data at mid-latitudes (i.e., the regions of interest here) and the variations in radiation seem to be generally of an acceptable magnitude. However, deviations could be expected at other latitudes or at locations with particular meteorological circumstances.

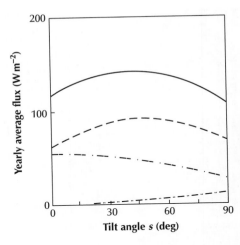

FIGURE 3.16 Components of yearly average fluxes on a vertical, south-facing surface. Key: —— total short-wavelength radiation; – – – direct radiation $D_{s,\gamma}$; –·—·— scattered radiation $d_{s,\gamma}$; —··—··— reflected radiation $R_{s,\gamma}$ (in all cases $\gamma = 0$). The calculation is based on the Danish reference year, $\phi = 56\,°N$, and an assumed albedo of 0.2 (0.9 with snow cover).

Looking for comparison at some of the empirical data for inclined surfaces, Fig. 3.16 gives, for a specific location, the variation of the yearly average fluxes with the tilt angle s, still for a south-facing surface ($\gamma = 0$). The variation of monthly averages of the total flux with tilt angle for a south-facing surface is given in Fig. 3.17.

According to Fig. 3.16, the maximum yearly short-wavelength radiation in Denmark ($\phi = 56\,°N$) is obtained on a south-facing surface tilted about $40\,°$, but

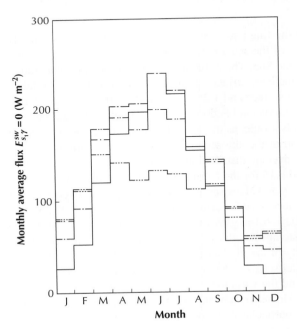

FIGURE 3.17 Monthly average of total short-wavelength radiation on inclined, south-facing surfaces, based on the Danish reference year, $\phi = 56\,°N$. Key: —— tilt angle $s = 0$; —·— $s = 30°$; —···— $s = 60°$; —····— $s = 90°$.

the maximum is broad. The maximum direct average flux is obtained at a tilt angle closer to ϕ, which is clear because at the top of the atmosphere the maximum would be attained for s equal to the latitude plus or minus the Sun's declination, the extremes of which are about $\pm23°$ at summer and winter solstices. From Fig. 3.17, one can see that the most constant radiation over the year is obtained for $s = 90°$. The December maximum value is for $s = 75°$, but the solar radiation on an $s = 90°$ surface is only 1%–2% smaller.

3.1.4 Long-wavelength Radiation

As suggested by Figs. 2.20 and 2.26, the long-wavelength radiation reaching a plane situated at the Earth's surface may be quite substantial, in fact, exceeding the short-wavelength radiation when averaged over 24 h. However, the outgoing long-wavelength radiation is (again on average) still larger, so that the net long-wavelength flux is directed away from the surface of the Earth. The ground and most objects at the Earth's surface emit long-wavelength (lw) radiation approximately as a blackbody, while the long-wavelength radiation from the atmosphere often deviates substantially from any blackbody spectral distribution.

In general, the emission and absorption of radiation by a body may be described by the spectral and directional emittance, $\varepsilon_\lambda(\Omega)$, and the corresponding absorptance, $\alpha_\lambda(\Omega)$. These quantities are related to $e(\nu)$ and $\kappa(\nu)$ of section 2.1.1, which may also depend on the direction Ω, by

$$e(\nu,\Omega) = \varepsilon_\lambda(\Omega)\, dF^{Planck}/d\nu, \quad \kappa(\nu,\Omega) = \alpha_\lambda(\Omega),$$

for corresponding values of frequency ν and wavelength λ.

Thus, $\varepsilon_\lambda(\Omega)$ is the emission of the body, as a fraction of the blackbody radiation (2.3) for emission in the direction Ω, and $\alpha_\lambda(\Omega)$ is the fraction of incident flux, again from the direction specified by Ω, which is absorbed by the body.

Where a body is in complete thermodynamic equilibrium with its surroundings (or more specifically those with which it exchanges radiation), Kirchhoff's law is valid (see, for example, Siegel and Howell, 1972),

$$\varepsilon_\lambda(\Omega) = \alpha_\lambda(\Omega). \tag{3.16}$$

However, since both $\varepsilon_\lambda(\Omega)$ and $\alpha_\lambda(\Omega)$ are properties of the body, they are independent of the surroundings, and hence (3.16) is also valid away from equilibrium. This does not apply if wavelength averages are considered. Then $\varepsilon(\Omega)$ is still a property of the surface, but $\alpha(\Omega)$ depends on the spectral composition of the incoming radiation, as follows from the definition of $\alpha(\Omega)$,

$$\alpha(\Omega) = \int_\lambda \alpha_\lambda(\Omega) S_\lambda^{inc}(\Omega) d\lambda \Big/ \int_\lambda S_\lambda^{inc}(\Omega) d\lambda.$$

Further averaging over directions yields the hemispherical emittance ε and absorptance α, the former still being a property of the surface, the latter not.

Consider now a surface of temperature T_s and orientation angles (s, γ), which emit long-wavelength radiation in such a way that the hemispherical energy flux is

$$E_-^{lw,\,emission} = \varepsilon^{lw} \sigma T_s^4, \tag{3.17}$$

where σ is Stefan's constant (5.7×10^{-8} W m^{-2} K^{-4}). For pure blackbody radiation from a surface in thermal equilibrium with the temperature T_s (forming an isolated system, e.g., a cavity emitter), ε^{lw} equals unity.

The assumption that $\varepsilon_\lambda = \varepsilon$ (emissivity independent of wavelength) is referred to as the *gray-body* assumption. It implies that the energy absorbed from the environment can be described by a single absorptivity $\alpha^{lw} = \varepsilon^{lw}$. Further, if the long-wavelength radiation from the environment is described as blackbody radiation corresponding to an effective temperature T_e, the incoming radiation absorbed by the surface may be written

$$E_+^{lw,abs.} = \varepsilon^{lw} \sigma T_e^4. \tag{3.18}$$

If the part of the environmental flux reflected by the surface is denoted $R\ (= \rho^{lw}\ \sigma T_e^4)$, then the total incoming long-wavelength radiation flux is $E_+^{lw} = R + E_+^{lw,abs.}$, and the total outgoing flux is $E_-^{lw} = R + E_-^{lw,emission}$, and thus the net long-wavelength flux is

$$E^{lw} = \varepsilon^{lw} \sigma (T_e^4 - T_s^4). \tag{3.19}$$

The reflection from the environment of temperature T_e back onto the surface of temperature T_s has not been considered, implying an assumption regarding the "smallness" of the surface considered as compared with the effective surface of the environment. Most surfaces that may be contemplated at the surface of the Earth (water, ice, grass, clay, glass, concrete, paints, etc.) have long-wavelength emissivities ε^{lw} close to unity (typically about 0.95). Materials like iron and aluminum with nonpolished surfaces have low long-wavelength emissivities (about 0.2), but often their temperature T_s is higher than the average temperature at the Earth's surface (due to high absorptivity for short-wavelength radiation), so that (3.19) may still be a fair approximation, if T_s is chosen as an average temperature of physical surfaces. If the surface is part of an energy-collecting device, a performance evaluation will require the use of the actual temperature T_s with its variations (cf. Chapter 4).

The deviations of the long-wavelength radiation received from the environment from that of a blackbody are more serious, leading both to a change in wavelength dependence and to a nonisotropic directional dependence. Decisive in determining the characteristics of this radiation component is the average distance, for each direction, to the point at which the long-wavelength radiation is emitted. Since the main absorbers in the long-wavelength frequency region are water vapor and CO_2 (cf. Fig. 2.40), and since the atmospheric content of water vapor is the most variable of these, then one may expect the long-wavelength flux to be primarily a function of the water vapor content m_v (Kondratyev and Podolskaya, 1953).

At wavelengths with few absorption bands, the points of emission may be several kilometers away, and owing to the temperature variation through the atmosphere (see, for example, Figs. 2.31 and 2.32), one expects this component of T_e to be 20–30 K below the ambient temperature T_a at the surface. As humidity increases, the average emission distance diminishes, and the effective temperature T_e becomes closer to T_a. The temperature of the surface itself, T_s, is equal to or larger than T_a, depending on the absorptive properties of the surface. This is also true for physical surfaces in the surroundings, and therefore the "environment" seen by an inclined surface generally comprises partly a fraction of the sky with an effective temperature T_e below T_a and partly a fraction of the ground, possibly with various other structures, which has an effective temperature of T_e above T_a.

Empirical Long-wavelength Evidence for Inclined Surfaces

Figure 3.18 shows the directional dependence of long-wavelength radiation from the environment, averaged over wavelengths, for very dry and humid atmospheres. The measurements on which the figure is based (Oetjen et al., 1960) show that the blackbody approximation is only valid for directions close to the horizon (presumably implying a short average distance to points of emission), whereas the spectral intensity exhibits deeper and deeper minima, in particular around $\lambda = 10^{-5}$ m, as the direction approaches zenith. Relative to the blackbody radiation at ambient temperature, T_a, the directional flux in Fig. 3.18 starts at unity at the horizon, but drops to 79% and 56%, respectively, for the humid (Florida Beach region) and the dry atmospheres (Colorado mountainous region). If an effective blackbody temperature is ascribed, although the spectral distributions are not Planckian, the reduced T_e is 94.3% and 86.5% of T_a, corresponding to temperatures 27.5 and 37.9 K below ambient temperature. Although it is not clear whether they are in fact, Meinel and Meinel (1976) suggest that the two curves in Fig. 3.18 be used as limiting cases, supplemented with blackbody emissions from ground and surrounding structures, which are characterized by ε^{lw} between 1.0 and 1.1, for the calculation of net long-wavelength radiation on inclined surfaces.

FIGURE 3.18 Variation in incoming long-wavelength radiation with height above the horizon (flat ground). The scale on the left is the ratio between the effective temperature of the long-wavelength radiation received and the ambient air temperature, whereas the scale on the right gives the average flux relative to that of blackbody radiation at ambient temperature. The two curves represent measurements for a very humid and a very dry atmosphere; performed by Oetjen *et al.* (1960) and quoted by Meinel and Meinel (1976).

In calculations of the performance of flat-plate solar collectors it has been customary to use an effective environmental temperature about 6 K less than the ambient temperature T_a (Duffie and Beckman, 1974; Meinel and Meinel, 1976), but since the surfaces considered in this context are usually placed with tilt angles in the range 45°–90°, a fraction of the hemisphere "seen" by the surface will be the ground (and buildings, trees, etc.). Thus, the effective temperature of the long-wavelength radiation received will be of the form

$$T_e = C\, T_{e,atmosphere} + (1-C)T_{e,ground}, \tag{3.20}$$

where C is at most $\cos^2(s/2)$, corresponding to the case of an infinitely extended, horizontal plane in front of the inclined surface considered. For this reason T_e will generally not be as much below T_a as indicated by Fig. 3.18, but it is hard to see how it could be as high as $T_a - 6$, since $T_{e,ground}$ is rarely more than a few degrees above T_a. Silverstein (1976) estimates values of $T_e - T_a$ equal to -20 K for horizontal surfaces and -7 K for vertical surfaces.

Figure 3.19 shows measured values of the ratio of net long-wavelength radiation (3.21) on a black-painted inclined surface to that on the same surface in horizontal position. The quantity varying with angle of inclination is the effective blackbody temperature (which could alternatively be calculated on the basis of curves like those given in Fig. 3.18), $T_{e,s,\gamma}$, so that Fig. 3.19 can be interpreted as giving the ratio

$$(T_{e,s,\gamma}^4 - T_s^4)/(T_{e,0,0}^4 - T_s^4).$$

No significant differences are found between the data points for $\gamma = 0°, \pm 90°$, and 180°, which are available for each value of the tilt angle s. The data are in agreement with the vertical to horizontal temperature difference $T_{e,s=90°} - T_{e,0}$

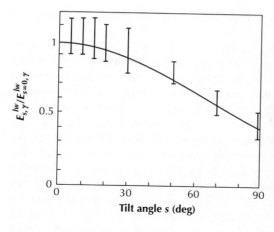

FIGURE 3.19 Ratio of net long-wavelength radiation on inclined surfaces and the corresponding flux for a horizontal surface. The error bars represent the spread of measured values. Although measurements for different azimuth angles γ have been aggregated, the position of data points within the error bars is not noticeably correlated with γ. The curve has been calculated with assumptions mentioned in the text. (Based on Kondratyev and Fedorova, 1976.)

found by Silverstein and the calculation of Kondratyev and Podolskaya (1953) for an assumed atmospheric water mixing ratio of about 0.02 m^3 m^{-2} vertical column. However, the accuracy is limited, and it would be valuable to expand the experimental activity in order to arrive at more precise determinations of the effective environmental temperature under different conditions.

3.1.5 Variability of Solar Radiation

The fact that the amount of solar radiation received at a given location at the Earth's surface varies with time is implicit in several of the data discussed in the preceding sections. The variability and part-time absence of solar radiation due the Earth's rotation (diurnal cycle) and orbital motion (seasonal cycle, depending on latitude) are well known and simple to describe, and so the

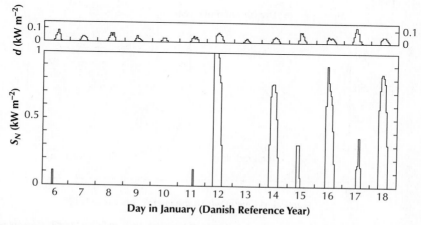

FIGURE 3.20 Hourly values of the normal incidence flux, S_N, and the scattered flux, d, on a horizontal plane for the Danish reference year, $\phi = 56\,°N$. Thirteen consecutive winter days are shown.

emphasis here is on a description of the less simple influence of the state of the atmosphere, cloud cover, etc.

Most of the data presented in sections 3.1.1–3.1.3 are in the form of averages over substantial lengths of time (e.g., a month), although some results are discussed in terms of instantaneous values, such as intensity as a function of solar zenith angle. In order to demonstrate the time structure of radiation quantities more clearly, Fig. 3.20 shows hourly averages of normal incidence radiation, S_N, as well as scattered radiation on a horizontal surface, d, hour by hour over a 13-day period of the Danish reference year (latitude 56 °N). For this period in January, several days without any direct (or normal incidence) radiation appear consecutively, and the scattered radiation is quite low. On the other hand, very clear days occasionally occur in winter, as witnessed by the normal flux received on 12 January in Fig. 3.20. Fluctuations within each hour are also present, as can be seen from the example shown in Fig. 3.21. The data are for two consecutive days at a latitude of about 39 °N (Goddard Space Flight Center in Maryland; Thekaekara, 1976), collected at 4-s intervals. The figure conveys the picture of intense fluctuations, particularly on a partially cloudy day.

FIGURE 3.21 Two days of continuous recording of total short-wavelength flux on a horizontal plane. (Based on Thekaekara, 1976.)

FIGURE 3.22a One-year power duration curves of total (*left side*) and direct (*right side*) short-wavelength radiation on south-facing surfaces of three different inclinations, based on the Danish reference year, $\phi = 56\,°N$. Fig. 3.22*b*: next page.

Geographical Distribution of Solar Power

The geographical distribution of average incoming (short-wave) solar radiation on a horizontal plane is shown in Figs. 3.1*a–d* for each of the four seasons. The data exhibit considerable dependence on conditions of cloud cover and other variables influencing the disposition of radiation on its way from the top to the bottom of the atmosphere. These data form the basis for further analysis in terms of suitability of the power flux for energy conversion in thermal and electricity-producing devices (Chapter 4). Methods for estimating solar radiation on inclined surfaces from the horizontal data are introduced in Chapter 6, along with an appraisal of the fraction of the solar resource that may be considered of practical use after consideration of environmental and area use constraints.

Power Duration Curves

The accumulated frequency distribution is called a *power duration curve*, since it gives the fraction of time during which the energy flux exceeds a given value E, as a function of E. Figure 3.22*a* gives power duration curves for total and direct radiation on a horizontal surface and two southward inclined surfaces for a location at latitude $\phi = 56\,°N$ (Danish reference year, cf. European Commission, 1985). Since data for an entire year have been used, it is not surprising that the energy flux is non-zero for approximately 50% of the time. The largest fluxes, above 800 W m^{-2}, are obtained for only a few hours a year, with the largest number of hours with high incident flux being for a surface inclined about 45° (cf. Fig. 3.16). The shape of the power duration curve may be used directly to predict the performance of a solar energy device, if this is known to be sensitive only to fluxes above a certain minimum value, say,

FIGURE 3.22b One-year power duration curves of normal incidence radiation alone, and of scattered radiation alone, based on the Danish reference year ($\phi = 56\,°N$). The normal incidence curve would correspond to the radiation received by a fully tracking instrument. The curve for scattered radiation is for a horizontal surface.

300 W m^{-2}. The power duration curves for direct radiation alone are shown on the right. They are relevant for devices sensitive only to direct radiation, such as most focusing collectors.

Figure 3.22b gives the power duration curves for normal incidence radiation, S_N, and for scattered radiation on a horizontal plane, d. The normal incidence curve is of interest for fully tracking devices, that is, devices that are being continuously moved in order to face the direction of the Sun. By comparing Fig. 3.22b with Fig. 3.22a, it can be seen that the normal incidence curve lies substantially above the direct radiation curve for an optimum but fixed inclination ($s = 45°$). The power duration curve for normal incidence radiation is not above the curve for total radiation at $s = 45°$, but it would be if the scattered radiation received by the tracking plane normal to the direction of the Sun were added to the normal incidence (direct by definition) radiation. The power duration curve for scattered radiation alone is also shown in Fig. 3.22b for a horizontal plane. The maximum scattered flux is about 350 W m^{-2}, much higher than the fluxes received during the winter days shown in Fig. 3.20.

3.2 WIND

It follows from the discussion in section 2.4.1 that the kinetic energy content of the atmosphere, on average, equals about seven days of kinetic energy production or dissipation, also assuming average rates. Utilizing wind energy means installing a device that converts part of the kinetic energy in the atmosphere to, say, useful meachnical energy, i.e., the device draws primarily on the energy stored in the atmospheric circulation and not on the energy flow into the general circulation (1200 TW according to Fig. 2.90). This is also clear from the fact that the production of kinetic energy, i.e., the conversion of available potential energy into kinetic energy, according to the general scheme of Fig. 2.54, does not take place to any appreciable extent near the Earth's surface. Newell et $al.$ (1969) give height–latitude distributions of the production of zonal kinetic energy for different seasons. According to these distributions, the regions of greatest kinetic energy, the zonal jet-streams at mid-latitudes and a height of about 12 km (see Fig. 2.49), are maintained from conversion of zonal available potential energy in the very same regions.

One of the most important problems to resolve in connection with any future large-scale energy extraction from the surface boundary region of the atmosphere's wind energy storage is that of the nature of the mechanism for restoring the kinetic energy reservoir in the presence of man-made energy extraction devices. To a substantial extent, this mechanism is known from the study of natural processes by which energy is removed from the kinetic energy storage in a given region of the atmosphere. The store of energy is replenished by the aforementioned conversion processes, by which kinetic energy is formed from the much larger reservoir of available potential energy (four times larger than the kinetic energy reservoir, according to section 2.4.1). The available potential energy is created by temperature and pressure differences, which in turn are formed by the solar radiation flux and the associated heat fluxes.

3.2.1 Wind Velocities

The Horizontal Wind Profile

The horizontal components of the wind velocity are typically two orders of magnitude larger than the vertical component, but the horizontal components may vary a great deal with height under the influence of frictional and impact forces on the ground (Fig. 2.52) and the geostrophic wind above, which is governed by the Earth's rotation (see section 2.3.1). A simple model for the variation of horizontal velocity with height in the lower (Prandtl) planetary boundary layer, assuming an adiabatic lapse rate (i.e., temperature gradient, cf. Eq. 2.32) and a flat, homogeneous surface, is given in (2.34).

The zero point of the velocity profile is often left as a parameter, so that (2.34) is replaced by

$$u = \kappa^{-1} (\tau/\rho)^{1/2} \log((z + z_0 - d_0)/z_0) \tag{3.21}$$

The zero displacement, d_0, allows the zero point of the profile to be determined independently of the roughness length z_0, both being properties of the surface. This is convenient, for example, for woods and cities with roughness lengths from about 0.3 m to several meters, for which the velocity profile "starts" at an elevation (of maybe 10 or 50 m) above the ground. For fixed z_0 and d_0, a family of profiles is described by (3.21), depending only on the parameter $(\tau/\rho)^{1/2}$, called the friction velocity. For a given location, it is essentially determined by the geostrophic wind speed, still assuming an adiabatic ("neutral") temperature profile.

If the atmosphere is not neutral, the model leading to (3.21) has to be modified. In fact, (2.34) may be viewed as a special case in a more general theory, due to Monin and Obukhov (1954) (see also Monin and Yaglom, 1965), according to which the wind speed may be expressed as

$$u(z) = \kappa^{-1} (\tau/r)^{1/2} (f(z/L) - f(z_0/L)).$$

Only in the case of an adiabatic lapse rate is the function f equal to a logarithm. This case is characterized by $z \ll L$, where the parameter L describing the atmosphere's degree of stability may be written

$$L = c_P \rho (\tau/\rho)^{3/2} T_0/(g\kappa Q_0). \tag{3.22}$$

L is called the Monin–Obukhov length, g is the acceleration of gravity, and T_0 and Q_0 are the temperature and heat flux at the surface. Thus, L is a measure of the transport of heat near the surface, which again determines the stability of the atmosphere. If the heat flux is sufficiently large and directed from ground to atmosphere (Q_0 and thus L negative), then transport of heat and matter is likely to be highly convective and unstable, so that there will be rapid mixing and therefore less variation of wind speed with height. On the other hand, if Q_0 and L are positive, the atmosphere becomes stable, mixing processes become weak, and stratification becomes pronounced, corresponding to almost laminar motion with increasing wind speed upward.

The connection between L and the temperature gradient becomes clear in the convective case, in which one can prove that z/L equals the Richardson number,

$$Ri = g\frac{1}{\theta}\frac{\partial\theta}{\partial z}\bigg/\left(\frac{\partial u}{\partial z}\right)^2, \tag{3.23}$$

where θ is the potential temperature (2.52). From the definition,

$$\frac{\partial\theta}{\partial z} = \frac{\theta}{T}\left(\frac{\partial T}{\partial z} - \frac{c_P - c_V}{c_P}\frac{T}{P}\frac{\partial P}{\partial z}\right),$$

which according to (2.32) is a direct measure of the deviation from adiabaticity. The adiabatic atmosphere is characterized by $Ri = 0$. The identity of z/L and Ri is easy to derive from (2.29), (2.31), and (2.33), but the linear relationship $l = \kappa z$ (2.33) may not be valid in general. Still, the identity may be maintained, if the potential temperature in (3.6) is regarded as a suitably defined average.

Figure 3.23 gives an example of calculated velocity profiles for neutral, stable, and unstable atmospheres, assuming fixed values of z_0 (0.01 m, a value typical of grass surfaces) and $(\tau/\rho)^{1/2} = 0.5$ m s^{-1}. (Based on Frost, 1975.) Many meteorological stations keep records of stability class, based on either observed turbulence, Ri, or simply $\partial T/\partial z$. Thus, it is often possible with a certain amount of confidence to extrapolate wind speed measurements taken at one height (e.g., the meteorological standard of 10 m) to other heights that are interesting from the point of view of wind energy extraction.

The wind profiles also show that a wind energy conversion device spanning several tens of meters and placed in the planetary boundary layer is likely to experience substantially different wind velocities at different ends of the device (e.g., rotor blades).

The simple parametrizations of wind profiles considered above can be expected to be valid only over flat and homogeneous terrain. The presence of obstacles, such as hills, vegetation of varying height, and building structures, may greatly alter the profiles and will often create regions of strong turbulence, where no simple average velocity distribution will give an adequate description. If an obstacle like a hilltop is sufficiently smooth, however, and the atmospheric lapse rate neutral or stable, the possibility exists of increasing the wind speed in a regular fashion at a given height where an energy collecting device may be placed. An example of this is shown in Fig. 3.24, based on a calculation by Frost et al. (1974). The elliptical hill shape extends infinitely in the direction perpendicular to the plane depicted. However, most natural hill shapes do not make a site more suitable for energy extraction. Another possibility of enhanced wind flow at certain heights exists in connection with certain shapes of valleys, which may act like a shroud to concentrate the intensity of the wind field, while still preserving its laminar features (see, for example, Frost, 1975). Again, if the shape of the valley is not perfect, regions of strong turbulence may be created and the suitability for energy extraction diminishes.

FIGURE 3.23 Wind speed profiles (i.e., variation with height) for a flat terrain characterized by a roughness length of 0.01 m and for a friction velocity of $(\tau/\rho)^{1/2} = 0.5$ m s^{-1}. The three different curves correspond to an atmosphere that is neutral (has an adiabatic lapse rate), stable, or unstable. The stability is expressed in terms of the Monin–Obukhov length. (Based on Frost, 1975.)

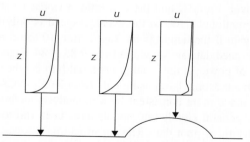

FIGURE 3.24 Wind speed profiles for a wind directed perpendicular to an elliptically shaped hill for two locations upwind, plus at the top of the hill. The profile far away from the disturbance corresponds to a surface of roughness length 0.005 m, a friction velocity of 0.4 m s^{-1}, and a neutral atmosphere. (Based on Frost et al., 1974.)

Only in relatively few cases can the local topography be used to enhance the kinetic energy of the wind at heights suitable for extraction. In the majority of cases, the optimum site selection will be one of flat terrain and smoothest possible surface in the directions of the prevailing winds. Since the roughness length over water is typically of the order 10^{-3} m, the best sites for wind energy conversion will usually be coastal sites with large fetch regions over water.

Wind Speed Data

Average wind speeds give only a very rough idea of the power in the wind (which is proportional to u^3) or the kinetic energy in the wind (which is proportional to u^2), owing to the fluctuating part of the velocity, which makes the average of the cube of u different from the cube of the average, etc. Whether the extra power of the positive speed excursions can be made useful depends on the response of the particular energy conversion device to such fluctuations.

For many locations, only average wind speed data are available, and since it is usually true in a qualitative way that the power that can be extracted increases with increasing average wind speed, a few features of average wind speed behavior are described.

In comparing wind speed data for different sites, the general wind speed–height relation (such as the ones shown in Fig. 3.23) should be kept in mind, along with the fact that roughness length, average friction velocity, and statistics of the occurrence of different stability classes are also site-dependent quantities.

Figure 3.25 shows the seasonal variation of wind speed, based on monthly average values, for selected locations. Except for Risø and Toronto, the sites represent near optimum wind conditions for their respective regions. Not all parts of the Earth are as favored with winds as the ones in Fig. 3.25. Figure 3.26 shows the variation of average wind speed throughout the day at two shoreline sites and one maritime site in Singapore. The average wind speed at the latter site is slightly over 3 m s^{-1}, and there is little seasonal variation. At the two land-based observational stations, the average speed is only around 2 m s^{-1}, with very little gain if the height is increased from 10 to 65 m.

Global wind speed data are shown in Figs. 2.83–2.84. Figures 3.27 and 3.28 show the levels of power in the wind at potential hub height for wind turbines (about 70 m), on a seasonal base, using data from NCEP-NCAR (1998). These data have been made more consistent by a re-analysis method (Kalnay *et al.*, 1996), in which general circulation models have been run to improve predictions in areas of little or poor data. The construction of power in the wind estimates from wind speed data is explained in section 6.3.1. It involves an averaging of circulation model data from height levels of 1000 and 925 mb (expressed as pressure levels), plus an empirical way of relating the average of the third power of the wind speed (the power) to the third power of the average wind speed. This procedure necessarily yields very approximate results. The main observation is that the highest power levels are found over oceans, including some regions close to shorelines. This points to the possibility of obtaining viable extraction of wind power from offshore plants located at suitable shallow depths. It is also seen that seasonal variations are significant, particularly over large continents. A new, more accurate method of determining offshore potential wind power production is presented in section 6.3.1.

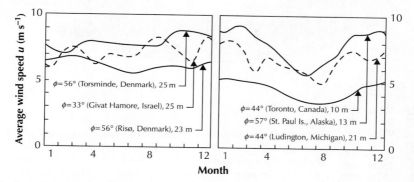

FIGURE 3.25 Seasonal variation of average wind speed for selected locations and heights. Torsminde data for 1961 from Jensen (1962), Givat Hamore data Nov. 1959 to Oct. 1960 from Frenkiel (1964), Risø 1958–1967 data from Petersen (1974), Toronto 1955–1972 data from Brown and Warne (1975), St. Paul Island 1943–1971 data from Wentink (1976), and the Ludington 1970, 1972 data from Asmussen (1975).

FIGURE 3.26 Variations of average wind speed with hour of the day for three Singapore stations. The data cover the period 1970–1974. (Based on Nathan *et al.*, 1976.)

(a)

Energy flow
W m⁻²

■	1600 to 6000
■	800 to 1600
■	600 to 800
■	400 to 600
■	300 to 400
■	250 to 300
■	200 to 250
■	160 to 200
■	120 to 160
■	80 to 120
■	40 to 80
■	0 to 40

FIGURE 3.27a,b Maps of wind power regimes for January and April 1997, based on NCEP-NCAR (1998). The power levels are estimated for a height of 70 m above ground, presently a typical hub height for wind turbines. The method of estimation is explained in section 6.3.1. [These and following area-based geographical information system (GIS) maps are from Sørensen and Meibom (1998), copyright B. Sørensen.]

(b)

(a)

Energy flow
W m⁻²

1600 to 6000
800 to 1600
600 to 800
400 to 600
300 to 400
250 to 300
200 to 250
160 to 200
120 to 160
80 to 120
40 to 80
0 to 40

FIGURE 3.28a,b Maps of wind power regimes for July and October 1997, based on NCEP-NCAR (1998). The power levels are estimated for a height of 70 m above ground, presently a typical hub height for wind turbines. The method of estimation is explained in section 6.3.1.

(b)

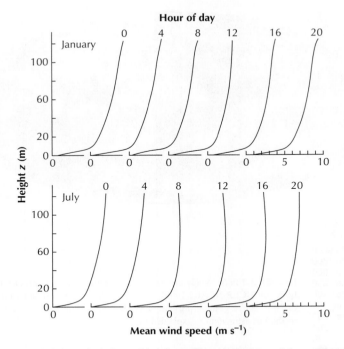

FIGURE 3.29 Variation of wind speed height profile with the hour of day at Risø, Denmark (56 °N). Each curve has been averaged over all data from the period 1958–1967 for one winter or summer month. (Based on data from Petersen, 1974.)

FIGURE 3.30 Annual average wind speed *vs.* height profile at latitudes 50–56 °N. The lower part of the curve is based on Risø data (Petersen, 1974), and the upper part is based on a set of curves derived from German measurements (Hütter, 1976). The one matching the lower part of the curve has been selected. Also indicated (*dashed line*) is an average height profile of the air density, used in the construction of Figs. 3.31–3.33.

Figure 3.29 shows an example of measured wind profiles as a function of the hour of the day, averaged over a summer and a winter month. The tendency to instability due to turbulent heat transfer is seen during summer day hours (cf. the discussion above). Changes in wind direction with height have been measured (e.g., by Petersen, 1974). In the upper part of the planetary boundary layer, this is simply the Ekman spiral (cf. section 2.3.2).

The continuation of the velocity profile above the Prandtl boundary layer may be important in connection with studies of the mechanisms of restoring kinetic energy lost to an energy extraction device (e.g., investigations of optimum spacing between individual devices). Extraction of energy in the upper troposphere does not appear attractive at the present technological level, but the possibility cannot be excluded. Figure 3.30 shows the 10-year average wind profile at Risø ($\phi = 56°$) matched to a profile stretching into the stratosphere, based on German measurements. The average wind speed near the tropopause compares well with the zonal means indicated in Fig. 2.49.

Height Scaling and Approximate Frequency Distribution of Wind

In practical applications, some approximate relations are often employed in simple wind power estimates. One describes the scaling of wind speeds with height, used, for example, to estimate speed u at a wind turbine hub height z from measured data at another height z_1. It is based on (3.21):

$$u(z) = u(z_1) \log((z - z_1)/z_0), \tag{3.24}$$

where the parameter z_0 describes the local roughness of the surface. However, (3.24) is valid only for stable atmospheres. Another often-used approximation is the Weibull distribution of wind speeds v over time at a given location (f is the fraction of time in a given speed interval; Sørensen, 1986):

$$f(v) = (k/c)\,(v/c)^{k-1} exp(-(v/c)^k), \tag{3.25}$$

where the parameter k is around 2 and c around 8, both with considerable variations from site to site.

3.2.2 Kinetic Energy in the Wind

In analogy to Fig. 3.30, the height variation of the kinetic energy density is obtained in the form shown in Fig. 3.31. The density variation used in (2.55) is the one indicated in Fig. 3.30, and the hourly data from Risø, Denmark, have been used to construct the kinetic energy at heights of 7–123 m, the average values of which form the lower part of the curve in Fig. 3.31. The fluctuations in wind speed diminish with height (cf. Fig. 3.36), and at $z = 123$ m the difference between the annual average of u^2 and the square of the average u is only about 1%. For this reason, the average wind speeds of Fig. 3.30 were used directly to form the average kinetic energy densities for the upper part of the figure. The kinetic energy density peaks some 3 km below the peak wind speed, owing to the diminishing air density. Fig. 3.31*b* illustrates, on a non-logarithmic scale, the bend of the curve at a height of 1–2 km, where the near-logarithmic increase of wind speed with height is replaced by a much stronger increase.

FIGURE 3.31*a,b*   Annual average height profile of kinetic energy density at latitudes 50°–56 °N. At low height, the curve is based on hourly data from Risø (Petersen, 1974), at larger heights calculated from the wind speeds of Fig. 3.30, neglecting any fluctuations. Fig. 3.31*b* is the same on two linear-scale segments.

FIGURE 3.32   Accumulated kinetic energy below a given height, for latitudes around 50°–56 °N, obtained by integration of the curve in Fig. 3.31.

The kinetic energy density curve of Fig. 3.31 may readily be integrated to give the accumulated amount of kinetic energy below a given height, as shown in Fig. 3.32. This curve illustrates the advantage of extending the sampling area of a wind energy-collecting device into the higher regions of the atmosphere. The asymptotic limit of the accumulation curve is close to 2×10^6 J m^{-2} of vertical column. This order of magnitude corresponds to that of the estimate given in Fig. 2.54, of 8×10^5 J m^{-2} as a global average of zonal kinetic energy, plus a similar amount of eddy kinetic energy. The eddy kinetic energy, which is primarily that of large-scale eddies (any deviation from zonal means), may, in part, be included in the curve based on Fig. 3.29, which gives mean wind speed and not zonal mean. Since the data are for latitude 50°–56 °N, it may also be assumed that the kinetic energy is above the global average, judging from the latitude distribution of zonal winds given in Fig. 2.49.

3.2.3 Power in the Wind

The energy flux passing through an arbitrarily oriented surface exposed to the wind is obtained by multiplying the kinetic energy density by $v \cdot n$, where v is the wind velocity and n is the normal to the surface. The energy flux (power) may then be written

$$E = 1/2 \, \rho \, u^3 \cos \theta, \tag{3.26a}$$

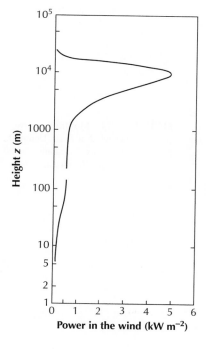

FIGURE 3.33   Annual average height profile of power in the wind, at latitudes 50°–56 °N. The lower part of the curve has been calculated from the Risø data of Petersen (1974), and the upper part is from the wind speeds of Fig. 3.30, neglecting any fluctuations.

where θ is the angle between v and $\mathbf{n}$. If the vertical component of the wind velocity can be neglected, as well as the short-term fluctuations [termed v' according to (2.23)], then the flux through a vertical plane perpendicular to the direction of the wind becomes

$$E = 1/2\,\overline{\rho}\,(V^*)^3, \tag{3.26b}$$

where V^* is the average, horizontal wind velocity corresponding to a given time-averaging interval (2.21). If the wind speed or direction cannot be regarded as constant over the time interval under consideration, then an average flux (average power) may be defined by

$$E = \frac{1}{\Delta t}\int_{t_1}^{t_1+\Delta t} \tfrac{1}{2}\rho v^3 \cos\theta\, dt_2,$$

where both v and θ (and in principle ρ) may depend on the time integrand t_2. A situation often met in practice is one of an energy-collecting device, which is able to follow some of the changes in wind direction, but not the very rapid ones. Such a "yaw" mechanism can be built into the prescription for defining θ, including the effect of a finite response time (i.e., so that the energy-collecting surface is being moved toward the direction the wind had slightly earlier).

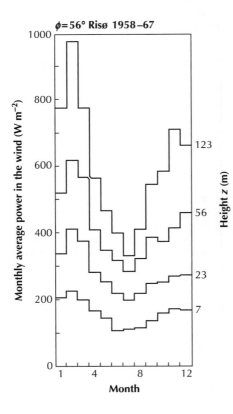

FIGURE 3.34 Monthly average power in the wind at Risø, Denmark, for different heights. Calculation based on 10 years of data (Petersen, 1974).

Most wind data are available in a form where short-term variations have been smoothed out. For example, the Risø data used in the lower part of Fig. 3.30 are 10-min averages centered on every hour. If such data are used to construct figures for the power in the wind, as in Figs. 3.33–3.34 and the duration curves in Figs. 3.39–3.40, then two sources of error are introduced. One is that random excursions away from the average wind speed will imply a level of power that on average is larger than that obtained from the average wind speed [owing to the cubic dependence in (3.24)]. The other is that, owing to changes in wind direction, the flux calculated from the average wind speed will not be passing through the same surface during the entire averaging period, but rather through surfaces of varying orientation. This second effect will tend to make the predicted power level an overestimate for any surface of fixed orientation.

Jensen (1962) overcame both of these difficulties by basing his measurements of power in the wind on a statistical ensemble of instantaneous measurements and by mounting unidirectional sensors on a yawing device that was chosen to respond to changing wind directions in a manner similar to that of the wind energy generators being considered. However, it is not believed that monthly average power or power duration curves will be substantially affected by using average data like those from Risø (note that the ergodic hypothesis in section 2.3.3, relating the time averages to the statistical ensemble means, is not necessarily valid if the time averages are formed only over one 10-min interval for each hour).

In the upper part of Fig. 3.33, the power has been calculated from the average wind speeds of Fig. 3.30, using (3.25). The seasonal variations at lower heights are shown in Fig. 3.34. It is clear that the amount of seasonal variation increases with height. This indicates that the upper part of Fig. 3.33 may well underestimate the average of summer and winter energy fluxes.

Additional discussion of the power in the wind, at different geographical locations, is found below in section 3.2.4. The overall global distribution of power in the wind at different seasons is shown in Fig. 3.27.

3.2.4 Variability in Wind Power

An example of short-term variations in wind speed at low height is given in Fig. 3.35. These fluctuations correspond to the region of frequencies above the spectral gap of Fig. 2.109. The occurrences of wind "gusts," during which the wind speed may double or drop to half the original value over a fraction of a second, are clearly of importance for the design of wind energy converters. On the other hand, the comparison made in Fig. 3.35 between two simultaneous measurements at distances separated horizontally by 90 m shows that little spatial correlation is present between the short-term fluctuations. Such fluctuations would thus be smoothed out by a wind energy conversion system, which comprises an array of separate units dispersed over a sufficiently large area.

The trends in amplitudes of diurnal and yearly wind speed variations are displayed in Fig. 3.36, as functions of height (Petersen, 1974). Such amplitudes are generally site-dependent, as one can deduce, for example, from Figs. 3.26 and 3.28 for diurnal variations and from Fig. 3.25 for seasonal variations. The diurnal amplitude in Fig. 3.26 diminishes with height, while the yearly amplitude increases with height. This is probably a general phenomenon when the wind approaches its geostrophic level, but the altitude dependence may depend on geographical position and local topography. At some locations, the diurnal cycle shows up as a 24-h peak in a Fourier decomposition of the wind speed data. This is the case at Risø, Denmark, as seen from Fig. 3.37, while the peak is much weaker at the lower height used in Fig. 2.109. The growth in seasonal amplitude with height is presumably determined by processes taking place at greater height (cf. Figs. 2.49 and 2.50) as well as by seasonal variations in atmospheric stability, etc.

FIGURE 3.35   Short-term variation in wind speed, from simultaneous records at two locations 90 m apart (measuring height probably in the range of 5 to 10 m). (Based on Banas and Sullivan, 1976.)

FIGURE 3.36   Height dependence of diurnal and yearly amplitude of oscillation of the wind speed at Risø. The average estimate is based on a Fourier decomposition of the data with no smoothing (Petersen, 1974).

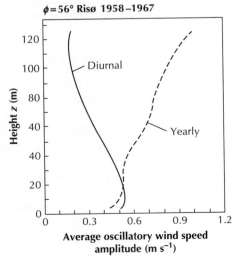

$\phi = 56°$ Risø 1958–1967

The wind speed variance spectrum (defined in section 2.5.2 in connection with Fig. 2.109) shown in Fig. 2.42 covers a frequency interval between the yearly period and the spectral gap. In addition to the 24-h periodicity, the amplitude of which diminishes with increasing height, the figure exhibits a group of spectral peaks with periods in the range 3–10 days. At the selected height of 56 m, the 4-day peak is the most pronounced, but moving up to 123 m, the peak period around 8 days is more marked (Petersen, 1974). It is likely that these peaks correspond to the passage time of typical meso-scale front and pressure systems.

In analyzing the variability of wind speed and power during a certain period (e.g., month or a year), the measured data are conveniently arranged in the forms of frequency distributions and power duration curves, much in the same manner as discussed in section 3.1.5. Figure 3.38 gives the 1-year frequency distribution of wind speeds at two Danish locations for a height of about 50 m. The wind speed frequency distribution (dashed curve) at Gedser (near the Baltic Sea) has two maxima, probably associated with winds from the sea and winds approaching from the land (of greater roughness). However, the corresponding frequency distribution of power (top curve) does not exhibit the lower peak, as a result of the cubic dependence on wind speed. At the Risø site, only one pronounced maximum is present in the wind speed distribution. Several irregularities in the power distribution (which are not preserved from year to year) bear witness to irregular landscapes with different roughness lengths, in different directions from the meteorological tower.

FIGURE 3.37 Variance spectrum of horizontal wind speeds at Risø ($\phi = 56°$ N) at a height of 56 m (based on 10 years of observation; Petersen, 1974). The spectrum is smoothed by using frequency intervals of finite length (cf. Fig. 2.109 of section 2.5.2, where similar data for a height of 7 m are given).

Despite the quite different appearance of the wind speed frequency distribution, the power distribution for the two Danish sites peaks at roughly the same wind speed, between 10 and 11 m s^{-1}.

Power Duration Curves

On the basis of the frequency distributions of the wind speeds (or alternatively that of power in the wind), power duration curves can be constructed giving the percentage of time when the power exceeds a given value. Figures 3.39 and 3.40 give a number of such curves, based on periods of a year or more. In Fig. 3.39, power duration curves are given for four U.S. sites that have been used or are being considered for wind energy conversion and for one of the very-low-wind Singapore sites.

Height 50 m, Gedser 1960/61   Height 56 m, Risø 1961

FIGURE 3.38
Frequency distribution of wind speed (*right-hand scale*) and of power for a height of about 50 m at two Danish sites. The middle curves give frequency distributions for the output of a typical wind energy generator (Sørensen, 1978a).

In Fig. 3.40, power duration curves are given for the two different Danish sites considered in Fig. 3.38, as well as for a site on the west coast of Sweden, at three different heights. The three curves for the same site have a similar shape, but in general the shape of the power duration curves in Figs. 3.39 and 3.40 depends on location. Although the Swedish Ringhals curves have non-negligible time percentages with very large power, the Danish Gedser site has the advantage of a greater frequency of medium power levels. This is not independent of conversion efficiency, which typically has a maximum as function of wind speed.

FIGURE 3.39 One-year duration curves of power in the wind, for a number of locations and heights [Swanson *et al.*, 1975 (Plum Brook and Grandpa's Knob); Coste, 1976 (Waterford); Asmussen, 1975 (Ludington); Nathan *et al.*, 1976 (Changi)].

FIGURE 3.40 One-year duration curves of power in the wind at Scandinavian sites (including different heights at the Ringhals site). [Based on data from Ljungström, 1975 (Ringhals); Petersen, 1974 (Risø); Jensen, 1962 (Gedser).]

3.3 WATER FLOWS AND RESERVOIRS, WAVES, AND TIDES

3.3.1 Ocean Currents

The maximum speed in the center of the Gulf Stream is about 2 m s^{-1}, corresponding to an energy density $1/2\rho_w V_w^2 = 2$ kJ m^{-3} and a power of $1/2 \rho_w V_w^3 = 4$ kW m^{-2}. This power level, for example, approaches that of wave power at reasonably good sites, taken as power per meter of wave crest rather than per square meter perpendicular to the flow, as used in the above case. However, high average speed of currents combined with stable direction is found only in a few places. Figure 3.41 shows the isotachs in a cross-section of the Gulf Stream, derived from a single set of measurements (in June 1938; Neumann, 1956). The maximum current speed is found at a depth of 100–200 m, some 300 km from the coast. The isotachs expand and become less regular further north, when the Gulf Stream moves further away from the coast (Niiler, 1977).

FIGURE 3.41   Contours of equal speed (in 10^{-2} m s^{-1}) along a cross-section through the Gulf Stream from Bermuda to the U.S. coast. The measurements were performed over a limited period of time (in June 1938, by Neumann, 1956), but they are consistent with other measurements in the same region (e.g., Florida Strait measurements by Brooks and Niiler, 1977).

Even for a strong current like the Gulf Stream, the compass direction at the surface has its most frequent value only slightly over 50% of the time (Royal Dutch Meteorological Institute, as quoted by Neumann and Pierson, 1966), and the power will, on average, deviate from that evaluated on the basis of average current speeds (as is the case for wind or waves) owing to the inequality of $<V^3>$ and $<V>^3$.

The geographical distribution of the strongest surface currents is indicated in Fig. 3.42. The surface currents in the Atlantic Ocean, along with the horizontal currents at three different depths, are sketched in Fig. 3.43. The current speeds indicated are generally decreasing downward, and the preferred directions are not the same at different depths. The apparent "collision" of oppositely directed currents, e.g., along the continental shelf of Central America at a depth of around 4 km, conceals the vertical motion that takes place in most of these cases. Figure 3.44 gives a vertical cross-section with outlined current directions. The figure shows how the oppositely directed water masses "slide" above and below each other. The coldest water is flowing along the bottom, while the warmer water from the North Atlantic is sliding above the cold water from the Antarctic.

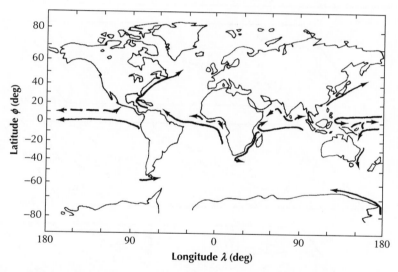

FIGURE 3.42 Indication of the location of strong surface currents (approximately defined as having average speeds above 0.5 m s^{-1}). Currents drawn with dashed lines are characterized by seasonal changes in direction. (Based on Defant, 1961.)

Variability in Current Power

Although the general features of circulation in the open ocean are described in section 2.3.2, the particular topography of coastal regions may have an important influence on currents. As an example, water forced through a narrow strait may acquire substantial speeds. The strait Storebælt ("Great Belt") between two Danish isles, which provides an outlet from the Baltic Sea, may serve as an illustration. It is not extremely narrow (roughly 20 km at the measurement site), but narrow enough to exhibit only two current directions, separated by 180°. The currents may seem fairly steady, except for periods when the direction is changing between north-going and south-going velocities, but when the energy flux is calculated from the third powers of the current speeds, the fluctuations turn out to be substantial. Figure 3.45, which gives the power at 3-hour intervals during two weeks of January 1972 clearly illustrates this.

FIGURE 3.43   Indication of average horizontal current speeds (in 10^{-2} m s^{-1}) in the Atlantic Ocean for different depths. (Based on Defant, 1961.)

Figure 3.46 shows, again for the Halskov Rev position in Storebælt, the variation in current speed with the hour of the day, based on 1-month averages. A 12-hour periodicity may be discerned, at least during January. This period is smaller than the one likely to be found in the open sea due to the motion of water particles in stationary circles under the influence of the Coriolis force [see (2.62)], having a period equal to 12 h divided by sin ϕ (ϕ is the latitude).

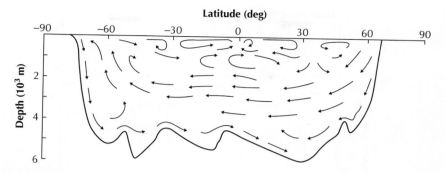

FIGURE 3.44 Indication of average current directions within a vertical cross-section of the Atlantic Ocean. (Based on Neumann and Pierson, 1966.)

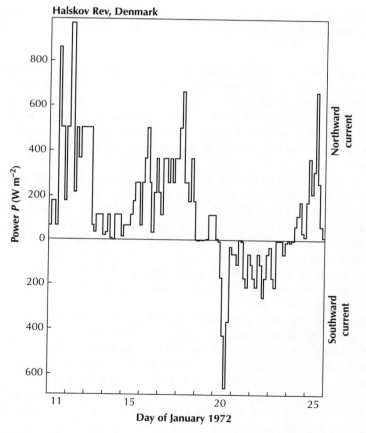

FIGURE 3.45 Power of surface current, based on observations made at 3-hour intervals, for Halskov Rev, Denmark, during a 15-day period in 1972 (the strait is narrow enough to exhibit only two opposite current directions). (Based on Danish Meteorological Institute, 1973.)

FIGURE 3.46   Dependence of average current speed on the hour of the day, for a summer and a winter month, for Halskov Rev, Denmark. (Based on Danish Meteorological Institute, 1973.)

In Fig. 3.47, the frequency distributions of current speed and power are shown for a summer and a winter month. These curves are useful in estimating the performance of an energy extraction device, and they can be used, for example, to construct the power duration curve of the current motion, as shown in Fig. 3.48. This is the power duration curve of the currents themselves. That of an energy-extracting device will have to be folded with the efficiency function of the device.

The peak in the frequency distribution of power is at a lower current speed for July than for January, and the average power is 93 W m^{-2} in July, as compared with 207 W m^{-2} in January (the corresponding kinetic energy densities are 138 and 247 J m^{-3}). This indicates that the fluctuations around the average values have a substantial effect on the energy and, in particular, on the power, because from the average current speeds, 0.46 m s^{-1} (July) and 0.65 m s^{-1} (January), the calculated kinetic energy densities would have been 108 and 211 J m^{-3}, and the power would have taken the values 50 and 134 W m^{-2}.

Few locations, whether in coastal regions or in open oceans, have higher average current speeds than the Danish location considered in Figs. 3.45–3.48, so that average power levels in the range 100–200 W m^{-2} are likely to be more representative than the 4000 W m^{-2} found (at least at the time of the measurement reported in Fig. 3.41) in the core of the Gulf Stream. This means that at many locations the power in currents is no greater than that found in the wind at quite low heights (cf. Fig. 3.34). Also, the seasonal variations and fluctuations are similar, which is not unexpected for wind-driven currents. For the currents at greater depths this may not hold true, partly because of the smoothing due to a long turnover time and partly

FIGURE 3.47 Frequency distributions of current speeds and power, based on a summer and a winter month (*solid* and *dashed curves*, respectively), for Halskov Rev, Denmark. The data (taken from Danish Meteorological Institute, 1973) have been smoothed in calculating the distributions. The monthly average speed and power are indicated on the figure (in parentheses).

because not all the deep-sea motion is wind driven, but may also be associated with temperature and salinity gradients, as discussed in section 2.3.2.

The power duration curves in Fig. 3.48 may be compared to those of wind power shown in Figs. 3.39 and 3.40. The fraction of time in which the monthly average power is available in the Halskov Rev surface current is 0.3 in both January and July. The overall average power of about 150 W m^{-2} is available for about 45% of the time in January, but only 17% of the time in July.

3.3.2 River Flows, Hydropower, and Elevated Water Storage

The kinetic energy of water flowing in rivers or other streams constitutes an energy source very similar to that of ocean currents. However, rather than being primarily wind driven or caused by differences in the state of the water masses themselves, the river flows are part of the hydrological cycle depicted in Fig. 2.65. Water vapor evaporated into the atmosphere is transported and eventually condensed. It reaches the ground as water or ice, at the elevation of the particular location. Thus, the primary form of energy is potential. In the case of ice and snow, a melting process (using solar energy) is usually necessary before the potential energy of elevation can start to transform into kinetic energy. The origin of many streams and rivers is precisely the ice-melting process, although their flows are joined by ground water flows along their subsequent routes. The area from which a given river derives its input of surface run-off, melt-off, and ground water is called its *drainage basin*.

The flow of water in a river may be regulated by means of dam building, if suitable reservoirs exist or can be formed. In this way, the potential energy of water stored at an elevation can be transformed into kinetic energy (e.g., driving a turbine) at times most convenient for utilization.

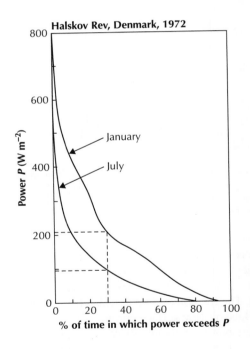

FIGURE 3.48 Power duration curves for power in currents at Halskov Rev, Denmark, based on a summer and a winter month of 1972. *Thin, dashed lines* indicate the monthly average power and the percentage of time during which it is available.

An estimate of the potential hydro-energy of a given geographical region could in principle be obtained by hypothetically assuming that all precipitation was retained at the altitude of the local terrain and multiplying the gravitational potential mg by the height above sea level. According to Fig. 2.65, the annual precipitation over land amounts to about 1.1×10^{17} kg of water, and taking the average elevation of the land area as 840 m (Sverdrup *et al.*, 1942), the annually accumulated potential energy would amount to 9×10^{20} J, corresponding to a mean energy flux (hydropower) of 2.9×10^{13} W.

Collection of precipitation is not usually performed as part of hydropower utilization; instead, the natural processes associated with soil moisture and vegetation are allowed to proceed, leading to a considerable re-evaporation and some transfer to deeper-lying ground water, which eventually reaches the oceans without passing through the rivers (see Fig. 2.65). The actual run-off from rivers and overground run-off from polar ice caps comprise, on average, only about 0.36 of the precipitation over land, and the height determining the potential energy may be lower than the average land altitude, namely, that given by the height at which the water enters a river flow (from ground water or directly). The minimum size of stream or river that can be considered useful for energy extraction is, of course, a matter of technology, but these general considerations would seem to place an upper limit on the hydro-energy available of roughly 3×10^{20} J y^{-1}, corresponding to a power average below 10^{13} W (10 TW).

If, instead of using average precipitation and evaporation rates together with average elevation, the geographical variations of these quantities are included, the result is also close to 3×10^{20} J y^{-1} or 10^{13} W. These figures are derived from the integral over all land areas,

$$\frac{dW^{pot}}{dt} = \int_{land\ area} (r - e)gz\,dA,$$

where r and e are the rates of precipitation and evaporation (mass of water, per unit of area and time), g is the gravitational acceleration, and z is the height above sea level. The observed annual mean precipitation and evaporation rates quoted by Holloway and Manabe (1971) were used in the numerical estimate.

Geographical Distribution of Hydropower Resources

A different estimate of hydropower potential is furnished by counts of actual rivers with known or assumed water transport and falling height. According to such an estimate by the World Energy Conference (1974, 1995), the installed or installable hydro-generation capacity resource at average flow conditions may, in principle, amount to 1.2×10^{12} W, for both large and small installations down to "micro-hydro" installations of around 1 MW. On the other hand, it is unlikely that environmental and other considerations will allow the utilization of all the water resources included in the estimate. The World Energy Conference (1995) estimates 626 GW as a realistic reserve (including an already installed capacity producing, on average, 70 GW).

Figure 3.49 gives an idea of the geographical distribution of the hydropower resources on a national basis. The largest remaining resources are in South America. The figures correspond to average flow conditions, and the seasonal variations in flow are very different for different regions. For example, in Zaire, practically all the reserves would be available year round, whereas in the United States, only 30% can be counted on during 95% of the year.

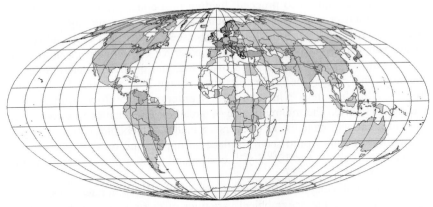

Energy flow
W m^{-2}

5	to 10
2	to 5
1	to 2
0.5	to 1
0.2	to 0.5
0.1	to 0.2
0.05	to 0.1
0.02	to 0.05
0.01	to 0.02
0.005	to 0.01
0.002	to 0.005
0.0001 to	0.002
all others	

FIGURE 3.49 Hydropower potential average production (W m^{-2}), including existing plants, plants under construction, and planned and contemplated installations, both for large- and small-scale power schemes. (Based on data from World Energy Council, 1995.)

FIGURE 3.50 Seasonal variation in the power associated with the water flow into Norwegian hydropower reservoirs for a typical year (1985) and a "dry year" (1990). (Based on Andersen, 1997; Meibom *et al.*, 1999.)

Figure 3.50 gives seasonal variations (for 2 years) in the flow into the existing hydropower reservoirs in Norway, a country where the primary filling of reservoirs is associated with the melting of snow and ice during the late spring and early summer months.

Environmental Impact

The environmental impact of nonregulated hydro-generation of power is mainly interference with the migration of fish and other biota across the turbine area, but the building of dams in connection with large hydro facilities may have an even more profound influence on the ecology of the region, in addition to introducing accident risks. Large reservoirs have caused serious destruction of natural landscapes and dislocation of populations living in areas to be flooded as reservoirs. There are ways to avoid some of these problems. For example, in Swizerland, modular construction, where the water is cascaded through several smaller reservoirs, has made a substantial reduction in the area modified. Furthermore, reservoirs need not be constructed in direct connection with the generating plants, but can be separate installations placed in optimum locations, with a two-way turbine that uses excess electric production from other regions to pump water up into a high-lying reservoir. When other generating facilities cannot meet demand, the water is then released back through the turbine. This means that, although the water cycle may be unchanged on an annual average basis, considerable seasonal modifications of the hydrological cycle may be involved. The influence of such modifications on the vegetation and climate of the region below the reservoir, which would otherwise receive a water flow at a different time, has to be studied in each individual case. The upper region, as well, may experience modifications—for example, owing to increased evaporation in the presence of a full reservoir.

Although the modifications are local, their influence on the ecosystems can have serious consequences for man. A notable example is the Aswan Dam in Egypt, which has allowed aquatic snails to migrate from the Nile delta to upstream areas. The snails can carry parasites causing schistosomiasis, and the disease has actually spread from the delta region to Upper Egypt since the building of the dam (Hayes, 1977).

It is unlikely that hydropower utilization will ever be able to produce changes in the seasonal hydrological cycle, which could have global consequences, but no detailed investigation has yet been made. Such a study could resemble the investigation of the influence of deforestation, shown in Fig. 2.97.

3.3.3 Ocean Waves

The order of magnitude of the total energy in wave motion is about 10^{-3} of the total kinetic energy in the atmospheric wind systems, according to the rough estimate made in section 2.4.1 in connection with Fig. 2.90. The wave energy of about 10 kJ m^{-2} found as an annual average in the North Atlantic Ocean corresponds to the accumulated wind energy up to a height of about 200 m, according to Fig. 3.32. This implies that, although the amount of energy stored in waves is much smaller than the amount stored in wind, wave energy may still be equivalent to the height-integrated fraction of wind energy accessible for practical use, at least at the current level of technology.

From the tentative estimates in section 2.4.1, the average turnover time for the energy in wave motion in the open ocean may be of the order of a few days. This time is consistent with average dissipation mechanisms, such as internal friction and weak wave–wave interactions, plus shoreline dissipation modes. The input of energy by the wind, on the other hand, seems to be an intermittent process that, for extended intervals of time, involves only slow transfer of energy between waves and turbulent wind components, or vice versa, and between wind and wave fields propagating in different directions (any angle from 0 to 2π). However, large amounts of energy may be transferred from wind to waves during short periods of time (i.e., "short periods" compared with the average turnover time). This suggests that the energy storage in waves may be varying more smoothly than the storage in wind (both waves and wind represent short-term stored solar energy, rather than primary energy flows, as discussed in connection with Fig. 2.90). As mentioned in section 2.3.2, the wave fields exhibit short-term fluctuations, which may be regarded as random. On a medium time scale, the characteristics of the creation and dissipation mechanisms may make wave energy a more "dependable" energy source than wind energy, but on a seasonal scale, the variations in wind and wave energy are expected to follow each other (cf. the discussion topic III.2 in Part III).

Wave Spectra

The energy spectrum $F(k)$ of a random wave field is defined by (2.79). Since the wavelength (or wave number k) is very difficult to observe directly, it is convenient instead to express the spectrum in terms of the frequency,

$$v = \omega(k)/2\pi = 1/T.$$

The frequency is obtained from the period, T, which for a harmonic wave equals the zero-crossing period, i.e., the time interval between successive passages of the wave surface through the zero (average) height, in the same direction. The spectral distribution of energy, or *energy spectrum* when expressed in terms of frequency, $F_1(v) = 2\pi F_1(\omega)$, is usually normalized to the total energy (Barnett and Kenyon, 1975), whereas the wave number-dependent spectrum $F(k)$, defined in section 2.3.2, is normalized to the average potential energy. Thus,

$$\int F_1(\omega)\, d\omega = W^{total} = 2 \int F(k)\mathrm{d}k.$$

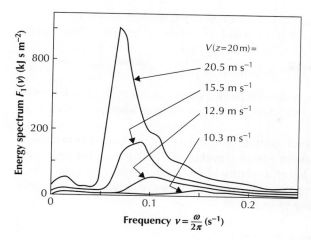

FIGURE 3.51   Energy spectrum of waves estimated to be "fully developed" for the Atlantic Ocean (data have been grouped according to the wind speed at a height of 20 m). (Based on Moskowitz, 1964.)

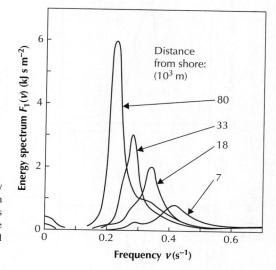

FIGURE 3.52   Fetch-limited energy spectrum of waves in the southern part of the North Sea. The wind is blowing from the continental shore (westward from Helgoland). (Based on Hasselmann *et al.*, 1973.)

Figure 3.51 shows a set of measured energy spectra, F_1, based on data from the Atlantic Ocean (Moskowitz, 1964). The wave fields selected were judged to correspond to fully developed waves, and data corresponding to the same wind speed at a height of 20 m were averaged in order to provide the spectra shown in the figure. The figure shows that the spectral intensity increases, and the frequency corresponding to the peak intensity decreases, with increasing wind speed.

Based on the similarity theory of Monin and Obukhov (1954), Kitaigorodskii (1970) suggested that $F_1(\omega)\, g^2 \rho_w^{-1} V^{-5}$ (V being the wind speed) would be a universal function of $\omega\, V\, g^{-1}$, with both quantities being dimensionless. Based on the data shown in Fig. 3.51, which approximately satisfy Kitaigorodskii's hypothesis, Moskowitz (1964) suggested the following analytical form for the energy spectrum of fully developed gravity waves,

$$F_1(\omega) = 8.1 \times 10^{-3} \rho_w g^3 \omega^{-5} \exp(-0.74 \, (V\{z = 20\,\mathrm{m}\}\, \omega/g)^{-4}). \qquad (3.27)$$

The usefulness of this relation is limited by its sensitivity to the wind speed at a single height and by the difficulty of determining whether a given wave field is fully developed.

If the wave field is "fetch-limited," i.e., if the wind has been able to act over only a limited length, then the energy spectrum will peak at a higher frequency, and the intensity will be lower, as shown in Fig. 3.52. Hasselmann *et al.* (1973) have generalized (3.27) to situations in which the wave field is not necessarily fully developed, according to such data.

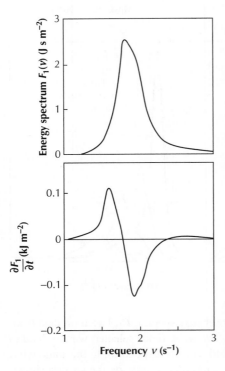

FIGURE 3.53 Simultaneous spectra of energy (*top*) and the time derivative of the energy spectrum. (Based on laboratory experiments by Mitsuyasu, 1968.)

The position of the spectral peak will move downward as a function of time owing to nonlinear wave–wave interactions, i.e., interactions between different spectral components of the wave field, as discussed in connection with (2.81) (Hasselmann, 1962). This behavior is clearly seen in the laboratory experiments of Mitsuyasu (1968), from which Fig. 3.53 shows an example of the time derivative of the energy spectrum, $\partial F_1(\nu)/\partial t$. Energy is transferred from middle to smaller frequencies. In other experiments, some transfer is also taking place in the direction of larger frequencies. Such a transfer is barely present in Fig. 3.53. The shape of the rate-of-transfer curve is similar to the one found in later experiments for real oceanic conditions (Hasselmann *et al.*, 1973), and both are in substantial agreement with the nonlinear theory of Hasselmann.

Observations of ocean waves often exist in the form of statistics on the occurrences of given combinations of wave height and period. The wave height may be expressed as the significant height, H_s, or the root mean square displacement of the wave surface, H_{rms}. For a harmonic wave, these quantities are related to the surface amplitude, a, by

$$a^2 = 2\,(H_{rms})^2 = H_s^2/8.$$

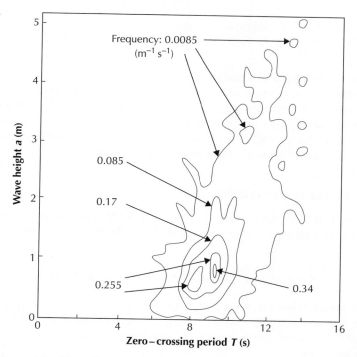

FIGURE 3.54 Frequency distribution of wave heights and zero-crossing periods for Station India (59 °N, 19 °W) in the North Atlantic. [Based on 1 year of observation by Draper and Squire (1967), quoted by Salter, 1974.]

The period is usually given by the observed zero-crossing period. Figures 3.54 and 3.55 give examples of such measured frequency distributions, based on a year's observation at the North Atlantic Station "India" (59 °N, 19 °W, the data quoted by Salter, 1974) and at the North Sea Station "Vyl" (55.5 °N, 5.0 °E, based on data discussed in Problem III.2). The probability of finding

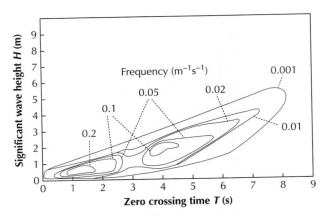

FIGURE 3.55   Frequency distribution of significant wave heights and zero-crossing periods for Station Vyl (55 °N, 5 °E) in the North Sea. (Based on 1 year of observation by Nielsen, 1977.)

periods between 8 and 10 s and wave heights, a, between 0.3 and 1.5 m is quite substantial at Station India, of the order of 0.3. Also, the probability of being within a band of zero-crossing periods with a width of about 2 s and a center that moves slowly upward with increasing wave height is nearly unity (the odd values characterizing the contour lines of Fig. 3.54 are due to the original definition of sampling intervals in terms of feet and H_s).

3.3.4 Power in the Waves

The power of a harmonic wave, i.e., the energy flux per unit length of wave crest, passing through a plane perpendicular to the direction of propagation, is from (2.84)

$$P = \rho_w \, g \, U_w \, a^2 / 4 = \rho_w \, g^2 \, T \, a^2 / (8\pi) = \rho_w \, g^2 a^2 / (4\omega). \qquad (3.28)$$

For the spectral distribution of energy given by the function F_1 (section 3.3.1), each component must be multiplied by the group velocity $\partial \omega \, (k) / \partial \, k$. Taking $\omega = (gk)^{1/2}$ for ocean gravity waves, the group velocity becomes $g/(2\omega)$, and the power becomes

$$P = \int \frac{\partial \omega}{\partial k} F_1(\omega) d\omega = \frac{1}{2} g \int \frac{F_1(\omega)}{\omega} d\omega. \qquad (3.29)$$

Based on observed energy spectra, F_1, as a function of frequency or period $T = 2\pi/\omega$, the power distribution [(3.29) before integration] and total power may be constructed. Figure 3.56 gives the power distribution at the North Atlantic station also considered in Fig. 3.54. Using data for an extended period, the average distribution (labeled "year") was calculated and, after extraction of data for the periods December–February and June–August, the curves labeled *winter* and *summer* were constructed (Mollison *et al.*, 1976).

Compared with Fig. 3.34, for example, Fig. 3.56 shows that the seasonal variations in wave power at the open ocean site are substantial and equivalent to the seasonal variations in wind power at considerable height (maybe around 100 m, considering that the roughness length over the ocean is smaller than at the continental site considered in Fig. 3.36).

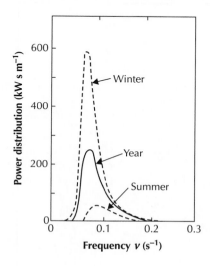

FIGURE 3.56 Frequency distribution of wave power, based on 1 year of observations (*solid line*) or summer or winter data only (*dashed lines*), for Station India (59 °N, 19 °W) in the North Atlantic. The yearly average power is 91 kW m^{-1}. (Based on Mollison *et al.*, 1976.)

FIGURE 3.57 Annual average wave power (in kW m^{-1}) for selected sites. (Based on United Kingdom Energy Technology Support Unit, 1976.)

Figure 3.57 summarizes some of the data available on the geographical distribution of wave power. The figures given are yearly mean power at a specific location, and most of the sites chosen are fairly near to a coast, although still in open ocean. The proximity of a shore is important for energy extraction from wave motion. Whether such a condition will continue to apply will depend on the development of suitable methods for long-range energy transfer (Chapter 5).

FIGURE 3.58   Contours of estimated equal annual average wave power in the North Sea, adjacent straits, and a part of the North Atlantic Ocean (unit: kW m^{-1}). (Based on United Kingdom Energy Technology Support Unit, 1976.)

In Fig. 3.58, a more detailed map of wave power availability for the north European region is shown, based on the iso-power lines estimated in initial assessments of wave power in the United Kingdom, supplemented with estimates based on data for the waters surrounding Denmark. The rapid decrease in power after passing the Hebrides in approaching the Scottish coast from the Atlantic and also after moving southward through the North Sea is noteworthy.

The variability of wave power may be described in terms similar to those used for wind energy. Figure 3.59 shows the power duration curves for Station India in the North Atlantic, based on all year or the summer or winter periods only, as in Fig. 3.56. Again, the occurrence of periods of low and high power depends markedly on seasonal changes.

Waves in a Climatic Context

As in the case of wind energy, little is known about the possible impact of large-scale energy extraction from wave power. One may argue that the total amount of energy involved is so small compared to the energy exchanged by atmospheric processes that any climatic consequence is unlikely. However, the exchange of latent and sensible heat, as well as matter, between the ocean and the atmosphere may, to a large extent, depend on the presence of waves. In particular, the rates of transfer may be high in the presence of breaking waves, and the extraction of energy from wave motion may prevent waves from developing a wave profile to the point of breaking. Thus, a study of the environmental implications of wave energy utilization, which seems to lend itself very naturally to computer simulation techniques, should be undertaken in connection with proposed energy extraction schemes. As suggested in previous sections, this is likely to be required for any large-scale use of renewable energy, despite the intuitive feeling that such energy sources are "nonpolluting." Yet the limited evidence available, mostly deriving from analogies to natural processes, does suggest that the results of more detailed analyses will be

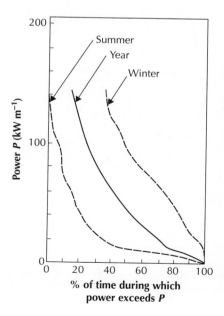

FIGURE 3.59 Time duration curves for wave power at Station India (59 °N, 19 °W) in the North Atlantic for the whole year (*solid line*) or only summer or winter (*dashed lines*). (Based on Mollison *et al.*, 1976.)

that renewable energy flows and stores can be utilized in quantities exceeding present technological capability, without worry about environmental or general climatic disturbances.

3.3.5 Tides

The average rate of dissipation of tidal energy, as estimated from the slowing down of the Earth's rotation, is about 3×10^{12} W. Of this, about a third can be accounted for by frictional losses in shallow sea regions, bays, and estuaries, according to Munk and MacDonald (1960).

In order to understand the concentration of tidal energy in certain coastal regions, a dynamic picture of water motion under the influence of tidal forces must be considered. The equations of motion for the oceans, (2.61) and (2.62), must be generalized to include the acceleration due to tidal attraction, that is, F^{tidal}/m where an approximate expression for the tidal force is given in (2.68). Numerical solutions to these equations (see, for example, Nihoul, 1977) show the existence of a complicated pattern of interfering tidal waves, exhibiting in some points zero amplitude (nodes) and in other regions deviations from the average water level far exceeding the "equilibrium tides" of a fraction of a meter (section 2.5.3). These features are in agreement with observed tides, an example of which is shown in Fig. 3.60. Newer data based on satellite measurements can be followed in near-real-time on the Internet (NASA, 2004).

FIGURE 3.60   Tidal range H (difference between highest and lowest level in m) of the semi-diurnal lunar tides in the North Sea and adjacent straits. (Based on Kiilerich, 1965; Cavanagh *et al.*, 1993.)

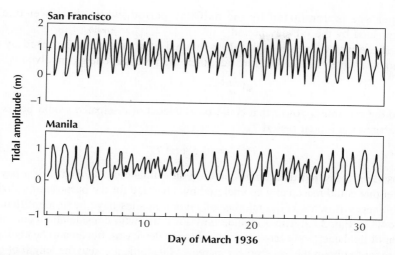

FIGURE 3.61 Examples of the time-development of tidal amplitudes at two different locations for the same month of 1936. (Based on Defant, 1961.)

The enhancement of tidal amplitudes in certain bays and inlets can be understood in terms of resonant waves. Approximating the inlet by a canal of constant depth h, the phase velocity of the tidal waves is $U_t = (gh)^{1/2}$ (Wehausen and Laitone, 1960), and the wavelength is

$$\lambda_t = T_t U_t,$$

where T_t is the period. For the most important tidal wave, T_t equals half a lunar day, and the condition for resonance may be written

$$L = i \lambda_t /4 = 0.25\, i\, T_t\, g^{1/2}\, h^{1/2}, \tag{3.30}$$

where i is an integer, so that the length L of the inlet is a multiple of a quarter wavelength. For $i = 1$, the resonance condition becomes $L = 34\,973\, h^{1/2}$ (L and h in meters). Bays and inlets satisfying this condition are likely to have high tidal ranges, with the range being the difference between highest and lowest water level. An example of this is the Severn inlet near Bristol in the United Kingdom, as seen from Fig. 3.60. Cavanagh, Clarke, and Price (1993) estimate the total European potential to be 54 GW or about 100 TWh y^{-1}, of which 90% is in France and the United Kingdom.

As discussed in section 2.5.3, the tides at a given location cannot be expected to have a simple periodicity, but rather are characterized by a superposition of components with different periods, the most important of which is equal to a half or full lunar or solar day. As a function of time, the relative phases of the different components change, leading to regularly changing amplitudes, of which two different patterns are shown in Fig. 3.61. The upper one is dominated by the half-day period, while the lower one is dominated by the full-day period.

If the water level at high tide, averaged over an area A, is raised by an amount H over the level at low tide, then the potential energy involved is

$$W^{tidal} = \rho_w H A g H,$$

and the maximum power that could be extracted or dissipated would be, as an average over a tidal period T_t,

$$P^{tidal} = \rho_w g A H^2 T_t^{-1}. \tag{3.31}$$

Based on measured tidal ranges, and on an estimate of the area A of a bay or inlet that could reasonably be enclosed by a barrage for the purpose of utilizing the energy flow (3.31), a number of favorable sites have been identified, as shown in Fig. 3.62. These include only localities with considerable concentration of tidal energy, because if the tidal range decreases, the area to be enclosed in order to obtain the same power increases quadratically, and the length of barrage will have to be correspondingly greater. For the same reason, sites of adequate tidal range, but no suitable bay, which could be enclosed by a reasonably small length of barrage, have been excluded. Of course, the term *reasonable* rests on an economic judgment that may be valid only under given circumstances. It is estimated that 2–3 GW may be extracted in Europe, half at costs in the range 10–20 euro-cents (or U.S. cents) per kWh (Cavanagh

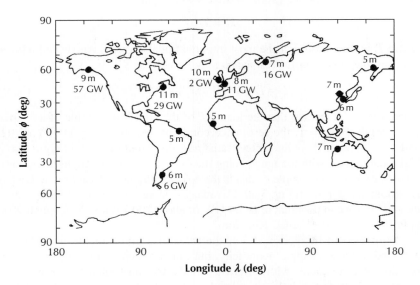

FIGURE 3.62 Tidal range for selected locations and the estimated average tidal power for each bay in a number of cases where utilization has been proposed. (Based on King Hubbert, 1969; Gray and Gashus, 1972; Sorensen and MacLennan, 1974.)

et al., 1993, using costing methodology of Baker, 1987), and 20–50 GW in North America (Sorensen and MacLennan, 1974; Bay of Fundy Tidal Power Review Board, 1977).

Environmental impacts may arise from utilization of tidal power. When the La Rance tidal plant was built in the 1960s, the upper estuary was drained for water over 2 years, a procedure that would hardly be considered environmentally acceptable today. Alternative building methods using caissons or diaphragms exist, but in all cases the construction times are long and careful measures have to be taken to protect the biosphere (e.g., when stirring up mud from the estuary seabed). Generally, the coastal environment is affected by the building and operation of tidal power plants, both during construction and to a lesser extent during operation, depending on the layout (fish bypasses, etc., as known from hydropower schemes). Some fish species may be killed in the turbines, and the interference with the periodic motion of bottom sand may lead to permanent siltation problems (and it has at La Rance).

The total estimated power of about 120 GW at the best sites throughout the world may be larger than what can be considered economical, but smaller than the amount of tidal energy actually available. It is still 12% of the above-mentioned estimate of the total dissipation of tidal energy in the vicinity of land, and it is unlikely that all the coastal sites yielding a total of 1000 GW would be suitable for utilization, so the order of magnitude is likely to be correct. The 2000 GW remaining [relative to the tidal power derived from astronomical data (Munk and MacDonald, 1960)] presumably becomes dissipated in the open ocean.

The maximal tidal power believed to be accessible, as well as the total resource estimate, is about 10% of the corresponding figures for hydropower, discussed in section 3.3.2. Tidal variations are coupled to river run-off and sea-level rise due to global greenhouse warming (Miller and Douglas, 2004).

3.4   HEAT FLOWS, RESERVOIRS, AND OTHER SOURCES OF ENERGY

A large fraction of incoming solar radiation is stored as heat near the surface of the Earth. According to Fig. 2.20, on average 47% is absorbed by the oceans and continents. The more detailed picture in Fig. 2.90 shows that 38% is absorbed by the oceans, 9% by the continents, and 24% by the atmosphere. Chapter 2 deals with some of the ways in which this energy could be distributed and eventually dissipated. Previous sections deal with the incoming radiation itself and with the kinetic energy in atmospheric and oceanic motion derived from the solar input by a number of physical processes. Storage of potential energy is also considered in connection with the processes of lifting water to a higher elevation, either in waves or in evaporated water, which may later condense and precipitate at a higher topographical level. Still, the energy involved in such kinetic and potential energy-creating processes is much less than the latent and sensible heat fluxes associated with the evaporation and condensation processes themselves and with the conversion of short-wavelength radiation to stored heat. The heat will be re-radiated as long-wavelength radiation in situations of equilibrium, but the temperature of the medium responsible for the absorption will rise, and in some cases the temperature gradients between the absorbing regions and other regions (deep soil, deep ocean, etc.), which do not themselves absorb solar radiation, cause significant heat flows.

Utilization of heat flows and stored heat may be direct if the desired temperature of use is no higher than that of the flow or storage. If this is not so, two reservoirs of different temperature may be used to establish a thermodynamic cycle yielding a certain amount of work, limited by the second law of thermodynamics. An alternative conversion scheme makes use of the heat pump principle by expending work added from the outside. Such conversion methods are considered in Chapter 4, while the focus here is on identifying the heat sources that are most suitable for utilization. The solar energy stores and flows are surveyed in section 3.4.1, whereas section 3.4.2 deals with the energy stored in the interior of the Earth and the corresponding geothermal flows.

3.4.1 Solar-derived Heat Sources

The ability of land and sea to act as a solar energy absorber and further as an energy store, to a degree determined by the local heat capacity and the processes that may carry heat away, is of profound importance for the entire biosphere. For example, food intake accounts only for 25%–30% of man's total acquisition of energy during a summer day in central Europe (Budyko, 1974). The rest is provided by the body's absorption of solar energy. The biological processes that characterize the present biosphere have vanishing rates if the temperature departs from a rather narrow range of about 270–320 K. Present forms of life thus depend on the greenhouse effect's maintaining temperatures at the Earth's surface in this range, at least during a fraction of the year (the growing season), and it is hard to imagine life on the frozen "white Earth" (see section 2.4.2) that would result if the absorption processes were minimized.

Utilization of heat stores and flows may then be defined as uses in addition to the benefits of the "natural" thermal regime of the Earth, but often no sharp division can be drawn between "natural" and "artificial" uses. For this reason, an assessment of the "magnitude" of the resource is somewhat arbitrary, and in each case it must be specified what is included in the resource base.

Total energy absorption rates are indicated in Fig. 2.90, and it is clear that the oceans are by far the most significant energy accumulators. The distributions of yearly average temperatures along cross-sections of the main oceans are shown in Figs. 2.67–2.69.

The potential of a given heat storage for heat pump usage depends on two temperatures: that of the storage and the required usage temperature. This implies that no energy "amount" can be associated with the energy source, and a discussion of heat pump energy conversion is therefore deferred to Chapter 4. An indication of possible storage temperatures may be derived from Figs. 2.67–2.69 for the oceans, Fig. 2.103 for a continental site, and Fig. 2.32 for the atmosphere.

For utilization without addition of high-grade mechanical or electric energy, two reservoirs of differing temperatures must be identified. In the oceans, Figs. 2.67–2.69 indicate the presence of a number of temperature gradients, most noticeable at latitudes below 50° and depths below 1000–2000 m. Near the Equator, the temperature gradients are largest over the first few hundred meters, and they are seasonally stable, in contrast to the gradients in regions away from the Equator, which are largest during summer and approach zero during winter. At still higher latitudes, there may be ice cover during a part or all of the year. The temperature at the lower boundary of the ice, i.e., at the water–ice interface, is almost constantly equal to 271.2 K (the freezing point of seawater), and in this case a stable temperature difference can be expected between the water and the atmosphere, particularly during winter.

It follows from Fig. 2.90 that over half the solar energy absorbed by the oceans is used to evaporate water. Of the remaining part, some will be transferred to the atmosphere as sensible heat convection, but most will eventually be re-radiated to the atmosphere (and maybe to space) as long-wavelength radiation. The length of time during which the absorbed energy will stay in the oceans before being used in one of the above ways, determines the temperature regimes. At the ocean surface, residence times of a fraction of a day are associated with diurnal temperature variations, while residence times normally increase with depth and reach values of several hundred years in the deepest regions (section 2.3.2).

If oceanic temperature gradients are used to extract energy from the surface region, a cooling of this region may result, unless the currents in the region are such that the cooling primarily takes place elsewhere. A prediction of the climatic impact of heat extraction from the oceans thus demands a dynamic investigation like the one offered by the general circulation models discussed in section 2.3.2. The energy extraction would be simulated by an additional source term in (2.65), and attention should be paid to the energy loss from the ocean's surface by evaporation, which would decrease at the places with lowered temperatures. Extraction of energy from the oceans may for this reason lead to an increase of the total downward energy flux through the surface, as noted by Zener (1973). In addition to possible climatic effects, oceanic energy extraction may have an ecological effect, partly because of the temperature change and partly because of other changes introduced by the mixing processes associated with at least some energy conversion schemes (e.g., changing the distribution of nutrients in the water). Assuming a slight downward increase in the concentration of carbon compounds in seawater, it has also been suggested that artificial upwelling of deep seawater (used as cooling water in some of the conversion schemes discussed in Chapter 4) could increase the CO_2 transfer from ocean to atmosphere (Williams, 1975).

The Power in Ocean Thermal Gradients

In order to isolate the important parameters for assessing the potential for oceanic heat conversion, Fig. 3.63 gives a closer view of the temperature gradients in a number of cases with large and stable gradients. For locations at the Equator, seasonal changes are minute, owing to the small variations in solar input. For locations in the Florida Strait, the profiles are stable, except for the upper 50–100 m, owing to the Gulf Stream transport of warm water originating in tropical regions. The heat energy content relative to some reference point of temperature T_{ref} is $c_V(T - T_{ref})$, but only a fraction of this can be converted to mechanical work, and the direct use of water at temperatures of about 25 °C above ambient is limited, at least in the region near the Equator. Therefore, in aiming at production of mechanical work or electricity, the heat value must be multiplied by the thermodynamic efficiency (cf. Chapter 4), i.e., the maximum fraction of the heat at temperature T that can be converted to work. This efficiency, which would be obtained by a hypothetical conversion device operating in accordance with an ideal Carnot cycle, is (cf. section 4.1.2)

$$\eta_{Carnot} = (T - T_{ref})/T, \tag{3.32}$$

where the temperatures should be in K.

Taking the reference temperature (the "cold side") as $T_{ref} = 6\,°C$, corresponding to depths of 380 m (Gulf Stream, 10 km from coast), 680 m (Gulf Stream, 50 km from coast), 660 m (Atlantic Ocean), 630 m (Pacific Ocean), and 1000 m (Indian Ocean) (see Fig. 3.63), one obtains the maximum energy that can be extracted from each cubic meter of water at temperature T (as a function of depth), in the form of mechanical work or electricity,

$$\eta_{Carnot}\, c_V\, (T - T_{ref}).$$

This quantity is shown in Fig. 3.64 for the sites considered in the preceding figure. It is seen that, owing to quadratic dependence on the temperature difference, the work density is high only for the upper 100 m of the tropical oceans. Near the coast of Florida, the work density drops more slowly with depth, owing to the presence of the core of the warm Gulf Stream current (cf. Fig. 3.50). Below the current, the work density drops with a different slope, or rather

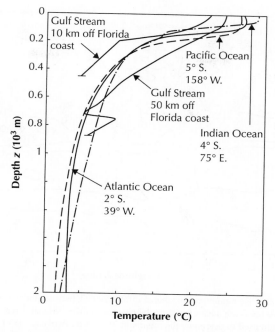

FIGURE 3.63 Temperature profiles for equatorial sites in the major oceans, as well as for the Gulf Stream at the Florida Strait. The profiles are derived from time-limited observations and may not represent annual averages. (Based on Neumann and Pierson, 1966; Sverdrup *et al.*, 1942.)

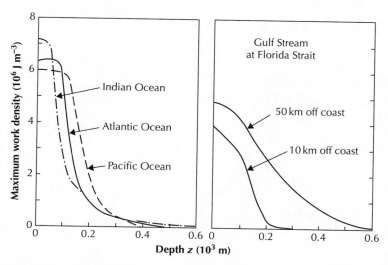

FIGURE 3.64 Maximum work density for the locations in Fig. 3.63, corresponding to an ideal Carnot machine extracting work from the temperature difference between the temperature at a given depth and a cold reservoir at constant temperature 6 °C (the depth at which such cooling water might be collected can be found in Fig. 3.63).

FIGURE 3.65   Average minimum temperature difference (°C) between the sea surface and a depth of 200 m. *Average minimum* means that a smoothed curve for the seasonal variations in temperature difference has been used to determine the minimum. (Based on Putnam, 1953.)

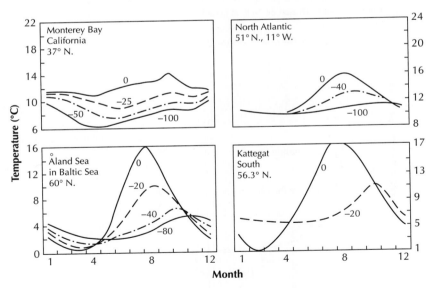

FIGURE 3.66   Trends of seasonal variations in water temperature at different depths (indicated in meters below the surface) for selected locations in open ocean and near shore. (Based on Sverdrup *et al.*, 1942; Neumann and Pierson, 1966; Danish Meteorological Institute, 1973.)

different slopes, because of the rather complex pattern of weaker currents and countercurrents of different temperatures.

The power that could be extracted is obtained by multiplying the work density by the transfer coefficient of the conversion device, that is, the volume of water from which a maximum energy prescribed by the Carnot value can be extracted per second. This quantity is also a function of T and T_{ref}, and losses relative to the ideal Carnot process are included in terms of a larger "effective" volume necessary for extracting a given amount of work. If the device were an ideal Carnot machine and the amount of water processed were determined by the natural flow through the device, then at a flow rate of 1 m s^{-1} (typical of the Gulf Stream current) the power extracted would be 3–4 MW m^{-2} facing the current flow direction, according to Fig. 3.64. In practice, as the treatment in Chapter 4 shows, only a fraction of this power will be obtained.

An indication of the global distribution of locations with large ocean temperature gradients may be obtained from Fig. 3.65, which gives the average temperature difference between the surface water and the water at a depth of 200 m for the time of the year when this difference is smallest. As mentioned above, seasonal variation is very small in the equatorial region, whereas the summer temperature difference is far greater than the winter temperature difference at mid- and high latitudes, as seen from Fig. 3.66.

The availability of currents that could promote flow through a temperature-gradient-utilizing device may be estimated from the material in section 3.3.1, but it should be borne in mind that the currents are warm only after passage through the equatorial region. It then follows from Fig. 3.42 that most currents on the Northern Hemisphere are warm, while most currents in the Southern Hemisphere originate in the Antarctic region and are cold.

Temperature Gradients in Upper Soil and Air

The temperature gradient due to absorption of solar radiation in continental soil or rock has a diurnal and a seasonal component, as discussed in section 2.3.2. Transport processes in soil or rock are primarily by conduction rather than by mass motion (ground water does move a little, but its speeds are negligible compared with ocean currents), which implies that the regions that are heated are much smaller (typically depths of up to 0–7 m in the diurnal cycle and up to 15 m in the seasonal cycle). The seasonal cycle may be practically absent for locations in the neighborhood of the Equator. An example of the yearly variation at fairly high latitude is given in Fig. 2.103. The picture of the diurnal cycle is very similar except for scale, as follows from the simple model given by (2.40)–(2.42). Therefore, the potential for extracting mechanical work or electricity from solar-derived soil or rock temperature gradients is very small. Direct use of the heat is usually excluded as well, because the temperatures in soil or rock are mostly similar to or slightly lower than that of the air. An exception may be dry rock, which can reach temperatures substantially above the ambient

FIGURE 3.67 Variations of air temperature for two heights over ground and two months (January and July), as a function of the hour of the day. Ten years of observations from the meteorological tower at Risø, Denmark (Petersen, 1974) have been used to form the average trends.

after some hours of strong solar exposure. Extraction of heat from soil or rock by means of the heat pump principle is possible and is discussed in the next chapter.

Temperature gradients in the atmosphere are rather small but of stable sign (except for the layer near the ground) up until the tropopause (see Fig. 2.29). At higher altitudes, the gradient changes sign a few times (Fig. 2.28). The temperature gradient over the first few hundred meters above ground is largely determined by the sign of the net total radiation flux, which exhibits a daily cycle (cf. Fig. 2.1) characterized by downward net flux during daylight and upward net flux at night. An example of the corresponding diurnal variation in temperature at heights of 7 and 123 m is shown in Fig. 3.67 for January and July at a location 56 °N.

The density of air also decreases with height (Fig. 2.28), and the heat capacity is small (e.g., compared with that of water), so the potential for energy conversion based on atmospheric heat is small per unit of volume. Energy production using the atmospheric temperature gradient would require the passage of very large quantities of air through an energy extraction device, and the cold air inlet would have to be placed at an elevation kilometers above the warm air inlet. However, heat pump use of air near ground level is possible, just as it is for solar energy stored in the upper soil.

3.4.2 Geothermal Flows and Stored Energy

Heat is created in some parts of the interior of the Earth: as a result of radioactive disintegration of atomic nuclei. In addition, the material of the Earth is in the process of cooling down from an initial high temperature or as a result of heat carried away after being released within the interior by condensation and possibly other physical and chemical processes.

Regions of Particularly High Heat Flow

Superimposed on a smoothly varying heat flow from the interior of the Earth toward the surface are several abnormal flow regions. The subsurface temperature exhibits variations associated with the magnitude of the heat flow and heat capacity—or, in general, the thermal properties—of the material. The presence of water (or steam) and other molten material is important for the transportation and concentration of heat. Until now, the vapor-dominated concentrations of geothermal energy have attracted the most attention as potential energy extraction sources. However, the occurrence of hot springs or underground steam reservoirs is limited to a very few locations. Reservoirs containing superheated water (brine) are probably more common, but are hard to detect from general geological data.

Some reservoirs attract special interest owing to chemical processes in brine–methane mixtures that release heat and increase the pressure, while the conductivity of such mixtures is low (so-called *geo-pressurized systems*, cf. Rowley, 1977). Dry rock accumulation of heat is probably the most common type of geothermal heat reservoir (i.e., storage with temperature above average), but it does not lend itself so directly to utilization because of the difficulties in establishing sufficiently large heat transfer surfaces. A number of high-temperature reservoirs are present in volcanic systems in the form of lava pools and magma chambers.

Whereas the smoothly varying average heat flow from the interior of the Earth can be considered a renewable energy resource (see below), the reservoirs of abnormal temperature are not necessarily large in relation to possible usage and do not get renewed at rates comparable to possible extraction rates. It is estimated that electric power generation by use of geothermal steam at the locations where such steam is available will be possible for only about 50 years. The total amount of geothermal heat stored in water or steam to a depth of 10 000 m is estimated to be 4×10^{21} J, of which some $(1-2) \times 10^{20}$ J is capable of generating steam of above 200 °C (World Energy Conference, 1974). The same source estimates the total amount of energy stored in dry rocks to a depth of 10 km as around 10^{27} J. Steam of temperatures above 200 °C represents an average power of 240×10^{9} W for a period of 50 years.

It follows from the above that most of the abnormal geothermal heat reservoirs must be considered nonrenewable resources on the same footing as fossil and fissile deposits. However, the average geothermal heat flow also corresponds to temperature gradients between the surface and reachable interior of the Earth, which can be used for energy extraction, at least if suitable transfer mechanisms can be established. These may be associated with water flow in regions of high permeability, so that water that has been cooled by an energy extraction device may regain the temperature of the surrounding layer in a relatively short time.

The Origin of Geothermal Heat

The radioactive elements mainly responsible for geothermal heat production at present are ^{235}U (decay rate 9.7×10^{-10} y^{-1}), ^{238}U (decay rate 1.5×10^{-10} y^{-1}), ^{232}Th (decay rate 5.0×10^{-11} y^{-1}), and ^{40}K (decay rate 5.3×10^{-10} y^{-1}). These isotopes are present in different concentrations in different geological formations. They are more abundant in the granite-containing continental shields than in the ocean floors. Most of the material containing the radioactive elements is concentrated in the upper part of the Earth's crust. In the lower half of the crust (total depth around 40 km), radiogenic heat production is believed to be fairly constant, at a value of about 2×10^{-7} Wm^{-3} (Pollack and Chapman, 1977). The rate of radiogenic heat production at the top of the continental crust is typically at least ten times higher, but it decreases with depth and reaches the lower crust value approximately halfway through the crust. The amount of radioactivity in the mantle (occupying the volume between the core and the crust) and in the core (the radius of which is about half the total radius) is poorly known, but models indicate that very little radioactive heat production takes place in the mantle or core.

From the radioactive decay of the above-mentioned isotopes, it can be deduced that the temperature of an average piece of rock near the surface of the crust must have decreased some 900 °C during the last 4.5×10^9 years (i.e., since the Sun entered the main burning sequence; Goguel, 1976). At present, radiogenic heat production accounts for an estimated 40% of the continental average heat flow at the surface. The rest, as well as most of the heat flow at the oceanic bottom, may then be due to cooling associated with expenditure of stored heat.

In order to assess the nature and origin of the heat stored in the interior of the Earth, the creation and evolution of the planet must be modeled. As discussed in section 2.1.1, the presence of heavy elements in the Earth's crust would be explained if the material forming the crust was originally produced during one or more supernovae outbursts. It is also consistent to assume that this material was formed over a prolonged period, with the last contribution occurring some 10^8 years before the condensation of the material forming the Earth (Schramm, 1974).

A plausible model of the formation of the planetary system assumes an initial nebula of dust and gases (including the supernova-produced heavy elements), the temperature of which would increase toward the center. The Sun might then be formed by gravitational collapse (cf. section 2.1.1). The matter not incorporated into the "protosun" would slowly cool, and parts of it would condense into definite chemical compounds. The planets would be formed by gravitational accretion of matter, at about the same time as, or slightly earlier than, the formation of the Sun (Reeves, 1975).

One hypothesis is that the temperature at a given distance from the proto-sun would be constant during the formation of planets like the Earth (*equilibrium condensation model*; see, for example, Lewis, 1974). As temperature declines, a sequence of condensation processes occurs: at 1600 K, oxides of calcium, aluminum, etc.; at 1300 K, nickel–iron alloys; at 1200 K, enstatite ($MgSiO_3$); at 1000 K, alkali-metal-containing minerals (feldspar, etc.); at 680 K, troilite (FeS). The remaining iron would be progressively oxidized, and, at about 275 K, water would be formed. The assumption that the Earth was formed at a constant temperature of about 600 K implies that it would have a composition characteristic of the condensation state at that temperature.

Other models can be formulated that would also be relatively consistent with the knowledge of the (initial) composition of the Earth's interior. If, for example, the planet formation process was slow compared to the cooling of the nebula, then different layers of the planet would have compositions characteristic of different temperatures and different stages of condensation.

It is also possible that the creation of the Sun and the organization of the primordial nebula did not follow the scheme outlined above, but were the results of more violent events, such as the close passage of a supernova star.

If the assumption of a constant temperature of about 600 K during the formation of the Earth is accepted, then the subsequent differentiation into crust, mantle, and core (each of the latter two having two subdivisions) is a matter of gravitational settling, provided that the interior was initially in the fluid phase or that it became molten as a result of the decay of radioactive isotopes (which, as mentioned, must have been much more abundant at that time, partly because of the exponential decay of the isotopes present today and partly owing to now-absent short-lived isotopes). The crust would have formed very quickly as a result of radiational cooling, but the presence of the crust would then have strongly suppressed further heat loss from the interior. The central region may have been colder than the mantle during the period of core formation by gravitational settling (Vollmer, 1977).

Gravitational settling itself is a process by which gravitational (potential) energy is transformed into other energy forms. It is believed that most of this energy emerged as heat, and only a smaller fraction as chemical energy or elastic energy. If the differentiation process took place over a short time interval, the release of 26×10^{30} J would have heated material of heat capacity similar to that of crustal rocks to a temperature of about 5000 K. If the differentiation was a prolonged process, then the accompanying heat would, to a substantial extent, have been radiated into space, and the temperature would never have been very high (Goguel, 1976). The present temperature increases from about 300 K at the surface to perhaps 4000 K in the center of the core. During the past, convective processes as well as formation of steam plumes in the mantle are likely to have been more intense than at present. The present heat flow between the mantle and the crust is about 25×10^{12} W, while the heat flow at the surface of the crust is about 30×10^{12} W (Chapman and Pollack, 1975; Pollack and Chapman, 1977).

If heat conduction and convective processes of all scales can be approximately described by the diffusion equation (2.28), supplemented with a heat source term describing the radioactive decay processes, in analogy to the general transport equation (2.44), then the equation describing the cooling of the Earth may be written

$$dT/dt = \mathrm{div}(K \, \mathrm{grad} \, T) + S/\rho, \qquad (3.33)$$

where K is the effective diffusion coefficient and S is the radiogenic heat production divided by the local heat capacity C. The rate of heat flow is given by

$$E^{sens} = \lambda \, \partial T / \partial r, \qquad (3.34)$$

where the thermal conductivity λ equals KC, according to (2.39). If the temperature distribution depends only on the radial co-ordinate r, and K is regarded as a constant, (3.33) reduces to

$$dT/dt = K(\partial^2 T/\partial r^2 + 2 \, r^{-1} \partial \, T/\partial r) + S/\rho. \qquad (3.35)$$

The simplified version is likely to be useful only for limited regions. The diffusion coefficient K is on average of the order of 10^{-6} m^2 s^{-1} in the crust and mantle, but increases in regions allowing transfer by convection. The thermal conductivity λ of a material depends on temperature and pressure. It averages about 3 W m^{-1} K^{-1} in crust and mantle, but may be much higher in the metallic regions of the core. The total amount of heat stored in the Earth (relative to the average surface temperature) may be estimated from $\int_{volume} \lambda \, K^{-1} \, (T - T_s) \, d\mathbf{x}$, which has been evaluated as 4×10^{30} J (World Energy Conference, 1974). Using the average heat outflow of 3×10^{13} W, the relative loss of stored energy presently amounts to 2.4×10^{-10} per year. This is the basis for treating the geothermal heat flow on the same footing as the truly renewable energy resources. Also, in the case of solar energy, the energy flow will not remain unchanged for periods of the order billions of years, but will slowly increase, as discussed in sections 2.1.1 and 2.4.2, until the Sun leaves the main sequence of stars (at which time the radiation will increase dramatically).

Distribution of the Smoothly Varying Part of the Heat Flow

The important quantities for evaluating the potential of a given region for utilization of the geothermal heat flow are the magnitude of the flow and its accessibility.

The temperature gradient and heat flow are usually large in young geological formations and, in particular, at the mid-oceanic ridges, which serve as spreading centers for the mass motion associated with continental drift. In the continental shields, the gradient and flow are much smaller, as illustrated in Fig. 3.68. Along sections from the mid-oceanic ridges to the continental shields,

FIGURE 3.68 Temperature as a function of depth for a young and an old geological formation, representing average extremes of temperature profiles due to the smoothly varying part of the geothermal heat flow. The young formation (*dashed line*) corresponds to a mid-ocean ridge, and the old formation (*solid line*) corresponds to a Precambrian continental shield. The kink in the solid line may be associated with the onset of convective processes in the mantle. (Based on a model calculation, adjusted to observations, by MacGregor and Basu, 1974.)

the gradient may gradually change from one to the other of the two types depicted in Fig. 3.68. In some regions the temperature gradients change much more abruptly (e.g., in going from the Fenno-Scandian Shield in Norway to the Danish Embayment; cf. Balling, 1976), implying in some cases a high heat flux for land areas (e.g., northwestern Mexico, cf. Smith, 1974).

Maps of the geographical distribution of heat flow, both at the surface of the crust and at the surface of the mantle, have been prepared by Chapman and Pollack (1975; Pollack and Chapman, 1977) by supplementing available data with estimates based on tectonic settling and, for oceanic regions, the age of the ocean floor. The results, shown in Figs. 3.69a and b, are contours of a representation in terms of spherical harmonic functions of latitude and longitude, $Y_{lm}(\phi, \lambda)$, of maximum degree $l = 12$. The advantage of this type of analysis is that oscillations of wavelength smaller than about 3000 km will be suppressed, so that the maps are presumably describing the smoothly varying average flow without perturbations from abnormal flow regions. A comparison between the calculation shown in Fig. 3.69a and one in which the model predictions have also been used in regions where measurements do exist, has convinced Chapman and Pollack that little change will result from future improvements in the data coverage. The map showing the mantle flow, Fig. 3.69b, is obtained by subtracting the contribution from the radiogenic heat production in the crust from the surface flow map. To do this, Pollack and Chapman have used a model for the continental regions in which the heat production decreases exponentially from its local value at the surface of the crust,

$$S = S_s \exp(-z/b),$$

where $b = 8.5$ km (if not measured), until it reaches the constant value $CS = 2.1 \times 10^{-7}$ W m^{-3} assumed to prevail throughout the lower crust. For oceanic regions, the difference between mantle and surface heat flows is estimated on the basis of the cooling of the oceanic crust (assumed to behave like a 6.5-km thick layer of basalt) with a fixed lower boundary temperature of 1200 °C. The time during which such cooling has taken place (the "age" of the sea floor) is determined from the present surface heat flow (Pollack and Chapman, 1977).

Mantle heat flow is very regular, with low continental values and increasing flow when approaching the oceanic ridges, particularly in the southern Pacific Ocean. The surface heat flow is considerably more irregular, in a manner determined by the composition of the crust at a given location, but the mean fluctuation around the average value of 5.9×10^{-2} W m^{-2} is only 22%. The surface heat flow has similar values if evaluated separately for continents and oceans

FIGURE 3.69a   Surface heat flow contours (in 10^{-3} W m^{-2}) for the smoothly varying part of the geothermal flux, calculated on the basis of available data supplemented with a theoretical model. From H. Pollack and D. Chapman (1977), *Earth and Planetary Science Letters* **34**, 174–184, copyright by Elsevier Scientific Publ. Co., Amsterdam.

FIGURE 3.69b   Heat flow contours at the top of the mantle (10^{-3} W m^{-2}), calculated from the surface fluxes shown in Fig. 3.69a by subtracting the contribution originating in the crust, for which a model has been constructed (see text). From H. Pollack and D. Chapman (1977), *Earth and Planetary Science Letters* **34**, 174–184, copyright by Elsevier Scientific Publ. Co., Amsterdam.

(5.3 and 6.2×10^{-2} W m^{-2}, respectively), while the contrast is larger for the upper mantle flow (2.8 and 5.7×10^{-2} W m^{-2}, respectively).

The usefulness of a given flow for purposes of energy extraction depends on the possibilities for establishing a heat transfer zone of sufficient flux. As mentioned earlier, the most appealing heat transfer method is by circulating water. In this case, the rate of extraction depends on the permeability of the geological material (defined as the fluid speed through the material, for a specified pressure gradient and fluid viscosity). The pressure is also of importance in determining whether hot water produced at a depth will rise to the top of a drill-hole unaided, or will have to be pumped to the surface. High permeability is likely to be found in sand-containing sedimentary deposits, while granite and gneiss rock have very low porosity and permeability. Schemes for hydraulic fractionation or granulation caused by detonation of explosives have been proposed, with the purpose of establishing a heat transfer area large enough that vast areas of hot rock formations could be accessed for extraction of geothermal energy, for example, by circulating water through man-made regions of porous material.

3.4.3 Ocean Thermal and Salinity Gradients

The preceding parts of this chapter are concerned with renewable energy flows and stores based on solar radiation, gravitation between celestial bodies, the mechanical energy involved in atmospheric and oceanic circulation, as well as a number of (sensible or latent) heat sources, the origin of which were solar or geological. It is believed that the most important of such sources are listed, but some significant sources may have been omitted.

In the description of heat sources, in particular, the list is far from complete, considering that any difference in temperature constitutes a potential source of energy. It may not be practical to attempt to extract energy from the difference in air or soil temperature between the Arctic and Equatorial regions, but the corresponding difference in water temperature is, in fact, the basis for the temperature gradients in some oceanic regions (discussed in section 3.4.2), owing to water transport by currents. One case in which a stable temperature difference might be usable in ways similar to those proposed for ocean temperature gradients is offered by the difference between the Arctic air temperature and the water temperature at the base of the ice sheets covering Arctic oceans. The average air temperature is about −25 °C for the period October–March, but the temperature at the ice–water interface is necessarily close to 0 °C (although not quite zero because of salinity- and pressure-induced depreciation of the freezing point, cf. Fig. 3.66). Other ocean thermal energy conversion (OTEC) proposals have focused on the temperature difference between warm currents, such as the Gulf Stream, and surrounding, colder bodies of water. The temperature difference is small and a thermodynamic engine working between such temperatures will have exceedingly low efficiency. In addition, tampering with currents like the Gulf Stream could have large environmental effects on the regions depending on heat delivery from warm currents (such as Northern Europe in the case of the Gulf Stream).

It is natural to ask whether other energy forms, such as electromagnetic, earthquake, chemical (of which bioenergy is treated in section 3.5), or nuclear energy, can be considered as energy resources and to what extent they are renewable. Examples of such energy forms are discussed in sections 3.4.4 and 3.4.5.

Useful chemical energy may be defined as energy that can be released through exothermic chemical reactions. In general, chemical energy is associated with chemical bonds between electrons, atoms, and molecules. The bonds may involve overlapping electron wavefunctions of two or more atoms, attraction between ionized atoms or molecules, and long-range electromagnetic fields created by the motion of charged particles (notably electrons). In all cases, the physical interaction involved is the Coulomb force. Examples of chemical energy connected with molecular bonds are given in section 3.5 (bioenergy sources, including fossil fuels).

The organization of atoms or molecules in regular lattice structures represents another manifestation of chemical bonds. Some substances possess different crystalline forms that may exist under given external conditions. In addition to the possibility of different solid phases, phase changes associated with transitions among solid, liquid, and gas phases all represent different levels of latent energy.

Salinity Differences

Solutions represent another form of chemical energy, relative to the pure solvent. The free energy of a substance with components $i = 1, 2, \ldots$, there being n_i mol of the ith component, may be written

$$G = \sum_i n_i \mu_i, \tag{3.36}$$

where μ_i is the *chemical potential* of component i. For a solution, μ_i can be expressed in the form (see, for example, Maron and Prutton, 1959)

$$\mu_i = \mu_i^0 + \mathscr{R} T \log (f_i x_i), \tag{3.37}$$

where $\mathscr{R}$ is the gas constant (8.3 J K^{-1} mol^{-1}), T is the temperature (K), and $x_i = n_i/(\Sigma_j n_j)$ the mole fraction. μ_i^0 is the chemical potential that would correspond to $x_i = 1$ at the given pressure P and temperature T, and f_i is the *activity coefficient*, an empirical constant that approaches unity for *ideal solutions*, an example of which is the solvent of a very dilute solution (whereas, in general, f_i cannot be expected to approach unity for the dissolved component of a dilute solution).

It follows from (3.36) and (3.37) that a solution represents a lower chemical energy than the pure solvent. The most common solution present in large amounts on the Earth is saline ocean water. Relative to this, pure or fresh water, such as river run-off, represents an elevated energy level. In addition, there are salinity differences within the oceans, as shown in Figs. 2.67–2.69.

Taking the average ocean salinity as about 33×10^{-3} (mass fraction), and regarding this entirely as ionized NaCl, $n_{Na+} = n_{Cl-}$ becomes about 0.56×10^3 mol and $n_{water} = 53.7 \times 10^3$ mol, considering a volume of 1 m³. The chemical potential of ocean water, μ, relative to that of fresh water, μ^0, is then from (3.37)

$$\mu - \mu^0 = \mathscr{R}\, T \log x_{water} \approx -2\, \mathscr{R}\, T\, n_{Na+}/n_{water}.$$

Consider now a membrane that is permeable for pure water but impermeable for salt (i.e., for Na⁺ and Cl⁻ ions) as indicated in Fig. 3.70. On one side of the membrane, there is pure (fresh) water, on the other side saline (ocean) water. Fresh water will flow through the membrane, trying to equalize the chemical potentials μ^0 and μ initially prevailing on each side. If the ocean can be considered as infinite and being rapidly mixed, then n_{Na+} will remain fixed, also in the vicinity of the membrane. In this case each m³ of fresh water penetrating the membrane and becoming mixed will release an amount of energy, which from (3.49) is

$$\delta G = \sum_i (n_i\, \delta\mu_i + \mu_i\, \delta n_i) \approx n_{water}(\mu^0 - \mu) \approx 2\mathscr{R}Tn_{Na+}. \tag{3.38}$$

FIGURE 3.70 Schematic picture of an osmotic pump. In order to mix the fresh water penetrating the semipermeable membrane toward the right and to maintain the salinity in the salt-water compartment, new saline water would have to be pumped into the salt-water compartment, and water motion near the membrane would have to be ensured.

For a temperature $T \approx 285$ K (considered fixed), $\delta G \approx 2.65 \times 10^6$ J. The power corresponding to a fresh-water flow of 1 m³ s⁻¹ is thus 2.65×10^6 W (cf. Norman, 1974). The worldwide run-off of about 4×10^{13} m³ y⁻¹ (Fig. 2.65) would thus correspond to an average power of around 3×10^{12} W.

The arrangement schematically shown in Fig. 3.70 is called an osmotic pump. The flow of pure water into the tube will ideally raise the water level in the tube, until the pressure of the water-head balances the force derived from the difference in chemical energy. The magnitude of this *osmotic pressure*, P^{osm}, relative to the atmospheric pressure P_0 on the fresh-water surface, is found from the thermodynamic relation

$$V dP - S dT = \sum_i n_i \, d\mu_i,$$

where V is the volume, S is the entropy, and T is the temperature. Assuming that the process will not change the temperature (i.e., considering the ocean a large reservoir of fixed temperature), insertion of (3.38) yields

$$P^{osm} = \delta P \approx n_{water} \, V^{-1} \delta \mu_{water} \approx 2 \, \mathcal{R} \, T \, n_{Na+} \, V^{-1}. \qquad (3.39)$$

Inserting the numerical values of the example above, $P^{osm} = 2.65 \times 10^6$ N m^{-2}, corresponding to a water-head some 250 m above the fresh-water surface. If the assumption of fixed mole fraction of salt in the tube is to be realized, it would presumably be necessary to pump saline water into the tube. The energy spent for pumping, however, would be mostly recoverable, since it also adds to the height of the water-head, which may be used to generate electricity as in a hydropower plant.

An alternative way of releasing the free energy difference between solutions and pure solvents is possible when the dissolved substance is ionized (the solution is then called electrolytic). In this case, direct conversion to electricity is possible, as further discussed in Chapter 4.

3.4.4 Nuclear Energy

Atomic nuclei (consisting of protons and neutrons) form the bulk of the mass of matter on Earth as well as in the known part of the present universe (cf. section 2.1.1). A nucleus is usually understood to be a bound system of Z protons and N neutrons (unbound systems, resonances, may be formed under laboratory conditions and they are observed in cosmic ray showers for short periods of time). Such a nucleus contains an amount of nuclear binding energy, given by

$$E_{Z,N} - (NM_n + ZM_p) c^2 = -B,$$

with the difference between the actual energy $E_{Z,N}$ of the bound system and the energy corresponding to the sum of the masses of the protons and neutrons if these were separated from each other. It thus costs energy to separate all the nucleons, and this is due to the attractive nature of most nuclear forces.

However, if the binding energy B (i.e., the above energy difference with opposite sign) is evaluated per nucleon, one obtains a maximum around ^{56}Fe (cf. section 2.1.1), with lower values for both lighter and heavier nuclei. Figure 3.71a shows the trends of $-B/A$, where $A = Z + N$ is the nucleon number. For each A, only the most tightly bound nucleus $(Z, A - Z)$ has been included, and only doubly even nuclei have been included (if Z or N is odd, the binding energy is about 1 MeV lower).

This implies that nuclear material away from the iron region could gain binding energy (and thereby release nuclear energy) if the protons and neutrons could be restructured to form ^{56}Fe. The reason why this does not happen spontaneously for all matter on Earth is that potential barriers separate the present states of nuclei from the state of the most stable nucleus and that very few nuclei are able to penetrate these barriers at the temperatures prevailing on Earth.

FIGURE 3.71a Trends of nuclear binding energy of bound nuclei (taken per nucleon) relative to a hypothetical situation in which all nucleons are separated (see text). (Based on Bohr and Mottelson, 1969.)

A few nuclei do spontaneously transform to more tightly bound systems (the natural radioactive nuclei mentioned in section 3.4.2), but the rate at which they penetrate the corresponding "barriers" is low, since otherwise they would no longer be present on Earth now, some 5×10^9 years after they were formed. As mentioned in section 3.4.2, these nuclei are responsible for 40% of the average heat flow at the surface of continents and contribute almost nothing to the heat flow at the ocean floors.

As schematically illustrated in Fig. 3.71b, the barrier that a nucleus of $A \approx 240$ must penetrate in order to fission is typically a very small fraction (2%–3%) of the energy released by the fission process. This barrier has to be penetrated by quantum tunneling. The width of the barrier depends on the initial state of the fissioning nucleus. It is smaller if the nucleus is in an excited state rather than in its ground state. Some heavy isotopes, such as ^{235}U, easily absorb neutrons, forming a compound system with a dramatically reduced fission barrier and hence with a dramatically increased fission probability. This process is called *induced fission*, and it implies that by adding a very small amount of energy to a "fissile" element like ^{235}U, by slow-neutron bombardment, a fission energy of some 200 MeV can be released, mostly in the form of kinetic energy of the products, but a few percent usually occurring as delayed radioactivity of unstable fragments. An example of binary fission, i.e., with two end-nuclei plus a number of excess neutrons (which can induce further fission reactions), is

$$^{235}_{92}U + n \rightarrow {}^{142}_{54}Xe + {}^{92}_{38}Sr + 2n.$$

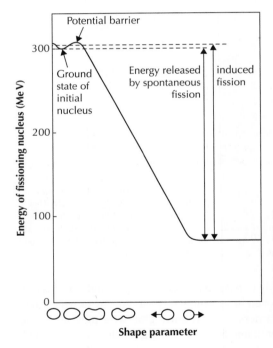

FIGURE 3.71*b*   Schematic view of a nuclear fission process described by a "one-dimensional" shape parameter (axially symmetric shapes are indicated along the lower boundary of the figure). The zero point of the energy scale on the ordinate corresponds to the energy of a (fractional) number of ^{56}Fe nuclei with mass identical to that of the fissioning nucleus.

The probability of finding an asymmetrical mass distribution of the fragments is often larger than the probability of symmetric fission. The reason why further fission processes yielding nuclei in the $^{56}_{26}$Fe region do not occur is the high barrier against fission for nuclei with $Z^2/A \lesssim 30$ and the low probability of direct fission of, say, uranium into three fragments closer to the iron region (plus a larger number of neutrons).

The amount of recoverable fissile material in the Earth's crust would correspond to a fission-energy of about 10^{22} J (Ion, 1975). However, heavy elements other than those fissioning under slow-neutron bombardment can be made fissile by bombardment with more energetic neutrons or some other suitable nuclear reaction, spending an amount of energy much smaller than that which may later be released by fission. An example is $^{238}_{92}$U, which may absorb a neutron and emit two electrons, thus forming $^{239}_{94}$Pu, which is a fissile element, capable of absorbing neutrons to form $^{240}_{94}$Pu, the fission cross-section of which is appreciable. Including resources by which fissile material may be "bred" in this way, a higher value may be attributed to the recoverable energy resource associated with nuclear fission. This value depends both on the availability of resource material, such as ^{238}U (the isotope ratio ^{238}U to ^{235}U is practically the same for all geological formations of terrestrial origin), and on the possibility of constructing devices with a high "breeding ratio." Some estimates indicate a 60-fold increase over the energy value of the "non-breeder" resource (World Energy Conference, 1974).

As indicated in Fig. 3.71b, the energy gain per nucleon that could be released by fusion of light elements is several times larger than that released by fission of heavy elements. A number of possible fusion processes are mentioned in section 2.1.1. These reactions take place in stars at high temperature. On Earth, they have been demonstrated in the form of explosive weapons, with the temperature and pressure requirements being provided by explosive fission reactions in a blanket around the elements to undergo fusion. Controlled energy production by fusion reactions such as ^{2}H + ^{3}H or ^{2}H + ^{2}H is under investigation, but at present the necessary temperatures and confinement requirements have not been met. Theoretically, the fusion processes of elements with mass below the iron region are, of course, highly exenergetic, and energy must be added only to ensure a sufficient collision frequency. In practice, it may take a while to realize devices for energy extraction for which the net energy gain is at all positive.

In theory, the nuclear fusion energy resources are much larger than the fission resources, but in any case only a fraction of the total nuclear energy on Earth (relative to an all-iron state) could ever become energy resources on a habitable planet. In principle, the nuclear energy resources are clearly nonrenewable, and the recoverable amounts of fissionable elements do not appear large compared with the possible requirements of man during the kind of time span for which he may hope to inhabit the planet Earth. However, it cannot be denied that the fusion resources might sustain man's energy expenditure for a long time, if, for example, the reaction chain entirely based on naturally occurring deuterium could be realized [cf. the cosmological reactions occurring some 300 s after the singularity in the Big Bang theory (section 2.1.1), at temperatures between 10^9 and 10^8 K]:

$$\,^2_1H + \,^2_1H \;\rightarrow\; \,^1_1H + \,^3_1H + 3.25\,\text{MeV},$$

$$\,^2_1H + \,^2_1H \;\rightarrow\; \,^1_0n + \,^3_2He + 4.0\,\text{MeV},$$

$$\,^2_1H + \,^3_1H \;\rightarrow\; \,^1_0n + \,^4_2He + 17.6\,\text{MeV},$$

$$\,^2_1H + \,^3_2He \;\rightarrow\; \,^1_1H + \,^4_2He + 18.3\,\text{MeV}.$$

The abundance of ^2H in seawater is about 34×10^{-6} (mass fraction). The potential nuclear energy released by one of the deuterium–to–helium fusion chains is thus 10^{13} J m^{-3} of seawater, or over 10^{31} J for all the oceans (an energy storage equivalent to the entire thermal energy stored in the interior of the Earth, cf. section 3.4.2).

The prime environmental concern about utilization of fission or fusion energy is the inherent change in the radioactive environment. The fragments formed by fission reactions cover a wide range of nuclear isotopes, most of which are unstable and emit nuclear radiation of gamma or particle type (see, for example, Holdren, 1974). Also, fusion devices are likely to produce large amounts of radioactivity, because the high-energy particles present in the reaction region may escape and experience frequent collisions with the materials forming the "walls," the confinement, and thereby induce nuclear reactions. The range of radioactive elements formed by fusion reactions can be partially optimized from an environmental point of view (minimizing the production of the most biologically hazardous isotopes) by choosing appropriate materials for the confinement, but the choice of materials is also limited by the temperature and stability requirements (see, for example, Post and Ribe, 1974).

Large-scale implementation of fission- or fusion-based energy conversion schemes will raise the question of whether it will be possible safely to manage and confine the radioactive "wastes" that, if released to the general environment, could cause acute accidents and long-range alterations in the radiological environment to which man is presently adapted. These dangers exist equally for the use of nuclear reactions in explosive weapons, in addition to the destructive effect of the explosion itself, and in their case no attempt is made to confine or to control the radioactive material.

In addition to possible health hazards, the use of nuclear energy may also have climatic consequences, such as those associated with enhanced heat flow (during nuclear war; cf. Feld, 1976) or with the routine emissions of radioactive material (e.g., ^{85}Kr from the uranium fuel cycle; cf. Boeck, 1976).

In summary, nuclear energy production based on existing once-through reactors constitutes a footnote in history, given the very limited amounts of fissile resources available for this mode of operation, comparable at most to those of oil. Nuclear fusion research has been ongoing for more than 50 years, so far with little success. Commercialization is still predicted to happen some 50 years into the future, just as it was at any earlier stage.

3.4.5 Atmospheric Electricity

The Earth is surrounded by a magnetic field, and from a height of about 80 km a substantial fraction of atmospheric gases is in an ionized state (cf. section 2.3.1). It is thus clear that electromagnetic forces play an important role in the upper atmosphere. Manifestations of electromagnetic energy in the lower parts of the atmosphere are well known in the form of lightning. Speculations on the possibility of extracting some of the electrical energy contained in a thunderstorm have appeared. An estimation of the amounts of energy involved requires knowledge of the electrical properties of the atmosphere.

In Fig. 3.72, calculated values of the electrical conductivity of the atmosphere are shown as a function of height for the lowest 200 km (Webb, 1976). The conductivity is low in the troposphere, increases through the stratosphere, and increases very sharply upon entering the ionosphere. Strong horizontal current systems are found in the region of high conductivity at a height of around 100 km (the "dynamo currents"). Vertical currents directed toward the ground are 6–7 orders of magnitude smaller for situations of fair weather, averaging about 10^{-12} A m^{-2} in the troposphere. Winds may play a role in generating the horizontal currents, but the energy input for the strong dynamo currents observed at mid-latitudes is believed to derive mainly from absorption of ultraviolet solar radiation by ozone (Webb, 1976; see also Fig. 2.20).

FIGURE 3.72 Electric conductivity of the atmosphere at noon on a clear day as a function of height (at about 10^5 m height, a substantial night-time reduction in λ takes place). (Based on Webb, 1976.)

The fair weather downward current of 10^{-12} A m^{-2} builds up a negative charge on the surface of the Earth, the total value of which on average is around 10^5 C. Locally, in the thunderstorm regions, the current is reversed and much more intense. Together, the currents constitute a closed circuit, the flow through which is 500–1000 A, but in such a way that the downward path is dispersed over most of the atmosphere, while the return current in the form of lightning is concentrated in a few regions (apart from being intermittent, causing time variations in the charge at the Earth's surface). The power P associated with the downward current may be found from

$$P = \int I^2 \, \lambda^{-1} \, dz,$$

where I ($\approx 10^{-12}$ A m^{-2}) is the average current and $\lambda(z)$ is the conductivity function (e.g., Fig. 3.80 at noon), or more simply from

$$P = IV = I\,Q/C,$$

where C is the capacitance of the Earth relative to infinity (V the corresponding potential difference) and Q ($\approx 10^5$ C) is its charge. The capacitance of a sphere with the Earth's radius r_s (6.4×10^6 m) is $C = 4\pi\,\varepsilon_0 r_s = 7 \times 10^{-4}$ F (ε_0 being the dielectric constant for vacuum), and thus the average power becomes $P \approx 6 \times 10^{-5}$ W m^{-2}.

The energy stored in the charge Q of the Earth's surface, relative to a situation in which the charges were all moved to infinity, is

$$W_\infty^{electric} = 1/2\,Q^2/C \approx 3 \times 10^{12} \text{ J}.$$

In "practice," the charges could only be moved up to the ionosphere (potential difference less than 10^6 V), reducing the value of the stored energy by almost two orders of magnitude. In any case, the charge of the Earth, as well as the tropospheric current system, constitutes energy sources of very small magnitude, even if the more concentrated return flux of lightning could be utilized (multiplying the average power estimate by the area of the Earth, one obtains a total power of 3×10^{10} W).

3.5 BIOLOGICAL CONVERSION AND STORES

The amount of solar energy stored in the biosphere in the form of standing crop biomass (plants, animals) is roughly 1.5×10^{22} J (see section 2.4.1), and the average rate of production is 1.33×10^{14} W or 0.26 W m^{-2}. The residing biomass per unit area is small in oceans and large in tropical forests. The average rate of production on the land surface is 7.6×10^{13} W or 0.51 W m^{-2} (Odum, 1972), and over 0.999 of the standing crop is found on land.

Still larger amounts of dead organic matter are present on the Earth in various degrees of fossilization (see Fig. 2.92, expressed in carbon mass units rather than in energy units; 1 kg carbon roughly corresponds to 41.8×10^6 J). Thus, fossil deposits (varieties of coal, oil, and natural gas) are estimated to be of the order of 6×10^{23} J. This may be an underestimate, in view of the floating distinction between low-concentration energy resources (peat, shale, tar sand, etc.) and nonspecific carbon-containing deposits. A maximum of about 1×10^{23} J is believed to be recoverable with technology that can be imagined at present (cf. section 2.4.2), suggesting that fossil energy resources are not very large compared with the annual biomass production. On the other hand, converting fresh biomass into useful energy requires more sophisticated methods than are generally required to convert the most pure and concentrated fossil fuel deposits. For bioenergy sources that have been in use for a long time, such as firewood and straw, an important requirement is often drying, which may be achieved by direct solar radiation.

It should not be forgotten that plants and derived forms of biomass serve man in essential ways other than as potential energy sources, namely, as food, as raw material for construction, and—in the case of green plants—as producers of atmospheric oxygen. The food aspect is partially an indirect energy usage, because man and other animals convert the energy stored in plants by metabolic processes that furnish energy required for their life processes. Biomass serving as food also acts as a source of mineral nutrients, vitamins, etc., which are required for reasons other than energy.

Assessment of the possible use of bioenergy for purposes other than those associated with life processes must take into account both biomass's food role and other tasks presently performed by vegetation and animal stock, such as prevention of soil erosion, conservation of diversity of species and balance of ecological systems. Only energy uses that can be implemented in harmony with (maybe even in combination with) these other requirements can be considered acceptable. Although plant systems convert only a small fraction of incident solar radiation, plans for energy usage should also consider the requirements necessary to avoid any adverse climatic impact. The management of extended areas of vegetation has the potential to interfere with climate, for example, as a result of the strong influence of vegetation on the water cycle (soil moisture, evaporation, etc.). Illustrations of such relations and actual climatic changes caused by removal of vegetation, overgrazing, etc., are given in section 2.4.2.

Before the question of plant productivity in different geographical regions and under different conditions is considered, the biochemistry of photosynthetic processes in green plants is briefly outlined.

3.5.1 Photosynthesis

Mechanism of Green Plant Photosynthesis

The cells of green plants contain a large number of chloroplasts in their cytoplasm. The interior of a chloroplast contains a liquid, the stroma, rich in dissolved proteins, in which floats a network of double membranes, the thylakoids. The thylakoid membranes form a closed space containing chlorophyll molecules in a range of slightly different forms, as well as specific proteins important for photosynthesis. In these membranes, photo-induced dissociation of water takes place,

$$2H_2O + \text{light} \rightarrow 4e^- + 4H^+ + O_2, \tag{3.40}$$

and the electrons are transferred, through a number of intermediaries, from the internal to the external side of the membrane. The transport system is depicted in Fig. 3.73, where the ordinate represents the redox potential [the difference in redox potential between two states roughly equals the amount of energy that is

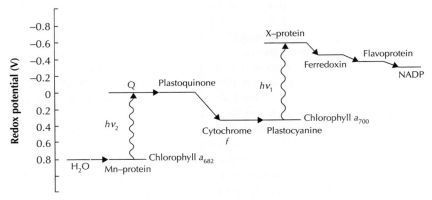

FIGURE 3.73 Model of the pathway of electron transfer in green plant photosynthesis. (Based on Trebst, 1974.)

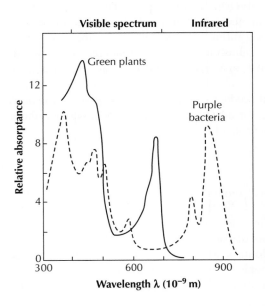

FIGURE 3.74 Spectrum of relative absorptance for a green plant and for a purple bacterium. (Based on Clayton, 1965.)

released or (if negative) that has to be added, in order to go from one state to the other].

The first step in the reaction is the absorption of solar radiation by a particular chlorophyll pigment (a_{680}). The process involves the catalytic action of an enzyme, identified as a manganese complex, that is supposed to trap water molecules and become ionized as a result of the solar energy absorption. Thereby, the molecules (of unknown structure) denoted Q in Fig. 3.73 become negatively charged, i.e., electrons have been transferred from the water–manganese complex to the Q-molecules of considerably more negative redox potential (Kok et al., 1975; Sørensen, 2005).

The photosensitive material involved in this step is called *photosystem II*. It contains the chlorophyll pigment a_{680} responsible for the electron transfer to the Q-molecules, where 680 is the approximate wavelength (in 10^{-9} m), for which the light absorptance peaks. It is believed that the initial absorption of solar radiation may be affected by any of the chlorophyll pigments present in the thylakoid and that a number of energy transfers take place before the particular pigment that will transport an electron from H_2O to Q receives the energy (Seliger and McElroy, 1965).

The gross absorptance as a function of wavelength is shown in Fig. 3.74 for green plants and, for comparison, a purple bacterium. The green plant spectrum exhibits a peak just above 400×10^{-9} m and another one a little below 700×10^{-9} m. The bacterial spectrum is quite different, with a pronounced peak in the infrared region around 900×10^{-9} m. The positions of the peaks are somewhat different for the different types of chlorophyll (a, b, c,...), and they also move if the chlorophyll is isolated from its cellular surroundings. For a given chlorophyll type, different pigments with different absorption peak positions result from incorporation into different chemical environments (notably, bonding to proteins).

Returning to Fig. 3.73, the electrons are transferred from the plastoquinone to cytochrome f by a redox reaction, which transfers a corresponding number of protons (H^+) across the thylakoid membrane. The electrons are further transferred to plastocyanine, which leaves them to be transported to another molecule of unknown structure, the X-protein. The energy required for this process is provided by a second photo-excitation step, in what is called *photosystem I*, by means of the chlorophyll pigment a_{700} (spectral peak approximately at 700×10^{-9} m).

The electrons are now delivered to the outside region, where they are picked up by ferredoxin, a small protein molecule situated at the outer side of the thylakoid membrane system. Ferredoxin may enter a number of different reactions; a very important one transfers the electrons via flavoprotein to nicotinamide–adenine dinucleotide phosphate (NADP), which reacts with protons to form $NADPH_2$, the basis for carbon dioxide assimilation. The protons formed by (3.40) and by the cytochrome f–plastoquinone reduction are both formed inside the membrane, but they penetrate the thylakoid membrane by a diffusion process, regulated by reaction (3.42) described below. The $NADPH_2$-forming reaction may then be written

$$2NADP + 4e^- + 4H^+ \rightarrow 2NADPH_2, \tag{3.41}$$

which describes the possible fate of electrons and protons formed by process (3.40), after transport by the sequence shown in Fig. 3.73 and proton diffusion through the membrane.

The transport of protons from outside the thylakoid membrane to its inside space, by means of the plastoquinone–cytochrome f redox cycle, creates an acidity (pH) gradient, which in turn furnishes the energy necessary for the phosphorylation of adenosine diphosphate (ADP) into adenosine triphosphate (ATP),

$$ADP^{3-} + H^+ + HPO_4^{2-} \rightarrow ATP^{4-} + H_2O. \tag{3.42}$$

This process, which involves the action of an enzyme, stores energy in the form of ATP (4.8×10^{-20} J per molecule), which may later be used to fuel energy-demanding processes, such as in the chloroplast stroma or in the cytoplasm outside the chloroplast (see, for example, Douce and Joyard, 1977). It is the analogue of the energy-stocking processes taking place in any aerobic plant or animal cells (i.e., cells using oxygen), as a result of the degradation of food (saccharides, lipids, and proteins), and associated with expenditure of oxygen (the respiratory chain, the Krebs cycle; see, for example, Volfin, 1971). The membrane system involved in the case of food metabolism is the mitochondrion.

By means of the energy stored in ATP and the high reducing potential of $NADPH_2$, the CO_2 assimilation process may take place in the stroma of the chloroplasts independently of the presence or absence of solar radiation. The carbon atoms are incorporated into glyceraldehyde 3-phosphate, which forms the basis for synthesis of glucose and starch. The gross formula for the process leading to the synthesis of glyceraldehyde 3-phosphate in the chloroplasts (the Benson-Bassham-Calvin cycle; cf. Douce and Joyard, 1977) is

$$3CO_2 + 9ATP^{4-} + 6NADPH_2 + 5H_2O \rightarrow$$
$$POCH_2CH(OH)CHO^{2-} + 9ADP^{3-} + 8HPO_4^{2-} + 6NADP + 9H^+. \tag{3.43}$$

Reactions (3.40) to (3.43) may be summarized,

$$3CO_2 + 2H_2O + HPO_4^{2-} + light \rightarrow POCH_2CH(OH)CHO^{2-} + 3O_2. \tag{3.44}$$

Going one step further and including the synthesis of glucose, the classical equation of photosynthesis is obtained,

$$6H_2O + 6CO_2 + 4.66 \times 10^{-18} J \rightarrow C_6H_{12}O_6 + 6O_2, \tag{3.45}$$

where 4.66×10^{-18} J is the net energy added by solar radiation.

Details of Photosynthetic Processes

Figure 3.75 shows the components of the photosystems residing along the thylakoid membranes. Insights into the structure of the membrane and component architecture have recently been obtained by atomic force microscopy (Bahatyrova *et al.*, 2004). The space inside the membranes is called the *lumen*, while on the outside there is a fluid, the *stroma*, which is rich in dissolved proteins. The whole assembly is in many cases enclosed within another membrane defining the chloroplasts, entities floating within the cytoplasm of living cells, while in some bacteria there is no outer barrier: they are called "cell-free."

FIGURE 3.75 Schematic picture of the components of the photosynthetic system in green plants and cyanobacteria. See text for explanation (Sørensen, 2005). (Compiled with use of Protein Data Bank IDs 1IZL (Kamiya and Shen, 2003), 1UM3 (Kurisu *et al.*, 2003), and 1JBO (Jordan *et al.*, 2001).)

Although some of the basic processes of photosynthesis have been inferred long ago (see Trebst, 1974), the molecular structures of the main components have only recently been identified. The reason is that the huge molecular systems have to be crystallized in order for spectroscopy, such as X-ray diffraction, to be used to identify atomic positions. Crystallization is achieved by a gelling agent, typically a lipid. Early attempts either damaged the structure of the system of interest, or used a gelling agent that showed up in the X-rays in a way difficult to separate from the atoms of interest. Very slow progress finally resulted in discovery of suitable agents and improvement of the resolution, from an initially more than 5 nm down to approximately 0.8 nm by 1998 and then recently to about 0.2 nm. The improved resolution allows most of the structure to be identified, although not the precise position (e.g., of hydrogen atoms). In the future, it is expected that the actual splitting of water may be followed in time-slice experiments under pulsed illumination.

Figure 3.75 places the main photosystems along the thylakoid membrane, shown schematically but with the three systems exhibiting extrusions on the luminal or stromal side according to experimental findings. The splitting of water by solar radiation takes place in photosystem II (PS II), with hydrogen emerging in the form of pqH_2, where pq is plastoquinone. Cytochrome b_6f (which resembles the mitochondrial cytochrome bc_1 in animals) transfers the energy in pqH_2 to plastocyanin (pc) and recycles pq to PS II. The plastocyanin migrates on the luminal side of the membrane to photosystem I (PS I), where more trapping of sunlight is needed to transfer the energy to the stromal side of the membrane, where it is picked up by ferredoxin (fd). Ferredoxin is the basis for the transformation in the stroma of NADP to $NADPH_2$, which is able to assimilate CO_2 by the Benson-Bassham-Calvin cycle (see (3.43)) and thereby form sugar-containing material, such as glucose and starch. However, a few organisms are capable of alternatively producing molecular hydrogen rather than $NADPH_2$. Selection of the hydrogen-producing process could possibly be managed by genetic engineering. In order to highlight the points where intervention may be possible, the individual processes are described in a little more detail.

The structure of PS II, which is similar for different plants and bacteria, helps to reveal the mechanisms of light trapping, water splitting, and transfer between sites. Figures 3.76a and b, based on a thermophilic cyanobacterium (*Thermosynechococcus vulcanus*) show the overall structure of PS II, seen from the side, as in Fig. 3.75, and from the top.[*] The structure is that of a dimer with two nearly identical parts. If one part is removed and pivoted to orient the "inner" surface (where the interface between the two parts was) directly at the viewer, the remaining monomer is as shown in Fig. 3.77, with a number of amino acid chains indicated as solid helices drawn over the molecular structure. In Fig. 3.78, the view is tilted in order to better display the reaction center details shown in Fig. 3.79. At the 0.37 nm resolution of the X-ray data used, 72 chlorophylls are identified in the PS II dimer. The zoomed monomer picture shows the four central chlorophyll rings (the experiment did not identify the side chains that allow the chlorophylls to attach to the membrane, cf. Fig. 3.80) and the unique manganese cluster of four atoms (there are another four in the other monomer). Manganese serves as a catalyst for the water splitting, as is discussed below. The many chlorophylls are part of an antenna system capable of absorbing solar radiation at the lower wavelength (higher energy) of the two peaks shown in Fig. 3.75 for green plants. The side chains shown in Fig. 3.80 and the long molecules of β-carotene probably also contribute to light absorption, with a subsequent transfer of energy from the initial chlorophyll molecules capturing radiation to the ones located in the reaction center region near the Mn atoms, as indicated in Fig. 3.79.

---

[*]The amino acid chains and helices in diagrams of molecular structures are displayed schematically (using the software package MSI (2000) for all diagrams of this kind).

(a)

FIGURE 3.76a,b Overall view of the structure of photosystem II (*a*: seen from side, with stroma at top and lumen below; *b*: seen from top, i.e., looking down from stroma perpendicular to the view in *a*). (Based on Protein Data Bank ID 1IZL; Kamiya and Shen, 2003.) Helices of amino acid chains are shown (cf. Fig. 3.77).

(b)

FIGURE 3.77 Protein arrangements in one monomer of photosystem II. A is the protein chain with the second quinone acceptor pq_B, B is a protein active in primary light harvesting, C and D are the subunits CP43 and D2, E and F are the cytochrome $b559$ α and β helices, K is the protein K of the reaction center, and V is cytochrome $b550$. The position of the four Mn^{2+} atoms of the reaction center is indicated. (Based on Protein Data Bank ID 1IZL; Kamiya and Shen, 2003.)

The present understanding of the working of PS II is that solar radiation is absorbed by the antenna system, including the large number of chlorophyll molecules, such as those of region B in Fig. 3.77, and possibly some of the other molecules, including, as mentioned, the β-carotenes. When absorbed, a light quantum of wavelength $\lambda = 680$ nm delivers an energy of

$$E = hc/\lambda = 2.9 \times 10^{-19}\, \text{J} = 1.8\, \text{eV}, \qquad (3.46)$$

where c is the vacuum speed of light (3×10^8 m s^{-1}) and h is Planck's constant (6.6×10^{-34} Js). The excited chlorophyll state is a collective state involving electrons in all parts of the central ring structure (not the central Mg atom in particular, as demonstrated by experiments substituting other elements in this position). The excited state may decay and re-emit the energy as a photon (fluorescence), but the probability of transferring the energy to a neighboring chlorophyll molecule is higher. In this way, the energy is handed from one chlorophyll molecule to another until it reaches one of the central chlorophyll-a molecules pointed out in Fig. 3.79b. The arrangement in this region favors a particular set of reactions, summarized in Fig. 3.81.

FIGURE 3.78 Left-hand side (as in Fig. 3.77) of photosystem II, but as seen from an angle of about 60° left rotation relative to that of Fig. 3.77 and from about 30° below the horizontal view of Fig. 3.77. To the left, amino acid chains are shown, while they are omitted on the right-hand side. The squares roughly indicate the part enlarged in Fig. 3.79. (Based on Protein Data Bank ID 1IZL; Kamiya and Shen, 2003.)

By transferring an electron to pheophytin (one of the two molecules shown in Fig. 3.79b), the donating chlorophyll-a is left in a positively charged state. The pheophytin passes on the electron to a plastoquinone, from which it is passed to a second plastoquinone believed to be better able to form pcH_2 and move around, after taking two protons from two of the processes shown in Fig. 3.81, and subsequently migrate from PS II to the cytochrome b_6f complex for the next step shown in Fig. 3.75. The plastoquinone structure is shown in Fig. 3.82, taken from a recent study of the cytochrome b_6f complex (Kurisu *et al.*, 2003); only the ring of one of the pq's is seen in the PS II X-ray studies of Kamiya and Shen (2003) or Zouni *et al.* (2001). Considerable attention has been paid to identification of the processes by which hydrogen ions are actually removed from water molecules, using 4 quanta of energy from the central chlorophyll absorbers. It has long been suspected that the Mn cluster serves to facilitate such processes, but first, energy has to get from the excited and positively charged chlorophyll molecule into the neighborhood of the Mn atoms. The Mn atoms have double positive charges and the cluster is believed to be held in place by weak forces tying them to some of the amino acids of the nearby A-chain (Fig. 3.79a). The chemical jargon is that the Mn atoms are being "co-ordinated" by the polypeptides of the $D1$ subunit. The amino acid most likely to donate an electron to the chl^+ ring is the ring structure of a tyrosine molecule identified as $tyr161$ (using the label numbering of the recent experiments quoted). The most recent X-ray experiment has a resolution of 0.35 nm, so there is considerable room for additional structures (and atoms) to be identified as resolution approaches typical atomic distances (around 0.1 nm).

(a)

FIGURE 3.79a   Reaction center of one-half of photosystem II, with amino acid chains shown in *a*, including two very close to the Mn cluster at the bottom middle. Amino acid chains are removed in the otherwise identical view *b* (*below*). H atoms are not shown. (Based on Protein Data Bank ID 1S5L; Ferreira *et al.*, 2004.)

(b)

FIGURE 3.79b   shows the same area as Fig. 3.79a, with amino acid backbones and side chains left out except tyrosines (labeled *tyr*), which have a special role in water splitting. The four chlorophyll-*a*'s (sometimes denoted chl_{D1}, $P_{680,D1}$, $P_{680,D2}$, and chl_{D2}) finally receiving the captured light energy are indicated (*chl*), as are the two pheophytin rings (*phe*) and a β-carotene that are part of the antenna system. The two middle-left chlorophyll rings and one pheophytin ring, together with the left tyrosine amino acid (sometimes denoted *tyr*161 or Y_Z), belong to subunit *D1*, the right-hand ones to *D2*. A non-heme Fe^{3+} atom is enlarged at the top, and the Mn cluster below has 4 O and 4 Mn atoms identified, with one Mn away from the center molecule. A Ca^{2+} atom is suggested as part of the cluster, and a molecule with the structure H_xCO_3 is nearby. Not all side chains of the ring molecules have been identified.

FIGURE 3.80 Molecular configurations of chlorophyll-*a* ($C_{55}H_{72}N_4O_5Mg$), pheophytin ($C_{55}H_{74}N_4O_5$), and tyrosine ($C_9H_{11}NO_3$). For clarity, the three-dimensional structure is not followed in detail. The surrounding environment would influence it. The tail components of the chlorophyll and pheophytin molecules serve to stabilize the placement of the molecules in the thylakoid membrane. Chlorophyll-*b*, which is found among the chlorophylls of the antenna system in some photosystems, differs from *chl a* by having one CH_3 group replaced by CHO.

In Fig. 3.81, all four chlorophyll molecules, having received solar energy, transfer an electron via pheophytin to plastoquinone-*A* and take one from a tyrosine molecule. In the first set of two reactions (top and bottom in Fig. 3.81), a hydrogen atom in a neighboring water molecule is made to transfer an electron to the tyrosine(-161), presumably under the influence of the Mn atoms. In the second set of reactions, the electron transfer from plastoquinone-*A* to an already negatively charged plastoquinone-*B* makes this doubly charged, which may also help to rip an electron from the second hydrogen atom in the water molecule. There are now two H^+ ions and a pq_B molecule with a charge of -2, which combine to form pqH_2. The subscript *B* is now dropped, as it is regarded as an indication of the site of reaction and therefore is irrelevant for a plastoquinone that can move along the membrane to the following process step (in the cytochrome b_6f complex). Finally, the two oxygen atoms left behind by the process will form an oxygen molecule, possibly again under the wings of the Mn catalyst. This understanding of water splitting is consistent with recent structural data, but the matter cannot yet be considered completely settled. Early theoretical explanations use sets of oxides MnO_x (with x = 2 or more) that change their charge or charge co-ordination state relative to the trapped water molecules, in consort with the one-by-one stripping and release of four protons (H^+) and four electrons (e^-) (Kok *et al.*, 1970; Hoganson and Babcock, 1997).

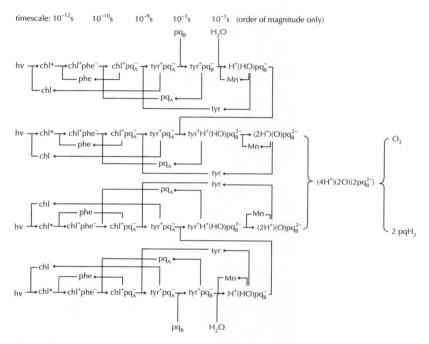

FIGURE 3.81 Photosystem II reaction center conversions of the energy furnished by 4 light quanta ($h\nu$) originally captured in the antenna system, under the catalyzing influence of the cluster of four Mn^{2+} atoms. Order of magnitude reaction times are listed at the top (Barber, 2002). The notation used is dictated by chemical balancing and does not in itself imply nearness of any pair of reagents, although some reaction mechanisms would require such physical nearness.

The scheme in Fig. 3.81 also allows an interpretation of where the plastoquinones at sites *A* and *B* are on equal footing: one is formed by the first reaction line in Fig. 3.81. If the second reaction involves the same pheophytin as the first, it will follow the reaction scheme in line 2 of Fig. 3.81; otherwise, it will follow the scheme in line 4. The remaining two steps are now fixed, provided that the plastoquinones have little probability for harboring more than two negative charges. With the positions of most atoms determined, it should be possible to perform more realistic theoretical calculations of the reaction processes in the near future and thereby verify the status of the proposed reaction paths.

Plastoquinone

Heme
(protoporphyrin
with Fe center)

FIGURE 3.82 Structure of plastoquinone ($C_{53}H_{80}O_2$) and a heme ($C_{34}H_{32}N_4O_4Fe$) as they appear in cytochrome b_6f.

Once the pqH_2 reaches cytochrome b_6f and gets close to a protoporphyrin heme molecule, the two protons are released and plastoquinone is recycled to the PS II. Figure 3.83 shows the overall structure of the cyanobacterium *Mastigocladus laminosus* cytochrome b_6f (Kurisu *et al.*, 2003), which, together with the structure derived from the alga *Chlamydomonas reinhardtii* (Stroebel *et al.*, 2003), is the most recent element in completion of the photosystem structure. Similar functions are performed in animal mitochondria by cytochrome bc_1 (Iwata *et al.*, 1998) and in purple bacteria by cytochrome c_2 (Camara-Artigas *et al.*, 2001). Figure 3.84 gives details of the b_6f components, including three iron-containing hemes in cytochrome b_6 and one in cytochrome f, plus the inorganic Fe_2S_2 molecule serving as an intermediary between the two cytochrome components.

Although as judged from Fig. 3.84 and the blow-up in Fig. 3.85, the Fe_2S_2 molecule seems too far away to be important, in fact the Rieske protein on which it sits can be turned around the hinge connecting it to the membrane part of the structure and thereby bring the Fe_2S_2 molecule close to the hemes b_L (*L* for luminal side) and f. The two hemes b_H and b_S closer to the stromal side are adjacent to a plastoquinone molecule (*PL9* in Fig. 3.85) and are thought to accomplish the recycling of pq after the hydrogen ions (protons) have been separated.

FIGURE 3.83   Overall structure of the dimer cytochrome b_6f, as measured by X-ray scattering for the thermophilic cyanobacterium *Mastigocladus laminosus*. Its position relative to the thylakoid membrane is indicated in Fig. 3.75. Resolution: 0.3 nm. (Based on Protein Data Bank ID 1UM3; Kurisu *et al.*, 2003.)

The path via heme f is taken by electrons on their way to the large plastocyanin (*pc*) molecule occupying a cavity on the luminal side before it moves with the extra electrons to PS I. The rather large *pc* molecule shown in Fig. 3.86 is from the blue-green alga (another word for cyanobacterium) *Anabaena variabilis* (Badsberg *et al.*, 1996). Smaller units with only one central Cu atom have been identified, for example, in the plant *Arabidopsis*, where the role of *pc* as an electron donor to PS I has been established by measurements indicating a large negative surface electrostatic charge (Molina-Heredia *et al.*, 2003).

Figure 3.85 also shows a few other molecules identified: TDS ($C_{25}H_{38}O_5$) and OPC ($C_{45}H_{87}NO_8P$), as well as more distant molecules of chlorophyll a and β-carotene. Amino acids from the protein chains (indicated in Fig. 3.85a) help to keep the heme molecules in place and also may facilitate the electron transfer. Transfer of electrons to the luminal side *pc* with corresponding positively charged hydrogen ions implies the creation of a proton gradient across

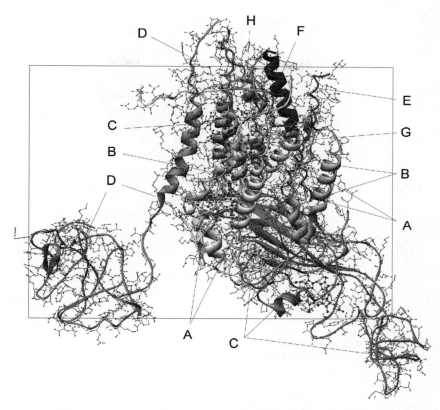

FIGURE 3.84 Shown is one-half of the cytochrome $b_6 f$ complex, rotated about 40° horizontally and moved about 20° upward relative to the view in Fig. 3.83. The structure at the lower right (C) is cytochrome f and the structure at the lower left (D) is called the Rieske protein. The upper structure (A) is cytochrome b_6. The other chains identified are a subunit (B) and four proteins (E–H). The area within the rectangle is enlarged in Fig. 3.85. (Based on Protein Data Bank ID 1UM3; Kurisu *et al.*, 2003.)

the membrane. This is where most of the energy transferred from PS II goes, and this energy may be stored for later use by the organism by the process

$$\text{ADP}^{3-} + \text{H}^+ + \text{HPO}_4^{2-} \rightarrow ATP^{4-} + \text{H}_2\text{O}, \qquad (3.47)$$

where ADP and ATP are adenosine di- and triphosphate (structure shown in Fig. 3.87). ATP is formed on the stromal side of the membrane, implying a transport of protons across the membrane.

As the next step in photosynthesis, PS I, shown in Fig. 3.88, receives the negatively charged plastocyanin $(pc)^{-1}$ from cytochrome f, but has to spend energy in transferring electrons across to the stromal side, where they are picked up by ferredoxin for use either in molecular hydrogen formation or for the CO_2-assimilating processes of the organism. In order to provide the energy required, PS I contains an antenna system of chlorophyll molecules capable of capturing

(a)

FIGURE 3.85a,b (without chains) Molecules relevant for the reaction path in cytochrome b_6f. See text for explanation. (Protein Data Bank ID 1UM3; Kurisu *et al.*, 2003.)

(b)

FIGURE 3.86 Structure of plastocyanin from the blue-green alga *Anabaena variabilis*, with amino acid chains and central Cu atoms indicated. (Based on Protein Data Bank ID 1NIN; Badsberg *et al.*, 1996.)

FIGURE 3.87 Molecular structure of $(H_3)ADP = (C_{10}H_{15}N_5O_{10}P_2)$ and $(H_4)ATP = (C_{10}H_{16}N_5O_{13}P_3)$.

further solar radiation, albeit at a slightly lower energy than that characterizing the central capture in PS II, corresponding now to wavelengths of about 700 nm.

Electrons are brought into PS I from the luminal side, either by plastocyanine *pc* or, in some cyanobacteria, by cytochrome c_6. It has been suggested that plants developed the route with *pc* and its Cu core at a time in their evolution

FIGURE 3.88   Structure of one photosystem I monomer (a total of three are at approximately 120° spacing) from the thermophilic cyanobacterium *Synechococcus elongatus*. A and B are two chlorophyll antennae, and C is the docking site for ferredoxin, harboring two of the three iron-sulfur centers (cf. Fig. 3.89, seen from an angle 40° to the right). Letters F to M are membrane-spanning protein subsystems, while D, E, and partly X extend upward into the stromal side. Resolution: 0.25 nm. (Based on Protein Data Bank ID 1JBO; Jordan *et al.*, 2001.)

when Fe was less available in the areas of plant growth (Rosa *et al.*, 2002). Referring to Fig. 3.88, a cavity is seen on the luminal (lower) side, associated with two horizontal helical pieces of subunit *F*. This cavity, quite close to the chlorophylls *chl-A*1 and *chl-B*1 at the beginning of the electron transfer chain (Fig. 3.90), is supposed to be the docking site of *cyt-b$_6$* or *pc* (Fromme *et al.*, 2003). The somewhat surprising presence of a Ca^{2+} atom on the luminal side (Fig. 3.89*a*) may have something to do with co-ordination of the three monomers that make up the total PS I (Fromme *et al.*, 2001). The three antenna systems, each with two branches (A and B in Fig. 3.88), consist of a total of 270 chlorophylls (mostly type *a*, cf. Fig. 3.80), to which come the 3 × 6 central ones (Fig. 3.90). This very large number (also compared with the 72 *chl*'s in PS II) is dictated by the fact that PS I must collect solar energy for all the requirements of the organism, whether plant or bacterium.

The energy captured from solar energy is transferred from *chl* to *chl* in the antennas A and B, until it reaches the central pair [*chl-A*1, *chl-B*1] (Fig. 3.90). These two chlorophylls (usually denoted P700) are slightly different in shape and it is not known if both participate in the same manner. An electron is quickly (about 1 ps) passed along one of the vertical paths to *chl-A*3 or *chl-B*3. It is not known if it passes through *chl-B*2 or *chl-A*2, or if these are parts of the primary collection site P700. In a following rather fast step (some 30 ps), the electron is passed to either *phy-A* or *phy-B*, from where somewhat slower processes (20–500 ns) carry the electron successively to the three Fe_4S_4 molecules *FeS-X*, *FeS-A*, and *FeS-B*. The surrounding amino acid chains play an important role in these transfers (Chitnis, 2001; Brettel and Leibl, 2001). Although charge recombination can occur, the primary electron donor is normally left in a positively charged state $P700^+$ throughout the successive electron transfers to *FeS-B*. The $P700^+$ (presumably *chl-A*1 or *chl-B*1) is brought back to its neutral state by the electron brought into PS I by *pc* (or *cyt-c_6*) and is then ready for handling the next electron transfer from lumen to stroma.

Unlike PS II, where four light quanta were required in a co-ordinated fashion, due to the energetic requirements for water splitting, PS I may proceed with just a single light quantum of 700 nm or 2.8×10^{-19} J (1.7 eV). However, this observation does not invite any simple explanation for the dual structure of the central energy transfer chains A and B starting from P700. It could perhaps be that a redundancy was built in during evolution and its use later abandoned.

While the protein chain C surrounding the iron-sulfur clusters (Fig. 3.89) is similar in structure to ferredoxin, it needs a real ferredoxin protein docked on the stromal side to transfer the electron from *FeS-B* away from PS I. The subsystems C, D, and E on the stromal side of PS I (Fig. 3.88) allow docking not only of the fairly small ferredoxin system (an example of which is shown in Fig. 3.91—there are variations in structure between species, such as plants and cyanobacteria), but also of the larger flavodoxin protein (Fromme *et al.*, 2003), an example of which is shown in Fig. 3.92.

After being brought to the stromal side by the succession of photosystems, a number of electrons are carried by the proteins ferredoxin and flavodoxin. Furthermore, protons (positively charged hydrogen ions) have been separated and moved across the thylakoid membrane (thereby creating an H^+ gradient from lumen to stroma). This makes room for two types of recombination processes on the stromal side. One is the "normal" reaction in plants and most bacteria:

$$NADP + 2e^- + 2H^+ \rightarrow NADPH_2, \tag{3.48}$$

where $NADPH_2$ combines with ATP^{4-} from (3.47) and water to assimilate CO_2 from the atmosphere for the synthesis of sugars and other molecules needed for the organism in question (see Sørensen, 2005). The second possibility is to induce (e.g., by genetic manipulation) the plant to produce and export molecular hydrogen by some process equivalent to

$$2e^- + 2H^+ \rightarrow H_2, \tag{3.49}$$

(a)

FIGURE 3.89a,b Photosystem I side (*a*) and top (*b*) views with protein chains omitted except for chain C on the stromal side near the three Fe_4S_4 clusters. A Ca atom is indicated on the luminal side. (Protein Data Bank ID 1JBO; Jordan *et al.*, 2001.)

(b)

FIGURE 3.90 Detail of photosystem I, seen from the side, with the division between units A and B (see Fig. 3.88) in the middle. Chlorophyll molecules (*chl*) other than the six central ones involved in transfer of energy to the three Fe_4S_4 clusters (FeS) are omitted. Chlorophylls, phylloquinones (*phy*, also known as vitamin K), and iron-sulfur clusters are further indexed with the subunit label A, B, or X (cf. Fig. 3.88), and the chlorophylls are shown with an index number 1 to 3. Other identified molecules include LHG ($C_{38}H_{75}O_{10}P$), LMG ($C_{45}H_{86}O_{10}$), and a number of β-carotenes (BCR). The scattered dots are oxygen atoms of water molecules (no hydrogens are shown). (Based on Protein Data Bank ID 1JBO; Jordan *et al.*, 2001.)

FIGURE 3.91 Ferredoxin from the cyanobacterium *Anabaena* PCC 7119. Each branch has an Fe_2S_2 cluster. Proteins are shown schematically as ribbons, and side chains are shown as atoms. Surrounding dots are oxygen atoms from water molecules (hydrogens not shown). (Based on Protein Data Bank ID 1CZP; Morales *et al.*, 1999.)

FIGURE 3.92   Flavodoxin from *Anabaena* PCC 7119. (Based on Protein Data Bank ID 1OBO; Lostao *et al.*, 2003.)

where, just as in the similar inorganic case, an "agent" (in the inorganic case, a catalyst) is needed for promoting the reaction. In the organic case, the agent may be hydrogenase.

Efficiency of Conversion

The basis for energy utilization has traditionally been the undifferentiated biomass produced. Other possibilities would include the direct dissociation of water by the action of sunlight (photolysis), or the formation of hydrogen rather than $NADPH_2$, after the initial oxygen formation (3.40) and electron transport to ferredoxin (Benemann and Weare, 1974; Mitsui and Kumazawa, 1977),

$$4e^- + 4H^+ \rightarrow 2H_2. \tag{3.50}$$

In both cases the problem is to control the recombination process,

$$2H_2 + O_2 \rightarrow 2H_2O + 9.47 \times 10^{-19} \, J, \tag{3.51}$$

so that the energy-releasing process takes place where desired and not immediately on the formation of hydrogen. Plants accomplish this by means of the thylakoid membrane and later by the chloroplast external membrane. Man-made processes may attempt to copy the membrane principle in various ways (Broda, 1975; Calvin, 1974, 1977; cf. Chapter 4), but it is not clear that present suggestions will allow separation of hydrogen and oxygen on a large scale. It may be advantageous to use green plants to perform the first step (3.40) and to transport electrons and protons (hydrogen ions) through their membranes; to prevent the energy from being too deeply trapped in organic matter, a strong reducing agent could be added to ferredoxin, as well as suitable enzymes, in order to accomplish the reaction (3.50) at this stage.

The maximum theoretical efficiency of this process is the ratio between the heat release 9.47×10^{-19} J in (3.51) and the solar energy input required. The latter depends on the absorption properties of the plant and its chlorophyll molecules (cf. Fig. 3.74), as well as on the number of light quanta required for each molecular transformation (3.40). The energy, E, of each light quantum of wavelength λ may be found from

$$\lambda E_\lambda = hc = 1.986 \times 10^{-25} \, \text{Jm}.$$

The minimum number of quanta required to transport one electron, as depicted in Fig. 3.73, is 2, one of wavelength 680×10^{-9} m and the other of 700×10^{-9} m (although these need not be the quanta originally absorbed). Since (3.40) requires the transport of four electrons, the minimum requirement would be 8 quanta with a total energy of 2.3×10^{-18} J. Experimental estimates of the number of quanta needed typically give values between 8 and 10.

The efficiency of the photosynthetic process containing only steps (3.40) and (3.50) may be written

$$\eta' = \eta_\lambda \, \eta_{geom} \, \eta_{chem}, \tag{3.52}$$

where η_λ is the fraction of the frequencies in the solar spectrum (depending on cloud cover, etc.) that is useful in the photosynthetic process, η_{geom} is the geometrical efficiency of passing the incoming radiation to the chlorophyll sites (depending on penetration depth in leaves, on reflectance from outer and inner surfaces, and on absorption by other constituents of the leaves), and η_{chem} is the efficiency of the photochemical reactions, the maximum value of which is given by

$$\eta_{chem} \leq 9.47 \times 10^{-19}/2.30 \times 10^{-18} = 0.41.$$

This efficiency pertains to the amount of internal heat produced by (3.51). Only a part of it can be converted into useful work. This part is obtained in (3.51) by replacing the enthalpy change 9.47×10^{-19} J with the change in free energy, $\Delta G = 7.87 \times 10^{-19}$ J (cf. section 4.1.1). This way, the efficiency of the photochemical reaction becomes

$$\eta_{chem,free} \leq 0.34.$$

The efficiency η_λ associated with the chlorophyll absorption spectrum typically lies in the range 0.4–0.5 (Berezin and Varfolomeev, 1976), and the geometrical efficiency η_{geom} may be around 0.8 (for the leaf of a green plant, not including the reduction associated with the penetration of radiation through other vegetation, e.g., in a forest environment).

The overall maximum efficiency of about $\eta' \approx 0.14$ found here for a hypothetical hydrogen (H_2)-producing system is valid also for actual green plants that assimilate CO_2. Over extended periods of time, the biomass production efficiency will have to incorporate still another factor, η_{resp}, expressing the respiration losses associated with the life cycle of the plant,

$$\eta = \eta' \eta_{resp}. \tag{3.53}$$

The respirative energy losses emerge as heat and evaporated water, at rates depending on temperature and wind speed (Gates, 1968). The value of η_{resp} is 0.4–0.5 for land plants and somewhat larger for aquatic plants and algae.

Actual plants may get close to the theoretical maximum efficiency, although the average plant does not. For the blue-green alga *Anacystis nidulans*, Goedheer and Hammans (1975) report an energy efficiency of $\eta_{chem} \approx 0.30$, based on 36 h of growth, including 6 h of irradiation, with a generous supply of nitrogen and other nutrients, as well as a CO_2-enriched atmosphere (i.e., 73% of the maximum energy efficiency calculated above).

It is estimated that, on average, each CO_2 molecule in the atmosphere becomes assimilated in a plant once every 200 years and that each O_2 molecule in the atmosphere is "renewed" through a plant once every 2000 years (Seliger and McElroy, 1965).

Bacterial Photosynthesis

Several bacteria use solar radiation to dissociate a compound of the general form H_2X, with a net reaction scheme of the form

$$CO_2 + 2H_2X + \text{light} \rightarrow (CH_2O) + H_2O + 2X \tag{3.54}$$

(Van Niel, 1941). Here (CH_2O) should be understood not as free formaldehyde, but as part of a general carbohydrate compound in analogy to the more precise equation (3.44). Actually, (3.54) was proposed to be valid for both green plant and bacterial photosynthesis, but there is no detailed analogy, since bacterial photosynthesis has been found to take place in a single step, resembling PS I of the green plant two-step process. Other photo-induced reactions do take place in bacteria, connected with ATP formation, which in this case is not a side-product of the primary photosynthetic process.

The compound H_2X may be H_2S (sulfur bacteria), ethanol C_2H_5OH (fermentation bacteria), etc. Most photosynthetic bacteria are capable of absorbing light in the infrared region (wavelength $800–1000 \times 10^{-9}$ m). The role that the NADP–NADPH$_2$ cycle (3.41) and (3.43) plays for green plant photosynthesis is played by NAD (nicotinamide-adenine dinucleotide)–NADH$_2$ for photosynthetic bacteria. The redox potential of NADH$_2$ is more or less the same as that of the initial compounds, e.g., H_2S, so practically no energy is stored in the process of bacterial photosynthesis (Hind and Olson, 1968).

3.5.2 Productivity in Different Environments

Ecological Systems

The gross primary production of a plant or, in general, of an ecological system is the rate at which solar energy is assimilated, i.e., the total amount of energy produced by photosynthesis. The net primary production, on the other hand, is the difference between gross primary production and respiration. Respiration involves an increase in redox potential (oxidation), either by consumption of oxygen [aerobic respiration, the net result being equivalent to (3.51)] or by the action of some other agent of oxidation (anaerobic respiration; if the oxidant is an organic compound, the respiration process is called fermentation).

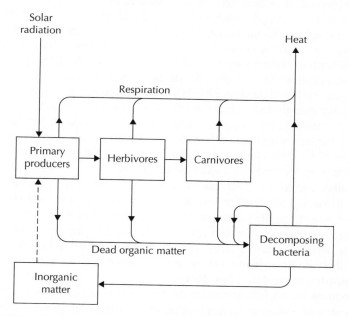

FIGURE 3.93 Model of an ecological system exhibiting flows of energy (*solid lines*, which in most cases also indicate the flow of organic matter) and of inorganic matter (*dashed lines*, which includes CO_2, nutrients, etc.). Exchange of heat between compartments and their environment is not shown, neither is the possible presence of photosynthetic bacteria in the system. Some of the flows may be "delayed," e.g., man's harvesting of food crops and wood, which may go into storage rather than being consumed immediately (i.e., transferred to the next compartment along the chain, herbivores in this example).

Primary producers are part of an ecological system. Figure 3.93 gives a schematic view of the energy and matter flow through such a system. The primary chain comprises the primary producers capable of carrying out photosynthesis, plant-eating organisms (herbivores), and a chain of successive carnivorous predators (carnivores), some of which may also eat plants (as man does). Each compartment in the chain is called a *trophic level*, and photosynthetic organisms are called *autotrophs*, while the consuming organisms further along the chain are called heterotrophs. Over short periods of time, withering and death of autotrophs can be neglected, and the net primary production available to the heterotrophic part of the community equals the gross primary production less the respiration of the autotrophs. Over longer periods, respiration, predation, and death must be considered in order to describe the amounts of biomass in each compartment (*standing crop*). Generally, biomass diminishes along the chain, but owing to the different average lifetimes, which are often longest for the highest trophic levels of the food chain, the biomass at a given time may be maximal for, say, the second or third member of the chain. Also, the seasonal dependence of primary production in most geographical regions, in conjunction with the short lifetimes of the autotrophs, leads to biomass distribution with a much increased long-range stability in the higher levels. The stability depends on the age of the ecosystem. A young system has fewer species and is more sensitive to external disturbances (climate variations, immigration of new predator species, etc.), which may destroy the food basis for the higher trophic levels. Old ecosystems are characterized by higher diversity and, therefore, are better equipped to circumvent sudden changes in external conditions, at least for a while.

Dead organic matter may precipitate toward the sea floor or detrital layer of the land surface, where an environment suitable for sustaining a number of decomposer bacteria evolves, so that nutrients are returned to the inorganic form necessary for renewed uptake by autotrophs.

It follows from Fig. 3.93 that the flows of energy that could be used for energy extraction serving human society diminish along the food chain. This does not negate the possibility that the concentration of energy flow may be higher and easier to utilize at a level higher than the primary producers. For example, livestock leave a large fraction of their respiration heat and manure within their sheds, a spatial region much smaller than the size of a grazing range that might yield the same energy value.

Important factors in determining the net productivity of a given autotrophic system are solar radiation, temperature regime, water availability, climate in general, and access to carbon dioxide and nutrients, as well as the nature of the entire ecosystem, density and structure of plants in the neighborhood, and predation and harvesting by heterotrophs (including man). Growth in biomass would be exponential if optimal conditions for photosynthesis could be maintained indefinitely. Actual growth is stopped when one or more of the conditions can no longer be met; factors responsible for halting the exponential growth are called *limiting factors*. A mature ecosystem may reach a stable situation in which the net production (and hence growth) of the community as a whole is zero.

Limiting Factors

The effective area of a plant that is exposed to solar radiation depends on shadow effects caused by other vegetation and on the shape and arrangements of leaves, or, more generally, of light-absorbing pigments. This, together with the loss caused by scattering and absorption on the leaf surface or other parts of the plant not involved in light accumulation, accounts for the value of the maximum geometrical efficiency in (3.52)—around 0.8. The effective area of fully developed terrestrial plant systems (e.g., forests, erect-leaved cereal crops) is large enough to intercept practically all incident radiation (Loomis and Gerakis, 1975), and most of the losses in these systems are due to reflection.

The availability of solar radiation itself is, of course, very important. (The geographical distribution of radiation is discussed in connection with Fig. 2.24 and in section 3.1.) The penetration of radiation through depths of water is illustrated in Fig. 2.30. The corresponding distributions of productivity for typical coastal and oceanic regions of the North Atlantic are shown in Fig. 3.94. Total autotrophic plus heterotrophic biomass at these sites was found to be 167×10^3 J m^{-2} (coastal region) and 8×10^3 J m^{-2} (open ocean) (Odum, 1972).

FIGURE 3.94 Net primary production as function of depth for two locations in the North Atlantic. The corresponding average biomasses (standing crops) are 167 kJ m^{-2} (coastal region) and 8 kJ m^{-2} (open ocean). (Based on Odum, 1972.)

Excessive solar intensity can diminish efficiency, if the chlorophyll pigments are unable to absorb it all or if the radiation destroys the chlorophyll. Also, the state of the photosynthetic systems as well as of the further energy conversion and transport system of the plant may play a role. Efficiency may be lowered by damage or aging of such components (Wassink, 1975).

If changes do not occur too rapidly, many plants are capable of a measure of adaptation to external conditions of light and humidity, etc. The pathway of carbon synthesis described in section 3.6.1 involves a molecule (3.40) with three carbon atoms (C_3-pathway). Some plants are able to follow a different pathway, in which four carbon atoms are synthesized (C_4-pathway), notably as malic acid ($HO_2CCH:CHCO_2H$). In this case, less CO_2 is needed for synthesis of, for example, sugar, so that C_4-plants may grow more rapidly than C_3-plants, at least in some environments.

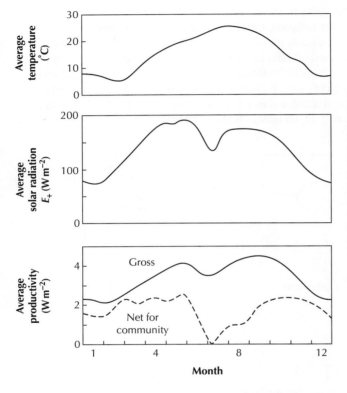

FIGURE 3.95   Trends of average temperature, solar radiation, and gross and net production of biomass for an evergreen oak (*quercus acuta*) forest community at Minamata, Kyushu, Japan. The curves have been constructed from data collected in 1971/1972, and the conversion factor 1 kg dry matter = 18.8 × 10^6 J has been used. (Based on Kira, 1975.)

Temperature plays an important role: the temperature of both air and soil for terrestial plants and the temperature of water for aquatic plants. The life cycles of aquatic plankton and algae are relatively short, and for some species, biomass production is confined to a few weeks of the year. Many terrestrial plants are dormant during cold periods, and they are able to withstand low temperatures during winter. If, however, sudden frost occurs in the middle of a growth period, their tolerance is often low and damage may result. For each species, the number of days during the year with temperatures above certain values may be defined as the *growing season*, a concept that has been used extensively in estimates of average productivity. A phenomenological treatment of the influence of thermal conditions on plant and animal life may be found in the work of Budyko (1974).

Figure 3.95 shows the trends in gross primary production of an evergreen oak forest, along with temperature and average solar radiation. The dashed line gives the net production of the community as a whole (trees and canopy), i.e., the surplus or growth rate, which is high most of the year (except in summer), with peaks in spring and autumn. This is an example of a growing ecosystem. If human intervention can be achieved without changing the trophic relationships within the community, it would appear that a harvest (energy extraction) corresponding to the net production could be sustained, leaving a stable ecosystem with no growth. Man would enter as an additional loop in the system depicted in Fig. 3.93, and the ecosystem would provide a truly renewable energy source.

When the standing crop biomass is small, only a small fraction of solar radiation is intercepted. Thus, productivity initially increases with increasing biomass until a turning point of optimal growth conditions is reached, as seen in Fig. 3.96. After this point, productivity slowly decreases toward equilibrium (zero).

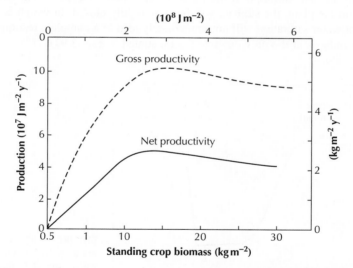

FIGURE 3.96 Trends of gross and net productivity as functions of standing crop biomass for a fir species (*Abies sachalinensis*). (Based on Kira, 1975.)

Water flows through all plants at average rates much higher than the rate at which water is dissociated by the first photosynthetic step (3.40). Extra water is required for the oxidation of the manganese enzyme involved in this process, as well as for the cyclic ADT–ATP and NADP–NADPH$_2$ processes, but these amounts of water are not consumed. The same is true for other metabolic processes in the plant, and water also performs physical tasks, such as keeping the surfaces of membranes wet and transporting dissolved substances.

The overall photosynthetic process (3.45) requires one water molecule for each carbon atom assimilated, or 1.5 kg of water for each 1 kg of carbon or for each 42×10^6 J of biomass, corresponding to 2–3 kg of dry matter. The wet weight is typically 4–10 times the dry weight, implying that the total water content of the plant is at least four times the amount fixed by photosynthesis (and typically is much higher).

The transport of water through a terrestrial plant is achieved by a free-energy gradient, often referred to as the *water potential* [cf. (3.36) and Canny (1977)]. The corresponding force may thus be more general than that associated with osmotic pressure. Loss of water by transpiration can be viewed as a local decrease in water potential near the area where water is leaving the plant. This means that a potential gradient has been established that will attract water from other parts of the plant. Therefore, the potential gradient moves through the plant and builds up in the root system, which then attracts water from the surrounding soil. If the water that can be drawn from the soil is insufficient, the plant will try to decrease its transpiration by contracting the stomata (openings at leaf surfaces). If the water deficit produces a strong and prolonged stress on the plant, the stomata will become tightly closed, in which case subsequent water saturation will not immediately lead to a complete resumption of photosynthesis at the previous rate (as illustrated in Fig. 3.97).

FIGURE 3.97 Effect on productivity of withholding water for 2.5 days. The experiment was performed on sunflower leaves under conditions of generous light. (Based on Slavik, 1975.)

The turnover of water associated with a biomass production of 1 kg of dry matter is quoted as lying in the range 200–900 kg of water, with values around 500 kg being typical for grass and cereal crops (Geiger, 1961; cf. also Penman, 1970, which quotes 400 kg for food crops).

As an upper limit on the flow of water available to terrestrial plants, one may take the total land precipitation rate of 1.1×10^{17} kg of water per year (Fig. 2.65). Assuming the plants' requirement to be 500 kg of water per kg of dry matter per year, one deduces a maximum sustainable production of 2.2×10^{14} kg of dry matter per year, or, in energy units, 4.0×10^{21} J y^{-1} (1.3×10^{14} W).

Since the total terrestrial gross primary production is presently 7.6×10^{13} W (Odum, 1972), and the net primary production may amount to about half this value, it is clear that a large fraction of the evapotranspiration (6.7×10^{16} kg of water per year according to Fig. 2.65) over land is due to plants. Accordingly, a substantial increase in terrestrial plant production would have to involve increased recycling of water on time scales shorter than 1 year. One might consider achieving this by artificial irrigation, using some of the water run-off from land areas (3.9×10^{16} kg of water per year according to Fig. 2.65). However, as Fig. 2.98 shows, about 70% of the water diverted to food crops is not immediately returned to run-off streams. In view of the many other important functions performed by the rivers and streams responsible for run-off, this suggests that there may be little room for expanding terrestrial biomass production on a worldwide scale if fresh-water resources alone are used for irrigation.

Therefore, schemes for ensuring quick recycling of water added from any source are as important as expansion of the irrigation potential, and furthermore, really significant increases in biomass production must involve oceanic water resources, either by marine biomass production or by use of (eventually distilled) seawater for irrigation. Some terrestrial plants actually accept seawater irrigation, with some reduction in conversion efficiency (Epstein and Norlyn, 1977).

If the availability of CO_2 is the only limiting factor, it is reasonable to assume that productivity is proportional to the amount of CO_2 entering the plant. The concentration X of CO_2 at the photosynthetic sites may be related to the atmospheric CO_2 concentration X_{atm}, expressed as a linear relation

$$X = f X_{atm},$$

where f depends on properties of the plant (inside pressure, opening of stomata, etc.), as well as on the CO_2 eddy diffusion parameter in the air surrounding the plant. The atmospheric concentration X_{atm} decreases with height when no photosynthesis takes place (night condition), as shown, for example, in Fig. 2.33. However, during daytime in low wind conditions, the CO_2 concentration exhibits a minimum halfway through the canopy of vegetation (Saeki, 1975).

If the proportionality factor between production P and CO_2 internal concentration X is g, when no other limiting factors are encountered, then the rate of primary production as function of incident radiation, E_+, may be written

$$\frac{1}{P} = \frac{1}{\eta' E_+} + \frac{1}{gX}, \tag{3.55}$$

where η' (3.52) is the efficiency at low incident radiation, when light is the only limiting factor. In (3.55), temperature requirements are assumed to be satisfied, and the water and nutrient supply are assumed to be adequate.

The state of the photosynthetic system within the plant may give rise to different efficiencies η_{chem} under different conditions. Basically, this can be understood in terms of competing pathways for the chain of molecular reactions involved. In the electron transport chain, Fig. 3.73, the Q to plastoquinone electron transport in PS II may have to compete with fluorescent de-excitation of the Q-molecule. A similar de-excitation of the X-protein may occur in PS I. In both cases, the energy captured by the chlorophyll pigments will be unavailable to biomass production. In summary, under less-than-optimal conditions, more than 8 light quanta are required for each CO_2 molecule assimilated (the excess energy being lost as heat or radiation).

In aquatic environments, the rate of photosynthesis may depend on such factors as salinity and pH value (acidity). It has been suggested that the decreased productivity of coniferous forests in Sweden is associated with increasing acidity of precipitation (and hence of soil) (Swedish Ministries of Foreign Affairs and of Agriculture, 1971).

Last, but not least, the role of nutrients as a limiting factor should be mentioned. Deprivation of nutrients decreases and eventually halts production. Transfer of plants to nutrient-free environments has shown effects of decreasing magnitude resulting from deprivation of nitrogen, phosphorus, and potassium, in that order.

It follows from Figs. 2.92 and 2.93 that an average of 1.4×10^{-3} kg of nitrogen is fixed for each kg of carbon fixed in the terrestrial biosphere, in which there is about 0.0343 kg of N (kg of C)$^{-1}$ (the standing crop biomass shows a different relationship between N and C, owing to different turnover times). The amount of nitrogen with which the soil must be enriched in order to increase productivity, if N is a limiting factor, may be much larger, owing to the nature of the pathways of uptake. Delwicke (1970) quotes experiments in which 0.76 kg of N was needed in order to increase production by 1 kg of dry matter.

In evaluating the net energy yield of cultivated land or sea, nonsolar energy subsidies must also be considered. These may be in the form of machinery and fuels for running the machinery (plowing and harvesting tools). In the present day practice of terrestrial agriculture, the largest energy subsidy in areas of intense farming is in the manufacture of fertilizers. In regions demanding irrigation, water pumping may expend substantial amounts of energy, whereas the amount of machinery employed typically is large only in regions of intense farming, so that the corresponding energy subsidy remains a fraction of that spent on fertilizer. If little machinery is used, more manpower is needed, combined with power provided by draught animals. Also, transport of harvested crops costs energy, particularly for remote fishing, and, in the case of food crops, in industrialized regions processing and packaging may account for the largest part of the total energy subsidy. Food derived from animals also involves less efficient utilization of primary energy, as discussed in connection with Fig. 3.93.

In the case of "energy crops," the size of energy subsidies in fertilizers may be of particular interest, in order to determine whether the extra primary production, i.e., solar energy utilization, exceeds the energy subsidy. Using present-day manufacturing techniques, including the mining of phosphate rock, the energy needed to produce 1 kg of nitrogen in the form of fertilizer is about 10^8 J, the energy needed for 1 kg of phosphorus is about 1.4×10^7 J, and the energy for 1 kg of potassium is about 9×10^6 J (Steinhart and Steinhart, 1974; Blaxter, 1974). As an example of fertilizer usage in intense farming practice, the average use of fertilizer in Denmark in 1965–1966 amounted to 0.0114 kg of N per m^2, 0.003 58 kg of P per m^2, and 0.0109 kg of K per m^2 (Danish Statistical Survey, 1968), which, with the above conversion figures, correspond to a total subsidy of 1.3×10^6 J m^{-2} y^{-1} (0.04 W m^{-2}). The harvested yield averaged 0.5 kg of dry matter per m^2 and year. For the most common crop (barley), the net primary production would be about twice the amount harvested, or 2×10^7 J m^{-2} y^{-1}, and the fertilizer energy subsidy would thus be 6.5%. By 1971, fertilizer-related energy consumption had risen by 23%, but the harvest yield was unchanged per unit area (Danish Statistical Survey, 1972). This implies that, because of this farming practice, nutrients suitable for uptake are being depleted from the soil so rapidly that increasing amounts of fertilizer must be added in order to maintain a constant productivity.

A proposed scheme for "energy plantations" yielding biomass at the rate of about 15×10^7 J m^{-2} y^{-1} (e.g., based on sunflower) under (southern) U.S. conditions anticipates the use of 4×10^6 J m^{-2} y^{-1} for fertilizers and 2×10^6 J m^{-2} y^{-1} for irrigation and other machinery, a total energy subsidy of 5% (Alich and Inman, 1976).

Productivity Data

Data on productivity of different species under different climatic conditions are plentiful, but often hard to summarize because of the differences in techniques used and in ways of representing the results. Most controlled experiments are performed under optimal conditions, and they do not reflect the average productivity of the particular plant and geographical region. On the other hand, statistical information for entire regions or countries does not usually convey the exact conditions of growth. In many cases, the productivity of a given crop depends on the history of previous uses of the land. For natural ecosystems, other factors make it difficult to compare data from different sources. Conditions are not always sufficiently well described by climatic zone and solar radiation data. For instance, the geology of the upper soil may vary over very short distances. Altogether, it is not reasonable to give more than broad ranges or limits inside which the productivity of a given plant in a given environment will normally lie, not excluding exceptional values outside the range given.

In order to partially eliminate the dependence on the amount of solar radiation, Fig. 3.98 indicates ranges for the overall efficiency (3.53), representing the ratio between net primary production and total solar radiation. In this way, for example, forests in different climatic regions (boreal coniferous, deciduous, temperate broad-leaved evergreen, and subtropical or tropical rain forests) are represented by a fairly narrow range of efficiencies, despite large variations in absolute productivity.

The range of efficiencies of natural plants goes from practically nothing to a little over 2%, the highest values being reached by tropical rain forests and algal cultures of coral reefs. Cultivated crops (terrestrial or marine) may reach 4%–5% under optimal conditions and nutrient subsidy. The optimum efficiency from the theoretical discussion of equations (3.52) and (3.53) is roughly

$$\eta = \eta_\lambda \, \eta_{geom} \, \eta_{chem} \, \eta_{resp} \lesssim 0.5 \times 0.8 \times 0.4 \times 0.6 \approx 0.1. \qquad (3.56)$$

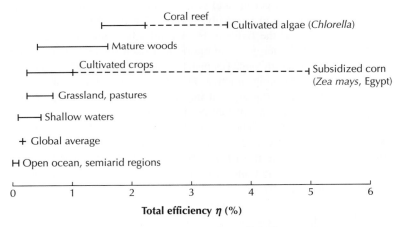

FIGURE 3.98 Intervals of typical values of total photosynthetic efficiency for different plants and communities (*solid lines*), as well as intervals of possible improvement for subsidized and optimized cultures (*dashed lines*). For nonperennial plants, the efficiency is an average over the growing season. (Constructed on the basis of information from Kira, 1975; Caldwell, 1975; Loomis and Gerakis, 1975.)

Each of the conditions is difficult to achieve in practice, at least for extended periods. As mentioned earlier, mature ecosystems are characterized by diversity, which may not be reconcilable with maximum efficiency of primary production, and young ecosystems, such as nonperennial crops, need a growing period (cf. Fig. 3.96) before they reach full productivity, and thus their average efficiency on an area basis and over prolonged periods of time cannot be optimal.

Figure 3.99 gives a recent estimate of the geographical distribution of potential biomass production. The data are net primary production data derived from the *Terrestrial Ecosystem Model* (TEM) of the Woods Hole group (Melillo and Helfrich, 1998; Raich *et al.*, 1991; Melillo *et al.*, 1993; McGuire *et al.*, 1997; Sellers *et al.*, 1997). The assumption is that a mature ecosystem of natural vegetation has developed, and the model takes into account solar radiation, moisture, and temperature, as well as access to water and nutrients. Not included here is global warming (increased CO_2), which could induce increased primary production in a fairly complex pattern and change the borders of natural vegetation zones, sometimes by several hundred kilometers (IPCC, 1996b). Use of these biomass data for energy scenarios is illustrated in Chapter 6. Seasonal variations of actual biomass land coverage, as deduced from satellite-based sensors, are regularly posted on the Internet (NOAA, 1998).

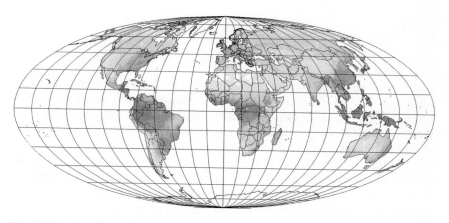

Energy linear scale
W m^{-2}

■	2 to 2.2
■	1.8 to 2
■	1.6 to 1.8
■	1.4 to 1.6
■	1.2 to 1.4
■	1 to 1.2
■	0.8 to 1
■	0.6 to 0.8
■	0.4 to 0.6
■	0.2 to 0.4
■	0.01 to 0.2
	all others

FIGURE 3.99 Annual average energy content in W m^{-2} of potential net biomass production in mature ecosystems. (Based on Melillo and Helfrich, 1998.)

PART I: MINI PROJECTS, DISCUSSION ISSUES, AND EXERCISES

I.1

Find out what has happened since the year 2000 (see in Table 1.1) to the amounts of renewable energy used, based upon available statistical data, either globally or for your region or country.

I.2

Order the prices given in Chapter 1.1 according to the type of energy (heat, fuels for industry or transportation, electricity) and the type of customers (wholesale, retail). Try to find out what is happening to these prices today (e.g., using annually updated sources, such as OECD, IEA, or Danish Energy Agency data). Have the relative prices changed?

I.3

Sustainability is a word often used in political statements. Discuss the difference between *renewability* and sustainability in the energy field. Examples could be wind power (where you can dismount the turbines should you no longer need them) and reservoir-based hydropower (which entails irreversible modifications of large flooded areas). Furthermore, is the sustainability requirement to be used only on the supply side? One might propose sustainability on the demand side to mean that the demand in question could be satisfied for the entire world population (should they so desire) without the environmental side-effects usually included in the definition of sustainability. Is it sustainable to drive large BMW cars using renewable hydrogen? Is it sustainable to build glass houses as long as the cooling and ventilation system is powered by wind? Discuss the argument that, because there is a ceiling on each renewable energy flow, every proposed demand should be evaluated and ranked, with the possibility of disregarding perceived needs that turn out to be poorly ranked.

I.4

How could one design instruments for measurement of the individual components of the net radiation flux, e.g., direct, scattered, and reflected short-wavelength radiation, and upward- and downward-directed long-wavelength radiation? (This question is probably most interesting for those who are not acquainted with such instruments, and who might wish to compare their proposals with instruments actually in use, cf. Thekaekara, 1976; Robinson, 1966; Sellers, 1965; Meinel and Meinel, 1976.)

I.5

Discuss the formulation of approximate equations for the flow of water in rivers (cf., for example, Arnason, 1971).

I.6

Discuss the determinism of climate.

I.7

Discuss tides in the solid Earth and compare their properties with those in the oceans and those in the atmosphere. Construct, for instance, a simple model of a uniform interior of the Earth, a complete cover of ocean with fixed depth, and a uniform atmosphere of some effective height, and assume the densities for each of the three components to be constant (cf., for example, Tomaschek, 1957; Kertz, 1957).

I.8

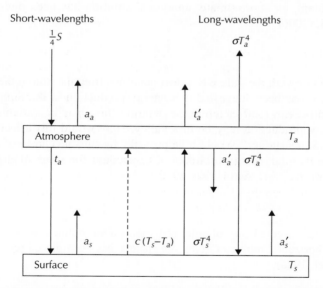

FIGURE I.1 Crude model of the Earth–atmosphere system. The average incoming radiation is $S/4$, and the symbols a and t represent fractions being reflected or transmitted at the location indicated. Multiple reflections are not shown, but can be calculated. The constant σ in Stefan–Boltzmann's law equals 5.7×10^{-8} W m^{-2} K^{-4}.

Construct a simple one-dimensional model of the greenhouse effect, containing the Earth's surface temperature T_s (variable), albedo a_s for short-wavelength radiation, and a'_s for long-wavelength radiation; and an idealized atmosphere visualized as a layer (Fig. I.1) of uniform temperature T_a (variable), short-wavelength albedo a_a, transmission coefficient t_a, long-wavelength albedo a'_a, and transmission coefficient t'_a. Both the surface and the atmosphere emit long-wavelength radiation as blackbodies. All fluxes are treated as vertical.

With a definite assumption regarding the nonradiative transfer between surface and atmosphere, such as assuming the flux to be proportional to $(T_s - T_a)$, a set of parameters may be set up that reproduce the main features of the observed Earth–atmosphere system (cf. Fig. 2.20). Now questions can be asked concerning the sensitivity of the model to the parameters entering into it. How does the surface temperature change if a_a is changed by $\pm 10\%$ (changing the albedo of the surface)? If the present energy use of human society were to be covered by solar collectors of average efficiency 0.2, and a_s was changed from 0.15 to zero for the area covered by these collectors, then how much would the total albedo a_s of the Earth's surface become diminished? How does the surface temperature change if a_a is changed by $\pm 10\%$ (e.g., due to presence of aerosols)? How does the surface temperature change if a'_a is changed by $\pm 10\%$ (e.g., due to the presence of CO_2)?

Although a computer-based model may offer the easiest approach to solving this problem, an approximate, analytical solution has been developed by Foong, S. (2006).

I.9

In connection with the failure to detect neutrinos from the Sun in the expected quantity, it has been suggested that energy production in the Sun's core by nuclear processes (and therefore the neutrino flux) may be variable and that the present production may be in a valley. How would this affect radiation from the Sun's surface? Would it be possible, by suitable choice of a periodic variation in solar energy production, to account for some of the climatic changes of the past? (See Ulrich, 1975.)

I.10

Discuss the reflected flux of solar radiation on an inclined plane, in front of which a horizontal mirror of given dimensions has been placed (cf., for example, Seitel, 1975).

I.11

What data on solar radiation, wind, waves, etc., are available in your region? (For example, check with the local meteorological service.)

I.12

Discuss the cooling of the Earth in terms of simplified models, such as: (a) a uniform sphere of constant heat capacity and diffusion coefficient, initially at a uniform temperature; (b) adding radiogenic heat production in the crust, by constraining the surface temperature to be the sum of two terms, one entirely due to the heat flow from the interior, and the other term exponentially decaying, such that the drop in 4.5×10^9 years becomes 900 K.

Use present surface heat flow and temperature (average values) as boundary conditions, and discuss the long-range development of thermal conditions in the interior of the Earth (cf., for example, Goguel, 1976). Recent models suggest a core temperature of nearly 4000 K (Hofmeister, 1999).

I.13

Consider a compartment model for biological energy production and transfer within a closed ecosystem of the general form shown in Fig. 3.73. Assume, for instance, that the limiting factor governing the rate of primary production is solar radiation, so that the production becomes proportional to a schematic solar input, constant or with only seasonal sinusoidal variations [cf. (3.47)]. Think of a food chain in which you may find data on the rates and biomass levels appearing in Fig. 3.73 or some modification of it, and try to determine conditions for stability (no net production in the community as a whole).

If you have a computer at hand, you might try to set up a numerical simulation model for the system, assuming each rate (time derivative of a compartment level) to be a linear combination of all the compartment levels (some coefficients may be zero, of course). In this case you may also be able to follow the growth phase of the system, and attempt to judge whether the parameter values chosen are reasonable (cf., for example, Odum, 1972, where examples of parameter values may also be found; Patten, 1971, 1972; Sørensen, 1975a).

I.14

Use current biomass production data (available on the Internet) mentioned in connection with Fig. 3.79 to estimate possible bioenergy sources present in your region and their seasonal distribution. Does the seasonality have implications for energy use? Compare your results with the model considered in Chapter 6.

Renewable Energy Technologies

Chapter 4

The Energy Conversion Processes

4.1 GENERAL PRINCIPLES

A large number of energy conversion processes take place in nature, some of which are described in Chapters 2 and 3. Man is capable of performing additional energy conversion processes by means of various devices he has invented. Such devices may be classified according to the type of construction used, according to the underlying physical or chemical principle, or according to the forms of energy appearing before and after the action of the device. This chapter surveys methods that may be suitable for the conversion of renewable energy flows or stored energy. A discussion of general conversion principles is made below, followed by an outline of engineering design details for specific energy conversion devices, ordered according to the energy form being converted and the energy form obtained. The collection is necessarily incomplete and involves judgment about the importance of various devices.

4.1.1 Basic Principles of Energy Conversion

Table 4.1 lists some examples of energy conversion processes or devices currently in use or contemplated, organized according to the energy form emerging after the conversion. In several cases, more than one energy form emerges as a result of the action of the device; for example, heat in addition to one of the other energy forms listed. Many devices also perform several energy conversion steps, rather than the single ones given in the table. A power plant, for example, may perform the conversion process chain between the energy forms: chemical → heat → mechanical → electrical. Diagonal transformations are also possible, such as conversion of mechanical energy into mechanical energy (potential energy of elevated fluid → kinetic energy of flowing fluid → rotational energy of turbine) or of heat into heat at a lower temperature (convection, conduction). A process in which the only change is heat transfer from a lower to a higher temperature is forbidden by the second law of thermodynamics. Such transfer can be established if at the same time some high-quality energy is degraded,

Renewable Energy, Fourth Edition

Initial Energy	Converted Energy Form				
Form	Chemical	Radiant	Electrical	Mechanical	Heat
Nuclear					Reactor
Chemical			Fuel cell, battery discharge		Burner, boiler
Radiant	Photolysis		Photovoltaic cell		Absorber
Electrical	Electrolysis, battery charging	Lamp, laser		Electric motor	Resistance, heat pump
Mechanical			Electric generator, MHD	Turbines	Friction, churning
Heat			Thermionic & thermoelectric generators	Thermodynamic engines	Convector, radiator, heat pipe

TABLE 4.1 Examples of energy conversion processes listed according to the initial energy form and one particular converted energy form (the one primarily wanted).

for example, by a heat pump (which is listed as a converter of electrical into heat energy in Table 4.1, but is further discussed in section 4.2.3).

The efficiency with which a given conversion process can be carried out, that is, the ratio between the output of the desired energy form and the energy input, depends on the physical and chemical laws governing the process. For heat engines, which convert heat into work or vice versa, the description of thermodynamic theory may be used in order to avoid a complicated description on the molecular level (which is, of course, possible, for example, on the basis of statistical assumptions). According to thermodynamic theory (again, the second law), no heat engine can have an efficiency higher than that of a reversible Carnot process, which is depicted in Fig. 4.1, in terms of different sets of thermodynamic state variables,

$$(P, V) = (\text{pressure, volume}),$$

$$(T, S) = (\text{absolute temperature, entropy}),$$

and

$$(H, S) = (\text{enthalpy, entropy}).$$

Entropy is defined in (1.1), apart from an arbitrary constant fixed by the third law of thermodynamics (Nernst's law), which states that S can be taken as zero at zero absolute temperature ($T = 0$). Enthalpy H is defined by

$$H = U + PV, \tag{4.1}$$

FIGURE 4.1   The cyclic Carnot process in different representations. Traversing the cycle in the direction $1 \to 2 \to 3 \to 4$ leads to the conversion of a certain amount of heat into work (see text for details).

in terms of P, V, and the internal energy U of the system. According to the first law of thermodynamics, U is a state variable given by

$$\Delta U = \int dQ + \int dW, \qquad (4.2)$$

in terms of the amounts of heat and work added to the system [Q and W are not state variables, and the individual integrals in (4.2) depend on the paths of integration]. Equation (4.2) determines U up to an arbitrary constant, the zero point of the energy scale. Using definition (1.1),

$$dQ = T \, dS$$

and

$$dW = -P \, dV,$$

both of which are valid only for reversible processes. The following relations are found among the differentials:

$$dU = T \, dS - P \, dV,$$
$$dH = T \, dS + V \, dP. \qquad (4.3)$$

These relations are often assumed to have general validity.

If chemical reactions occur in the system, additional terms $\mu_i \, dn_i$ should be added on the right-hand side of both relations (4.3), in terms of the chemical potentials, which are discussed briefly in section 3.5.3.

For a cyclic process like the one shown in Fig. 4.1, $\int dU = 0$ upon returning to the initial locus in one of the diagrams, and thus, according to (4.3), $\int T \, dS = \int P \, dV$. This means that the area enclosed by the path of the cyclic process in either the (P, V)- or the (T, S)-diagram equals the work $-W$ performed by the system during one cycle (in the direction of increasing numbers in Fig. 4.1).

The amount of heat added to the system during the isothermal process 2-3 is $\Delta Q_{23} = T(S_3 - S_2)$, if the constant temperature is denoted T. The heat added in the other isothermal process, 4-1, at a temperature T_{ref}, is $\Delta Q_{41} = -T_{ref} (S_3 - S_2)$. It follows from the (T, S)-diagram that $\Delta Q_{23} + \Delta Q_{41} = -W$.

The efficiency by which the Carnot process converts heat available at temperature T into work, when a reference temperature of T_{ref} is available, is then

$$\eta = \frac{-W}{\Delta Q_{23}} = \frac{T - T_{ref}}{T}. \qquad (4.4)$$

The Carnot cycle (Fig. 4.1) consists of four steps: 1-2, adiabatic compression (no heat exchange with the surroundings, i.e., $dQ = 0$ and $dS = 0$); 2-3, heat drawn reversibly from the surroundings at constant temperature (the amount of heat transfer ΔQ_{23} is given by the area enclosed by the path 2-3-5-6-2 in the (T, S)-diagram); 3-4, adiabatic expansion; and 4-1, heat given away to the surroundings by a reversible process at constant temperature [$|\Delta Q_{41}|$ is equal to the area of the path 4-5-6-1-4 in the (T, S)-diagram].

The (H, S)-diagram is an example of a representation in which energy differences can be read directly on the ordinate, rather than being represented by an area.

Long periods of time are required to perform the steps involved in the Carnot cycle in a way that approaches reversibility. Because time is important for human enterprises (the goal of the energy conversion process being power, rather than just an amount of energy), irreversible processes are deliberately introduced into the thermodynamic cycles of actual conversion devices. The thermodynamics of irreversible processes are described below using a practical approximation, which is referred to in several of the examples to follow. Readers without specific interest in the thermodynamic description may go lightly over the formulae.

Irreversible Thermodynamics

The degree of irreversibility is measured in terms of the rate of energy dissipation,

$$D = T\, dS/dt, \qquad (4.5)$$

where dS/dt is the entropy production of the system while it is held at the constant temperature T (i.e., T may be thought of as the temperature of a large heat reservoir, with which the system is in contact). In order to describe the nature of the dissipation process, the concept of *free energy* is introduced (cf. Prigogine, 1968; Callen, 1960).

The free energy, G, of a system is defined as the maximum work that can be drawn from the system under conditions where the exchange of work is the only interaction between the system and its surroundings. A system of this kind is said to be in thermodynamic equilibrium if its free energy is zero.

Consider now a system divided into two subsystems, a small one with extensive variables (i.e., variables proportional to the size of the system) U, S, V, etc., and a large one with intensive variables T_{ref}, P_{ref}, etc., which is initially in thermodynamic equilibrium. The terms *small system* and *large system* are meant to imply that the intensive variables of the large system (but not its extensive variables U_{ref}, S_{ref}, etc.) can be regarded as constant, regardless of the processes by which the entire system approaches equilibrium.

This implies that the intensive variables of the small system, which may not even be defined during the process, approach those of the large system when the combined system approaches equilibrium. The free energy, or maximum work, is found by considering a reversible process between the initial state and the equilibrium. It equals the difference between the initial internal energy, $U_{init} = U + U_{ref}$, and the final internal energy, U_{eq}, or it may be written (all in terms of initial state variables) as

$$G = U - T_{ref} S + P_{ref} V, \tag{4.6}$$

plus terms of the form $\Sigma \mu_{i,ref} n_i$ if chemical reactions are involved, and similar generalizations in case of electromagnetic interactions, etc.

If the entire system is closed, it develops spontaneously toward equilibrium through internal, irreversible processes, with a rate of free energy change

$$\frac{dG}{dt} = \frac{d}{dt}(U_{init} - U_{eq}(t)) = \left(\frac{\partial}{\partial S(t)} U_{eq}(t)\right) \frac{dS(t)}{dt},$$

assuming that the entropy is the only variable. $S(t)$ is the entropy at time t of the entire system, and $U_{eq}(t)$ is the internal energy that would be possessed by a hypothetical equilibrium state defined by the actual state variables at time t, i.e., $S(t)$, etc. For any of these equilibrium states, $\partial U_{eq}(t)/\partial S(t)$ equals T_{ref} according to (4.3), and by comparison with (4.5) it is seen that the rate of dissipation can be identified with the loss of free energy, as well as with the increase in entropy,

$$D = -dG/dt = T_{ref}\, dS(t)/dt. \tag{4.7}$$

For systems met in practice, there are often constraints preventing the system from reaching the absolute equilibrium state of zero free energy. For instance, the small system considered above may be separated from the large one by walls that keep the volume V constant. In such cases, the available free energy (i.e., the maximum amount of useful work that can be extracted) becomes the absolute amount of free energy, (4.6), minus the free energy of the relative equilibrium that the combined system can be made to approach in the presence of the constraint. If the extensive variables in the constrained equilibrium state are denoted U^0, S^0, V^0, etc., then the available free energy becomes

$$\Delta G = (U - U^0) - T_{ref}(S - S^0) + P_{ref}(V - V^0), \tag{4.8}$$

eventually with the additions involving chemical potentials, etc. In (4.6) or (4.8), G is called the Gibbs potential. If the small system is constrained by walls, so that the volume cannot be changed, the free energy reduces to the Helmholtz potential $U - TS$, and if the small system is constrained so that it is incapable of exchanging heat, the free energy reduces to the enthalpy (4.1).

The corresponding forms of (4.8) give the maximum work that can be obtained from a thermodynamic system with the given constraints.

A description of the course of an actual process as a function of time requires knowledge of *equations of motion* for the extensive variables, i.e., equations that relate the currents, such as

$$J_s = dS/dt \text{ (entropy flow rate)} \quad \text{or} \quad J_Q = dQ/dt \text{ (heat flow rate)},$$
$$J_m = dm/dt \text{ (mass flow rate)} \quad \text{or} \quad J_\theta = d\theta/dt \text{ (angular velocity)}, \qquad (4.9)$$
$$J_q = dq/dt = I \text{ (charge flow rate or electrical current)}, \text{ etc.}$$

to the (generalized) forces of the system. As a first approximation, the relation between the currents and the forces may be taken as linear (Onsager, 1931),

$$J_i = \sum_j L_{ij} F_j. \qquad (4.10)$$

The direction of each flow component is J_i/J_i. The arbitrariness in choosing the generalized forces is reduced by requiring, as did Onsager, that the dissipation be given by

$$D = -dG/dt = \sum_i J_i \cdot F_i. \qquad (4.11)$$

Examples of the linear relationships (4.10) are Ohm's law, stating that the electric current J_q is proportional to the gradient of the electric potential ($F_q \propto$ grad ϕ), and Fourier's law (3.34) for heat conduction or diffusion, stating that the heat flow rate $E^{sens} = J_Q$ is proportional to the gradient of the temperature.

Considering the isothermal expansion process required in the Carnot cycle (Fig. 4.1), heat must be flowing to the system at a rate $J_Q = dQ/dt$, with $J_Q = LF_Q$ according to (4.10) in its simplest form. Using (4.11), the energy dissipation takes the form

$$D = T \, dS/dt = J_Q F_Q = L^{-1} J_Q^2.$$

For a finite time Δt, the entropy increase becomes

$$\Delta S = (dS/dt) \, \Delta t = (LT)^{-1} J_Q^2 \, \Delta t = (LT \, \Delta t)^{-1} (\Delta Q)^2,$$

so that in order to transfer a finite amount of heat ΔQ, the product $\Delta S \, \Delta t$ must equal the quantity $(LT)^{-1} (\Delta Q)^2$. For the process to approach reversibility, as the ideal Carnot cycle should, ΔS must approach zero, which is seen to imply that Δt approaches infinity. This qualifies the statement made in the beginning of this subsection that, in order to go through a thermodynamic engine cycle in a finite time, one has to give up reversibility and accept a finite amount of energy dissipation and an efficiency that is smaller than the ideal (4.4).

Efficiency of an Energy Conversion Device

A schematic picture of an energy conversion device is shown in Fig. 4.2, and it is sufficiently general to cover most types of converters in practical use (Angrist, 1976; Osterle, 1964). There is a mass flow into the device and another one out from it, as well as incoming and outgoing heat flow. The work output may be in the form of electric or rotating shaft power.

It may be assumed that the converter is in a steady state, implying that the incoming and outgoing mass flows are identical and that the entropy of the device itself is constant, i.e., that all entropy created is being carried away by the outgoing flows.

FIGURE 4.2 Schematic picture of an energy conversion device with a steady–state mass flow. The sign convention is different from the one used in (4.2), where all fluxes into the system were taken as positive.

From the first law of thermodynamics, the power extracted, E, equals the net energy input,

$$E = J_{Q,in} - J_{Q,out} + J_m(w_{in} - w_{out}). \tag{4.12}$$

The magnitude of the currents is given by (4.9), and their conventional signs may be inferred from Fig. 4.2. The specific energy content of the incoming mass flow, w_{in}, and of the outgoing mass flow, w_{out}, are the sums of potential energy, kinetic energy, and enthalpy. The significance of the enthalpy to represent the thermodynamic energy of a stationary flow is established by Bernoulli's theorem (Pippard, 1966). It states that for a stationary flow, if heat conduction can be neglected, the enthalpy is constant along a streamline. For the uniform mass flows assumed for the device in Fig. 4.2, the specific enthalpy, h, thus becomes a property of the flow, in analogy with the kinetic energy of motion and, for example, the geopotential energy,

$$w = w^{pot} + w^{kin} + h. \tag{4.13}$$

The power output may be written

$$E = -J_\theta \cdot F_\theta - J_q \cdot F_q, \tag{4.14}$$

with the magnitude of currents given by (4.9) and the generalized forces given by

$$\begin{aligned} F_\theta &= \int r \times dF_{mech}(r) && \text{(torque)}, \\ F_q &= -\text{grad}(\phi) && \text{(electric field)} \end{aligned} \tag{4.15}$$

corresponding to a mechanical torque and an electric potential gradient. The rate of entropy creation, i.e., the rate of entropy increase in the surroundings of the conversion device (as mentioned, the entropy inside the device is constant in the steady-state model), is

$$dS/dt = (T_{ref})^{-1} J_{Q,out} - T^{-1} J_{Q,in} + J_m(s_{m,out} - s_{m,in}),$$

where $s_{m,in}$ is the specific entropy of the mass (fluid, gas, etc.) flowing into the device, and $s_{m,out}$ is the specific entropy of the outgoing mass flow. $J_{Q,out}$ may be eliminated by use of (4.12), and the rate of dissipation obtained from (4.7),

$$\begin{aligned} D = T_{ref} \, dS/dt &= J_{Q,in}(1 - T_{ref}/T) + J_m(w_{in} - w_{out} - T_{ref}(s_{m,in} - s_{m,out})) - E \\ &= \max(E) - E. \end{aligned} \tag{4.16}$$

The maximum possible work (obtained for $dS/dt = 0$) is seen to consist of a Carnot term (closed cycle, i.e., no external flows) plus a term proportional to the mass flow. The dissipation (4.16) is brought in the Onsager form (4.11),

$$D = J_{Q,in} F_{Q,in} + J_m F_m + J_\theta \cdot F_\theta + J_q \cdot F_q, \tag{4.17}$$

by defining generalized forces

$$\begin{aligned} F_{Q,in} &= 1 - T_{ref}/T, \\ F_m &= w_{in} - w_{out} - T_{ref}(s_{m,in} - s_{m,out}) \end{aligned} \tag{4.18}$$

in addition to those of (4.15).

The efficiency with which the heat and mass flow into the device is converted to power is, in analogy to (4.4),

$$\eta = \frac{E}{J_{Q,in} + J_m w_{in}}, \tag{4.19}$$

where expression (4.16) may be inserted for E. This efficiency is sometimes referred to as the "first law efficiency," because it only deals with the amounts of energy input and output in the desired form and not with the "quality" of the energy input related to that of the energy output.

In order to include reference to the energy quality, in the sense of the second law of thermodynamics, account must be taken of the changes in entropy taking place in connection with the heat and mass flows through the conversion device. This is accomplished by the *second law efficiency*, which for power-generating devices is defined by

$$\eta^{(2.\,law)} = \frac{E}{\max(E)} = -\frac{\mathbf{J}_\theta \cdot \mathbf{F}_\theta + \mathbf{J}_q \cdot \mathbf{F}_q}{J_{Q,in}F_{Q,in} + J_m F_m}, \qquad (4.20)$$

where the second expression is valid specifically for the device considered in Fig. 4.2, while the first expression is of general applicability, when $\max(E)$ is taken as the maximum rate of work extraction permitted by the second law of thermodynamics. It should be noted that $\max(E)$ depends not only on the system and the controlled energy inputs, but also on the state of the surroundings.

Conversion devices for which the desired energy form is not work may be treated in a way analogous to the example in Fig. 4.2. In (4.17), no distinction is made between input and output of the different energy forms. Taking, for example, electrical power as input (sign change), output may be obtained in the form of heat or in the form of a mass stream. The efficiency expressions (4.19) and (4.20) must be altered, placing the actual input terms in the denominator and the actual output terms in the numerator. If the desired output energy form is denoted W, the second law efficiency can be written in the general form

$$h^{(2.\,law)} = W/\max(W). \qquad (4.21)$$

For conversion processes based on other principles than those considered in the thermodynamic description of phenomena, alternative efficiencies could be defined by (4.21), with $\max(W)$ calculated under consideration of the nonthermodynamic types of constraints. In such cases, the name "second law efficiency" would have to be modified.

4.1.2 Thermodynamic Engine Cycles

A number of thermodynamic cycles, i.e., (closed) paths in a representation of conjugate variables, have been demonstrated in practice. They offer examples of the compromises made in modifying the "prototype" Carnot cycle into a cycle that can be traversed in a finite amount of time. Each cycle can be used to convert heat into work, but in traditional uses the source of heat has mostly been the combustion of fuels, i.e., an initial energy conversion process, by which high-grade chemical energy is degraded to heat at a certain temperature, associated with a certain entropy production.

Figure 4.3 shows a number of engine cycles in (P, V)-, (T, S)-, and (H, S)-diagrams corresponding to Fig. 4.1.

The working substance of the Brayton cycle is a gas, which is adiabatically compressed in step 1-2 and expanded in step 3-4. The remaining two steps take place at constant pressure (isobars), and heat is added in step 2-3. The useful work is extracted during the adiabatic expansion 3-4, and the simple efficiency is thus equal to the enthalpy difference $H_3 - H_4$ divided by the total input $H_3 - H_1$. Examples of devices operating on the Brayton cycle are gas turbines and jet engines. In these cases, the cycle is usually not closed, since the gas is exhausted at point 4 and step 4-1 is thus absent. The somewhat contradictory name given to such processes is *open cycles*.

The Otto cycle, traditionally used in automobile engines, differs from the Brayton cycle in that steps 2-3 and 4-1 (if the cycle is closed) are carried out at constant volume (isochores) rather than at constant pressure.

The Diesel cycle (used in ship and truck engines, and, with the invention of the common-rail design, increasingly in passenger vehicles) has step 2-3 as isobar and step 4-1 as isochore, while the two remaining steps are approximately adiabatic. The actual design of conventional diesel engines may be found in engineering textbooks (e.g., Hütte, 1954), and new designs, such as the common-rail concept, are discussed in Sørensen (2006, 2010c).

Closer to the Carnot ideal is the Stirling cycle, involving two isochores (1-2 and 3-4) and two isotherms.

The Ericsson cycle has been developed with the purpose of using hot air as the working fluid. It consists of two isochores (2-3 and 4-1) and two curves between isotherms and adiabates (cf., for example, Meinel and Meinel, 1976).

The last cycle is the one used in thermal power plants, the Rankine cycle depicted in Fig. 4.3. It has a more complicated appearance owing to the presence of two phases of the working fluid. Step 1-2-3 describes the heating of the fluid to its boiling point. Step 3-4 corresponds to the evaporation of the fluid, with both fluid and gaseous phases being present. It is an isotherm as well as an isobar. Step 4-5 represents the superheating of the gas, followed by an adiabatic expansion step 5-7. These two steps are sometimes repeated one or more times, with the superheating taking place at gradually lowered pressure, after each step of expansion to saturation. Finally, step 7-1 again involves mixed phases with condensation at constant pressure and temperature. The condensation often does not start until a temperature below that of saturation is reached. Useful work is extracted during the expansion step 5-7, so the simple efficiency equals the enthalpy difference $H_5 - H_7$ divided by the total input $H_6 - H_1$. The second law efficiency is obtained by dividing the simple efficiency by the Carnot value (4.4), for $T = T_5$ and $T_{ref} = T_7$.

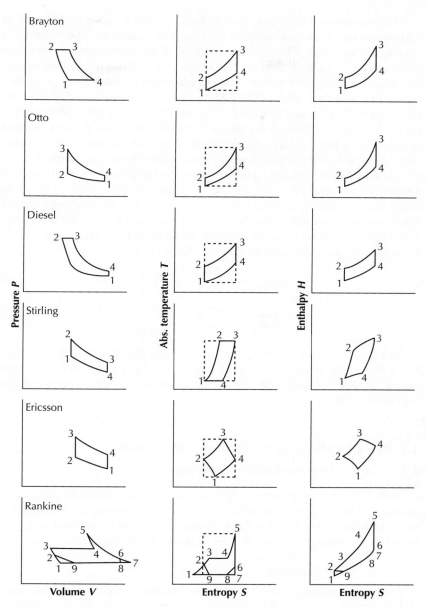

FIGURE 4.3 Examples of thermodynamic cycles in different representations. For comparison, the Carnot cycle is indicated in the (*P, S*)-diagram (*dashed lines*). Further descriptions of the individual cycles are given in the text (cf. also section 4.4.4 for an alternative version of the Ericsson cycle).

Thermodynamic cycles like those in Figs. 4.1 and 4.3 may be traversed in the opposite direction, thus using the work input to create a low temperature T_{ref} (cooling, refrigeration; T being the temperature of the surroundings) or to create a temperature T higher than that (T_{ref}) of the surroundings (heat pumping). In this case, step 7-5 of the Rankine cycle is a compression (8-6-5 if the gas experiences superheating). After cooling (5-4), the gas condenses at the constant temperature T (4-3), and the fluid is expanded, often by passage through a nozzle. The passage through the nozzle is considered to take place at constant enthalpy (2-9), but this step may be preceded by undercooling (3-2). Finally, step 9-8 (or 9-7) corresponds to evaporation at the constant temperature T_{ref}.

For a cooling device, the simple efficiency is the ratio of the heat removed from the surroundings, $H_7 - H_9$, and the work input, $H_5 - H_7$, whereas for a heat pump it is the ratio of the heat delivered, $H_5 - H_2$, and the work input. Such efficiencies are often called *coefficients of performance* (COP), and the second law efficiency may be found by dividing the COP by the corresponding quantity ε_{Carnot} for the ideal Carnot cycle (cf. Fig. 4.1),

$$\varepsilon_{Carnot}^{cooling} = \frac{\Delta Q_{14}}{W} = \frac{T_{ref}}{T - T_{ref}}, \qquad (4.22a)$$

$$\varepsilon_{Carnot}^{heatpump} = \frac{\Delta Q_{32}}{W} = \frac{T}{T - T_{ref}}. \qquad (4.22b)$$

In practice, the compression work $H_5 - H_7$ (for the Rankine cycle in Fig. 4.3) may be less than the energy input to the compressor, thus further reducing the COP and the second law efficiency, relative to the primary source of high-quality energy.

4.2 HEAT ENERGY CONVERSION PROCESSES

4.2.1 Direct Thermoelectric Conversion

If the high-quality energy form desired is electricity, and the initial energy is in the form of heat, there is a possibility of utilizing direct conversion processes, rather than first using a thermodynamic engine to create mechanical work followed by a second conversion step using an electricity generator.

Thermoelectric Generators

One direct conversion process makes use of the thermoelectric effect associated with heating the junction of two different conducting materials, e.g., metals or semiconductors. If a stable electric current, I, passes across the junction between two conductors A and B, in an arrangement of the type depicted in Fig. 4.4, then quantum electron theory requires that the Fermi energy level (which may be regarded as a chemical potential μ_i) is the same in the two materials ($\mu_A = \mu_B$). If the spectrum of electron quantum states is different in the two materials, the crossing of negatively charged electrons or positively charged "holes" (electron vacancies) will not preserve the statistical distribution of electrons around the Fermi level,

$$f(E) = (\exp((E - \mu_i)/kT) + 1)^{-1}, \tag{4.23}$$

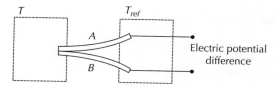

FIGURE 4.4 Schematic picture of a thermoelectric generator (thermocouple). The rods A and B are made of different materials (metals or better p- and n-type semiconductors).

with E being the electron energy and k being Boltzmann's constant. The altered distribution may imply a shift toward a lower or a higher temperature, such that the maintenance of the current may require addition or removal of heat. Correspondingly, heating the junction will increase or decrease the electric current. The first case represents a thermoelectric generator, and the voltage across the external connections (Fig. 4.4) receives a term in addition to the ohmic term associated with the internal resistance R_{int} of the rods A and B,

$$\Delta\phi = -IR_{int} + \int_{T_{ref}}^{T} \alpha \, dT',$$

The coefficient α is called the Seebeck coefficient. It is the sum of the Seebeck coefficients for the two materials A and B, and it may be expressed in terms of the quantum statistical properties of the materials (Angrist, 1976). If α is assumed independent of temperature in the range from T_{ref} to T, then the generalized electrical force (4.15) may be written

$$F_q = R_{int}J_q - \alpha T F_{Q,in}, \tag{4.24}$$

where J_q and $F_{q,in}$ are given in (4.9) and (4.18).

Considering the thermoelectric generator (Fig. 4.4) as a particular example of the conversion device shown in Fig. 4.2, with no mass flows, the dissipation (4.11) may be written

$$D = J_Q F_Q + J_q F_q.$$

In the linear approximation (4.10), the flows are of the form

$$J_Q = L_{QQ} F_Q + L_{Qq} F_q,$$

$$J_q = L_{qQ} F_Q + L_{qq} Fq,$$

with $L_{Qq} = L_{qQ}$ because of microscopic reversibility (Onsager, 1931). Considering F_Q and J_q (Carnot factor and electric current) as the "controllable" variables, one may solve for F_q and J_Q, obtaining F_q in the form (4.24) with $F_Q = F_{Q,in}$ and

$$L_{qq} = (R_{int})^{-1}; \qquad L_{qQ} = L_{Qq} = \alpha T / R_{int}.$$

The equation for J_Q takes the form

$$J_Q = CTF_Q + \alpha T J_q, \tag{4.25}$$

where the conductance C is given by

$$C = (L_{QQ} L_{qq} - L_{Qq} L_{qQ}) / (LqqT).$$

Using (4.24) and (4.25), the dissipation may be written

$$D = CTF_Q^2 + R_{int} J_q^2, \tag{4.26}$$

and the simple efficiency (4.19) may be written

$$\eta = \frac{-J_q F_q}{J_Q} = \frac{-J_q (R_{int} J_q - \alpha T F_Q)}{CTF_Q + \alpha T J_q}. \tag{4.27}$$

If the reservoir temperatures T and T_{ref} are maintained at a constant value, F_Q can be regarded as fixed, and the maximum efficiency can be found by variation of J_q. The efficiency (4.27) has an extremum at

$$J_q = \frac{CF_Q}{\alpha} \left(\left(1 + \frac{\alpha^2 T}{R_{int} C} \right)^{1/2} - 1 \right), \tag{4.28}$$

corresponding to a maximum value

$$\max(\eta) = F_Q \frac{(1 + ZT)^{1/2} - 1}{(1 + ZT)^{1/2} + 1}, \tag{4.29}$$

with $Z = \alpha^2 (R_{int} C)^{-1}$. Equation (4.29) is accurate only if the linear approximation (4.10) is valid. The maximum second law efficiency is obtained from (4.29), dividing by F_Q [cf. (4.20)].

The efficiencies are seen to increase with temperature, as well as with Z. Z is largest for certain materials (A and B in Fig. 4.4) of semiconductor structure and is small for metals as well as for insulators. Although R_{int} is small for metals and large for insulators, the same is true for the Seebeck coefficient α, which appears squared. C is larger for metals than for insulators. Together, these features combine to produce a peak in Z in the semiconductor region. Typical values of Z are about 2×10^{-3} (K)$^{-1}$ at $T = 300$ K (Angrist, 1976). The two materials A and B may be taken as a p-type and an n-type semiconductor, which have Seebeck coefficients of opposite signs, so that their contributions add coherently for a configuration of the kind shown in Fig. 4.4.

Thermionic Generators

Thermionic converters consist of two conductor plates separated by vacuum or plasma. The plates are maintained at different temperatures. One, the emitter, is at a temperature T large enough to allow a substantial emission of electrons into the space between the plates due to the thermal statistical spread in electron energy (4.23). The electrons (e.g., of a metal emitter) move in a potential field characterized by a barrier at the surface of the plate. The shape of this barrier is usually such that the probability of an electron penetrating it is small until a critical temperature, after which it increases rapidly ("red-glowing" metals). The other plate is maintained at a lower temperature T_{ref}. In order not to have a buildup of space charge between the emitter and the collector, atoms of a substance like cesium may be introduced in this area. These atoms become ionized near the hot emitter (they give away electrons to make up for the electron deficit in the emitter material), and, for a given cesium pressure, the positive ions exactly neutralize the space charges of the traveling electrons. At the collector surface, recombination of cesium ions takes place.

The layout of the emitter design must allow the transfer of large quantities of heat to a small area in order to maximize the electron current responsible for creating the electric voltage difference across the emitter–collector system, which may be utilized through an external load circuit. This heat transfer can be accomplished by a so-called *heat pipe*—a fluid-containing pipe that allows the fluid to evaporate in one chamber when heat is added. The vapor then travels to the other end of the pipe, condenses, and gives off the latent heat of evaporation to the surroundings, after which it returns to the first chamber through capillary channels, under the influence of surface tension forces.

The description of the thermionic generator in terms of the model converter shown in Fig. 4.2 is very similar to that of the thermoelectric generator. With the two temperatures T and T_{ref} defined above, the generalized force F_Q is defined. The electrical output current, J_q, is equal to the emitter current, provided that back-emission from the collector at temperature T_{ref} can be neglected and provided that the positive-ion current in the intermediate space is negligible in comparison with the electron current. If the space charges are saturated, the ratio between ion and electron currents is simply the inverse of the square root of the mass ratio, and the positive-ion current will be a fraction of a percent of the electron current. According to quantum statistics, the emission current (and hence J_q) may be written

$$J_Q = AT^2 \exp(-e\,\phi_e/(kT)), \qquad (4.30)$$

where ϕ_e is the electric potential of the emitter, $e\phi_e$ is the potential barrier of the surface in energy units, and A is a constant (Angrist, 1976). Neglecting heat conduction losses in plates and the intermediate space, as well as light emission, the heat $J_{Q,in}$ to be supplied to keep the emitter at the elevated temperature T equals the energy carried away by the electrons emitted,

$$J_{Q,in} = J_q(\phi_e + \delta + 2kT/e), \tag{4.31}$$

where the three terms in brackets represent the surface barrier, the barrier effectively seen by an electron due to the space charge in the intermediate space, and the original average kinetic energy of the electrons at temperature T (divided by e), respectively.

Finally, neglecting internal resistance in plates and wires, the generalized electrical force equals the difference between the potential ϕ_e and the corresponding potential for the collector ϕ_c,

$$-F_q = \phi_c - \phi_e, \tag{4.32}$$

with insertion of the above expressions (4.30) to (4.32). Alternatively, these expressions may be linearized in the form (4.10) and the efficiency calculated exactly as in the case of the thermoelectric device. It is clear, however, that a linear approximation to (4.30), for example, would be very poor.

4.2.2 Engine Conversion of Solar Energy

The conversion of heat to shaft power or electricity is generally achieved by one of the thermodynamic cycles, examples of which were shown in Fig. 4.3. The cycles may be closed, as in Fig. 4.3, or they may be "open," in that the working fluid is not recycled through the cooling step (4-1 in most of the cycles shown in Fig. 4.3). Instead, new fluid is added for the heating or compression stage, and "used" fluid is rejected after the expansion stage.

It should be kept in mind that thermodynamic cycles convert heat of temperature T into work plus some residual heat of temperature above the reference temperature T_{ref} (in the form of heated cooling fluid or rejected working fluid). Emphasis should therefore be placed on utilizing both the work and the "waste heat." This is done, for example, by co-generation of electricity and water for district heating.

The present chapter looks at a thermodynamic engine concept considered particularly suited for conversion of solar energy. Other examples of the use of thermodynamic cycles in the conversion of heat derived from solar collectors into work is given in sections 4.4.4 and 4.4.5. The dependence of the limiting Carnot efficiency on temperature is shown in Fig. 4.93 for selected values of a parameter describing the concentrating ability of the collector and its short-wavelength absorption to long-wavelength emission ratio. Section 4.4.5 describes devices aimed at converting solar heat into mechanical work for water pumping, while the devices of interest in section 4.4.4 convert heat from a solar concentrator into electricity.

Ericsson Hot-air Engine

The engines in the examples mentioned above were based on the Rankine or the Stirling cycle. It is also possible that the Ericsson cycle (which was actually invented for the purpose of solar energy conversion) will prove advantageous in some solar energy applications. It is based on a gas (usually air) as a working fluid and may have the layout shown in Fig. 4.5. In order to describe the cycle depicted in Fig. 4.3, the valves must be closed at definite times and the pistons must be detached from the rotating shaft (in contrast to the situation shown in Fig. 4.5), so that heat may be supplied at constant volume. In a different mode of operation, the valves open and close as soon as a pressure difference between the air on the two sides begins to develop. In this case, the heat is added at constant pressure, as in the Brayton cycle, and the piston movement is approximately at constant temperature, as in the Stirling cycle (this variant is not shown in Fig. 4.3).

Heating fluid inlet

Cooling fluid inlet

FIGURE 4.5   Example of an Ericsson hot-air engine.

The efficiency can easily be calculated for the latter version of the Ericsson cycle, for which the temperatures T_{up} and T_{low} in the compression and expansion piston cylinders are constant, in a steady situation. This implies that the air enters the heating chamber with temperature T_{low} and leaves it with temperature T_{up}. The heat exchanger equations (cf. section 4.2.3, but not assuming that T_3 is constant) take the form

$$-J_m^f \, C_p^f dT^f(x)/dx = h'(T^f(x) - T^g(x)),$$
$$J_m^g C_p^g dT^g(x)/dx = h'(T^f(x) - T^g(x)),$$

where the superscript g stands for the gas performing the thermodynamic cycle, f stands for the fluid leading heat to the heating chamber for heat exchange, and x increases from zero at the entrance to the heating chamber to a maximum value at the exit. C_p is a constant-pressure heat capacity per unit mass, and J_m is a mass flow rate. Both these and h', the heat exchange rate per unit length dx, are assumed constant, in which case the equations may be explicitly integrated to give

$$
\begin{aligned}
T_{c,in} &= T_{c,out} - \frac{J_m^g \, C_p^g}{J_m^f \, C_p^f} (T_{up} - T_{low}) \\
&= \frac{J_m^g C_p^g (1-H)}{J_m^g C_p^g + J_m^f C_p^f} T_{low} + \left(H + \frac{J_m^f C_p^f (1-H)}{J_m^g C_p^g + J_m^f C_p^f} \right) T_{c,out}.
\end{aligned}
\tag{4.33}
$$

Here $T_{c,out} = T$ is the temperature provided by the solar collector or other heat source, and $T_{c,in}$ is the temperature of the collector fluid when it leaves the heat exchanger of the heating chamber, to be re-circulated to the collector or to a heat storage connected to it. H is given by

$$H = \exp(-h((J_m^f C_p^f)^{-1} + (J_m^g C_p^g)^{-1})),$$

where $h = \int h' \, dx$. Two equations analogous to (4.33) may be written for the heat exchange in the cooling chamber of Fig. 4.5, relating the reject temperature $T_{r,in}$ and the temperature of the coolant at inlet, $T_{r,out} = T_{ref}$, to T_{low} and T_{up}. $T_{c,in}$ may then be eliminated from (4.33) and $T_{r,in}$ from the analogous equation, leaving two equations for determination of T_{up} and T_{low} as functions of known quantities, notably the temperature levels T and T_{ref}. The reason for not having to consider equations for the piston motion in order to determine all relevant temperatures is, of course, that the processes associated with the piston motion have been assumed to be isothermal.

The amounts of heat added, Q_{add}, and rejected, Q_{rej}, per cycle are

$$Q_{add} = mC_p^g(T_{up} - T_{low}) + n\mathscr{R} \, T_{up} \log(V_{max}/V_{min}), \tag{4.34a}$$

$$Q_{rej} = mC_p^g(T_{up} - T_{low}) + n\mathscr{R} \, T_{low}\log(V_{max}/V_{min}) + Q'_{rej}, \tag{4.34b}$$

where m is the mass of air involved in the cycle and n is the number of moles of air involved. $\mathscr{R}$ is the gas constant, and V_{min} and V_{max} are the minimum and maximum volumes occupied by the gas during the compression or expansion stages (for simplicity, the *compression ratio* V_{max}/V_{min} has been assumed to be the same for the two processes, although they take place in different cylinders). The ideal gas law has been assumed in calculating the relation between heat amount and work in (4.34), and Q'_{rej} represents heat losses not contributing to transfer of heat from working gas to coolant flow (piston friction, etc.). The efficiency is

$$\eta = (Q_{add} - Q_{rej})/Q_{add},$$

and the maximum efficiency that can be obtained with this version of the Ericsson engine is obtained for negligible Q'_{rej} and ideal heat exchangers providing $T_{up} = T$ and $T_{low} = T_{ref}$,

$$\max(\eta) = (1 - T_{ref}/T)\Big/\left(1 + \frac{mC_p^g}{n \mathscr{R} \log(V_{max}/V_{min})}(1 - T_{ref}/T)\right). \qquad (4.35)$$

The ideal Carnot efficiency may even be approached, if the second term in the denominator can be made small. (However, to make the compression ratio very large implies an increase in the length of time required per cycle, such that the rate of power production may actually go down, as discussed in section 4.1.1.) The power may be calculated by evaluating (4.34) per unit time instead of per cycle.

4.2.3 Heat Pumps

In some cases, it is possible to produce heat at precisely the temperature needed by primary conversion. However, often the initial temperature is lower or higher than required, in the latter case even considering losses in transmission and heat drop across heat exchangers. In such situations, appropriate temperatures are commonly obtained by mixing (if the heat is stored as sensible heat in a fluid like water, the water may be mixed with colder water, a procedure often used in connection with fossil-fuel burners). This procedure is wasteful in the context of the second law of thermodynamics, since the energy is, in the first place, produced with a higher quality than subsequently needed. In other words, the second law efficiency of conversion (2.20) is low, because there will be other schemes of conversion by which the primary energy can be made to produce a larger quantity of heat at the temperature needed at load. An extreme case of a "detour" is the conversion of heat to heat by first generating electricity by thermal conversion (as is done today in fossil power plants) and then degrading the electricity to heat of low temperature by passing a current through an ohmic resistance ("electric heaters"). However, there are better ways.

Heat Pump Operation

If heat of modest temperature is required, and a high-quality form of energy is available, a device is needed that can derive additional benefits from the high quality of the primary energy source. This can be achieved by using one of the thermodynamic cycles described in section 4.1.2, provided that a large reservoir of approximately constant temperature is available. The cycles (cf. Fig. 4.3) must be traversed "anti-clockwise," such that high-quality energy (electricity, mechanical shaft power, fuel combustion at high temperature, etc.) is added, and heat energy is thereby delivered at a temperature T higher than the temperature T_{ref} of the reference reservoir from which it is drawn. Most commonly, the Rankine cycle, with maximum efficiencies bounded by (4.22), is used (e.g., in an arrangement of the type shown in Fig. 4.6). The fluid of the closed cycle, which should have a liquid and a gaseous phase in the temperature interval of interest, may be a fluorochloromethane compound (which needs to be recycled owing to climatic effects caused if it is released to the atmosphere). The external circuits may contain an inexpensive fluid (e.g., water), and they may be omitted if it is practical to circulate the primary working fluid directly to the load area or to the reference reservoir.

The heat pump contains a compressor, which performs step 7-5 in the Rankine cycle depicted in Fig. 4.3, and a nozzle, which performs step 2-9. The intermediate steps are performed in the two heat exchangers, giving the working fluid the temperatures T_{up} and T_{low}, respectively. The equations for determining these temperatures are of the form of equation (4.37). There are four such equations, which must be supplemented by equations for the compressor and nozzle performance, in order to allow a determination of all the unknown temperatures indicated in Fig. 4.6, for given T_{ref}, given load, and a certain energy expenditure to the compressor. Losses in the compressor are in the form of heat, which in some cases can be credited to the load area.

FIGURE 4.6 Schematic picture of a heat pump.

An indication of the departures from the Carnot limit of the *coefficients of performance*, $\varepsilon^{heat\ pump}$, encountered in practice, is given in Fig. 4.7, as a function of the temperature difference $T_{up}-T_{low}$ at the heat pump and for selected values of T_{up}. In the interval of temperature differences covered, the $\varepsilon^{heat\ pump}$ is about 50% of the Carnot limit (4.22b), but it falls more and more below the Carnot value as the temperature difference decreases, although the absolute value of the coefficient of performance increases.

Several possibilities exist for the choice of the reference reservoir. Systems in use for space heating or space cooling (achieved by reversing the flow in the compressor and expansion-nozzle circuit) have utilized river water, lake water, and seawater, and air, as well as soil as reservoirs. The temperatures of such reservoirs are not entirely constant, and it must therefore be acceptable that the performance of heat pump systems will vary with time. The variations are damped if water or soil reservoirs at sufficient depth are used, because weather-related temperature variations disappear as one goes just a few meters down into the soil. Alternative types of reservoirs for use with heat pumps include city waste sites, livestock manure, ventilation air from households or from livestock barns (where the rate of air exchange has to be particularly high), and heat storage tanks connected to solar collectors.

With solar heating systems, the heat pump may be connected between the heat store and the load area (whenever the storage temperature is too low for direct circulation), or it may be connected between the heat store and the collector, so that the fluid let into the solar collector is cooled in order to improve the collector performance. Of course, a heat pump operating on its own reservoir (soil, air, etc.) may also provide the auxiliary heat for a solar heating system of capacity below the demand.

FIGURE 4.7 Measured coefficient of performance, $\varepsilon^{heat\ pump}$, for a heat pump with a semi-hermetic piston-type compressor (*solid lines*, based on Trenkowitz, 1969), and corresponding curves for ideal Carnot cycles.

The high-quality energy input to the compressor of a heat pump may also come from a renewable resource, e.g., by wind or solar energy conversion, either directly or via a utility grid carrying electricity generated by renewable energy resources. As for insulation materials, concern has been expressed over the use of CFC (Chloro-Fluoro-Carbon) gases in the processing or as a working fluid, and substitutes believed to have less negative impacts have been developed.

Heat Exchange

A situation like the one depicted in Fig. 4.8 is often encountered in energy supply systems. A fluid is passing through a reservoir of temperature T_3, thereby changing the fluid temperature from T_1 to T_2. In order to determine T_2 in terms of T_1 and T_3, in a situation where the change in T_3 is much smaller than the change from T_1 to T_2, the incremental temperature change in the fluid traveling a short distance dx through the pipe system is related to the amount of heat transferred to the reservoir, assumed to depend linearly on the temperature difference,

$$J_m C_p^{fluid}\, dT^{fluid}/dx = h'(T_3 - T^{fluid}).$$

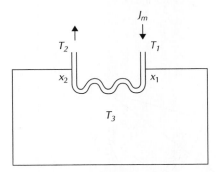

FIGURE 4.8   Heat exchanger, an idealized example of a well-mixed T_3-reservoir.

Integrating from T_1 at the inlet ($x = x_1$) gives

$$T^{fluid}(x) - T_3 = (T_1 - T_3)\exp(-h'(x - x_1)/(J_m C_p^{fluid})), \qquad (4.36)$$

where h' is the heat transfer per unit time from a unit length of the pipe for a temperature difference of one unit. The heat transfer coefficient for the entire heat exchanger is

$$h = \int_{x_1}^{x_2} h'\, dx,$$

which is sometimes written $h = U_h A_h$, with U_h being the transfer coefficient per unit area of pipe wall and A_h being the effective area of the heat exchanger. For $x = x_2$, (4.36) becomes (upon re-ordering)

$$(T_1 - T_2) = (T_1 - T_3)(1 - \exp(-h/(J_m C_p^{fluid}))). \tag{4.37}$$

4.2.4 Geothermal and Ocean-thermal Conversion

Geothermal heat sources have been utilized by means of thermodynamic engines (e.g., Brayton cycles), in cases where the geothermal heat has been in the form of steam (water vapor). In some regions, geothermal sources provide a mixture of water and steam, including suspended soil and rock particles, so that conventional turbines cannot be used. Work has been done on a special "brine screw" that can operate under such conditions (McKay and Sprankle, 1974).

FIGURE 4.9a,b Examples of the utilization of geothermal heat: (a) based on the presence of an aquifer; (b) based on a region of fractured rock.

However, in most regions the geothermal resources are in the form of heat-containing rock or sediments, with little possibility of direct use. If an aquifer passes through the region, it may collect heat from the surrounding layers and allow a substantial rate of heat extraction, for example, by means of two holes drilled from the surface to the aquifer and separated from each other, as indicated in Fig. 4.9a. Hot water (not developing much steam unless the aquifer lies very deep—several kilometers—or its temperature is exceptionally high) is pumped or rises by its own pressure to the surface at one hole and is re-injected through a second hole, in a closed cycle, in order to avoid pollution from various undesired chemical substances often contained in the aquifer water. The heat extracted from a heat exchanger may be used directly (e.g., as district heating; cf. Clot, 1977) or may generate electricity through one of the "low-temperature" thermodynamic cycles considered above in connection with solar collectors (Mock *et al.*, 1997).

If no aquifer is present to establish a "heat exchange surface" in the heat-containing rock, it may be feasible to create suitable fractures artificially (by explosives or induced pressure). An arrangement of this type is illustrated in Fig. 4.9b, counting on the fluid that is pumped down through one drilled hole to make its way through the fractured region of rock to the return drilled hole in such a way that continued heat extraction can be sustained. The heat transfer can be predicted only in highly idealized cases (see, for example, Gringarten *et al.*, 1975), which may not be realized as a result of the fairly uncontrolled methods of rock fractionation available.

One important result of the model calculations is that the heat extraction rate deemed necessary for practical applications is often higher than the geothermal flux into the region of extraction, so that the temperature of the extracted heat will drop. This nonsustainable use of geothermal energy is apparent in actual installations in New Zealand and Italy (where temperatures of extracted steam are dropping by something like 1 °C per year, the number being highly dependent on fracture distribution, rock structure, etc.).

Conversion of Ocean Thermal Energy

Downward gradients of temperature exist in most oceans, and they are particularly stable (i.e., without variations with time) in the tropical oceans. The utilization of such temperature gradients for electricity generation (e.g., by use of a Rankine cycle) has been considered several times since the first suggestions by d'Arsonval (1881).

The temperature differences available over the first 500–1000 m of water depth are only about 25 °C. For a closed Rankine cycle, with a working fluid (e.g., ammonia) that evaporates and condenses at convenient temperatures, placed near the ocean surface, colder water will have to be pumped through a pipe from the depth to a heat exchanger for condensation of the working fluid. Further, a warm water heat exchanger is required for evaporating the working fluid. If the heat exchange surface is such that, say, 5 °C is "lost" at each heat exchanger, the temperature difference available to the thermodynamic cycle is only 15 °C, corresponding to a limiting Carnot efficiency of roughly 0.05. For an actual engine, the efficiency is still lower, and from the power generated should subtract the power needed to pump hot and cold water through the heat exchangers and to pump cold water from its original depth to the converter level. It is expected that overall efficiencies around 0.02 may be achieved (cf., for example, McGowan, 1976).

In order to save energy to pump the hot water through the heat exchanger, it has been suggested that the converters be placed in strong currents, such as the Gulf Stream (Heronemus, 1975). Of course, the power extraction from ocean thermal gradients could have adverse environmental effects. This danger may be increased if ocean currents are incorporated into the scheme, because of the possible sensitivity of the currents to small perturbations, and because of the dependence of climatic zones on the course of currents like the Gulf Stream and the Kuro Shio. (Similar concerns are discussed in connection with global warming caused by increased anthropogenic injection of greenhouse gases into the atmosphere.)

Open thermodynamic cycles have also been suggested for conversion of ocean thermal energy (Claude, 1930; Beck, 1975; Zener and Fetkovich, 1975), for example, based on creating a rising mixture of water and steam bubbles or "foam"; the mixture is separated at a height above sea level, so that the water can be used to drive a turbine rotor.

If viable systems could be developed for conversion of ocean thermal energy, then there would be a number of other applications of such conversion devices in connection with other heat sources of a temperature little higher than that of the surroundings, especially when such heat sources can be regarded as "free." Examples are the reject or "waste" heat flows from other conversion devices operating at higher initial temperature differences, including fuel-based power plants.

4.3 MECHANICAL ENERGY CONVERSION PROCESSES

4.3.1 Basic Description of Flow-driven Converters

A turbine is a device delivering rotational shaft power on the basis of some other type of mechanical energy. If the temperature of the surroundings is regarded as fixed, the simple model in Fig. 4.2 allows the energy dissipation (4.17) to be written

$$D = J_m F_m + J_\theta F_\theta, \tag{4.38}$$

since, from (4.18), $F_{Q,in}$ is zero, and no electrical output has been considered in this conversion step. The output variables are the angular velocity of the shaft, J_θ (4.9), and the torque acting on the system, F_θ (4.15), while the input variables are the mass flow rate, J_m (4.9), and the generalized force F_m given in (4.18). The specific energy contents w_{in} and w_{out} are of the form (4.13), corresponding to, for example, the geopotential energy of a given water head,

$$w_{in}^{pot} = g\Delta z, \qquad w_{out}^{pot} = 0, \tag{4.39}$$

the kinetic energy of the working fluid,

$$w_{in}^{kin} = \tfrac{1}{2}u_{in}^2, \qquad w_{out}^{kin} = \tfrac{1}{2}u_{out}^2, \tag{4.40}$$

and the enthalpy connected with the pressure changes,

$$h_{in} = P_{in}/\rho_{in}, \qquad h_{out} = P_{out}/\rho_{out}, \tag{4.41}$$

where the internal energy term in (4.1), assumed constant, has been left out, and the specific volume has been expressed in terms of the fluid densities ρ_{in} and ρ_{out} at input and output.

If a linear model of the Onsager type (4.10) is adopted for J_m and J_θ and these equations are solved for J_m and F_θ, one obtains

$$J_m = L_{m\theta}J_\theta/L_{\theta\theta} + (L_{mm} - L_{m\theta}L_{\theta m}/L_{\theta\theta})F_m,$$

$$-F_\theta = -J_\theta/L_{\theta\theta} + L_{\theta m}F_m/L_{\theta\theta}. \tag{4.42}$$

The coefficients may be interpreted as follows: $L_{m\theta}/L_{\theta\theta}$ is the mass of fluid displaced by the turbine during one radian of revolution, $(L_{mm} - L_{m\theta}L_{\theta m}/L_{\theta\theta})$ is a "leakage factor" associated with fluid getting through the turbine without contributing to the shaft power, and finally, $L_{\theta\theta}^{-1}$ represents the friction losses. Insertion into (4.38) gives the linear approximation for the dissipation,

$$D = (L_{mm} - L_{m\theta}L_{\theta m}/L_{\theta\theta})(F_m)^2 + (J_\theta)^2/L_{\theta\theta}. \tag{4.43}$$

An ideal conversion process may be approached if no heat is exchanged with the surroundings, in which case (4.19) and (4.12) give the simple efficiency

$$\eta = (w_{in} - w_{out})/w_{in}. \tag{4.44}$$

The second law efficiency in this case is, from (4.20), (4.14), and (4.12),

$$\eta^{(2.\,law)} = (w_{in} - w_{out})/(w_{in} - w_{out} - T_{ref}(s_{m,in} - s_{m,out})). \tag{4.45}$$

The second law efficiency becomes unity if no entropy change takes place in the mass stream. The first law efficiency (4.44) may approach unity if only potential energy change of the form (4.39) is involved. In this case $w_{out} = 0$, and the fluid velocity, density, and pressure are the same before and after the turbine. Hydroelectric generators approach this limit if working entirely on a static water head. Overshot waterwheels may operate in this way, and so may the more advanced turbine constructions, if the potential to kinetic energy conversion (in penstocks) and pressure buildup (in the nozzle of a Pelton turbine and in the inlet tube of many Francis turbine installations) are regarded as "internal" to the device (cf. section 4.3.4). However, if there is a change in velocity or pressure across the converter, the analysis must take this into account, and it is no longer obvious whether the first law efficiency may approach unity or not.

Free-stream-flow Turbines

Consider, for example, a free stream flow passing horizontally through a converter. In this case, the potential energy (4.39) does not change and may be left out. The pressure may vary near the converting device, but far behind and far ahead of the device, the pressure is the same if the stream flow is free. Thus,

$$w = w^{kin} = 1/2\,(u_x^2 + u_y^2 + u_z^2) = 1/2\,\boldsymbol{u}\cdot\boldsymbol{u},$$

and

$$w_{in} - w_{out} = \tfrac{1}{2}\,(\boldsymbol{u}_{in} - \boldsymbol{u}_{out})\cdot(\boldsymbol{u}_{in} + \boldsymbol{u}_{out}). \tag{4.46}$$

This expression, and hence the efficiency, would be maximum if $\boldsymbol{u}_{out}$ could be made zero. However, the conservation of the mass flow J_m requires that $\boldsymbol{u}_{in}$ and $\boldsymbol{u}_{out}$ satisfy an additional relationship. For a pure, homogeneous streamline flow along the x-axis, the rate of mass flow is

$$J_m = \rho A_{in}\,u_{x,in} = \rho A_{out} u_{x,out}, \tag{4.47}$$

in terms of areas A_{in} and A_{out} enclosing the same streamlines, before and after the passage through the conversion device. In a more general situation, assuming rotational symmetry around the x-axis, the device may have induced a radial as well as a circular flow component. This situation is illustrated in Fig. 4.10. It is further discussed below, and the only case treated here is the simple one in which the radial and tangential components of the velocity field, u_r and u_t, which may be induced by the conversion device, can be neglected.

FIGURE 4.10 Schematic picture of a free-stream-flow converter or turbine. The incoming flow is a uniform streamline flow in the x-direction, while the outgoing flow is allowed to have a radial and a tangential component. The diagram indicates how a streamline may be transformed into an expanding helix by the device. The effective area of the converter, A, is defined in (4.49).

The axial force ("thrust") acting on the converter equals the momentum change,

$$F_x = J_m(u_{x,in} - u_{x,out}).\qquad(4.48)$$

If the flow velocity in the converter is denoted u, an effective area of conversion, A, may be defined by

$$J_m = \rho A u_x,\qquad(4.49)$$

according to the continuity equation (4.47). Dividing (4.48) by ρA, one obtains the specific energy transfer from the mass flow to the converter, within the conversion area A. This should equal the expression (4.46) for the change in specific energy, specialized to the case of homogeneous flows $\boldsymbol{u}_{in}$ and $\boldsymbol{u}_{out}$ along the x-axis,

$$u_x(u_{x,in} - u_{x,out}) = 1/2\,(u_{x,in} + u_{x,out})(u_{x,in} - u_{x,out})$$

or

$$u_x = 1/2\,(u_{x,in} + u_{x,out}).\qquad(4.50)$$

The physical principle behind this equality is simply energy conservation, and the assumptions so far have been the absence of heat exchange [so that the energy change (4.12) becomes proportional to the kinetic energy difference (4.46) and the absence of induced rotation (so that only x-components of the velocity need to be considered). On both sides of the converter, Bernoulli's equation is valid, stating that the specific energy is constant along a streamline (Bernoulli's equation for an irrotational flow is given in section 2.5.3, with $\phi =$ constant defining the streamlines). Far from the converter, the pressures are equal but the velocities are different, while the velocity just in front of or behind the converter may be taken as u_x, implying a pressure drop across the converter,

$$\Delta P = 1/2\,\rho(u_{x,in} + u_{x,out})(u_{x,in} - u_{x,out}).\qquad(4.51)$$

The area enclosing a given streamline field increases continuously across the converter at the same time as the fluid velocity continuously decreases. The pressure, on the other hand, rises above the ambient pressure in front of the converter, then discontinuously drops to a value below the ambient one, and finally increases toward the ambient pressure again, behind ("in the wake of") the converter.

It is customary (see, for example, Wilson and Lissaman, 1974) to define an *axial interference factor*, a, by

$$u_x = u_{x,in}(1 - a),\qquad(4.52)$$

in which case (4.50) implies that $u_{x,out} = u_{x,in}\,(1 - 2a)$. With this, the power output of the conversion device can be written

$$E = J_m(w_{in} - w_{out}) = \rho A\,(u_{x,in})^3\,2a\,(1 - a)^2,\qquad(4.53)$$

and the efficiency can be written

$$\eta = E/(J_m w_{in}) = 4a(1-a).$$ (4.54)

It is seen that the maximum value of η is unity, obtained for $a = 1/2$, corresponding to $u_{x,out} = 0$. The continuity equation (4.47) then implies an infinite area A_{out}, and it will clearly be difficult to defend the assumption of no induced radial motion.

In fact, for a free stream device of this type, the efficiency (4.54) is of little relevance since the input flux may not be independent of the details of the device. The input area A_{in}, from which streamlines would connect with a fixed converter area A, could conceivably be changed by altering the construction of the converter. It is therefore more appropriate to ask for the maximum power output for fixed A, as well as fixed input velocity $u_{x,in}$, this being equivalent to maximizing the *power coefficient* defined by

$$C_p = E/(1/2 \rho A \, (u_{x,in})^3) = 4a(1-a)^2.$$ (4.55)

The maximum value is obtained for $a = 1/3$, yielding $C_p = 16/27$ and $u_{x,out} = u_{x,in}/3$. The areas are $A_{in} = (1-a)A = 2/3\,A$ and $A_{out} = (1-a)A/(1-2a) = 2A$, so in this case it may be a reasonable approximation to neglect the radial velocity component in the far wake.

The maximum found above for C_p is only a true upper limit with the assumptions made. By discarding the assumption of irrotational flow, it becomes possible for the converter to induce a velocity field, for which $rot(u)$ is no longer zero. It has been shown that if the additional field is in the form of a vortex ring around the converter region, so that it does not contribute to the far wake, then it is possible to exceed the upper limit power coefficient 16/27 found above (cf. section 4.3.3).

Methods of Describing the Conversion of Wind Flows

Conversion of wind energy into linear motion of a body has been utilized extensively, particularly for transportation across water surfaces. A large sailing vessels of the type used in the 19th century would have converted wind energy at peak rates of a quarter of a megawatt or more.

The force on a sail or a wing (i.e., profiles of negligible or finite thickness) may be broken down into a component in the direction of the undisturbed wind (drag) and a component perpendicular to the undisturbed wind direction (lift). In referring to an undisturbed wind direction it is assumed that a uniform wind field is modified in a region around the sail or the wing, but that beyond a certain distance such modifications can be disregarded.

In order to determine the force components, Euler's equations (cf. section 2.5.2) may be used. If viscous and external forces are neglected, and the flow is assumed to be irrotational (so that Bernoulli's equation is valid) and steady (so that the time-derivative of the velocity potential vanishes, cf. section 2.5.3), then the force on a segment of the airfoil (sail or wing) may be written

$$\frac{\mathrm{d}\mathbf{F}}{\mathrm{d}z} = \oint_C P\mathbf{n}\,\mathrm{d}s = -\tfrac{1}{2}\rho \oint_C (\mathbf{v}\cdot\mathbf{v})\,\mathbf{n}\,\mathrm{d}s.$$

Here $\mathrm{d}z$ is the segment length (cf. Fig. 4.11), C is a closed contour containing the airfoil profile, $\mathbf{n}$ is a unit vector normal to the contour [in the (x, y)-plane], and $\mathrm{d}s$ is the path-length increment, directed along the tangent to the contour, still in the (x, y)-plane. Taking advantage of the fact that the wind velocity $\mathbf{v}$ approaches a homogeneous field W (assumed to be along the x-axis) far from the airfoil, the contour integral may be reduced and evaluated (e.g., along a circular path),

$$\mathrm{d}\mathbf{F}/\mathrm{d}z = \rho\, W\, \Gamma\, \mathbf{e}_y, \tag{4.56}$$

$$\Gamma = \oint_C \mathbf{v}\cdot\mathrm{d}s \approx \pi\, c\, W \sin\alpha. \tag{4.57}$$

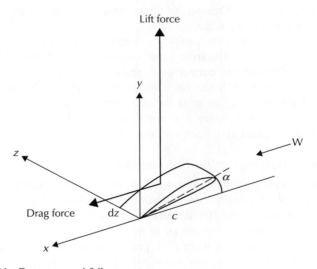

FIGURE 4.11   Forces on an airfoil segment.

Here $\mathbf{e}_y$ is a unit vector along the y-axis, c is the airfoil chord length, and α is the angle between the airfoil and W. In the evaluation of the circulation Γ, it has been assumed that the airfoil is thin and without curvature. In this case, c and α are well defined, but in general the circulation depends on the details of the profile, although an expression similar to the right-hand side of (4.57) is still valid as a first approximation, for some average chord length and angle of attack. Equation (4.56) is known as the theorem of Kutta (1902) and Joukowski (1906).

The expressions (4.56) and (4.57) are valid in a co-ordinate system fixed relative to the airfoil (Fig. 4.11), and if the airfoil is moving with a velocity U, the velocities v and W are to be interpreted as relative ones, so that

$$W = u_{in} - U, \qquad (4.58)$$

if the undisturbed wind velocity is u_{in}.

The assumption that viscous forces may be neglected is responsible for obtaining in (4.56) only a lift force, the drag force being zero. Primitive sailing vessels, as well as primitive windmills, have been primarily aimed at utilizing the drag force. It is possible, however, with suitably constructed airfoils, to make the lift force one or two orders of magnitude larger than the drag force and thereby effectively approach the limit where the viscous forces and hence the drag can be neglected. This usually requires careful "setting" of the airfoil, i.e., careful choice of the angle of attack, α, and, in order to study operation at arbitrary conditions, the drag component should be retained.

It is customary to describe the drag and lift forces on an airfoil of given shape as a function of α, in terms of two dimensionless constants, $C_D(\alpha)$ and $C_L(\alpha)$, defined by

$$dF_x/dz = 1/2\rho\, C_D W^2 c, \qquad (4.59a)$$

$$dF_y/dz = 1/2\rho\, C_L W^2 c. \qquad (4.59b)$$

The constants C_D and C_L are not quite independent of the size of the system, which is not unexpected since the viscous forces (friction) in air contribute most to turbulent motion on smaller scales (cf. the discussion in section 2.5.2). Introducing the Reynolds number,

$$\mathrm{Re} = Wc/\eta,$$

where η is the kinematic viscosity of air defined in section 2.5.2 as a measure of the ratio between "inertial" and "viscous" forces acting between airfoil and air, the α-dependence of C_D and C_L for fixed Re, as well as the Re-dependence for the value of α that gives the highest lift-to-drag ratio, $L/D = C_L/C_D$, may appear as shown in Figs. 4.12 and 4.13. The contours of these "high lift" profiles are indicated in Fig. 4.12.

Assuming that C_D, C_L, and W are constant over the area $A = \int c\, dz$ of the airfoil, the work done by a uniform (except in the vicinity of the airfoil) wind field u_{in} on a device (e.g., a ship) moving with a velocity U can be derived from (4.58) and (4.59),

$$E = F \cdot U.$$

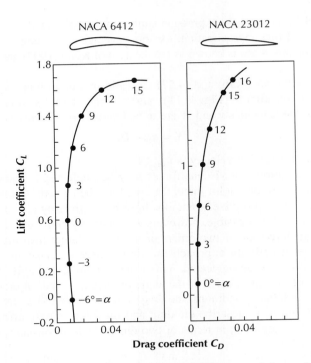

FIGURE 4.12 Lift and drag forces as a function of the angle of attack for two NACA airfoils (National Advisory Committee for Aeronautics; cf., for example, Betz, 1959). The Reynolds number is Re = 8 × 10⁶.

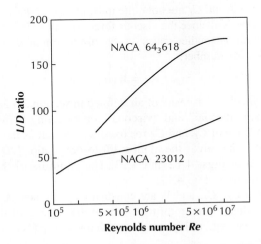

FIGURE 4.13 Reynolds number dependence of the lift-to-drag ratio, defined as the maximum value (as a function of the angle of attack) of the ratio between the lift and drag coefficients C_L and C_D. (Based on Hütter, 1977.)

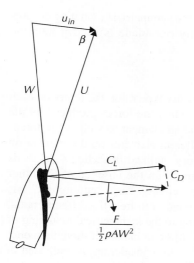

FIGURE 4.14 Velocity and force components for a sailing vessel.

The angle β between $\boldsymbol{u}_{in}$ and $\boldsymbol{U}$ (see Fig. 4.14) may be maintained by a rudder. The power coefficient (4.55) becomes

$$C_p = f(C_L \sin\beta - C_D(1 - \sin^2\beta + f^2 - 2f\cos\beta)^{\frac{1}{2}})(1 + f^2 - 2f\cos\beta)^{\frac{1}{2}},$$

with $f = U/u_{in}$. For $C_L = 0$, the maximum C_p is $4C_D/27$, obtained for $f = 1/3$ and $\beta = 0$, whereas the maximum C_p for high lift-to-drag ratios L/D is obtained for β close to $1/2\pi$ and f around $2C_L/(3C_D)$. In this case, the maximum C_p may exceed C_L by one to two orders of magnitude (Wilson and Lissaman, 1974).

It is quite difficult to maintain the high speeds U required for optimum performance in a linear motion of the airfoil, and it is natural to focus the attention on rotating devices, in case the desired energy form is shaft or electric power and not propulsion. Wind-driven propulsion in the past (mostly of ships at sea) has been restricted to U/u_{in} values far below the optimum region for high L/D airfoils (owing to friction against the water), and wind-driven propulsion on land or in the air has received little attention.

4.3.2 Propeller-type Converters

Propellers have been extensively used in aircraft to propel the air in a direction parallel to that of the propeller axis, thereby providing the necessary lift force on the airplane wings. Propeller-type rotors are similarly used for windmills, but here the motion of the air (i.e., the wind) makes the propeller, which should be placed with its axis parallel to the wind direction, rotate, thus providing the possibility of power extraction. The propeller consists of a number of blades that are evenly distributed around the axis (cf. Fig. 4.15), with each blade having an aerodynamic profile designed to produce a high lift force, as discussed in section 4.3.1. If there are two or more blades, the symmetrical mounting ensures a symmetrical mass distribution, but if only one blade is used it must be balanced by a counterweight.

Theory of Non-interacting Streamtubes

In order to describe the performance of a propeller-type rotor, the forces on each element of the blade must be calculated, including the forces produced by the direct action of the wind field on each individual element as well as the forces arising as a result of interactions between different elements on the same or on different blades. Since the simple airfoil theory outlined in section 4.3.1 deals only with the forces on a given blade segment, in the absence of the other ones and also without the inclusion of "edge effects" from "cutting out" this one segment, it is tempting as a first approximation to treat the different radial segments of each blade as independent. Owing to the symmetrical mounting of blades, the corresponding radial segments of different blades (if more than one) may be treated together for a uniform wind field, considering, as indicated in Fig. 4.15, an annulus-shaped *streamtube* of flow, bordered by streamlines intersecting the rotor plane at radial distances r and $r + dr$. The approximation of independent contributions from each streamtube, implying that radially induced velocities (u_r in Fig. 4.10) are neglected, allows the total shaft power to be expressed as

$$E = \int_0^R \frac{dE}{dr} dr,$$

where dE/dr depends only on the conditions of the streamtube at r. Similar sums of independent contributions can in the same order of approximation be used to describe other overall quantities, such as the axial force component T and the torque Q (equal to the power E divided by the angular velocity Ω of the propeller).

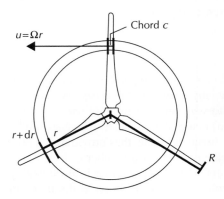

$u = \Omega r$

Chord c

$r + dr$ r

R

FIGURE 4.15   Definition of the streamtubes and blade integration variable for a propeller-type rotor.

In Fig. 4.16, a section of a wing profile (blade profile) is seen from a direction perpendicular to the cut. The distance of the blade segment from the axis of rotation is r, and its velocity $r\Omega$ is directed along the y-axis. This defines a co-ordinate system with a fixed x-axis along the rotor axis and moving y- and z-axes such that the blade segment is fixed relative to the co-ordinate system.

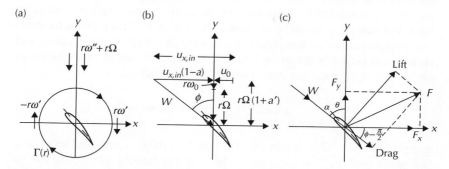

FIGURE 4.16a,b,c Velocity and force components of a rotating blade segment in a co-ordinate system following the rotation (which is in the direction of the y-axis): (a) determination of induced angular velocities; (b) determination of the direction of the apparent wind velocity; (c) determination of force components along x- and y-axes.

In order to utilize the method developed in section 4.3.1, the apparent wind velocity W to be used in (4.59) must be determined. It is given by (4.58) only if the velocity induced by, and remaining in the wake of, the device, $\boldsymbol{u}^{ind} = \boldsymbol{u}_{out} - \boldsymbol{u}_{in}$ (cf. Fig. 4.10), is negligible. Since the radial component u_r of $\boldsymbol{u}^{ind}$ has been neglected, $\boldsymbol{u}^{ind}$ has two components, one along the x-axis,

$$\boldsymbol{u}_x^{ind} = u_{x,out} - u_{x,in} = -2au_{x,in}$$

[cf. (4.52)], and one in the tangential direction,

$$u_t^{ind} = \omega_r^{ind} = 2a'\Omega\,r,$$

when expressed in a nonrotating co-ordinate system (the second equality defines a quantity a', called the *tangential interference factor*).

W is determined by the air velocity components in the rotor plane. From the momentum considerations underlying (4.52), the induced x-component u_0 in the rotor plane is seen to be

$$u_0 = 1/2u_x^{ind} = -au_{x,in}. \tag{4.60}$$

In order to determine $r\omega_0$, the induced velocity of rotation of the air in the rotor plane, one may use the following argument (Glauert, 1935). The induced rotation is partly due to the rotation of the wings and partly due to the non-zero circulation (4.57) around the blade segment profiles (see Fig. 4.16a). This circulation may be considered to imply an induced air velocity component of the same magnitude,

$|r\omega'|$, but with opposite direction in front of and behind the rotor plane. If the magnitude of the component induced by the wing rotation is called $r\omega''$ in the fixed co-ordinate system, it will be $r\omega'' + r\Omega$ and directed along the negative y-axis in the co-ordinate system following the blade. It has the same sign in front of and behind the rotor. The total y-component of the induced air velocity in front of the rotor is then $-(r\omega'' + r\Omega + r\omega')$, and the tangential component of the induced air velocity in the fixed co-ordinate system is $-(r\omega''- r\omega')$, still in front of the rotor. But here there should be no induced velocity at all. This follows, for example, from performing a closed line integral of the air velocity v along a circle with radius r and perpendicular to the x-axis. This integral should be zero because the air in front of the rotor has been assumed to be irrotational,

$$0 = \int_{circle\ area} \text{rot}\,v\,dA = \oint v \cdot ds = -2\pi r(r\omega''-r\omega'),$$

implying $\omega'' = \omega'$. This identity also fixes the total induced tangential air velocity behind the rotor, $r\omega^{ind} = r\omega'' + r\omega' = 2r\omega''$, and in the rotor plane (note that the circulation component $r\omega'$ is here perpendicular to the y-axis),

$$r\omega_0 = r\omega'' = 1/2 r\omega^{ind} = \Omega\,ra'. \tag{4.61}$$

The apparent wind velocity in the co-ordinate system moving with the blade segment, W, is now determined as seen in Fig. 4-16b. Its x-component is u_x, given by (4.52), and its y-component is obtained by taking (4.61) to the rotating co-ordinate system,

$$W_x = u_x = u_{x,in}(1 - a), \tag{4.62a}$$

$$W_y = -(r\omega_x + r\Omega) = -r\Omega\,(1 + a'). \tag{4.62b}$$

The lift and drag forces (Fig. 4.16c) are now obtained from (4.61) (except that the two components are no longer directed along the co-ordinate axes), with values of C_D, C_L, and c pertaining to the segments of wings at the streamtube intersecting the rotor plane at the distance r from the axis. As indicated in Figs. 4.16b and c, the angle of attack, α, is the difference between the angle ϕ, determined by

$$\tan\phi = -W_x/W_y, \tag{4.63}$$

and the pitch angle θ between blade and rotor plane,

$$\alpha = \phi - \theta. \tag{4.64}$$

Referred to the co-ordinate axes of Fig. 4.16, the force components for a single blade segment become

$$F_x = 1/2\rho\,cW^2(C_D(\alpha)\,\sin\phi + C_L(\alpha)\,\cos\phi), \tag{4.65a}$$

$$F_y = 1/2\rho\,cW^2(-C_D(\alpha)\,\cos\phi + C_L(\alpha)\,\sin\phi), \tag{4.65b}$$

and the axial force and torque contributions from a streamtube with B individual blades are given by

$$dT/dr = BF(r),$$ (4.66a)

$$dQ/dr = BrF_y(r).$$ (4.66b)

Combining (4.66) with equations expressing momentum and angular momentum conservation for each of the (assumed non-interacting) streamtubes, a closed set of equations is obtained. Equating the momentum change of the wind in the x-direction to the axial force on the rotor part intersected by an individual streamtube [i.e., (4.48) and (4.49) with $A = 2\pi r$], one obtains

$$dT/dr = -A\rho u_x u_x^{ind} = 2\pi r\rho u_{x,in}(1-a)\,2au_{x,in},$$ (4.67)

and similarly equating the torque on the streamtube rotor part to the change in angular momentum of the wind (from zero to $r \times u_t^{ind}$), one gets

$$dQ/dr = A\rho u_x u_t^{ind} = 2\pi r^2\rho u_{x,in}(1-a)2a'\Omega r.$$ (4.68)

Inserting $W = u_{x,in}(1-a)/\sin \phi$ or $W = r\Omega (1+a')/\cos \phi$ (cf. Fig. 4.16c) as necessary, one obtains a and a' expressed in terms of ϕ by combining (4.65) – (4.68). Since, on the other hand, ϕ depends on a and a' through (4.63) and (4.62), an iterative method of solution should be used. Once a consistent set of (a, a', ϕ)-values has been determined as function of r [using a given blade profile implying known values of θ, c, $C_D(\alpha)$, and $C_L(\alpha)$ as function of r], either (4.66) or (4.67) and (4.68) may be integrated over r to yield the total axial force, total torque, or total shaft power $E = \Omega Q$.

One may also determine the contribution of a single streamtube to the power coefficient (4.55),

$$C_p(r) = \frac{\Omega \, dQ/dr}{\frac{1}{2} \rho u_{x,in}^3 2\pi r} = 4a'(1-a)\left(\frac{r\Omega}{u_{x,in}}\right)^2.$$ (4.69)

The design of a rotor may utilize $C_p(r)$ for a given wind speed $u_{x,in}^{design}$ to optimize the choice of blade profile (C_D and C_L), pitch angle (θ), and solidity ($Bc/(\pi r)$) for given angular velocity of rotation (Ω). If the angular velocity Ω is not fixed (as it might be by use of a suitable asynchronous electrical generator, except at start and stop), a dynamic calculation involving $d\Omega/dt$ must be performed. Not all rotor parameters need to be fixed. For example, the pitch angles may be variable, by rotation of the entire wing around a radial axis, such that all pitch angles θ (r) may be modified by an additive constant θ_0. This type of regulation is useful in order to limit the C_p-drop when the wind speed moves away from the design value $u_{x.in}^{design}$. The expressions given above would then give the actual C_p for the actual setting of the pitch angles given by θ $(r)_{design} + \theta_0$ and the actual wind speed $u_{x,in}$, with all other parameters left unchanged.

   In introducing the streamtube expressions (4.67) and (4.68), it has been
assumed that the induced velocities, and thus a and a', are constant along
the circle periphery of radius r in the rotor plane. As the model used to estimate the magnitude of the induced velocities (Fig. 4.16a) is based on being
near to a rotor blade, it is expected that the average induced velocities in
the rotor plane, u_0 and $r\omega_0$, are smaller than the values calculated by the
above expressions, unless the solidity is very large. In practical applications,
it is customary to compensate by multiplying a and a' by a common factor, F,
less than unity and a function of B, r, and ϕ (see, for example, Wilson and
Lissaman, 1974).
   Furthermore, edge effects associated with the finite length R of the rotor
wings have been neglected, as have the "edge" effects at $r = 0$ due to the
presence of the axis, transmission machinery, etc. These effects may be
described in terms of *trailing vortices* shed from the blade tips and blade
roots and moving in helical orbits away from the rotor in its wake. Vorticity
is a name for the vector field rot$v$, and the vortices connected with the circulation around the blade profiles (Fig. 4.16a) are called *bound vorticity*. This
can "leave" the blade only at the tip or at the root. The removal of bound vorticity is equivalent to a loss of circulation I (4.57) and hence a reduction in
the lift force. Its dominant effect is on the tangential interference factor a'
in (4.68), and so it has the same form and may be treated on the same footing
as the corrections due to finite blade number. (Both are often referred to as *tip
losses*, since the correction due to finite blade number is usually appreciable
only near the blade tips, and they may be approximately described by the
factor F introduced above.) Other losses may be associated with the "tower
shadow," etc.

Model Behavior of Power Output and Matching to Load

A calculated overall C_p for a three-bladed, propeller-type wind energy converter
is shown in Fig. 4.17, as a function of the tip-speed ratio,

$$\lambda = \Omega R/u_{x,in},$$

and for different settings of the overall pitch angle, θ_0. It is clear from (4.69) and
the equations for determining a and a' that C_p depends on the angular
velocity Ω and the wind speed $u_{x,in}$ only through the ratio λ. Each blade has
been assumed to be an airfoil of the type NACA 23012 (cf. Fig. 4.12), with
chord c and twist angle θ changing along the blade from root to tip, as indicated
in Fig. 4.18. A tip-loss factor F has been included in the calculation of sets of
corresponding values of a, a', and ϕ for each radial station (each "streamtube"),
according to the Prandtl model described by Wilson and Lissaman (1974). The
dashed regions in Fig. 4.17 correspond to values of the product aF (corresponding to a in the expression without tip loss) larger than 0.5. Since the axial air
velocity in the wake is $u_{x,out} = u_{x,in}(1 - 2aF)$ [cf. (4.52) with $F = 1$], $aF > 0.5$

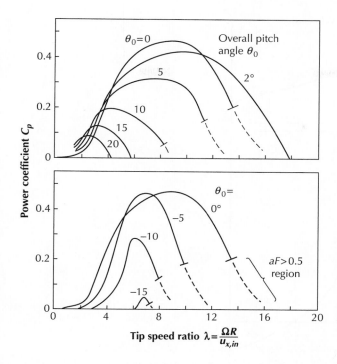

FIGURE 4.17 Dependence of power coefficient on tip-speed ratio for different settings of the overall pitch angle. The calculation is based on the wing geometry shown in Fig. 4.18 and the NACA 23012 profile of Fig. 4.12.

implies reversed or re-circulating flow behind the wind energy converter, a possibility that has not been included in the above description. The C_p values in the dashed regions may thus be inaccurate and presumably overestimated.

Figure 4.17 shows that, for negative settings of the overall pitch angle θ_0, the C_p distribution on λ values is narrow, whereas it is broad for positive θ_0 of modest size. For large positive θ_0, the important region of C_p moves to smaller λ values. The behavior of $C_p(\lambda)$ in the limit of λ approaching zero is important for operating the wind energy converter at small rotor angular velocities and, in particular, for determining whether the rotor will be self-starting, as discussed in more detail below.

For a given angular speed Ω of the rotor blades, the power coefficient curve specifies the fraction of the power in the wind that is converted, as a function of the wind speed $u_{x,in}$. Multiplying C_p by $1/2\ \rho A u_{x,in}^3$, where the rotor area is $A = \pi R^2$ (if "coning" is disregarded, i.e., the blades are assumed to be in the plane of rotation), the power transferred to the shaft can be obtained as a function of wind speed, assuming the converter to be oriented ("yawed") such that the rotor plane is perpendicular to the direction of the incoming wind. Figures 4.19 and 4.20 show such plots of power E, as functions of wind speed and overall pitch angle, for two definite angular velocities ($\Omega = 4.185$ and 2.222 rad s^{-1}, if the length of the wings is $R = 27$ m). For the range of wind speeds

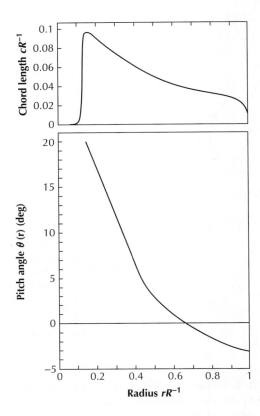

FIGURE 4.18 Chord variation and twist along the blades used for the calculations in this section.

encountered in the planetary boundary layer of the atmosphere (e.g., at heights between 25 and 100 m), a maximum power level around 4000 W per average square meter swept by the rotor (10 MW total) is reached for the device with the higher rotational speed (tip speed $R\Omega = 113$ ms^{-1}), whereas about 1000 W m^{-2} is reached by the device with the lower rotational velocity ($R\Omega = 60$ m s^{-1}).

The rotor has been designed to yield a maximum C_p at about $\lambda = 9$ (Fig. 4.17), corresponding to $u_{x,in} = 12.6$ and 6.6 m s^{-1} in the two cases. At wind speeds above these values, the total power varies less and less strongly, and for negative pitch angles it even starts to decrease with increasing wind speeds in the range characterizing "stormy weather." This phenomenon can be used to make the wind energy converter self-regulating, provided that the "flat" or "decreasing" power regions are suitably chosen and that the constancy of the angular velocity at operation is ensured by, for example, coupling the shaft to a suitable asynchronous electricity generator (a generator allowing only minute changes in rotational velocity, with the generator angular speed Ω_g being fixed relative to the shaft angular speed Ω, e.g., by a constant exchange ratio $n = \Omega_g/\Omega$ of a gearbox). Self-regulating wind energy converters with fixed

FIGURE 4.19 Power output of the wind energy converter described in Figs. 4.17 and 4.18, for a tip speed of 113 m s^{-1}, as a function of wind speed and overall pitch angle. The actual Reynolds number is in this case about 5×10^6 from blade root to tip, in reasonable consistency with the value used in the airfoil data of Fig. 4.12. Dashed lines indicate regions where a [defined in (4.62)] times the tip speed correction, F, exceeds 0.5 for some radial stations along the blades.

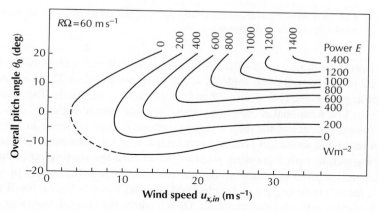

FIGURE 4.20 Same as Fig. 4.19, but for a tip speed of 60 ms^{-1}.

overall pitch angle θ_0 and constant angular velocity have long been used for AC electricity generation (Juul, 1961).

According to its definition, the torque $Q = E/\Omega$ can be written

$$Q = 1/2\rho\pi R^3 u_{x,in}^2 C_p(\lambda)/\lambda, \quad (4.70)$$

and Fig. 4.21 shows C_p/λ as a function of λ for the same design as the one considered in Fig. 4.17. For pitch angles θ_0 less than about $-3°$, C_p/λ is negative for $\lambda = 0$, whereas the value is positive if θ_0 is above the critical value of about $-3°$.

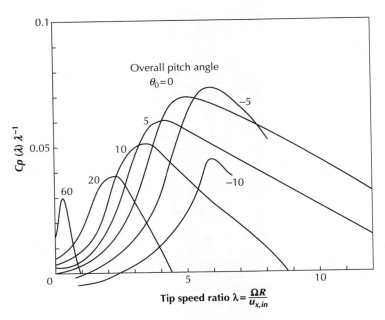

FIGURE 4.21   Variation of power coefficient over tip-speed ratio, as a function of tip-speed ratio, for the converter considered in the previous figures.

The corresponding values of the total torque (4.70), for a rotor with radius $R = 27$ m and overall pitch angle $\theta_0 = 0°$, are shown in Fig. 4.22, as a function of the angular velocity $\Omega = \lambda u_{x,in}/R$ and for different wind speeds. The small dip in Q for low rotational speeds disappears for higher pitch angles θ_0.

The dependence of the torque at $\Omega = 0$ on pitch angle is shown in Fig. 4.23 for a few wind speeds. Advantage can be taken of the substantial increase in starting torque with increasing pitch angle, in case the starting torque at the pitch angle desired at operating angular velocities is insufficient and provided the overall pitch angle can be changed. In that case, a high overall pitch angle is chosen to start the rotor from $\Omega = 0$, where the internal resistance (friction in bearings, gearbox, etc.) is large. When an angular speed Ω of a few degrees per second is reached, the pitch angle is diminished to a value close to the optimal one (otherwise the torque will pass a maximum and soon start to decrease again, as seen from, for example, the $\theta_0 = 60°$ curve in Fig. 4.21). Usually, the internal resistance diminishes as soon as Ω is non-zero. The internal resistance may be represented by an internal torque, $Q_0(\Omega)$, so that the torque available for some load, "external" to the wind power to shaft power converter (e.g., an electric generator), may be written

$$Q^{available} = Q(\Omega) - Q_0(\Omega).$$

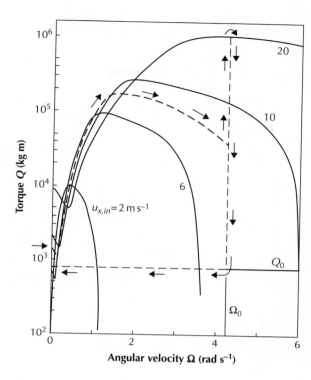

FIGURE 4.22 Torque variation with angular velocity for a wind energy converter of the type described in the previous figures, with radius $R = 27$ m and overall pitch angle $0°$. The dashed, vertical line represents operation at fixed angular velocity (e.g., with a synchronous generator). Q_0 illustrates the internal torque.

If the "load" is characterized by demanding a fixed rotational speed Ω (such as the asynchronous electric generator), it may be represented by a vertical line in the diagram shown in Fig. 4.22. As an example of the operation of a wind energy generator of this type, the dashed lines (with accompanying arrows) in Fig. 4.22 describe a situation with initial wind speed of 8 m s^{-1}, assuming that this provides enough torque to start the rotor at the fixed overall pitch angle $\theta_0 = 0°$ (i.e., the torque produced by the wind at 8 m s^{-1} and $\Omega = 0$ is above the internal torque Q_0, here assumed to be constant). The excess torque makes the angular velocity of the rotor, Ω, increase along the specific curve for $u_{x,in} = 8$ m s^{-1}, until the value Ω_0 characterizing the electric generator is reached. At this point, the load is connected and the wind energy converter begins to deliver power to the load area. If the wind speed later increases, the torque will increase along the vertical line at $\Omega = \Omega_0$ in Fig. 4.22, and if the wind speed decreases, the torque will decline along the same vertical line until it reaches the internal torque value Q_0. Then the angular velocity diminishes and the rotor is brought to a halt.

An alternative type of load may not require a constant angular velocity. Synchronous DC electric generators are of this type, providing an increasing power output with increasing angular velocity (assuming a fixed exchange

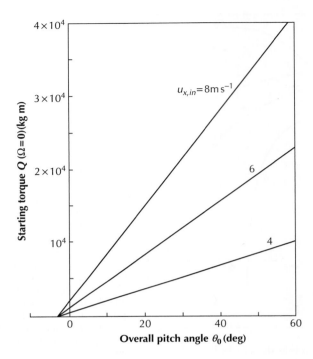

FIGURE 4.23 Starting torque as a function of overall pitch angle for the wind energy converter considered in the preceding figures.

ratio between rotor angular velocity Ω and generator angular velocity Ω_g). Instead of staying on a vertical line in the torque-versus-Ω diagram, for varying wind speed, the torque now varies along some fixed, monotonically increasing curve characterizing the generator. An optimal synchronous generator would be characterized by a $Q(\Omega)$ curve that for each wind speed $u_{x,in}$ corresponds to the value of $\Omega = \lambda u_{x,in}/R$ that provides the maximum power coefficient C_p (Fig. 4.17). This situation is illustrated in Fig. 4.24, with a set of dashed curves again indicating the torque variation for an initial wind speed of 8 m s^{-1}, followed by an increasing and later again decreasing wind speed. In this case, the $Q = Q_0$ limit is finally reached at a very low angular velocity, indicating that power is still delivered during the major part of the slowing-down process.

Non-uniform Wind Velocity

The velocity field of the wind may be non-uniform in time as well as in spatial distribution. The influence of time variations in wind speed on power output of a propeller-type wind energy converter is touched upon in the previous subsection, although a detailed investigation involving the actual time dependence of the angular velocity Ω is not included. In general, the direction of the wind velocity is also time dependent, and the conversion device should be able to successively align its rotor axis with the long-range trends in wind direction,

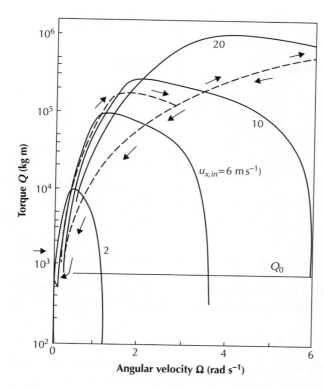

FIGURE 4.24 Same as Fig. 4.22, but here operation is illustrated for a load (e.g., asynchronous generator) that is optimized, i.e., corresponds to maximum C_p at each wind speed.

or suffer a power reduction that is not just the cosine to the angle β between the rotor axis and the wind direction (*yaw angle*), but involves calculating the performance of each blade segment for an apparent wind speed W and angle of attack α different from the ones previously used, and no longer axially symmetric. This means that the quantities W and α (or ϕ) are no longer the same inside a given annular streamtube, but they depend on the directional position of the segment considered (e.g., characterized by a rotational angle ψ in a plane perpendicular to the rotor axis).

Assuming that both the wind direction and the wind speed are functions of time and of height h (measured from ground level or from the lower boundary of the velocity profile z_0 discussed in section 2.5.2), but that the direction remains horizontal, then the situation will be as depicted in Fig. 4.25. The co-ordinate system (x_0, y_0, z_0) is fixed and has its origin in hub height h_0, where the rotor blades are fastened. Consider now a blade element at a distance r from the origin, on the ith blade. The projection of this position on to the (y_0, z_0)-plane is turned the angle ψ_i from vertical, where

$$\psi_i = 2\pi/i + \Omega t, \tag{4.71}$$

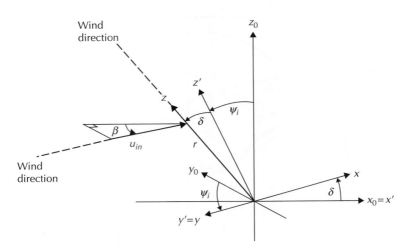

FIGURE 4.25   Definition of co-ordinate systems for a coning rotor in a non-uniform wind field.

at the time t. The coning angle δ is the angle between the blade and its projection onto the (y_0, z_0)-plane, and the height of the blade element above the ground is

$$h = h_0 + r \cos \delta \cos \psi_i, \qquad (4.72)$$

where h_0 is the hub height. The height h enters as a parameter in the wind speed $u_{in} = |\boldsymbol{u}_{in}(h, t)|$ and the yaw angle $\beta = \beta\,(h, t)$, in addition to time.

Now, an attempt can be made to copy the procedure used for a uniform wind velocity along the rotor axis, i.e., to evaluate the force components for an individual streamtube both by the momentum consideration and by the lift and drag approach of section 4.3.1. The individual streamtubes can no longer be taken as annuli, but must be of an area A_s (perpendicular to the wind direction) small enough to permit the neglect of variations in wind velocity over the area at a given time. It will still be assumed that the flows inside different streamtubes do not interact, and also, for simplicity, the streamtubes will be treated as "straight lines," i.e., not expanding or bending as they pass the region of the conversion device (as before, this situation arises when induced radial velocities are left out).

Consider first the "momentum equation" (4.48), with s denoting "in the streamwise direction" or "along the streamtube,"

$$F_s = -J_m u_s^{ind}.$$

This force has components along the x-, y-, and z-directions of the local co-ordinate system of a blade turned the angle ψ from vertical (cf. Fig. 4.69). The y-direction is tangential to the rotation of the blade element, and, in addition to the component of F_s, there may be an induced tangential velocity u_t^{ind}

(and force) of the same type as the one considered in the absence of yaw (in which case F_s is perpendicular to the y-axis). The total force components in the local co-ordinate system are thus

$$F_x = -J_m u_s^{ind} (\cos \beta \cos \delta + \sin \beta \sin \delta \sin \psi),$$
$$F_y = -J_m u_s^{ind} \sin \beta \cos \psi + J_m u_t^{ind}. \tag{4.73}$$

From the discussion of Fig. 4.10, (4.49), and (4.52),

$$J_m = A_s \rho u_s = A_s \rho u_{in}(1-a),$$
$$u_s^{ind} = -2a u_{in},$$

and, in analogy with (4.61), a tangential interference factor a' may be defined by

$$u_t^{ind} = 2a' r \Omega \cos \delta.$$

However, the other relation contained in (4.61), from which the induced tangential velocity in the rotor plane is half the one in the wake, cannot be derived in the same way without the assumption of axial symmetry. Instead, the variation in the induced tangential velocity as a function of rotational angle ψ is bound to lead to crossing of the helical wake strains and it will be difficult to maintain the assumption of non-interacting streamtubes. Here, the induced tangential velocity in the rotor plane will still be taken as $1/2 u_t^{ind}$, an assumption which at least gives reasonable results in the limiting case of a nearly uniform incident wind field and zero or very small yaw angle.

Secondly, the force components may be evaluated from (4.63)–(4.65) for each blade segment, defining the local co-ordinate system (x, y, z) as in Fig. 4.25 with the z-axis along the blade and the y-axis in the direction of the blade's rotational motion. The total force, averaged over a rotational period, is obtained by multiplying by the number of blades, B, and by the fraction of time each blade spends in the streamtube. Defining the streamtube dimensions by an increment dr in the z-direction and an increment $d\psi$ in the rotational angle ψ, each blade spends the time fraction $d\psi/(2\pi)$ in the streamtube, at constant angular velocity Ω. The force components are then

$$F_x = B(d\psi/2\pi)1/2\rho c W^2 (C_D(\alpha) \sin \phi + C_L(\alpha) \cos \phi) dr,$$
$$F_y = B(d\psi/2\pi)1/2\rho c W^2 (-C_D(\alpha) \cos \phi + C_L(\alpha) \sin \phi) dr, \tag{4.74}$$

in the notation of (4.65). The angles ϕ and α are given by (4.63) and (4.64), but the apparent velocity W is the vector difference between the streamwise velocity $u_{in} + 1/2 u_s^{ind}$ and the tangential velocity $r\Omega \cos \delta + 1/2 u_t^{ind}$, both taken in the rotor plane. From Fig. 4.25,

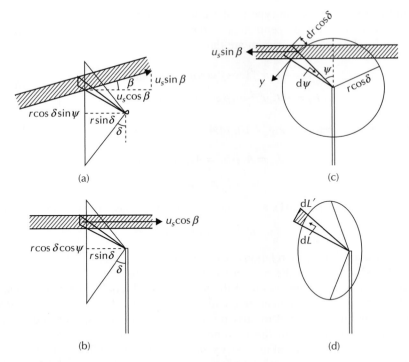

FIGURE 4.26a,b,c,d Streamtube definition (*hatched areas*) for a coning rotor in a non-uniform wind field: (*a*) view from top; (*b*) side view; (*c*) front view; (*d*) view along streamtube.

$$W_x = u_{in}(1-a)(\cos\beta\cos\delta + \sin\beta\sin\delta\sin\psi),$$
$$W_y = u_{in}(1-a)\sin\beta\cos\psi + r\Omega\cos\delta(1+a'), \qquad (4.75)$$
$$W_z = u_{in}(1-a)(-\cos\beta\sin\delta + \sin\beta\cos\delta\sin\psi).$$

The appropriate W^2 to insert into (4.74) is $W_x^2 + W_y^2$. Finally, the relation between the streamtube area A_s and the increments dr and $d\psi$ must be established. As indicated in Fig. 4.26, the streamtube is approximately a rectangle with sides dL and dL', given by

$$dL = r\cos\delta(\cos 2\beta\cos 2\psi + \sin 2\psi)^{1/2}d\psi,$$
$$dL'' = (\sin 2\delta(1 + \cos 2\beta) + \cos 2\delta(\cos 2\beta\sin 2\psi + \cos 2\psi))^{1/2}dr, \qquad (4.76)$$

and

$$A_s = dL\,dL'.$$

Now, for each streamtube, a and a' are obtained by equating the x- and y-components of (4.73) and (4.74) and using the auxiliary equations for

(W, ϕ) or (W_x, W_y). The total thrust and torque are obtained by integrating F_x and $r \cos \delta F_y$, over dr and $d\psi$ (i.e., over all streamtubes).

Restoration of Wind Profile in Wake, and Implications for Turbine Arrays

For a wind energy converter placed in the planetary boundary layer (i.e., in the lowest part of the Earth's atmosphere), the reduced wake wind speed in the streamwise direction, $u_{s,out}$, will not remain below the wind speed $u_{s,in}$ of the initial wind field, provided that this is not diminishing with time. The processes responsible for maintaining the general kinetic motion in the atmosphere (cf. section 2.3.1), making up for the kinetic energy lost by surface friction and other dissipative processes, will also act in the direction of restoring the initial wind profile (speed as function of height) in the wake of a power-extracting device by transferring energy from higher air layers (or from the "sides") to the partially depleted region (Sørensen, 1996d). The large amounts of energy available at greater altitude (cf. Fig. 3.31) make such processes possible almost everywhere at the Earth's surface and not just at those locations where new kinetic energy is predominantly being created (cf. Fig. 2.60).

In the near wake, the wind field is nonlaminar, owing to the induced tangential velocity component, u_t^{ind}, and owing to vorticity shed from the wing tips and the hub region (cf. discussion of Fig. 4.16 and (4.57)). It is then expected that these turbulent components gradually disappear further downstream in the wake, as a result of interactions with the random eddy motion of different scales present in the "unperturbed" wind field. "Disappear" here means "get distributed on a large number of individual degrees of freedom," so that no contributions to the time-averaged quantities considered [cf. (2.21)] remain. For typical operation of a propeller-type wind conversion device, such as the situations illustrated in Figs. 4.19 and 4.20, the tangential interference factor a' is small compared to the axial interference factor a, implying that the most visible effect of the passage of the wind field through the rotor region is the change in streamwise wind speed. Based on wind tunnel measurements, this is a function of r, the distance of the blade segment from the hub center, as illustrated in Fig. 4.27. The induced r-dependence of the axial velocity is seen to gradually smear out, although it is clearly visible at a distance of two rotor radii in the set-up studied.

Figure 4.28 suggests that, under average atmospheric conditions, the axial velocity, $u_{s,in}$, will be restored to better than 90% at about 10 rotor diameters behind the rotor plane and better than 80% at a distance of 5–6 rotor diameters behind the rotor plane, but restoration is rather strongly dependent on the amount of turbulence in the "undisturbed" wind field.

A second wind energy converter may be placed behind the first one, in the wind direction, at a location where the wind profile and magnitude are reasonably well restored. According to simplified investigations in wind tunnels (Fig. 4.28), supported by field measurements behind buildings, forests, and fences, a suitable

FIGURE 4.27   Wind tunnel results illustrating the distortion of the wind field along the radial direction (R is the rotor radius) at various distances from the rotor plane. (Based on Hütter, 1976.)

distance would seem to be 5–10 rotor diameters (increasing to over 20 rotor diameters if "complete restoration" is required). If there is a prevailing wind direction, the distance between conversion units perpendicular to this direction may be smaller (essentially determined by the induced radial velocities, which were neglected in the preceding subsections, but appear qualitatively in Fig. 4.10). If, on the other hand, several wind directions are important, and the converters

FIGURE 4.28   Wind tunnel results indicating the restoration at streamwise wind speed behind a rotor (placed at distance 0, the rotor being simulated by a gauze disc). The approaching wind has a logarithmic profile (*solid line*) or is uniform with approximately laminar flow (*dashed line*). (Based on Pelser, 1975.)

are designed to be able to "yaw against the wind," then the distance required by wake considerations should be kept in all directions.

More severe limitations may possibly be encountered with a larger array of converters, say, distributed over an extended area with average spacing X between units. Even if X is chosen so that the relative loss in streamwise wind speed is small from one unit to the next, the accumulated effect may be substantial, and the entire boundary layer circulation may become altered in such a way that the power extracted decreases more sharply than expected from the simple wake considerations. Thus, large-scale conversion of wind energy may even be capable of inducing local climatic changes.

A detailed investigation of the mutual effects of an extended array of wind energy converters and the general circulation on each other requires a combination of a model of the atmospheric motion, e.g., along the lines presented in section 2.3.1, with a suitable model of the disturbances induced in the wake of individual converters. In one of the first discussions of this problem, Templin (1976) considered the influence of an infinite two-dimensional array of wind energy converters with fixed average spacing on the boundary layer motion to be restricted to a change of the roughness length z_0 in the logarithmic expression (see section 2.5.1) for the wind profile, assumed to describe the wind approaching any converter in the array. The change in z_0 can be calculated from the stress τ^{ind} exerted by the converters on the wind, due to the axial force F_x in (4.65) or (4.67), which according to (4.51) can be written

$$\tau^{ind} = F_x/S = 1/2\rho(u_{s,in})^2\, 2(1-a)^2 A/S,$$

FIGURE 4.29 Calculated (see text) and measured power outputs from one row of turbines perpendicular to the wind direction, in a wind park located at Nørrekær Enge in Denmark. It consists of a total of five rows with a turbine spacing of 6 rotor diameters (Taylor *et al.*, 1993).

with S being the average ground surface area available for each converter and A/S being the "density parameter," equal to the ratio of rotor-swept area to ground area. For a quadratic array with regular spacing, S may be taken as X^2. According to section 2.5.1, $u_{x,in}$ taken at hub height h_0 can, in the case of a neutral atmosphere, be written

$$u_{x,in}(h_0) = \frac{1}{\kappa} \left(\frac{\tau^0 + \tau^{ind}}{\rho} \right)^{1/2} \log\left(\frac{h_0}{z_0'} \right),$$

where τ^0 is the stress in the absence of wind energy converters, and z_0' is the roughness length in the presence of the converters, which can now be determined from this equation and τ^{ind} from the previous one.

Figure 4.29 shows the results of a calculation for a finite array of converters (Taylor *et al.*, 1993) using a simple model with fixed loss fractions (Jensen, 1994). This model is incapable of reproducing the fast restoration of winds through the turbine array, presumably associated with the propagation of the enhanced wind regions created just outside the swept areas (as seen in Fig. 4.27). A three-dimensional fluid dynamics model is required for describing the details of array shadowing effects. Such calculations are in principle possible, but so far no convincing implementation has been presented. The problem is the very accurate description needed for the complex three-dimensional flows around the turbines and for volumes comprising the entire wind farm of maybe hundreds of turbines. This is an intermediate regime between the existing three-dimensional models for gross wind flow over complex terrain and the detailed models of flow around a single turbine used in calculations of aerodynamic stability.

Offshore Foundation Issues

The possibility of placing wind turbines offshore, typically in shallow waters of up to 20 m depth, relies on use of low-cost foundation methods developed earlier for harbor and oil-well uses. The best design depends on the material constituting the local water floor, as well as local hydrological conditions, including strength of currents and icing problems. Breakers are currently used to prevent ice from damaging the structure. The most common structure in place today is a concrete caisson, as shown in Fig. 4.30a, but in the interest of cost minimizing, the steel solutions shown in Fig. 4.30b,c have received increasing attention and the monopile (Fig. 4.30c) was selected for the recent Samsø wind farm in the middle of Denmark (Birch and Gormsen, 1999; Offshore Windenergy Europe, 2003). Employing the suction effect (Fig. 4.30b) may help cope with short-term gusting forces, while the general stability must be ensured by the properties of the overall structure itself. The possibility of offshore employment is of course relevant for all types of wind converters, including those described below in section 4.3.3.

FIGURE 4.30a,b,c Some foundation types for offshore wind turbine towers placed in shallow waters; (*a*) sand-filled caisson standing on the bottom, (*b*) suction buckets, (*c*) steel monopile. These foundation types are in use at locations with clay-till water floor material.

4.3.3 Cross-wind and Other Alternative Converter Concepts

Wind energy converters of the cross-wind type have the rotor axis perpendicular to the wind direction. The rotor axis may be horizontal, as in wheel-type converters (in analogy to waterwheels), or vertical, as in the panemones used in Iran and China. The blades (ranging from simple "paddles" to optimized airfoil sections) will be moving with and against the wind direction on alternative sides of the rotor axis, necessitating some way of emphasizing the forces acting on the blades on one side. Possible ways are simply to shield half of the swept area, as in the Persian panemones (Wulff, 1966); to curve the "paddles" so that the (drag) forces are smaller on the convex than on the concave side, as in the Savonius rotor (Savonius, 1931); or to use aerodynamically shaped wing blades producing high lift forces for wind incident on the "front edge," but small or inadequate forces for wind incident on the "back edge," as in the Darrieus rotor (cf. Fig. 4.31) and related concepts. Another possibility is to allow for changes in the pitch angle of each blade, as, for example, achieved by hinged vertical blades in the Chinese panemone type (Chen Li, 1951). In this case the blades on one side of a vertical axis have favorable pitch angles, while those on the other side have unfavorable settings. Apart from shielded ones, vertical axis cross-wind converters are omnidirectional, i.e., they accept any horizontal wind direction on equal footing.

Performance of a Darrieus-type Converter

The performance of a cross-wind converter, such as the Darrieus rotor shown in Fig. 4.31, may be calculated in a way similar to that used in section 4.3.2, deriving the unknown, induced velocities by comparing expressions of the forces in terms of lift and drag on the blade segments with expressions in terms of the momentum changes between incident wind and wake flow. Also, the flow may be divided into a number of streamtubes (assumed to be non-interacting), according to assumptions about the symmetries of the flow field. Figure 4.21 gives an example of the streamtube definitions for a two-bladed Darrieus rotor with angular velocity Ω and a rotational angle ψ describing the position of a given blade, according to (4.71). The blade chord, c, has been illustrated as constant, although it may actually be taken as varying to give the optimum performance for any blade segment at the distance r from the rotor axis. As in the propeller rotor case, it is not practical to extend the chord increase to the regions near the axis.

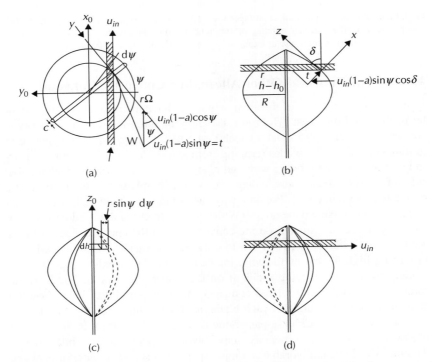

FIGURE 4.31a,b,c,d Streamtube definition (*hatched areas*) for two-bladed Darrieus rotor, and determination of apparent wind velocity, in the case of negligible cross-wind induced velocity ($u_{y0}^{ind} = 0$): (*a*) view from top; (*b*) view along tangent to blade motion; (*c*) view along streamtube; (*d*) view perpendicular to streamtube.

The bending of the blade, characterized by an angle δ (h) depending on the height h (the height of the rotor center is denoted h_0), may be taken as a troposkien curve (Blackwell and Reis, 1974), characterized by the absence of bending forces on the blades when they are rotating freely. Since the blade profiles encounter wind directions at both positive and negative forward angles, the profiles are often taken as symmetrical (e.g., NACA 00XX profiles).

Assuming, as in section 4.3.2, that the induced velocities in the rotor region are half of those in the wake, the streamtube expressions for momentum and angular momentum conservation analogous to (4.67) and (4.68) are

$$F_{x0} = -J_m u_s^{ind} = 2\rho A_s (u_{in})^2 (1 - a)a,$$
$$F_{y0} = J_m u_{c.w.}^{ind}, \tag{4.77}$$

where the axial interference factor a is defined as in (4.60), and where the streamtube area A_s corresponding to height and angular increments dh and dψ is (cf. Fig. 4.31)

$$A_s = r \sin \psi \, d\psi \, dh.$$

The cross-wind induced velocity $u_{c.w.}^{ind}$ is not of the form (4.61), since the tangent to the blade's rotational motion is not along the y_0-axis. The sign of $u_{c.w.}^{ind}$ will fluctuate with time, and for low chordal ratio c/R (R being the maximum value of r) it may be permitted to put F_{y0} equal to zero (Lissaman, 1976; Strickland, 1975). This approximation is made in the following. It is also assumed that the streamtube area does not change by passage through the rotor region and that individual streamtubes do not interact (these assumptions being the same as those made for the propeller-type rotor).

The forces along the instantaneous x- and y-axes due to the passage of the rotor blades at a fixed streamtube location (h, ψ) can be expressed in analogy to (4.74), averaged over one rotational period,

$$F_x = B(d\psi/2\pi)1/2\rho \, c \, W^2 (C_D \sin \phi + C_L \cos \phi) dh/\cos \delta, \tag{4.78a}$$

$$F_y = B(d\psi/2\pi)1/2\rho \, c \, W^2 (-C_D \cos \phi + C_L \sin \phi) dh/\cos \delta, \tag{4.78b}$$

where

$$\tan \phi = -W_x/W_y; \quad W^2 = W_x^2 + W_y^2$$

and (cf. Fig. 4.21)

$$W_x = u_{in}(1 - a) \sin \psi \, \cos \delta,$$
$$W_y = -r\Omega - u_{in}(1 - a) \cos \psi, \tag{4.79}$$
$$W_z = -u_{in}(1 - a) \sin \psi \sin \delta.$$

The angle of attack is still given by (4.64), with the pitch angle θ being the angle between the y-axis and the blade center chord line (cf. Fig. 4.16c).

FIGURE 4.32 Power co-efficient as a function of tip-speed ratio ("tip" means point furthest away from axis), for a two-bladed Darrieus rotor, with a chord ratio $c/R = 0.1$ and blade data corresponding to a symmetrical profile (NACA 0012 with Reynolds number Re $= 3 \times 10^6$). A stream-tube model of the type described in Fig. 4.31 has been used. (Based on Strickland, 1975.)

The force components (4.78) may be transformed to the fixed (x_0, y_0, z_0) co-ordinate system, yielding

$$F_{x0} = F_x \cos \delta \sin \psi + F_y \cos \psi, \qquad (4.80)$$

with the other components F_{y0} and F_{z0} being neglected due to the assumptions made. Using the auxiliary relations given, a may be determined by equating the two expressions (4.77) and (4.80) for F_{x0}.

After integration over dh and $d\psi$, the total torque Q and power coefficient C_p can be calculated. Figure 4.32 gives an example of the calculated C_p for a low-solidity, two-bladed Darrieus rotor with NACA 0012 blade profiles and a size corresponding to Reynolds number Re $= 3 \times 10^6$. The curve is similar to the ones obtained for propeller-type rotors (e.g., Fig. 4.17), but the maximum C_p is slightly lower. The reasons why this type of cross-wind converter cannot reach the maximum C_p of 16/27 derived from (4.55) (the "Betz limit") are associated with the fact that the blade orientation cannot remain optimal for all rotational angles ψ_i, as discussed (in terms of a simplified solution to the model presented above) by Lissaman (1976).

Figure 4.33 gives the torque as a function of angular velocity Ω for a small three-bladed Darrieus rotor. When compared with the corresponding curves for propeller-type rotors shown in Figs. 4.22 and 4.24 (or generally Fig. 4.21), it is evident that the torque at $\Omega = 0$ is zero for the Darrieus rotor, implying that it is not self-starting. For application with an electric grid or some other back-up system, this is no problem, since the auxiliary power needed to start the Darrieus rotor at the appropriate times is very small compared with the wind converter output, on a yearly average basis. However, for application as an isolated source of power (e.g., in rural areas), it is a disadvantage, and it has been suggested that a small Savonius rotor should be placed on the main rotor axis in order to provide the starting torque (Banas and Sullivan, 1975).

FIGURE 4.33 Torque as a function of angular velocity for a small three-bladed Darrieus rotor and at selected wind speeds. The experimental rotor had a radius $R = 2.25$ m, a chord ratio $c/R = 0.085$, and NACA 0012 blade profiles. (Based on Banas and Sullivan, 1975.)

Another feature of the Darrieus converter (as well as of some propeller-type converters), which is evident from Fig. 4.33, is that for application with a load of constant Ω (as in Fig. 4.22), there will be a self-regulating effect, in that the torque will rise with increasing wind speed only up to a certain wind speed. If the wind speed increases further, the torque will begin to decrease. For variable-Ω types of load (as in Fig. 4.24), the behavior of the curves in Fig. 4.33 implies that cases of irregular increase and decrease of torque, with increasing wind speed, can be expected.

Augmenters and Other "Advanced" Converters

In the preceding sections, it has been assumed that the induced velocities in the converter region were half of those in the wake. This is strictly true for situations where all cross-wind induced velocities (u_t and u_r) can be neglected, as shown in (4.50), but if suitable cross-wind velocities can be induced so that the total streamwise velocity in the converter region, u_x, exceeds the value of $-1/2(u_{x,in} + u_{x,out})$ by a positive amount δu_x^{ind}, then the Betz limit on the power coefficient, $C_p = 16/27$, may be exceeded,

$$u_x = 1/2(u_{x,in} + u_{x,out}) + \delta u_x^{ind}.$$

A condition for this to occur is that the extra induced streamwise velocity δu_x^{ind} does not contribute to the induced velocity in the distant wake, u_x^{ind}, which is given implicitly by the above form of u_x, since

$$u_{x,out} = u_{x,in} + u_x^{ind} = u_{x,in}(1 - 2a).$$

The streamtube flow at the converter, (4.49), is then

$$J_m = \rho A_s u_x = \rho A_s u_{x,in}(1 - a + \tilde{a}) \tag{4.81}$$

with $\tilde{a} = \delta u_x^{ind}/u_{x,in}$, and the power (4.53) and power coefficient (4.55) are replaced by

$$E = \rho A_s (u_{x,in})^3 \, 2a(1-a) \, (1-a+\tilde{a}),$$
$$C_p = 4a(1-a) \, (1-a+\tilde{a}),$$

(4.82)

where the streamtube area A_s equals the total converter area A, if the single-streamtube model is used.

Ducted Rotor

In order to create a positive increment $\tilde{a}u_{x,in}$ of axial velocity in the converter region, one may try to take advantage of the possibility of inducing a particular type of cross-wind velocity, which causes the streamtube area to contract in the converter region. If the streamtube cross-section is circular, this may be achieved by an induced radial outward force acting on the air, which again can be caused by the lift force of a wing section placed at the periphery of the circular converter area, as illustrated in Fig. 4.34.

Figure 4.34 compares a free propeller-type rotor (top), for which the streamtube area is nearly constant (as was actually assumed in section 4.3.2 and also for the Darrieus rotor) or expanding as a result of radially induced velocities, with a propeller rotor of the same dimensions, shrouded by a duct-shaped wing-profile. In this case the radial inward lift force F_L on the shroud corresponds (by momentum conservation) to a radial outward force on the air, which causes the streamtube to expand on both sides of the shrouded propeller; in other words, it causes the streamlines passing through the duct to define a streamtube, which contracts from an initial cross-section to reach a minimum area within the duct and which again expands in the wake of the converter. From (4.59), the magnitude of the lift force is

$$F_L^{duct} = 1/2\rho \, W_{duct}^2 \, c_{duct} \, C_L^{duct},$$

with the duct chord c_{duct} defined in Fig. 4.34 and W_{duct} related to the incoming wind speed $u_{x,in}$ by an axial interference factor a_{duct} for the duct, in analogy with the corresponding one for the rotor itself, (4.52),

$$W_{duct} = u_{x,in}(1 - a_{duct}).$$

F_L^{duct} is associated with a circulation Γ around the shroud profile (shown in Fig. 4.34), given by (4.57). The induced velocity δu_x^{ind} inside the duct appears in the velocity being integrated over in the circulation integral (4.57), and to a first approximation it may be assumed that the average induced velocity is simply proportional to the circulation (which in itself does not depend on the actual choice of the closed path around the profile),

$$\delta u_x^{ind} = k_{duct} \, \Gamma_{duct}/R,$$

(4.83)

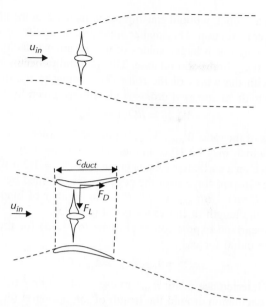

FIGURE 4.34 Peripheral streamlines for free stream (*top*) and ducted rotor (*below*).

where the radius of the duct, R (assumed to be similar to that of the inside propeller-type rotor), has been introduced because the path-length in the integral Γ_{duct} is proportional to R and the factor k_{duct} appearing in (4.83) is therefore reasonably independent of R. Writing $\Gamma_{duct} = (\delta\ W_{duct})^{-1}\ F_L^{duct}$ in analogy to (4.57), and introducing the relations found above,

$$\tilde{a}_{duct} = k_{duct}\, c_{duct}\, C_L^{duct}(1 - a_{duct})/2R. \qquad (4.84)$$

If the length of the duct, c_{duct}, is made comparable to or larger than R, and the other factors in (4.84) can be kept near to unity, it is seen from (4.82) that a power coefficient about unity or larger is possible. This advantage may, however, be outweighed by the much larger amounts of materials needed to build a ducted converter, relative to a free rotor.

Augmenters taking advantage of the lift forces on a suitably situated aerodynamic profile need not provide a fully surrounding duct around the simple rotor device. A vertical cylindrical tower structure (e.g., with a wing profile similar to that of the shroud in Fig. 4.34) may suffice to produce a reduction in the widths of the streamtubes relevant to an adjacently located rotor and thus may produce some enhancement of power extraction.

Rotor with Tip-vanes

A formally appealing design, shown in Fig. 4.35, places tip-vanes of modest dimensions on the rotor blade tips (van Holten, 1976). The idea is that the tip-vanes act like a duct, without causing much increase in the amount of materials needed in construction. The smaller areas producing lift forces are compensated for by having much larger values of the apparent velocity W_{vane} seen by the vanes than W_{duct} in the shroud case. This possibility occurs because the tip-vanes rotate with the wings of the rotor (in contrast to the duct), and hence experience air with an apparent cross-wind velocity given by

$$W_{vane} = R\Omega(1 - a_{vane}).$$

The magnitude of the ratio W_{vane}/W_{duct} is thus of the order of the tip-speed ratio $\lambda = R\Omega/u_{x,in}$, which may be 10 or higher (cf. section 4.3.2).

The lift force on a particular vane is given by $F_L^{vane} = 1/2\, \rho\, W_{vane}^2\, c_{vane}\, C_L^{vane}$, and the average inward force over the periphery is obtained by multiplying this expression by $(2\pi R)^{-1} Bb_{vane}$, where B is the number of blades (vanes), and where b_{vane} is the length of the vanes (see Fig. 4.35).

Using a linearized expression analogous to (4.83) for the axial velocity induced by the radial forces,

$$\delta u_x^{ind} = \tilde{a}_{vane}\, u_{x,in} = k_{vane}\, \Gamma/R, \tag{4.85}$$

the additional interference factor $\tilde{a}_{vane}$ to use in (4.82) may be calculated. Here Γ is the total circulation around the length of the tip-vanes (cf. Fig. 4.35), and not the circulation $\Gamma_{vane} = (\rho\, W_{vane})^{-1} F_L^{vane}$ in the plane of the vane lift and drag forces (left-hand side of Fig. 4.35). Therefore, Γ may be written

$$\Gamma = F_L^{vane} B\, b_{vane}/(2\pi R\rho\, w''),$$

i.e., it is equal to the average inward force divided by ρ times the average axial velocity at the tip-vane containing peripheral annulus,

$$w'' = u_{x,in}(1 + a'').$$

Expressions for a'' have been derived by van Holten (1976). By inserting the above expressions into (4.85), $\tilde{a} = \tilde{a}_{vane}$ is obtained in the form

$$\tilde{a} = k_{vane}(1 - a_{vane})^2 B\, c_{vane} b_{vane} \lambda^2\, C_L^{vane}/(4\pi R^2(1 + a'')). \tag{4.86}$$

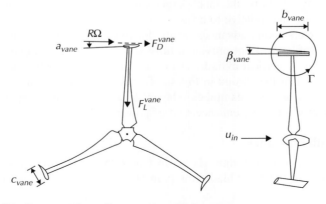

FIGURE 4.35 Geometry of propeller-type rotor with tip-vanes.

The part $(Bc_{vane}b_{vane}/\pi R^2)$ in the above expression is the ratio between the summed tip-vane area and the rotor-swept area. Taking this as 0.05 and [according to van Holten (1976), for $b_{vane}/R = 0.5$] $k_{vane}(1-a_{vane})^2/(1+a'')$ as 0.7, the tip-speed ratio λ as 10, and $C_L^{vane} = 1.5$, the resulting $\tilde{a}$ is 1.31 and the power coefficient according to the condition that maximizes (4.82) as function of a,

$$a = (2 + \tilde{a} - ((2 + \tilde{a})^2 - 3(1 + \tilde{a}))^{1/2}) = 0.43; \quad C_p = 1.85.$$

The drag forces F_D^{vane} induced by the tip-vanes (cf. Fig. 4.35) represent a power loss from the converter that has not been incorporated in the above treatment. According to van Holten (1976), in the numerical example studied above, this loss may reduce C_p to about 1.2. This is still twice the Betz limit for a tip-vane area that would equal the area of the propeller blades for a three-bladed rotor with average chord ratio $c/R = 0.05$. It is conceivable that the use of such tip-vanes to increase the power output would in some cases be preferable to achieving the same power increase by increasing the rotor dimensions.

Other Concepts

A large number of alternative devices for utilization of wind energy have been studied, in addition to those already discussed. The lift-producing profile exposed to the wind may be hollow, with holes through which inside air is driven out by the lift forces. Through a hollow tower structure, replacement air is then drawn to the wing profiles from sets of openings placed so that the air must pass one or more turbine propellers in order to reach the wings. The turbine propellers provide the power output, but the overall efficiency, based on the total dimensions of the device, is low (Hewson, 1975). The efficiency of devices that try to concentrate the wind energy before letting it reach a modest-size propeller may be improved by first converting the mainly axial flow of the wind into a

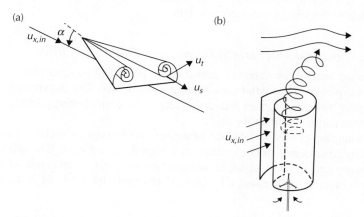

FIGURE 4.36a,b Augmenter concepts: (*a*) delta wing forming trailing vortices; (*b*) ducted vortex system. (Based on Sforza, 1976; Yen, 1976.)

flow with a large vorticity of a simple structure, such as a circular air motion. Such vortices are, for example, formed over the sharp edges of highly swept delta wings, as illustrated in Fig. 4.36a (Sforza, 1976). There are two positions along the baseline of the delta wing where propeller rotors can be placed in an environment with streamwise velocities u_s/u_{in} of 2–3 and tangential velocities u_t of the same order of magnitude as u_{in} (Sforza, 1976).

If the wake streamlines can be diffused over a large region, so that interaction with the wind field not contributing to the streamtubes passing through the converter can transfer energy to the slipstream motion and thereby increase the mass flow J_m through the converter, then further increase in power extraction can be expected. This is the case for the ducted system shown in Fig. 4.34 (lower part), but it may be combined with the vorticity concept described above to form a device of the general layout shown in Fig. 4.36b (Yen, 1976). Wind enters the vertical cylinder through a vertical slit and is forced to rotate by the inside cylinder walls. The vortex system created this way (an "artificial tornado") is pushed up through the cylinder by pressure forces and leaves the open cylinder top. Owing to the strong vorticity, the rotating air may retain its identity high up in the atmosphere, where its diffusion extracts energy from the strong winds expected at that height. This energy is supposed to be transferred down the "tornado," strengthening its vorticity and thus providing more power for the turbines placed between the "tornado" bottom and a number of air inlets at the tower foot, which replace air that has left the top of the cylinder (cf. Fig. 4.36b). Neither of these two constructions has found practical applications.

Conventional (or unconventional) wind energy converters may be placed on floating structures at sea (where the wind speeds are often higher than on land) or may be mounted on balloons (e.g., a pair of counterrotating propellers beside one another) in order to utilize the increasing wind speed usually found at elevations not accessible to ordinary building structures on the ground. In order to serve as a power source, the balloons must be guyed to a fixed point on the ground, with wires of sufficient strength.

Heat, Electrical/Mechanical Power, and Fuel Generation

The wind energy converters described in the preceding sections primarily convert the power in the wind into rotating shaft power. The conversion system generally includes a further conversion step if the desired energy form is different from that of the rotating shaft.

Examples of this are electric generators with fixed or variable rotational velocity, mentioned in connection with Figs 4.22 and 4.24. The other types of energy can, in most cases, be obtained by secondary conversion of electric power. In some such cases, the "quality" of electricity need not be as high as that usually maintained by utility grid systems, in respect to voltage fluctuations and variations in frequency in the (most widespread) case of alternating current (AC). For wind energy converters aimed at constant working angular velocity Ω, it is customary to use a gearbox and an induction-type generator. This maintains an AC frequency equal to that of the grid and constant to within about 1%. Alternatively, the correct frequency can be prescribed electronically. In both cases, reactive power is created (i.e., power that, like that of a condenser or coil, is phase shifted), which may be an advantage or disadvantage, depending on the loads on the grid.

For variable-frequency wind energy converters, the electric output would be from a synchronous generator and in the form of variable-frequency AC. This would have to be subjected to a time-dependent frequency conversion, and for arrays of wind turbines, phase mismatch would have to be avoided. Several schemes exist for achieving this, for example, semiconductor rectifying devices (thyristors), which first convert the variable frequency AC to DC (direct current) and then, in a second step, the DC to AC of the required fixed frequency.

If the desired energy form is heat, "low-quality" electricity may first be produced, and the heat may then be generated by leading the current through a high ohmic resistance. Better efficiency can be achieved if the electricity can be used to drive the compressor of a heat pump (see section 4.2.3), taking the required heat from a reservoir of temperature lower than the desired one. It is also possible to convert the shaft power more directly into heat. For example, the shaft power may drive a pump, pumping a viscous fluid through a nozzle, so that the pressure energy is converted into heat. Alternatively, the shaft rotation may be used to drive a "paddle" through a fluid, in such a way that large drag forces arise and the fluid is put into turbulent motion, gradually dissipating the kinetic energy into heat. If water is used as the fluid medium, the arrangement is called a *water-brake*.

Windmill shaft power has traditionally been used to perform mechanical work of various kinds, including flour milling, threshing, lifting, and pumping. Pumping of water (e.g., for irrigation purposes), with a pump connected to the rotating shaft, may be particularly suitable as an application of wind energy, since variable and intermittent power would, in most cases, be acceptable, as long as the average power supply, in the form of lifted water over an extended period of time, is sufficient.

In other cases, an auxiliary source of power may be needed, so that demand can be met at any time. For grid-based systems, this can be achieved by trade of energy (cf. scenarios described in Chapter 6). Demand matching can also be ensured if an energy storage facility of sufficient capacity is attached to the wind energy conversion system. A number of such storage facilities are mentioned in Chapter 5, and among them are the storage of energy in the form of fuels, such as hydrogen. Hydrogen may be produced, along with oxygen, by electrolysis of water, using electricity from the wind energy converter. The detailed working of these mechanisms over time and space is simulated in the scenarios outlined in Chapter 6 and related studies (Sørensen *et al.*, 2003; Sørensen, 2004a).

The primary interest may also be oxygen, for example, to be dissolved into the water of lakes that are deficient in oxygen (say, as a result of pollution), or to be used in connection with "ocean farming," where oxygen may be a limiting factor if nutrients are supplied in large quantities, e.g., by artificial upwelling. The oxygen may be supplied by wind energy converters, while the co-produced hydrogen may be used to produce or move the nutrients. Again, in this type of application, large power fluctuations may be acceptable.

4.3.4 Hydro and Tidal Energy Conversion

Electricity generation from water possessing potential, kinetic, or pressure energy (4.39)–(4.41) can be achieved by means of a turbine, the general theory of which is outlined in section 4.3.1. The design of the particular turbine to be used depends on whether there is a flow J_m through the device, which must be kept constant for continuity reasons, or whether it is possible to obtain zero fluid velocity after passage through the turbine.

The energy at the entrance of the turbine may be kinetic or pressure energy, causing the forces on the turbine blades to be a combination of "impulse" and "reaction" forces, which can be modified easily. If elevated water is allowed to "fall," its potential energy forms kinetic energy, or it may act as a pressure source through a water-filled tube connecting the elevated water source with the turbine placed below. Conversely, pressure energy may be transformed into kinetic energy by passage through a nozzle.

Typical classical turbine designs are illustrated in Fig. 4.37. For high specific-energy differences $w_{in}-w_{out}$ (large heads), the Pelton turbine, which is a high-speed variant of the simple undershot waterwheel, may be used. It has inflow through a nozzle, providing purely kinetic energy, and negligible w_{out} (if the reference point for potential energy is taken to correspond to the water level after passing the turbine). Also, the Francis turbine (Fig. 4.37b)

FIGURE 4.37a,b,c   Pelton (a), Francis (b), and Kaplan (c) water turbines.

is used with large water heads. Here the water is allowed to approach the entire rotor and is guided to obtain optimal angles of attack; the rotor moves owing to the reaction forces resulting from both the excess pressure at the entrance and the suction at the exit.

A third type of turbine, illustrated in Fig. 4.37c, can be used for low water heads. Here the rotor is a propeller, designed to obtain high angular speeds. Again, the angle of attack may be optimized by installation of guiding blades at the entrance to the rotor region. If the blade pitch angle is fixed, it is called a Nagler turbine. If it can be varied, it is called a Kaplan turbine.

Figure 4.38 gives examples of actual efficiencies for the types of turbines described above (Fabritz, 1954) as functions of the power level. The design point for these turbines is about 90% of the rated power (which is set to 100% on the figure), so the power levels below this point correspond to situations in which the water head is insufficient to provide the design power level.

Pelton and Francis turbines have been used in connection with river flows with rapid descent, including waterfalls, and, in many cases, construction of dams has provided a steady energy source throughout most of the year. The water is stored at an elevated level in natural or artificial reservoirs and is allowed to descend to the turbines only when needed. An example of natural variations inviting regulation is shown in Fig. 3.50. The operation of systems that include several sources of flow (springs, glacier or snow melt, rainy regions), several reservoirs of given capacity, a given number of turbine power stations, and a load of electricity usage that also varies with time, has been studied and optimized by simulation techniques (see, for example, Jamshidi and Mohseni, 1976).

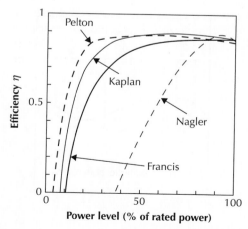

FIGURE 4.38 Efficiency of water turbines as a function of power level. (Based on Fabritz, 1954.)

For many years, hydropower has been the most widely used renewable source of electricity and also—among all types of power plants, including fossil and nuclear—the technology involving the largest power plants (rated at several gigawatts) and often gigantic artificial water reservoirs. The development of such schemes, while disregarding social and environmental problems, has given hydropower a negative reputation. In developing countries, thousands of people have been forcefully removed from their homes, with no compensation, to make room for flooded reservoirs that cause monumental environmental disruption (although this practice is not restricted to developing countries, as the examples of Tasmania and Norway show), and in some cases destroying price-less archaeological sites (e.g., Turkey). In recent decades, it has become clear that, in many cases, the environmental damage can be minimized, albeit at a higher cost. The discussion in section 3.4.2 mentions Swiss efforts to use cas-cading systems in order to do away with large reservoirs, accepting that the smaller reservoirs, located along the flow of water and designed to minimize local impacts, provide somewhat less regulation latitude. Today, in most socie-ties, full consideration of these concerns is a prerequisite for considering hydro-power a benign energy source.

Kaplan (or Nagler) turbines are used in connection with low water heads (e.g., the local community power plants in China; cf. *China Reconstructs*, 1975) and tidal plants (André, 1976), and they may be used if ocean currents are to be exploited. The 240 MW tidal plant at la Rance in France has turbines placed in a dam structure across an inlet, which serves as a reservoir for two-way operation of the turbines (filling and emptying of the reservoir). According to André (1976), the turbine efficiency is over 90%, but the turbines are only generating electricity during part of the day, according to the scheme outlined in Fig. 4.39. The times at which generation

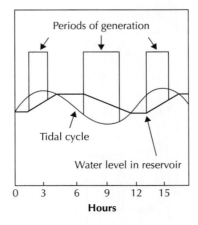

FIGURE 4.39 Operation of the tidal power plant at la Rance. (Based on André, 1976.)

is possible are determined by the tidal cycle, according to the simple scheme of operation, but modifications to better suit load variations are possible, e.g., by using the turbines to pump water into the reservoir at times when filling would not occur as a result of the tidal cycle itself. The installation has had several problems with siltation, causing it to be operated most of the time only for water flows in one direction, despite its design to accept water inflow from both sides.

4.3.5 Magneto-hydrodynamic Converters

For the turbines considered above, it is explicitly assumed that no heat was added. Other flow-type converters are designed to receive heat during the process. An example of this is the gas turbine, which is described in section 4.1.2 from the point of view of thermodynamics. The gas turbine (Brayton cycle) allows the conversion of heat into shaft power, but it may equally well be viewed as the successive conversion of heat into kinetic energy of flow and of the kinetic energy of flow into shaft power.

The magneto-hydrodynamic converter is another device converting heat into work, but delivering the work directly as electrical power without intermediate steps of mechanical shaft power. The advantage is not in avoiding the shaft power to electrical power conversion, which can be done with small losses, but rather in avoiding a construction with moving parts, thereby permitting higher working temperatures and higher efficiency. The heat added is used to ionize a gas, and this conducting gas ("plasma") is allowed to move through an expanding duct, upon which an external magnetic field B is acting. The motion of the gas is sustained by a pressure drop between the chamber where heat is added and the open end of the expanding duct. The charged particles of velocity u in the plasma are subjected to a Lorentz force

$$F = \rho_{el} u \times B, \tag{4.87}$$

where the direction of this induced force is perpendicular to B and u, but opposite for the positive atoms and for the negatively charged electrons. Since the mass of an electron is much smaller than that of an atom, the net induced current will be in the direction given by a negative value of ρ_{el}. Assuming a linear relationship between the induced current J_{ind} and the induced electric field $E_{ind} = F/\rho_{el} = u \times B$, the induced current may be written

$$J_{ind} = \sigma u \times B, \tag{4.88}$$

where σ is the electrical conductivity of the plasma. This outlines the mechanism by which the magneto-hydrodynamic (MHD) generator converts kinetic energy of moving charges into electrical power associated with the induced current J_{ind} across the turbine. A more detailed treatment must take into account the contributions to the force (4.87) on the charges, which arise from the induced velocity component J_{ind}/ρ_{el}, as well as the effect of variations (if any) in the flow velocity u through the generator stage (see, for example, Angrist, 1976).

The generator part of the MHD generator has an efficiency determined by the net power output after subtraction of the power needed for maintaining the magnetic field B. Only the gross power output can be considered as given by (4.12). Material considerations require that the turbine be cooled, so in addition to power output, there is a heat output in the form of a coolant flow, as well as the outgoing flow of cooled gas. The temperature of the out-flowing gas is still high (otherwise recombination of ions would inhibit the functioning of the converter), and the MHD stage is envisaged as being fol-lowed by one or more conventional turbine stages. It is believed that the total power generation in all stages could be made to exceed that of a conver-sion system based entirely on turbines with moving parts, for the same heat input.

Very high temperatures are required for the ionization to be accomplished thermally. The ionization process can be enhanced in various ways. One is to "seed" the gas with suitable metal dust (sodium, potassium, cesium, etc.), for which case working MHD machines operating at temperatures around 2500 K have been demonstrated (Hammond $et\ al.$, 1973). If the heat source is fossil fuel, and particularly if it is coal with high sulfur content, the seeding has the advantage of removing practically all the sulfur from the exhaust gases (the seeding metals are rather easily retrieved and must anyway be recycled for economic reasons).

4.3.6 Wave Energy Conversion

A large number of devices for converting wave energy to shaft power or com-pression energy have been suggested, and a few of them have been tested on a modest scale. Reviews of wave devices may be found, for example, in Leichman and Scobie (1975), Isaacs $et\ al.$ (1976), Slotta (1976), Clarke (1981), Sørensen (1999a), Thorpe (2001), CRES (2002), and DEA Wave Pro-gram (2002). An overview of possibilities is illustrated schematically in Fig. 4.40. Then, the technical details are given for two typical examples of actual devices: the oscillating water column device that has been in successful small-scale operation for many years in powering mid-sea buoys, and the Salter duck, which theoretically has a very high efficiency, but has not been successful in actual prototyping experiments. However, first some general remarks are necessary.

The resource evaluation in section 3.3.4 indicated that the most promising locations for a wave-utilization apparatus would be in the open ocean, rather than in coastal or shallow water regions. Yet all three device types have been researched: (a) shore-fixated devices for concentrating waves into a modestly

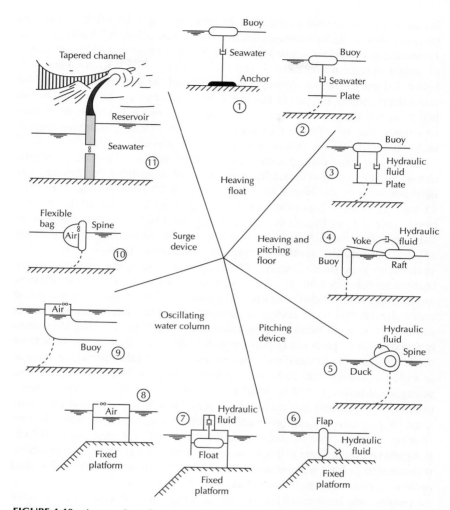

FIGURE 4.40 An overview of wave energy converter types (Hagerman and Heller, 1988; Falnes and Løvseth, 1991).

elevated reservoir, from which the water may drive conventional hydro-turbines; (b) near-shore devices making use of the oscillations of a water column or a float on top of the waves, relative to a structure standing at the sea floor; and (c) floating devices capturing energy by differential movements of different parts of the device.

As a first orientation toward the prospects of developing economically viable wave-power devices, a comparison with (e.g., offshore) wind power may be instructive. One may first look at the weight of the construction relative to its rated power. For an onshore wind turbine, this number is around

0.1 kg/W_{rated}, while adding the extra foundation weight for offshore turbines (except caisson in-fill) increases the number to just below 0.2 kg/W_{rated}. For 15 wave-power devices studied by the DEA Wave Program (2002), the range of weight to rated power ratios is from 0.4 to 15 kg/W_{rated}. The two numbers below 1.0 are for device concepts not tested and for use on shore or at low water depth, where the power resources are small anyway. So the conclusion is that the weight/power ratio is likely to be at least twice, but likely more than five times, that of offshore wind turbines, which to a first approximation is also a statement on the relative cost of the two concepts.

Using the same data, one may instead look at the ratio of actually produced power at a particular location and the weight of the device. For offshore wind in the North Sea, this is around 20 kWh y^{-1} kg^{-1}. For the 15 wave devices, values of 0.1 to 10 are found, by using for all devices the same wave data estimated for a location some 150 km west of the city of Esbjerg in Denmark.[*] Clearly, it is not reasonable to use data for a location 150 km out into the ocean for wave-power devices that must stand on the shore or at very shallow water. Omitting these cases, the resulting range reduces to 0.1–1.5 kWh y^{-1} kg^{-1}, or over 13 times less than for offshore wind in the same region. Again, the simplistic translation from weight to cost indicates that wave energy is economically unattractive, because at the same location wind power can be extracted at much lower cost. Also, there are no particular reasons to expect the distribution of weight on less expensive materials (concrete or steel) and more expensive materials (special mechanical and electric equipment) to be substantially different. It is also argued that, where wave power would be feasible, offshore wind power is also available, because it is the wind that creates the waves. Only at large water depths, say, over 20 m, where foundation would be problematic, might the wave devices floating on the surface be more attractive than wind. Yet there are no indications above that the cost of such mid-ocean wave-power extraction and cable transmission to land will be economically viable, unless all near-shore wind options have already been exploited.

Finally, one must consider the time-distribution of the power from wave devices. As seen from Figs. 3.56 and 3.59, wave power exhibits large variations with seasons. For example, in the North Sea, variations are large: Rambøll (1999) finds 6 times more average wave power in January than in June, where the corresponding factor for wind power is 2 (Sørensen, 2000a). Acceptance of wave power into grids serving electricity demands is thus going to be considerably more difficult than acceptance of wind, which roughly

---

[*]The wave data used are calculated from the 10 m wind data shown in Figs. 2.79–2.80 using a fetch model (Rambøll, 1999). As mentioned in Chapters 2 and 6, the 10 m data are not very reliable, and the grid is coarse, but comparison with the actual measurements at Vyl a few kilometers to the west (shown in Fig. 3.45) indicates that the error is well under 50% (however, the calculated data do not, for example, exhibit the two peaks shown in Fig. 3.55, but only one).

has the same seasonal variation as demand, at least on the Northern Hemisphere. Thus, in addition to an initial cost likely to be substantially higher than that of wind power, additional costs for energy storage or other supply–demand mismatch management must be considered.

Pneumatic Converter

The only wave energy conversion device that has been in considerable practical use, although on a fairly small scale, is the buoy of Masuda (1971), shown in schematic form in Fig. 4.41. Several similar wave-power devices exist, based on an oscillating water column driving an air turbine; in some cases they are shore-based and have only one chamber (see, for example, CRES, 2002). The buoy in Fig. 4.41 contains a tube extending downward, into which water can enter from the bottom, and a "double-action" air turbine, i.e., a turbine that turns the same way under the influence of both pressure and suction (as illustrated by the non-return valves).

Wave motion will cause the whole buoy to move up and down and thereby create an up-and-down motion of the water level in the center tube, which in turn produces the pressure increases and decreases that make the turbine in the upper air chamber rotate. Special nozzles may be added to increase the speed of the air impinging on the turbine blades.

For a simple sinusoidal wave motion, as described in section 2.5.3, but omitting the viscosity-dependent exponential factor, the variables σ_1 and Z, describing the water level in the center tube and the vertical displacement of the entire buoy, respectively, may also be expected to vary sinusoidally, but with phase delays and different amplitudes,

$$\sigma = a \cos (kx - \omega t),$$
$$\sigma_1 = \sigma_{10} \cos (kx - \omega t - \delta_\sigma),$$
$$Z = Z_0 \cos (kx - \omega t - \delta_Z),$$

with $\omega = kU_w$ in terms of the wave number k and phase velocity U_w (cf. section 2.5.3). The relative air displacement ρ_1 in the upper part of the center tube (the area of which is $A_1 = \pi R_1^2$) is now (see Fig. 4.41)

$$\rho_1 = \sigma_1 - Z.$$

Assuming the air flow to be incompressible, continuity requires that the relative air velocity $d\rho_2/dt$ in the upper tubes forming the air chamber (of area $A_2 = \pi R_2^2$) be

$$d\rho_2/dt = (A_1/A_2) \, d\rho_1/dt.$$

It is now possible to set up the equations of motion for σ_1 and Z, equating the accelerations $d^2\sigma_1/dt^2$ and d^2Z/dt^2 multiplied by the appropriate masses

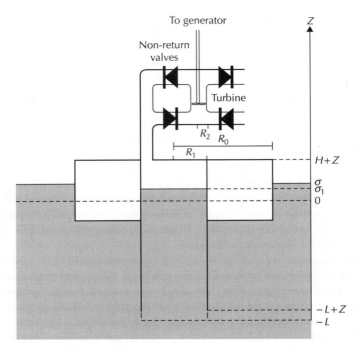

FIGURE 4.41   Masuda's pneumatic wave energy conversion device.

(of the water column in the center tube and of the entire device) to the sum of forces acting on the mass in question. These are buoyancy and pressure forces, as well as friction forces. The air displacement variable ρ_2 satisfies a similar equation involving the turbine blade reaction forces. The equations are coupled, but some of the mutual interaction can be incorporated as damping terms in each equation of motion. McCormick (1976) uses linear damping terms $-b(d\sigma_1/dt)$, etc., with empirically chosen damping constants, and drops other coupling terms, so that the determination of σ_1 and Z becomes independent. Yet the $d\rho_2/dt$ values determined from such solutions are in good agreement with measured ones, for small wave amplitudes a (or significant heights $H_S = 2 \times 2^{1/2}a$; cf. section 3.3.3) and wave periods $T = 2\pi/\omega$, which are rather close to the resonant values of the entire buoy, T_0, or to that of the center tube water column, T_1. Between these resonant periods, the agreement is less good, probably indicating that couplings are stronger between the resonances than near them, as might be expected.

The power transferred to the turbine shaft cannot be calculated from expression (4.12) because the flow is not steady. The mass flows into and out of the converter vary with time, and at a given time they are not equal. However, owing to the assumption that air is incompressible, there will be no buildup of compressed air energy storage inside the air chamber, and therefore, power can still be obtained as a product of a mass flux, i.e., that to the turbine,

$$J_m = \rho_a A_1 \, d\rho_1/dt$$

(ρ_a being the density of air), and a specific energy change Δw (neglecting heat dissipation). In addition to the terms (4.39)–(4.41) describing potential energy changes (unimportant for the air chamber), kinetic energy changes, and pressure/enthalpy changes, Δw will contain a term depending on the time variation of the air velocity $d\rho_2/dt$ at the turbine entrance,

$$\Delta w \approx \tfrac{1}{2}\left(\left(\frac{d\rho_1}{dt}\right)^2 - \left(\frac{d\rho_2}{dt}\right)^2\right) + \frac{\Delta P}{\rho_a} - \int\left(\frac{d^2\rho_2}{dt^2}\right)d\rho,$$

where ΔP is the difference between the air pressure in the air chamber (and upper center tube, since these are assumed equal) and in the free outside air, and ρ is the co-ordinate along a streamline through the air chamber (McCormick, 1976). Internal losses in the turbine have not been taken into account.

Figure 4.42 shows an example of the simple efficiency of conversion, η, given by the ratio of $J_m\Delta w$ and the power in the incident waves, which for sinusoidal waves is given by (3.28). Since expression (3.28) is the power per unit length (perpendicular to the direction of propagation, i.e., power integrated over depth z), it should be multiplied by a dimension of the device that characterizes the width of wave acceptance. As the entire buoy is set into vertical motion by the action of the waves, it is natural to take the overall diameter $2R_0$ (cf. Fig. 4.41) as this parameter, such that

$$\eta = 8\pi J_m \, \Delta w/(2R_0 \rho_w g^2 Ta^2) = 32\pi J_m \, \Delta w/(\rho_w g^2 THs^2 R_0). \qquad (4.89)$$

Efficiency curves as functions of the period T, for selected significant wave heights H_S, which are shown in Fig. 4.42, are based on a calculation by McCormick (1976) for a 3650 kg buoy with overall diameter $2R_0 = 2.44$ m and center tube diameter $2R_1 = 0.61$ m. For fixed H_S, η has two maxima, one just below the resonant period of the entire buoy in the absence of damping terms, $T_0 \approx 3$ s, and one near the air chamber resonant period, $T_1 \approx 5$ s. This implies that the device may have to be "tuned" to the periods in the wave spectra giving the largest power contributions (cf. Fig. 3.56), and that the efficiency will be poor in certain other intervals of T. According to the calculation based on solving the equations outlined above and with the assumptions made, the efficiency η becomes directly proportional to H_S. This can only be true for small amplitudes, and Fig. 4.42 shows that unit efficiency would be obtained for $H_S \approx 1.2$ m from this model, implying that nonlinear terms have to be kept in the model if the power obtained for waves of such heights is to be correctly calculated. McCormick (1976) also refers to experiments that suggest that the power actually obtained is smaller than that calculated.

FIGURE 4.42 Efficiency of a pneumatic wave energy converter as a function of wave period, calculated for two wave amplitudes. (Based on McCormick, 1976.)

Masuda has proposed placing a ring of pneumatic converters (floating on water of modest depth) such that the diameter of the ring is larger than the wavelength of the waves of interest. He claims that the transmission of wave power through such a ring barrier is small, indicating that most of the wave energy will be trapped (e.g., multiply reflected on the interior barriers of the ring) and thus eventually will be absorbed by the buoy and internal fluid motion in the center tubes. In other developments of this concept, the converters are moored to the sea floor or leaning toward a cliff on the shore (Clarke, 1981; Vindeløv, 1994; Thorpe, 2001).

Oscillating-vane Converter

Since water particles in a sinusoidal wave are moving in circular orbits, it may be expected that complete absorption of the wave energy is only possible with a device possessing two degrees of freedom, i.e., one allowing displacement in both vertical and horizontal directions. Indeed, it has been shown by Ogilvie (1963) that an immersed cylinder of a suitable radius, moving in a circular orbit around a fixed axis (parallel to its own axis) with a certain frequency, is capable of completely absorbing the energy of a wave incident perpendicular to the cylinder axis, i.e., with zero transmission and no wave motion "behind" the cylinder.

Complete absorption is equivalent to energy conversion at 100% efficiency, and Evans (1976) has shown that this is also possible for a half-immersed cylinder performing forced oscillations in two dimensions, around an equilibrium position, whereas the maximum efficiency is 50% for a device that is only capable of performing oscillations in one dimension.

The system may be described in terms of coupled equations of motion for the displacement co-ordinates X_i ($i = 1, 2$),

$$m \, \mathrm{d}^2 X_i / \mathrm{d}t^2 = -d_i \, \mathrm{d}X_i / \mathrm{d}t - k_i X_i + \sum_j F_{ij},$$

where m is the mass of the device, d_i is the damping coefficient in the ith direction, k_i correspondingly is the restoring force or "spring constant" (the vertical component of which may include the buoyancy forces), and F_{ij} is a matrix representing the complete hydrodynamic force on the body. It depends on the incident wave field and on the time derivatives of the co-ordinates X_i, the non-diagonal terms representing couplings of the motion in vertical and horizontal directions.

The average power absorbed over a wave period T is

$$E = \sum_i T^{-1} \int_0^T \frac{\mathrm{d}X_i}{\mathrm{d}t} \sum_j F_{ij} \mathrm{d}t, \tag{4.90}$$

and the simple efficiency η is obtained by inserting each of the force components in (4.90) as forces per unit length of cylinder, and dividing by the power in the waves (3.28). One may regard the following as design parameters: the damping parameters d_i, which depend on the power extraction system, the spring constants k_i and the mass distribution inside the cylinder, which determine the moments and hence influence the forces F_{ij}. There is one value of each d_i for a given radius R of the cylinder and a given wave period that must be chosen as a necessary condition for obtaining maximum efficiency. A second condition involves k_i, and if both can be fulfilled, the device is said to be *tuned*. In this case it will be able to reach maximum efficiency (0.5 or 1, for oscillations in one or two dimensions). Generally, the second condition can be fulfilled only in a certain interval of the ratio R/λ between the radius of the device and the wavelength $\lambda = gT^2/(2\pi)$.

Figure 4.43 shows the efficiency as a function of $2\pi R/\lambda$ for a half-immersed cylinder tuned at $2\pi R/\lambda_0 = 1$ (full line) and for a partially tuned case, in which the choice of damping coefficient corresponds to tuning at $2\pi R/\lambda_0 = 0.3$, but where the second requirement is not fulfilled. Such *partial tuning* offers a possibility of widening the range of wave periods accepted by the device.

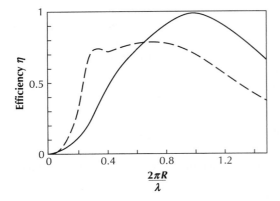

FIGURE 4.43 Efficiency of wave energy absorption by oscillating motion of a half-immersed cylinder of radius R. The calculation assumes parameters describing damping and coupling terms, such that the device becomes fully tuned at $2\pi R/\lambda = 1$ (*solid line*) or partially tuned at $2\pi R/\lambda = 0.3$ (*dashed line*). (Based on Evans, 1976.)

FIGURE 4.44 Salter's design of an oscillating-vane wave energy converter. The cross-section shows the backbone and axis of rotation as a small circle and the location of the center of gravity as a cross. The waves are incident from the right-hand side.

FIGURE 4.45 Efficiency of Salter's device based on model experiments (*solid line*) and extrapolations (*dashed line*). (Based on Mollison *et al.*, 1976.) See also the discussion topics and associated figures in Part III.

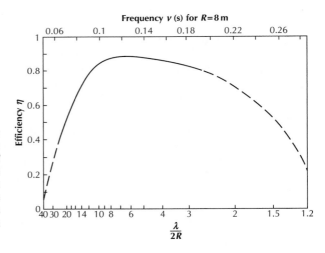

A system whose performance may approach the theoretical maximum for the cylinder with two degrees of freedom is the oscillating vane or cam of Salter (1974), illustrated in Fig. 4.44. Several structures of the cross-section indicated are supposed to be mounted on a common backbone providing the fixed axis around which the vanes can oscillate, with different vanes not necessarily moving in phase. The backbone has to be long in order to provide an approximate inertial frame, relative to which the oscillations can be utilized for power extraction, for example, by having the rocking motion create pressure pulses in a suitable fluid (contained in compression chambers between the backbone and oscillating structure). The necessity of such a backbone is also the weakness of the system, owing to the large bending forces along the structure, which must be accepted during storms.

The efficiency of wave power absorption for a Salter *cam* or *duck* of the type shown in Fig. 4.44 is indicated in Fig. 4.45, based on measurements on a model of modest scale ($R = 0.05$ m). Actual devices for use in, say, a good wave energy location in the North Atlantic (cf. Fig. 3.56) would need a radius of 8 m or more if tuned at the same value of $\lambda/(2\pi R)$ as the device used to construct Fig. 4.45 (Mollison *et al.*, 1976). Electricity generators of high efficiency are available for converting the oscillating-vane motion (the angular amplitude of which may be of the order of half a radian) to electric power, either directly from the mechanical energy in the rocking motion or via the compression chambers mentioned above, from which the fluid may be let to an electricity-producing turbine. The pulsed compression is not incompatible with continuous power generation (Korn, 1972). Later experiments have shown that the structural strength needed for the backbone is a problem. For all wave devices contemplated, placed in favorable locations at mid-ocean, transmission to shore is a problem for the economy of the scheme (cf. discussion topics after Part III).

4.4 SOLAR RADIATION CONVERSION

4.4.1 Photovoltaic Conversion

To facilitate discussion of the physics of photovoltaic conversion devices, here follows a short primer on energy bands in semiconductors. It is not required for following the subsequent sections (Photovoltaic devices, etc.).

The electrons in a solid move in a potential, which for crystalline forms is periodic, corresponding to the lattice structure of the material. The forces on a given electron are electromagnetic, comprising attractive interaction with the positively charged nuclei (the effective charge of which may be diminished by the screening effect of more tightly bound electrons) as well as repulsive interaction with other electrons. Owing to the small mass ratio between electrons and nuclei, the positions of the nuclei (R_i, $i = 1,2,3...$) may to a first approximation be regarded as fixed. This is the basis for saying that the electrons are moving in a *potential*. When an electron is close to a definite nucleus, its wavefunction may be approximated by one of the atomic wavefunctions $\psi_{i,n(i)}(r - R_i)$ for this isolated nucleus. One may therefore attempt to describe an arbitrary electron wavefunction as a linear combination of such atomic wavefunctions (*orbitals*),

$$\psi(r) = \sum_{i=1}^{N} \sum_{n(i)} c_{i,n(i)} \psi_{i,n(i)}(r - R_i). \tag{4.91}$$

Here N is the total number of atomic nuclei, and r is the position vector of the electron.

A simple example is a diatomic molecule $(N = 2)$ with only one important level in each atom,

$$\psi(r) = c_{1,n(1)} \, \psi_{1,n(1)}(r - R_1) + c_{2,n(2)} \, \psi_{2,n(2)}(r - R_2),$$

for which two normalized solutions exist, with opposite relative sign between the two terms. If the overlap

$$S = \int \psi_{1,n(1)}{}^*(r - R_1) \, \psi_{2,n(2)}(r - R_2) \, dr$$

(where the asterisk denotes complex conjugation) is non-zero, the energies of the two superposition states will be different from those ($W_{1,n(1)}$ and $W_{2,n(2)}$) of the atomic wavefunctions $\int \psi_{1,n(1)}$ and $\psi_{2,n(2)}$. The most tightly bound solution, $\psi_b(r)$, will correspond to an energy lower than the lowest of the original ones $W_{1,n(1)}$ and $W_{2,n(2)}$, while the other solution, $\psi_a(r)$, will correspond to an energy higher than the highest of the original ones (see, for example, Ballhausen and Gray, 1965).

If the energies W_b and W_a of the bonding solution $\psi_b(r)$ and the antibonding solution $\psi_a(r)$ are calculated for various relative positions $R = |R_1 - R_2|$, one may obtain a picture like that shown in Fig. 4.46 (approximately describing an ionized hydrogen molecule H_2^+).

If the number of (say, identical) nuclei, N, is larger than two, but still placed in a regular geometrical relationship characterized by a single distance R (e.g., the distance between atomic nuclei in a cubic lattice or a diamond-type lattice), the superposition-type wavefunction (4.91) still gives an energy picture similar to Fig. 4.46, but now with N curves, the highest and lowest of which may look like the two curves for $N = 2$.

For a solid crystal, N may be of the order of 10^{24}, so the region between the lowest energy state and the highest one will be filled out with energy levels that for all practical purposes can be regarded as continuous. Such a region is called an *energy band*. For each pair of overlapping atomic orbits, $n(1)$ and $n(2)$, there will be a set of linear combinations of wavefunctions in the $N = 2$ case, with bonding and antibonding characteristics. In the large-N case, each combination of overlapping atomic orbits will lead to an energy band, such that the overall picture of energy levels as a function of lattice characteristics (in the simple case, just R) will be one of a number energy bands inside which all electron energies are "allowed," plus the spaces between energy bands in which no allowed energy is present. The energy band structure may, for example, resemble that shown in Fig. 4.47.

If the distance between the nuclei is large, an electron "feels" only one nucleus (plus a number of tightly bound "inner" electrons, if the nucleus has $Z > 2$) at a time, and the energy spectrum is identical to that of an individual

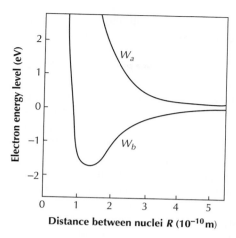

FIGURE 4.46 Electronic energy levels in a diatomic molecule as a function of interatomic distance (e.g., levels formed from $1s$ atomic orbitals in the H_2^+ molecule).

FIGURE 4.47 Energy band structure of a solid as a function of lattice constant (assumed to completely specify interatomic distances, as it does for crystals of diamond structure). Atomic physics labeling of orbits are indicated to the right. The band structure in the left-hand region of overlapping bands should be regarded as highly schematic. It may take different forms for different solids. The figure indicates how the overlapping $3s$ and $3p$ bands in silicon form the 1.14 eV energy gap further illustrated in Fig. 4.48.

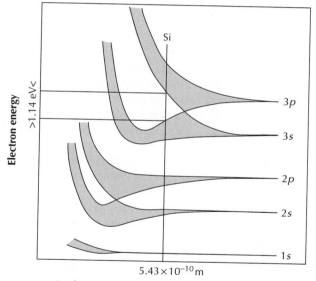

atom (right-hand side of Fig. 4.47). As the assumed distance between nuclei becomes smaller, energy bands develop and become wider. When the energy bands begin to overlap and cross (left-hand side of Fig. 4.47), the basis for the linear assumption (4.91) breaks down, at least in the form discussed so far. If the restriction that the bands are formed from one definite atomic orbital [so that there is no sum over $n(i)$ in (4.91)] is relaxed, it still may be possible to use an expression of the form (4.91) to describe some features of the region of overlapping bands.

As an example, it is possible to obtain the two separate bands in silicon (see Fig. 4.47) by first defining two suitable superpositions of $3s$ and $3p$ atomic orbits [corresponding to the summation over $n(i)$, this procedure being called *hybridization* in molecular theory], and then by applying the band theory [corresponding to the summation over i in (4.91)]. It is evident that, in this region near the band crossing point, it is possible to find energy gaps between adjacent bands that are much smaller than in the non-overlapping region.

The electron potentials initially represent a situation with one electron per atom being considered. The simplest approximation of many-electron states in one atom is to regard the electrons as independent of each other, except that they have to obey the Pauli exclusion principle, according to which there can be at most one electron in each quantum state. Because the electron has an intrinsic spin of $S = 1/2$, there are two spin states ("spin-up" and "spin-down") for each spatial orbit. Thus, the periodic system would be derived by first placing two electrons in the $1s$-orbit (cf. Fig. 4.47), then two in the $2s$-orbit, and then six in the $2p$-orbit (comprising three degenerate spatial orbits, because the orbital angular momentum is $L = 1$), etc.

A more accurate solution of the Schrödinger equation of quantum mechanics is necessary in order to include the effects of electron–electron interactions, which make the entire energy spectrum depend on electron number in addition to the smooth dependence on the charge of the nucleus (atomic number). In several cases, these interactions even influence the succession of orbits being filled in the periodic system.

Similar statements can be made for the energy bands in many-atom lattices. The spectrum of energy bands can be used to predict the filling systematics, as a function of the number of electrons per atom, but again the spectrum is not independent of the degree of filling (for this reason, no energy scale is provided in Fig. 4.47). Thus, a diagram like Fig. 4.47 cannot be used directly to predict the atomic distance (e.g., represented by the lattice constant) that will give the lowest total energy and hence the ground state of the system. This is in contrast to the H_2^+ molecule (Fig. 4.46), for which the minimum of the W_b curve determines the equilibrium distance between the nuclei.

Knowing the lattice structure and the lattice constant (the dimension of an elementary cube in the lattice), a vertical cut in a diagram of the type in Fig. 4.47 will give the allowed and forbidden energy values, with the reservations made above. Filling the proper number of electrons per atom into the energy bands, with reference to the Pauli principle, allows the energy of the latest added electron to be determined, as well as the lowest available energy level into which an additional electron added to the system may go.

If the electrons available exactly fill a band, and if there is a substantial distance to the next higher level, the material is an electrical *insulator* (no electrons can change state, i.e., they cannot "move"). If a band is partially filled, the material is a good *conductor* (e.g., a metal). The continuum of levels in the partially filled band allows the electrons to move throughout the lattice. If the distance between the highest filled band and an empty band is small, the material is called a *semiconductor*. At zero absolute temperature, a semiconductor is an insulator, but, because of the thermal energy spread at higher temperatures [given by (4.23) because electrons are Fermi particles], some electrons will be excited into the higher band (the *conduction band*). The conductance of silicon increases by a factor of 10^6 between 250 and 450 K.

The bands corresponding to $1s$, $2s$, and $2p$ atomic orbits are completely filled in Si, but then four electrons remain per atom. According to Fig. 4.47, the $3s$ and $3p$ bands are overlapping and allow two mixed bands to be constructed, with an intermediate gap of about 1 eV (1.6×10^{-5} J). The lower band (the *valence band*) can hold four electrons per atom, so this band will be completely full and the other one empty at zero temperature.

If a few of the atoms in a silicon lattice are replaced by an atom with higher Z (atomic number), e.g., phosphorus, the additional electrons associated with the impurity cannot be accommodated in the valence band, but will occupy a discrete level (the *donor level*) just below the conduction band (the energy depression being due to the larger attractive force from the atom of higher Z). A semiconductor material with this type of impurity is called *n*-type. The electrons in the donor-level are very easily excited into the conduction band. Adding *n*-type impurities makes the Fermi level [μ_i in (4.23)] move upward from the initial position approximately halfway between the valence and conduction bands.

Impurities with lower Z (e.g., Al in a Si-lattice) lead to electron vacancies or *holes*, which are associated with an energy slightly higher than the top of the valence band, again due to the Z-dependence of the attractive force. These *acceptor levels* easily transfer holes to the valence band, a process that may, of course, alternatively be described as the excitation of electrons from the valence band into the acceptor level. Semiconductor material with this type of impurity is called *p*-type. Holes can formally be treated as particles like electrons, but with positive charges and provided that the state of the pure semiconductor at zero temperature is taken as reference ("vacuum state" in quantum theory).

The energy diagrams of doped semiconductors of *n*- and *p*-types are sketched in Fig. 4.48. The distribution of electron densities on the energy bands in the valence and the conduction bands is indicated, for *n*- or *p*-doped and un-doped materials. For a more complete account of basic semiconductor theory, see, for example, Shockley (1950).

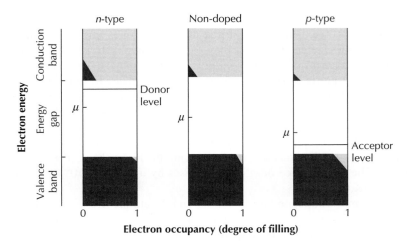

FIGURE 4.48   Energy band structure near the Fermi energy (μ) for a semiconductor material without impurities (*middle column*) or with *n*- or *p*-type doping (cf. text). The dark shading indicates the electron occupancy as a function of energy for a finite temperature (occupancy equal to unity corresponds to the maximum number of electrons at a given energy or energy interval that is consistent with the exclusion principle for Fermi-type particles).

Photovoltaic Devices

Conversion of radiant energy (light quanta) into electrical energy can be achieved with the use of semiconductor materials, for which the electron excitation caused by impinging light quanta has a strongly enhancing effect on the conductivity.

It is not sufficient, however, that electrons are excited and are able to move more freely, if there is no force to make them move. Such a force would arise from the presence of a gradient of electrical potential, such as the one found in a *p–n* junction of doped semiconductor materials (a *p–n* junction is a junction of a *p*-type and an *n*-type semiconductor, as further described below). A *p–n* junction provides an electrical field that will cause the electrons excited by radiation (such as solar) to move in the direction from the *p*-type to the *n*-type material and cause the vacancies (holes) left by the excited electrons to move in the opposite direction. If the electrons and holes reach the respective edges of the semiconductor material, the device is capable of delivering electrical power to an external circuit. The motion of electrons or holes receives competition from recombination processes (electrons being recaptured into vacancies), making such factors as overall dimensions and electron mobility in the material used important.

The p–n Junction

An essential constituent of photovoltaic cells is the *p–n* junction, a fundamental device in semiconductor physics based on the principles described in the short primer above. When a p-type and an n-type semiconductor are joined so that they acquire a common surface, they are said to form a p–n junction. This initially causes electrons to flow in the n to p direction because, as seen in Fig. 4.48, the electron density in the conduction band is higher in n-type than in p-type material and because the hole density in the valence band is higher in the p-type than in the n-type material (the electron flow in the valence band can also be described as a flow of positive holes in the direction p to n).

This electron flow builds up a surplus of positive charge in the n-type material and a surplus of negative charge in the p-type material, in the neighborhood of the junction (mainly restricted to distances from the junction of the order of the mean traveling distance before recombination of an electron or a hole in the respective materials). These surplus charges form a dipole layer, with an associated electrostatic potential difference, which will tend to hinder any further unidirectional electron flow. Finally, an equilibrium is reached in which the potential difference is such that no net transfer of electrons takes place.

Another way of stating the equilibrium condition is in terms of the Fermi energy [cf. Fig. 4.48 and (4.23)]. Originally, the Fermi energies of p- and n-type materials, μ_p and μ_n, are different, but at equilibrium $\mu_p = \mu_n$. This is illustrated in Fig. 4.49, and it is seen that the change in the relative positions of the conduction (or valence) bands in the two types of material must equal the electron charge, $-e$, times the equilibrium electrostatic potential.

The number of electrons in the conduction band may be written

$$n_c = \int_{E_c}^{E'_c} n'(E) f(E) \, dE, \tag{4.92}$$

where E_c and E'_c are the lower and upper energy limit of the conduction band, $n'(E)$ is the number of quantum states per unit energy interval (and, for example, per unit volume of material, if the electron number per unit volume is desired), and finally, $f(E)$ is the Fermi–Dirac distribution (4.23). If the electrons in the conduction band are regarded as free, elementary quantum mechanics gives (see, for example, Shockley, 1950)

$$n'(E) = 4\pi h^{-3} (2m)^{3/2} E^{1/2}, \tag{4.93}$$

where h is Planck's constant and m *is* the electron mass. The corrections for electrons moving in condensed matter, rather than being free, may to a first approximation be included by replacing the electron mass by an "effective" value.

If the Fermi energy is not close to the conduction band,

$$E_c - \mu >> kT,$$

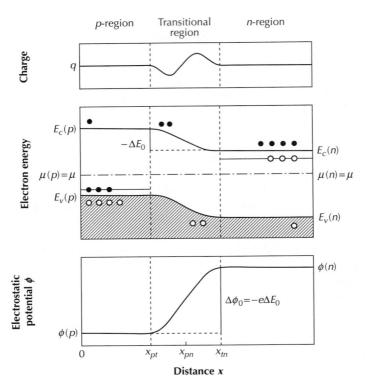

FIGURE 4.49 Schematic picture of the properties of a *p–n* junction in an equilibrium condition. The *x*-direction is perpendicular to the junction (all properties are assumed to be homogeneous in the *y*- and *z*-directions). The charge (*top*) is the sum of electron charges in the conduction band and positive hole charges in the valence band, plus charge excess or deficit associated with the acceptor and donor levels. In the electron energy diagram (*middle*), the abundance of minority charge carriers (*closed circles* for electrons, *open circles* for holes) is schematically illustrated. The properties are further discussed in the text.

the Fermi–Dirac distribution (4.23) may be replaced by the Boltzmann distribution,

$$f_B(E) = \exp(-(E - \mu)/kT). \qquad (4.94)$$

Evaluating the integral, (4.92) then gives an expression of the form

$$n_c = N_c \exp(-(E_c - \mu)/kT). \qquad (4.95)$$

The number of holes in the valence band is found in an analogous way,

$$n_v = N_v \exp(-(\mu - E_v)/kT), \qquad (4.96)$$

where E_v is the upper limit energy of the valence band.

The equilibrium currents in a p–n junction like the one illustrated in Fig. 4.49 can now be calculated. Considering first the electron currents in the conduction band, the electrons thermally excited into the conduction band in the p-region can freely flow into the n-type materials. The corresponding current, $I_0^-(p)$, may be considered proportional to the number of electrons in the conduction band in the p-region, $n_c(p)$, given by (4.95),

$$I_0^-(p) = \alpha N_c \exp(-(E_c(p) - \mu(p))/kT), \qquad (4.97)$$

where the constant α depends on electron mobility in the material and on the electrostatic potential gradient, grad ϕ. The electrons excited into the conduction band in the n-type region have to climb the potential barrier in order to move into the p-region. The fraction of electrons capable of doing this is given by a Boltzmann factor of the form (4.95), but with the additional energy barrier $\Delta E_0 = -\Delta\phi_0/e$ ($-e$ being the electron charge),

$$n_c(n) = N_c \exp(-(E_c(n) - \mu(n) - \Delta E_0)/kT).$$

Using $-\Delta E_0 = E_c(p) - E_c(n)$ (cf. Fig. 4.49) and considering the current $I_0^-(n)$ as being proportional to $n_c(n)$, the corresponding current may be written

$$I_0^-(n) = \alpha' N_c \exp(-(E_c(n) - \mu(n))/kT), \qquad (4.98)$$

where α' depends on the diffusion parameter and on the relative change in electron density, $n_c^{-1}\mathrm{grad}(n_c)$, considering the electron motion against the electrostatic potential as a diffusion process. The statistical mechanical condition for thermal equilibrium demands that $\alpha = -\alpha'$ (Einstein, 1905), so (4.97) and (4.98) show that the net electron current,

$$I_0^- = I_0^-(p) + I_0^-(n),$$

becomes zero precisely when

$$\mu(p) = \mu(n),$$

which is then the condition for thermal equilibrium. The same is true for the hole current,

$$I_0^+ = I_0^+(p) + I_0^+(n).$$

If an external voltage source is applied to the p–n junction in such a way that the n-type terminal receives an additional electrostatic potential $\Delta\phi_{ext}$ relative to the p-type terminal, then the junction is no longer in thermal equilibrium, and the Fermi energy in the p-region is no longer equal to that of the n-region, but satisfies

$$\mu(p) - \mu(n) = e^{-1}\Delta\phi_{ext} = \Delta E_{ext} \qquad (4.99)$$

if the Boltzmann distributions of electrons and of holes are to maintain their shapes in both p- and n-regions. Similarly, $E_c(p) - E_c(n) = -(\Delta E_0 + \Delta E_{ext})$, and assuming that the proportionality factors in (4.97) and (4.98) still bear the relationship $\alpha = -\alpha'$ in the presence of the external potential, the currents are connected by the expression

$$I^-(n) = -I^-(p) \exp(\Delta E_{ext}/kT).$$

The net electron current in the conduction band then becomes

$$I^- = I^-(n) + I^-(p) = -I^-(p) \left(\exp(\Delta E_{ext}/kT) - 1\right). \tag{4.100}$$

For a positive $\Delta\phi_{ext}$, the potential barrier that electrons in the n-region conduction band (see Fig. 4.49) have to climb increases, and the current $I^-(n)$ decreases exponentially (ΔE_{ext} negative, *reverse bias*). In this case, the net current I^- approaches a saturation value equal to $I^-(p)$, according to (4.100).

For negative $\Delta\phi_{ext}$, (positive ΔE_{ext}, *forward bias*), the current $I^-(n)$ increases exponentially with the external potential. In both cases, $I^-(p)$ is assumed to remain practically unchanged, when the external potential of one or the other sign is applied, considering that $I^-(p)$ is primarily limited by the number of electrons excited into the conduction band in the p-type material, a number that is assumed to be small in comparison with the conduction band electrons in the n-type material (cf. Figs. 4.48 and 4.49).

The contributions to the hole current, I^+, behave similarly to those of the electron current, and the total current I across a p–n junction with an external potential $\Delta\phi_{ext} = -e \Delta E_{ext}$ may be written

$$I = I^- + I^+ = -I(p) \left(\exp(\Delta E_{ext}/kT) - 1\right). \tag{4.101}$$

The relationship between current and potential is called the *characteristic* of the device, and the relation (4.101) for the p–n junction is illustrated in Fig. 4.50 by the curve labeled "no light." The constant saturation current $I(p)$ is sometimes referred to as the *dark current*.

Solar Cells

A p–n junction may be utilized to convert solar radiation energy into electric power. A solar cell is formed by shaping the junction in such a way that, for example, the p-type material can be reached by incident solar radiation, e.g., a thin layer of p-type material is placed on top of a piece of n-type semiconductor. In the dark and with no external voltage, the net current across the junction is zero, as is shown in the previous subsection, i.e., the intrinsic potential difference $\Delta\phi_0$ is unable to perform external work.

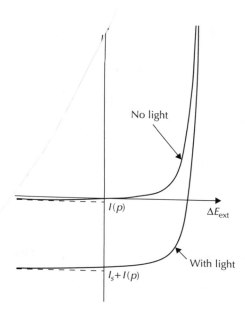

FIGURE 4.50 Characteristics (i.e., current as a function of external voltage) of a p–n junction, in the dark and with applied light. The magnitude of the short-circuit current, I_s, is a function of light intensity and spectral distribution.

However, when irradiated with light quanta of an energy $E_{light} = h\nu = hc/\lambda$ (h is Planck's constant, c is the velocity of light, and ν and λ are the frequency and wavelength of radiation), which is larger than the energy difference between the conduction and valence band for the p-type material,

$$E_{light} \geq E_c(p) - E_v(p),$$

then electrons may be photo-excited from the valence band into the conduction band. The absorption of light quanta produces as many holes in the valence band of the p-type material as electrons in the conduction band. Since in the dark there are many fewer electrons in the p-type conduction band than holes in the valence band, a dramatic increase in the number of conduction-band electrons can take place without significantly altering the number of holes in the valence band. If the excess electrons are sufficiently close to the junction to be able to reach it by diffusion before recombining with a hole, then the current in this direction exceeds $I_0^-(p)$ of (4.97) by an amount I_s, which is the net current through the junction in case of a short-circuited external connection from the n-type to the p-type material. The photo-induced current is not altered if there is a finite potential drop in the external circuit, since the relation between the current (4.101) and the external potential drop $e\,\Delta E_{ext}$ was derived with reference only to the changes in the n-region.

An alternative n–p type of solar cell may consist of a thin n-type layer exposed to solar radiation on top of a p-type base. In this case, the excess holes in the n-type valence band produce the photo-induced current I_s.

The total current in the case of light being absorbed in the p-type material and with an external potential drop is then

$$I = I_s - I(p)\,(\exp(-\Delta\,\phi_{ext}/kT) - 1). \qquad (4.102)$$

The short-circuit current I_s depends on the amount of incident light with frequencies sufficient to excite electrons into the conduction band, on the fraction of this light actually being absorbed, and on the conditions for transporting the excess electrons created in the conduction band, in competition with electron-hole recombination processes. I_s may be written as the sum of a conduction- and diffusion-type current, both related to the number of excess electrons in the conduction band, n_c^{ind}, induced by the absorption of light,

$$I_s = e(m_c E_e n_c^{ind} + k_c\, dn_c^{ind}/dx), \qquad (4.103)$$

where e is the numerical value of the electron charge (1.6×10^{-19} C), m_c is the mobility of conduction band electrons [e.g., 0.12 m^2 V^{-1} s^{-1} for silicon (Loferski, 1956), the dependence on the degree of doping being displayed in Fig. 4.51], E_e is the local electrical field, k_c is the diffusion constant [cf. (2.28) and (3.33); e.g., $k_c = 10^{-3}$ m^2 s^{-1} (Loferski, 1956)], and x is the depth below the solar cell surface, assumed to be the only significant coordinate (as in Fig. 4.49).

The excess electron number density, $n_c^{ind}(x)$, at a depth x, changes when additional electrons are photo-excited, when electrons are carried away from x by the current I_s, and when electrons recombine with holes,

$$\frac{\partial n_c^{ind}(x)}{\partial t} = \int \sigma(\nu) n_{ph}(\nu) \exp(-\sigma(\nu)x)\, d\nu + \frac{I}{e}\frac{\partial I_s}{\partial x} - n_c^{ind}(x)\frac{1}{\tau_c}. \qquad (4.104)$$

Here $\sigma(\nu)$ is the cross-section for absorption of light quanta (photons) in the p-type material, and $n_{ph}(\nu)$ is the number of photons at the cell surface ($x = 0$) per unit time and unit interval of frequency ν. The absorption cross-section is zero for photon energies below the semiconductor energy gap, $h\nu < E_c(p) - E_v(p)$, i.e., the material is transparent to such light. The most energetic light quanta in visible light could theoretically excite more than one electron per photon (e.g., 2–3 in Si with an energy gap slightly over 1 eV), but the probability for exciting just one electron to a higher energy is higher, and such a process is usually followed by a transfer of energy to other degrees of freedom (e.g., lattice vibrations and ultimately heat), as the excited electron approaches the lower part of the conduction band, or as the hole left by the electron de-excites from a deep level to the upper valence band. Thus, in practice, the quantum efficiency (number of electron–hole pairs per photon) hardly exceeds one.

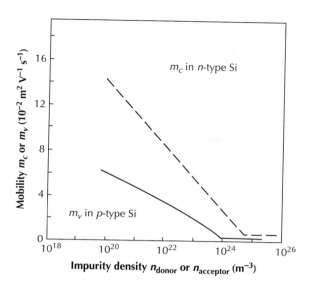

FIGURE 4.51 Mobility of minority carriers in Si at room temperature (about 300 K), extrapolated from measurements (Wolf, 1963). The mobility plotted is the *conduction mobility*, equal to the conductivity divided by the number of minority carriers and by the electron charge. The attenuation of the flow of actual carriers by recombination effects (trapping) is not considered.

The last parameter introduced in (4.104), τ_c, is the average lifetime of an electron excited into the conduction band, before recombination [τ_c may lie in the interval 10^{-11} to 10^{-7}, with 10^{-9} being a typical value (Wolf, 1963)]. The lifetime τ_c is connected to the cross-section for recombination, σ_c, and to the mean free path l_c of electrons in the conduction band by

$$l_c = \sigma_c^{-1} = v_c \tau_c N_a,$$

where v_c is the average thermal velocity of the electrons, $v_c = (2kT/m)^{1/2}$ (*m* being the electron mass, k being Boltzmann's constant, and T being the absolute temperature), and N_a is the number of recombination centers (*acceptor impurities*, cf. Fig. 4.48).

The boundary conditions for solving (4.104) may be taken as the absence of excess minority carriers (electrons or holes) at the junction $x = x_{pn}$,

$$n_c^{ind}(x_{pn}) = 0,$$

and a prescribed (material-dependent) excess electron gradient at the surface $x = 0$. This gradient, $(dn_c^{ind}/dx)|_{x=0}$, is often expressed in terms of a surface recombination velocity, s_c, through (4.103) by writing the left-hand side

$$I_s = s_c n_c^{ind}(0).$$

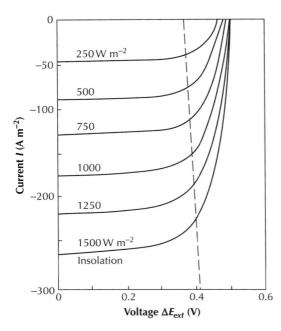

FIGURE 4.52 Characteristics of Cu_2S–CdS solar cell at 300 K for different intensities of incident radiation with typical clear-sky solar frequency distribution. The points of intersection with the dashed line give the voltage and current leading to maximum power output for given solar radiation. (Based on Shirland, 1966.)

Typical values of s_c are of the order of 10^3 m s^{-1} (Wolf, 1963, 1971). For n–p type solar cells, expressions analogous to the above can be used.

Once $n_c^{ind}(x)$ has been found, I_s can be calculated. Figure 4.50 shows an example of the total current through a p–n junction, as a function of applied voltage but for a fixed rate of incoming solar radiation on the p-type surface. The short-circuit current I_s increases linearly with intensity of light, if the spectral composition is kept constant, for the entire interval of intensities relevant for applications of solar cells at or near Earth. This is illustrated in Fig. 4.52 for a solar cell based on a p–n heterojunction, with the p-type material being Cu_2S and the n-type material CdS.

For an open-ended solar cell (i.e., no external circuit), the difference in electrical potential between the terminals, $V_{oc} = \Delta\phi_{ext}(I = 0)$, is obtained by putting I equal to zero in (4.102),

$$V_{oc} = kTe^{-1}(\log(I_s/I(p)) + 1). \qquad (4.105)$$

The amount of electrical power, E, delivered by the irradiated cell to the external circuit is obtained by multiplication of (4.102) by the external voltage,

$$E = (\Delta\phi_{ext})I = \Delta\phi_{ext}(I_s - I(p)(\exp(-e\Delta\phi_{ext}/kT) - 1)). \qquad (4.106)$$

From $\partial E/\partial (\Delta\phi_{ext}) = 0$, the external voltage V_{opt} may be found, which leads to the maximum value of power, E_{max}. In the situations of interest, V_{opt} is a slowly varying function of the amount of incident radiation, as illustrated by Fig. 4.52. The corresponding current may be denoted I_{opt}.

The efficiency of solar cell radiant-to-electrical energy conversion is the ratio of the power E delivered and the incident energy, denoted E_+^{sw}, as in sections 3.1 and 2.2.2 (eventually for a tilted orientation of the solar cell), $\eta = E/E_+^{sw}$. In terms of the flux of photons of given frequency incident on the solar cell [introduced in (4.104)], the non-reflected energy flux at the surface may be written (a is the albedo of the cell surface)

$$E_+^{sw}(1-a) = \int_0^\infty h\nu n_{ph}(\nu)\, d\nu, \qquad (4.107)$$

where h is Planck's constant. For a given semiconductor material, the maximum fraction of the energy (4.107) that can be absorbed is

$$\int_{h\nu = E_c(p) - E_v(p)}^\infty h\nu n_{ph}(\nu)\, d\nu.$$

The part of the integral from zero up to the energy gap (i.e., the part above a certain wavelength of light) constitutes a fundamental loss. The same can be said of the energy of each light quantum in excess of the semiconductor energy gap $E_c(p) - E_v(p)$, assuming a quantum efficiency of at most one, i.e., that all such quanta are indeed absorbed (which may not be true if their energy is, say, between the upper limit of the conduction band and the lower limit of the following band) and that all excess energy is spent in exciting lattice degrees of freedom (vibrational phonons) that do not contribute to the photovoltaic process. In that case, the energy flux available for photoconversion is only

$$(E_c(p) - E_v(p)) \int_{h\nu = E_c(p) - E_v(p)}^\infty n_{ph}(\nu)\, d\nu = E^{avail}. \qquad (4.108)$$

Further losses in addition to reflection and insufficient or excess photon energy may be associated with imperfections in the junction materials or in the current extraction system, causing heat formation or light re-emission rather than electrical power creation. Both heat creation (in the lattice) and re-radiation may take place in connection with the recombination of photo-excited electrons and holes. Since many of these processes are highly temperature dependent, the maximum efficiency that can be obtained in practice is also temperature dependent. Examples of maximum theoretical efficiencies, as well as those obtained in practice, are given in section 4.4.4.

Rather than being p- and n-doped materials of the same elemental semiconductor, the solar cell junction may be based on different materials (a *heterojunction*) or on a metal and a semiconductor (a *Schottky junction*).

Individual types of solar cells are discussed below, after presentation of some general remarks on the efficiency of photovoltaic cells.

Design of Photovoltaic Converters

A photovoltaic converter consists of a number of solar cells suitably connected, plus eventually some auxiliary equipment, such as focusing devices in front of the cells and tracking systems. The maximum efficiency of a solar cell is given by the ratio of the maximum power output (4.106) and the incident radiation flux,

$$\max(\eta) = \max(E)/E_{s,y}^{sw}. \tag{4.109}$$

The maximum efficiency is smaller than unity for a number of reasons. First, as discussed above, radiation of frequency below the semiconductor band gap is not absorbed. Second, according to (4.108), the excess energy of radiation with frequencies above the semiconductor band gap is not available to the photovoltaic conversion process. This loss would be small if the solar spectrum were peaked across the band gap, but most semiconductor gaps correspond to only a limited part of the broad solar spectrum (cf. Fig. 2.17).

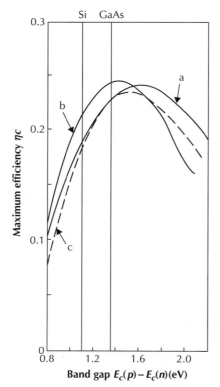

FIGURE 4.53 Early calculation of maximum efficiency for simple p-on-n solar cell; a outside the Earth's atmosphere ($E^{sw} = 1350$ W m^{-2}); b at the Earth's surface under standard conditions ($E^{sw} = 890$ W m^{-2}, air mass 1, water content 0.02 m^3 m^{-2}, and major absorption bands included in the calculation); c overcast condition ($E^{sw} = 120$ W m^{-2}). (Based on Loferski, 1956.)

Third, as seen, for example, in Fig. 4.52, the maximum power output is less than the maximum current times the maximum voltage. This reduction in voltage, necessary in order to get a finite current, is analogous to the requirement for a finite dissipation term in a thermodynamic engine in order to get energy out in a finite time (cf. section 4.1.1). Expression (4.106) does not fit measured power values in detail, and it has been suggested that a second exponential term be added in the current–voltage relation (4.102), of similar form but with $\Delta\phi_{ext}$ replaced by $1/2\Delta\phi_{ext}$ (Sah *et al.*, 1957). The origin of such a term is thermal generation and recombination of carriers in the vicinity of the junction. Fourth, the external potential (times the electron charge) has a maximum value (4.105), which is smaller than the semiconductor gap $E_c(p) - E_c(n)$, since it equals the difference in Fermi level between the *p*- and *n*-regions [cf. (4.99) and Fig. 4.48]. This loss may be diminished by increasing the number of impurities (and hence the number of minority carriers) in both the *p*- and *n*-region. However, if these are increased above a certain level, the increased recombination probability offsets the voltage gain.

Figure 4.53 shows an early example of calculated maximum efficiency as a function of the semiconductor band gap, including the above-mentioned losses, for radiation conditions corresponding to the top of the Earth's atmosphere (*a*), for a clear-sky day at the Earth's surface with average atmospheric absorption and scattering conditions (*b*), and for a cloud-covered sky (*c*). It is shown that for common solar cell materials, such as Si or GaAs, the maximum efficiency is larger for the spectral composition of clear-day solar radiation at ground level than for the solar spectrum not disturbed by the Earth's atmosphere. The performance for scattered solar radiation (the overcast-sky case) is not substantially impaired.

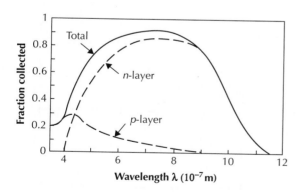

FIGURE 4.54 Spectral collection efficiency for a simple *p*-on-*n* solar cell. The curve labeled *total* is based on measurements, and the individual contributions from the *p*-layer (thickness 5×10^{-7} m) and the *n*-layer have been calculated. The total thickness is about 4.5×10^{-4} m. (Based on Wolf, 1963.)

The wavelength dependence of the collection efficiency (i.e., the efficiency including the first two loss terms discussed above, but not the last two) is shown in Fig. 4.54 for a simple silicon cell consisting of a thin (5×10^{-7} m) *p*-layer on top of an *n*-layer base (of thickness 4.5×10^{-4} m). The total efficiency curve is based on measurements, whereas the separate contributions from the *p*- and *n*-layers have been calculated (Wolf, 1963). Short wavelengths are absorbed in the *p*-layer and give rise to a current of electrons directed toward the junction, but the bulk of the solar wavelengths are not absorbed until the photons reach the *n*-layer base. They give rise to a hole current toward the junction. The efficiency is not independent of cell temperature, as indicated in Fig. 4.55. The silicon cells currently used exhibits an absolute loss in efficiency of 0.4%–0.5% for each °C of temperature rise.

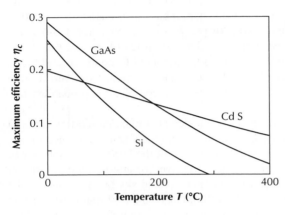

FIGURE 4.55 Calculated temperature dependence of solar cell efficiency. (Based on Wysocki and Rappaport, 1960.)

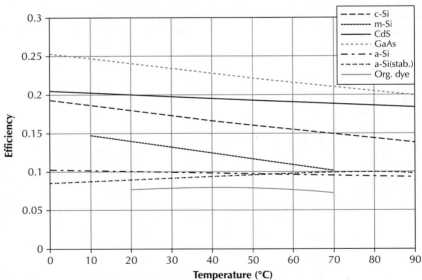

FIGURE 4.56 Solar cell efficiency as a function of operating temperature, normalized to a typical 25 °C efficiency for each cell type. [From Sørensen (2000b), based on Ricaud (1999), Dutta *et al.* (1992), Wysocky and Rappaport (1960), Yamamoto *et al.* (1999), Rijnberg *et al.* (1998).]

Figure 4.56 adds more recent data on the temperature dependence of different types of solar cells, based upon measurements.

The data shown in Fig. 4.56 have been normalized, for each type of solar cell, to a typical absolute efficiency for current commercial or near-commercial versions of the type of device in question. Early theoretical calculations (Wysocki and Rappaport, 1960) are largely confirmed by current measurements, and the mechanisms are thus well understood, at least for conventional photovoltaic devices. The temperature dependence is chiefly due to band-gap effects, which explains why the slope of the crystalline silicon (c-Si) and multicrystalline silicon (m-Si) are identical (Yamamoto et al., 1999). In other words, the grain boundaries do not give rise to additional temperature effects. Cd–S cells have a lower but still significant temperature gradient, whereas the temperature effect for amorphous silicon cells and organic dye-sensitized TiO_2 cells is very small.

The temperature effect is negative with increasing working temperature for all devices except two: organic cells show a maximum near 40 °C (Rijnberg et al., 1998) and the amorphous silicon (a-Si)–hydrogen cells show a reversal of temperature trends after annealing (Dutta et al., 1992). This positive temperature coefficient persists only until the undegraded efficiency is reached, and it requires annealing as opposed to light-soaking treatment, which causes the development of a stronger negative temperature coefficient. The modest temperature dependence is conveniently modeled by a power expansion of the efficiency,

$$\eta = \eta(298 \text{ K}) + a\,(T - 298 \text{ K}) + b\,(T - 298 \text{ K})^2. \qquad (4.110)$$

The operating temperature dependence of the solar-energy-to-electricity conversion efficiency suggests that cooling the cell by extracting heat may improve the electric performance of the cell and thereby pay for some of the extra expense of the heat extraction equipment. (This is further discussed in a case study in section 6.5.1.) Typical operating temperatures for uncooled cells are about 50 °C. Fig. 4.56 shows that improvement is indeed obtained for crystalline or multicrystalline silicon photovoltaic (PV) cells, for example, but not notably for dye-sensitized cells or amorphous PV cells. On the other hand, in order to make use of the heat, it should preferably be collected at higher temperatures, which would indicate that the best solutions are those with little operating temperature effect on the electricity yields. This trade-off is further illustrated in the simulation models of section 6.5.1.

In cells currently produced, each loss factor is carefully minimized, and the resulting efficiencies have increased over time, for all types of solar cells, as discussed below.

Monocrystalline Silicon Cells

The photovoltaic cell principles described above form the basis for monocrystalline cells, which are cells constructed from single crystals, usually in the form of ingots sliced into a number of cells.

A number of improvements have brought the cell efficiency of state-of-the-art monocrystalline silicon cells up to about 25%. The light capture is improved through trapping structures that minimize reflection in directions not benefiting the collection area and by back-side designs reflecting light rays back into the active areas (see, for example, Fig. 4.57). The doping degree is altered near electrodes (n^+ and p^+ areas), and a thin oxide layer further helps to prevent electrons from reaching the surface rather than the electrode (this process is termed *passivation*). Further, top electrodes may be buried, in order not to produce shadowing effects for the incoming light (Green, 1992).

Figure 4.58 shows the measured characteristics, i.e., current as a function of voltage, for a cell of the type shown in Fig. 4.57.

Simulation of the light-trapping and electron-transport processes in one, two, or three dimensions has helped in selecting the best geometry and degree of doping (Basore, 1991; Müller *et al.*, 1992). Figure 4.59 gives an example of one-dimensional simulation of the variation in cell performance as a function of doping degree for a cell of the kind shown in Fig. 4.58.

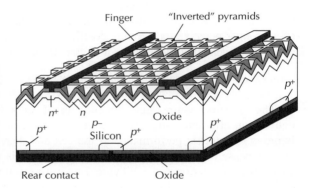

FIGURE 4.57 Structure of a monocrystalline silicon cell with passivated emitter and a locally diffused rear structure (PERL), used to obtain a 23% module efficiency. (From Green *et al.*, 1998, used with permission.)

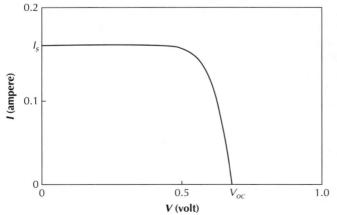

FIGURE 4.58 Current–voltage curve for a PERL cell similar to that of Fig. 4.57 (but with cell efficiency 22%), as measured by Sandia Laboratories (USA) at 1006 W m^{-2} airmass 1.5 simulated radiation, for a 4 cm^2 cell. Key findings are V_{oc} = 696 mV (4.105), I_s = 160 mA (4.103), and fill factor = 0.79 (Wenham *et al.*, 1995b). The fill factor is the ratio between the area under the curve and $I_s V_{oc}$.

FIGURE 4.59 Calculated efficiency as a function of doping parameters for a simplified silicon cell of the type depicted in Fig. 4.57. The one-dimensional finite-element model used is described in the text (Sørensen, 1994c).

The model takes into account the 10 μm facet depth and uses a curved one-dimensional path through the cell. The two most important doping parameters (impurities per unit volume) are the uniform bulk doping of the p-material and the n-doping at the front, assumed to fall off as an error function, thereby simulating the average behavior of near-electrode sites and sites away from electrodes. Transport and recombination are calculated in a finite-element model (Basore, 1991). The back-side doping is kept at 2×10^{19} cm^{-3}.

Multicrystalline Cells

Another technology for producing solar cells uses multicrystalline (sometimes referred to as *polycrystalline*) materials, instead of single-crystal materials. Multicrystalline materials consist of small domains or grains of crystalline material, randomly oriented relative to each other. The crystal grains in multicrystalline materials sustain conductivity in the same way that single crystals do, but the transport of electrons across grain boundaries induces losses, reduces conductivity, and thus makes the cells less efficient. On the other hand, they can be produced by simpler methods than those needed for monocrystals, e.g., by evaporating suitable coatings onto a substrate. This field is rapidly developing, as it is becoming possible to deposit only a few atomic layers onto a substrate, and with suitable techniques (such as using magnetic fields to align grains) it may soon be possible to form near-monocrystalline layers without having to grow crystals.

It was initially believed that the additional losses at grain boundaries would necessarily make the efficiency of multicrystalline cells substantially lower than what could be obtained by crystalline materials. Actually, the difference has narrowed as a result of better understanding of the options for optimizing performance of complex cell structures. One problem has been the damage inflicted upon multi-crystalline cells by attempting to copy to them some of the efficiency-improving techniques that have worked well for monocrystalline cells (surface texturing, rear passivation by oxide layers). Yet, etching of inverted pyramids on the surface of multicrystalline cells has improved efficiency considerably (Stock *et al.*, 1996), and, recently, less-damaging honeycomb texture patterns have brought the efficiency up to 20% (Zhao *et al.*, 1998). This trend is likely to induce the long-predicted change from expensive ingot-grown monocrystalline cell materials to multicrystalline materials more suitable for mass production and price-reduction efforts. However, development away from single-crystalline solar cell materials is slower than anticipated because of the higher maturity of the crystalline industry processes. Figure 4.60 shows the structure of a multicrystalline cell that is 20% efficient and absorbs more than 90% of incoming radiation. The advantages of thin-film multicrystalline solar cells over monocrystalline ones would seem to more than compensate for the remaining 5% efficiency difference. This does not exclude that crystalline and multicrystalline technologies will continue to co-exist in the marketplace for a while.

FIGURE 4.60 Texturing used to raise multicrystalline cell efficiency to 20% in a PERL-type cell. (From Zhao *et al.*, 1998, used with permission.)

Stacked Cells

Instead of basing a solar cell on just a single *p–n* junction, it is possible to stack several identical or different cells on top of each other. The use of different cell materials aims at capturing a wider range of frequencies than possible with a single junction. In this case materials of different band gaps will be stacked (Yazawa *et al.*, 1996; Takamoto *et al.*, 1997). In the case of stacked identical cells, the aim is to be able to use lower quality material (e.g., the mentioned thinly sprayed crystalline cells in contrast to ingot-grown ones) and still get an acceptable overall efficiency by stacking several layers of low individual efficiency (Wenham *et al.*, 1995a). This concept is illustrated in Fig. 4.61, where a calculation performed for low-quality multicrystalline silicon finds a broad maximum of efficiency for about six layers (Green, 1994).

Supporting substrate (or superstrate)

FIGURE 4.61 Concept of a multilayer thin-film silicon solar cell (Wenham *et al.*, 1995a).

FIGURE 4.62 Production steps for multiplayer cells: deposition, grooving, and metallization. (From Sproul *et al.*, 1995, used with permission.)

Figure 4.62 shows the industrial process that may be used for producing multilayer cells, using vapor deposition to deposit each layer and laser grooving to make room for electrodes. Module assembly uses automatic series connection achieved by the layout shown in Fig. 4.63. With six junctions of low-quality material (grain sizes above 3 μm), a cell efficiency of 15.2% has been measured (for a small specimen; Sproul et al., 1995). The dependence on the number of layers is weak for the optimum bulk doping concentration of about 10^{18} cm^{-3} and a total thickness of around 10 μm. Efforts to scale the processes up to industrial scale production levels have not succeeded so far, due to difficulties in controlling the transport of electrons from the layers of low-quality silicon to the electrodes.

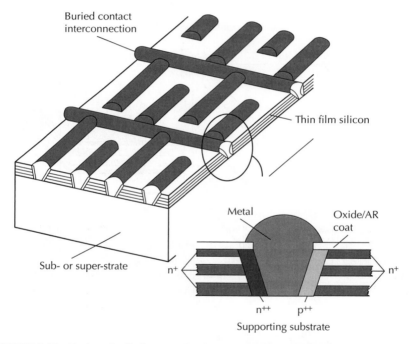

FIGURE 4.63   Design of cells for easy electrical interconnection between adjacent cells in a module. (From Wenham et al., 1995a, used with permission.)

Amorphous Cells

Amorphous semiconductor materials exhibit properties of interest for solar cell applications. While elemental amorphous silicon has a fairly uniform energy distribution of electron levels, composite materials have been constructed that exhibit a pronounced energy gap, i.e., an interval of energy essentially without any states, as in a crystal. Spear and Le Comber (1975) first produced such an amorphous material, which was later proved to be a silicon–hydrogen alloy, with the distribution of energy states shown in Fig. 4.64 as curve *a*. Ovshinsky (1978) produced a silicon–fluorine–hydrogen alloy with further depression of gap states (Fig. 4.64, curve *b*). The gap is about 1.6 eV wide and thus should be more favorable with respect to the solar energy spectrum than the 1.1 eV gap of crystalline silicon. Furthermore, doping (introduction of boron or phosphorus atoms) has proved possible, so that *p*- and *n*-type amorphous semiconductors can readily be made, and a certain amount of "engineering" of materials for exactly the desired gap structure, doping efficiency, conductivity, temperature sensitivity, and structural stability (lifetime) can be performed.

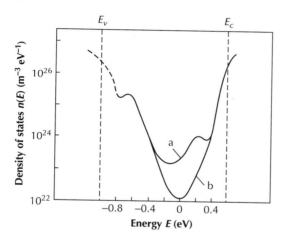

FIGURE 4.64 Density of electron states as a function of excitation energy for silicon-based amorphous materials: *a* silicon–hydrogen alloy; *b* silicon–fluorine–hydrogen alloy. (Based on Spear and Le Comber, 1975; Ovshinsky, 1978.)

A theoretical description of the gap occurrence and electron conduction in amorphous materials was first presented by Street and Mott (1975) and followed up by Fritsche (1977) and Pfister and Scher (1977). The basis is the occurrence in amorphous material of defects of particular affinity to attract or reject electrons (e.g., lone pair sites), and the transport of electrons is thought of as a quasi-random "hopping" between sites, some of which are capable of "trapping" an electron for a shorter or longer period of time. An abundance of broken or "dangling" bonds may give rise to serious trapping problems (low conductivity), and the success obtained by incorporating hydrogen seems due to its occupying and neutralizing such sites.

Not very long after the theoretical description of amorphous solar cells, Japanese scientists succeeded in creating cells that were suited for industrial production (Hamakawa *et al.*, 1981) and that soon found a market in powering calculators and similar small-scale devices, where the cell cost was relatively unimportant. A typical structure of a commercial amorphous cell is illustrated in Fig. 4.63. Band gaps in the range from 1.0 to 3.6 eV can be engineered with different silicon alloys (SiGe, Si, SiC), and such cells may be stacked to obtain a broader frequency acceptance (Ichikawa, 1993; Hamakawa, 1998).

However, the simplest version has just one type of material: an intrinsic layer of an a-Si:H compound is the main area of light absorption, and adjacent p- and n-type layers ensure the transport to the electrodes, the front one of which is made of a transparent material. The whole structure is less than 1 μm thick and is deposited onto a plastic backing material. Maximum efficiencies of around 13% have been demonstrated (Fig. 4.66), but one problem has persisted: because the structure of the material is without order, bombardment with light quanta may push atoms around, and the material degrades with time. Current practice is to degrade commercial cells before they leave the factory, thereby decreasing the efficiency by some 20%, but in return obtaining reasonable stability over a 10-year period under average solar radiation conditions (Sakai, 1993). Several layers of p-, i-, and n-layers may be stacked, and the highest efficiency is obtained by replacing the amorphous n-layers with a multicrystalline pure Si or silicon-compound layer (Ichikawa, 1993; Ma et al., 1995).

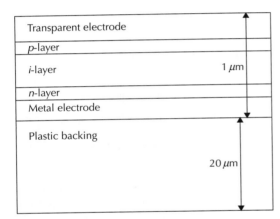

FIGURE 4.65 Structure of the most common a-Si cell. Several p- i- and n-layers may be stacked.

FIGURE 4.66 Efficiency development in a-Si solar cell for different cell areas (before degradation; Sakai, 1993).

Other Materials and Other Thin-film Cells

Use of materials from chemical groups III and V, such as GaAs, CdS, and CdTe, instead of silicon, allows better engineering of band gaps in crystalline solar cells to suit particular purposes and advances new properties suitable for certain tasks (notably space applications and use in concentrating collectors). Important considerations in selecting materials include temperature dependence, where crystalline silicon cell efficiency drops quickly with the increasing temperature likely to prevail (despite possible active cooling), for cells operating at high levels of solar radiation (see Figs. 4.55 and 4.56).

The GaAs band gap of 1.43 eV is well suited for the solar spectrum, and, with a tandem cell of GaAs plus $GaInP_2$, an efficiency of over 30% has been reached (Bertness et al., 1994; Deb, 1998). Currently, these cells are expensive and are mainly used in space. However, thin-film versions may be developed, as in the Si case, and they already have for CIS (copper–indium–diselenide) cells. The highest efficiency obtained so far is about 17% for a Cu (In,Ga) Se_2 structure (Tuttle et al., 1996).

Among a range of nontraditional designs of solar cells is the use of spherical droplets. This design grew out of a proposed re-use of scrap material from the microelectronics industry, and it may increase acceptance of light from different directions and reduce reflection that otherwise would have to be dealt with by surface texturing (Maag, 1993; Drewes, 2003; ATS, 2003).

Concentrating devices can be used for both thermal and electricity-producing systems. (They are treated together in section 4.4.4.) Naturally, a photovoltaic cell receiving concentrated sunlight will need to be able to function with the increased thermal load, normally leading to higher temperatures. The implications of this can be derived from the discussion of Figs. 4.55 and 4.56.

Module Construction

Individual cells based on monocrystalline materials typically have areas of 100 cm^2 (limited by techniques for growing ingots of monocrystalline material). Multicrystalline and photoelectrochemical cells, where semiconductor material is deposited on a backing template (usually glass), may have larger cell size, and amorphous cells have essentially no limitations. Amorphous cells have been produced on rolls of flexible plastic backing materials with widths of 1–2 m and rolls of any length. The same may also become possible for other thin-film types, such as spray-deposited multicrystalline materials.

It is customary to assemble cells into modules by a mixture of parallel and series connections, so that the resulting voltages become suitable for standard electricity-handling equipment, such as inverters transforming DC currents into AC currents of grid specifications and quality. Alternatively, microprocessor inverters may be integrated into each module or even into each cell in order to minimize transport losses. In recent years, specific inverters optimized for solar cell applications have been produced, with an inverter efficiency increase from 90% to about 98% (IEA, 1999).

Solar cell technology, then, is characterized by two main solar radiation conversion efficiencies: the efficiency of each cell and the efficiency of the entire module sold to the customer. The latter is currently about 5% lower than the former, notably because of phase mismatch between the individual cell current components, but this does not need to be so, and the difference between the two efficiencies is expected to diminish in the future.

Use of Residual Energy

Given that practical efficiencies of photovoltaic devices are 10%–30%, it is natural to think of putting the remaining solar energy to work. This has first been achieved for amorphous cells, which have been integrated into windowpanes, so that at least a part of the energy not absorbed is passed through to the room behind the pane. Of course, conductors and other nontransparent features reduce the transmittance somewhat. The same should be possible for other thin-film photovoltaic materials, including the emerging multicrystalline silicon cells.

Another possibility, particularly for PV panels not serving as windows, is to convert the solar energy not giving rise to electricity into heat (as some of it in actuality already is). One may think of cooling the modules of cells by running pipes filled with water or another suitable substance along their bottom side, carrying the heat to a thermal store, or alternatively by an air flow above the collector (but below a transparent cover layer). The energy available for this purpose is the incoming radiation energy minus the reflected and converted part. Reflection from the back of the semiconductor material, aimed at increasing the path-length of the photons in the material, could be chosen to optimize the value of the combined heat and power production, rather than only the power production. Examples of combined-cycle solar panels are modeled in section 6.5.1.

For concentrator cells, active cooling is needed in any case, because of the temperature dependence of the photovoltaic process (cf. Figs. 4.55 and 4.56).

The maximum electrical efficiency of a single junction photovoltaic device implied by semiconductor physics is around 40% [see discussion following (4.109)]. If reflections can be minimized and many different light absorption materials are stacked, the theoretical Carnot efficiency (4.4) for the temperature of the Sun relative to a reference temperature at the surface of the Earth, i.e., about 95%, may be approached. Considerations of power-flow optimization diminish this limit further. Honsberg et al. (2001) and Green (2001) find a thermodynamic value of some 87% for an infinite stack of cells, which can be used as a starting point for discussing the losses deriving from semiconductor physical arguments.

For a solar cell with electric efficiency below 40%, the possible associated heat gain is another 40% (cf. the simulation results in section 6.5.1).

4.4.2 Photo-electrochemical Conversion

Photo-electrochemistry is an area of confluence between solar cell technology, discussed here, and battery or fuel-cell technology, discussed in section 4.7. Organic solar cells are a special kind of photo-electrochemical (PEC) devices that try to take advantage of inexpensive organic materials, as opposed to the more expensive metals and doped semiconducting materials used in the photovoltaic devices discussed above. As discussed for biological photosynthesis in section 3.6, suitable organic materials can trap sunlight and convert radiation into other forms of energy. Copying this process has been attempted in various ways. Calvin (1974) considered a double membrane that would separate the ionized reactants of a photo-excitation process,

$$A + h\nu \to A^*; \quad A^* + B \to A^+ + B^-.$$

In addition, a transport system is needed to get the ions to an electrode. No practical version of this idea has been produced. The same is the case for a

FIGURE 4.67 Layout of photo-electrochemical solar cell with indication of energy levels (E_f is the Fermi level of the semiconductor material). Solar radiation is absorbed in a dye layer, creating an excited electron state, from which an electron is transferred to the semiconductor at left and replenished from the counter-electrode through a redox cycle in an electrolyte to the right (Sørensen, 2003a).

concept aimed at both removing CO_2 from the atmosphere and simultaneously producing methanol (Jensen and Sørensen, 1984, pp. 217–218). The absorption of solar radiation is used to fix atmospheric carbon dioxide to a ruthenium complex [Ru], which is then heated with water steam,

$$[Ru]CO_2 + 2H_2O + 713 \, kJ \, mol^{-1} \rightarrow [Ru] + CH_3OH + 1/2 \, O_2.$$

One scheme that has been realized is the attachment of a ruthenium complex as a dye to TiO_2, which may then transport electrons formed by photo-excitation in the Ru-complex to an electrode. The process is similar to the dye-excitation processes used in conventional photographic prints and was first proposed by Moser (1887). He called the dye substance enhancing the absorption of solar radiation above what can be achieved by the TiO_2 a "sensi-tizer." The Ru-complex is restored by a redox process in an electrolyte joining the other electrode, with use of a platinum catalyst, as indicated in Fig. 4.67. Because a monolayer of even a very efficiently absorbing dye will absorb less than 1% of incoming solar radiation, the dye-absorption layer is made three-dimensional by adhering the dye to a nanostructured TiO_2 network of nodules, as first demonstrated by Tsubomura et al. (1976). They also introduced a liquid electrolyte, capable of penetrating into the cavities in the nanostructured sensitized TiO_2 and providing the necessary contact for transfer of the replace-ment electron to the dye. The effective absorption surface may be increased by three orders of magnitude and provides an overall cell efficiency of about 10%. It should still be possible for the TiO_2 nodules to transfer the absorbed electron to the back electrode through a series of transport processes.

TiO₂
Anatase

FIGURE 4.68   Anatase structure (first determined by Horn et al., 1970) and unit cell (Sørensen, 2003a).

The material presently favored for the anode nanoparticles is TiO_2 in the form of anatase (Fig. 4.68). Compared to other forms of titanium dioxide (rutile and brookite), anatase better accommodates the dye molecules and forms nodules rather than flakes. The large-side dimension of the unit cell shown in Fig. 4.68 is about 0.2 nm. Several other semiconductor materials have been investigated, but so far anatase has shown the best overall properties and is used fairly universally. Electron transport through the anatase layers follows conventional solid-state physics, except for the issue of nodule coherence. A simple modeling effort has successfully described the transport as a random walk, rather than hopping (Nelson et al., 2001).

On the cathode side, a redox couple is used to supply the electron to replace the one being excited in the dye and transferred to the anatase before it decays back to the dye ground state. The electrolyte is typically acetonitrile (C_2H_3N), and the redox couple is iodine/tri-iodine (I^-/I_3^-), which has been used almost exclusively since the work of Tsubomura et al. (1976). This does not seem ideal, since the difference between the anatase Fermi level and the I^-/I_3^- chemical potential, which determines the cell's open-circuit voltage, is only about 0.9 eV, as compared with typical dye-excitation energies of 1.8 eV. Many efforts have been directed at finding more appropriate redox shuttle systems, but so far none has shown overall properties making it preferable to the I^-/I_3^- couple (Wolfbauer, 1999). Electrolyte and redox couple integrity and lifetimes are of concern. A comparison to batteries is appropriate, and battery lifetimes are rarely as long as desired for solar cells that may be incorporated directly into building components and structures.

A further energy loss takes place at the cathode, where application of a catalyst is required in order to obtain the desired rate of electron transfer from electrode to electrolyte. As in batteries and fuel cells, traditionally preferred catalysts are based on platinum, but alternatives are under investigation. Generally speaking, the use of liquid electrolytes and catalysts is undesirable, and the much slower electron transfer through the electrolyte and its redox couple (as compared with semiconductor transport) is likely to be the overall limiting factor for current in the device. However, the reason for still choosing a liquid electrolyte is also obvious. The cell is produced by deposition of anatase layers on the anode and subsequent annealing, processes requiring temperatures well over 100 °C. The dye is then applied, either by soaking or by flushing, creating the huge intrinsic surface for solar collection. Typical dye melting points are 80 °C–100 °C, so applying a second semiconductor material (if one with appropriate properties could be found) from the other side at appropriate temperatures would destroy the cell. More gentle application not requiring high temperatures is not likely to allow the surface contact area between dye and semiconductor to be equally large on both sides.

An alternative is to find another material not requiring high temperatures for penetrating into the cavity structure of the initial semiconductor plus dye layers. Possible candidates are conducting polymers or the ion-carrying polymers used in fuel cells. Actual 2%–3% energy conversion efficiency has been obtained with two types of polymer systems. One uses a gel network polymer as electrolyte (Ren et al., 2001). The other is a type of plastic solar cell, where the known ability of ^{60}C molecules to absorb solar radiation (Sariciftci et al., 1992) is used to create a fairly large absorption area of ^{60}C sensitizer imbedded in a suitable polymer (Shaheen et al., 2001; Yu et al., 1995).

The choice of sensitizer is ideally based upon fulfillment of requirements that include at least the following:

- high absorption capability over the range of spectral frequencies characteristic of sunlight
- energetically suitable excited states
- good attachment to semiconductor nanoparticles that ensures rapid electron transfer (in competition with de-excitation and back-transfer from semiconductor surface to dye sensitizer)
- easy acceptance of the replacement electron from electrolyte
- dye lifetime consistent with stipulated device life.

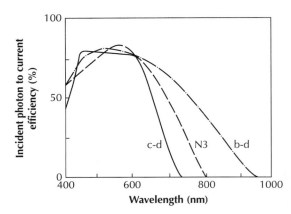

FIGURE 4.69 Spectral sensitivity: c-d indicates coumarin derivative; N3 and b-d (black dye) indicate ruthenium complexes. (Based upon Hara *et al.*, 2001; Nazeeruddin *et al.*, 2001.)

The search for optimized sensitizers has usually focused on a particular family of molecules. For example, one group (O'Regan and Grätzel, 1991; Nazeeruddin *et al.*, 1993, 2001; Shklover *et al.*, 1998) has looked at metal complexes based on ruthenium polypyridines, meticulously synthesizing one variant after the other, adding rings, thiocyanate ligands, and carboxylate groups in different combinations. The size of the molecule, in combination with its excitation spectrum, determines the frequencies of solar radiation that can be absorbed and the associated cross-sections. The *black dye* (1 ruthenium atom, 3 pyridine rings, 3 thiocyanate ligands, and 3 carboxylate groups) has led to the currently highest overall conversion efficiency of 10% for laboratory cells (area about 10^{-4} m^2). In comparison, an efficiency of 5% is claimed for a large cell (on the order of 1 m^2) that was briefly in industrial production (STI, 2002). An earlier favorite was *N3 dye* (Ru, 2 bipyridine rings, 2 thiocyanate ligands, and 4 carboxylate groups), which is particularly able to transfer an excited electron to an anatase surface, a condition that has been attributed to its attachment to the anatase surface by two carboxylate binding sites at approximately the same spacing as the "indents" in one anatase surface. However, the dye's light absorption stops below 800 nm, implying a smaller efficiency for many potential real-life collector sites. Figure 4.69 compares spectral sensitivities of the two dyes just mentioned, plus the coumarin-derivative organic dye considered in the following.

FIGURE 4.70 The structure of N3 ($RuS_2O_8N_6C_{26}H_{24}$, *left*) and b-d ($RuS_3O_6N_6C_{21}H_{22}$, *right*) ruthenium sensitizers synthesized by Nazeeruddin *et al.* (1993; 2001). A modest structure optimization has been performed (Sørensen, 2004b).

Figure 4.70 gives the molecular structure of the purple N3 and the black ruthenium dye (Nazeeruddin *et al.* 1993, 2001). Several variants have been studied (see, for example, Shklover *et al.*, 1998; Zakeeruddin *et al.*, 1997).

The structure of modified sensitizer molecules is often roughly given by applying the known structures of the components plus some general rules of thumb. Measurements of spectra, such as the nuclear magnetic resonance (NMR) spectra, help determine the positions of specific atoms (e.g., hydrogen atoms), but not always in a unique way. The addition of new features to a dye molecule in order to enhance solar absorption or to aid electron transfer out of the molecule may give rise to structures not amenable to simple guesses, e.g., due to isomerism or other reasons, such as energy surfaces in configuration space with more than one minimum.

Quantum mechanical modeling leads to theoretical predictions of likely molecular structures, as outcomes of optimization studies. These involve following a path of steepest descent in the potential energy surface, eventually using second-order derivatives in order to stabilize the sizes of jumps made for each of the iterations (Sørensen, 2003a).[*]

---

[*]So-called *ab initio* calculations in quantum chemistry involve solving the Schrödinger equation under several simplifying assumptions: nuclear motion is neglected, the basis functions used are linear combinations of Gaussian functions for each atom and far fewer than needed for completeness. Interactions are first treated by the Hartree-Fock method, implying that each electron is assumed to move in the mean field from all the other particles. Higher electron correlations are added in a perturbative way or by the phenomenological density functional theory.

FIGURE 4.71 Optimized structure of coumarin-derivative organic dye (gross formula $H_{26}C_{25}N_2O_4$) synthesized by Hara *et al.* (2001) and yielding overall efficiencies of around 6% when used in PEC solar cells (Sørensen, 2003a).

Figure 4.71 shows the result of a theoretical optimization (Sørensen, 2001a; 2003a) for a modification of a coumarin dye that is part of a family of molecules studied by Hara *et al.* (2001) as potential inexpensive, purely organic candidates for dyes to use with TiO_2 semiconductor nanoparticles in solar cells.

Figure 4.72 shows the molecular orbits near the Fermi level for the coumarin-derivative dye, obtained from a quantum mechanical Hartree-Fock self-consistent field calculation of the best ground-state configuration [using the software of Frisch *et al.* (1998) and including a basis of 641 molecular states, or about 3 for each physical orbit]. It is seen that moving an electron from the highest occupied orbit (HOMO) to the lowest unoccupied orbit (LUMO) involves reducing the electron density in the coumarin part of the molecule and increasing it along, and particularly at the end of, the "arm" attached to the molecule. However, at this level of approximation, the energy difference (LUMO minus HOMO) is still nearly 8 eV, as opposed to an experimental value of 1.8 eV (Hara *et al.*, 2001). The second unoccupied orbit (Fig. 4.72, *bottom*) is quite different.

In order to estimate more realistically the energy of the first excited state of the coumarin-derivative dye molecule, a number of calculations have been performed (Sørensen, 2003a). Interactions not included in the SCF ground state calculation may be added in a time-dependent Hartree-Foch (TDHF) calculation for one or more excited states (Casida *et al.*, 1998), and further improvement is obtained by performing the TDHF calculation on top of ground-state calculations including further interaction, such as including

FIGURE 4.72 Electron density of HOMO (*top*), LUMO (*middle*), and second unoccupied molecular orbital (*bottom*) for a coumarin-derivative dye based on self-consistent field (SCF) calculation (Sørensen, 2001a, 2003a).

exchange forces in a density functional method (Kohn and Sham, 1965; Becke, 1993). Figure 4.73 shows the excitation energy of the lowest spin-0 and spin-1 excited states using successive improvements in the sophistication of the interactions included and in the number of basis states used for the calculations (Sørensen, 2003b).

The fact that the first excited state comes down from an initial 8 eV molecular orbital energy difference to the observed 1.8 eV indicates that it must contain a substantial amount of correlations, or, in other words, that this excitation must comprise collective involvement of a large number of electrons.

This is borne out in Fig. 4.74, showing the electron density difference between the first singlet excited state of the coumarin-derivative, and the ground state, using the large-basis TDHF calculation. The excited state is

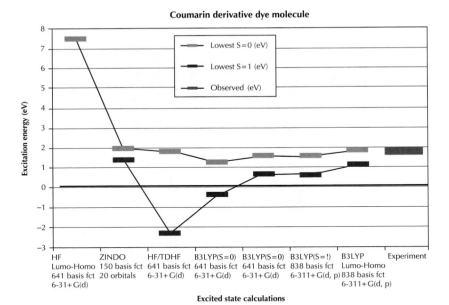

FIGURE 4.73 Measured (*right-hand side*) and calculated excitation energies of first excited states of spin 0 or 1 for the coumarin-derivative molecule shown in Fig. 4.71. The abscissa indicates acronyms for the type of calculation made, with complexity increasing from left to right (except the last, which, by comparison with the first column, shows the correlations built into the molecular orbits calculated with inclusion of exchange forces). Two of the calculations give unphysical states of negative energy, indicating the fragile nature of the approximations that effectively use different electron interactions for the excited-state calculation than for the ground-state calculation (because, at a realistic level of approximation, the same approach will not yield both). The problem is solved in the subsequent columns by using basis states more appropriate for the spin-1 calculations (Sørensen, 2003b).

made up of a dozen significant molecular orbital (MO) excitation pairs. The density difference shows the expected effect, already surmised from the Hartree-Fock (HF) ground-state MOs, that the excitation moves electron density from the coumarin core to the peripheral arm added to the molecule. It would then be natural to assume that this is where the transfer to the anatase surface, to which the dye adheres, takes place. This interpretation is supported by the large dipole moment found in the calculations (13.7 debyes).

The precise attachment of the dye to the anatase surface might be investigated by a combined optimization of the dye plus a chunk of semiconductor, with the distance and rotation angles of the dye relative to the surface as parameters (cf. Fig. 4.75). A recent study of a similar material has revealed the nature of the surface distortion and surface Ti and O molecules (Erdman *et al.*, 2002). The distortion penetrates only one layer down, in contrast to the over

FIGURE 4.74 Calculated electron density difference between the first excited singlet and ground state of coumarin-derivative dye (*top*: positive values enhanced; *below*: negative values enhanced; Sørensen, 2001a, 2003a).

10-nm-thick space charge regions in surfaces of solids. The conclusion from the present calculation is that both dye and surface are modified, with the dye "arm" being bent to better attach to the surface, and the surface atoms to accommodate the dye particle (but even in the absence of the dye, the surface of the lattice structure would be different from the regular interior used as starting point for the optimization).

A further "handle" on the quantum chemical calculations is the comparison of measured spectra (IR, NMR, etc.) with those predicted by the calculations. Understanding of complex molecules has progressed rapidly as a result of the combination of spectral studies in the laboratory with quantum model calculations.

The photo-electrochemical dye and nanostructure technique has several applications beyond the formation of solar cells. Among them are smart windows (Bechinger *et al.*, 1996; Granqvist *et al.*, 1998; STI, 2002), energy storage (Hauch *et al.*, 2001; Krasovec *et al.*, 2001), environmental monitors (Kamat and Vinodgopal, 1998; Yartym *et al.*, 2001), hydrogen production (Khaselev and Turner, 1998; Luzzi, 1999; Mathew *et al.*, 2001), computer and TV screens also suitable for outdoor use (using the dye as light emitter rather than absorber; Rajeswaran *et al.*, 2000; Tang, 2001; Müller *et al.*, 2003), and three-dimensional data storage (Dwayne-Miller *et al.*, 2001).

FIGURE 4.75 Attachment of coumarin-derivative dye to and modification of anatase surface. *Left:* initial guess, *right:* optimized (the depth of TiO$_2$ surface layer distortion cannot be determined by modeling just the top layer; Sørensen, 2003a, 2004b).

4.4.3 Solar Thermal Conversion

Heat Generation

Conversion of solar energy to heat requires a light-absorbing material, a *collector*, which is able to distribute the absorbed radiant energy over internal degrees of freedom associated with kinetic energy of motion at the molecular level (e.g., lattice vibrations in the case of a solid). The Earth and its atmosphere are examples of such collectors, as discussed in sections 2.2 and 2.3. Absorption of solar energy will raise the temperature of the collector or transfer energy to a reservoir, if the collector is connected to one. The collector will also emit radiation, and it may lose heat energy by conduction and convection processes. The frequency spectrum of the emitted radiation will correspond to the Planck spectrum (2.3) for the collector temperature T_c, if the collector is in a state allowing the definition of a thermodynamic temperature.

Man-made collectors may try to achieve a large absorption by minimizing reflection and transmission and to achieve small losses, e.g., by operating the collector at temperatures not much above ambient air temperatures or, if higher load temperatures are required, by reducing the heat loss rates by suitable transparent covers and insulation.

One may distinguish between *passive* and *active* systems, according to whether energy is specifically added (pumps, etc.) in order to bring the collector heat gain to the load areas. A passive system need not be characterized by the absence of definite heat flow paths between collectors and load areas, but such flows should be natural, i.e., they should not depend on other energy inputs provided by man. There may be borderline cases in which the term *natural circulation* would be difficult to define.

factors for translating the scatterometer data to potential wind speeds at turbine hub heights to vary rapidly with location. Comparing the new blended data with the 1998 re-analysis data, one finds no compelling reasons to alter the 1.3 scaling-factor value over land in estimating regional wind-power. One might have suspected that larger roughness would lead to a larger scaling factor, but the blended data for continental locations are dominated by the circulation-model data, and the "effective height" associated with them (for the lowest model level) is likely to be considerably larger than the 3 m found for the ocean data.

For offshore locations, the spatial resolution is quite acceptable due to the absence of abrupt roughness variations. Variations in power production over the year are considered in planning offshore wind parks. Averages similar to Fig. 6.30, but for the wind speeds rather than turbine output power, have been published (Jet Propulsion Laboratory, 2007; Risien and Chelton, 2006), but they are not suited for wind turbine power estimates, due to the nonlinear power curve that excludes deriving accurate average power levels from average wind speeds.

Figure 6.30 emphasizes, first, the huge advantage of producing wind-power over water, not only in mid-ocean locations but also at shore-regions of several continents. However, it also indicates that a number of coastal sites are less suitable, notably sites on the western side of continents near the Equator and on both sides of Equatorial sites south and southeast of the Asian continent. Closer inspection of shorelines farther away from the Equator shows notable differences in the generally large wind potentials, with a number of very-high-potential coast-lines existing on all continents, but some of them not yet considered for energy exploitation. Of course, additional factors may influence suitability for construct-ing power facilities, such as whether the water depth increases more or less rapidly away from the shore. Further, a number of inland water systems stand out as highly suitable for wind-power, including the Great Lakes in North America, the Baltic Sea and parts of the Mediterranean Sea in Europe, the Caspian and Aral Seas near the Europe-Asia border, and Lake Victoria in Africa.

To address wind turbine operational issues, such as the need for power exchange between regions or storage (Sørensen, 2008c), the blended data offer 6-hour calculations of both the expected power output and the direction from which the turbine must accept winds. Figure 6.31 shows an example of such a sequence of 6-hour data, for a region in the North Atlantic Ocean. The data are for January 26. There is a storm blowing from the southwest at the Florida coast and turning to the west as it reaches the British Isles and Scandinavia. These countries and Germany are seen to be capable of producing large amounts of power, while France, Central Europe, and the Mediterranean countries (except for a few south-facing coasts) need to import power or use alternative solutions. Such examples indicate some modes of application made possible by the quality of data derived from the scatterometer satellite measurements.

FIGURE 6.31   Plot of wind directions (*top*) and calculated power production (*bottom*) (W m^{-2} of turbine-swept area) based on satellite data for the North Atlantic Ocean and adjacent land areas, on January 26, 2000, at 18 h GMT. The abrupt changes from maximum to zero power output (*white squares*) are due to the assumption that, for safety reasons, wind turbines are shut down if the wind speed exceeds 25 m s^{-1}.

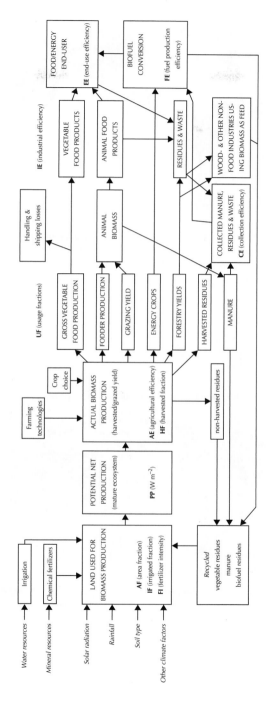

FIGURE 6.32   Overview of model used for describing the agricultural and silvicultural sectors in a scenario for future integrated food and energy production.

Area fraction

■	0.9	to 1
■	0.8	to 0.9
■	0.7	to 0.8
■	0.6	to 0.7
■	0.5	to 0.6
■	0.4	to 0.5
■	0.3	to 0.4
■	0.2	to 0.3
■	0.1	to 0.2
■	0.001	to 0.1
	all others	

FIGURE 6.33   The fraction of area used as cropland and cropland in rotation (taken as a fraction of each $0.5° \times 0.5°$ grid cell used for these purposes, on the basis of the classification in the U.S. Geological Survey, 1997).

FIGURE 6.34   The fraction of area used as rangeland (cf. Fig. 6.33).

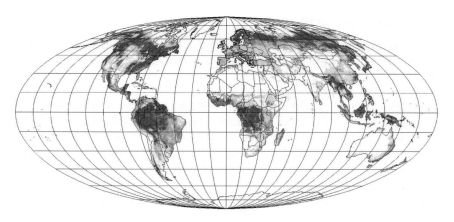

Area fraction   **FIGURE 6.35**   The fraction of area occupied by forest (cf. Fig. 6.33).

- 0.9   to 1
- 0.8   to 0.9
- 0.7   to 0.8
- 0.6   to 0.7
- 0.5   to 0.6
- 0.4   to 0.5
- 0.3   to 0.4
- 0.2   to 0.3
- 0.1   to 0.2
- 0.001 to 0.1
- all others   **FIGURE 6.36**   The fraction of area considered marginal land (cf. Fig. 6.33).

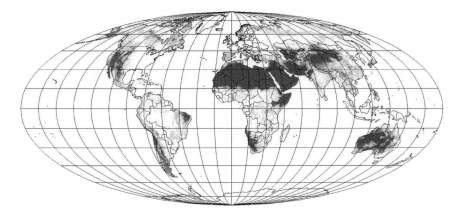

6.3.3 Food Production

The general model used in the scenario model to describe the biomass sector is shown in Fig. 6.32. It is a refinement of a model developed earlier (Sørensen et al., 1994; Sørensen, 1994a). Each part of the model is discussed below, and numerical assumptions are stated.

In the 2050 scenario, the land area used for food crops is considered to be the same as now. This includes the cropland area fraction (Fig. 6.34; AF in Fig. 6.33) and, for grazing, rangeland as well (Fig. 6.35). Marginal land (Fig. 6.36) is not used for food crops. Some marginal land is used today for grazing in a less-intensive way, in contrast to the use of cropland in rotation for occasional grazing. In some areas (e.g., Africa), crop cultivation on the cropland fraction is little intensive, and present yields strongly reflect the agricultural practices of each region. As an indication of the potential biomass production in these areas, the model shown in Fig. 6.33 uses calculated net primary production data from the *Terrestrial Ecosystem Model (TEM)* of the Woods Hole group (Melillo and Helfrich, 1998; model evolution and description in Raich et al., 1991; Melillo et al., 1993; McGuire et al., 1997; Sellers et al., 1997), shown in Fig. 3.99. Global warming may induce increased primary production in a fairly complex pattern and the borders of natural vegetation zones are expected to change, sometimes by several hundred kilometers (IPCC, 1996a).

Global warming–induced changes in area fractions are not included in the model, assuming that farming practice is able to gradually change and replace the crops cultivated in response to altered conditions, which are, in any case, long-term changes compared to the lifespans of annual crops. The model does not specify which crops will be cultivated at a given location, but simply assumes productivity consistent with growing crops suited for the conditions. The TEM data are for a mature ecosystem, taking into account natural water, humidity, and nutrient constraints, along with solar radiation and temperature conditions. Annual crops are likely to have smaller yields because of only partial ground cover during part of the year and the corresponding smaller capture of radiation. On the other hand, the crops selected for cultivation may be favorably adapted to the conditions and therefore may give higher yields than the natural vegetation at the location. Furthermore, irrigation may prevent yield losses in dry periods, and application of chemical fertilizers may improve overall yields.

It is conceivable that ecological concern will dictate restricted use of these techniques and will lead to increased use of organic agricultural principles, which are currently used in about 10% of agricultural areas in Europe and a few other places. The 2050 scenario assumes what is called *integrated agriculture* (Sørensen et al., 1994), a concept that includes a ban on pesticide use and that encourages use of recycled vegetable residues and animal manure for nutrient restoration, but that does not exclude biological pest control and limited use of chemical fertilizers. The yield losses associated with integrated agriculture are under 10% of current chemical-based agriculture yields, according to experience in Northern Europe.

On cultivated land (including grazing land and managed forests) in regions like Denmark (modest solar radiation and good soil and water access), the average annual biomass production is 0.62 W m^{-2} (of which 0.3 W m^{-2} are cereal crops; Sørensen et al., 1994). This equals the value for a grid cell in Denmark given in the TEM database for mature natural productivity. In Southern Europe, current production is about half (Nielsen and Sørensen, 1998), while the TEM database gives a slightly higher value than for Denmark. The reasons for this are less-intensive agricultural practice and water limitations for the growth pattern of the crops cultivated (limitations that would be less severe for a mature ecosystem). Therefore, it seems reasonable to use the TEM in the scenario as a proxy for cultivation yields, provided that one assumes improved farming techniques by the year 2050 and assumes that irrigation and chemical fertilizers are used when necessary. This is precisely the assumption stated above as the basis for the scenario. The model then uses the net natural primary production data of the TEM globally, but without adding further increases on the basis of irrigation (which in dry regions can double agricultural output) and use of chemical fertilizers (which can provide a further doubling, if the soil is poor in nutrients or nutrients are not returned to the fields). In other words, the disadvantage of going from mature vegetation to annual crops is offset by the advantage of reduced limiting factors related to water and nutrients. In Fig. 6.32, this means disregarding the irrigation and fertilizer parameters IF and FI and proceeding with the potential production PP taken from the TEM database.

The TEM global biomass production estimates for PP shown in Fig. 3.99 are expressed in energy units (1 g carbon per year is equal to a rate of energy production of 0.00133 W).

Currently, in Denmark only about 10% of this energy is contained in food consumed domestically, which indicates that there is room for altered management of the system, by diverting residues to energy extraction and later returning nutrients to the fields. One may also note that the current system is based on high meat consumption and the associated emphasis on raising livestock and, in the Danish case, export. Even the modest change in the vegetable to animal food consumption ratio that is assumed in the demand scenario shown in Figs. 6.5 and 6.6 shows that it is possible globally to divert substantial amounts of biomass to energy purposes without jeopardizing the ability to provide food for a growing world population.

It is not assumed that the intensive agricultural practices of Northern Europe will have been adopted globally by year the 2050. The agricultural efficiency factor AE in Fig. 6.32 is taken as unity only for regions 1 and 2 (cf. Table 6.3). For Africa (region 6) it is taken as 0.4, and for the remaining regions it is taken as 0.7. The fraction of the biomass production actually harvested is taken globally as $HF = 0.4$. The remaining fraction consists of roots and residues plowed under in order to provide natural fertilization for the following growth season.

For land areas classified as cropland, the assumed distribution of uses is given in Table 6.7, based on considerations of cropland scarcity and traditional practices in livestock raising. The low animal fodder value in Africa reflects the fact that the African tradition is to raise livestock on rangeland, not to use cropland to provide fodder.

Region	AE (Cropland)	HF	UF (Vegetable Food)	UF (Fodder)	UF (Energy Crops)
1	1	0.4	0.4	0.5	0.1
2	1	0.4	0.4	0.5	0.1
3	0.7	0.4	0.5	0.5	0
4	0.7	0.4	0.4	0.5	0.1
5	0.7	0.4	0.7	0.3	0
6	0.4	0.4	0.8	0.2	0

TABLE 6.7   Parameters used for cropland biomass production in the 2050 scenario. (Regions are defined in Table 6.2; see also Fig. 6.32 for explanation of abbreviations.)

The amounts of vegetable-type food that can potentially be produced and delivered to end-users in this scenario can now be calculated on an area basis, assuming the losses in going from vegetable food produced to vegetable food delivered as 25% [$IE(veg.\ products) = 0.75$ for vegetable food products in Fig. 6.23],

Delivered vegetable food =

$$AF(cropl.) \times PP[\mathrm{W\ m^{-2}}] \times AE \times HF \times UF(veg.\ food) \times IE(veg.\ products),$$

where AF and PP depend on the precise geographic location and the others only on region. The calculated distribution of vegetable food delivered to consumers is shown in Fig. 6.37.

For food from animals, such as meat, milk, and eggs, the average efficiency in transforming biomass to delivered animal products is assumed to be $IE(animal\ products) = 0.15$, a value reflecting a typical average of a smaller efficiency for meat production and a higher one for milk and eggs (Sørensen et al., 1994). The amount of animal-based food using cropland-derived fodder and delivered to the consumer is thus

Delivered animal food (1) =

$$AF(cropl.) \times PP[\mathrm{W\ m^{-2}}] \times AE \times HF \times UF(fodder) \times IE(\text{animal products}).$$

The distribution of potential animal food deliveries based on livestock being fed fodder produced on cropland is shown in Fig. 6.38.

The other part of animal food is from animals grazing on rangeland, where we shall assume that livestock grazes HF = 0.4 of the biomass production per unit of area, and put AE = 1. The use of rangeland is assumed to be 50% for grazing and 50% for other purposes (such as energy crops or no commercial use). Thus, the utilization factor is taken as $UF(grazing)$ = 0.5:

Delivered animal food (2) =

$$AF(rangeland) \times PP[\text{W m}^{-2}] \times HF \times UF(grazing) \times IE(\text{animal production}).$$

The distribution of potential animal food deliveries to end-users based on rangeland grazing is shown in Fig. 6.39. The ratio of the two contributions (crop feeding and rangeland grazing) is determined by the area set aside for each. The resulting fraction of animal food derived from rangeland grazing is 37%, in terms of energy content.

The efficiency of the end-user's making use of delivered food, denoted EE in Fig. 6.32, is for all bioenergy routes included in the definition of gross demand in section 6.2.3.

6.3.4 Biofuel Production

A number of fuels may be produced from biomass and residues derived from vegetable and animal production or from forestry and dedicated energy crops. They range from fuels for direct combustion over biogas (mainly methane mixed with carbon dioxide) to liquid biofuels, such as ethanol and methanol, or gaseous fuels, such as synthesis gas (a mixture of mainly carbon monoxide and hydrogen, also being an intermediate step in producing methanol) or pure hydrogen. The production of biofuels by thermochemical processes is based on high-temperature gasification and various cleaning and transformation processes (Jensen and Sørensen, 1984; Nielsen and Sørensen, 1998).

Whether biofuel production is by thermal or biological processes, the expected conversion efficiency is of the order of FE = 50% (cf. Fig. 6.32). This is to be compounded with a factor describing the ability of the biofuel industry to collect the necessary feedstock. This collection-efficiency factor, which we call CF, describes the efficiency in collecting biomass for industrial uses. For vegetable foods, it is assumed that CF *(veg. waste)* = 25% of the gross production is available for energy production [some of this would come from food-industry losses of $(1 - IE(veg. prod.))$ = 25%, and some from the subsequent household losses of 30%; cf. sections 6.2.3 and 6.3.3]. The overall yield of biofuels from vegetable crops is then

Biofuels from vegetable foodcrops =

$$AF(cropl.) \times PP[\text{W m}^{-2}] \times AE \times HF \times UF(veg. food) \times CF(veg. waste) \times FE.$$

Of course, manure is available only when livestock are kept in stables or similar areas that allow easy collection. The model assumes that grazing animals leave manure in the fields and that it is not collected (although it could be in some cases), but that animals being fed fodder from crops will be in situations where collection of manure is feasible. Furthermore, although the 85% of animal biomass not ending up in food products will be used both to maintain the metabolism of livestock and the process of producing manure, it will also contain a fraction that may be used directly for fuel production (e.g., slaughterhouse wastes). Combined with manure, this is considered to amount to $CF(anim.) = 0.6$, giving, for the fodder-to-animal route to biofuels:

Biofuels from manure and other animal residues =

$$AF(cropl.) \times PP[\text{W m}^{-2}] \times AE \times H \times UF(fodder) \times (1 - IE(anim.\ prod.))$$
$$\times CF(anim.) \times FE$$

The possibility of producing biofuels from forestry residues (either scrap derived from the wood industry or residues collected as part of forest management) may be described by a factor $CF(forestry) = 0.3$, defined as a percentage of the total forest biomass production. This is the fraction collected, expressed in energy units. For managed forests, it depends on the fraction of wood suitable for the wood manufacturing industry (furniture, etc.), which depends on the tree type. Adding wood scrap from industry and discarded wooden items, as well as residue from forest management, would in many regions exceed 30%. However, starting from an enumeration of all forests, including the rainforests and other preservation-worthy forest areas that are not proposed to be touched, and considering only managed forests that deliver to wood industries, 30% is probably a maximum for the year 2050. The forest-residue-to-biofuel route is then

Biofuels from forest management =

$$AF(forestland) \times PP[\text{W m}^{-2}] \times HF \times CF(forestry) \times FE.$$

The potential amounts of biofuels that could be derived from forestry are shown in Fig. 6.40, and the sum of the three routes to biofuels described above are given on an area basis in Fig. 6.41. This may be called *decentralized fuel production*, although forestry may not be seen as entirely decentral in nature. However, forestry is an ongoing activity that is distinct from producing biofuels from land used exclusively for energy crops. Two energy-crop routes are considered.

While the biomass production on rangeland does not give rise to biofuel production because manure from grazing livestock is not collected, some of the rangeland may be suited for cultivation of dedicated energy crops. Because only 50% of rangeland is assumed to be used for grazing, one may consider the remaining 50% a potentially exploitable area for energy purposes, in the scenario versions where centralized energy schemes are considered acceptable [i.e., $UF(rangeland\ energy\ crops) = 0.5$]. For cropland, it is also assumed that a maximum of 10% is set aside for energy crops in areas with generous resources, such as Western Europe and the Americas [see the *UF(cropland energy crops)* values of 0.1 or 0 in Table 6.7]. The potential biofuel production from these areas is

Biofuels from energy crops on cropland =

$$AF(cropl.) \times PP[\mathrm{W\ m^{-2}}] \times AE \times HF \times UF(cropland\ energy\ crops) \times FE,$$

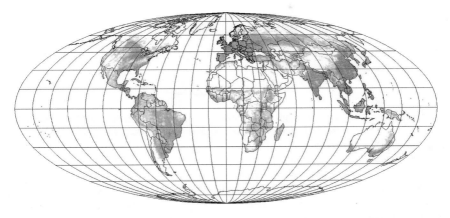

Energy flow
W m⁻²

5	to	10
2	to	5
1	to	2
0.5	to	1
0.2	to	0.5
0.1	to	0.2
0.05	to	0.1
0.02	to	0.05
0.01	to	0.02
0.005	to	0.01
0.002	to	0.005
0.0001	to	0.002
all others		

FIGURE 6.37 Potential delivery to final consumer of vegetable-based food from cropland production, expressed in units of energy flow (W m⁻²).

FIGURE 6.38 Potential delivery to final consumer of animal-based food from livestock being fed with fodder grown on cropland, in energy units.

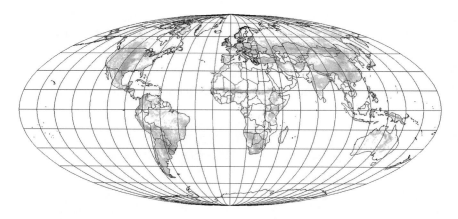

Energy flow
W m⁻²

5	to 10
2	to 5
1	to 2
0.5	to 1
0.2	to 0.5
0.1	to 0.2
0.05	to 0.1
0.02	to 0.05
0.01	to 0.02
0.005	to 0.01
0.002	to 0.005
0.0001 to	0.002

all others

FIGURE 6.39   Potential delivery to final consumer of animal-based food from livestock grazing on rangeland, expressed in units of energy flow (W m⁻²).

FIGURE 6.40   Potential delivery to final consumer of biofuels based on forestry residues and wood waste, expressed in energy units.

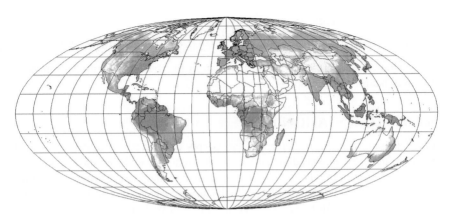

Energy flow
W m⁻²

5	to 10
2	to 5
1	to 2
0.5	to 1
0.2	to 0.5
0.1	to 0.2
0.05	to 0.1
0.02	to 0.05
0.01	to 0.02
0.005	to 0.01
0.002	to 0.005
0.0001	to 0.002
all others	

FIGURE 6.41 Potential delivery to final consumer of all "decentralized" biofuels based on forestry, agricultural residues, manure, and household waste, in units of energy flow (W m⁻²).

FIGURE 6.42 Potential delivery to final consumer of "centralized" biofuels based on dedicated energy crops grown on part of rangeland and minor parts of cropland, in energy units.

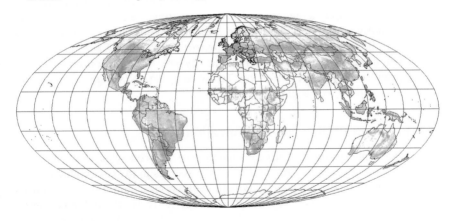

Biofuels from energy crops on rangeland =

$$AF(rangeland) \times PP[\text{W m}^{-2}] \times HF \times UF(rangeland\ energy\ crops) \times FE.$$

The value of FE is 0.5. The area distribution for the two potential sources of biofuels from energy crops is shown in Fig. 6.42.

6.4 IMPLEMENTATION ISSUES

Once demand has been determined and supply options mapped, there may still be more than one intermediate conversion system that will join the supply-and-demand requirements. They could be selected according to the criterion of minimal economic, environmental, and social impact (according to the evaluation methods presented in Chapter 7). Assuming this is done, there is a complete system in the hands of the planner, ready to be checked for consistency. Consistency is in part *internal consistency*, meaning that the system functions smoothly at all times and for all consumers, and in part consistency in the sense that the system condition can be reached from the present situation. Implementation studies usually end up quantifying a number of decisions that have to be made, at specified moments in time, in order for the development from the current state to the scenario state to happen.

Implementation evaluations have been made by first checking the technical consistency of the scenario system in the world assumed at the time envisaged for the scenario to be realized and subsequently specifying the path that has to be taken to get there (Sørensen, 1975b to 2008b and 2008c). An alternative method for an implementation study is *backcasting*, i.e., starting at the final scenario and working backward to the present, making notes of the decisions needing to be taken along the way (Robinson, 1982; 1988).

A few general remarks on intermediary system optimization and scenario consistency checking are made below, followed by several sections of examples on smaller and larger geographic scales.

6.4.1 System Choice and Optimization

A number of physical constraints are imposed on a given energy system, including the availability of different renewable energy sources in the geographic regions of interest; the characteristics of different system components, such as response time for changing the level of energy delivery; the delivery energy form, which may not be uniquely given by the nature of the load; and so on. Fig. 6.43 provides some possible components of energy systems based on renewable resources. More details are in the scenarios presented in sections 6.5 to 6.7.

The rated capacity of each component of the energy supply system will generally be determined by the maximum load expected to be demanded, with some adjustments made if the system includes stores and energy-exchange facilities, adapting system performance according to the expected time distribution of source fluxes and loads. Situations may arise in which not all incident energy can be converted, because the demand is already met and the stores (or those for a given energy form) are full. In this case, as well as in situations where different parts of the system would each be capable of satisfying the load, system operational problems arise. Regulation procedures would then have to be defined that could determine which converter to tune down or stop in cases of system overflow and which stores should be depleted first when more than one is available to serve a given load at a given time. Also, strategies for exchange of energy between regions with different load structures (or time zones) offer possibilities for avoiding over- or underflow (Meibom *et al.*, 1999). Because the system is nonlinear, there is no simple way of optimizing it, although relative minima in cost or other parameters may be sought (Sørensen, 1999c).

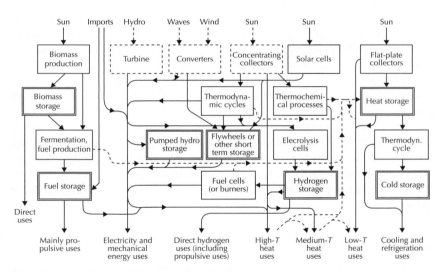

FIGURE 6.43   Example of the components of a supply–demand matching energy system based on renewable sources, with primary and secondary conversion processes, storage, and transmission.

6.4.2 Consistency of Simulation

The consistency of a given scenario is checked by performing a simulation, typically with 1-hour time steps, through 1 or more years. Realistic variations in demand are introduced, although the actual demands at a future scenario year are of course not known. Typically, one would take the present data with temporal variations and scale it to the scenario assumption for the total energy demanded, using the energy forms considered. More subtle extrapolations can be made if the scenario contains prescriptions for normative changes in society, such as altered consumption habits or work relations.

Taking note of the system performance at each step and allowing use of storage reservoirs and import/export options, one checks that the are no misses in supply–demand matching and adjusts the size of the system components downward or adds more (when possible), if there is unused or insufficient capacity. This constitutes the consistency check, and if system expansion beyond what can be achieved with the sources designated as potentially available is not required, then the system is labeled as realistic.

Ideally, the implementation study would also require consistency checks at least at selected stages along the path of establishing the system, taking into account phase-out of existing equipment (with consideration of the economic effects of too early retirement).

There may be other things to keep track of than just the technical performance of the system. If, for instance, the purpose of the scenario is to avoid greenhouse warming, then the reduction of CO_2 emissions along the timeline of implementation should be calculated. If a motivation for the scenario is to allow more energy-related decisions to become decentralized or even to reach the level of individual personal choice, then these are qualities to monitor. Decentralization may involve a choice between large, centralized primary energy production plants (which are also possible for renewable energy— think of desert photovoltaic systems or gigawatt offshore wind farms, not to mention dammed hydro stations) or smaller units placed closer to the users. At the extreme, every household may fully control its energy supply, e.g., by reversible fuel-cell installations capable of producing vehicle fuel and conversion of rooftop photovoltaic power to hydrogen or hydrogen to electricity, as needed.

6.5 LOCAL SYSTEMS

Renewable energy conversion systems or combinations of systems for use in different regions and in differently structured societies will very probably have to be different. The contribution of a single type of conversion in the system may depend on local resource availability. In the early phase, co-existence of renewable and nonrenewable components is likely and it is important to make them work together. Often this offers the advantage that fossil-fuel systems can alleviate the need for dedicated energy stores for the renewable components, until the renewable energy contribution grows large. Solar thermal collectors would work well in some regions of the world, but, in other regions, it may be difficult to reach 100% coverage of heat needs during winter. The examples given in this section deal with some of these problems, which are of immediate interest for current implementations of energy supply systems, by investigating the relation between fractional load coverage and various system parameters characterizing the size and complexity of the system.

6.5.1 Solar Heat or Heat-and-electricity Systems

The examples described in this subsection are based on flat-plate solar collector devices (cf. section 4.4.3) in various configurations, and the system simulations use the database for the Danish Reference Year, which is summarized in section 3.1.3 (e.g., in Fig. 3.17). It is clear that use of the same systems at latitudes with higher levels of solar radiation will improve performance and makes it possible to cover loads with smaller and less expensive systems. On the other hand, the simulation of solar energy systems at fairly high latitudes (56 °N for the Danish reference data) raises a number of design problems, which may show up more clearly than for latitudes with higher and more stable solar input and smaller heating loads, but which may still be of general interest. In any case, the need for heating is higher at higher latitudes, making the optimization of systems for such use particularly relevant.

Systems involving solar thermal collection into a heat store (cf. Fig. 4.84) have a complex dynamic behavior, due to the direct influence on absorber performance of heat-storage temperature, collector fluid inlet temperature, and speed of circulation. Indirectly, the collector circuit is influenced by heat-storage temperature of load patterns governing the heat exchange between store and load circuits. If a photovoltaic part is present (thus forming a hybrid photovoltaic-thermal—PVT—system), its performance depends on the complex temperature behavior only through the temperature dependence of power production efficiency (cf. Fig. 4.56), which in any case is much weaker than that of the thermal collection part.

Model Description

The system modeled (using the software package NSES, 2001) is a building with solar collectors, heat stores (which could be communal, i.e., common for several buildings or even part of a district heating system), and load patterns consisting of time profiles of the use of electricity, space heating, and hot water. The design of the building has to be specified in order to calculate heat losses and energy gains through passive features (such as windows), as does the activities in the building relevant for energy flows [occupancy, use of electric appliances and lighting, opening and closing of shutters (if any) in front of windows, etc.]. The external parameters include ambient temperature, wind speed, and solar radiation.

In case both thermal and electric energy are produced, the thermal absorber could be the back plate of a photovoltaic collector sheet, or the thermal collection could be from the air flow over the PV panel but below a transparent cover. In the first case, the heat-transfer fluid would typically be water, and, in the second, case it would be air. Heat collection in front of the PV panel could have negative effects, such as more hours with dew formation on the cover glass. Heat collection behind the PV panel could be lowered, if reflective back layers are used in the PV cell design to increase light capture probability by prolonging the radiation path-length within the cell.

The simulations reported below use the model described in section 4.4.3 for solar thermal collection and heat exchange in the collector–storage circuit. Details of the radiation data used in the numerical examples and the load description entering the modeling of the storage–load circuit are given in the following subsections.

Solar Radiation Data Used

Radiation data are taken from a stream of reference year data for a position 56 °N and 10 °E (cf. section 3.1.3). The data are collections of hourly global and diffuse radiation covering a year and exhibiting variations typical of what is contained in 30-year meteorological time-series. A reference year (of which several have been constructed, covering cities in Europe and the United States) is put together by pieces of actual data from the 30-year data source (30 years is the selected time span for defining a "climate"). Typical segment length is 14 days, in order to preserve the short-term correlations in the data arising from continuity and the typical passage time of weather fronts. Thus, the data are supposed to have both the typical short-term correlations and the long-term variability that can ensure realistic modeling of solar system performances over the lifetime of the system in question.

The most recent[*] Danish reference year data set employed is illustrated in Figs. 6.44 and 6.45 and shows the hourly Danish 1996 reference year profiles

---

[*]The Danish reference year data set was first established in 1974 (Andersen et al., 1974). This set was not based on a full 30-year underlying database and had some unrealistic features, including too much solar radiation in spring. International projects launched by the International Energy Agency and the European Commission's R&D program tried to establish a common prescription for the reference year sets of participating countries. A Danish reference year data set from this effort is included in the European publication (Lund et al., 1985). However, not all problems were resolved, including the mentioned unrealistic feature of the data set, which reappeared in the 1985 data. In recent years, new efforts have been made to obtain a more consistent tool for building calculations, and at least the 1996 Danish part of this effort is now free from obvious inconsistencies (DTI, 2000). There are other objections that could be made, as the new data sets are now primarily directed at small solar collector systems aimed only at providing hot water during summer periods. For example, the snow cover present in early data sets is now omitted, making it difficult to predict reflected light received by collectors with tilt angles near 90°, which in Denmark would be very relevant for winter collection.

for ambient temperature and solar radiation incident on a south-facing plane tilted 45°.

The collection of short-wavelength radiation by the solar panel takes place in a system comprising one or several cover layers followed by one or more absorbing layers. If present, the photovoltaic modules constitute one layer. Part of the absorbed radiation is transformed into an electric current, while the rest is transformed into long-wavelength radiation, which is either converted into useful heat or lost to the surroundings.

FIGURE 6.44   Danish 1996 reference year profile of ambient temperatures.

In contemporary thermal solar collectors, a heat-transfer fluid is passed over or under the usually building-integrated solar collector. The connection with the rest of the system (heat-store and building loads), illustrated in Fig. 4.84, usually contains two circuits, called the collector and the load circuit, both connected to the heat store by heat exchangers.

The heat-transfer fluid is typically air or water. Use of water in PVT systems entails a requirement for sealed paths and protection against corrosion, just as for thermal solar collectors. Although the use of air as a transfer fluid is easier, the low heat capacity of air and low energy transfer rates to and from air limit its applicability.

Thermal collection in front of the solar cell (but below a cover glass) may alter the efficiency of the PV conversion in a negative way, whereas collection behind the solar cell should pose few problems once it has been ensured that the PV and thermal plate layers are well connected, so

FIGURE 6.45 Danish 1996 reference year profile of 45 °S solar radiation over the year.

that heat transfer (conduction) is high. Incoming solar radiation is made useful for electricity production, is transformed into heat taken up by components of the collector, or leaves the cell through front or rear walls. A reflective rear layer may be used at the back of the PV module. It causes the light to pass through the solar cell twice, with a greater overall chance of capture. On the other hand, this excludes the possibility for light to pass through the PV layers and only to be absorbed in an underlying thermal collector layer.

Heat Load of an Individual House

The heat load of a building is fixed by specifying the indoor temperature of the building and the hot-water need as a function of time. The indoor temperature may need to be fixed, or it may be allowed to decrease during the night and when the house is unoccupied. For the purpose of simulation, the hot-water needs may be considered to be identical on every day of the year and concentrated in certain hours, such as a morning peak plus a period of moderate use in late afternoon and throughout the evening (including clothes washing, dish washing, baths, etc.). The amount of heat required to keep the indoor temperature at the prescribed value does not, on the other hand, remain constant throughout the year, but equals the heat losses through doors, windows, walls, floor, and roof, as well as losses associated with air exchange, which may be regulated by a ventilation system or may be entirely "natural," i.e., associated with crevices and the like in building materials and assembly sites.

The heat losses through building surfaces depend on the difference between inside and outside temperature and may be approximated by a linear expression

$$h = (T_L - T_a) \sum_i A_i U_i, \tag{6.2}$$

where T_L is the indoor ("load-") temperature, T_a is the outside temperature, and A_i and U_i are corresponding areas and specific heat-loss rates for different parts of the external building surface (walls, windows, roof, floor, etc.). The floor against the ground may be poorly described by the expression with T_a, and the heat-loss expression should perhaps contain a particular term describing the floor loss, being proportional to the difference between the load and soil temperatures, rather than the difference between the load and air temperatures. However, the heat-loss rates U_i are in any case only approximately constant, and they may have a further dependence on temperature (e.g., if radiative transfer is involved) and on outside wind speed, moisture, etc., as discussed above in connection with the collecting absorber.

An average contemporary one-family detached dwelling located in the temperate climatic zone may be characterized by wall and ceiling losses amounting to around 57 W °C^{-1} (U_{wall} in W m^{-2} °C^{-1} times wall and ceiling total area) and ground floor losses of 18 W °C^{-1} (U_{floor} often slightly lower than U_{wall}). With 20 m^2 window area and U_{window} = 2 W m^{-2} °C^{-1} (two layers of glazing; with intermediate space filled with low-conduction gas, the value may be reduced to half), the total "conduction" loss becomes 115 W °C^{-1}. If the house has 1½ stories (cf. sketch in Fig. 6.46), the ground floor area may be about 100 m^2 and the combined floor area above 150 m^2 (depending on how low ceiling heights are accepted as contributing to floor area for the upper floor). This corresponds to an inside volume of about 330 m^3, and if the rate of air exchange is 0.75 per hour, then V_{air} = 250 m^3 of air per hour will have to be heated from T_a to T_L (reduction is possible by letting the outgoing air preheat the incoming air in a heat exchanger),

$$(A_i U_i)_{air\ exchange} = V_{air} C_v^{air} \approx 87\ W\ °C^{-1},$$

since C_v^{air} = 0.347 Wh m^{-3} °C^{-1}. For the reference calculation, V_{air} is taken as 250 m^3 h^{-1} only during the period 16:00 to 08:00 and as 125 m^3 h^{-1} between 08:00 and 16:00 (assuming lower occupancy during these hours and a programmable ventilation system). Measurements have suggested that the air-exchange rate in naturally ventilated Danish dwellings (individual houses as well as apartments) may be lower than required for reasons of health (Collet et al., 1976). In a sample of 81 dwellings, a range of air-exchange rates between 0.15 and 1.3 times per hour was found, with the average value 0.62 times per hour.

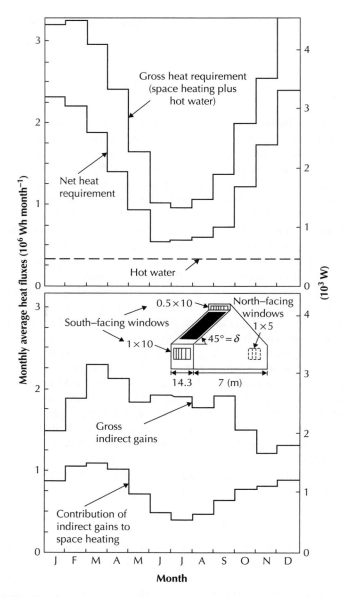

FIGURE 6.46 Heat-load data for a one-family dwelling, in terms of monthly averages of heat fluxes. The layout and dimensions of the house are indicated as an inset.

The minimum ventilation requirement has been quoted as 8 m³ h⁻¹/cap (Ross and Williams, 1975), but it may be more if the ventilation air is polluted by, for example, automobile exhaust or if tobacco smoking, open-air cooking, fireplace

1. **Building heat losses**

   Loss rate through surface (except windows) 75 W °C^{-1}.

   Loss rate through windows (no shutters) 40 W °C^{-1}.

   Ventilation air 250 m^3 h^{-1} from 16:00 to 08:00, 125 m^3 h^{-1} from 08:00 to 16:00.

   Minimum inlet temperature required for space heating load, 28 °C.

   Indoor temperature 20 °C from 07:00 to 09:00 and from 16:00 to 24:00, 19 °C from 09:00 to 16:00 and 16 °C from 0:00 to 07:00.

   Hot water usage: 50 kg from 07:00 to 08:00, 25 kg h^{-1} from 16:00 to 18:00.

   Required hot water temperature: 42 °C above an assumed input temperature of 8 °C.

2. **Indirect heat gains**

   Persons: 100 W from 22:00 to 07:00, 200 W from 07:00 to 09:00 and from 16:00 to 22:00, and 0 W from 09:00 to 16:00.

   Equipment: 60 W from 00:00 to 07:00, 600 W from 07:00 to 08:00 and from 23:00 to 24:00, 120 W from 08:00 to 16:00, 1200 W from 16:00 to 17:00 and from 18:00 to 21:00, 1800 W from 17:00 to 18:00, and 900 W from 21:00 to 24:00.

   Lighting: 100 W from 07:30 to 08:30 and from 15:30 to 00:30, unless global solar radiation exceeds 50 W m^{-2}, zero during other hours.

   Gains from 13 m^2 vertically mounted windows, of which 8 m^2 are facing south, the rest north. Glazing: three layers of glass with refractive index 1.526, extinction co-efficient times glass thickness (the product xL in $P^{t.a.}$) is 0.05. An absorptance $\alpha^{sw} = 0.85$ is used when working with $P^{t.a}$ to specify the rooms behind the windows.

3. **Flat-plate solar collector**

   Area 2, 6 or 40 m^2, tilt angle 45°, azimuth 0° and geographical latitude 55.69 °N. Danish Reference Year 1996 version used.

   Albedo of ground in front of collectors, 0.2.

   Absorbing plates: short-wavelength absorptance $\alpha^{sw} = 0.9$, reflection 0.1, long-wavelength emittance $\varepsilon^{lw} = 0.11$, full transfer between PV and heat absorbing plates.

   Efficiency of electricity production by PV layers: 0.15−0.0005 (T_c−20 °C).

   Cover system: one layer of glass (refractive index 1.526, xL in (4.81) equal to 0.05, emittance $\varepsilon^{lw} = 0.88$).

   Heat capacity of collector: 1.4 + 0.8 times number of glass layers (Wh m^{-2} °C^{-1}).

   Collector back heat loss coefficient: 0.2 W °C^{-1} m^{-2}.

   Collector circuit flow rate times fluid heat capacity, $J_m C_p = 41.8$ W °C^{-1} per m^2 of collector area.

4. **Storage facility**

   Capacity of heat storage given in terms of sensible heat storage in a container holding 0.1, 0.2 or 40 m^3 of water, the latter usually divided into 2 sections of 20 m^3.

   Storage loss coefficient, 0.2 or 1.2 W °C^{-1} (surroundings at 8 °C for larger stores).

   Heat exchange rate between collector circuit and storage (100 W °C^{-1} per m^2 of collector).

TABLE 6.8 Reference data for simulation of solar system for a Danish one-family house.

combustion, or similar practices take place on the premises. The value quoted would roughly correspond to the smallest values encountered in the Danish measurements, assuming that a family with 4–5 members occupies the dwelling. Although this minimum air exchange generously provides the necessary oxygen and removes CO_2 from the air the family breathes, it may not be sufficient to ensure health and well-being. Indeed, when "dust" or other particulate matter is abundantly injected into the air (e.g., from carpets or from activities taking place in the building), then increased ventilation is definitely required. Analyses of particulate matter from the air of private dwellings show the largest component is often pieces of human skin tissue. It should also be kept in mind that many building materials release gases (some of which have adverse effects on health) in substantial quantities, for at least the first few years after installation.

The heat required to compensate for the building's heat losses may be distributed by a device (radiative and/or convective) with a heat-transfer area chosen in such a way that any expected load can be met. In earlier individual or district heating fuel-based systems, it was customary to work with high inlet temperatures (steam at $100\,°C$, water at $80\,°C$). This was lowered to about $60\,°C$ after the first oil crisis in 1973. The minimum working inlet temperature would be about $45\,°C$ for a water radiator system, whereas distribution based on air circulation may employ heat supply of temperature around $28\,°C$. The latter choice is made in the present calculations, partly because it is suitable for solar systems aiming at coverage during as large a fraction of the year as possible and partly because efforts to decrease building heat losses through better insulation will make the capital costs of a water distribution system less appealing. For each of the heat exchangers, a temperature loss of $8\,°C$ is assumed.

Hot water is assumed to be required at a temperature $42\,°C$ above the inlet (waterworks) temperature, which in Denmark is taken as $8\,°C$.

The metabolism of persons present within the house releases heat, corresponding to roughly 125 W per person as an average net heat transfer to the surroundings. The changing occupancy of the house, as well as the lower rate of metabolism during sleep and the higher one during periods of high activity, are modeled by a variation in the gains from persons, assumed to be of an identical pattern every day of the year, as specified in Table 6.8. Also the use of appliances that convert electricity or another form of energy (cookers, washing machines, dishwashers, lights, amplifiers, and other electronic equipment, etc.) contributes to indirect heat gains, since the various conversions in most cases result in the release of the entire energy in the form of heat inside the house. The contribution from cooking is reflected in the distribution of the heat releases from appliances over different hours of the day, the assumed pattern being also given in Table 6.8. The gain from lighting, assumed to be dominated by highly efficient bulbs, is assumed to be 100 W during 1 hour in the morning and 8 hours in the evening, except when the outside total solar radiation exceeds $50\,W\,m^{-2}$. Again, the use of more efficient equipment, including low electricity-consuming light bulbs, does have the effect of diminishing indirect contribution to space heating with time.

However, a potentially very large contribution to the gross indirect gains of the house considered is due to solar radiation transmitted through the windows and absorbed in the interior. The magnitude depends on the design of the building, which of course is often different in cold climates, as opposed to regions where shielding against solar heat is desired. Here, most of the window area has been assumed to be on the southern side of the building, and, on average, the gain through windows facing south far exceeds the heat losses due to the temperature difference between inside and outside air. The gains are calculated as if the window glazing were the cover system of a solar collector and as if the inside of the room were acting as an absorber plate of short-wavelength absorptance $\alpha_{sw} = 0.85$ (due to the small fraction of transmitted light that will be multiply reflected on indoor objects and will be retransmitted to the outside through window areas). Indirect gains occurring at times when they are not needed are assumed lost. Often, surplus gains (or parts of them) would be stored as sensible heat in the wall materials, etc., thus potentially providing storage from day to night, but if it is desirable to avoid temperatures higher than the stipulated ones, surpluses may be disposed of by venting. Active storage by air heat exchangers working with a heat store is not considered.

The solar system modeling passes through the steps described above, with the temperature increase in the collector circuit and store being determined iteratively as described, and assuming that the fluid in this circuit is pumped through the collector only when the temperature of the collector has reached that of the energy storage (the collected heat is used to achieve this during a "morning start-up" period) and only if a positive amount of energy can be gained by operating the collector.

A number of simplifying assumptions have been made in formulating the model used, which include the neglect of reduced transparency of the collector cover in case of dirt or dust on the cover surface, the neglect of edge losses from the collector and heat losses associated with special geometry (e.g., where the piping enters and leaves the collector sections) and also of the detailed temperature distribution across the collector, from bottom to top, which may not be adequately described by the averaging procedure used. Another set of complications not included in the simulation model are those associated with possible condensation of water on the outer or inner surfaces of the collector's cover system, icing conditions, and snow layers accumulating on the top surface of the collector. It is assumed that water condensation conditions, which are expected to occur, especially on some mornings, simply prolong the morning start-up time somewhat, while the collected solar energy provides the evaporation heat as well as the heat required for raising the temperature of the collector. Sample runs of the model with different assumptions regarding the amount of heat required for start-up have revealed that these amounts of heat are anyway very small compared with the total gain, when averaged over the year.

For the purpose of the simulation, the year is divided into 12 identical months, each 730 h long. The monthly averages given below are all based on these artificial months, rather than on calendar months. The parameters for the solar heat system are summarized in Table 6.8. Most of them are kept fixed, and the variations performed primarily relate to collector size, storage volume, and types of operation.

Heat-pump Systems

The model described above calculates the heat loads not covered by the solar system and evaluates the auxiliary heat required, relative to heat of insufficient temperature from the solar heat stores. For a system with both electricity and heat outputs, it is natural to consider adding a heat pump, using the solar heat store as its cold side.

FIGURE 6.47 Schematic solar heat system with heat pump that may be switched between collector and load side. The main operational modes correspond to opening the following switches indicated on the figure: mode 1: c_1, i_1, u_1, and l_1 (simple operation); mode 2: c_2, e_2, k_1, i_2, u_1, and l_1 (collector cooling); mode 3: c_1, i_1, u_2, e_1, k_2, and l_2 (heat pumping to load) (Sørensen, 1976).

Theoretically, a heat pump may be added to the solar energy system in different ways. Two of these are illustrated in Fig. 6.47. In one, the heat pump is situated between the storage and the collector, in order to increase the temperature difference between inlet and outlet fluid, $T_{c,out} - T_{c,in}$, and thereby improve the efficiency of the collector. In the other mode, the heat pump is inserted between the storage and the load areas, so that it can be used to "boost" the load circuit temperature difference, $T_{L,in} - T_{L,out}$, whenever the storage temperature is below the required minimum value of $T_{L,in}$. The heat pump could use as its low-temperature reservoir the soil or air of the surroundings, rather than the solar storage facility, and thus be an independent energy source only related to the solar heating system by the controls, which determine the times of providing auxiliary heat. The merits of such a system, comparing solar thermal collection with solar electric plus heat pump, are discussed in Sørensen (2000b). The option of using the heat pump to cool the collector fluid has been tested and found unsuitable, because it improves efficiency of collection only when the storage temperature (which in the absence of the heat pump would determine the collector inlet temperature) is high, and in this case the collector output temperature is mostly lower than the store temperature, leading to no heat transfer (although $T_{c,out} - T_{c,in}$ is increased by lowering $T_{c,in}$, $T_{c,out}$ may not be high enough, i.e. above T_s).

For the mode 3 operation considered here as an option, the heat pump compressor may be set into action on the condition that the heat store temperature is below a certain value (e.g. 50 °C for hot water, 28 °C for space heating). Additional heat exchange losses occur between heat pump and storage (again assumed to entail a temperature difference of 8 °C, i.e. the heat pump effectively operates between $T_{up} = T_{load} + 8$ °C and $T_{low} = T_s - 8$ °C). The coefficient of performance (COP = heat output over electricity input) is taken by the empirical relation (cf. Fig. 4.7, all temperatures in K)

$$\text{COP}^{-1} = (1.8 + 0.00025(T_{up} - T_{low} - 50)^2 + 1000/(T_{up} - T_{low})^3)(T_{up} - T_{low})/T_{up}.$$

Typical COP's for the temperature lifts considered are in the range of 2 to 5.

Software Validation and Results of Simulations

Three examples of hybrid PVT system simulations are presented in the following, with 2, 6, and 40 m^2 of solar collectors, and heat stores ranging from a very small 0.1 m^3 water tank to one of 40 m^3. For the largest system, the variant of communal storage common for a number of buildings is considered. Finally, the possibility of supplying auxiliary heat from a heat pump is discussed. However, first the software must be validated.

The NSES software has been tested against an experimentally investigated system similar to the one described in section 6.5.1 (Table 6.8), but making the store so large (10 000 m^3) in the calculation, that the collector inlet temperature could be regarded as a constant 30 °C. A special set of meteorological input data was used, with ambient temperature constantly at 20 °C and wind speed at 5 m s^{-1}, and with solar radiation stepping through a range from zero to maximum magnitude (formally, this was taken as diffuse radiation, because in that way the program would not attempt to calculate incident angle variations for the sample hours). Parameters deviating from the values of Table 6.8 are given in Table 6.9. The resulting solar thermal efficiencies, parameterized in terms of $T^* = (T_{c,in} - T_e)/E_{c+}^{sw}$, are shown in Fig. 6.48.

Solar collector area:	1 m^2
Short-wavelength absorption of PV panel:	0.93
Reflectance of PV panel:	0.07
Long-wavelength emittance of PV panel:	1.00
Plate temperature:	30 °C
PV efficiency at plate temperature 20 °C:	0.08
Circuit flow times fluid heat capacity:	83.6 W/m^2/K
Back heat loss coefficient:	1 W/m^2/K
Albedo in front of collector:	0
Capacity of heat store:	10 000 m^3

TABLE 6.9 Parameters changed from the values given in Table 6.8 for the test run of a Racell PVT panel tested empirically.

The most important parameters influencing the normalized efficiency given in Fig. 6.48 are the plate absorptance, emittance, and back thermal loss. These parameters quoted in Table 6.9 are based on the manufacturer's specifications rather than measured (Bosanac *et al.*, 2001). In cases where they can be compared, such as the test shown in Fig. 6.48, NSES gives results similar but not identical to those of the program TRNSYS, which has been in extensive use for many years (Duffie and Beckman, 1991). The water-flow parameter indicated in Table 6.9 is fairly high, and the resulting heat exchange is less perfect than that assumed in the simulations below, in keeping with the experimental set-up and the temperature rise found. The back thermal loss is also taken somewhat higher (a factor of 2) than what is indicated by the type of and thickness of insulation. This is because U_{back} is supposed to describe all thermal losses from the collector, including the edge losses, which may be substantial for a small collector like the one tested (actual area 2.2 m^2). Finally, the plate emittance would be expected to be lower than the absorptance (as they would be equal for a blackbody and the emittance would be lower for a selectively coated surface). The measurements also leave some doubt about the influence of electricity co-production, as the panel's quoted 8% electric efficiency does not match the <5% difference between measurements with and without electricity collection. In the calculation shown in Fig. 6.48, the emittance is left at unity, but it should be said that the slope of the curve diminishes rapidly with falling ε_p^{lw}. Although the experimental errors are not known, the agreement appears to be good, both between theory and experiments and between the two software packages. The methodology used in both software packages is highly dependent on *ad hoc* parametrizations derived from other experiments, and they do not give exactly the same results for identical parameters.

FIGURE 6.48 NSES efficiency results for test case (*solid line*) compared with measured data (◆). The effective temperature used is $T^* = (T_{c,in} - T_e)/E_{c+}^{sw}$.

Small PVT System Simulation

The smallest system simulated is similar to the Racell prototype used for program verification: the PV collecting panel is 2 m², and thermal collection takes place under the PV panel by tubes attached to, or part of, a thermally conducting plate. The thermal storage tank is a mere 100-liter water container, with heat exchangers to collector and load circuits. The photovoltaic collection efficiency is taken as 15%, representing typical contemporary commercial silicon modules, and only a linear term is included in the thermal performance degradation of the PV efficiency (cf. Fig. 4.56). The only simulation parameter not discussed above is the initial temperature, which for consistency has to be taken in such a way that it resembles the temperature at the end of the simulation period, which here is an entire year. This is not important for small stores, as they have no long-term memory.

The resulting monthly average behavior is shown in Fig. 6.49, and Figs. 6.50–6.51 show the hour-by-hour behavior of storage temperature and the requirement for auxiliary heat.

The top two panels in Fig. 6.49 show the monthly collection of heat and electricity and compare it with demands. The heat diagram also shows the indirect heat gains from persons, activities, and windows and the auxiliary heat needed in order to cover the total hot water and heating needs. Evidently, the solar system contributes little heat or electricity during winter months. During summer, solar electricity production is still far below building demands, whereas solar heat production covers up to two-thirds of demands. Percentages relative to loads are shown in the middle line of panels; for heat, percentages relative to the load remaining after application of auxiliary heat input are also shown.

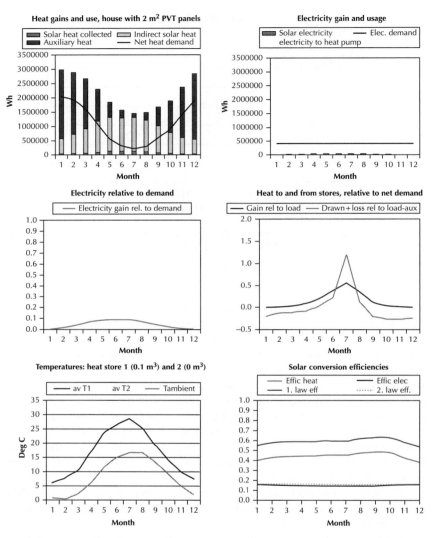

FIGURE 6.49 Monthly simulation results for a small PVT system placed in Denmark. See discussion in text.

The bottom line of diagrams shows the average temperatures of the water tank and of the ambient air and the efficiencies of electric and thermal conversion, plus first and second law thermodynamic efficiencies. The electric efficiency deviates a little from 15% in a downward direction, due to the thermal effect during summer months. The thermal efficiency remains >40% year-round; the first law efficiency is the sum of electric and thermal efficiencies; and the second law efficiency is only slightly greater than the electric efficiency, due to the modest temperature rises accomplished by the solar system.

FIGURE 6.50 Hourly storage temperatures from NSES simulation of a small PVT system.

The simulated hourly course of storage temperatures given in Fig. 6.50 shows that the solar system contributes to space heating during the period from April through September (storage temperature above 28 °C). The store temperature rarely exceeds 50 °C, so some auxiliary energy is needed for hot water year-round, as is shown in Fig. 6.49. This could be partially avoided by lowering the flow rate of the water passed through the collector circuit, with some associated decrease in thermal efficiency. Figure 6.51 gives the temporal distribution of auxiliary heat added in order to cover the demand at all times; there are some days even in winter where the system is able to cover the space heating need.

FIGURE 6.51 Hourly auxiliary heat requirements from NSES simulation of a small PVT system.

Medium-size PVT System Simulation

The medium system considered has 6 m² of PVT collectors, but still a very modest heat store (200 liters), because long-term storage is not intended, given the collector size, solar radiation conditions at the location (Denmark), and load. As the middle left entry in Fig. 6.53 indicates, it is now possible to cover >30% of the electricity load during the months of May, June, and July. Also for heat, the coverage in summer is high (roughly 100% in July), due to the higher summer storage temperature (see Fig. 6.55) relative to the small system considered in the previous subsection. From May to October 1, the storage temperature oscillates between 15 °C and 80 °C.

The 15% electricity production makes the heat-collection efficiency (bottom line, Fig. 6.53) almost constantly 15% below that of a pure thermal system of the same size, the monthly simulation results of which are shown in Fig. 6.52. During summer, the system without electricity production has a higher average store temperature (in July, 60 °C as compared to 50 °C), whereas during winter months there is no difference. The thermal efficiency of the PVT system is, on average, 35%, with small variations: during spring and autumn, 40% is reached, whereas during the summer, a minimum of 30% coincides with the peak store temperature.

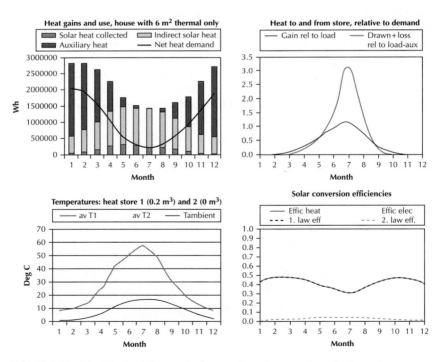

FIGURE 6.52 Monthly simulation results for a medium-size, pure thermal solar system.

Figure 6.54 shows the solar thermal gain for each hour of the year and also the energy gain by the heat store, which is a little less, due to incomplete transfer through the heat exchangers (solar absorber to working fluid and working fluid to store). Figure 6.55 gives the hourly store temperatures, which are more appropriate for the heating applications than those of the small system considered in Fig. 6.50. As a consequence, the need for auxiliary heat,

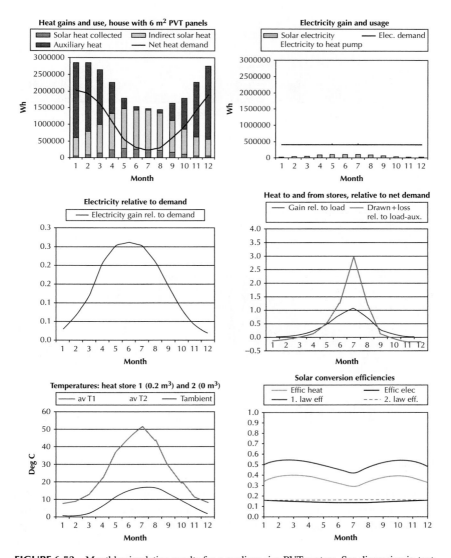

FIGURE 6.53 Monthly simulation results for a medium-size PVT system. See discussion in text.

FIGURE 6.54   Hourly solar heat gains from simulation of a medium-size PVT system.

FIGURE 6.55   Hourly storage temperatures from simulation of a medium-size PVT system.

shown in Fig. 6.57, is reduced, particularly during the summer. Figure 6.56 shows the net heat drawn from the store to the load circuits (hot water and air duct heating), including heat losses through the store insulation. During winter, the net heat drawn is seen to be often negative. This is when auxiliary energy is used to satisfy the load. The auxiliary energy brings the amount of

FIGURE 6.56 Net heat drawn hourly from store from simulation of a medium-size PVT system.

FIGURE 6.57 Hourly auxiliary heat requirements from simulation of a medium-size PVT system.

hot water needed up to the required temperature and likewise the amount of air circulated between the heat store and load areas. The return air and water (after the load heat exchange) may in winter have temperatures above that of the store and thus actually help to increase the store temperature. The implied "net solar gain to store" is therefore negative, as nonsolar heat is used to heat the store.

Large PVT System Simulation

The large system considered has 40 m^2 of PVT collectors and 40 m^3 of water storage. This may be divided into two hemispherical parts with a common interface, to simulate stratification. The input to the collector is from the lower store, and the heat to load is extracted from the upper part, provided it has a temperature higher than the lower one. The water passing through the solar collector is delivered at the top of the upper store, where a heat-exchanger spiral is continuing down to the lower store. This means that if the temperature of the water heated by the solar collector exceeds that of the upper store, it delivers energy to that store and reaches the lower store at a diminished temperature, possibly allowing it to pass further heat to the lower part. This is not meant as a substitute for a more detailed simulation of stratification, which is not done here.

The input parameters are as in Table 6.8, but with two storage compartments of 20 m^3 of water and heat-loss coefficients (U-values) of 0.55. The heat loss from the stores is diminished to about a tenth of the value that would be proper for a single store with 20–30 cm of insulation, because it is supposed that some 100 buildings share a common store, in which case the surface-to-volume advantage allows this kind of reduction (cf. the simulations made in Sørensen 1979; 2000a). The program has been instructed to perform a 1% mixing of the upper and lower store for each iteration step (here 1 h), as a crude model of the stratification of a single storage volume. The resulting average store temperatures are indicated in the lower line of Fig. 6.58 and in more detail in Fig. 6.60. It is seen that the mixing does not allow temperature differences between top and bottom to persist for very long.

The top diagrams of Fig. 6.58 show that electricity needs are covered from late March to early September, but that solar heat is only able to satisfy demand without assistance from late March to the end of August. There is a strong asymmetry between the behavior in spring and autumn, which is directly caused by the temperature variations of the store and thus the collector circuit inlet temperature. The store temperature remains high until November, which causes the thermal solar collection to perform poorly, while, on the other hand, the store is capable of satisfying heat demands from heat collected earlier, as shown in the middle, right-hand diagram: there is no need for auxiliary heat from May to November. This proves that the store is large enough to act as a seasonal store, but, on the other hand, there is a very negative effect on solar thermal collection during the later months of the year. This is borne out very clearly by the efficiency curves on the lower right.

The hourly solar gain shown in Fig. 6.59 supports this interpretation, showing a clear asymmetry between spring and autumn, with a very poor performance toward the end of the year. The hourly course of storage temperature (1 is lower store, 2 is upper) is shown in Fig. 6.60. It peaks in August and has its minimum value in February; that is, it is displaced by 1–2 months relative to the solar radiation. Due to the size of the store, there are no rapid variations like those for the store temperature of the smaller systems shown in Figs. 6.50 and 6.55.

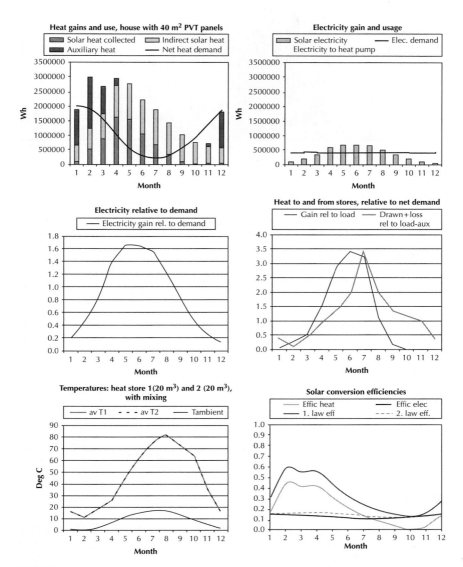

FIGURE 6.58 Monthly NSES simulation results for a large PVT system. See text for details.

FIGURE 6.59 Hourly solar gains from simulation of a large PVT system.

FIGURE 6.60 Hourly storage temperatures for NSES simulation of a large PVT system.

The net heat drawn from the (two) stores is given in Fig. 6.61. The negative values explained above are now confined to a few weeks in January and February. Finally, Fig. 6.62 gives the necessary auxiliary heat inputs, which are still substantial during winter, although the solar heating system now satisfies all loads from May to November.

FIGURE 6.61 Net heat drawn hourly from the two stores of the large PVT system.

FIGURE 6.62 Hourly auxiliary heat requirements for the large PVT system.

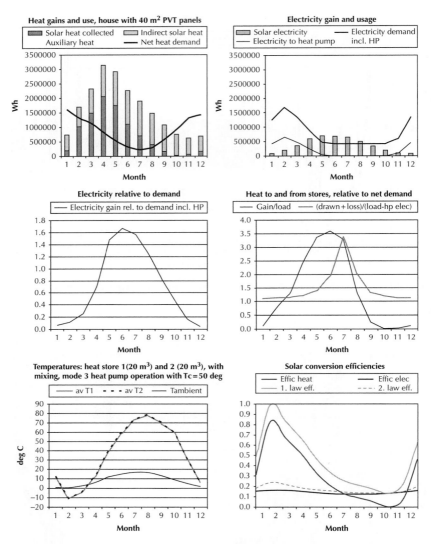

FIGURE 6.63   Monthly simulation results for the large PVT system with heat pump operating in mode 3 and with control temperature 50 °C (see text for details).

PVT Systems with Heat Pump

The final set of calculations adds a heat pump operating in mode 3 (see the discussion associated with Fig. 6.47) to the large PVT system. In mode 3, the heat pump operates on the heat store as its lower temperature reservoir and delivers heat at appropriate temperatures for covering space heating and hot water needs.

Setting the control temperature of the heat pump to 50 °C, it is possible to deliver all auxiliary heat from the heat-pump system. The monthly system performance curves are shown in Fig. 6.63. In the electricity diagram (upper right), the electric energy inputs to the heat pump are indicated. At the peak, they increase the electricity demand by nearly 70%.

The store temperature curves (lower left) show that during February and March the heat pump cools the store down to a temperature below the ambient one. As a consequence, the apparent total efficiency (lower right) exceeds 100%, because heat from the surroundings is drawn into the heat store. Figure 6.64 shows how the net energy drawn from the store is now always positive, and, in order to make the distinction between hot water and space heating coverage easier, the space heating demand is shown separately in Fig. 6.65.

FIGURE 6.64 Net heat drawn hourly from stores for a large PVT system with heat pump operating in mode 3 and with control temperature 50 °C. No other auxiliary heat is needed.

The store temperatures below 0 °C are not quite realistic, partly because a water store has been assumed, so that large quantities of antifreeze additives are required (in addition to the small amounts used for the fluid in the collector circuit, due to occasional operation of the collector in situations of solar radiation and low temperatures), and partly because having a large cold store may increase the need for electric energy to the heat pump (if the store is within the building) or it may make district heating transmission more difficult (frost in pumps, etc.).

For these reasons, an alternative heat-pump run is made, where the control temperature is set at 29 °C. This means that the heat pump will take care of all auxiliary energy for space heating, but not for hot water. The summary results are shown in Fig. 6.66. The store temperature now stays above 7 °C (also in the detailed hourly output), and there is no longer any anomaly in the efficiency curves. The figure shows (in the upper right diagram) that the electric energy to the heat pump now is modest, but (in the upper left diagram) that some direct auxiliary heat has to be provided during winter months.

FIGURE 6.65　Hourly space heating demand (used for all simulation cases).

6.5.2 Wind Electricity Systems

Figures 6.67 and 6.68 give the variation in wind-power production for a large portion of the currently installed wind turbines in Denmark. Such data can be used for system simulation and inquiry into questions of surplus and deficit management through import/export or through storage. It is seen that, for turbines on land, maximum output peaks over four times the average are found during winter months, whereas for offshore plants, the maximum is only 2.3 times the average. This may indicate greater resource stability over water, and it may be that the power curve of the turbines used offshore is not approximating the power curve leading to maximum annual production as well as the power curve for turbines used on land. This design choice is illustrated in Fig. 6.70. Figure 6.69 shows the power duration curves for wind turbines placed on land and offshore in Denmark, as well as a combined curve for the onshore/offshore mix of a particular scenario for future wind coverage (Sørensen *et al.*, 2001). The latter is seen to produce power during every hour of the year.

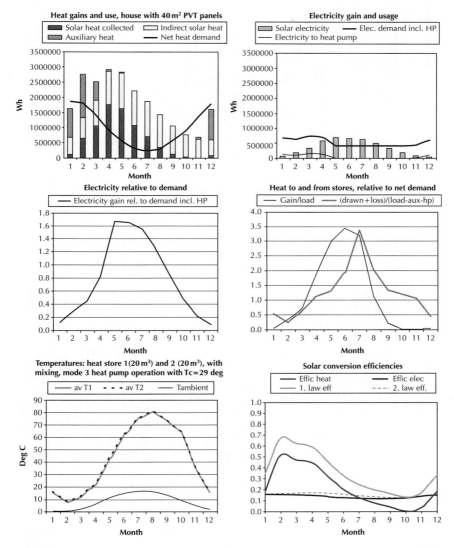

FIGURE 6.66 Monthly NSES simulation results for a large PVT system with heat pump operating in mode 3 and with control temperature 29 °C (see text for details).

FIGURE 6.67   Hourly variation in wind-power production during the year 2000 for some 3000 Danish wind turbines placed on land. The overall annual average power output is normalized to one (Ravn, 2001; Eltra/Elkraft, 2001).

FIGURE 6.68   Hourly variation in wind-power production during the year 2000, normalized to annual average, for about 100 Danish wind turbines placed offshore (Ravn, 2001; Eltra/Elkraft, 2001).

FIGURE 6.69 Power duration curves for wind-power produced at a single site on land or offshore in Denmark and for a combined scenario for future generation on many sites, with substantial expansion of the offshore wind contribution, but use of only onshore sites already hosting a wind turbine today (Sørensen *et al.*, 2001).

An example of how the power distribution of a propeller-type wind energy converter depends on design is shown in Fig. 6.70. The highest annual power production is obtained with design (a) and not design (c), which peaks at an average wind speed of 10 m s^{-1}, due to the asymmetry of the curves. The assumed hub height is 56 m, and 1961 data from Risø are used (cf. Fig. 3.34). Several of the calculations described in this section (taken from Sørensen 1976c, 1977, 1978a,b) use these data and agree remarkably with newer results, notably because hub heights of 50–60 m are the actual standard for present-day commercial wind turbines.

Wind-power System without Energy Storage

Power duration curves (or "time duration curves for power") can be constructed using detailed hour-by-hour simulation, or even without it, provided that the wind-speed data exist in the form of a frequency distribution (such as the dashed curves in Fig. 3.38). Once a storage facility or an auxiliary conversion system has to be operated together with wind energy converters, the system simulation requires the retention of the time sequences of the meteorological data.

For a system without energy storage, there may be an advantage in choosing a power coefficient curve that peaks well below the peak of the frequency distribution of the power in the wind, rather than just maximizing annual production.

FIGURE 6.70   Power coefficients (percent of power in the wind) are given as a function of wind speed for four typical designs of wind energy generators (cf. Fig. 4.17), with the corresponding power output (*below*). The 1961 average power production at Risø, Denmark (height 56 m) is given in square brackets for each type of design (Sørensen, 1978a).

Figures 6.69 and 6.71 give examples of power duration curves that illustrate the advantage of dispersing the locations of turbines. In Fig. 6.69, the substantial smoothing obtained for the Danish sites currently in use, supplemented with continued build-up at offshore locations, is seen to significantly stretch the power duration curve. A similar conclusion can be drawn from Fig. 6.71, in this case for 18 locations in Denmark, Ireland, and Germany. If the wind-speed frequency distribution function is denoted $h(u) = df(u)/du$, where u is the scalar wind speed and $f(u)$ is the accumulated frequency of wind speeds less than or equal to u, then the average power production is given by

$$E = \int \tfrac{1}{2} \rho \, C_p(u) u^3 h(u) \, du \qquad (6.3)$$

for a wind energy converter with sufficient yawing capability to make the cos θ factor in (3.26) insignificantly different from unity during all (or nearly all) the time covered by the integral in (6.3). The power duration curve is then

$$F(E) = \int \delta(\tfrac{1}{2}\, \rho\, C_p(u)u^3 \geq E)h(u)\, du, \tag{6.4}$$

where the δ function is unity if the inequality in its argument is fulfilled, and zero otherwise. If the data are in the form of a time-series, the integration may be made directly over dt (t is then time, in units of the total interval being integrated over) rather than over $h(u)du$. $F(E)$ gives the fraction of time [i.e., of the time period being integrated over in (6.4)], during which the power generation from the wind energy converter exceeds the value E.

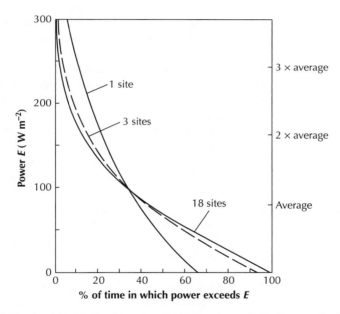

FIGURE 6.71 Annual power duration curves for wind energy systems with converters at 1, 3, or 18 different sites, based on 1973 data from West German weather stations, taken at the height 10 m but extrapolated to 72 m. The assumed power coefficient curve is similar to type (b) in Fig. 6.70. (Based on Molly, 1977.)

The coupling of just three units, at different locations in Germany (typical distances about 240 km), is seen in Fig. 6.71 to imply practically as much improvement as the coupling of 18 units. The reason is that the maximum distances within Germany are still not sufficient to involve the frequent occurrence of entirely different wind regimes within the area, and thus the general meteorological conditions are often the same for all the converters except for minor differences, which can be taken advantage of by selecting three different locations. In order to take advantage of more different meteorological front system patterns, distances of 1000 km or more are required. A study of the synergy of 60 European sites made by Giebel (1999) still finds variations in wind-power production substantially at variance with demand, indicating a fundamental need for back-up or storage.

Regulation of a Fuel-based Back-up System

Consider now an existing fuel-based electricity-producing system, to which a number of wind energy converters are added with the purpose of saving fuel. It is conceivable that this will increase the requirements for regulation of the fuel-based units and lead to an increased number of starts and stops for the fuel-based units. Since most of these units will not be able to start delivering power to the grid at short notice, the decision to initiate the start-up procedure for a given unit must be taken some time in advance, based on expectations concerning the load and the fraction of load that the wind energy converters may be able to cover.

The time required for start-up of a conventional fuel-based unit is associated with preheating the boiler and turbine materials and thus depends on the time elapsed since the last usage. Traditionally, for start-up times less than 7 h, the start is called a "warm start." Start-up times are generally diminishing due to new technology, so variations between plants of different age is often large, with quoted start-up times being 20–60 min for coal-fired base and inter-mediate-load units, 1–7 h for pressurized-water fission reactors, and 5–6 h for boiling-water fission reactors (Nordel, 1975). For fission reactors, "warm starts" assume start from an under-critical reactor with control rods in, and operating pressure. If the fuel-based units have cooled completely since the last usage, start-up times are 0.6–6 h for large coal plants, 18 h for a pressurized-water reactor, and 14 h (to reach 90% of the rated power) to 48 h (full power) for a boiling-water reactor. These start-up times are representative, and actual times may be smaller for some newer plants.

Now the question is to what extent the power production of the wind energy converters can be predicted in advance by time intervals correspond-ing to the different start-up times for the relevant fuel-based units. Two types of prediction may be considered, one being actual forecasts based on (mental or mathematical) models of the dynamics of general circulation and the other being associated with analysis of the statistical behavior of wind patterns, possibly including correlations that make simple methods of extrapolation permissible.

Standard meteorological forecasting, as practiced routinely by meteorologi-cal services throughout the world, may be very useful in predicting storms and thus in identifying periods of very high wind-power production, with advance notice usually the same as that required by the quoted start-up times for fuel-based units. However, forecasts at present are usually unable to specify very precisely the expected wind speed, if this is in the critical range of 5–10 m s^{-1} covering the very steep part of the converter performance curve from zero to quite high production (cf. Fig. 6.70).

Therefore, it may be worth taking advantage of the correlations present in sequences of wind production data and in wind-speed data, such as those implied by the spectral decomposition (see, for example, Fig. 3.37). The continuity of the general circulation and the nature of the equations describing the atmospheric motion (section 2.3.1) combine to produce sequences of wind energy production figures that exhibit regularities of seasonal and diurnal scale as well as regularities of 3–5 days' duration, associated with the passage of mesoscale front systems over a given geographical location. The following sections describe models based on trend extrapolation of power production sequences, models adding elements of stochastic fluctuations, and models using physical forecasting based on climate modeling.

Simple Extrapolation Model

Figure 6.72 shows a sequence of hourly, sequential production data for a wind energy converter [power coefficient (a) in Fig. 6.70] placed at the Risø turbine test site 56 m above the ground. After a period of insufficient wind (insufficient to start the wind energy converter), the average production rises, goes through a peak, and decreases again in a matter of about 4 days. It then follows a period of small and fluctuating production. Also, during the 4 days of large production, hourly fluctuations are superimposed on the general trend. However, it is evident that a power prediction simply saying that the production will be high, modest, or near zero on a given day, if the power production is high, modest, or low, respectively, on the previous day, will be correct quite often, and, similarly, the production during a given hour is strongly correlated with the production for the previous hour.

An extrapolation model may then be tried that bases the decision to initiate start-up procedures for fuel-based units on some (possibly weighted) average of the known wind energy production data for the time period immediately before.

For simplicity, the fuel-based supply system is assumed to consist of a number of identical units, each with a maximum power generation that is a tenth of the average load, E_{av}, for the entire grid of the utility system considered. It is assumed that the units can be regulated down to a third of their rated power, and that some are operated as base-load plants (i.e., at maximum power whenever possible), while others are operated as intermediate-load plants (i.e., at most times allowing for regulation in both upward and downward direction). The start-up time for the base-load units is assumed to be 6 h, and, once such a unit has been started, it will not be stopped for at least 24 h. The start-up

time assumed for the intermediate-load units is 1 h. For simplicity, the load is
factorized in the form

$$E_{load} = E_{av}C_1C_2,   \quad\quad (6.5)$$

where E_{av} is the annual average load, C_1 is the seasonally varying monthly
averages (such as those shown in Fig. 6.13), and C_2 is the average diurnal
variations (such as those shown in Fig. 6.15). Both C_1 and C_2 are normalized
so that the annual average implied by (6.5) is E_{av}. The approximation involved
in (6.5) implies neglect of seasonal variations in diurnal load profiles, as well as
weekday/holiday differences, and is used only for the purpose of illustration.
The actual hourly data are used in the forecasting subsection below.

FIGURE 6.72   Hourly output from wind energy system without (*top*) and with short-term storage
(*bottom*) for a 10-day period in January 1961. The wind energy converters are assumed to be of type
(a) (Fig. 6.70), 56 m Risø data are used, and the 24 h energy storage is assumed to be loss-free and
operated in a base-load mode (Sørensen, 1978a).

In the first set of calculations to be discussed, the extrapolation of wind-power production is based on two averages of past production. One is the average production of the preceding 24 h, A_{24}, which is used once every 24 h in conjunction with the minimum value of the expected load (6.5) for the following day to decide on the number n_{base} of base-load fuel-fired units to start. If this number is different from the number already in use, the change will become effective 6 h after the decision is made, according to the assumed base-load start-up time. Thus,

$$n_{base} = \left[\frac{E_{av}C_1 \min(C_2) - A_{24}}{0.1\, E_{av}} \right],$$

where the square bracket denotes integer truncation.

The other wind-power average used in the regulation of the utility system is the average wind energy production in the previous 4 h, A_4, which is used every hour in conjunction with the expected load for the following hour to make a trial estimate of the required number of intermediate-load units, n_{int}^{trial},

$$n_{int}^{trial} = \text{integer} \left(\frac{E_{av}C_1C_2(t + 1\text{h}) - A_4}{0.1\, E_{av}} \right) - n_{base} + 2,$$

where the operation "integer" rounds to the closest integer number. This trial value is two units larger than the minimum value indicated by the predicted wind-power production, in order to make room for regulation of the intermediate-load units in both directions if the wind-power prediction should turn out to be inaccurate. The value of additional units is chosen as 2, because this is the number that, for the Danish power system, gives the smallest number of start-ups per year for the fuel-based units, in the absence of wind energy contributions. If the trial value, n_{int}^{trial}, of intermediate-load units for the following hour exceeds the number already in use, a number of start-ups are initiated, so that the number of units actually operating 1 h later will be $n_{int} = n_{int}^{trial}$. If the number indicated by n_{int}^{trial} is smaller than the one already operating, it is not necessarily implemented; rather, the units already in operation are regulated downward. If it can be calculated that the regulation of all operating units down to the technical minimum level (assumed to be a third of the rated power) is insufficient, then one unit is stopped. It has been found advantageous on an annual basis not to stop more than one unit each consecutive hour, even if the difference between the minimum power production of the operating units and the expected demand will be lost. This is because of the uncertainty in the wind-power predictions, which in any case leads to losses of energy from the fuel-based units every time the wind-power production exceeds expectation by an amount that is too large to be counterbalanced by regulating the fuel-based units down to their technical minimum level (in this situation it is also permitted to regulate the base-load units). Not stopping too many of the intermediate-load units makes it easier to cope with an unexpected occurrence of a wind-power production below the predicted one during the following hour.

If the wind-power deficit is so large that, for a given hour, it cannot be covered even by regulating all fuel-based units in operation to maximum power level, then peak-load units must be called in. They can be fuel-based units with very short start-up times (some gas turbines and diesel units), hydro-power units (with reservoirs), or "imports", i.e., transfers from other utility systems through (eventually international) grid connections. [The role of international exchange in regulating a wind-power system has been investigated by Sørensen (1981a) and by Meibom *et al.* (1997, 1999).] In a sufficiently large utility pool, there will always be units somewhere in the system that can be regulated upward at short notice. Another option is, of course, dedicated energy storage (including again hydro reservoirs).

All of these peak-load possibilities are in use within currently operated utility systems, and they are used to cover peaks in the expected load variation as well as unexpected load increases, and to make up for power production from suddenly disabled units ("outage").

Summary of Simulation Results

Based on the 1-year simulation of the fuel-based system described above, with an installed capacity I_{wind} of wind energy converters added, a number of quantities are calculated. The installed wind energy capacity I_{wind} is defined as the ratio between annual wind energy production from the number of converters erected, E_{wind}^{prod} given by (6.3), and the average load,

$$I_{wind} = E_{wind}^{prod}/E_{av}. \tag{6.6}$$

The quantities recorded include the number of starts of the base- and intermediate-load units, the average and maximum number of such units in operation, and the average power delivered by each type of unit. For the peak-load facilities, a record is kept of the number of hours per year in which peak units are called and the average and the maximum power furnished by such units. Finally, the average power covered by wind is calculated. It may differ from the average production I_{wind} by the wind energy converters because wind-produced power is lost if it exceeds the load minus the minimum power delivered by the fuel-based units. If the annual average power lost is denoted E_{lost}, the fraction of the load actually covered by wind energy, C_{wind}, may be written as

$$C_{wind} = \frac{E_{wind}^{prod} - E^{lost}}{E_{av}}. \tag{6.7}$$

Figure 6.73 shows the average number of base- and intermediate-load unit start-up operations per unit and per year, as a function of I_{wind}. Also shown is the average number of base- and intermediate-load units in operation. The maximum number of base-load units operating is six, and the maximum number of base- plus intermediate-load units is 15.

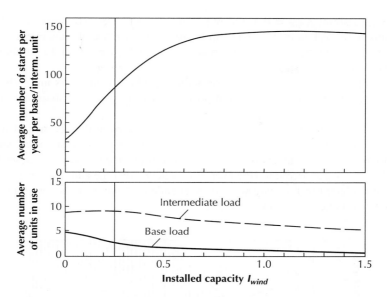

FIGURE 6.73 Start-up requirements and average number of fuel-based units in operation, as a function of installed wind capacity in a combined system without storage possibilities (Sørensen, 1978b).

FIGURE 6.74 The number of hours requiring utilization of peak units (or imports) and the maximum amount of peak power required during the year for a combined fuel-plus-wind electricity-producing system without storage possibilities (Sørensen, 1978b).

Up to about $I_{wind} = 0.3$, the wind energy converters are seen to replace mainly base-load units. The number of starts in the absence of wind energy converters is, on average, 35 per unit, rising to 98 for $I_{wind} = 0.3$, and saturating at about 144.

In Fig. 6.74, the number of hours during which peak units are called is given, together with the maximum power drawn from such units in any hour of the year.

In the simulation, both of these quantities are zero in the absence of wind energy converters. This is an unrealistic feature of the model, since actual fuel-based utility systems do use peak-load facilities for the purposes described in the previous subsection. However, it is believed that the additional peak-load usage excluded in this way will be independent of whether the system comprises wind energy converters or not.

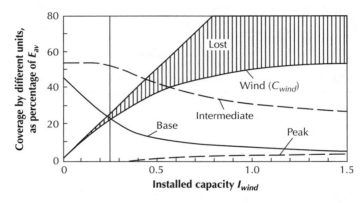

FIGURE 6.75　Fractional coverage of the annual average power required (total annual load) by wind and three types of fuel-based units for a combined system without storage. The wind energy production not used to cover loads is indicated by the shaded region "lost." (From Sørensen, 1978b.)

The number of peak unit calls associated with inappropriate wind predictions rises with installed wind capacity. For $I_{wind} = 0.25$, it amounts to about 140 h per year or 1.6% of the hours, with a maximum peak power requirement that is about 40% of E_{av}. The asymptotic values for large I_{wind} are 1000 h per year (11.4%) and 155% of E_{av}.

The total amount of energy delivered by the peak units is not very large, as seen in Fig. 6.75. For I_{wind} up to about 0.4, it is completely negligible, and it never exceeds a few percent of the total annual load for any I_{wind}. This is considered important because the cost of fuels for peak-load units is typically much higher than for base- or intermediate-load units.

In Fig. 6.75, the fractional coverage by wind, base, intermediate, and peak units is shown as a function of I_{wind}, and the fraction of produced wind energy that is lost, $I_{wind}-C_{wind}$, is indicated as a shaded area. Up to an installed wind energy capacity of about $I_{wind} = 0.1$, the lost fraction is negligible, and up to I_{wind} around 0.25, it constitutes only a few percent. For larger I_{wind}, the lost fraction increases rapidly, and the wind coverage, C_{wind}, actually approaches a maximum value of about 60%. Figure 6.75 also indicates how the wind energy converters first replace base-load units and affect the coverage by intermediate-load units only for larger I_{wind}.

Stochastic Model

The regulation procedures described are quite rigorous, and one may ask how much the system performance will be altered if the prescriptions are changed. By modifying the prescription for choosing n_{base}, the relative usage of base- and intermediate-load units may be altered, and by modifying the prescription for choosing n_{int}, the relative usage of intermediate- and peak-load units can be changed. The prescriptions chosen for n_{base} and n_{int} give the smallest total number of starts of base plus intermediate units for zero wind contribution ($I_{wind} = 0$).

Also amenable to modification is the extrapolation procedure used to predict future wind energy production. Rather than using plain averages over the preceding 4 and 24 h (A_4 and A_{24}), one may try using a weighted average, which would emphasize the most recent hours more strongly. A number of such weighted averages were tried, one of which is illustrated in Fig. 6.76. Here A_{24} was replaced by $(0.8A_{24} + 0.2A_4)$ in the expression for n_{base}, and A_4 was replaced by $(0.8A_4 + 0.2A_1)$, A_1 being the wind energy production for the preceding hour, in the expression for n_{int}^{trial}. The resulting changes are very modest, for this as well as for other combinations tried, and no clear improvements over the simple averages are obtained in any of the cases.

One may then ask if knowing the recent history of wind-power production is at all important for regulating the fuel-based back-up system. To demonstrate the nonrandom nature of the hour-to-hour variation in wind-power production, the predictive model based on previous average production was replaced by a stochastic model, which used the power production at a randomly selected time of the year, each time a prediction was required.

The basis for this procedure is a power duration curve [such as the ones in Fig. 6.69; however, for the present case, the Risø curve obtained by combining Fig. 3.40 and Fig. 6.70(a) was used]. A random number in the interval from zero to one (0%–100%) was generated, and the power-level prediction was taken as the power E from the power duration curve, which corresponds to the time fraction given by the random number. In this way, the stochastic model is consistent with the correct power duration curve for the location considered and with the correct annual average power. The procedure outlined was used to predict the power generation from the wind, to replace the averages A_{24} and A_4 in the expressions for n_{base} and n_{int}^{trial}.

The results of the stochastic model (c) are compared to those of the extra-polation type model (a) in Fig. 6.76. The most dramatic difference is in the number of start-ups required for the fuel-based units, indicating that a substantial advantage is achieved in regulating the fuel-based system by using a good predictive model for the wind-power contribution and that the extrapolation model serves this purpose much better than the model based on random predictions.

FIGURE 6.76 Excerpts from Figs. 6.73–6.75 (*curves labeled a*), compared to two other calculations of the performance of a combined fuel/wind energy system. *Dashed curves* (*b*) represent a calculation where the extrapolation-type wind energy predictions are based on previous average production data with a higher weighting of the most recent hours; *heavy lines* (*c*) represent a calculation where the wind-power predictions are not extrapolated, but randomly chosen within a stochastic model described in the text (Sørensen, 1978b).

Some differences are also seen in the requirements for peak units, which are smaller for the extrapolation type model up to $I_{wind} = 0.3$ and then larger. The latter behavior is probably connected with the lowering of the total wind coverage (bottom right in the figure) when using the stochastic model. This means that the number of fuel-based units started will more often be wrong, and when it is too high, more of the wind-produced power will be lost, the wind coverage will be lower, and the need for auxiliary peak power will be less frequent. The reason why these situations outnumber those of too few fuel-based units started, and hence an increased need for peak power, is associated with the form chosen for the n_{int} regulating procedure. If this is the explanation for the lowering of the number of peak unit calls in the stochastic model for I_{wind} above 0.3 (upper right in Fig. 6.76), then the fact that the number of peak unit calls is lower for the extrapolation-type model than for the stochastic model in the region of I_{wind} below 0.3 implies that the extrapolation model yields incorrect predictions less often.

Economic evaluation of the requirements for additional start-ups, increased utilization of peak-load facilities, etc., of course, depends on the relative cost of such operations in the total cost calculation. For a particular choice of all such parameters, as well as fuel costs and the capital cost of the wind energy converters, a maximum installed wind capacity I_{wind} can be derived from the point of view of economic viability (cf. Chapter 7).

Wind-forecasting Model

For the reasons explained in Chapter 2, general weather forecasts are inaccurate for periods longer than a few days, with the exact period depending on the strength of coupling between large-scale and chaotic atmospheric motion. For short periods, say under 12 h, the extrapolation methods described above in connection with time delays in starting up back-up power plants are expected to perform as well as meteorological forecasts based on general circulation models. Thus, the role of wind forecasting for power systems with wind-power is in the intermediate time range of 1–2 days. The interest in this interval has recently increased, due to establishment of power pools, where surpluses and deficits in power production are auctioned at a pool comprising several international power producers interconnected by transmission lines of suitable capacity. The Nordic Power Pool currently functions with a 36 h delay between bidding and power delivery. Therefore, the possibility of profitably selling surplus wind-power or buying power in case of insufficient wind depends on the ability to predict wind-power production 36 h into the future.

The quality of wind-power predictions, even for as little as 12 h, using the circulation models currently used in 1- to 5-day forecasts, is illustrated in Fig. 6.77. It clearly reflects the problem caused by nonlinearity, as explained above. Yet, for the 36 h forecasts needed for bidding at the power pool, the average error of the forecasts of wind-power production (at Danish wind farms) is only 60% of that of extrapolation values (Meibom et al., 1997; 1999). The effect that this has on pool bidding is illustrated in Figs. 6.78 and 6.79 for a model in which the Danish power system is 100% based on wind-power, while the other Nordic countries primarily use hydropower, as today. Figure 6.69 shows the amount of wind-power buying and selling, after covering their own needs, for Eastern and Western Denmark. Fig. 6.79

similarly gives the power duration curve for deviations of actual wind-power delivery from the bid submitted 36 h earlier. In terms of cost, using current trading prices at the Nordic Pool, incorrect bids will imply a monetary penalty amounting to 20% of the revenue from pool trade (this depends on a facility called the adjustment market, where additional bids for short-notice "regulation power" are used to avoid any mismatch between supply and demand). No other energy storage systems could presently provide back-up for wind-power at such a low cost.

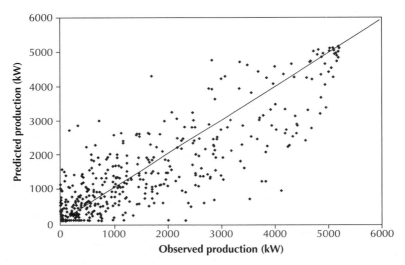

FIGURE 6.77   Correlation between 12 h meteorological forecasts and observed wind-power production at Danish wind turbine park. (Based on Landberg *et al.*, 1999.)

Wind Systems with Short-term Storage

All of the storage types mentioned in section 5.3 may be utilized as short-term stores in connection with wind energy systems. The purpose of short-term stores may be to smooth out the hourly fluctuations (the fluctuations on a minute or sub-minute scale being smoothed simply by having the system consist of several units of modest spacing). A back-up system must still be available if the extended periods of insufficient wind (typical duration one or a few weeks) are not to give rise to uncovered demands.

If the inclusion of short-term stores is to be effective in reducing the use of fuel-based peak-load facilities, it is important to choose storage types with short access times. It is also preferable for economic reasons to have high storage-cycle efficiency. This may be achieved for flywheel storage, pumped hydro storage, and superconducting storage (see Table 5.5).

A wind energy system including storage may be operated in different ways, such as:

(i) *Base-load operation*. The system attempts to produce a constant power E_0. If the production is above E_0 and the storage is not full, the surplus is delivered to the storage. If the production is below E_0 and the storage is not empty, the complementary power is drawn from the storage. If the production is above E_0 and the storage is full, energy will be lost. If the production is below E_0 and the storage is empty, the deficit has to be covered by auxiliary (e.g., fuel-based) generating systems.

FIGURE 6.78   Annual duration curves for trade of Danish wind-power at the Nordic Power Pool for East (DK-E) and West Denmark (DK-W) (Meibom *et al.*, 1997, 1999).

FIGURE 6.79   Annual duration curve for adjustment in Danish wind-power delivery to Nordic Pool, relative to bid made 36 h in advance. Notation as in Fig. 6.78.

(ii) *Load-following operation.* The wind energy system attempts to follow the load (6.3), i.e., to produce power at a level E proportional to E_{load},

$$E = E_0 C_1 C_2.$$

If production by the wind energy converters is above the actual value of E and the storage is not full, the surplus is delivered to the storage. If the production is below E and the storage is not empty, the difference is drawn from the storage. Energy loss or need for auxiliary energy may arise as in (i).

(iii) *Peak-load operation.* Outside the (somehow defined) time periods of peak load, all wind-produced energy is transferred to the storage facility. During the peak-load periods, a prescribed (e.g., constant) amount of power is drawn from the wind energy converters and their storage. Surplus production (if the storage is full) during off-peak hours may be transferred to the grid to be used to replace base or intermediate-load energy, if the regulation of the corresponding (fuel-based) units allows. A deficit during peak hours may, of course, occur. This system has two advantages. One is that the power to be used in a given peak-load period (say, 2 h) can be collected during a longer period (say, 10 h), because the probability of having adequate wind during part of this period is higher than the probability of having adequate wind during the peak hour, when the energy is to be used. The other advantage is that the wind energy in this mode of operation will be replacing fuel for peak-load units, i.e., usually the most expensive type of fuel. With respect to the first claimed advantage, the weight of the argument depends on the rating of the wind energy converters relative to their planned contribution to covering the peak load. If the wind energy converters chosen are very small, they would have to operate during most of the off-peak hours to provide the storage with sufficient energy. If they are very large, the advantage mentioned will probably arise, but, on the other hand, surplus energy that cannot be contained in the storage will be produced more often. It all depends, then, on the relative costs of the wind energy converters and the storage facilities. If the latter are low, a storage large enough to hold the produced energy at any time may be added, but if the storage costs are high relative to the costs of installing more converters, the particular advantage of an extended accumulation period may be decreased.

(iv) *Other modes of operation.* Depending, among other things, on the installed wind energy capacity relative to the total load and on the load distribution, other schemes of operating a wind energy system, including short-term storage, may be preferable. For example, some fraction of the wind energy production may always be transferred to the storage, with the rest going directly to the grid. This may increase the probability that the storage will not be empty when needed. Alternatively, wind energy production above a certain fixed level may be transferred to the storage, but the system will still attempt to follow the actual load. There is also the possibility of having a certain fraction of the wind energy production in a load-following mode, but not all of it. Clearly, optimization problems of this type can be assessed by performing computer simulations.

Results of Simulation Studies

If short-term storage is added to a wind energy system in the base-load operation mode, the hourly sequence of power deliveries from the system may change as indicated in Fig. 6.72. On a yearly basis, the modifications may be summarized in the form of a modified power duration curve (compare Fig. 6.80 with the single-turbine curves in Fig. 6.69), where some of the "excess" power production occupying levels above average has been moved over to the region below the average level to fill up part of what would otherwise represent a power deficit.

The results of Fig. 6.80 have been calculated on the basis of the 1961 Risø data (height 56 m), using the power coefficient curve (a) in Fig. 6.70, in an hour-by-hour simulation of system performance. The storage size is kept fixed

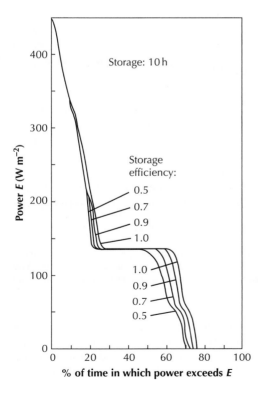

FIGURE 6.80   Annual power duration curves for a wind energy system with 10 h storage capacity $[t_A(E_0)]$, as a function of the storage-cycle efficiency. The average wind-power production is E_0. The calculation is for a wind energy converter of type (a) (Fig. 6.70), operated in base-load mode, and using 1961 Risø data at the height 56 m (Sørensen, 1978a).

at 10 h in Fig. 6.80 and is defined as the number of hours of average production (from the wind energy converters) to which the maximum energy content of the storage facility is equivalent. Thus, the storage capacity W_s may be written as

$$W_s = E_0 t_s, \qquad (6.8)$$

where W_s is in energy units and t_s is the storage capacity in hours. E_0 is the power level that the wind energy system in base-load operation mode aims at satisfying (for other modes of operation, E_0 is the average power aimed at).

Storage-cycle efficiencies lower than unity are considered in Fig. 6.80, assuming for simplicity that the entire energy loss takes place when energy is drawn from storage. Thus, an energy amount W/η_s must be removed from the storage in order to deliver the energy amount W, where η_s is the storage-cycle efficiency.

The effect of the storage is to increase the fraction of time during which the power level E_0 (or power levels $0 < E \le E_0$) is available, causing the fractions of time during which the power level exceeds E_0 to decrease. The availability of a given power level E is the time fraction $t_A(E)$, during which the power level exceeds E, and Fig. 6.81 shows the availability of the base-load power level, $t_A(E_0)$, for the case of a loss-free storage (full line), as a function of storage capacity.

FIGURE 6.81 Power availability as a function of storage capacity $t_A(E_0)$ for a loss-free storage, wind energy converters of type (a) (Fig. 6.70) with average power production E_0 and 1961 Risø data. The system is operated in a base-load (*solid line*) or a load-following mode (*dashed line*) (Sørensen, 1978a).

Thus, modest storage leads to a substantial improvement in availability, but improvements above availabilities of 75%–80% come much more slowly and require storage sizes that can no longer be characterized as short-term stores. However, the availabilities that can be achieved with short-term stores would be considered acceptable by many utility systems presently operating, owing to system structure and grid connections.

Another way to characterize the effect of adding a storage facility is to calculate the decrease in the average fluctuation σ of the system output E away from E_0,

$$\sigma = \left(N^{-1} \sum_{i=1}^{N} \left(E(t_i) - E_0 \right)^2 \right)^{1/2}, \tag{6.9}$$

where t_i is the time associated with the ith time interval in the data sequence of N intervals. The annual fluctuation is given in Fig. 6.82 as a function of storage capacity t_s, again for a loss-free storage.

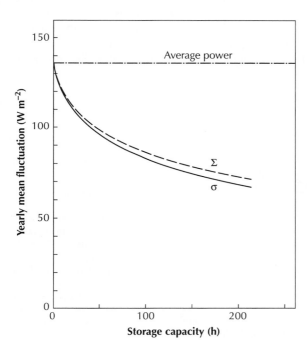

FIGURE 6.82 Annual average power fluctuations as a function of storage capacity $t_A(E_0)$ for a loss-free storage, wind energy converters of type (a) (Fig. 6.70) and 1961 Risø 56 m data. The system is operated in base-load mode (*solid line*) or in a load-following mode (*dashed curve*) (Sørensen, 1978a).

In the hourly simulation method, it is easy to keep track of the number of times that the operation of the storage switches from receiving to delivering power, or vice versa. This number is particularly relevant to those storage facilities whose lifetimes are approximately proportional to the number of charge/discharge cycles. Figure 6.83 gives the annual number of charge/discharge cycles, as a function of storage capacity, for the same assumptions as those made in Fig. 6.82.

All these ways of illustrating the performance of a store indicate that the improvement per unit of storage capacity is high at first, but that each additional storage unit above $t_s \approx 70$ h gives a very small return.

If the wind energy system including short-term storage is operated in a load-following mode, the storage size may still be characterized by t_s derived from (6.8), but, in calculating fluctuations (called $\sum$ for convenience of distinction), E_0 in (6.9) should be replaced by $E_0 C_1 C_2$, and, in constructing a time duration curve, the relevant question is whether the power delivered by the system exceeds $E_0 C_1 C_2$. Therefore, the power duration curve may be replaced by a curve plotting the percentage of time during which the difference $E - E_0 C_1 C_2$ between the power produced at a given time, E, and the load-following power level aimed at, exceeds a given value ΔE. Such curves are given in Fig. 6.84 for an "average load" E_0 equal to the annual average power production from the wind energy converters and assuming a loss-free store. The effect of augmenting storage capacity t_s is to prolong the fraction of time during which the power difference defined above exceeds the value zero, i.e., the fraction of time during which the load aimed at can indeed be covered.

FIGURE 6.83 Number of annual charge/discharge cycles of energy storage for the systems considered in Fig. 6.82. Again, the system is operated in base-load mode (*solid line*) or in a load-following mode (*dashed curve*) (Sørensen, 1978a).

The results of load-following operation are also indicated in Figs. 6.81–6.83 (dashed lines), which show that the wind energy system actually performs better in the load-following mode than in the base-load mode, when a non-zero storage capacity is added.

Finally, Fig. 6.85 gives some results of operating a wind energy system with short-term storage in the peak-load mode, based on a simulation calculation performed by Coste (1976). The storage facility is imagined as an advanced battery type of cycle efficiency 70%. The system is aimed at delivering a constant peak-load power level, E_{peak}, but only during certain hours of the day and only on weekdays. The remaining time is used to accumulate energy in the storage whenever the wind energy converters are productive. If the storage units become filled, surplus generation may be delivered to the grid as off-peak power (at reduced credits). Figure 6.85 gives the storage capacity needed in this mode of operation in order to achieve an availability of 80%, as a function of the peak-load time fraction, i.e., the ratio between the accumulated time designed as peak-load periods and the total time period considered (1 year). The lack of continuity in the calculated points is due to dependence on the actual position of the peak-load hours during the day. It is determined by the load variation (cf. Fig. 6.14) and represents an increasing amount of "peak shaving" as the peak-load time fraction increases. This influences the length of accumulation times between peak-load periods (of which there are generally two, unevenly spaced over the day), which again is responsible for some of the discontinuity. The rest is due to the correlation between wind-power production

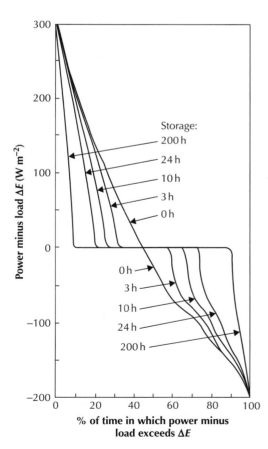

FIGURE 6.84 Annual time duration curve of the difference between the power produced by a wind energy system operated in the load-following mode and the actual load (or the fixed fraction E_0/E_{av} of it that it is aimed at covering) for a wind energy system with average power production E_0 and a loss-free storage of the capacity $t_A(E_0)$ indicated. Wind energy data from Risø (1961, 56 m) are used together with the power coefficient of type (a) in Fig. 6.70 (Sørensen, 1978a).

and time of the day [cf. the spectrum shown in Fig. 3.37 or the average hourly variation, of which an example is given in Sørensen (1978a)], again in conjunction with the uneven distribution of the peak-load periods over the day.

If the storage size is measured in terms of hours of peak-load power E_{peak}, the storage size required to reach an 80% availability increases with the peak-load time fraction, but, relative to the capacity of the wind energy converters, the trend is decreasing or constant. For small peak-load time fractions, the converter capacity is "small" compared to the storage size, because the accumulation time is much longer than the "delivery" time. In the opposite extreme, for a peak-load time fraction near unity, the system performs as in the base-load operation mode, and the storage size given as hours of E_{peak} coincides with those given in the previous figures in terms of E_0.

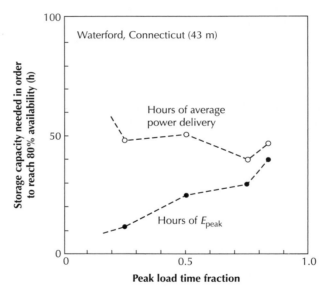

FIGURE 6.85 Storage capacity needed in order to reach 80% availability for a wind energy system in peak-load operational mode: $t_A(E_{peak})$ and $t_A(E_0)$ as a function of peak-load time fraction. The assumed wind converter design has an average power production of 57 W m^{-2} at Waterford, Connecticut (cf. Fig. 3.39), and the storage facility has a cycle efficiency of 70%. The average power production (related to the number of installed converters) is slightly above the average power delivery E_0 (which simply equals E_{peak} times the peak-load time fraction). (Based on Coste, 1976.)

Wind-power Systems with Long-term Storage

Long-term storage is of interest in connection with wind energy systems aiming at covering a high fraction of the actual load. In this case, it is not possible to rely entirely on fuel-based back-up units to provide auxiliary power, and the wind energy system itself has to approach autonomy (to a degree determined by grid connections and the nature of the utility systems to which such connections exist).

The mode of operation used for wind energy systems that include long-term storage is therefore usually the load-following mode. A simple system of this sort uses pumped hydro storage (eventually sharing this storage system with a hydropower system, as mentioned earlier), in which case high storage-cycle efficiency can be obtained. The performance of such a system would be an extrapolation of the trends shown in Fig. 6.81.

In many cases, storage options using pumped hydro would not be available and other schemes would have to be considered (e.g., among those discussed in section 5.3). The best candidate for long-term storage and versatility in use is presently thought to be hydrogen storage (Gregory and Pangborn, 1976; Lotker, 1974; Bockris, 1975; Sørensen *et al.*, 2003). The cycle efficiency is fairly low (cf. Table 5.5) for electricity regeneration by means of fuel cells (and still lower for combustion), but the fuel-cell units could be fairly small and could be placed in a decentralized fashion, so that the "waste heat" could be utilized conveniently (Sørensen, 1978a; Sørensen *et al.*, 2003). The production of hydrogen would also provide a fuel that could become important in the transportation sector. The use of hydrogen for short-term storage has also been considered (Fernandes, 1974), but, provided that other short-term stores of higher cycle efficiency are available, a more viable

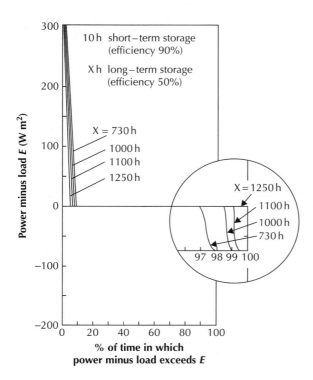

FIGURE 6.86 Annual power duration curve for a system comprising both short- and long-term storage facilities (with cycle efficiencies 0.9 and 0.5, respectively). Type (a) converters (Fig. 6.59) and 1961 Risø 56-meter data are used, and the average wind energy production (i.e., the number of converters installed) is 36% above the average load E_{av} (Sørensen, 1978a).

system could be constructed by using storage types with high cycle efficiency for short-term storage and hydrogen for long-term storage only.

An example of the performance of a wind energy system of this kind is given in Fig. 6.86, using the simulation model described previously. A 10-hour storage (of average production E_0, which has to exceed average load E_{av} owing to the storage losses) with 90% cycle efficiency is supplemented by long-term storage of assumed efficiency 50% (e.g., electrolysis-hydrogen-fuel cells), and the capacity of the long-term storage is varied until the minimum value that will enable the system to cover the total load at any time of the year is found. This solution, characterized by a power-minus-load duration curve that is positive 100% of the time, is seen to require a long-term storage of capacity $t_A(E_{av}) = 1250$ h. If the availability requirement is reduced to 97% of the time, the long-term storage size may be reduced to 730 h (1 month). Hydrogen storage in underground caverns or aquifers, which satisfies these requirements, is available in many locations (Sørensen *et al.*, 2001, 2003; Sørensen, 2004a).

6.6 REGIONAL SYSTEMS

In this section, scenarios for use of renewable energy are assessed on a regional basis. This means that the appraisal of energy demands and renewable energy supply options can be done on a regional basis, often with cultural similarities that make it easier to agree on the needs leading to energy demands. Furthermore, the possibility of exchanging energy between countries in the same region will alter the options for optimizing the system, as collaborative management of the system can reduce the need for each country in the region to install enough renewable energy to satisfy the highest demand. Generally, it will be cheaper to rely on the largest renewable energy sources in each country and to trade energy, either because a surplus capacity is fortuitously located in a country different from the one needing the energy or because the load peaks occur at different times, so that collaboration smoothes the necessary production in time.

Three examples are given: the Northern European region, which is special due to its high availability of (existing) hydro power; the North American region; and the Mediterranean region. For all three regions, a detailed assessment of available renewable energy supply options is made, with more accurate methods than the ones used for the global models of section 6.7. For one region, the Northern European one, a complete energy system scenario is constructed, showing in detail the advantages of cooperating, with shared use of energy stores and exchange of power or fuels (biofuels or renewably produced hydrogen), whenever time-related performance can benefit from imports or exports. The importance of the trade options is illustrated by first designing an energy system for each country as if it were an island that has to solve all its energy-system problems by itself. However, this is not the case today, because exchange of power and fuels takes place regularly between countries.

6.6.1 Regional Scenario Construction

Section 6.2 presents considerations in the construction of demand scenarios for use in scenarios for the future. This general category is termed *precursor scenarios*. To simplify the variety of scenario combinations, here it is assumed that the highest-efficiency intermediary-conversion scenario goes together with the highest-efficiency end-use scenario, and similarly that the lowest-efficiency choice applies both to intermediate and end-use conversion. Below is an outline of three such scenarios, corresponding to the range of behaviors anticipated in the precursor scenarios. The single load-scenario described in section 6.2.3 is used in the global scenarios discussed in Chapter 6.7. The regional scenarios discussed below in sections 6.6.2 to 6.6.4 choose the middle of the three end-use scenario alternatives made explicit below. It is interesting that, in some cases, new developments have made the middle-efficiency scenario closer to what 10 years ago would have been considered the highest-efficiency assumption.

Three Efficiency Scenarios

Biologically Acceptable Surroundings

The fixed relationship between indoor temperature T and space heating or cooling requirement P (for power) assumed in section 6.2.3 for 60 m^2 floor space per person (in work, home, and leisure heated or cooled spaces) was

$$P = c \times \Delta T \text{ with } c = 36 \text{ W/cap/°C}.$$

This assumption corresponds to use of the best technology available in 1980. For the 2050–2060 scenarios, this is taken as the *lowest, unregulated-efficiency scenario*, realized by leaving the development to builders. Using instead the best technology available in Denmark by 2005, as a result of repeated but modest tightening of building codes, gives $c = 24$ W/cap/°C. This lightly regulated development constitutes the *middle-efficiency scenario*, actually used in the examples of sections 6.6.2 to 6.6.5. The *highest-efficiency scenario* assumes a further technology improvement to take place and to be employed between now and the scenario year, yielding $c = 18$ W/cap/°C.

Using the middle scenario requires delivery of space heating and cooling on an area base (for GIS calculations) at the rates of

$$P_{heating} = 24 \times d \times (16\,°C - T),$$

$$P_{cooling} = 24 \times d \times (T - 24\,°C),$$

as it is assumed that the dependence of the comfort temperature zone on outdoor temperature exhibits a ±4 °C interval due to the flexible influence of indoor activities, as affected by body heat and clothing. The factor d is the population density, people per unit area (cap m^{-2}), and P is thus expressed in W m^{-2}.

Food, Health, and Security

In accord with section 6.2.3, an average end-user energy requirement of 120 W cap^{-1} for food, 18 W cap^{-1} for food storage, and small contributions for water supply, institutions, and security (police, military), are assumed to be included here or in the building and production energy figures. Only hot water for hygiene and clothes washing is different in the scenarios, amounting to an average of 100 W cap^{-1} in the high-efficiency scenario, 142 W cap^{-1} in the intermediate scenario used below (which is the same as used in the 1998 study; Sørensen et al., 2001), and 200 W cap^{-1} in the unregulated-efficiency scenario.

Human Relations

Energy use for human relations and leisure activities at the end-user translated in the 1998 study to an average 100 W cap^{-1} and is chiefly electricity for lighting, telecommunications, electronic entertainment, and computing. Not anticipated at the time was the magnitude of demand increase: for instance, CRT computer and television screens were phased out and were replaced by flat screens of similar size that use less than 10% of the energy CRTs used. In reality, much of the energy gain was diminished by a perceived demand for larger screens (25-inch TV screens were replaced by 50-inch flat screens, greatly affecting the reduction in energy use) and more screens (most dwellings already approach one screen per room). One positive development is reduced equipment energy use when idling (in stand-by mode): sensors can switch off equipment (to stand-by) whenever there are no people in a particular room. That would return the energy use to one screen per capita. For computers (which have already replaced music and video players) this might also be the case, except for those that perform special services while unattended (burglary surveillance, satellite recording, and possibly plain computing, at least in the case of this scientist!).

Consequently, the three demand scenarios used here have electricity use for relations set at 100, 200, and 300 W cap^{-1}, typically derived from people's three times higher rates of energy use during the 8 hours a day when they are not at work, commuting, or sleeping. Again, the middle-efficiency scenario is used in the regional scenarios. Currently, in affluent countries, such average end-use is a little above 100 W cap^{-1}.

Heat energy for leisure-activities venues is assumed to be incorporated into the building's heat use, while indirect uses for equipment, freight, and services are incorporated into the human activities energy use.

Transportation for human relations includes visits to family and friends, theaters, and concert venues, as well as vacations. When shopping is included, annual use today averages a total of about 10 000 km y^{-1} cap^{-1} (Sørensen, 2008b), with both short and long trips included. There is a large spread, and the scenarios assume 8000, 16 000, and 24 000 km y^{-1} cap^{-1} for non-work-related travel. Distances traveled are translated into energy units by using 0.04 W/(km y^{-1}) for personal transport and 0.1 W/(ton-km y^{-1})

for freight transport. These values correspond to state-of-the-art vehicles used at present.

Human Activities

Nonleisure activities are what humans do in exchange for food and leisure provision. They include the devices involved (vehicles, washing machines, and computers, and so on), which have to be manufactured. Furthermore, manufacturing requires a range of raw materials that have to be obtained (e.g., by mining). Energy is one such requirement. Section 6.2.3 lists a range of such indirect demands. Because there are many ways to arrive at a particular product, the manufacturing techniques and inputs change with time. Innovation can follow different directions and is at times very rapid and thus difficult to predict in a scenario context. Technological change also has a nontechnical component: for example, when we change our behavior to avoid diminishing animal diversity, or to avoid pesticides in our food intake and skin care products by choosing ecological (organic) produce or products. The negative effects of certain energy systems can also induce altered preferences, such as selecting "green electricity" rather than nuclear or coal electricity (a choice given by several power utilities). All these considerations should be reflected in the range of human activity scenarios offered.

Indirect activities for food and leisure equipment production, for distribution and sale of products, as well as for building and infrastructure construction and consequently materials provision, are estimated in section 6.2.3 to comprise some 60 W cap^{-1} of mainly high-quality energy (electric or mechanical energy), plus 3000 ton-km y^{-1} cap^{-1} of goods transportation plus 10 000 km y^{-1} cap^{-1} of work-related passenger transportation (e.g., in the service sector). These are all to be included in the activities energy use. The transportation figures translate into 400 W cap^{-1} average power use for person-transport and 300 W cap^{-1} for freight transportation. These numbers are taken as describing the highest-efficiency scenario.

The middle-efficiency scenario assumes 120 W cap^{-1} for agriculture and industry (for countries concentrating on information technologies, cf. Fig. 6.1), 600 W cap^{-1} for work-related transport (commuting included), and 400 W cap^{-1} for freight transportation, and the unregulated-efficiency scenario, 180 W cap^{-1}, 1000 W cap^{-1}, and 500 W cap^{-1}, respectively. The high values for business-related personal transportation are caused by insistence on personal-contact meetings instead of video conferences. The high values for transportation in the unregulated scenario imply a similar high use of time, which may be undesirable to business and industry unless travel time can be used productively (e.g., in trains and planes). People already complain about time wasted in personal car travel (such as commuting to and from work). One solution could be to make the employer responsible for transporting employees to and from work (presumably leading to selection of energy-efficient solutions) and including commuting time as work hours (because otherwise the employer could impose on employees' leisure time by choosing slow but cheap transportation solutions).

6.6.2 Mediterranean Region

Solar energy resources for the Northern Hemisphere are presented in Figs. 3.14 and 3.15, using 6-hour data for the year 2000. Based on the full 6-hour data, solar energy potentials for the Mediterranean region can be constructed, making the conversion process modeling described in section 4.4. In the Southern Hemisphere, appropriate tilt angles for solar collectors would be north-facing rather than south-facing. For those locations where measurements have been performed, the model seems to agree with data at mid-latitudes (i.e., the regions of interest here) and the variations in radiation seem to be generally of an acceptable magnitude. However, deviations could be expected at other latitudes or at locations with particular meteorological circumstances, due to the simplistic tilt-angle transformations.

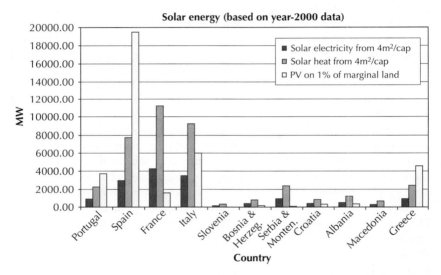

FIGURE 6.87 Country totals (in MW) for annual average solar electricity and heat-production rates from building-integrated collectors and stand-alone panels on marginal land in north-shore Mediterranean countries (year-2000 names used).

Figure 6.87 gives an estimate of practical available solar energy potentials for the countries on the northern Mediterranean coast, and Fig. 6.88 gives similar estimates for the southern coast. For building-integrated use, it is assumed that an average of 4 m^2 of collectors placed on roofs or facades of suitable orientation (close to south-facing and without other shadowing structures) would be available per capita. This area can provide solar power at an assumed photovoltaic panel efficiency of 15% and heat at an assumed 40% efficiency, allowing the same 4 m^2 to serve both purposes. For centralized solar power plants, the available area is taken as 1% of the marginal land in the country. Population densities and land area uses are given in Figs. 6.1, 6.2, and Figs. 6.33–6.36. The centralized solar cell panels are placed in fixed horizontal arrays, which, according to the distributions underlying Figs. 3.14 and 3.15, work as well as tilted panels. However, if concentrators are used, it is necessary to add some amount of tracking. Furthermore, desert areas contained in several African Mediterranean countries provide such

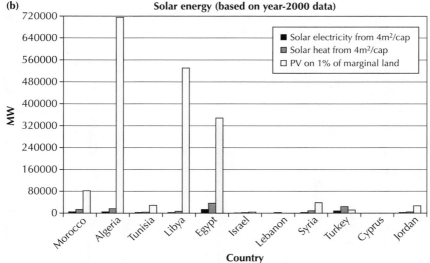

FIGURE 6.88a,b Country totals (in MW) for annual average solar electricity and heat-production rates from building-integrated collectors and stand-alone panels on marginal land in southern- and eastern-shore Mediterranean countries. Panel (*b*) is on a scale suited for central PV production potentials.

high levels of potential solar power that it is shown on a separate scale in Fig. 6.88*b*. This clearly constitutes a potential source for power export, if suitable transmission can be established.

For comparison, Figs. 6.89 and 6.90 show the scenario heating and cooling demands in the same countries, using the methodology of section 6.2 and

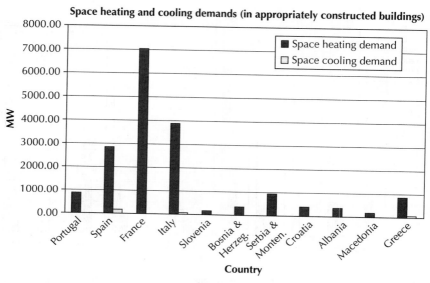

FIGURE 6.89   Country totals (in MW) for annual average solar heating and cooling demands (in middle-efficiency scenario), for northern-shore Mediterranean countries.

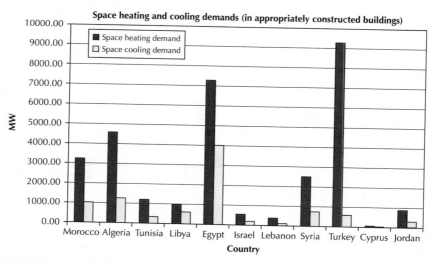

FIGURE 6.90   Country totals (in MW) for annual average solar heating and cooling demands (in middle-efficiency scenario), for southern- and eastern-shore Mediterranean countries.

middle scenarios for consumption development (see further discussion below about the northern European scenario).

FIGURE 6.91 Time-series comparison of space-cooling demand and rooftop PV electricity supply (MW) for Spain (annual totals given in Figs. 6.87 and 6.89).

FIGURE 6.92 Time-series comparison of space-heating demand and rooftop solar-panel heat production (MW) for France (annual totals given in Figs. 6.87 and 6.89).

The temporal behavior of the 6-hour time-series is illustrated in Figs. 6.91 and 6.92. In Fig. 6.91, the potential power production from building-integrated solar panels in Spain is compared to the demand for space cooling in properly designed buildings. Space cooling is assumed to be required if the temperature exceeds 24 °C, as described in section 6.6.1. This fairly low energy expenditure for space cooling is valid for energy-efficient buildings (high insulation, controlled ventilation). Solar energy is well correlated with space cooling and the rooftop collection is more than sufficient for keeping the temperature below the limit selected.

In Fig. 6.92, the thermal energy yield of 4 m^2 cap^{-1} solar collectors in France is compared to the demands for space heating, again for energy-efficient buildings. The assumption is that space heating is required if the temperature is below 16 °C, again with details given in section 6.6.1. Here there is an anti-correlation between solar energy collection and demand, and although the annually collectable solar energy is adequate, heat stores of several month's capacity must be added to make a solar heating system able to cover demand year-round. It should also be mentioned that the functioning of solar heat collectors depends sensitively on the temperature of the inlet fluid, and therefore there is an expected decreased efficiency in late summer and possibly a deficiency in temperature of the outlet fluid during winter, relative to the requirement of the heat-distribution system (worst for radiator heating, best for floor or air heating). During the summer months, additional heat collection for other uses than space heating is possible. Since the electricity production from the building-integrated solar PVT cells is assumed to be 15/40 of the heat production, it is clear that the rooftop solar collectors are far from capable

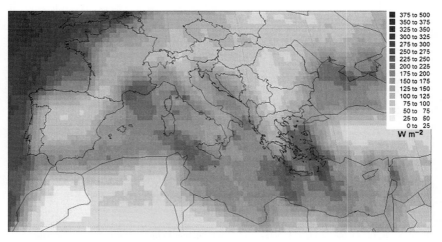

FIGURE 6.93   Annual average potential wind-power production (W per m^2 swept), for the Mediterranean region (based on Fig. 6.30).

of satisfying total power needs. The electricity that may be derived from solar panels on marginal land (Fig. 6.87) is insignificant in France, but in Spain it could contribute to general electricity demands.

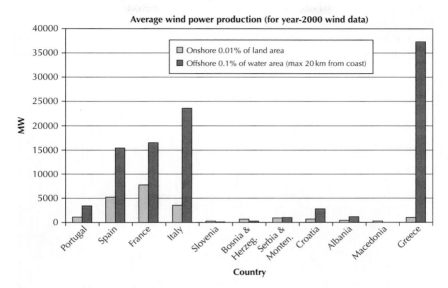

FIGURE 6.94 Country totals (in MW) of average wind-power production potentials for northern Mediterranean region. The ratios of swept and ground areas are given.

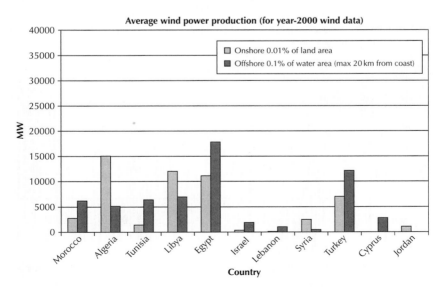

FIGURE 6.95 Country totals (in MW) of average wind-power production potentials for the southern and eastern Mediterranean regions. The ratios of swept and ground areas are given.

Wind resources could be used by decentralized placement of wind turbines on agricultural land, or more centralized placement on marginal land or offshore in fairly shallow waters that allow conventional foundation work (section 6.3.2). Figure 6.30 shows the potential wind-power production globally. The Mediterranean results are shown on a larger scale in Fig. 6.93. There are several areas of high potential, particularly in the eastern part of the region. Figure 6.94 shows the totals for each country along the northern Mediterranean shores, and

FIGURE 6.96a,b   Time-series comparison of potential on- and offshore wind-power production (MW) for Portugal (*a*) and Greece (*b*).

Fig. 6.95 shows totals for the southern and eastern shores. It is assumed that 0.01% of the land area is suitable for wind production (representing the rotor-swept area and giving generous room for placing the turbines so that production is not lowered by interference between individual turbines), and that 0.1% of the sea area may be used without interference with other uses, such as fishing, shipping routes, or military training. Only areas with water depth below 20 m are considered.

Countries near the Atlantic Ocean have substantial potentials, also inland, but in the eastern Mediterranean, Greece and the Turkish west coast are characterized by very large offshore potentials but little mainland onshore wind. This is illustrated by the time-series shown in Figs. 6.96a and b for Portugal and Greece. The combined wind potential for the total region is large enough to offer coverage of a substantial part of the electricity demand, together with smaller additions by photovoltaic power. The potential wind-power estimates are generally larger than the rooftop PV estimates, which should not be surprising, since wind farms on agricultural and marginal land are included. These locations also make the solar potential large, but it is restricted to areas without other uses, because solar cells cover the area while wind turbines have small footprints and allow agriculture on the same area.

FIGURE 6.97 Vision of Mediterranean power grid connections enabling export of Saharan solar energy to Europe (Nielsen and Sørensen, 1996).

The total renewable electricity production in the European Mediterranean countries (with the restricted assumptions on PV areas and number of wind sites, 0.01% of land area being actually less dense than the present penetration in Denmark) is not sufficient to cover the power demands of the middle scenario. However, the African Mediterranean countries have a huge surplus due to the potential of desert-located solar farms on just 1% of the land. This has caused several investigators to propose power exchange through one of the three cost-effective routes (Fig. 6.97). Political issues in some of the countries involved have so far prevented these ideas from being seriously pursued.

Biomass resources may be divided into resources associated with agricultural activity, with forestry, and with aquaculture. In each case, the primary products, such as food and timber, should get first priority. However, residues not used for these purposes can be used for a variety of other purposes, including energy. This has long been the case for simple combustion of biomass, but considerations of recycling nutrients to the soil or using biomass as industrial feedstock for a range of products may make it preferable to avoid simple burning, which has associated negative impacts from emissions to the atmosphere.

Biomass resources are estimated from the model of plant growth described in section 3.5.2, expressed as net primary production (NPP). The maximum amount of biomass that can be used sustainably for energy (as harvested crops, wood, etc., but recycling nutrients when possible), considering variations in solar radiation, precipitation, and nutrients as function of geographic location, is shown in Fig. 3.99. Figure 6.98 shows a variant where water is freely available and thus no longer a limiting factor. This gives the biomass production that would be possible with sufficient irrigation. Of course, it cannot be globally realized because many regions have water limitations that preclude full irrigation.

Based on the global models shown in Figs. 3.99 and 6.98, the annual biofuel potential in energy units is given in Fig. 6.99 and 6.100 for the Mediterranean

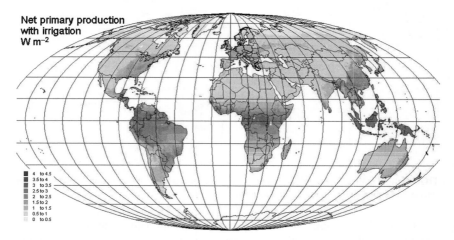

FIGURE 6.98 Net primary production (W m^{-2}) for model used in Fig. 3.99, but without constraints on water access, i.e., admitting irrigation as needed, disregarding issues of water availability, lowering of ground water table, and possible desertification.

region, with the following assumptions: Irrigation is not included, although some regions in the area considered presently use artificial addition of water, but nutrient recycling has been assumed, although it is not now fully practiced in the Mediterranean region (and administration of chemical fertilizer is used in some regions). Of the total primary production modeled, only half is taken as available for energy production. This means that, in the agricultural sector, considerably more than food is left to other uses (under 10% of the biomass energy yield ends up used as food, the rest being residues, such as straw or waste from processing and final use in households, all of which could theoretically be recycled to the fields). In forestry, 50% of NPP used for energy means that uses for timber, paper, and pulp are not affected, although sustainable wood usage would produce a similar amount of discarded wood products that could be recycled. The biofuel production efficiency is assumed to be 50% (of the raw material, hence 25% of the primary biomass produced), which is consistent with the 30%–50% efficiency expected for upcoming plants producing ethanol, methanol, biodiesel, or other fuels from biomass residues (the existing unsustainable use of food grains for producing biofuels in many cases has an efficiency greater than 50%).

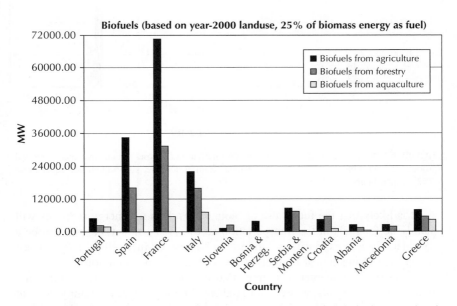

FIGURE 6.99   Northern Mediterranean countries' annual averages of biofuel production (MW) based on feedstock of residues from agriculture, silviculture, and aquaculture (25% of total biomass yields).

As shown in Fig. 6.99, France, particularly with its large food production (partially for export), has a similarly large energy potential for biofuels produced from agricultural waste, and Spain and Italy have a considerable potential as well. Some potential is also available in the Balkan region, and the same countries have an important potential derived from forestry residues. There is an amount of potential energy production from aquaculture, the largest being in Italy. The future will show how much of the potential water sites will be used for additional food production and how much for other aquaculture applications. The time variations in biomass supply are not particularly relevant, as biomass can be stored before or after conversion to fuels. As Fig. 6.100 shows, the soil is not amenable to large crop yields on the African side of the Mediterranean. Turkey, however, has substantial agricultural output.

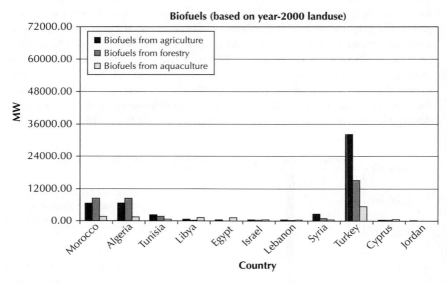

FIGURE 6.100 Southern- and eastern-shore Mediterranean countries' annual averages of biofuel production (MW) based on feedstock of residues from agriculture, silviculture, and aquaculture (25% of total biomass yields).

It is likely that the future use of biomass residues for energy purposes will mainly be for fuels in the transportation sector. Here, biomass appears to be a considerably less expensive solution than battery- or fuel-cell-driven vehicles. However, combustion of biofuels still contributes to air pollution.

Other renewable energy sources that may play some role are geothermal energy; ocean currents, waves, and gradients in temperature or salinity; and, of course, hydro power, for which no more large installations are expected to achieve environmental acceptability, but for which there may still be a potential for smaller installations. The results can be used to construct full scenarios matching the renewable energy supply with expected future demands, and to identify the opportunities for trade of energy between the regions or countries. Such trade is very important because it can eliminate the need for energy storage to deal with the intermittency of some renewable energy sources, and because, as the preliminaries above show, there are areas in Europe that do not seem to have sufficient renewable energy sources domestically to support the level of population envisaged and their energy demands. An example of trade solutions (in the north of Europe) is treated in section 6.6.4. Here, basic supply–demand matching is carried out for one of the Mediterranean countries, Greece. The time-series of assumed demands, distributed on energy qualities, are depicted in Fig. 6.101.

FIGURE 6.101   Greece: Scenario for energy demands around 2050 (in PJ^{-1}).

FIGURE 6.102   Greece: Time-series of rooftop solar thermal coverage of low-temperature heat demands around 2050 (in PJ^{-1}).

Figure 6.102 shows that a substantial part of the low-temperature heat demand can be covered by rooftop solar thermal collectors. The balance would be covered by heat pumps or by waste heat from building-size fuel cells, if they become viable by the scenario time, about 2050.

FIGURE 6.103   Greece: Time-series of the disposition of PV/wind power production on demands around 2050 (in PJ^{-1}). From the bottom one finds electricity to heap pumps and furnaces (hidden), direct electricity loads, and electricity to fuel cells for hydrogen production, and, at top, surplus electricity that may be exported.

FIGURE 6.104   Greece: Time-series of biofuel and hydrogen coverage of transportation demands around 2050 (in PJ^{-1}). Dips signal insufficient power for hydrogen production by PV/wind (which could be handled by imports or larger hydrogen stores).

Figure 6.103 shows the scenario disposition of electricity produced by wind or photovoltaic cells. A fluctuating but quite large amount of power is available for export. Finally, Fig. 6.104 shows the coverage of transportation energy demand. Roughly half is covered by biofuels, and half by hydrogen in fuel-cell vehicles (more efficiency, hence lower energy demand). Only nominal hydrogen stores have been included in the model. They would have to be increased to avoid occasionally insufficient hydrogen production. The surplus power to feed the stores is available, as seen in Fig. 6.103.

6.6.3 North America

The North American solar radiation resource can be derived from Figs. 3.14 and 3.15, the wind resources from Fig. 6.30 (from which the blow-up shown in Fig. 6.105 is derived), and the biomass potential from Fig. 3.99 (blow-up shown in Fig. 6.106).

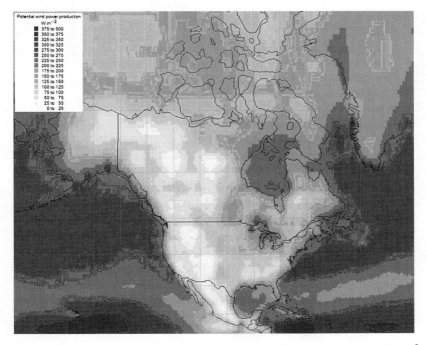

FIGURE 6.105 Potential annual average wind-power production in North America (W m^{-2}). Shown in Magellan coordinate projection (Sørensen, 2007).

The potential wind power that can be derived from very modest areas, preferably near-shore and offshore locations, is summarized in Fig. 6.107. The potential for biomass utilization is, according to Fig. 6.106, largest in southern

Mexico, the eastern United States, and a narrow coastal strip along the West
Coast. If artificial irrigation is added, as it is today, the Midwest potential is
considerably increased. Other renewable sources include hydro (assumed to
stay at current level) and smaller sources, such as geothermal, wave, and tidal.

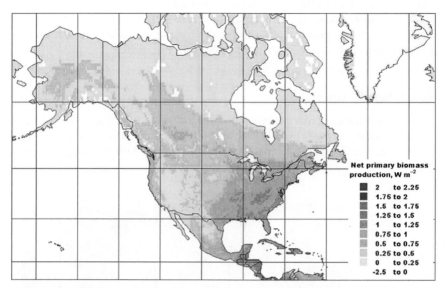

FIGURE 6.106   Potential annual biomass production in North America, in W m^{-2} (Sørensen, 2007).

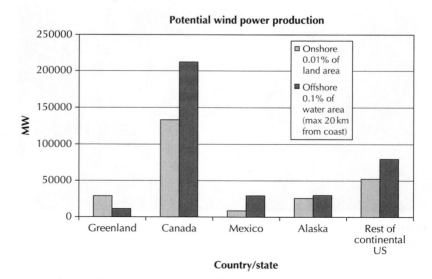

FIGURE 6.107   Practical potential of wind resources in North America, in MW (Sørensen, 2007).

Preliminary simulations for the United States (demand time-series illustrated in Fig. 6.108) show that the direct generation of electricity by hydro, sun, and wind, based on the practical potentials identified above, is insufficient for covering electricity demands, high-temperature process heat, and generating hydrogen for a part of the vehicle fleet. The most obvious candidate for supplying the deficit is Canada, which has a large power surplus. However, the generation of liquid fuels from biomass residues derived from agricultural and forestry residues and possibly by additional aquaculture activities (e.g., using the waters between offshore wind turbines) more than covers the need for such fuels and allows the United States to be self-sufficient, provided that the best available technology is used to achieve a high standard of energy conversion efficiency for all types of conversion. If hydrogen fuel-cell vehicles are more attractive than biofuel-operated ones, an alternative would be to import the necessary electric power for electrolysis from neighboring countries, notably Canada. The alternative of creating hydrogen from biomass residues is also available, although it is perhaps less attractive than using liquid biofuels.

FIGURE 6.108 Energy demands in the scenario for 2050, for the continental United States, except Alaska (PJ y^{-1}). The distribution on energy qualities can be done in different ways. Here, liquid fuels are used in the transportation sector, and electricity is used for all purposes other than low-temperature heat.

6.6.4 Northern Europe

A very detailed study, mapping energy supply and demand on a geographical grid of 0.5 km × 0.5 km, has been made for Denmark for an energy-efficient 2050 scenario. It is described in Sørensen (2005). Here an update of that study is looked at, expanding the region considered to the Nordic countries (except Iceland) and Germany. The new study uses the methodology described in section 6.6.1, choosing the middle-efficiency scenario as in section 6.6.2. The region is interesting because of the large amount of hydro with seasonal reservoirs present in Norway and Sweden, and because of the much higher population and energy-use density in Germany and Denmark, compared to the other three countries. Denmark has generous supplies of wind and agricultural energy, Sweden and Finland similar supplies of forestry energy, and Norway and Finland have particularly good supplies of wind energy. Germany has reasonable agricultural assets, but its wind resources are only on the northern shores. Some solar energy can be derived, particularly in the south, but there remains a large deficit in electricity provision. However, as the study concludes, the surplus of energy in the Nordic countries is more than twice as large as the deficit in Germany, implying that German supply and demand can be balanced either by importing electricity from wind or hydro, or by importing less electricity and more biofuels, eventually with some final conversions (e.g., to hydrogen) in Germany.

One purpose of the new study (Sørensen *et al.*, 2008; Sørensen, 2008c) was to compare the alternative supply scenarios based on hydrogen and fuel cells, or on biofuels, as described in Sørensen (2008b). Here, the variant assuming that, in the transportation sector, liquid biofuels will be successful and hydrogen-fueled fuel cells will not, is illustrated. Figures 6.109*a-e* show the 2060 energy demands for the five countries, according to the middle-efficiency scenario. Differences in the demand profiles reflect different climatic conditions and different industry structures in the countries.

FIGURE 6.109a Danish energy demands in 2060 scenario (PJ y^{-1}). "Electricity" is dedicated electricity loads and the transportation fuels are 50% liquid, 50% gaseous.

(b)

(c)

FIGURE 6.109*b,c*   Swedish (*b*) and Norwegian (*c*) energy demands in 2060 scenario (PJ y^{-1}).

In contrast to Denmark, the other countries have substantial heavy industry demanding high-temperature heat and electricity, particularly in Norway. Despite its small size, Denmark has relatively higher transportation demands than the other countries, possibly reflecting the different industry structure. Space-heating needs show the expected dependence on degree-days of required heating of buildings.

Generally speaking, the Northern European renewable energy supply options contain a very high percentage of electricity provision. This implies that high-temperature heat for industry would most conveniently be supplied

FIGURE 6.109*d,e*   Finnish (*d*) and German (*e*) energy demands in 2060 scenario (PJ y^{-1}) (Sørensen, 2008c).

by electricity, that low-temperature heat not derived from waste heat from other conversion processes would be supplied by electric heat pumps, and that electric vehicles are seen as an attractive option, if prices can be brought sufficiently down. Figure 6.110 (Nordel, 2005) indicates the options available in a system with hydropower based on large amounts of seasonal storage in the form of elevated water, which are predictable at the start of the season.

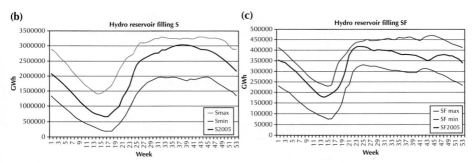

FIGURE 6.110a,b,c Weekly filling (GWh stored) of Nordic hydro reservoirs aggregated on a country basis: *a*: Norway; *b*: Sweden; *c*: Finland. Data for 2005 are shown, together with maximum and minimum over 20 years.

Wind conditions are excellent at nearly all locations near the Atlantic or Baltic coastlines. Figure 6.111 shows annual averages on a per capita basis. Relative to its size, the Danish wind potential is impressive, but the other Nordic countries, especially Norway, have large wind resources, notably offshore. On the other hand, several Norwegian offshore locations may be difficult to exploit (the criterion for inclusion in this study was a distance from the coast of less than 20 km, rather than a water-depth limit), primarily due to difficult conditions for transmission ashore and further to load centers. On the other hand, Norwegian winds are much more consistent than those of Sweden and Finland, and these are again more consistent than those in Denmark, as shown by the annual power duration curves in Fig. 6.112. The evaluation of the power duration curves assumes that the identified potential is fully used. It places turbines on 0.1% of the offshore areas closer than 20 km to the coast, and on 0.01% of the inland areas (0.02% for Denmark, due to its lower fraction of inaccessible mountainous areas). In Norway, half the average power output is available more than 95% of the time, in Sweden and Finland, 85% of the time, and in Denmark and Germany, just under 80% of the time.

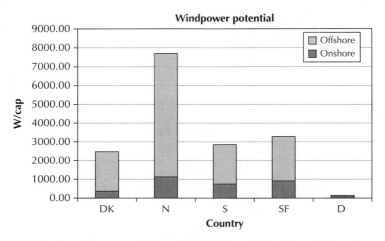

FIGURE 6.111   Potential average wind-power production, Northern Europe (W cap^{-1}).

FIGURE 6.112   Power duration curves for wind in each Northern European country.

The Nordic countries can handle any remaining mismatch between supply and demand by using small amounts of the energy stored in the hydro reservoirs. Denmark could avoid using hydro stores situated in neighboring countries, should it wish to: The existing underground storage in caverns, presently used for natural gas, could store hydrogen produced from excess wind, and electricity could be regenerated when needed, using fuel cells or Carnot cycle power plants. As shown in Fig. 6.113, only a small store is required (less that the existing ones). Table 6.10 summarizes the renewable energy sources available in Northern Europe, for the assumed level of use.

FIGURE 6.113 Filling variations through year in a small (1.37 PJ) Danish hydrogen store.

Country:	DK	N	S	SF	D
Wind onshore	64	167	201	147	157
Wind offshore	358	974	579	391	177
Biofuels from agriculture	241	51	111	49	1993
Biofuels from forestry	58	523	1670	1180	892
Biofuels from aquaculture	153	223	320	205	108
Hydro	-	510	263	49	27
Solar PVT electricity	-	-	-	-	129
Solar PVT heat	-	-	-	-	275

TABLE 6.10 Summary of potential renewable energy supply available for use in the Northern European countries considered (in PJ y^{-1}). PVT is combined photovoltaic and thermal collectors.

The year-2060 Nordic energy scenario first examines demand coverage in each country by indigenous resources, and then examines what potential remains for export. For all countries except Germany, full coverage of demands is possible without import (this does not say that economic optimization would not make it advantageous to trade more than strictly necessary).

FIGURE 6.114   German import options (PJ y^{-1}). Electricity is given for each 6 hours, fuels only for each month.

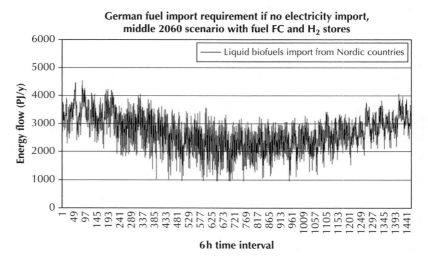

FIGURE 6.115   German biofuel deficit, per 6-hour interval, if the needs of the transportation sector are to be covered by biofuels (PJ y^{-1}). On average, the deficit equals import requirements.

The import options available for Germany are depicted in the rather messy Fig. 6.114. There are total import options corresponding to more than twice the German requirement, so the mix of sources to import can be selected freely. A likely outcome would be to import any electricity needed for direct or industrial heat use and to let the transport sector import be determined by the relative success in developing second-generation biofuel vehicles, electric or hydrogen fuel-cell vehicles, or some hybrid between them. Figure 6.115 shows the German biofuel deficit as function of time. Of course, actual import need not follow this time pattern, because biofuels can be stored (e.g., at filling stations). Indigenously produced German biofuels suffice to satisfy heating needs in boilers, if this is the scenario chosen (Fig. 6.116). If the alternative high electricity import is selected, heating not provided by solar thermal collectors would be provided by heat pumps (details in Sørensen, 2008c).

FIGURE 6.116   Coverage of German low-temperature heat in the 2060 scenario variant with only biofuel import (Sørensen, 2008c).

6.7   A GLOBAL ENERGY SCENARIO

In section 6.2, a demand scenario is developed for the year 2050, and, in section 6.3, estimates are made of potential energy utilization from the most important renewable energy sources for the same period of time, based on Sørensen and Meibom (1998; 2000). Satellite data used were rather simple compared to the data used in section 6.6, but the later calculations largely support the simplifications used as being reasonably accurate.

The supply–demand matching first uses only the renewable energy that is estimated to be available locally; subsequently, if necessary, more centralized production, such as offshore wind farms or photovoltaic farms on marginal land, are used. The study was probably the first to accept energy in food as part of the renewable energy concept. The reason is of course that some renewable energy sources have a double role, in both energy and nutrition, a possibility not applicable to fossil fuels.

The vegetable part of food production is shown in Fig. 6.117, which contains comparisons of local supply and demand, with Fig. 6.117a showing the amount of surplus for the geographical grid cells where supply exceeds demand, and Fig. 6.117b the amount of deficit for the local cells where demand exceeds supply. Regional sums are given in Table 6.11. It follows that, on average, year-2050 worldwide supply exceeds demand by 35%. This gives room for variations in crop harvests and thus variations in food production from year to year, as well as for differences in eating habits.

Like today, in year 2050 there is surplus vegetable food production in the Americas and Western Europe (regions 1, 2, and 4, cf. Table 6.5) and by the year 2050 also in region 3 (including Russia), owing to proposed substantial improvements in agricultural practices assumed for this region. Region 5 (including China and India) will be barely self-sufficient by the year 2050, whereas Africa (region 6) will have a deficit that must be covered by imports. In the scenario, Africa is the only region that by 2050 is in a development situation, where it may offer labor at lower expense than the other regions, and, thus, there will be the possibility of paying for food imports by industrial revenues, provided that an education policy is pursued that will give the workforce the necessary skills. In addition to interregional exchange, upon closer inspection, Fig. 6.117 allows a deduction of the requirements for transport of vegetable food within regions, e.g., from farming areas to cities.

The scenario assumptions for interregional trade in food are summarized in Fig. 6.132, where the regional exports have been selected from the surpluses available.

For animal-based food from either rangeland or fodder-fed animals, the surpluses and deficits are shown in Fig. 6.118. In regions 1 to 4, the picture is similar to that for vegetable foods, with surpluses, but in regions 5 and 6 there are deficits. This is due to the increase in the meat and milk fractions of diets assumed for Asia (Table 6.5), but the amounts are easily covered by imports from other regions, as indicated in Figs. 6.129 and 6.130. Overall, the animal food supply exceeds demand by 27%, which again is considered adequate in view of additional losses. Variations between years are smaller than for primary crops (because of the storage functions performed by livestock), but a reserve is required due to epidemics of animal diseases.

(a)

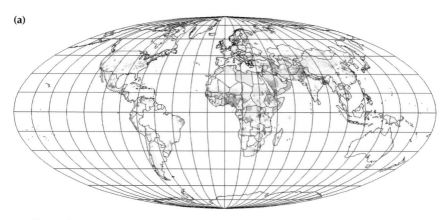

Energy flow
W m⁻²

■	5	to 10
■	2	to 5
■	1	to 2
■	0.5	to 1
■	0.2	to 0.5
■	0.1	to 0.2
■	0.05	to 0.1
■	0.02	to 0.05
■	0.01	to 0.02
■	0.005	to 0.01
■	0.002	to 0.005
■	0.0001	to 0.002
■	all others	

FIGURE 6.117a,b Local surplus (*a, scale left*: annual average supply minus demand in W m⁻² is shown if positive) and deficit (*b, scale right*: annual average demand minus supply in W m⁻² is shown if positive) of vegetable-based food supply relative to demand, on an area basis, for 2050 scenario. (From Sørensen and Meibom, 1998; 2000.)

Energy flow
W m⁻²

■	-0.002 to	-0.0001
■	-0.005 to	-0.002
■	-0.01 to	-0.005
■	-0.02 to	-0.01
■	-0.05 to	-0.02
■	-0.1 to	-0.05
■	-0.2 to	-0.1
■	-0.5 to	-0.2
■	-1 to	-0.5
■	-2 to	-1
■	-5 to	-2
■	all others	

(b)

(a)

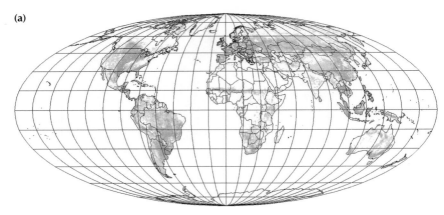

Energy flow
W m⁻²

5	to 10
2	to 5
1	to 2
0.5	to 1
0.2	to 0.5
0.1	to 0.2
0.05	to 0.1
0.02	to 0.05
0.01	to 0.02
0.005	to 0.01
0.002	to 0.005
0.0001	to 0.002
all others	

Energy flow
W m⁻²

-0.002 to -0.0001
-0.005 to -0.002
-0.01 to -0.005
-0.02 to -0.01
-0.05 to -0.02
-0.1 to -0.05
-0.2 to -0.1
-0.5 to -0.2
-1 to -0.5
-2 to -1
-5 to -2
all others

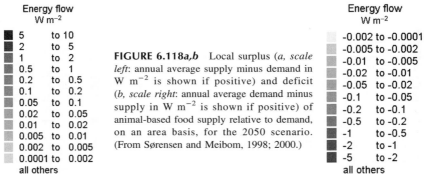

FIGURE 6.118a,b Local surplus (*a, scale left*: annual average supply minus demand in W m⁻² is shown if positive) and deficit (*b, scale right*: annual average demand minus supply in W m⁻² is shown if positive) of animal-based food supply relative to demand, on an area basis, for the 2050 scenario. (From Sørensen and Meibom, 1998; 2000.)

(b)

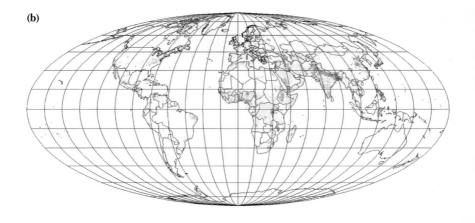

Region (cf. Table 6.5): 1	2	3	4	5	6	Total	Unit
Vegetable food	108	113	91	192	40	−140	402	GW
Animal food	43	31	16	91	−63	−27	91	GW

TABLE 6.11 Balance of potential regional food supply and demand in 2050 scenario.

The regional energy sources employed to satisfy energy demands beyond food energy are listed in Table 6.12. For hydropower, existing plants and those already under construction are kept, but no additional plants are allowed to be built, in consideration of their environmental impacts. Untapped hydro resources are primarily in South America (cf. Fig. 3.49).

Region (cf. Table 6.5)	1	2	3	4	5	6	Total	Unit
Total food balance (as in Table 6.10)	151	144	107	283	−23	−167	493	GW
Total biofuels used	250	190	300	640	327	192	1899	GW
Of which decentralized biofuels	250	166	236	640	295	192	1779	GW
Balance: total biofuels minus use for transportation	136	158	146	277	392	61	1170	GW
Hydropower	80	70	50	120	110	10	440	GW
Decentralized solar power	27	50	67	109	189	23	465	GW
Centralized solar power	40	20	114	19	400	100	693	GW
Decentralized wind power	20	37	80	100	100	10	347	GW
Onshore wind parks	12	10	20	38	46	0	126	GW
Offshore wind power	0	40	0	0	13	0	53	GW
Balance: other energy, annual average	100	−63	139	288	−660	217	22	GW
Balance: other energy, Jan.	74	−120	169	306	−890	214	−250	GW
Balance: other energy, Apr.	87	−51	72	243	−710	200	−160	GW
Balance: other energy, July	129	−17	180	352	−410	233	461	GW
Balance: other energy, Oct.	110	−64	129	252	−610	220	35	GW

TABLE 6.12 Assumed energy supply (after storage conversion cycles and transmission but before interregional imports/exports) in the 2050 scenario, and corresponding supply–demand balances.

The sources proposed to be exploited in Table 6.12 are shown in Table 6.13 as fractions of the potential resources that they constitute. For region 4, current rainforest areas are not exploited but are left as preservation areas.

Region (cf. Table 6.5)	1	2	3	4	5	6	Total
Decentralized biofuels	0.91	0.98	0.96	0.93	1.00	0.92	0.94
Centralized biofuels	0	0.12	0.64	0	0	0	0.21
Hydropower	0.88	0.82	0.95	0.48	0.90	0.67	0.71
Decentralized solar power	0.54	0.93	0.71	0.76	0.99	0.20	0.72
Centralized solar power	0.013	0.003	0.014	0.004	0.041	0.005	0.013
Decentralized wind power	0.56	0.85	0.45	0.75	1.00	0.21	0.64
Onshore wind parks	0.75	0.068	0.27	0.66	1.00	0	0.21
Offshore wind power	0	0.52	0	0	0.70	0	0.16

TABLE 6.13   Fraction of potential resources used in the 2050 scenario.

The resulting energy system is very robust because there are additional renewable resources to use if the assumptions behind the scenario should turn out not to be right. This modesty is of course necessary, owing to the variations between seasons and years of renewable energy production, neither of which are, or can be, modeled precisely. Table 6.12 indicates substantial variations in supply–demand balances between seasons. Stores will have to handle the mismatch, and a detailed simulation is needed to ensure that it works, as was done for the regional scenarios of sections 6.5 and 6.6.

The area-based distributions of supply–demand balancing corresponding to the totals given in Table 6.12 are given in Fig. 6.119 for transportation biofuels and in Figs. 6.120–6.124 for stationary use of high-quality renewable energy sources. Figures 6.125–6.131 show block diagrams of the regional and global flows through the energy system from source to end-use.

The main routes of import and export are depicted in Fig. 6.132. The amount of centralized energy used in the scenario is still modest compared to the potential listed in section 6.3, even when new large hydro installations or rainforest conversion are renounced. The PV and wind energy export options arising for desert land give such areas new opportunities for development and, in some cases, make up for current oil-export revenues lost. Also the large biofuel exports considered for South America give credibility to the assumed substantial economic development of this continent.

While it is not surprising that the most densely populated regions should have difficulty being self-sufficient in energy provision from renewable energy (Fig. 6.132), just as they usually are with the energy sources presently used, it is a little surprising to see how large a surplus of energy is offered for purchase by the regions with surplus renewable energy sources. This has implications for economic development (as discussed in Chapter 7). The magnitude of these surpluses is illustrated in Table 6.13, which gives the percentages of estimated available resources actually used in the scenario.

(a)

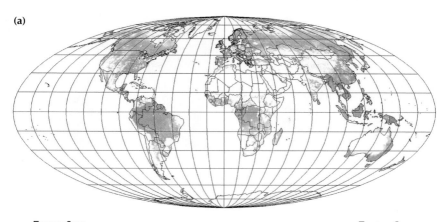

Energy flow
W m⁻²

■	5	to 10
	2	to 5
	1	to 2
	0.5	to 1
	0.2	to 0.5
	0.1	to 0.2
	0.05	to 0.1
	0.02	to 0.05
	0.01	to 0.02
	0.005	to 0.01
	0.002	to 0.005
	0.0001	to 0.002
	all others	

Energy flow
W m⁻²

	-0.002 to -0.0001
	-0.005 to -0.002
	-0.01 to -0.005
	-0.02 to -0.01
	-0.05 to -0.02
	-0.1 to -0.05
	-0.2 to -0.1
	-0.5 to -0.2
■	-1 to -0.5
■	-2 to -1
■	-5 to -2
	all others

FIGURE 6.119a,b Local surplus (*a, scale left:* annual average supply minus demand in W m⁻² is shown if positive) and deficit (*b, scale right:* annual average demand minus supply in W m⁻² is shown if positive) of transportation biofuel supply (such as methanol) relative to transportation demand, on an area basis, according to the 2050 scenario.

(b)

(a)

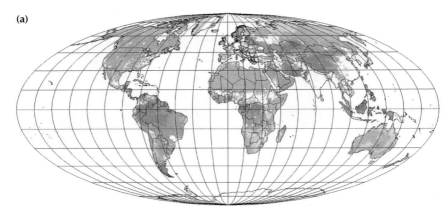

Energy flow
W m⁻²

5	to 10
2	to 5
1	to 2
0.5	to 1
0.2	to 0.5
0.1	to 0.2
0.05	to 0.1
0.02	to 0.05
0.01	to 0.02
0.005	to 0.01
0.002	to 0.005
0.0001 to	0.002
all others	

FIGURE 6.120a,b January local surplus (*a, scale left:* average supply minus demand in W m⁻² is shown if positive) and deficit (*b, scale right:* average demand minus supply in W m⁻² is shown if positive) of electricity and biofuels for stationary use (such as biogas or hydrogen) relative to demand, on an area basis, according to the 2050 scenario.

Energy flow
W m⁻²

-0.002 to -0.0001	
-0.005 to -0.002	
-0.01 to -0.005	
-0.02 to -0.01	
-0.05 to -0.02	
-0.1 to -0.05	
-0.2 to -0.1	
-0.5 to -0.2	
-1	to -0.5
-2	to -1
-5	to -2
all others	

(b)

(a)

Energy flow
W m^{-2}

5	to 10
2	to 5
1	to 2
0.5	to 1
0.2	to 0.5
0.1	to 0.2
0.05	to 0.1
0.02	to 0.05
0.01	to 0.02
0.005	to 0.01
0.002	to 0.005
0.0001 to	0.002
all others	

FIGURE 6.121a,b   April local surplus (*a, scale left:* average supply minus demand in W m^{-2} is shown if positive) and deficit (*b, scale right:* average demand minus supply in W m^{-2} is shown if positive) of electricity and biofuels for stationary use (such as biogas or hydrogen) relative to demand, on an area basis, according to the 2050 scenario.

Energy flow
W m^{-2}

-0.002 to -0.0001
-0.005 to -0.002
-0.01 to -0.005
-0.02 to -0.01
-0.05 to -0.02
-0.1   to -0.05
-0.2   to -0.1
-0.5   to -0.2
-1   to -0.5
-2   to -1
-5   to -2
all others

(b)

(a)

Energy flow
W m⁻²

5	to 10
2	to 5
1	to 2
0.5	to 1
0.2	to 0.5
0.1	to 0.2
0.05	to 0.1
0.02	to 0.05
0.01	to 0.02
0.005	to 0.01
0.002	to 0.005
0.0001 to	0.002

all others

FIGURE 6.122a,b July local surplus (*a, scale left:* average supply minus demand in W m⁻² is shown if positive) and deficit (*b, scale right:* average demand minus supply in W m⁻² is shown if positive) of electricity and biofuels for stationary use (such as biogas or hydrogen) relative to demand, on an area basis, according to the 2050 scenario.

Energy flow
W m⁻²

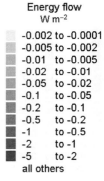

-0.002 to -0.0001
-0.005 to -0.002
-0.01 to -0.005
-0.02 to -0.01
-0.05 to -0.02
-0.1 to -0.05
-0.2 to -0.1
-0.5 to -0.2
-1 to -0.5
-2 to -1
-5 to -2

all others

(b)

(a)

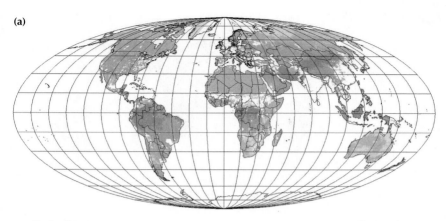

Energy flow
W m⁻²

5	to	10
2	to	5
1	to	2
0.5	to	1
0.2	to	0.5
0.1	to	0.2
0.05	to	0.1
0.02	to	0.05
0.01	to	0.02
0.005	to	0.01
0.002	to	0.005
0.0001	to	0.002
all others		

Energy flow
W m⁻²

-0.002	to	-0.0001
-0.005	to	-0.002
-0.01	to	-0.005
-0.02	to	-0.01
-0.05	to	-0.02
-0.1	to	-0.05
-0.2	to	-0.1
-0.5	to	-0.2
-1	to	-0.5
-2	to	-1
-5	to	-2
all others		

FIGURE 6.123a,b October local surplus (*a, scale left:* average supply minus demand in W m⁻² is shown if positive) and deficit (*b, scale right:* average demand minus supply in W m⁻² is shown if positive) of electricity and biofuels for stationary use (such as biogas or hydrogen) relative to demand, on an area basis, according to the 2050 scenario.

(b)

(a)

Energy flow
W m⁻²

5	to 10
2	to 5
1	to 2
0.5	to 1
0.2	to 0.5
0.1	to 0.2
0.05	to 0.1
0.02	to 0.05
0.01	to 0.02
0.005	to 0.01
0.002	to 0.005
0.0001 to	0.002

all others

FIGURE 6.124a,b   Annual average local surplus (*a, scale left:* average supply minus demand in W m⁻² is shown if positive) and deficit (*b, scale right:* average demand minus supply in W m⁻² is shown if positive) of electricity and biofuels for stationary use (such as biogas or hydrogen) relative to demand, on an area basis, according to the 2050 scenario.

Energy flow
W m⁻²

-0.002 to	-0.0001
-0.005 to	-0.002
-0.01 to	-0.005
-0.02 to	-0.01
-0.05 to	-0.02
-0.1 to	-0.05
-0.2 to	-0.1
-0.5 to	-0.2
-1	to -0.5
-2	to -1
-5	to -2

all others

(b)

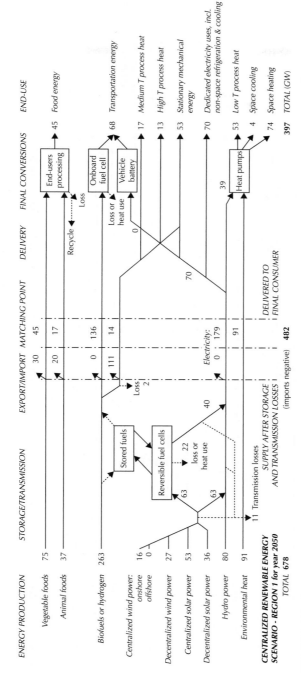

FIGURE 6.125 Region 1, 2050 scenario, with indication of import and export of energy (the unit used for energy flows is GW or GWy y^{-1}).

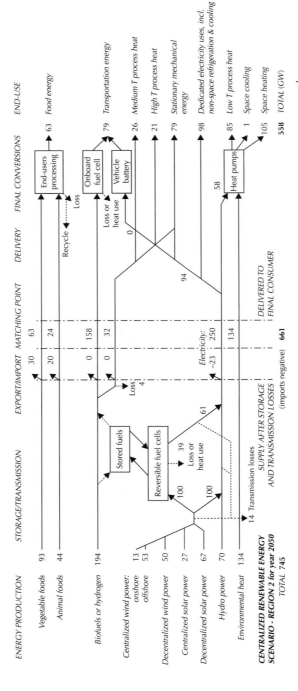

FIGURE 6.126 Region 2, 2050 scenario, with indication of import and export of energy (the unit used for energy flows is GW or GWy y⁻¹).

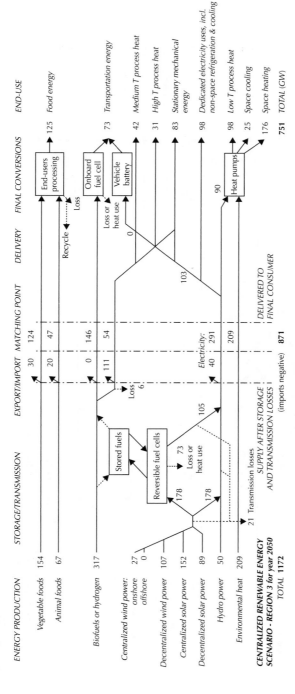

FIGURE 6.127   Region 3, 2050 scenario, with indication of import and export of energy (the unit used for energy flows is GW or GWy y^{-1}).

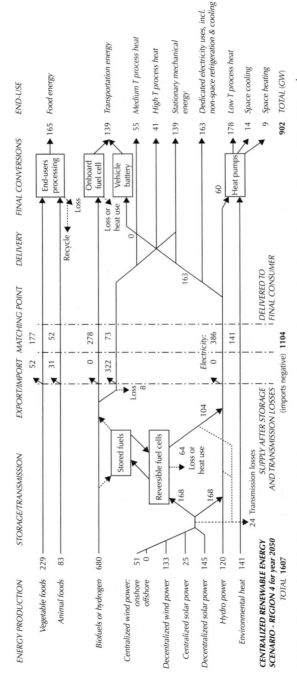

FIGURE 6.128 Region 4, 2050 scenario, with indication of import and export of energy (the unit used for energy flows is GW or GWy y⁻¹).

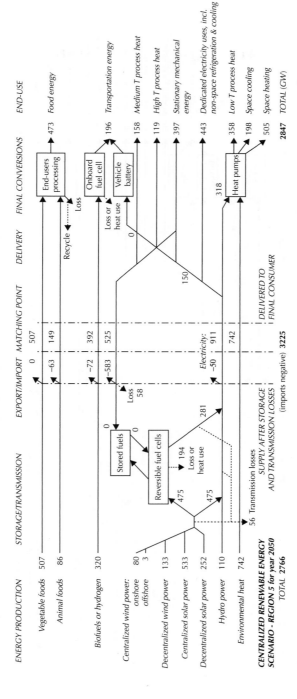

FIGURE 6.129 Region 5, 2050 scenario, with indication of import and export of energy (the unit used for energy flows is GW or GWy y⁻¹).

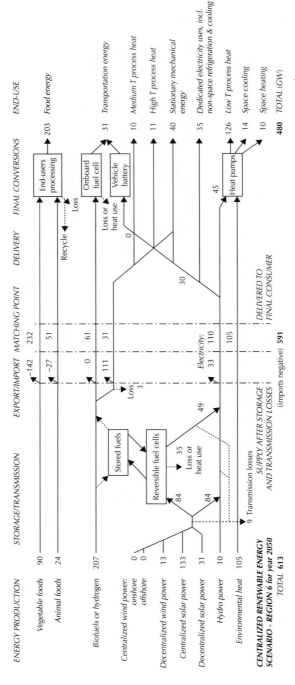

FIGURE 6.130   Region 6, 2050 scenario, with indication of import and export of energy (the unit used for energy flows is GW or GWy y^{-1}).

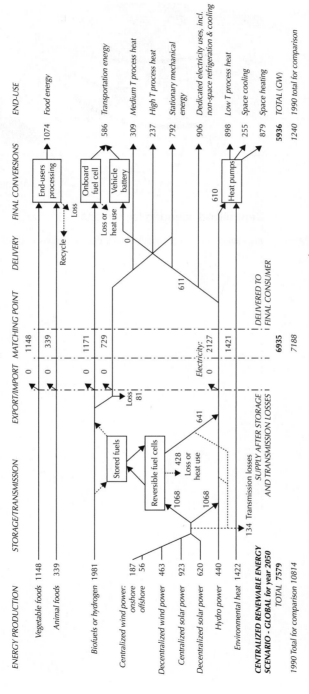

FIGURE 6.131 Overview of the 2050 scenario (the unit used for energy flows is GW or GWy y^{-1}).

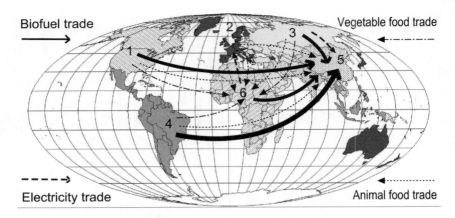

Centralized scenario for regions 1-6

FIGURE 6.132   Regional patterns of energy trade between the six regions in the 2050 scenario.

Chapter 7

Socioeconomic Assessment

7.1 SOCIAL AND ECONOMIC FRAMEWORK

The choice of energy systems, like any other feature of major social organizations, is a product of historical development, of prevailing social value systems, and often of the role of influential individuals from various layers of society: political decision-makers, industrialists, or intellectual figures.

In this chapter, the tools available for analysis and comparative assessment of energy systems are presented. Because they are not independent of social preferences, one should expect to find that different tools are available, catering to different positions in the social debate. However, when presented systematically, the effects of differences in assumptions become transparent, and the outcome is to show how different results are due not just to uncertainty but to specific differences in underlying choices. This also forces the users of these methods to specify their normative positions.

The chapter presents both simple methods for including the direct economic impacts of an energy system and more advanced methods that include, or at least keep track of, indirect and non-monetary impacts.

7.1.1 Social Values and the Introduction of Monetary Economy

Social value systems are by no means laws of nature. They differ from one culture to another, and they change with time. They can be influenced substantially by conscious or unconscious propaganda (from teachers, advertisements, spiritual or political indoctrinators, etc.) and by willingness to imitate (thus making it possible for value changes to be caused by small groups within society, e.g., the rich, the eloquent, the pious, or the well-educated).

However, more fundamental social values, associated with basic needs, are determined by the biology of human beings and thus are less subject to modification. In addition, there are values increasingly linked to the type of society in question. Some such values may be basic, but, owing to the adaptability of human beings, conditions that violate these values or "needs" may persist without apparent dissatisfaction, at least not on a conscious level. Examples of social values of this type are human interaction in a broad sense, meaningful activities ("work"), and stimulating physical surroundings.

This view of basic values is used in building the demand scenarios for energy described in section 6.2. The present world is, to a large extent, characterized by widespread deprivation of even basic social needs: people suffer from hunger, inadequate shelter, and ill health; they are offered unsatisfying work instead of meaningful activities; and they are surrounded by a polluted environment, ugly varieties of urbanization, etc.

Current societies are either managed for the benefit of a dictator or an oligarchy, or they herald democracy but then emphasize the single goal of maximizing the gross national product, which is then effectively taken to represent all the desired needs in a single number (see, for example, Mishan, 1969; *gross national product* is defined as the added value of all activities or transactions that can be expressed in monetary terms). It is then assumed that the "economic growth" represented by an increasing gross national product improves social values, such as access to food, housing, and a range of consumer goods, the production and sale of which form the basis for economic growth. Measures aimed at reducing the side-effects of production and material consumption are sometimes implemented, causing definite improvements in environmental quality, health, and working conditions, but in the end are often subordinated to the primary goal of economic growth.

The emphasis on economic figures is associated with the use of monetary units in discussions of social issues, with the attendant risk that needs that cannot be assigned a monetary value will be ignored.

In principle, any measure of value could be used as a foundation for economic relations between people and nations. The important thing is the way in which monetary values are assigned and used. Historically, the price of products has been influenced by the bargaining strength of sellers and buyers, constituting a *market*. Even intangible quantities, such as currency exchange rates, have been allowed to be determined by markets. One view is that this paves the way for less-developed economies, because low valuation of their products and labor will make them more competitive, until they "catch up." In practice, this unfortunately also means accepting environmentally flawed production methods in developing economies. A framework for more appropriate valuation of products and services is life-cycle assessment, which includes indirect effects. Life-cycle assessment allows the assignment of monetary values to environmental impacts and depletion of resources in a manner consistent with their long-term social value. These challenges are taken up in section 7.3 on life-cycle analysis, after a miniature crash-course in basic economic theory and remarks on governance at different levels of social organization.

7.1.2 Economic Theory

Economic theories propose to identify quantitative relations between variables describing economic behavior in society, based on observations of the past. Only part of society is amenable to such quantification, and assignment of value is not an objective enterprise. The normative positions of economists range from believing that everything can be monetized to considerably more modest claims. Because external conditions, technology, knowledge, and preferences can all change with time, economic rules extracted from past experience are not necessarily valid for the future, and economic theory therefore cannot be used for guiding national or international planning and policy, because they depend on perceptions of future conditions. However, economic theory is often used to make precisely such forecasts. They are adequately termed "business-as-usual" forecasts, because they describe what would happen only if political decisions, technical innovation, lifestyles, or natural phenomena don't deviate from those of the past.

Economic theories usually deal with a homogeneous society, often a nation, containing a private business sector as well as a public service sector. There is clearly a need to consider interactions between such individual "economies" through foreign trade and other contacts, which is another subject for economic studies. Simple economic theories neglect the effect of the size of individual enterprises and the size of each production unit, both of which may have an influence on the time required for changing the level of production, as well as on prices.

Economic theories may be regarded as tools for answering "if–then" questions on the basis of past experience. The parameters used in formulating economic theories are obtained from short- or long-term data series for past societies. The parameters could reflect past changes by incorporating trends in a functional form, and the theory would then predict a dynamic change constituting a nonlinear extrapolation of history.

Future conditions are fundamentally uncertain for two reasons. One is the possibility of changing attitudes (e.g., to environmental quality) and political prescriptions (which are the result of "free will"); the other is forced changes in conditions (resource depletion, appearance of new diseases, climate variations causing new ice ages, etc.). The political task of planning therefore essentially requires techniques for "planning under uncertainty."

A number of features of an economy are highly dependent on the organization of society. The presence of an influential public sector is characteristic of a substantial number of societies today, but not long ago the industrial sectors in some countries were largely unaffected by government, and the rapid growth of industrialization in the Western world about 200 years ago (and somewhat later in several other regions) was influenced by the transition from feudal societies, with local concentrations of power, to nations with increasing amounts of power concentrated in the hands of a central government. In the industrial sector, an increasing division has taken place between the economy of the enterprise itself and that of the owners (stockholders or the state) and the employees. On the other hand, similar divisions used to be rare in agricultural and small-scale commercial enterprises (farms, family shops, etc.).

There are important formal differences between the ways that current economic theories view the production process. In a socialist economy, the power to produce goods is entirely attributed to the workers, i.e., only labor can produce a "surplus value" or profit, and machinery and raw materials are expenses on the same footing. In a capitalist economy, machinery and other equipment are considered capable of producing a profit, implying that if one machine operated by one worker is producing the same output as 100 workers, then 99% of the profit is attributed to the machine, in contrast to the socialist economy, in which all of the profit would be attributed to the single worker operating the machine—formally increasing his productivity a hundred-fold from that before the machines were introduced. Alternatively formulated, in a capitalist economy, only capital is capable of producing profit, and labor is on the same footing as raw materials and machinery: commodities bought at the lowest possible prices by an optimal allocation of capital to these expenses.

It is, of course, only of theoretical interest to discuss which of the production factors are assigned the power to produce profit, but it is of decisive practical interest to determine how the profits are actually shared between labor and capital, i.e., how the profits of a given enterprise are distributed among workers and employees on one side and owners (capital providers) on the other side. As in socialist economies, the state may be considered to collectively represent the workers and may administer some of (or all of) their profit shares, but, in many non-socialist economies, public spending includes a number of direct or indirect supports to the capitalists, as well as public goods to be shared by all members of society, and a certain amount of recycling of capital back to the enterprises. In mixed economies, which have developed from the basic capitalist ones, the net distribution of profits may involve one portion's being pre-distributed to benefit society as a whole, with only the residual profit placed at the disposal of the capitalists.

The *long-term prices* of goods produced in the industrial sector are, in most cases today, determined by the full-cost rule or mark-up procedure, counting the total production costs and adding a standard profit (percentage). In order to change prices, changes in production method, wages, or costs of raw materials and duties must take place, or the *fixprice* method may be temporarily abandoned. In a fixprice economy, equalization of output (supply) and demand is accomplished by regulating output up or down when supply falls short of, or exceeds, demand.

In other sectors, such as the agricultural sector, profit often declines with increasing production, and pricing is not done by the mark-up procedure, but instead according to a *flexprice* method, in which equalization of supply and demand is achieved by putting the price up if demand exceeds supply and down if supply exceeds demand. The flexprice system is also in use (in capitalist economies) for determining *short-term* price variations of industrial products.

In a socialist economy, it is possible to maintain fixed prices on essential goods, such as food, despite fluctuations in the cost of the raw materials due to resource scarcity or abundance, by shifting funds from other sectors (e.g., to import food, if local production fails), but such procedures are, of course, only applicable if the scarce resources are physically available at a higher cost. The danger of this approach is that it invites inefficient use of assets.

Production Planning

The organization of the production processes in a number of interrelated industries that buy and supply materials and goods among themselves may be described in terms of a *production function* Y (see, for example, Morishima, 1976), giving a measure of net total output,

$$Y = Y(\{M_i, L_i, R_i\}).$$

Here M_i, L_i, and R_i stand for the machinery, labor, and raw materials required by the ith process, and the net output z_j of one of the goods entering somewhere in the hierarchy of production processes may be assumed to be given by a linear expression,

$$z_j = \sum_i b_{ji} x_i, \tag{7.1}$$

where x_i is the rate of the ith process and $\{b_{ji}\}$ is a set of coefficients. The entire collection of process rates, $\{x_i\}$, constitutes a *production plan*. It is seen that the conglomerate production function Y involves a weighting of the individual goods (index j), which will normally be described in different units. Equation (7.1) describes both inputs and outputs of goods, prescribing that those z_j that are inputs are given a negative sign and those that are outputs are given a positive sign.

A production plan aiming at maximizing the total output, as specified by Y, for given inputs, makes little sense, unless the weighting of different goods is properly chosen. The choice of this weighting may be taken to transform the measure of output into a measure of profit. If the monetary value (price) assigned to each good z_j is denoted p_j (in monetary units per unit quantity), then the net profit, y, may be written

$$y = \sum_j p_j z_j. \tag{7.2}$$

The production plan $\{x_i\}$ may now be chosen, through (7.1) and (7.2), so as to maximize the total profit y. The method may be generalized to include all sectors of society, including the public sector.

Allocation of the net total profit to the individual processes that may be performed by different enterprises (with different owners in a capitalist economy) presents another problem. A "rational" (but not necessarily socially acceptable) distribution might be to return the total profit to the individual processes, in the same ratios as the contributions of the processes to the total profit. The prices of the individual goods defined in this way are called *shadow prices*. Thus, if all the prices p_j appearing in (7.2) are taken to be the shadow prices, then the net total profit y becomes zero. In practice, this would rarely be the case, and even if individual prices within the production hierarchy are at one time identical to the shadow prices, then changes in, for example, raw material costs or improved methods in some of the processes will tend to make y non-zero and will lead to frequent adjustments of shadow prices.

This behavior emphasizes the static nature of the above methodology. No dynamic development of production and demand is included, and changes over time can be dealt with only in a "quasi-static" manner by evaluating the optimal production plan, the pricing, and the profit distribution each time the external conditions change and by neglecting time delays in implementing the planning modifications. Dynamic descriptions of the time development of non-static economies have been formulated in qualitative terms, e.g., in the historical analysis of Marx (1859) and in the approach of Schumpeter (1961).

The quasi-static theory is particularly unsuited for describing economies with a long planning horizon and considerations of resource depletion or environmental quality. By assigning a monetary value to such *externalities* or *external diseconomies*, a dynamic simulation calculation can be carried out, defining the basic economic rules and including time delays in implementing changes, delayed health effects from pollution, and the time constants for significant depletion of nonrenewable resources, either physical depletion or depletion of raw materials in a given price range. Rather than planning based on instantaneous values of profit, planning would be based on suitably weighted sums of profits during the entire length of the planning horizon (see, for example, Sørensen, 1976a). The determination of values to be ascribed to externality costs is discussed in section 7.3 on life-cycle analysis.

Furthermore, the applicability of the simulation approach is limited because of the uncertainty inherent in time integration over very long time intervals, using economic parameters that are certainly going to be modified during the course of time or replaced by completely different concepts in response to changes in external conditions as well as major changes in the organization and value systems of societies. Examples of problems that defy description in terms of conventional economic theories with use of simulation techniques may be the socioeconomic costs of possible climatic impact from human (profit-seeking) activities, the global contamination resulting from nuclear warfare, and the more subtle aspects of passing radioactive waste to future generations. Such problems, it would seem, should be discussed in a general framework of social values, including the values associated with not restricting the possibilities for future generations to make their own choices by restricting their access to natural resources or by forcing them to spend a considerable part of their time on our leftover problems rather than on their own.

The quantification of, for example, environmental values and their inclusion in economic planning still suffer from the lack of a suitable theoretical framework. It has been suggested that contingent evaluation should be employed, implying that funds are allocated to environmental purposes (reducing pollution from given production processes, setting aside land for recreational purposes, etc.) to an extent determined by the willingness of people (salary earners) to accept the reduction in salary that corresponds to the different allocation of a given amount of profit (see, for example, Hjalte *et al.*, 1977; Freeman *et al.*, 1973). It is clearly difficult to include long-term considerations in this type of decision process. In many cases, the adverse environmental effects of present activities will not show up for many years, or our knowledge is insufficient to justify statements that adverse effects will indeed occur. Only recently have public debates about such issues started to take place before the activity in question has begun. Thus, the discussion has often been about halting an ongoing activity for environmental purposes, and the argument against doing so has been loss of investments already made. It is therefore important to evaluate environmental problems of production or consumption in the planning phase, which implies that, if environmental costs can be evaluated, they can be directly included in the planning prescription [such as (7.1) and (7.2)] as a cost of production on the same level as the raw materials and machinery (Sørensen, 1976a).

Another method that is in fairly widespread use in present economies is to implement government regulations, often in the form of threshold values for pollution levels (emissions, concentrations, etc.) that should not be exceeded. If, then, for a given production process, the emission of pollutants is proportional to the output, the regulation implies a government-prescribed ceiling on the output z_j of a given product in (7.2). This is similar to the ceilings that could be prescribed in cases of a scarce resource (e.g., cereal after a year of poor harvest) or other rationing requirement (e.g., during wartime). Such an environmental policy will stimulate development of new production processes with less pollution per unit of output, so that the output of goods may still be increased without violating the pollution limits.

The problem is, however, that government-enforced limits on pollution are usually made stricter with time, because they were initially chosen too leniently and because additional long-term adverse effects keep appearing (an example is radiation exposure standards). Thus, the socioeconomic effects of environmental offenses are constantly underestimated in this approach, and industries and consumers are indirectly stimulated to pollute up to the maximum permitted at any given time. By contrast, the approach whereby environmental impact must be considered earlier, in the production-planning phase, and whereby governments may determine the cost of polluting to be at such a high value that long-term effects are also included, has the effect of encouraging firms and individuals to keep pollution at the lowest possible level.

Distribution Problems

Classical market theory assumes completely free competition between a large number of small firms. There are no monopolies and no government intervention in the economy. Extension of this theory to include, for example, a public sector and other features of present societies, without changing the basic assumption, is called *neoclassical theory*. According to this view, the distribution of production on the firms, and the distribution of demand on the goods produced, may attain equilibrium, with balanced prices and wages determined from the flexprice mechanism (this is called a *Walras equilibrium*).

In neoclassical theory, it is assumed that unemployment is a temporary phenomenon that occurs when the economy is not in equilibrium. Based on observations in times of actual economic crises occurring in capitalist countries, Keynes (1936) suggested that in reality this was not so, for a number of reasons. Many sectors of the "late capitalist" economies were of the fixprice category rather than the flexprice one, wages had a tendency to be rigid against downward adjustment, the price mechanism was not functioning effectively (e.g., owing to the presence of monopolies), and the relation between prices and effective demand was therefore not following the Walras equilibrium law. Since prices are rigid in the fixprice economy, full employment can be obtained only when demand and investments are sufficiently high, and Keynes suggested that governments could help stimulate demand by lowering taxes and that they could help increase economic activity by increasing public spending (causing a "multiplier effect" due to increased demand on consumer goods) or by making it more attractive (for capitalists) to invest their money in means of production. The distinction between consumer goods and capital goods is important, since unemployment may exist together with a high, unsatisfied demand for consumer goods if capital investments are insufficient (which again is connected with the expected rates of interest and inflation). This is a feature of the recent (2008+) financial crisis, which was caused by allowing nonproductive sectors, such as the financial sector, to attract a very high fraction of the total investments in society (in itself detrimental) and then to grossly mismanage the assets.

In a socialist economy, allocation of resources and investments in means of production is made according to an overall plan (formulated collectively or by representatives), whereas in a capitalist society the members of the capitalist class make individual decisions on the size of investment and the types of production in which to invest. Then it may well be that the highest profit is in the production of goods that are undesirable to society but are still demanded by customers if sales efforts are backed by aggressive advertising campaigns. It is in any case not contradictory that there may be a demand for such goods, if more desirable goods are not offered or are too expensive for most consumers. Yet relatively little is done, even in present mixed-economy societies, to influence the quality of the use of capital, although governments do try to influence the level of investments.

Unemployment is by definition absent in a socialist society. In a capitalist society, the effect of introducing new technology that can replace labor is often to create unemployment. For this reason, new technology is often fiercely opposed by workers and unions, even if the benefits of the machinery include relieving workers from the hardest work, avoiding direct human contact with dangerous or toxic substances, etc. In such cases, the rules of capitalist economy clearly oppose improvements in social values.

Actual Pricing Policies

The mechanism of price adjustments actually used in a given economy is basically a matter of convention, i.e., it cannot be predicted by a single theory, but theories can be formulated that reflect some features of the price mechanism actually found in a given society at a given time. A mechanism characteristic of present growth-oriented economies (both socialist and capitalist) is to adjust prices in such a way that excess profit per unit of good is reduced, whenever an excess demand exists or can be created, in order to be able to increase output (Keynes–Leontief mechanism).

Important parameters in an empirical description of these relations are the *price flexibility*,

$$\eta_i = \frac{\Delta p_i/p_i}{\delta d_i/d_i},\tag{7.3}$$

and the demand elasticity,

$$\varepsilon_i = -\frac{\Delta d_i/d_i}{\Delta p_i/p_i},\tag{7.4}$$

of the ith good or sector. Here p_i and d_i represent price and demand, and δd_i is an exogenous change in demand, whereas Δd_i is the change in demand resulting from a change in price from p_i to $p_i + \Delta p_i$. The change in output, Δz_j, being indicated by an exogenously produced change in demand, Δd_i, is

$$\Delta z_i = \Delta d_i + \delta d_i = (1 - \eta_i \varepsilon_i)\delta d_i.\tag{7.5}$$

The use of the indicators η_i and d_i for predicting the results of changing prices or demand is equivalent to a quasi-static theory, whereas a dynamic theory would require η_i and ε_i to be dynamic variables coupled to all other parameters describing the economy through a set of nonlinear, coupled differential equations (i.e., including feedback mechanisms left out in simple extrapolative approaches).

The large number of assumptions, mostly of historical origin, underlying the present theoretical description of pricing policy (actual or "ideal") does suggest that comparative cost evaluations may constitute an inadequate or at least incomplete basis for making major decisions with social implications. Indeed, many important political and business decisions are made in bold disregard of economic theory. This is relevant for the discussion of energy systems. A statement like, one system is 10% cheaper than another one, would be seen in a new light in cases where the combination of ingredients (raw materials, labor, environmental impact, availability of capital) determining the prices of the two systems are different. There might be a choice between changing the production process to make a desired system competitive or, alternatively, making society change the rules by which it chooses to weight the cost of different production factors.

7.1.3 Direct Cost and Inflation

In many economies, changes in prices have taken place that neither are associated with changes in the value of the goods in question nor reflect increased difficulties in obtaining raw materials, etc. Such changes may be called inflation (negative inflation is also called deflation). If they affect all goods, they merely amount to changing the unit of monetary measure and thus would have no effect on the economy, were it not for the fact that the rate of inflation cannot be precisely predicted. In negotiated labor markets, inflation is a way to make up for labor cost increases considered too high by industry. However, people and institutions dealing with money cannot simply correct their interest rate for inflation, but have to estimate the likely inflation over an entire investment depreciation time in order, for example, to offer a loan on fixed interest terms.

In order to reduce the chance of losing money due to an incorrect inflation estimate, it is likely that the interest rate demanded will be higher than necessary, which again has a profound influence on the economy. It makes the relative cost of capital as a production factor too high, and for a (private or public) customer considering a choice between alternative purchases (e.g., of energy systems performing the same task), a bias in the direction of favoring the least capital-intensive system results. Governments in many capitalist countries try to compensate (and sometimes overcompensate) for the adverse effects of unreasonably high interest rates by offering tax credits for interest paid.

Often inflation is taken to have a somewhat extended meaning. Individual nations publish indices of prices, giving the relative change in the cost of labor, raw materials, etc., with time. Each of the indices is derived as a weighted average over different sectors or commodities. To the extent that the composition of these averages represents the same quantity, the price indices thus include not only inflation, as defined above, but also real changes (e.g., due to increased difficulty of extracting nonrenewable resources, increased living standard, etc.). If the rate of inflation is taken to include all price changes, then the use of the price indices offers one way of comparing the cost of goods produced at different times by correcting for inflation in the extended sense, i.e., by transforming all costs to monetary values pertaining to a fixed point in time.

If the annual inflation of prices relevant to a given problem is denoted i (fraction per year), and the market interest is r_m (fraction per year), then the effective interest rate r is given by

$$(1+r) = (1+r_m)/(1+i). \tag{7.6}$$

Other production costs, such as the cost of raw materials, machinery, and labor, may be similarly corrected, and if (7.6) and the analogous relations are applied repeatedly to transform all costs to the levels of a given year, then the calculation is said to be "in fixed prices."

The difficulty in carrying out a fixed price calculation is again that the inflation rate i may differ from year to year, and it has to be estimated for future years if the calculation involves future interest payments, fuels, or working expenses, such as operation and maintenance of equipment.

If all changes in prices from year to year are small, (7.6) may be approximated by $r \approx r_m - i$.

7.1.4 Interest and Present Value

Private and Social Interest Rate, Intergenerational Interest

The concept of interest is associated with placing a higher value on current possession of assets than on promised future possession. Using the monetary value system, a positive interest rate thus reflects the assumption that it is better to possess a certain sum of money today than it is to be assured of having it in 1 year or in 20 years from today. The numerical value of the interest rate indicates how much better it is to have the money today, and if the interest has been corrected for the expected inflation, it is a pure measure of the advantage given to the present.

For the individual, who has a limited life span, this attitude seems very reasonable. But for a society, it is difficult to see a justification for rating the present higher than the future. New generations of people will constitute the society in the future, and placing a positive interest on the measures of value implies that the "present value" (7.7 below) of an expense to be paid in the future is lower than that of one to be paid now. This has a number of implications.

Nonrenewable resources left for use by future generations are ascribed a value relatively smaller than those used immediately. This places an energy system requiring an amount of nonrenewable fuels to be converted every year in a more favorable position than a renewable energy system demanding a large capital investment now.

Part of the potential pollution created by fuel-based energy conversion, as well as by other industrial processes, is in the form of wastes, which may be either treated (changed into more acceptable substances), diluted, and spread in the environment, or concentrated and kept, i.e., stored, either for good or for later treatment, assuming that future generations will develop more advanced and appropriate methods of treatment or final disposal. A positive interest rate tends to make it more attractive, when possible, to leave the wastes to future management, because the present value of even a realistic evaluation of the costs of future treatment is low.

For instance, any finite cost of "cleaning-up" some environmental side-effect of a technology (e.g., nuclear waste) can be accommodated as a negligible increase in the capital cost of the technology, if the clean-up can be postponed sufficiently far into the future. This is because the clean-up costs will be reduced by the factor $(1 + r)^n$ if the payment can wait n years (cf. the discussion in section 7.1.2 of proposed economic theories for dealing with environmental side-effects, in connection with "production planning").

It may then be suggested that a zero interest rate be used in matters of long-term planning seen in the interest of society as a whole. The social values of future generations, whatever they may be, would, in this way, be given a weight equal to those of the present society. In any case, changing from the market interest rate to one corrected for inflation represents a considerable step away from the interest (in both senses) of private capitalist enterprises and toward that of society as a whole. As mentioned above, the interest rate corrected for inflation according to historical values of market interest and inflation, and averaged over several decades, amounts to a few percent per year. However, when considering present or presently planned activities that could have an impact several generations into the future (e.g., resource depletion, environmental impact, including possible climatic changes), then even an interest level of a few percent would appear to be too high to take into proper consideration the welfare of future inhabitants of Earth. For such activities, the use of a social interest rate equal to zero in present value calculations does provide one way of giving the future a fair weight in the discussions underlying present planning decisions.

One might argue that two facts speak in favor of a positive social interest rate: (i) the capital stock, in terms of buildings and other physical structures, that society passes to future generations is generally increasing with time, and (ii) technological progress may lead to future solutions that are better and cheaper than those we would use to deal with today's problems, such as waste disposal. The latter point is the reason why decommissioned nuclear reactors and high-level waste are left to future generations to deal with. However, there are also reasons in favor of a negative social interest rate: environmental standards have become increasingly more strict with time, and regulatory limits for injection of pollutants into both humans (through food) and the environment have been lowered as function of time, often as a result of new knowledge revealing negative effects of substances previously considered safe. The problems left to the future may thus appear larger then, causing a negative contribution to social interest, as it will be more expensive to solve the problems according to future standards. The use of a zero social interest rate may well be the best choice, given the uncertainties.

However, it is one thing to use a zero interest rate in the present value calculations employed to compare alternative future strategies, but quite another thing to obtain capital for the planned investments. If the capital is not available and has to be procured by loans, it is unlikely that the loans will be offered at zero interest rate. This applies to private individuals in a capitalist society and to nations accustomed to financing certain demanding projects by foreign loans.

Such borrowing is feasible only if surpluses of capital are available in some economies and if excess demand for capital is available in others. Assuming that this is not the case, or that a certain region or nation is determined to avoid dependence on outside economies, then the question of financing is simply one of allocation of the capital assets created within the economy.

If a limited amount of capital is available to such a society, it does not matter if the actual interest level is zero or not. In any case, the choices between projects involving capital investments, which cannot all be carried out because of the limited available capital, can then be made on the basis of zero interest comparisons. This makes it possible to maximize long-term social values, at the expense of internal interest for individual enterprises. Social or "state" intervention is necessary in order to ensure a capital distribution according to social interest comparisons, rather than according to classical economic thinking, in which scarcity of money resources is directly reflected in a higher "cost of money," i.e., positive interest rate.

The opposite situation, that too much capital is available, would lead to a low interest rate, as has been seen during the first decade of the 21st century and is still prevailing. A low short-term market interest signals that an insufficient number of quality proposals for investment are being generated in society, and this may explain (but not justify) why financial institutions during this period made unsecured loans to unworthy purposes, such as excess luxury consumption and substandard industrial or commercial product ideas. Declining quality of education in the richest countries may be invoked as an explanation for the lack of worthy new investment ideas, or it may be a result of the many technological advances during the preceding decades having generated larger-than-usual profits, i.e., capital to be reinvested.

Net Present Value Calculation

The costs of energy conversion may be considered a sum of the capital costs of the conversion devices, operation and maintenance costs, plus fuel costs if there are any (e.g., conversion of nonrenewable fuels or commercially available renewable fuels, such as wood and other biomass products). Some of these expenses have to be paid throughout the operating lifetime of the equipment and some have to be paid during the construction period, or the money for the capital costs may be borrowed and paid back in instalments.

Thus, different energy systems would be characterized by different distributions of payments throughout the operating lifetime or at least during a conventionally chosen depreciation time. In order to compare such costs, they must be referred to a common point in time or otherwise made commensurable. A series of yearly payments may be referred to a given point in time by means of the *present value* concept. The present value is defined by

$$P = \sum_{i=1}^{n} p_i (1+r)^{-1}, \tag{7.7}$$

where $\{p_i\}$ are the annual payments and n is the number of years considered. The interest rate r is assumed to be constant and is ascribed only once every year (rather than continuously), according to present practice. The interest is paid *post numerando*, i.e., the first payment of interest would be after the first year, $i = 1$. Thus, P is simply the amount of money that one must have at year zero in order to be able to make the required payments for each of the following years, from the original sum plus the accumulated interest earned, if all remaining capital is earning interest at the annual rate r.

If the payments p_i are equal during all the years, (7.7) reduces to

$$P = p_{(i)} + (1+r)^{-1}((1+r)^{-n}-1)/((1+r)^{-1}-1),$$

and if the payments increase from year to year, by the same percentage, $p_i = (1 + e)p_{i-1}$ for $i = 2, \dots, n$, then the present value is

$$P = p_0(1+r)^{-1}\frac{((1+e)/(1+r))^n-1}{(1+e)/(1+r)-1}. \tag{7.8}$$

Here e is the rate of price escalation (above the inflation rate if the average inflation is corrected for in the choice of interest r, etc.) and p_0 is the (fictive) payment at year zero, corresponding to a first payment at year one equal to $p_0(1 + e)$.

The present value of capital cost C paid at year zero is $P = C$, whether the sum is paid in cash or as n annual instalments based on the same interest rate as used in the definition of the present value. For an annuity-type loan, the total annual instalment A (interest plus back-payment) is constant and given by

$$A = C(1+r)\frac{(1+r)^{-1}-1}{(1+r)^{-n}-1}, \tag{7.9}$$

which, combined with the above expression for the present value, proves that $P = C$. When r approaches zero, the annuity A approaches C/n.

The present value of running expenses, such as fuels or operation/maintenance, is given by (7.7), or by (7.8), if cost increases yearly by a constant fraction. It is usually necessary to assign different escalation rates e for different fuels and other raw materials; for labor of local origin, skilled or unskilled; and for imported parts or other special services. Only rarely will the mix of ingredients behave like the mix of goods and services defining the average rate of inflation or price index (in this case, prices will follow inflation and the escalation will be zero). Thus, for a given energy system, the present value will normally consist of a capital cost C plus a number of terms that are of the form (7.7) or (7.8) and that generally depend on future prices that can at best be estimated on the basis of existing economic planning and expectations.

7.1.5 Cost Profiles and Break-even Prices

If successive present values of all expenditures in connection with, say, an energy system are added from year zero to n, an *accumulated present value* is obtained. As a function of n, it increases monotonically. It is illustrated in Fig. 7.1 by two different systems. One is a system in which the only expenses are fuels and the related handling. Its accumulated present value starts at zero and increases along a curve that bends downward owing to the effect of the finite interest level in diminishing the present value of future payments. If fuel prices increase, the curve shows an overall tendency to bend upward. The other system, which may be thought of as a renewable energy conversion system, has a high initial present value reflecting the capital cost of the conversion equipment, but no fuel costs, and therefore only a slightly increasing accumulated present value due to working costs.

The accumulated present value offers a way of comparing the costs of different energy supply systems over a period of time. Curves like those of Fig. 7.1 give a measure of total costs in commensurable units, as a function of the time horizon of the comparison. If the planning horizon is taken as 25 years (e.g., the life expected for the conversion equipment), then the total cost of the renewable energy system in the figure is lower than that of buying fuel to produce the same amount of energy, whereas if the planning horizon is set at 15 years (the standard depreciation period adopted by some utility companies), then there is roughly a break-even system.

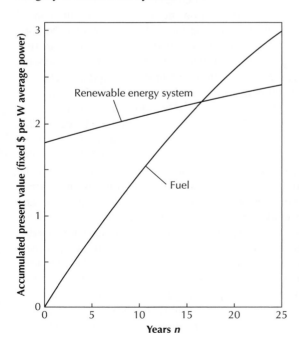

FIGURE 7.1 Accumulated present value to the year n of total costs for a renewable energy system and for converting fuel in an existing system, as a function of n. The monetary unit is denoted "$." The $-values indicated at the ordinate would correspond to typical wind turbine costs and typical coal prices as experienced during the late 20th century in Europe.

The accumulated present value may also be used to define the *break-even* capital cost of a new energy system, relative to an existing system. In this case, the depreciation time n used in expressions such as (7.8) is kept fixed, but the capital cost of the new system, C, is considered a variable. The break-even cost, $C(b.e.)$, is then defined as the lowest capital cost C for which the total present value cost (accumulated to the year n) of the new system is no greater than that of the existing system, including all relevant costs for both systems, such as the (known) capital cost of the existing system (if the new system is aimed at replacing its capacity), and the costs of fuels, operation, and maintenance. If the depreciation time selected for the comparison is taken as the physical lifetime of the equipment, situations may arise in which different periods n should be used for the two systems being compared.

If the construction period is extended, and systems with different construction times are to be compared, the distribution of capital costs over the construction period is simply treated as the running payments and is referred to the common year (year 0) by means of (7.7).

The example shown in Fig. 7.1 assumes a given capital cost C of the renewable energy system. For a depreciation period of 25 years, it is seen that the break-even cost relative to the fuel (that would be saved by installing the renewable energy system) is higher than the assumed cost, $C(b.e.) > C$, implying that the renewable system could be more expensive and still be competitive according to the break-even criterion. For n below 15 years, the cost C is above the break-even value, and the system's cost will have to be reduced in order to make the system competitive according to given assumptions regarding fuel cost, etc.

The production cost of energy depends on the distribution of the capital costs over the depreciation time. Whether the method of financing is by loan taking, by use of equity, or by a combination of the two, the cost can (but does not have to) be distributed over different years according to the scheme of annuity payments or according to a return demanded on capital assets.

Assume first that the capital is raised on the basis of an annuity loan or an equivalent scheme with annual instalments given by an expression of the form (7.9), where the interest r is to be taken as the market interest rate r. This implies that no correction for inflation is made and that the annual instalments A are of equal denomination in current monetary values. The left-hand side of Fig. 7.2 illustrates the energy production costs according to this scheme for the two alternatives considered in Fig. 7.1. The rate of inflation has been assumed to be 10% per year, the rate of interest corrected for inflation is 3% per year, and thus according to (7.6) the market interest $r \approx 13.3\%$ per year.

If these same energy costs are corrected for inflation, so that they correspond to values in fixed year-0 monetary units, the results are as shown in the right-hand side of Fig. 7.2 (solid lines). The fuel cost is constant in fixed prices (assumption), but the cost of the renewable energy declines rapidly, especially during the first few years (although it is increasing in current prices, as shown to the left).

This behavior illustrates a barrier that the annuity loan system poses against any capital-intensive system. Although the renewable energy system is competitive according to the accumulated present value criterion (when considered over 25 years, as in Fig. 7.1), the energy cost during the first year is about 50% higher than that of the fuel it replaces, and only after the sixth year does it become lower. From a private economy point of view, it may not be possible to afford the expenses of the first 5 years, although in the long run the system is the less expensive one.

The example also illustrates the fallacy of comparing costs of different (energy) systems by just quoting the payments (running plus instalments on loans) for the first year of operation, as is sometimes done.

If cost comparisons are to be made in terms of energy costs, an improved prescription is to quote the average cost during the entire depreciation period (or lifetime) considered. This is meaningful only if fixed prices are used, but then it is equivalent to the present value method, in the sense that a system that is competitive according to one criterion will also be according to the other.

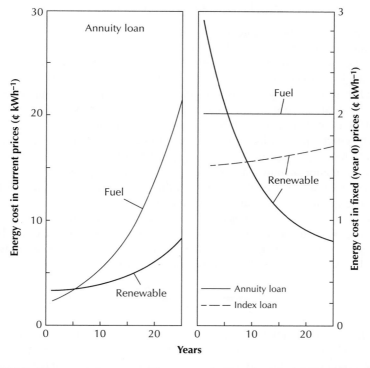

FIGURE 7.2 Energy costs for a renewable energy or fuel-based system (cf. Fig. 7.1), excluding cost of transmission. *Solid lines* correspond to financing through annuity loans, and the *dashed line* (for the renewable energy system) corresponds to financing by index loan, required to give the same 25-year accumulated present value of payments as the annuity loan. The left-hand part of the figure is in current prices, whereas the right-hand side is in fixed prices, assuming an annual rate of inflation equal to 10%. Note the difference in scale.

The creation of loan conditions that would eliminate the adverse effect of inflation on the back-payment profile has been considered (Hvelplund, 1975). The idea is to abandon the constancy (in current prices) of the annual instalments and instead modify each payment according to the current price index. If this procedure is followed in such a way that the accumulated present value of all the payments during the depreciation period equals that of the annuity loan, then both types of instalment are given by expression (7.9), but for the index loan, the interest r is to be corrected for inflation, and A is given in fixed year-0 prices. The index-regulated instalments in current prices are then

$$A_m(j) = (1+i)^j C(1+r)((1+r)^{-1}-1)/((1+r)^{-n}-1),$$

for the ith year, assuming the inflation to be constant and equal to the fraction i per year. For the annuity loan, the instalments in current prices are

$$A_m(j) = C(1+r_m)((1+r_m)^{-1}-1)/((1+r_m)^{-n}-1).$$

In fixed prices, both expressions should be divided by $(1+i)^j$. For the example considered in Fig. 7.2, both payment schemes are illustrated on the right-hand side, with the index loan given by a dashed line. It shows that the system with the lowest cost in terms of present value is also lowest in annual energy cost, for each of the years, and in fixed as well as in current prices. Price-index-regulated loans may not have not been used in practice, but a variety of variable-interest loans have been in fairly widespread use.

The energy prices discussed so far are to be considered production prices, for different types of financing. The price actually charged for energy is, of course, a matter of policy by the company or state selling the energy. They could choose to distribute the prices over the time interval in a manner different from the distribution of actual production costs.

In practical uses of the break-even technique for comparing energy systems, it is customary to add a sensitivity analysis, where the sensitivity of the break-even price, $C(b.e.)$, to variations in the ith parameter, π_i, may, when π_I is non-zero, be expressed as a relative sensitivity, s_i, defined by

$$s_i = \frac{\pi_i}{C(b.e.)} \frac{\partial C(b.e.)}{\partial \pi_i}. \tag{7.10}$$

Close to the reference parameters, this linear expression may be used (evaluated with the reference parameters inserted) to predict the dependence of the break-even price on the different parameters. During the early stages of considering wind as a novel energy source, after the 1973/1974 oil-supply crisis, such estimates played an important role (Sørensen, 1975c). Sophistications, such as the consideration of start-up costs if drops in wind production force utilities to employ fossil power plants on stand-by, were also considered (Sørensen, 1979,

2004a), and the procedures were extended to include the use of energy stores. In this case, the structure of the system cost C_{system} may be specified as

$$C_{system} = C_{converter} + C_{s,in} + C_s + C_{s,out},$$

where $C_{s,in}$ is the cost of the equipment transferring surplus energy to the storage (e.g., electrolysis unit for hydrogen storage), C_s is the cost of the storage itself (e.g., the hydrogen containers), and $C_{s,out}$ is the cost of retrieving the energy as electricity (e.g., fuel-cell unit in the hydrogen example). For some types of storage, the equipment for inserting and withdrawing energy from the storage is the same (the two-way turbines of a pumped hydro installation, the electro-motor/generator of a flywheel storage), and in some cases all three components may be physically combined (some battery types). If the three costs are distinct, only C_s will be expected to increase with the storage capacity. $C_{s,in}$ corresponds to a device rated at the maximum surplus power expected from the renewable energy converter, and $C_{s,out}$ corresponds to a device rated at the maximum power deficit that is expected between current wind power production and load.

Working entirely in prices referred to a fixed moment in time (i.e., corrected for inflation), the terms necessary for calculating the break-even price of a system with storage may be written

$$\frac{C_{system}(b.e.)}{8.76E_{system}^{net}} + \sum_i p_{system,i} N(n,r,e_i) = \sum_j \frac{A_j C_{alt,j}}{8.76E_{alt,j}^{net}} + \sum_j p_{alt,j} N(n,r,e_j). \quad (7.11)$$

The expression is designed to express every term in monetary value for the net production of one kWh of energy every year during the depreciation period of n years. In order to ensure this, the costs C must be in, say, dollars or euros for a given size system (e.g., expressed as m^2 swept by a wind energy generator or as kW rated power), and the energy E must be in watts of net power production for the same size system. The annual payments p must be in units of dollars (or other currency as used in the other terms) per kWh of net energy production. The subscript alt represents the alternative (e.g., fuel-based) system being compared with, and $p_{system,i}$ for the (renewable energy) system includes terms giving the cost of operation and maintenance of both converter and storage and terms describing eventual penalties in terms of increased costs of regulating the fuel-based back-up system, whereas the $p_{alt,j}$ on the alternative system side includes operation and maintenance related to both equipment and fuels, as well as terms giving the fuel costs themselves, for each of the units forming the alternative system. The capacity credit of the renewable energy system is split on each of the alternative units j, with a fraction A_j specifying the capacity of a particular unit (or unit type, such as base, intermediate, or peak unit), which is being replaced. The present value factor $N(n,r,e_i)$ is given as in the right-hand side of (7.8), where n is the depreciation time in years, r is the annual interest rate, and e_i is the annual price escalation above inflation, for the ith component of payments (fuels, raw materials, labor, etc.) (Sørensen, 1978a).

7.1.6 Indirect Economic Considerations

The term *indirect economics* may be understood to relate to those social values (costs or benefits) that are not, or cannot be, evaluated in monetary units. Life-cycle assessment (section 7.3) is an attempt to evaluate indirect influences from, for example, resource depletion or environmental impacts and, when possible, to quantify them.

The effort required in order to extract nonrenewable resources is generally expected to increase as easily accessible deposits become depleted and less and less accessible sources have to be exploited. This tendency is counteracted by the development of more ingenious extraction methods, but if effort is measured in terms of the energy required for extraction of a given raw material (e.g., bringing a certain amount of composite rock to the surface and breaking the physical or chemical bonds by which the desired material is attached to other rock materials), then there is a well-defined minimum effort which has to be provided.

For a number of raw materials, methods exist or could be developed for recycling, i.e., using discarded production goods as a resource basis rather than materials extracted from the natural surroundings. This would seem obviously advantageous if the concentration of the desired material is higher in the scrap material than in available natural deposits and if the collection and extraction costs for the recycled material are lower than the extraction and transportation costs for the natural resource. A question to be raised in this connection is whether it might be advantageous to base decisions about recycling not on instantaneous costs of recycling versus new resource extraction, but instead on the present value cost of using either method for an extended period of time (e.g., the planning horizon discussed earlier).

The recycling decision would not have to be taken in advance if the scrap resources are kept for later use, waiting for the process to be profitable. However, in many cases, a choice is being made between recycling and further dispersion or dilution of the waste materials to a form and concentration that would be virtually impossible to use later as a resource base. Thus, resource depletion contributes to the indirect economy of a given production process and should be considered if the planning horizon is extended.

Of all the nonrenewable resources, energy resources play a special role, partly because they cannot be recycled after the energy conversion has taken place (combustion, nuclear transformation) and partly because of their part in the extraction of other raw materials and manufacturing processes.

Energy Analysis

For energy conversion systems, the energy accounting is of vital interest because the *net energy* delivered from the system during its physical lifetime is equal to its energy production minus the energy inputs into the materials forming the equipment, its construction, and its maintenance. Different energy supply systems producing the same gross energy output may differ in regard to energy inputs, so that a cost comparison based on net energy outputs may not agree with a comparison based on gross output.

The energy accounting may be done by assigning an *energy value* to a given commodity, which is equal to the energy used in its production and in preparing the raw materials and intermediate components involved in the commodity, all the way back to the resource extraction, but no further. Thus, nuclear, chemical, or other energy already present in the resource in its "natural" surroundings is not counted. It is clear that energy values obtained in this way are highly dependent on the techniques used for resource extraction and manufacture. Energy analysis along these lines has served as a precursor for full life-cycle analyses of energy systems.

For goods that may be used for energy purposes (e.g., fuels), the "natural" energy content is sometimes included in the energy value. There is also some ambiguity in deciding on the accounting procedure for some energy inputs from "natural" sources. Solar energy input into, for example, agricultural products is usually counted along with energy inputs through fertilizers, soil conditioning, and harvesting machinery, and (if applicable) active heating of greenhouses, but it is debatable what fraction of long-wavelength or latent heat inputs should be counted (e.g., for the greenhouse cultures), which may imply modifications of several of the net energy exchanges with the environment. Clearly, the usefulness of applying energy analysis to different problems depends heavily on agreeing on the definitions and prescriptions that will be useful in the context in which the energy value concept is to be applied.

Early energy analyses were attempted even before sufficient data had become available (Chapman, 1974; Slesser, 1975; von Hippel *et al.*, 1975).

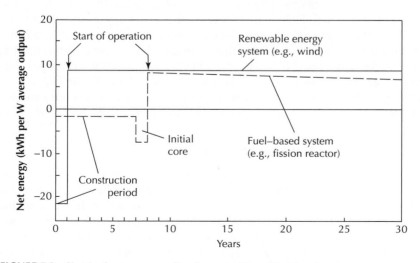

FIGURE 7.3 Sketch of net energy profiles for renewable and fuel-based systems.

Still, they were able to make specific suggestions, such as not building wave energy converters entirely of steel (Musgrove, 1976). More recently, a large number of energy systems have been analyzed for their net energy production (see, for example, the collection in Nieuwlaar and Alsema, 1997).

It should be said, in anticipation of the life-cycle approach discussed in section 7.3, that, if all impacts of energy systems are included in a comparative assessment, then there is no need to perform a specific net energy analysis. Whether a solar-based energy device uses 10% or 50% of the originally captured solar radiation internally is of no consequence as long as the overall impacts are acceptable (including those caused by having to use a larger collection area to cover a given demand).

In order to illustrate that net energy is also characterized by time profiles that may differ for different systems, Fig. 7.3 sketches the expected time profiles for the net energy production of two different types of energy conversion systems. They are derived from cost estimates in Sørensen (1976d). For both systems, the initial energy "investment" is recovered over 2–3 years of energy production.

The prolonged construction period of some energy systems does present a problem if such systems are introduced on a large scale over a short period of time. The extended delay between the commencement of energy expenditures and energy production makes the construction period a potentially serious drain of energy (Chapman, 1974). This problem is less serious for modular systems, which have short construction periods for each individual unit.

7.2 SCALE OF ANALYSIS

Traditional energy analysis often uses entities like nations for aggregate presentation of energy data. However, many features of an energy system are not captured by this approach, and many social and other indirect effects are better described from a perspective other than the national perspective. In this treatise, an area-based or person-based approach has been preferred wherever possible, that is, giving energy resources and loads per unit of area or per capita. One advantage of this *energy use density* approach is that it allows subsequent discussion on any scale, from local to national to regional to global.

7.2.1 Local and National Economy

Local communities constitute the basis for any working democracy. It is on the local level that politics directly interfere with daily life and thus the local level is most stimulating for conducting a continued political debate. It is also on the local level that groups of citizens have defied national decision-makers and done what they considered right, at least as long as no laws were broken. An example in Denmark is the erection of some 4000 wind turbines by individuals or small owner groups (nonprofit companies called *co-ops* or *guilds*) during the late 1970s, in response to a national debate over introduction of nuclear energy, an option favored at the time by the government and 99% of the parliament members.

Local interest may be seen as a preference for decentralized solutions, as opposed to passing control to a large utility company or governments situated "far away." Many owners of detached houses prefer having their own heating system, whether an oil or gas boiler in the basement or solar panels on the roof, rather than receiving heat through district heating lines. Currently, a similar interest is growing for combined electric power and heat-generating systems (using either natural gas or PVT), which are seen as precursors for reversible fuel-cell systems capable of producing both power and heat, and also for disposing of excess power from photovoltaic or wind systems (that the citizen may be a co-owner of) by producing hydrogen for use in future vehicles. Even today, plug-in hybrid diesel-plus-battery vehicles are getting attention. These are all signs that people place a high value on having as much control as possible over their lives and specifically over their energy provision, even at a premium cost.

Citizens want influence on physical planning locally and on regional infrastructure. On the other hand, they are influenced by infrastructure decisions imposed upon them, such as the creation of dispersed individual housing and large shopping centers located at a substantial distance from living areas, which contribute to a large energy demand for individual transportation, such as using automobiles (because distances appear too large for walking or bicycling, and population densities are too small for providing collective transportation). The increased distances between home and workplace that have emerged in most industrialized regions have been made possible only by access to cheap oil products used as energy for commuting. Initially, the distances were seen as a positive development, because homes and children were separated from polluted cities and industrial workplaces, but today, pollution is unacceptable and therefore has been reduced, except for that from automobile exhaust, which is strangely regarded as a necessary evil.

Some city and area planners dare go against the stream: No vehicle emissions are allowed in inner cities. If energy efficiency were considered sufficiently important, individual houses with large heating requirements (due to large surface area and more wind exposure) would be valued less than more densely placed dwelling units. The latter, along with associated service facilities and perhaps also office and production structures, would be placed in groups along public transportation routes. The ideal structure for such a society would still be fairly decentralized, because the minimization of transportation distances would still require many small centers, each possessing facilities for service and recreation.

It is sometimes said that the introduction of large energy conversion units and extensive transmission systems will promote centralization of activities as well as of control, while small conversion units (e.g., based on renewable energy) would promote societies of decentralized structure, with emphasis on regional development. This view obviously is not entirely substantiated by any observed effects of the size of conversion plants. An array of 100 wind energy converters or a solar farm may have rated capacities of magnitudes similar to those of large fuel-based power plants, and in the case of wind energy, for example, it would not make much sense to disperse the array of converters from a wind-optimal site to the actual load areas, which would typically be characterized by poor wind conditions. Furthermore, many renewable energy systems greatly benefit from the existence of power grids and other energy transmission systems, allowing alternative supply during periods of insufficient local renewable energy, instead of requiring dedicated local energy stores. However, as renewable energy like wind power becomes economically attractive, large power companies tend to take over the erection and ownership of turbines, previously in the hands by consumer guilds, and blur the market structure by tariff manipulations (such as shifting delivered power rates to fixed connection rates).

Some renewable energy technologies depend mainly on materials and skills that are found or may be developed locally in most regions. They are then well placed to encourage regional development in a balanced way, in contrast to technologies that depend heavily on very specialized technology that is available in only a few places. Generally, the structure of research and development programs related to individual energy conversion and transmission technologies play an important indirect economic role. In some cases, society will subsidize a particular technology by paying its research and development costs; in other situations, private enterprises pay these costs and include them in the final selling price of the energy conversion equipment. Another indirect effect is the influence of R&D efforts, as well as the types of skills required to produce and run specific (energy) technological systems, on preferences and directions in education and choice of profession. For example, emphasis on either nuclear engineering or ecology in the educational system may influence political decisions on selection of energy options or military strategy, or the other way round.

National Economy

In mixed or socialist economies, a production may be maintained even if it yields a loss. One reason may be to maintain the capability of supplying important products when external conditions change (e.g., maintaining food production capacity even in periods when imported foods are cheaper) or to maintain a regional dispersion of activity that might be lost if economic activities were selected strictly according to direct profit considerations. This may apply to regions within a nation or to nations within larger entities with a certain level of coordination.

Because energy supply is important to societies, they are likely to be willing to transfer considerable funds from other sectors to ensure security of energy supplies and national control over energy sources and systems.

International trade of raw materials and intermediate and final products has now reached very substantial dimensions. Often, quite similar goods are exchanged between nations in different parts of the world, with the only visible result being increased demand for transportation. Of course, all such activities contribute to the gross national product, which has been increasing rapidly (even after correction for inflation) over recent decades in most countries in the world.

A measure of a nation's success in international trade is its *balance of foreign payments*, i.e., the net sum of money in comparable units that is received by the nation from all other nations. Since the worldwide sum of foreign payment balances is by definition zero, not all nations can have a positive balance of payments, but they could, of course, all be zero.

An important factor in foreign trade is each dealer's willingness to accept payments in a given currency. To some extent, currencies are simply traded like other commodities, using basically the flexprice method for fixing exchange rates. It is also customary in international trade that governments are called upon to provide some form of guarantee that payments will be made. In international finance, the "trade of money" has led to an increasing amount of foreign loans, given both to individual enterprises and to local and national governments. A positive balance of current foreign payments may thus indicate not that a surplus of goods has been exported, but that a large number of loans have been obtained, which will later have to be returned in instalments, including interest. In addition to the balance of current payments, the *balance of foreign debts* must also be considered in order to assess the status of an individual nation. The origin of international finance is, of course, the accumulation of capital in some countries, capital that has to be invested somewhere in order to yield a positive interest to its owners.

Assuming that all nations try to have a zero or positive balance of foreign payments, their choice of, for example, energy systems would be made in accordance with this objective. For this reason, direct economic evaluation should be complemented by a list of the *import values* of all expenses (equipment, maintenance, and fuels, if any) for the energy supply systems considered. Import value is defined as the fraction of the expense that is related to imported goods or services (and which usually has to be paid in foreign currency). The import value of different energy technologies depends on methods of construction and on the choice of firms responsible for the individual enterprises. It is therefore heavily dependent on local and time-dependent factors, but is nevertheless a quantity of considerable importance for national planning. In many cases, the immediate flexibility of the import value of a given system is less than that implied by the above discussion, because of the time lags involved in forming a national industry in a new field or, depending on the number of restrictions placed on foreign trade, because of the existence of well-established foreign enterprises with which competition from new firms would be hard to establish. Examples of calculations of import values for different energy technologies, valid for the small nation of Denmark, may be found in Blegaa *et al.* (1976).

Employment

In a comparison of alternative investments, governments may evaluate the influence of each investment on the employment situation. If the social benefits from the different investments are similar, the one with the highest employment requirement would be chosen in a situation of unemployment, and the one with lowest employment requirement would be chosen in a situation of labor scarcity. It is not surprising that decision-makers make this evaluation in capitalist economies, but the employment impact of different technologies may also be considered in socialist or "welfare" economies, even if the concepts of unemployment or over-employment are not officially used. *Hidden unemployment* occurs if there is not enough meaningful work available for distribution among the work force, and, of course, if the work that is available is shared, the workload on the individual may be higher than desired in times of rapid expansion.

The employment situation relevant for a given (energy) system may be described in terms of an *employment factor*, defined as the direct employment (as measured in numbers of man-years or man-hours) associated with a given sum of money spent on the system. In addition, there may be an indirect employment factor associated with the activities initiated on the basis of the profits and payments made to workers (directly and when they purchase commodities), assuming that they would otherwise be unemployed. In a socialist economy, this "multiplier effect" is directly associated with the surplus value created, and in a (mixed) capitalist economy it is divided into one part being at the disposal of the workers (salary minus unemployment compensation, after taxes), one part at the disposal of the capitalists (net profit after taxation), and finally, one part administered by the public (the taxes).

Not all the employment is necessarily local, and it may be relevant in a national evaluation to disregard employment created outside the country. If this is done, the employment factor becomes closely related to the import value, in fact proportional to 1 minus the import fraction, if employment factors and import values are assumed to be the same for all sectors of the economy, to a first approximation.

If the purpose is to satisfy an expected energy demand, it may be more relevant to compare employment for different systems supplying the same average amount of energy, rather than comparing employment for systems with the same cost (the employment factor per unit investment defined above). It should also be kept in mind that if the real end-use can be obtained with reduced energy expenditure, then this has the same value (both for the individual and for society) as satisfying the demand by addition of new conversion units. Thus, as long as the final service rendered is unaltered, an energy unit saved by conservation is worth as much as an additional energy unit produced. This has the implication for the direct cost comparison that investments in energy conservation efforts should be considered together with investments in new supply systems and that conservation should be given priority, as long as the cost of saving one energy unit is lower than that of producing one. As mentioned in the previous sections, such a priority system should not exclude long-term considerations, including the time delays in the implementation of conservation measures as well as new systems, but it would appear that many conservation efforts not undertaken at present are less costly than any proposed alternative of additional supply systems.

For the employment considerations, the employment factor of schemes for improving conversion efficiency, as well as that of other energy conservation measures that do not change the end-use social values, should be compared with those of energy supply systems. Examples of both kinds may be found (e.g., in Blegaa *et al.*, 1976; Hvelplund *et al.*, 1983). They derive total employment factors, which are fairly independent of the technology considered (all of which are highly industrialized, but on different scales), and are roughly equal to 35 man-years per million 1978 US\$ spent. The national employment is obtained by multiplying this number by the fraction not imported.

As mentioned earlier, employment or "work" is in itself not a goal of society, and unemployment can always be eliminated by a distribution procedure that simply shares the available work. However, meaningful work for each individual may well be a social goal that can be pursued by proper incorporation of the employment implications of alternative systems in the planning considerations.

Use of Subsidies for Introducing "Appropriate Technology"

In capitalist economies with an influential public sector concerned with social priorities (i.e., a mixed economy), the allocation policy of the government may be directed to support individuals or firms that wish to introduce the technological solutions judged most appropriate from the point of view of society, even though other solutions would be more attractive in a private economic assessment.

A number of subsidy methods are available, including direct subsidy either to manufacturers or to customers, tax credits, loan guarantees, offers of special loans with favorable conditions, and, specifically for fluctuating energy generation, favorable utility buy-back schemes for energy not used by the owner of the production equipment.

Subsidies can be used for a limited period to speed up the initial introduction of appropriate technology, assuming that in a mature phase the technology can survive on its own. However, the discussion in the previous sections regarding the arbitrariness of costs included in a direct economic assessment makes it likely that some solutions may be very attractive to society and yet unable to compete in a market ruled by direct costs alone. Here redistribution subsidies can be used, or governments can decide to change the rules of price fixation (using environmental taxation or regulatory commands regarding choice of technology, as is done in building regulations for reasons of safety). In both cases, democratic support or at least tolerance (as with taxes) for these actions must be present.

Government allocation of funds generally offers a way of changing the distribution of profits in a given direction. The funds available to a government, in part, derive from taxes that are collected according to a certain distribution key (rates of taxation usually differ for different income levels and are different for individual taxpayers and enterprises). In recent years, taxation has been used to deal with externality costs, i.e., taxes have been directly related to environmental costs to society (e.g., carbon taxes in the case of global warming).

Since renewable energy technologies help relieve the burden of continued exploitation of nonrenewable resources, with the associated environmental problems, they may be particularly eligible "appropriate technologies" that should be supported and, if necessary, given government subsidies. Obvious conditions for the renewable energy technologies to fulfill are that they do not themselves create environmental problems comparable to those of the replaced technologies and that the construction and operation of the renewable energy systems do not involve increased exploitation of nonrenewable resources. The need to consider the environmental impact of large-scale renewable energy utilization is discussed earlier (e.g., for hydro in Chapter 3), but, clearly, renewable energy extraction offers the possibility of limiting climatic impacts in a way that can be planned and controlled, and the operation of renewable energy systems usually does not involve pollution of the kind associated with fuel-based energy conversion (biofuels need consideration in this respect).

The resource demand of renewable energy systems may be divided into a part comprising materials that can be recycled and another part comprising nonrenewable inputs, including fuels. In most cases, the renewable energy system will itself supply these inputs after an initial period of market penetration.

From the national government's point of view, if indirect economic considerations point to renewable energy sources as an optimal long-term solution, then they are worthy of support and possibly subsidy, in case short-term economic considerations give a less favorable result than the long-term considerations. In situations of unemployment, governments of nations with mixed "welfare" economies providing unemployment benefits could consider replacing the unemployment benefit with subsidies to build renewable energy systems, without serious resource implications (perhaps even with a positive influence on the balance of foreign payments, if the renewable energy systems are replacing imported fuels) and without having to increase taxes. Historically, unemployment periods and financial crises have offered opportunities for positive restructuring of infrastructure in society. Clearly, utilization of such opportunities entails a danger of worsening national debt accounts if the measures do not live up to their promise.

In those cases where the energy supply is in the hands of the government (e.g., decided by a department of energy), it can simply choose the appropriate solutions.

If, on the other hand, the investments and choices of energy supply system are made by individual citizens (as is often the case for residential heating systems) or by various enterprises (industries with their own energy supplies, electricity-selling utility companies, which increasingly are private companies), then the government has to use either compulsory regulations or taxation/subsidies to influence the choice of systems.

If technological improvements demanding large-scale sales to be viable are the limiting factors, direct government subsidy to customers for a limited period of time may increase the market sufficiently. Direct subsidies to producers are often refused by industrial enterprises operating in competitive markets. On the other hand, placement of large orders by the government (e.g., energy systems for public buildings, etc.) may have the desired effect. If no obvious mass production benefits in terms of reduced prices are discernible, but the technology is still socially desirable despite needing a long depreciation period to obtain break-even with competing systems, then direct subsidies are not likely to be of much help. However, the government may offer, or arrange access to, loans with flexible payback schemes. This means that a private investor can borrow money for the renewable energy system in question and pay back according to the index loan curve (illustrated by the dashed line on the right-hand side of Fig. 7.2), rather than according to the annuity loans common at present (heavy, solid line in Fig. 7.2). Direct subsidies to customers, in contrast, may cause prices to rise, which of course is not the intention.

A widely accepted form of government subsidy is to provide research and basic development efforts. A problem revealed by looking at the history of government involvement in science and technology R&D is that of dual-purpose support, for example, providing the energy industry with highly developed technology transferred from other sectors, such as technology developed for military or space applications. It is not clear that such transferred technology is optimal or even appropriate for use in the energy sector.

One important planning parameter is the amount of uncertainty associated with the evaluation of a given technology that is considered for introduction into the energy supply system. This uncertainty may be economic or physical. Systems based on future fuel supplies have important economic uncertainties associated with the future development of fuel prices. Renewable energy systems, on the other hand, are usually characterized by having most of the lifetime costs concentrated in the initial investment, implying that, once the construction is over, there is no uncertainty in energy cost. On the other hand, such systems may have physical uncertainties, for example, to do with length of life and maintenance costs, particularly if the technology is new and totally unlike technologies that have been used long enough in other fields for the relevant experience to be gathered.

As an example of the sensitivity of choosing the right subsidy policy, one may look at the introduction of wind power in various countries since 1975. After the 1973/1974 oil supply crisis, many countries offered subsidies for introduction of renewable energy, including wind turbines, but without reaching the desired effect. In several countries, subsidies were given directly to manufacturers, while in other countries, the subsidies were given to the purchasers of wind turbines. In most cases, wind turbine prices did not decline as a result of industry subsidy. One exception was Denmark, where the subsidy (initially 30% of the cost, declining with time to zero) was given to the customers, but only for approved technologies, where the approval was based on the manufacturer's being able to make a plausible argument (not prove!) that his technology had, or would soon have, a payback time under 8 years. This scheme was implemented by the Danish government following recommendations from an independent working group (ATV, 1975, 1976). At the time, neither wind power nor the other new renewable energy technologies could achieve an 8-year payback time, but the manufacturers provided calculations suggesting the plausibility of getting there in the near future, and once they were engaged in selling units receiving the subsidy, they had a strong incentive to lower prices to the promised level. There was no way the subsidy could just be added to the sales price, and the following years saw an effort to improve the technology and streamline the production process that made Danish wind turbines increasingly viable and that helped create a strong export market, in addition to a healthy home market. Other countries with a more conventional subsidy policy did not succeed in creating a viable wind industry, and it is only in recent years that the Danish wind industry has seen serious competition.

7.2.2 Regional and Global Economy

At present, there are vast discrepancies between different parts of the world, and sometimes between different regions within a nation, with respect to degree of industrialization and richness of natural resources, as well as the extent to which basic human needs are satisfied, human rights are honored, and material goods are available. The goals of different societies ought to include the coverage of basic needs (cf. section 6.2) but otherwise may differ greatly.

It is important to recognize that some societies may set goals that appear reasonable when seen in isolation, but which actually exclude the fulfillment of (perhaps even basic) goals in other societies. An example of this is the exploitation, by representatives of wealthy industrialized countries, of natural resources found in countries where basic needs are unmet and where the profits are not shared. In the same way, if all the reserves of arable land were located in only some countries, while other countries were unable to provide basic food for their populations, one might ask whether it would be reasonable for the land-rich countries to set aside their surplus arable land for recreational purposes.

In principle, countries with many unmet needs may sell their natural resources to other nations valuing them (e.g., fuels), and in this way obtain capital to use for satisfying their own needs. However, not all countries in need actually have such resources, or the prices they obtain for their resources may not be high enough to cover their needs. It may also be that exporting a resource is in immediate conflict with the use of the same resource within the country (e.g., corn for exported for biofuels instead of being used for food), or that the resources exported would have been vital in pursuing developments within the country over a longer time period.

Despite the intense international trade in raw materials and produced goods (industrial or agricultural), it appears that inequities in the availability of development options are increasing, implying that the mechanisms of (more or less free) competitive trade are inadequate in the pursuit of global goals. Furthermore, the conditions of trade are traditionally fixed in a fairly unilateral way by the "strong" industrial nations [e.g., through the World Trade Organization (WTO)]. The countries with unmet basic needs, but without excess natural resources to sell, are forced to offer primarily cheap labor and goods made particularly competitive by low wages (e.g., textiles). However, strong nations have repeatedly enforced limits on the quantities of such goods to be imported, and, as a result, the rate of improving conditions in the "weak" countries has been slow. Unmet basic needs are still widespread and have increased as a result of population growth's offsetting the positive effect of trade, international aid, and increasing industrialization.

The present globalization trend is connected to inappropriate energy pricing. Not paying for environmental and climatic damage (WTO has forced international freight and passenger transport to be free of taxation, including compensation for damage caused) has increased unnecessary flows of goods and components between continents. Globalization has also increased awareness in underprivileged countries of the lifestyles in rich countries, creating a desire for imitating them, despite sober calculations indicating that this cannot be compatible with either present population size, population growth, or natural resources available. Globalization also blurs the expectation that national goals and the social values upon which they are based can, and probably should, be different for different regions. It would be wrong to attempt to push all nations through the same kind of development (and mistakes) as traversed by the most "advanced" nations, and it would be wrong if the advanced nations try to dump the technologies they no longer support themselves in poorer parts of the world.

The reflections made above are relevant in discussions of "appropriate" technology in the energy field. The energy technologies most likely to provide early benefits in rural areas are necessarily decentralized, as are most renewable energy sources. Furthermore, at least some of the renewable energy technologies are consistent with using the local skills and raw materials present almost everywhere.

In highly industrialized nations, the most urgent priorities from a global point of view are stopping those activities that constitute a drain on the development possibilities in other parts of the world. This probably involves more prudent use of nonrenewable resources, increased recycling, and higher efficiency with respect to end-use. Use of renewable resources (e.g., for energy production) and securing increased efficiency in conversion to end-use are both examples of policies that are consistent with a global perspective. A transitional period of mixed fuel and renewable energy-based supply systems is, of course, required.

Without doubt, the most important reason for adopting a course with a strong emphasis on producing equitable development options for every nation or region is the desirability of reducing any factors likely to lead to conflict. Globalization may also lead to sharpened ideological (e.g., religious) conflicts. Increased information regarding what other people do and say may benefit mutual understanding, but increased migration and creation of mixed cultures can have the opposite effect. At present, both affluent nations and some poor nations are spending large sums on preparing themselves for military confrontations. As weapons arsenals of incredible sizes have already transformed the Earth into a global minefield, no planning considerations should ignore their possible effect on enhancing or damping the tensions in the world.

7.2.3 An Example: Privatization of the Energy Industry

In recent years, several countries have debated about the structure of their energy industry, starting with the electricity utilities. In many countries, electricity utilities are state enterprises or semi-public concessioned companies, i.e., companies with selling prices determined by the government (typically through an "electricity commission") and with local city or provincial administrations as majority shareholders. Privatization has allowed state and local governments to sell off these assets at a substantial one-time profit, but at the expense of losing detailed control over prices and supply security. The term *deregulation* has been used, although, as detailed below, it is not appropriate. The first countries/states to carry through privatization of the public electricity utilities were the United Kingdom and California, followed by Norway and then several other countries.

The outcome of the early privatizations has been a general deterioration of the electric utility sector, leading to poorer service, often higher prices, and, in most cases, a sharp decline in development efforts and new investments, making the power plant stock increasingly outdated and prone to failure, causing serious power blackouts. More recent privatizations have attempted to avoid some of these problems by additional legislation, e.g., imposing on the utilities a *public service obligation* (PSO) in the form of a fixed kWh tax earmarked for research and development of new technology relevant to the sector, and by requiring a fixed percentage of power to be produced by nonpolluting renewable energy sources. Alternatively, in some countries, renewable energy is being offered to environmentally concerned customers at a price above the price for pollution-creating power. Since one initial claim has been that privatization would lower consumer prices, efforts have been made to invite this outcome. Key among these is the auction, or "pool," concept, where power producers bid for the contract to supply power for periods ranging from an hour to a couple of days, and where long-term bids are handled analogously to futures options in a stock market.

This has led to a division of the power industry into producers, transmitters, and distributors, where typically the producers are privatized and the net-providers are still semi-public concessioned companies (because it is regarded as improper to create competition in power networking, since it would not be rational to have several independent grids in the same region—a problem familiar to privatized fixed or mobile telephone grids). One company is then given the concession to be the grid-balancing administrator, i.e., to reconcile the possibly lowest bids received from producers into a pattern compatible with the grid transmission capabilities available, and to correct mismatch or fall-out situations by drawing on producers having offered bids for emergency power (to be provided with no advance notice). This means that the balancing administrator can use any available transmission facilities regardless of ownership and that the grid owners will receive compensation according to set tariffs. This also allows customers to enter into contractual agreements with specific producers (bypassing the pool system) and yet be assured of transmission through third-party grid sections. Given the limited capacity of any grid, this clearly entails some intricate work to be done by the grid's responsible agent, making the best use of available transmission capacity by dispatching power from the producers on contract to the end-users [who of course will receive power from nearby producers through their grid, while contractual producers farther away deliver a corresponding amount of power into the grid for delivery to clients closer to them (cf. system modeling by Meibom et al., 1999)].

The result is the creation of huge power systems, comprising all of North America or continental Europe, but still limited by the capacity of transmission between countries or regions. Had the transmission capacity been infinite and the grid failure rate zero, this interconnection would, of course, lead to a highly stable and reliable electricity supply system, but, given the existing finite ceilings on transfer, in combination with the free choice of suppliers, the result may well be the opposite. The issue is very complex, and although simulations of adverse situations have been attempted, at the present time there is no complete answer to the question of whether the very massive blackouts experienced in various parts of the world are connected to the liberalization of trading options.

The final branch of the industry, the distribution from endpoints of transmission to individual customers, clearly also should be kept on public hands, since no competition can be created (and nobody would care to have ten power lines laid down for each customer, just as privatization of phone service has not led to several telephone lines sent to the same house). Strangely, in a number of countries, these obvious observations have been ignored and distribution has been fully privatized. Typically, the power-producing companies acquire the distribution grid and use it to raise consumer prices and their own profits, shifting electricity price components from production (where there is competition) to distribution (where competition cannot be established).

Some countries have supplemented privatization with legislation aimed at reducing pollution. This could be in the form of ceilings on emissions of SO_2, NO_x, particles, and CO_2, which may be forewarned to become lowered at intervals in time. It is then up to the power industry to adjust its generation mix so that the ceilings are never exceeded. In some cases, these schemes are supplemented with issuance of tradable permits for a given amount of pollution. The idea is that polluting power plants can trade a given amount of pollution permits among them, leading ideally to spending money on pollution abatement where it is most effective. In practice, the half-decent power plants of more developed regions can continue to pollute at a constant level, while the power plants of less developed regions, which would likely have been phased out in any case, get an added incentive to close down (as is clearly the case in Western and Eastern Europe). The argument against tradable permits is that one could economically assist less developed regions in replacing their most polluting equipment without, at the same time, halting the progress of pollution reduction in the regions most capable of carrying through a more advanced reduction.

The preceding paragraphs underline the inappropriateness of the term deregulation. In fact, the privatized energy sector needs a substantial increase in regulation to handle all the economic problems incurred nationally or globally.

It is interesting that the wave of privatization proposals occurs in countries where the electricity supply has been fully established. Think for a moment what would have happened if these ideological schemes had been implemented 100 years ago, when electricity systems were initially being established. The cost of providing electricity was obviously lowest in densely populated areas, such as city centers, and much higher in rural areas. On the one hand, this would have made it uneconomical to supply electricity outside cities through grids. On the other hand, it would have offered better conditions for local generation by renewable energy sources. The struggle between wind power and coal-based power, which went on in certain European countries up to the 1920s, when coal power finally "won," could have had the opposite outcome. However, we would have lost the benefits offered by the power grids, in terms of stable supply despite failure of some power units and of making optimum use of the different peak power demands in different regions to reduce the required rated capacity. The actual historical development of power supply can be characterized as one of solidarity, where city customers subsidized rural dwellers through the policy of maintaining a similar price for all electricity customers, in contrast to current deregulated systems with different prices for different customers according to actual delivery costs and preferences for polluting or nonpolluting production methods. It should be acknowledged that early 20th century prices were not fully equalized, because a higher price was charged for power on isolated islands, causing wind power to prevail as the energy source there for a while.

The final concern is about the implication of societies' losing control of essential areas. Energy supply is clearly an area of high strategic importance. If this sector belongs to private companies, it may be sold to foreign investors with agendas incompatible with the interests of a particular country. The foreign owner may elect supply options that are at variance with local customers' preferences regarding accident risks, supply security, environmental impacts, and so on. Again, the only possible solution to such concerns is to surround the energy supply sector with so much regulation that the owners of the production and grid system do not have any room left to maneuver—in which case it would be simpler to retain public ownership.

7.3 LIFE-CYCLE ANALYSIS

The abbreviation LCA is used for both life-cycle analysis and life-cycle assessment. However, they are two different concepts: life-cycle analysis is the scientific and technical analysis of impacts associated with a product or a system, while life-cycle assessment is the political evaluation based upon the analysis.

The need for incorporating study of environmental impacts in all assessment work performed in our societies, from consumer product evaluation to long-term planning decisions, is increasingly being accepted. Energy systems were among the first to be subjected to LCA, in an attempt to identify environmental and social impacts (such as effects on health), or, in other words, to include in the analysis impacts that have not traditionally been reflected in prices paid in the marketplace. This focuses on the sometimes huge difference between direct cost and full cost, including what are termed externalities: the social costs that are not incorporated in market prices. It may be seen as the role of societies and their governments to make sure that the indirect costs are not neglected in consumer choices or decision-making processes related to planning in a society. The precise way in which externalities are included will depend on political preferences. Possible avenues range from taxation to legislative regulation.

Life-cycle analysis is a tool to assist planners and decision-makers in performing assessments of external costs. The LCA method aims at assessing all direct and indirect impacts of a technology, whether a product, an industrial plant, a system, or an entire sector of society. LCA incorporates impacts over time, including impacts derived from materials or facilities used to manufacture tools and equipment for the process under study, and it includes final disposal of equipment and materials, whether it is re-use, recycling, or waste disposal. The two important characteristics of LCA are

- Inclusion of "cradle-to-grave" impacts
- Inclusion of indirect impacts imbedded in materials and equipment

The ideas behind LCA were developed during the 1970s and initially were called "total assessment," "including externalities," or "least cost planning." Some of the first applications of LCA were in the energy field, including both individual energy technologies and entire energy supply systems. It was soon realized that procurement of all the required data was a difficult problem. As a result, emphasis turned toward LCA of individual products, where the data handling seemed more manageable. However, it is still a very open-ended process, because manufacture of, say, a milk container requires both materials and energy and to assess the impacts of the energy input calls for LCA of the energy supply system. Only as the gathering of relevant data has been ongoing for a considerable time has it become possible to perform credible LCAs.

In recent years, product LCA has been promoted by organizations like SETAC (Consoli *et al.*, 1993), and several applications have appeared (e.g., Mekel and Huppes, 1990; Pommer *et al.*, 1991; Johnson *et al.*, 1994; DATV, 1995). Site- and technology-specific LCA of energy systems has been addressed by the European Commission (1995f) and by other recent projects (Petersen, 1991; Inaba *et al.*, 1993; Kato *et al.*, 1993; Meyer *et al.*, 1994; Sørensen and Watt, 1993; Yasukawa *et al.*, 1996; Sørensen, 1994b, 1995a, 1996c; Kuemmel *et al.*, 1997). Methodological issues have been addressed by Baumgartner (1993), Sørensen (1993b, 1995b, 1996b, 1997b), and Engelenburg and Nieuwlaar (1993), and systemwide considerations for energy have been addressed by Knöepfel (1993), Kuemmel *et al.* (1997), and Sørensen (1997c), the latter with emphasis on greenhouse gas emission impacts.

7.3.1 Methodology of Life-cycle Analysis

The first consideration in formulating a life-cycle assessment strategy is to formulate the purpose of the analysis. Several uses may be contemplated:

(*a*) To determine impacts from different ways of producing the same product
(*b*) To determine impacts from different products serving the same purpose
(*c*) To determine all impacts from a sector of the economy, e.g., the energy sector
(*d*) To determine all impacts from the entire social system and its activities

If the purpose is either (*a*) or (*b*), the analysis is called a product LCA, whereas purposes (*c*) and (*d*) define a systems LCA. Here we concentrate on studies pertaining to (*c*), but there are borderline cases, such as the analysis of power produced by a particular power plant, with its upstream and downstream processes, or the analysis of building insulation, with its inputs of insulation materials and installation work. In such cases, we talk about a single chain of energy conversions based on site- and technology-specific components. The course of the actual investigation may thus employ different types of analysis:

(*A*) Chain analysis (with side-chains)
(*B*) System-level analysis (each device treated separately)
(*C*) Partial system analysis (e.g., confined to energy sector).

Energy conversion chain

FIGURE 7.4 Generic energy chain (Sørensen, 1995c).

In a chain analysis (*A*), impacts include those of the main chain (Fig. 7.4) as well as impacts caused by provision of sideline inputs (Fig. 7.5), which are allocated to the chain investigated by the fraction of production entering the chain. For example, if equipment used in a chain, such as an oil refinery, is provided by a manufacturer who sells 20% of his production to the oil refinery, then 20% of each of his impacts (environmental, social) are allocated to the oil refinery. Each physical component of the chain undergoes a number of life-cycle phases, from construction activities through the period of operation and concurrent maintenance, evolving into stages of major repairs or dismantling as part of final decommissioning. Each stage has inputs of materials, energy, and labor, and outputs of pollutants and useful components. Impacts are thus positive or negative: the positive impacts are generally the benefits of the activity, notably the products or services associated with energy use, while the negative impacts are a range of environmental and social impacts. Skill of operating the components of a system is a determining factor for the magnitude of impacts, as is of course the technology used and the structure of the society receiving the impacts.

An energy system is a complete system for generating, distributing, and supplying energy to a range of end-users, defined by some domain of demand, e.g., specified by type, by geographical coverage, or by recipient profile. Examples were shown in section 6.4. Physically, the system components occupying the center of Fig. 7.5 would be facilities for extracting or collecting energy, for importing or exporting energy, for converting energy from one form to another, for transporting and distributing energy, and, finally, for transforming the energy into a useful product or service, as indicated in Fig. 7.4. Products and services are the quantities in demand in the human society. They obviously change with time and according to development stage of a given society.

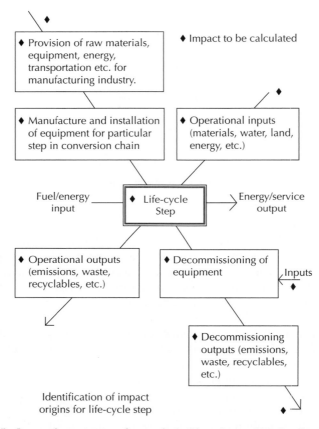

FIGURE 7.5   Input and output streams for a particular life-cycle step. (Based on Sørensen, 1997a.)

In the system level analysis (*B*), the impacts from each device in the system are calculated separately and summed up in the end. For example, the direct impacts from running the oil refinery are calculated, and the direct impacts of the equipment manufacturer are calculated, as well as any inputs he may receive from other actors. At the end, summing up all the calculated impacts will provide a true total without double counting.

Analyses made so far using the LCA method have mostly been chain calculations (*A*) and partial system analyses (*C*), in which the energy sector is treated directly, while other sectors of the economy are treated indirectly. In the latter case, it is necessary to calculate impacts individually according to system-level analysis (*B*) for the components of the energy system itself, whereas equipment input from other sectors is to be evaluated as product, with their imbedded impacts from other actors included. A double-counting problem arises in this type of analysis when the imbedded impacts include energy use by one of the manufacturers or other actors not treated explicitly. If all steps of the energy conversion chain are treated directly, the impacts from such inputs should be excluded from the indirect side-chains outside the energy sector. In many cases, this distinction is easy to make, because the impacts found in the litera-ture are normally divided into direct and indirect, and the indirect ones are dis-tributed over their origin, such as from energy use or other material use. There are, however, cases in which the data do not allow a precise splitting of the energy and non-energy inputs. In such cases, one has to estimate this split judgmentally.

Figure 7.6 illustrates the double-counting problem: if a chain LCA is made for each chain identified in Fig. 7.6*b*, there will be double-counting of both some direct and some indirect impacts. The solution for simple double-counting of the major chains is to calculate impacts for each compartment in Fig. 7.6*a* and then sum them up, but, for the indirect impacts, one has to make sure that there is no hidden double-counting. This may, in principle, be accomplished by including only the fraction of impacts corresponding to the input reaching the compartment focused upon (as in Fig. 7.4), but some of these inputs may involve energy that emerges as output from the energy system (i.e., Fig. 7.6*a*) itself. In other words, if the entire energy system is included in the analysis, one should simply omit energy-related impacts from the indirect side-chains. If only a partial energy system is being analyzed, one would still have to include impacts from other parts not explic-itly included within the system.

Treatment of Import and Export

Many of the available data sources include indirect impacts based on an assumed mix of materials and labor input, taking into account the specific loca-tion of production units, mines, refineries, etc. This means that an attempt has been made to retrace the origin of all ingredients in a specific product (such as electricity) in a bottom-up approach, which is suited for comparison of different ways of furnishing the product in question (e.g., comparing wind, coal, and nuclear electricity).

In the case of the analysis of an entire energy system, the data for specific sites and technology, as well as specific countries from which to import, are not the best suited. Especially for the future scenarios, it seems improper to use data based on the current location of mines, refineries, and other installa-tions. One may, of course, average over many different sets of data, in order to obtain average or "generic" data, but the selection of future energy systems

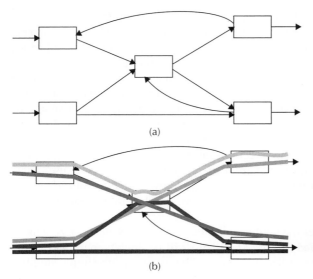

FIGURE 7.6   Energy system (*a*) with forward chains indicated (*b*) (Kuemmel *et al.*, 1997).

should not depend sensitively on where, for example, our utilities choose to purchase their coal this particular year, and therefore a different approach has to be found.

One consistent methodology is to consider the energy system of each country being studied a part of the national economy, so that, if that country chooses to produce more energy than needed domestically in order to export it, this constitutes an economic activity no different from, say, Denmark's producing more LEGO® blocks than can be used by Danish children. Exports are an integral part of the economy of a small country like Denmark, because it cannot and does not produce every item needed in its society and thus must export some goods in order to be able to import other ones that it needs. Seen in this way, it is "their own fault" if manufacturing the goods they export turns out to have more environmental side-effects than the imports have in their countries of origin, and the total evaluation of impacts should simply include those generated as part of the economy of the country investigated, whether for domestic consumption or export. Similarly, the impacts of goods imported should be excluded from the evaluation for the importing country, except, of course, for impacts arising during the operation or use of the imported items in the country (for example, burning of imported coal).

A consistent methodology is thus to include all impacts of energy production and use within the country considered, but to exclude impacts inherent in imported materials and energy. If this is done for a Danish scenario (Kuemmel *et al.*, 1997), the impact calculation related to energy for the present system turns out not to differ greatly from one based on the impacts at the place of production, because Denmark is about neutral with respect to energy imports and exports (this varies from year to year, but currently, oil and gas exports roughly balance coal imports). However, for other countries, or for future systems, this could be very different, because the impacts of renewable systems are chiefly from manufacture of conversion equipment, which consists of concrete, metals, and other materials, of which 30%–50% may be imported.

The argument against confining externality calculations to impacts originating within the country is that this could lead to purchase of the most environmentally problematic parts of the system from countries paying less attention to the environment than we claim to do. Countries in the early stages of industrial development have a tendency to pay less attention to environmental degradation, making room for what is negatively termed "export of polluting activities to the Third World." The counterargument is that this is better for the countries involved than no development, and that, when they reach a certain level of industrialization, they will start to concern themselves with the environmental impacts (examples are Singapore and Hong Kong). Unfortunately, this does not seem to be universally valid, and the implication of environmental neglect in China, India, and countries in South America extends outside their borders, as in the case of global warming, which is not confined to the country of origin or its close regional neighbors. Still, from a methodological point of view, the confinement of LCA impacts to those associated with activities in one country does provide a fair picture of the cost of any chosen path of development, for industrializing as well as for highly industrialized countries.

The problem is that a large body of existing data is based on the other methodology, where each energy form is viewed as a product, and impacts are included for the actual pathway from mining through refining and transformation to conversion, transmission, and use, with the indirect impacts calculated where they occur, that is, in different countries. In actuality, the difficulty in obtaining data from some of the countries involved in the early stages of the energy pathway has forced many externality studies to use data "as if" the mining and refining stages had occurred in the country of use. For example, the European Commission study (1995c) uses coal mined in Germany or England, based on the impacts of mining in these countries, rather than the less well-known impacts associated with coal mining in major coal-exporting countries. It has been pointed out that this approach to energy externalities can make the LCA too uncertain to be used meaningfully in decision processes (Schmidt et al., 1994). Recent product LCAs looking at soft-drink bottles and cans clearly suffer from this deficiency, making very specific assumptions regarding the place of production of aluminum and glass and of the type of energy inputs to these processes (UMIP, 1996). If the can manufacturer chooses to import aluminum from Tasmania or the United States instead of, say, from Norway, the balance between the two types of packing may tip the other way.

(a) Counting impacts imbedded in imports,
     neglecting impacts in exported good and services

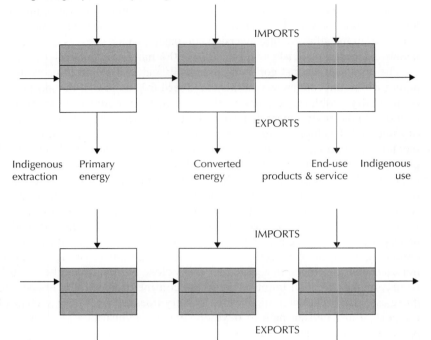

(b) Counting all impacts generated within economy, ignoring impacts from imports

FIGURE 7.7   Two consistent ways of treating imports and exports in LCA (Kuemmel *et al.*, 1997).

The two types of analysis are illustrated in Fig. 7.7. The method in common use for product LCA, where imbedded impacts are traced through the economies of different countries, not only seems problematic when used in any practical example, but also is unsuited for the system analysis that decision-makers demand. In contrast, one may use a conventional economic approach, where import and export are specific features of a given economy, deliberately chosen in consideration of the assets of the country, and calculate LCA impacts only for the economy in question in order to be consistent.

What to Include in an LCA

The history of LCA has taken two distinct paths. One is associated with the energy LCAs, developed from chain analysis without imbedded impacts to analysis including such impacts, as regards both environmental and social impacts. The history and state of the art of energy LCA are described in Sørensen (1993b, 1996b). The main ingredients suggested for inclusion in an energy

LCA are listed below. The other path is associated with product LCA and has been pushed by the food and chemical industries, particularly through the SETAC (Consoli *et al.*, 1993). SETAC tends to consider only environmental impacts, neglecting most social impacts. The same is true for the prescriptions of standardization institutions (ISO, 1997, 1999), which mainly focus on establishing a common inventory list of harmful substances to consider in LCA work. Both types of LCA were discussed methodologically around 1970, but only recently have they been transferred to credible calculations, because of the deficiencies in available data and incompleteness. Many absurd LCA conclusions were published from the late 1960s to the late 1980s. The present stance is much more careful, realizing that LCA is not, and cannot be made, a routine screening method for products or energy systems, but has to remain an attempt to furnish more information to the political decision-maker than has previously been available. The decision process will be of a higher quality if these broader impacts are considered, but the technique is never going to be a computerized decision tool capable of replacing political debate before decisions are made. This is also evident from the incommensurability of different impacts, which cannot always be meaningfully brought to a common scale of units. To emphasize this view of the scope of LCA, this section lists impacts to consider, without claiming that the list is inclusive or complete.

The impacts contemplated for assessment reflect to some extent the issues that at a given moment in time have been identified as important in a given society. It is therefore possible that the list will be modified with time and that some societies will add new concerns to the list. However, the following groups of impacts, a summary of which is listed in Table 7.1, constitute a fairly comprehensive list of impacts considered in most studies made to date (Sørensen, 1993b):

- Economic impacts, such as impacts on owners' economy and on national economy, including questions of balance of foreign payments and employment.

These impacts aim at the direct economy reflected in market prices and costs. All other impacts can be said to constitute indirect costs or externalities, the latter if they are not included in prices through, for example, environmental taxes. Economics is basically a way of allocating resources. When economic assessment is applied to an energy system, the different payment times of different expenses have to be taken into account, e.g., by discounting individual costs to present values. This again gives rise to different economic evaluations for an individual, an enterprise, a nation, and some imaginary global stakeholder. One possible way of dealing with these issues is to apply different sets of interest rates for the different actors and, in some cases, even different interest rates for short-term costs and for long-term, intergenerational costs, for the same actor. The ingredients in this kind of economic evaluation are the separate private economy and national economy accounts, often made in the past. The national economy evaluation includes factors like import fraction (balance of foreign payments), employment impact (i.e., distribution between labor and nonlabor costs), and more subtle components, such as regional economic impacts. Impact evaluations must pay particular attention to imports and exports, as many of the indirect impacts often will not be included in trade prices or their presence or absence will be unknown.

a.	Economic impacts, such as impacts on owners' economy and on national economy, including questions of balance of foreign payments and employment.
b.	Environmental impacts, such as land use; noise; visual impact; local, regional, and global pollution of soil, water, air, and biota; impacts on climate.
c.	Social impacts, related to satisfaction of needs, impacts on health and work environment, risk and impact of large accidents.
d.	Security and resilience, including supply security, safety against misuse and terrorist actions, as well as sensitivity to system failures, planning uncertainties, and changes in future criteria for impact assessment.
e.	Developmental and political impacts, such as degree of consistency with goals of a given society, effects of control requirements and institutions, openness to decentralization, and democratic participation.

TABLE 7.1 Impacts to be considered in life-cycle analysis of energy systems. (Based on Sørensen, 1996f.)

- Environmental impacts, such as land use; noise; visual impact; local pollution of soil, water, air, and biota; regional and global pollution; and other impacts on the Earth–atmosphere system, such as climatic change.

Environmental impacts include a very wide range of impacts on the natural environment, including the atmosphere, hydrosphere, lithosphere, and biosphere, usually with human society left out (but included in evaluation of social impacts; see below). Impacts may be classified as local, regional, and global. At the resource extraction stage, in addition to the impacts associated with extraction, there is the impact of resource depletion.

In many evaluations, the resource efficiency issue of energy use in resource extraction is treated in conjunction with energy use further along the energy conversion chain, including energy used to manufacture and operate production equipment. The resulting figure is often expressed as an energy payback time, which is reasonable because the sole purpose of the system is to produce energy, and thus it would be unacceptable if energy inputs exceeded outputs. In practice, the level of energy input over output that is acceptable depends on the overall cost and should be adequately represented by the other impacts, which presumably would become large compared with the benefits if energy inputs approached outputs. In other words, energy payback time is a secondary indicator, which should not itself be included in the assessment, when the primary indicators of positive and negative impacts are sufficiently well estimated. Also, issues of the quality of the environment, as affected by anthropogenic factors, should be included here. They include noise, smell, and visual impacts associated with the cycles in the energy activity. Other concerns could be the preservation of natural flora and fauna.

It is necessary to distinguish between impacts on the natural ecosystems and those affecting human well-being or health. Although human societies are, of course, part of the natural ecosystem, it is convenient and often necessary to treat some impacts on human societies separately, which is done under "social impacts" below. However, often a pollutant is first injected into the natural environment and later finds its way to humans, e.g., by inhalation or through food and water. In such cases, the evaluation of health impacts involves a number of calculation steps (dispersal, dose–response relation) that naturally have to be carried out in order.

- Social impacts, related to satisfaction of needs, impacts on health and work environment, risks, impact of large accidents.

Social impacts include the impacts from using the energy provided, which means the positive impacts that derive from services and products associated with the energy use (usually with other inputs as well) and the negative impacts associated with the energy end-use conversion. Furthermore, social impacts derive from each step in the energy production, conversion, and transmission chain. Examples are health impacts, work environment, job satisfaction, and risk, including the risk of large accidents. It is often useful to distinguish between occupational impacts and impacts to the general public. Many of these impacts involve transfer of pollutants first to the general environment and then to human society, where each transfer requires separate investigation, as stated above. This is true both for releases during normal operation of the facilities in question and for accidents. Clearly, the accident part is a basic risk problem that involves estimating probabilities of accidental events of increasing magnitude.

- Security impacts, including both supply security and also safety against misuse, terrorist actions, etc.

Security can be understood in different ways. One is supply security, and another is the security of energy installations and materials against theft, sabotage, and hostage situations. Both are relevant in a life-cycle analysis of an energy system. Supply security is a very important issue, especially for energy systems depending on fuels unevenly spread over the planet. Indeed, some of the most threatening crises in energy supply have been related to supply security (1973/1974 oil-supply withdrawal, 1991 Gulf War).

- Resilience, i.e., sensitivity to system failures, planning uncertainties, and future changes in criteria for impact assessment.

Resilience is also a concept with two interpretations. One is technical resilience, including fault resistance and parallelism, e.g., in providing more than one transmission route between important energy supply and use locations. Another is a more broadly defined resilience against planning errors (e.g., resulting from a misjudgment of resources, fuel price developments, or future demand development). A trickier, self-referencing issue is resilience against errors in impact assessment, assuming that the impact assessment is used to make energy policy choices. All the resilience issues are connected to certain features of the system choice and layout, including modularity, unit size, and transmission strategy. The resilience questions may well be formulated in terms of risk.

- Development impacts (e.g., consistency of a product or technology with the goals of a given society).

Energy systems may exert an influence on the direction of development a society will take or may be compatible with one development goal and not with another goal. The goals could be decentralization, concentration on knowledge-based business rather than heavy industry, etc. For so-called developing countries, clear goals usually include satisfying basic needs, furthering education, and raising standards. Goals of industrialized nations are often more difficult to identify.

- Political impacts, such as effects of control requirements and openness to decentralization in both physical and decision-making terms.

There is a geopolitical dimension to the above issues: development or political goals calling for import of fuels for energy may imply increased competition for scarce resources, an impact that may be evaluated in terms of increasing cost expectations or in terms of increasing political unrest (more "energy wars"). The political issue also has a local component, pertaining to the amount of freedom local societies have to choose their own solutions, possibly different from the solutions selected by neighboring local areas.

Qualitative or Quantitative Estimates of Impacts

There is a consensus that one should try to quantify as much as possible in any impact assessment. However, items for discussion arise in the handling of those impacts that cannot be quantified (and later for those quantifiable impacts that prove to be hard to monetize). One historically common approach is to ignore impacts that cannot be quantified. Alternatively, one may clearly mark the presence of such impacts and, in any quantitative summary, add a warning that the numbers given for impacts cannot be summed up to a total, as some impacts are missing. As, for example, Ottinger (1991) points out, the danger is that policymakers will still ignore the warnings and use the partial sums as if they were totals. Hopefully, this is an underestimation of the capacities of decision-makers, as their task is precisely to make decisions in areas where only part of the consequences are known at any given time and where most quantitative data are uncertain. If this were not the case, there would be no need for decision-makers, as the calculated total impact values would directly point to the best alternative. We return to some of these issues in section 7.3.2 below, where we discuss ways of presenting the results of LCAs.

Treatment of Risk-related Impacts and Accidents in LCA

Of the impacts listed above, some involve an element of risk. Risk may be defined as a possible impact occurring or causing damage only with a finite probability (for example, being hit by a meteor, or developing lung disease as a result of air pollution). The insurance industry uses a more narrow definition requiring the event to be *sudden*, i.e., excluding the damage caused by general air pollution. In LCA, all types of impacts have to be considered, but the treatment may depend on whether they are *certain* or *stochastic* in nature and, in the latter case, whether they are insurance risks or not.

As regards the health impacts associated with dispersal of air pollutants followed by ingestion and an application of a dose–response function describing processes in the human body (possibly leading to illness or death), one can use established probabilistic relationships, provided population densities and distributions are such that the use of statistical data is appropriate. A similar approach can be taken for the risk of accidents occurring fairly frequently, such as automobile accidents, but a different situation occurs for low-probability accidents with high levels of associated damage (e.g., nuclear accidents).

The treatment of high-damage, low-probability accident risks needs additional considerations. The standard risk assessment used in the aircraft industry, for example, consists of applying fault-tree analysis or event-tree analysis to trace accident probabilities forward from initiating events or backward from final outcomes. The idea is that each step in the evaluation involves a known type of failure associated with a specific piece of equipment and that the probability for failure of such a component should be known from experience. The combined probability is thus the sum of products of partial event probabilities for each step along a series of identified pathways. It is important to realize that the purpose of this evaluation is to improve design by pointing out the areas where improved design is likely to pay off.

Clearly, unanticipated event chains cannot be included. In areas like aircraft safety, one is aware that the total accident probability consists of one part made up of anticipated event trees and one part made up of unanticipated events. The purpose of the design efforts is clearly to make those accidents that can be predicted by the fault-tree analysis (and thus may be said to constitute "built-in" weaknesses of design) small compared with the unanticipated accidents, for which no defense is possible, except to learn from actual experiences and ideally move event chains including common mode failures from the "unanticipated" category into the "anticipated," where engineering design efforts may be addressed. This procedure has led to declining aircraft accident rates, while the ratio between unanticipated and anticipated events has stayed at approximately 10:1.

The term *probability* is often used loosely, with no proof of a common, underlying statistical distribution. For example, technological change makes the empirical data different from the outcome of a large number of identical experiments. This is equally true in cases of oil spills or nuclear accidents, for which the empirical data are necessarily weak, owing to the low frequency of catastrophic events (albeit compounded with potentially large consequences). Here the term *probability* is really out of place and if used should be interpreted as just "a frequency indicator."

Choosing the Context

When the purpose of the LCA is to obtain generic energy technology and systems evaluations (e.g., as inputs into planning and policy debates), one should try to avoid using data depending too strongly on the specific site selected for the installation. Still, available studies may depend strongly on location (e.g., special dispersal features in mountainous terrain) or a specific population distribution (presence of high-density settlements downstream relative to emissions from the energy installation studied). For policy uses, these special situations should be avoided if possible and be left for the later, detailed plant-siting phase to eliminate unsuitable locations. This is not normally a problem if the total planning area is sufficiently diverse.

Pure emission data often depend only on the physical characteristics of a given facility (power plant stack heights, quality of electrostatic filters, sulfate scrubbers, nitrogen oxide treatment facilities, etc.) and not on the site. However, dispersion models are, of course, site dependent, but general concentration versus distance relations can usually be derived in model calculations avoiding any special features of sites. The dose commitment will necessarily depend on population distribution, while the dose–response relationship should not depend on it. As a result, in many cases a generic assessment can be performed with only a few adjustable parameters left in the calculation, such as the population density distribution, which may be replaced with average densities for an extended region.

The approach outlined above will serve only as a template for assessing new energy systems, as the technology must be specified and usually would involve a comparison between different new state-of-the-art technologies. If the impact of the existing energy system in a given nation or region has to be evaluated, the diversity of the technologies in place must be included in the analysis, which would ideally have to proceed as a site- and technology-specific analysis for each piece of equipment.

In generic assessments, not only do technology and population distributions have to be fixed, but also a number of features characterizing the surrounding society will have to be assumed, in as much as they may influence the valuation of the calculated impacts (and in some cases also the physical evaluation, e.g., as regards society's preparedness for handling major accidents, which may influence the impact assessment in essential ways).

Aggregation Issues

Because of the importance of aggregation issues, both for data definition and for calculation of impacts, this topic is discussed in more detail. There are at least four dimensions of aggregation that play a role in impact assessments:

- Aggregation over technologies
- Aggregation over sites
- Aggregation over time
- Aggregation over social settings

The most disaggregated studies done today are termed "bottom-up" studies of a specific technology located at a specific site. Since the impacts will continue over the lifetime of the installation, and possibly longer (e.g., radioactive contamination), there is certainly an aggregation over time involved in stating the impacts in compact form. The better studies attempt to display impacts as a function of time, e.g., as short-, medium-, and long-term effects. However, even this approach may not catch important concerns, as it will typically aggregate over social settings, assuming them to be inert as a function of time. This is, of course, never the case in reality, and in recent centuries, the development with time of societies has been very rapid, entailing also rapid changes in social perceptions of a given impact. For example, the importance presently accorded to environmental damage was absent just a few decades ago, and over the next decades there are bound to be issues that society will be concerned about, but which currently are considered marginal by broad sections of society.

Aggregation over social settings also has a precise meaning for a given instance. For example, the impacts of a nuclear accident will greatly depend on the response of the society. Will there be heroic firemen, as in Chernobyl, who will sacrifice their own lives in order to diminish the consequences of the accident? Has the population been properly informed about what to do in case of an accident (going indoors, closing and opening windows at appropriate times, etc.)? Have there been drills of evacuation procedures? For the Chernobyl accident in 1986, the answer was no; in Sweden today, it would be yes. A study making assumptions on accident mitigation effects must be in accordance with the make-up of the society for which the analysis is being performed. Again, the uncertainty in estimating changes in social context over time is obvious.

Aggregation over sites implies that peculiarities in topography (leading perhaps to irregular dispersal of airborne pollutants) are not considered, and that variations in population density around the energy installation studied will be disregarded. This may be a sensible approach in a planning phase, where the actual location of the installation may not have been selected. It also gives more weight to the technologies themselves, making this approach suitable for generic planning choices between classes of technology (e.g., nuclear, fossil, renewable). Of course, once actual installations are to be built, new site-specific analyses will have to be done in order to determine the best location.

In most cases, aggregation over technologies would not make sense. However, in a particular case, where, for example, the existing stock of power plants in a region is to be assessed, something like technology aggregation may play a role. For example, one might use average technology for the impact analysis, rather than performing multiple calculations for specific installations involving both the most advanced and the most outdated technology. Thus, in order to assess the total impact of an existing energy system, one might aggregate over coal-fired power stations built at different times, with differences in efficiency and cleaning technologies being averaged. On the other hand, if the purpose is to make cost-benefit analyses of various sulfur- and nitrogen-cleaning technologies, each plant has to be treated separately.

In a strict sense, aggregation is clearly not allowed in any case, because the impacts that play a role never depend linearly or in simple ways on assumptions of technology, topography, population distribution, and so on. One should, in principle, treat all installations individually and make the desired averages on the basis of the actual data. This may sound obvious, but in most cases it is also unachievable, because available data are always incomplete and so is the characterization of social settings over the time periods needed for a complete assessment. As regards the preferences and concerns of future societies, or the impacts of current releases in the future (such as climatic impacts), one will always have to do some indirect analysis, involving aggregation and assumptions on future societies (using, for example, the scenario method described in Chapter 6).

One may conclude that some aggregation is always required, but that the level of aggregation must depend on the purpose of the assessment. In line with the general characterization given above, one may discern the following specific purposes for conducting an LCA:

- Licensing of a particular installation
- Energy system assessment
- Assistance with energy planning and policy efforts.

For licensing of a particular installation along a fuel chain or for a renewable energy system, clearly a site- and technology-specific analysis has to be performed, making use of actual data for physical pathways and populations at risk (as well as corresponding data for impacts on ecosystems, etc.). For the assessment of a particular energy system, the elements of full chain from mining or extraction through refining, treatment plants, and transportation to power plants, transmission, and final use must be considered separately, as they would typically involve different locations. A complication in this respect is that, for a fuel-based system, it is highly probable that over the lifetime of the installation fuel would be purchased from different vendors, and the fuel would often come from many geographical areas with widely different extraction methods and impacts (e.g., Middle East versus North Sea oil or gas, German or Bolivian coal mines, open-pit coal extraction in Australia, and so on). Future prices and environmental regulations will determine the change in fuel mix over the lifetime of the installation, and any specific assumptions may turn out to be invalid.

For the planning type of assessment, in most industrialized nations it would be normal to consider only state-of-the-art technology, although even in some advanced countries there is a reluctance to applying known and available environmental cleaning options (currently for particle, SO_2, and NO_x emissions, in the future probably also for CO_2 sequestering or other removal of greenhouse gases). In developing countries, there is even more of a tendency to ignore available but costly environmental mitigation options. In some cases, the level of sophistication selected for a given technology may depend on the intended site (e.g., close to or far away from population centers). Another issue is maintenance policies. The lifetime of a given installation depends sensitively on the willingness to spend money on maintenance, and the level of spending opted for is a matter to be considered in the planning decisions. The following list enumerates some of the issues involved (Sørensen, 1993b):

Technology and organization
- Type and scale of technology
- Age of technology
- Maintenance state and policy
- Matching technology with the level of skills available
- Management and control set-up

Natural setting
- Topography, vegetation, location of waterways, water table, etc.
- Climatic regime: temperature, solar radiation, wind conditions, water currents (if applicable), cloud cover, precipitation patterns, air stability, atmospheric particle content

Social setting
- Scale and diversity of society
- Development stage and goals
- Types of government, institutions, and infrastructure

Human setting
- Values and attitudes, goals of individuals
- Level of participation, level of decentralization of decision-making

Impact assessments suitable for addressing these issues involve the construction of scenarios for future society in order to have a reference frame for discussing social impacts. Because the scenario method has normative components, in most cases it would be best to consider more than one scenario, spanning important positions in the social debate occurring in the society in question.

Another issue is the emergence of new technologies that may play a role during the planning period considered. Most scenarios of future societies involve an assumption that new technologies will come into use, based on current research and development. However, actual development is likely to involve new technologies that were not anticipated at the time the assessment was made. It is possible, to some extent, to analyze scenarios for sensitivity to such new technologies, as well as for sensitivity to possible errors in other scenario assumptions. This makes it possible to distinguish between future scenarios that are resilient, i.e., do not become totally invalidated by changes in assumptions, and those that depend strongly on the assumptions made.

In the case of energy technologies, it is equally important to consider the uncertainty of demand assumptions and assumptions about supply technologies. The demand may vary according to social preferences, as well as due to the emergence of new end-use technologies that may provide the same or better services with less energy input. It is therefore essential that the entire energy chain be looked at, not just to the energy delivered but also to the non-energy service derived. No one demands energy, but we demand transportation, air conditioning, computing, entertainment, and so on.

The discussion of aggregation issues clearly points to the dilemma of impact analyses: those answers that would be most useful in the political context often are answers that can be given only with large uncertainty. This places the final responsibility in the hands of the political decision-maker, who has to weigh the impacts associated with different solutions and in that process to take the uncertainties into account (e.g., choosing a more expensive solution because it has less uncertainty). But this is, of course, what decision-making is about.

Social Context

The social context in which a given energy system is placed may have profound consequences for a life-cycle assessment of the energy system. The social context influences both the nature and the magnitude of impacts to be considered. Key aspects of describing the social context of an energy system are the natural setting, the social setting, and the human setting.

The natural setting has to do with geography and demography. Geography may force people to settle in definite patterns, which may influence the impact of pollution of air, waterways, and soils. In other words, these impacts will not be the same for the same type of energy equipment if it is placed in different geographical settings. The patterns of releases and dispersal are different, and the chance of affecting the population is also different, say, for a city placed in a narrow valley as compared with one situated on an extended plain.

The social setting includes many features of a society: its stage of development, its scale and diversity, its institutional infrastructure, and its type of government. Many of the social factors are important determinants in the selection of an energy system for a particular society, and they are equally important for determining the way that operation of the system is conducted, as well as the way in which society deals with various impacts. This may pertain to the distribution of positive implications of the energy system, but it may also relate to the actions taken in the case of negative impacts (e.g., the way society deals with a major accident in the energy sector).

The human setting involves the values and attitudes of individual members of society. They are important in individuals' choices between different types of end-use technology, and of course also in people's opinion about energy planning and energy future toward which they would like their society to move. In democratic societies, the role of attitudes is to influence the political debate, either by making certain technological choices attractive to decision-makers or by protesting against choices about to be made by governments or political assemblies, thereby expressing the lack of public support for such decisions. Examples of both kinds of political influence are numerous.

The processes are further complicated by feedback mechanisms, such as media attention and interest groups' attempts to influence attitudes in the general population.

Data related to social setting should be used in the impact calculation. Health impacts of energy systems depend on the age and health of members of the society in question, social impacts depend on the social structure, and environmental impacts may depend on the settlement type, geography, and climate of the region in question.

Most countries have statistics pertaining to these kinds of issues, but it is rare to see them used in connection with energy impact analyses. It is therefore likely that an effort is required to juxtapose all the relevant types of data, but in principle it can be done with available tools.

More difficult is the question of incorporating the values and attitudes of the members of a given society in the assessment. Available studies are often made differently in different societies, and in any case it is unlikely that the impacts associated with the values and attitudes of a society can be expressed in terms of numbers that can be compared to economic figures and the other data characterizing the energy system.

In other words, one would have to accept that impacts must be described in different phrases or units and that not all of them can be numerically compared. This should not imply that some impacts should *a priori* be given a smaller weight. In fact, what the social evaluation is all about is to discuss in political terms those issues that do not lend themselves to a straightforward numerical evaluation.

The influence of media coverage, which in many societies plays an important role in shaping political perceptions and their feedback on values and attitudes, has been studied by Stolwijk and Canny (1991), and the influence of protest movements and public hearings have been studied by Gerlach (1987) and Gale (1987) (cf. also the general introductory chapter in Shubik, 1991). The role of institutions has been studied by Lau (1987), by Hooker and van Hulst (1980), and by Wynne (1984).

Monetizing Issues

The desire to use common units for as many impacts in an LCA as possible is, of course, aimed at facilitating the job of a decision-maker wanting to make a comparison between different solutions. However, it is important that this procedure does not further marginalize the impacts that cannot be quantified or that seem to resist monetizing efforts. The basic question is really whether the further uncertainty introduced by monetizing offsets the benefit of being able to use common units.

Monetizing may be accomplished by expressing damage in monetary terms or by substituting the cost of reducing the emissions to some threshold value (avoidance cost). Damage costs may be obtained from health impacts by counting hospitalizations and workday salaries lost, replanting cost of dead forests, restoration of historic buildings damaged by acid rain, and so on. Accidental human death may, for example, be replaced by the life insurance cost.

Unavailability of data on monetization has led to the alternative approach of interviewing cross-sections of the affected population on the amount of money they would be willing to pay to avoid a specific impact or to monitor their actual investments (contingency evaluations, such as hedonic pricing, revealed preferences, or willingness to pay). Such measures may change from day to day, depending on exposure to random bits of information (whether true or false), and also depend strongly on the income at the respondents' disposal, as well as competing expenses of perhaps more tangible nature. Should the monetized value of losing a human life (the "statistical value of life," SVL, discussed below) be reduced by the fraction of people actually taking out life insurance? Should it be allowed to take different values in societies of different affluence?

All of the monetizing methods mentioned are clearly deficient: the damage caused by not including a (political) weighting of different issues (e.g., weighting immediate impacts against impacts occurring in the future), and the contingency evaluation by doing so on a wrong basis (being influenced by people's knowledge of the issues, by their available assets, etc.). The best alternative may be to avoid monetizing entirely by using a multivariate analysis, e.g., by presenting an entire impact profile to decision-makers, in the original units and with a time sequence indicating when each impact is believed to occur, and then inviting a true political debate on the proper weighting of the different issues. However, the use of monetary values to discuss alternative policies is so common in current societies that it may seem a pity not to use this established framework wherever possible. It is also a fact that many impacts can meaningfully be described in monetary terms, so the challenge is to make sure that the remaining ones are treated adequately and do not "drop out" of the decision process.

The translation of impacts from physical terms (number of health effects, amount of building damage, number of people affected by noise, etc.) to monetary terms (US\$/PJ, DKr/kWh, etc.) is proposed in studies like the European Commission project (1995a,b) to be carried out by a study of the affected population's willingness to pay (WTP) for avoiding the impacts. This means that the study does not aim at estimating the cost to society, but rather the sum of costs inflicted on individual citizens. The concept of WTP was introduced by Starr (1969). The WTP concept has a number of inherent problems, some of which are:

- Interview studies may lead people to quote higher amounts than they would pay in an actual case.
- The resulting WTPs will depend on disposable income.
- The resulting WTPs will depend on the level of knowledge of the mechanism by which the impacts in question work.

The outcome of an actual development governed by the WTP principle may be inconsistent with agreed social goals of equity and fairness, because it may lead to polluting installations being built in the poorest areas.

The accidental deaths associated with energy provision turn out in most studies of individual energy supply chains, such as the European Commission study, to be the most significant impact, fairly independently of details in the monetizing procedure selected. We therefore deal with choice of the monetized value of an additional death caused by the energy system in a little more detail below. At this point, it should only be said that our preference is to work with a monetized damage reflecting the full LCA cost of energy to society, rather than the cost to selected individual citizens.

Statistical Value of Life

In calculating externalities, a European Commission (EC) project uses the value of 2.6 M euro (about 3 million US\$) to monetize the loss of a life for all the European energy chains considered (European Commission, 1995a–f). This value is based on a survey of three types of data:

- Willingness to accept a higher risk of death, as revealed by salary increases in risky jobs as compared with salaries for similar jobs with smaller risk
- Contingency valuation studies, i.e., interviews aimed at getting statements of WTP for risks of death
- Actual expenditures paid to reduce risk of loss of life (e.g., purchase of automobile air bags, antismoking medication, etc.)

Compensations paid by European governments to families of civil servants dying in connection with their job were also considered in the European Commission study. The scatter in data reviewed ranged from 0.3 to 16 M euros per death. Outside Western Europe, the study recommends to use the purchase parity translation (i.e., same purchasing power) of the statistical value of life (SVL) used in the European case studies.

A feeling for the SVL can be obtained by considering the salary lost by accidental death. If one assumes that the accidental death on average occurs at the middle of working life and then calculates the total salary that would have been earned during the remaining time to retirement, one would, in Denmark, get a little over 20 years multiplied by the average salary for the high seniority part of a work career, amounting to between 300 000 and 400 000 DKr per year, or around 8 M DKr (some 1.25 M US\$ or 1.1 M euro). If this were paid to an individual, it should be corrected for interest earned by giving the corresponding present value of annual payments of about 60 k euro/y over 20 years. However, when calculating a cost to society, it may be argued that no discount should take place, because society does not set money aside for future salary payments.

Two other arguments might be considered. One is that, in times of unemployment, the social value of a person fit to work may be less than the potential salary. Accepting this kind of argument implies that the outcome of technology choices in a society would depend on the ability of that society to distribute the available amount of work fairly (the total amount of salaries involved is not fixed, because salaries are influenced by the level of unemployment). The other argument is that the members of a society have a value to that society above their ability to work. If this were not the case, a society would not provide health services that prolong people's lives beyond retirement age. A judgment on the merits of these arguments would lead to the conclusion that the SVL for society is more than the 1.1 M euros estimated above, but not how much more. One could say that the European Commission study's value of 2.6 M euros represents a fairly generous estimate of nontangible values to society of its members and that a lower value may be easier to defend. However, as stated above, the EC estimate has an entirely different basis, representing an individual SVL rather than one seen from the point of view of society. Instead, the conclusion may be that it is reassuring that two so different approaches do not lead to more different values.

One further consideration is that not all deaths associated with, say, Danish use of energy take place in Denmark. If coal is imported from Bolivia, coal-mining deaths would occur there, and the question arises whether a smaller value of life should be used in that case, reflecting the lower salary earnings in Bolivia (and perhaps a smaller concern by society). This would easily be stamped as a colonial view, and the EC study effectively opted to use the same SVL no matter where in the world the death occurs (this is achieved by assuming that, for example, all coal comes from mines in Germany or the United Kingdom, whereas in reality Europe imports coal from many different parts of the world).

The global equity problem is one reason why the concept of SVL has been attacked. Another is the ethical problem of putting a value on a human life. The reply to the latter may be that SVL is just a poorly chosen name selected in the attempt to give the political decision-making process a clear signal (read: in monetary units) regarding the importance of including consideration of accidental death in decisions on energy system choice and siting. The debate about the use of SVL was taken up by Grubb in connection with the greenhouse-warming issue (Grubb, 1996), using arguments similar to those given above.

For the examples of externality and LCA results to be presented below, a monetized SVL value of 2.6 M euro/death has been used. The discussion above suggests that if this SVL is on the high side, it is so by at most a factor of 2.

Since impacts from energy devices occur throughout the lifetime of the equipment and possibly after decommissioning, one point to discuss is whether expenses occurring in the future should be discounted. This is discussed in section 7.1 in connection with the concept of a social interest rate.

Chain Calculations

This section outlines the use of LCA to perform what is termed a chain calculation above. The procedure consists in following the chain of conversion steps leading to a given energy output, as illustrated in Fig. 7.4, but considering input from, and outputs to, side-chains, as exemplified in Fig. 7.5. Chain LCAs are the equivalent of product LCAs and usually involve specific assumptions about the technology used in each step along the chain. For example, the immediate LCA concerns may be emissions and waste from particular devices in the chain. However, before these can be translated into actual damage figures, one has to follow their trajectories through the environment and their uptake by human beings, as well as the processes in the human body possibly leading to health damage. The method generally applied to this problem is called the *pathway method*.

The pathway method consists of calculating, for each step in the life cycle, the emissions and other impacts directly released or caused by that life-cycle step, then tracing the fate of the direct impact through the natural and human ecosystems, e.g., by applying a dispersion model to the emissions in order to determine the concentration of pollutants in space and time. The final step is to determine the impacts on humans, on society, or on the ecosystem, using for instance dose–response relationships between intake of harmful substances and health effects that have been established separately. The structure of a pathway is indicated in Fig. 7.8.

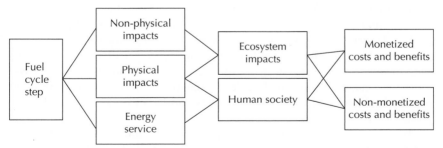

E.g.: Emissions → Dispersal → Level/concentration → Dose-response → Valuation → Cost

FIGURE 7.8 Illustration of the pathway method (Sørensen, 1995c).

Consider as an example electricity produced by a fossil-fuel power station using coal (Fig. 7.9). The first step would be the mining of coal, which may emit dust and cause health problems for miners, then follows cleaning, drying, and transportation of the coal, spending energy like oil for transportation by ship (here the impacts from using oil have to be incorporated). The next step is storage and combustion of the coal in the boiler of the power station, leading to partial releases of particulate matter, sulfur dioxide, and nitrogen oxides through the stack. These emissions would then have to be traced by a dispersion model, which would be used to calculate air concentrations at different distances and in different directions away from the stack. Based upon these concentrations, inhalation amounts and the health effects of these substances in the human body are obtained by using the relation between dose (exposure) and effect, taken from some other study or World Health Organization databases. Other pathways should also be considered, for instance, pollutants washed out and deposited by rain and subsequently taken up by plants, such as vegetables and cereals that may later find their way to humans and cause health problems.

For each life-cycle step, the indirect impacts associated with the chain of equipment used to produce any necessary installation, the equipment used to produce the factories producing the primary equipment, and so on, have to be assessed, together with the stream of impacts occurring during operation of the equipment both for the life-cycle step itself and its predecessors (cf. Fig. 7.5). The same is true for the technology used to handle the equipment employed in the life-cycle step, after it has been decommissioned, in another chain of discarding, recycling, and re-use of the materials involved.

In the coal power example, the costs of respiratory diseases associated with particulates inhaled may be established from hospitalization and lost workday statistics, and the cancers induced by sulfur and nitrogen oxides may be similarly monetized, using insurance payments as a proxy for deaths caused by these agents.

Generally, the initiating step in calculating chain impacts may be in the form of emissions (e.g., of chemical or radioactive substances) from the installation to the atmosphere, releases of similar substances to other environmental reservoirs, visual impacts, or noise. Other impacts would be from inputs to the fuel-cycle step (water, energy, and materials like chalk for scrubbers). Basic emission data are routinely collected for many power plants, whereas the data for other conversion steps are often more difficult to obtain. Of course, emission data, e.g., from road vehicles, may be available in some form, but are rarely distributed over driving modes and location (at release), as one would need in most assessment work.

Based on the releases, the next step, calculating the dispersal in the ecosphere, may exploit available atmospheric or aquatic dispersion models. In the case of radioactivity, decay and transformation have to be considered. For airborne pollutants, the concentration in the atmosphere is used to calculate deposition (using dry deposition models, deposition by precipitation scavenging or after adsorption or absorption of the pollutants by water droplets). As a result, the distribution of pollutants (possibly transformed from their original form, e.g., from sulfur dioxide to sulfate aerosols) in air and on ground or in water bodies will result, normally given as a function of time, because further physical processes may move the pollutants down through the soil (eventually reaching ground water or aquifers) or again into the atmosphere (e.g., as dust).

Given the concentration of dispersed pollutants as function of place and time, the next step along the pathway is to establish the impact on human society, such as by human ingestion of the pollutant. Quite extended areas may have to be considered, both for normal releases from fossil-fuel power plants and for nuclear plant accidents (typically, the area considered includes a distance from the energy installation of a thousand kilometers or more,

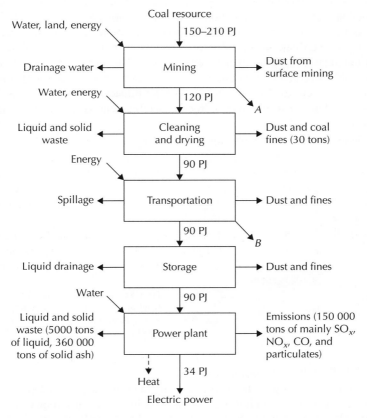

FIGURE 7.9 Coal-based electricity chain (Sørensen, 1993b). Modern plants reduce the power-plant emissions indicated, by use of filters.

according to the European Commission study, 1995c). Along with the negative impacts there is, of course, the positive impact derived from the energy delivered. In the end, these are the impacts that will have to be weighed against each other. Finally, one may attempt to assist the comparison by translating the dose–responses (primarily given as number of cancers, deaths, workdays lost, and so on) into monetary values. This translation of many units into one should only be done if the additional uncertainty introduced by monetizing is not so large that the comparison is weakened (see Fig. 7.10). In any case, some impacts are likely to remain that cannot meaningfully be expressed in monetary terms. The impacts pertaining to a given step in the chain of energy conversions or transport may be divided into those characterizing normal operation and those arising in case of accidents. In reality, the borderlines between often occurring problems during routine operation, mishaps of varying degree of seriousness, and accidents of different size are fairly blurry and may be described in terms of declining frequency for various magnitudes of problems.

To a considerable extent, the pathways of impact development are similar for routine and accidental situations involving injuries and other local effects, such as those connected with ingestion or inhalation of pollutants and with the release and dispersal of substances causing a nuisance where they reach inhabited areas, croplands, or recreational areas. The analysis of these transfers involves identifying all important pathways from the responsible component of the energy system to the final recipient of the impact, such as a person developing a mortal illness, possibly with considerable delays, such as late cancers.

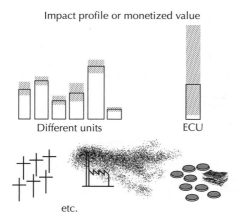

FIGURE 7.10 Multivariate versus monetized presentation of LCA results ("ECU" was a forerunner of the euro; Sørensen, 1996b).

Matrix Calculations

A full systemic LCA calculation proceeds by the same steps as the chain calculations, but without summing over the indirect impact contributions from sidechains. All devices have to be treated independently, and only at the end may one sum impacts over the totality of devices in the system in order to obtain systemwide results. In practice, the difference between chain and system LCAs in work effort is not great, because in chain LCA calculations each device is usually treated separately before the chain totals are evaluated. There are exceptions, however, where previous results for energy devices in chains that already include indirect impacts are used. In such cases, the chain results cannot immediately be used as part of a systemwide evaluation.

Particular issues arise when LCA evaluations are made for systems considered for future implementation: there may be substitutions between human labor and machinery, linking the analysis to models of employment and leisure activities. In order to find all the impacts, vital parts of the economic transactions of society have to be studied. The total energy system comprises conversion equipment and transmission lines or transportation, as well as end-use devices converting energy to the desired services or products. Demand modeling involves consideration of the development of society beyond the energy sector.

More factual is the relation between inputs and outputs of a given device, which may be highly nonlinear but in most cases are given by a deterministic relationship. Exceptions are, for example, combined heat-and-power plants, where the same fuel input may be producing a range of different proportions between heat and electricity. This gives rise to an optimization problem for the operator of the plant (who will have to consider fluctuating demands along with different dispatch options involving different costs). A strategy for operation is required in this case before the system LCA can proceed. But once the actual mode of operation is identified, the determination of inputs and outputs is, of course, unique. The impact assessment then has to trace where the inputs came from and keep track of where the outputs are going in order to determine which devices need to be included in the analysis. For each device, total impacts have to be determined, and the cases where successive transfers may lead back to devices elsewhere in the system can be dealt with by setting up a matrix of all transfers between devices belonging to the energy system. In what corresponds to an economic input–output model, the items associated with positive or negative impacts must be kept track of, such that all impacts belonging to the system under consideration in the life-cycle analysis will be covered.

Once this is done, the impact assessment itself involves summation of impacts over all devices in the system, as well as integration over time and space, or just a determination of the distribution of impacts over time and space. As in the chain studies, the spatial part involves use of dispersal models or compartment transfer models (Sørensen, 1992b), while the temporal part involves charting the presence of offensive agents (pollutants, global warming inducers, etc.) as a function of time for each located device as well as a determination of the impacts (health impacts, global warming trends, and so on) with their associated further time delays. This can be a substantial effort, as it has to be done for each device in the system, or at least for each category of devices (an example of strong aggregation is shown in Fig. 6.83).

The first step is similar to what is done in the chain analysis, i.e., running a dispersion model that uses emissions from point or area sources as input (for each device in the system) and thus calculating air concentration and land deposition as function of place and time. For example, the RAINS model is used to calculate SO_2 dispersal on the basis of long-term average meteorological data, aggregated with the help of a large number of trajectory calculations (Alcamo et al., 1990; Hordijk, 1991; Amann and Dhoondia, 1994).

Based on the output from the dispersion model, ingestion rates and other uptake routes are again used to calculate human intake of pollutants through breathing, skin, etc., and a model of disposition in the human body, with emphasis on accumulation in organs and rates of excretion, is used. The resulting concentration for each substance and its relevant depository organs is via a dose–response function used to calculate the morbidity and mortality arising from the human uptake of pollution. It is customary to use a linear dose–response function extending down to (0,0) in cases where measurements only give information on effects for high doses. This is done in the European Commission study, based on a precautionary point of view as well as theoretical considerations supporting the linear relation (European Commission, 1995a–f). The alternative, assuming a threshold below which there is no effect, is often used in regulatory schemes, usually as a result of industry pressure rather than scientific evidence.

Systemwide calculations are often interpreted as comprising only those components that are directly related to energy conversion. Sometimes this borderline is difficult to determine; for example, transportation vehicles cause traffic and thus link to all the problems of transportation infrastructure. A general approach would be to treat all components of the energy system proper according to the system approach, but to treat links into the larger societal institutions and transactions as in the chain LCA. In this way, the overwhelming prospect of a detailed modeling all of society is avoided, and yet the double-counting problem is minimized because energy loops do not occur (although loops of other materials may exist in the chains extending outside the energy sector).

Marginal Versus Systemic Change

Studies like those described above make the assumption that the energy installations considered are marginal additions to the existing system. For instance, one particular coal-fired power plant is added to a system, which is otherwise unchanged. This implies that in calculating indirect impact, the whole set-up of society and its industrial capacity is the current one. Such an approach would not be valid in a systemic study of a scenario for the future system. This scenario will be imbedded into a society that may be very different from the present society with respect to energy flows, industry structure, and social habits. In studying the future impacts from manufacturing, e.g., photovoltaic panels, the process energy input will not come from the present mix of mainly fossil-fuel power plants, but will have to reflect the actual structure of the future energy system assumed.

Evidently, this systemic approach gives results very different from those emerging from treating the same future energy system as the result of a series of marginal changes from the present and it may thus affect the determination of an optimum system solution (e.g., if the energy input to a future renewable energy installation is higher than that of a competing fossil-fuel installation, then the marginal evaluation based on current fossil-fuel power supply will be less favorable than one based on a future scenario of non-fossil-fuel energy supply).

One workable alternative to the marginal assumption, in case different forms of energy supply have to be compared to each other without being part of an overall scenario, is to consider each system as autonomous, i.e., to assume for the photovoltaic power plant that the energy for manufacture comes from similar photovoltaic plants. This makes the impact evaluation self-contained, and the assumption is generally fair, e.g., if the power for site-specific work mostly comes from nearby installations, rather than from the national average system. Because renewable systems like wind turbines and solar plants are of small unit size, the gradual establishment of a sizeable capacity could indeed be seen as involving energy use based on the previously installed plants of the same kind. This suggestion may not always apply; in some cases, energy inputs have to be in forms different from the one associated with the installation studied.

7.3.2 Communicating with Decision-makers

Because the purpose of LCA is to facilitate decision-making, some thought should be given to the way the results of an analysis are presented to the target group. This underlies, of course, the quest for monetizing all impacts: it is believed that decision-makers understand monetary impacts better than physical ones and that qualitative descriptions have little say in policy. From a scientific point of view, the dividing line goes between qualitative and quantitative impact statements. That the quantifiable impacts cannot all be expressed in the same unit is intuitively clear: numbers of cancer deaths, loss of agricultural crops, acid rain damage to Greek temples, and traffic noise are fundamentally impacts expressed in "different units." The translation into monetary values, however it is done, loses part of the message. This is why monetizing should be used only if it does not significantly increase uncertainty, which means that the decision-makers should not be exposed to the monetizing simplification unless it preserves their possibility of making a fair assessment.

If those impacts that can be quantified are kept in different units, a question arises about how they can be presented to the decision-maker in a form facilitating their use. The common answer is to use a multivariate approach, where, as indicated on the left side of Fig. 7.10, each category of impact is presented in its own units. Figure 7.11 expands on one methodology for multivariate presentation (Sørensen, 1993b), suggesting the use of what is called an *impact profile*. The idea of the profile is that each particular type of impact is evaluated in the same way for different systems. Thus, the magnitudes indicated by the profile are no more subjective than the monetized values, although they cannot be summed across different impact categories. Clearly those impacts that can be meaningfully monetized should be, but the impact profile actually gives much more information, as it tells the decision-maker whether two energy solutions have the same type (i.e., the same profile) of impacts or whether the profile is different and thus makes it necessary for the decision-maker to assign weights to different kinds of impacts (e.g., comparing greenhouse-warming impacts of a fossil-fuel system with noise impacts of wind turbines). The assignment of weights to different impact categories is the central political input into the decision process.

Evaluation of impacts:

Impact item	Fundamentally unacceptable	Strongly negative	Neutral	Strongly positive	Unconditionally acceptable	Weight
a				●		1.0
b			●			0.8
c				●		1.0
d		●				1.0
e		●				0.2
f			●			1.0
g					●	0.9
h				●		1.0
i				●		1.0
j		●				0.9
k	●					0.9
l		●				1.0
m	●					1.0
n			●			0.3
o				●		1.0
p			●			0.0

Average impact:

Absolutism:

FIGURE 7.11   Layout of multivariate impact assessment scheme (Sørensen, 1982, 1993b).

The impact profile approach also makes it a little easier to handle qualitative impacts that may only allow a description in words, because such impacts can often be characterized vaguely as "small," "medium," or "large," a classification that can be rendered in the profiles and compared for different energy systems. Hence, the advantage of the profile method is that the decision-maker sees both the bars representing monetized values and the adjacent bars describing the outcome of a qualitative assessment. Thus, the chance of over-looking important impacts is diminished. In any case, as mentioned, the multi-variate profile approach gives the decision-maker more information than a single monetary value. A further point that may play a role in decision-making is the presence of value systems making certain impacts "unconditionally acceptable" or "unconditionally unacceptable." Such absolute constraints can be accommodated in the assignment of weights (zero or infinite), as indicated in Fig. 7.11. Figure 7.12 shows an example of a profile of the impacts from two energy installations, including both positive and negative impacts (details are given in Sørensen, 1994b).

Section 7.3.3 below gives several examples of monetized impacts based on recent LCA studies. It is evident that, although the derivation of each single impact figure requires a large effort, the result still may involve substantial uncertainty. The analyses presented in the following section also show that

FIGURE 7.12   LCA impact profiles for coal and wind power chains (Sørensen, 1994b).

the largest uncertainties are often found for the most important impacts, such as nuclear accidents and greenhouse warming. Clearly, there is a general need to improve data by collecting information pertinent to these analyses. This need is most clearly met by doing site- and technology-specific studies. As regards indirect inputs, national input–output data are often based upon statistical aggregation choices failing to align with the needs for characterizing transactions relevant for the energy sector. Furthermore, there are usually gaps in data availability. One conclusion from these observations is that there is a need to be able to present qualitative and quantitative impacts to a decision-maker in such a way that the weight and importance of each item become clear, despite uncertainties and possibly different units used. The multivariate presentation tools invite the decision-maker to employ multi-criteria assessment.

The difficulties encountered in presenting the results of externality studies and life-cycle analyses in a form suited for the political decision-making process may be partly offset by the advantages of bringing into the debate the many impacts often disregarded (which is, of course, the core definition of "externalities," meaning issues not included in the market prices). It may be fair to say that LCAs and the imbedded risk assessments will hardly ever become a routine method of computerized assessment, but they may continue to serve a useful purpose by focusing and sharpening the debate involved in any decision-making process and ideally help increase the quality of the basic information upon which a final decision is taken, whether on starting to manufacture a given new product or to arrange a sector of society (such as the energy sector) in one or another way.

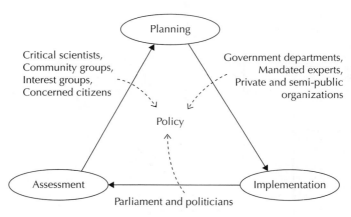

FIGURE 7.13   The actor triangle, a model of democratic planning, decision-making, and continued assessment (Sørensen, 1993a).

Finally, Fig. 7.13 indicates how decision-making is a continuous process, involving planning, implementation, and assessment in a cyclic fashion, with the assessment of actual experiences leading to adjustments of plans or, in some cases, to entirely new planning.

7.3.3 Application of Life-cycle Analysis

Examples of LCA are presented in the following sections. Each is important in its own right, but they also illustrate the span of issues that one may encounter in different applications. The first is greenhouse-warming impacts associated with emissions of substances like carbon dioxide. It involves global dispersal of the emitted substance, followed by a subtle range of impacts stretching far into the future and depending on a number of poorly known factors, for instance, the ability of nations in central Africa to cope with vector-borne diseases like malaria. The uncertainties in arriving at any quantitative impact measure are discussed.

The second example is that of chain calculations for various power plants using fuels or renewable energy. It illustrates the state of the art of such calculations, which are indeed most developed in the case of electricity provision chains. Life-cycle impact estimates for the emerging hydrogen and fuel-cell technologies may be found in Sørensen (2004a).

The third case presents the impacts of road traffic, which are included to show an example of a much more complex LCA chain, with interactions between the energy-related impacts and impacts that would normally be ascribed to other sectors in the economy of a society.

In a practical approach to life-cycle analysis and assessment, one typically goes through the following steps:

- Establishing an inventory list, i.e., identifying and categorizing materials and processes inherent in the production, use, and final disposal of the product or system.
- Performing an impact analysis, i.e., describing the impacts on environment, health, etc. for each item in the inventory.
- Damage assessment, translating the physical impacts into damage figures (in monetary units where possible).
- Overall assessment, identifying the most crucial impact pathways.
- Mitigation proposals, suggesting ways of avoiding damage, e.g., by use of alternative materials or processes.
- Regulatory proposals, describing how norms and regulative legislation or prescriptions for manufacture and use can decrease damage.

One may say that the first two items constitute the life-cycle analysis, the two next items the life-cycle assessment, and the two final ones, optional political undertakings.

LCA of Greenhouse Gas Emissions

The injection of carbon dioxide and other greenhouse gases, such as methane, nitrous oxide, and chlorofluorocarbons, into the atmosphere changes the disposition of incoming solar radiation and outgoing heat radiation, leading to an enhancement of the natural greenhouse effect. Modeling of the Earth's climate system in order to determine the long-term effect of greenhouse gas emissions has taken place over the last 40 years with models of increasing sophistication and detail. Still, there are many mechanisms in the interaction between clouds and minor constituents of the atmosphere, as well as coupling to oceanic motion, that are only modeled in a crude form. For instance, the effect of sulfur dioxide, which is emitted from burning fossil fuel and in the atmosphere becomes transformed into small particles (aerosols) affecting the radiation balance, has been realistically modeled only over the last couple of years. Because of the direct health and acid-rain impacts of SO_2, and because SO_2 is much easier to remove than CO_2, emissions of SO_2 are being curbed in many countries. As the residence time of SO_2 in the atmosphere is about a week, in contrast to the 80–120 years for CO_2, climate models have to be performed dynamically over periods of 50–100 years (cf. Chapter 2).

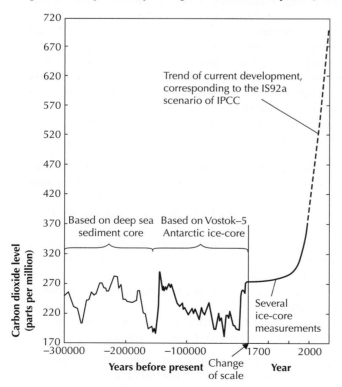

FIGURE 7.14 History of atmospheric CO_2 levels based on ice-core data. (Based upon Sørensen, 1991; Emiliani, 1978; Berger, 1982; Barnola *et al.*, 1987; Neftel *et al.*, 1985; Friedli *et al.*, 1986; Siegenthaler and Oeschger, 1987; IPCC, 1996a.)

Figure 7.14 shows measured values of CO_2 concentrations in the lower atmosphere over the last 300 000 years. During ice-age cycles, systematic variations between 190 and 280 ppm took place, but the unprecedented increase that has taken place since about 1800 is primarily due to combustion of fossil fuels, with additional contributions from changing land use, including felling of tropical forests. If current emission trends continue, the atmospheric CO_2 concentration will have doubled around the mid-21st century, relative to the pre-industrial value. The excess CO_2 corresponds to slightly over half of anthropogenic emissions, which is in accordance with the models of the overall carbon cycle (IPCC, 1996b). The ice-core data upon which the historical part of Fig. 7.14 is based also allow the trends for other greenhouse gases to be established. The findings for methane are similar to those for CO_2, whereas there is too much scatter for N_2O to allow strong conclusions. For the CFC gases, which are being phased out in certain sectors, less than 40 years of data are available. Both CO_2 and methane concentrations show regular seasonal variations, as well as a distinct asymmetry between the Northern and the Southern Hemispheres (Sørensen, 1992b).

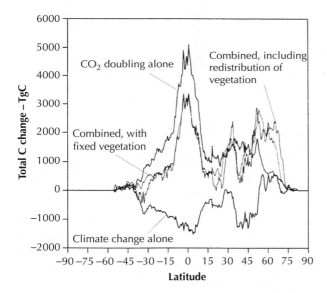

FIGURE 7.15 Simulated changes in equilibrium soil-stored carbon as a function of latitude (units: 1012 g carbon per half degree latitude), with individual contributions indicated. (Based upon Melillo *et al.*, 1993 and IPCC, 1996b.)

Despite the incompleteness of climate models, they are deemed realistic enough to permit the further calculation of impacts due to the climate changes predicted for various scenarios of future greenhouse gas emissions. This is true for the 1996 IPCC assessment work used in the LCA work here, and there is little change in the two subsequent IPCC reports.

It should be mentioned that the discussion above pertains only to the influence of greenhouse gases on the physical earth–ocean–atmosphere system. There is a subtle interaction with biological activity in each compartment, with and without anthropogenic interference. Some of the mechanisms identified by the IPCC (1996b, Chapter 9), in addition to the temperature changes discussed in Chapter 2, are: changed vegetation cover and agricultural practices influence carbon balance and amounts of carbon stored in soils; wetlands may dry out owing to temperature increases, thereby reducing methane release; and albedo changes may influence the radiation balance. Figure 7.15 shows how carbon storage is increased owing to the combined effects of higher levels of atmospheric carbon (leading to enhanced plant growth) and changed vegetation zones, as predicted by the climate models (affecting moisture, precipitation, and temperature). Comprehensive modeling of the combined effect of all the identified effects has not yet been performed.

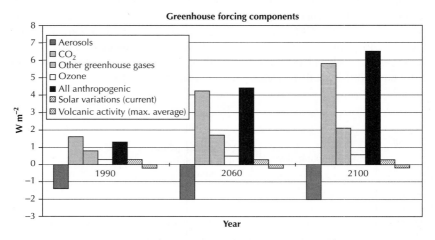

FIGURE 7.16 Components of present and future (geographically averaged) greenhouse forcing, according to the IS92a scenario of the IPCC. Based on IPCC, 1996b, p. 81 and p. 320 for the year 1990, p. 321 for 2060—the year that CO_2 concentrations have doubled—and, finally, pp. 320–321 for the year 2100, with future natural forcing components taken at 1990 values (Kuemmel et al., 1997).

The current anthropogenic greenhouse forcing, i.e., the net excess flow of energy into the atmosphere taking into account anthropogenic effects since the late 18th century, is estimated at 1.3 W m^{-2}, as a balance between a twice as big greenhouse gas contribution and a negative contribution from sulfate aerosols (IPCC, 1996b, p. 320). The further increase in the concentration of CO_2 and other greenhouse gases, together with at least a stabilization in current emissions leading to the creation of sulfate aerosols, as assumed in the IPCC business-as-usual scenario (called IS92a), is estimated to raise the radiative forcing to some 6–7 W m^{-2} by year 2100 (IPCC, 1996b, p. 320). For the IS92a scenario, the doubling of CO_2 occurs around the year 2060, where the forcing is estimated to be 4.4 W m^{-2}. The estimates are summarized in Fig. 7.16, which also gives the non-anthropogenic contributions from presently observed variations in solar radiation (somewhat uncertain) and from volcanic activity (very irregular variations with time).

It should be noted that it is not strictly allowable to make a "sum" of all anthropogenic forcing components. For example, aerosols currently have a quite different geographical distribution from that of carbon dioxide, and the combined effect is far more complex than the simple difference between the two absolute values.

Greenhouse-warming Impacts

Several studies have tried to estimate the damage that may occur as a consequence of the enhanced greenhouse effect. Early studies have concentrated on the damage that may occur in industrialized countries or specifically in North America (Cline, 1992; Frankhauser, 1990; Tol, 1995; Nordhaus, 1994; summaries may be found in IPCC, 1996c). Exceptions are the top-down studies of Hohmeyer (1988) and Ottinger (1991), using rather uninformed guesses regarding starvation deaths from climate change.

The IPCC Second Assessment Report (IPCC, 1996b) summarizes information on the expected physical damage caused by global warming published up to 1994. Similar reports have been published more recently (IPCC, 2007), with essentially the same message. Based on the 1996 work, a monetary evaluation using assumptions similar to those of the European Commission study has been made (cf. Sørensen, 1997c). It should be said that the energy-use scenarios upon which the impact studies were based are typically the reference scenario of the 1992 IPCC report (business-as-usual increase in energy consumption and doubling of CO_2; IPCC, 1992) or earlier work, and the physical warming assumptions are those of either the IPCC 1994 study (IPCC, 1995) or, in most cases, earlier studies. The teams for the 1995 IPCC impact assessments were provided with the IPCC climate-model results (seasonal temperatures, precipitation, and soil moisture) for three global models (IPCC, 1994), which, however, were all calculations made without inclusion of the sulfur aerosol contributions. Impacts could possibly be overestimated, although in many cases they involve uncertainties in precisely defining the warming that will lead to the effect discussed, which probably overshadows the effects of improvements in the basic climate modeling.

The estimation of impacts on agriculture and forestry particularly depends on the change in natural vegetation zones and local changes in soil moisture and temperatures. Although precise location of the zones within a few hundred kilometers may not be essential, their relative magnitudes would directly affect the impact assessment; as a result, the confidence we can attach to the impacts predicted is not particularly high. In addition, there is, as mentioned in section 7.3, a factor-of-two uncertainty in the monetary valuation used for accidental loss of life. One reason for making these crude impact estimations is to find out whether the mitigation or avoidance costs are likely to be lower than the damage costs for not taking any action.

The model presented here uses the 2060 situation as an average representing the entire 21st century. It is clear that valuation studies must be made on the basis of impact studies already available and that impact studies must be made on the basis of earlier climate-model results, which again are based on existing scenarios. The possibility of inconsistency is evident.

Impact Description	Ref.	Valuation* H (T$)	Valuation* L (T$)
Additional heatwave deaths (doubling, additional 0.1 M y^{-1}, valued at 3 M US$ each)	a	30	20
Fires due to additional dry spells (doubling)	b	1	0.5
Loss of hardwood and fuelwood yields (20% relative to 1980 production)	c, d, e	4	1
Increase in skin cancer due to UV radiation	b	3	3
Additional asthma and allergy cases (increase in pollen and air pollution due to warming)	b	3	1.5
Financial impact of doubling extreme events	b	3	0.8
Increase in insect attacks on livestock and humans	b	?	?
Food production (failure to adopt new crops, increased pests and insect attacks, production loss = 10%–30%; population at risk increasing from present 640 M to 700–1000 M), additional deaths from starvation due to crop loss (100 M additional deaths, chiefly in developing countries)	f, g	300	0
Deaths connected with migration caused by additional droughts or floods (50 M deaths, the affected population being over 300 M)	b	150	0
Increased mortality and morbidity due to malaria (presently 2400 M at risk and 400 M affected; increase due to warming = 100 M cases in tropical and partly in subtropical regions, a potential 7-fold increase in temperate regions assumed to be curbed)	b, h	300	75
Increased mortality and morbidity due to onchocerciasis (presently 120 M at risk and 17.5 M affected; increase and spread due to warming = 10 M additional cases, primarily in developing countries)	b, i	30	5
Increased mortality and morbidity due to schistosomiasis (presently 600 M at risk and 200 M cases; increase due to warming = 25%, in developing countries)	b	150	20
Increased mortality and morbidity due to dengue (presently 1800 M at risk and 20 M cases; increase due to warming = 25%, in developing countries)	b	15	0
Other effects of sanitation and fresh-water problems connected with salt water intrusion, droughts, floods, and migration (developing countries)	b	50	0
Loss of tourism, socioeconomic problems, loss of species, ecosystem damage, etc.	b	?	?
Total of valuated impacts (order of magnitude)		**1000**	**100**

TABLE 7.2 Estimated global warming impacts during the entire 21st century, for IPCC reference case (CO_2 doubling) (unit: 10^{12} US$). H: high SVL of 3 M $ used throughout. L: SVL put at zero for Third World (Sørensen, 1997c).

Key to references: Base reference IPCC (1996a), specific references in the table are:
[a]Kalkstein and Smoyer (1993)
[b]McMichael et al., in IPCC (1996a), Chapter 18
[c]Zuidema et al. (1994)
[d]Kirschbaum et al., in IPCC (1996a), Chapter 1
[e]Solomon et al., in IPCC (1996a), Chapter 15
[f]Reilly et al., in IPCC (1996a), Chapter 13 and Summary for Policymakers
[g]Rosenzweig et al. (1993); Parry and Rosenzweig (1993)
[h]Martens et al. (1994)
[i]Walsh et al. (1981); Mills (1995)
*The valuations involve further estimates and should be regarded as very uncertain.

Table 7.2 gives a list of impacts identified by the IPCC Second Assessment Report (IPCC, 1996a), in terms of the additional number of people exposed to a range of risks due to the average warming, sea-level rise, and additional extreme events predicted by the climate studies reviewed by the (earlier) IPCC studies. Because many impacts involve human deaths, the results scale practically linearly with the statistical values of life (SVL) used. This means that a choice of a value different from the chosen 3 M US$ (which is close to the 2.6 M euro used by the European Commission study discussed above) can be accommodated as an overall factor.

The individual entries in Table 7.2 are based on the sources indicated, with monetization of impacts evaluated in two different ways. The first valuation column uses the high SVL value of 3 M US$ globally, while the second takes the SVL to be zero for developing countries, in order to display the geographical differences of impacts. Here follows a brief explanation of the impacts included:

Heat-wave deaths occur in major cities due to the heat island effect, possibly combined with urban pollution. The doubling estimated by Kalkstein and Smoyer (1993) is mostly due to increased occurrence at mid-latitudes (an inner-city temperature rise of 2.5 °C is assumed), and thus two-thirds of the increased mortality is assumed to happen in industrialized countries. A case study for New York City (Kalkstein, 1993) finds an increased mortality of 4×10^{-5} over a 5- to 10-day period. The present annual New York rate of heat-wave deaths is 320, and Kalkstein has collected similar numbers for a number of large cities around the world experiencing days with temperatures above 33 °C. The estimated doubling of incidence will thus imply an additional order of magnitude of 10^5 heat-wave deaths, annually and globally, valued at 30 T US$ over the 100-year period. Uncertainties are related to possible acclimatization effects and from increased populations in large cities expected over the 21st century.

The doubling of fires causes mainly economic loss, assumed to be evenly distributed between developed and developing countries, whereas the 20% loss of hardwood and fuelwood yields is predicted to follow a complex geographical pattern (Kirschbaum *et al.*, IPCC, 1996a, Chapter 1), but with the highest losses occurring in tropical regions (Solomon *et al.*, IPCC, 1996a, Chapter 15). It is assumed that 75% of the economic losses occur in developing countries. The increased number of cases of skin cancer, attributable to an assumed 10% loss of ozone, occurs mainly at higher latitudes (IPCC, 1996a, Ch. 18).

Additional allergy cases would be associated with increased levels of pollen and air pollution due to heating and would occur predominantly at lower latitudes, whereas asthma incidence is highest in humid climates and is expected to be enhanced at higher latitudes (IPCC, 1996a, Chapter 18). These impacts are assumed to be equally divided between developed and developing regions. Owing to an expected shortfall of hardwood supply relative to demand during the 21st century, the actual economic value may be considerably higher than the estimate given in Table 7.2. The financial loss associated with a predicted doubling of extreme events (droughts and floods, IPCC, 1996a,

Chapter 18) is assumed to occur with 75% of the incidences in developing countries. The predicted increased frequency of insect bites is not valued, but could have an economic impact, e.g., on livestock production.

One major issue is the impact of climate change on agricultural production. Earlier evaluations (e.g., Hohmeyer, 1988) found food shortage to be the greenhouse impact of highest importance. However, the 1995 IPCC assessment suggests that, in developed countries, farmers will be able to adapt crop choices to the slowly changing climate, and that the impacts will be entirely in the Third World, where farmers will lack the skills needed to adapt. The estimated production loss amounts to 10%–30% of current global production (Reilly *et al.*, IPCC, 1996a, Chapter 13), increasing the number of people exposed to risk of hunger from the present 640 million to somewhere between 700 and 1000 million (Parry and Rosenzweig, 1993). There are also unexploited possibilities for increasing crop yields in developing countries, so the outcome will depend on many factors, including speed of technology transfer and development. Assuming the lower estimate of 100 million additional starvation deaths during the 21st century, one arrives at the 300 T US$ given in Table 7.2, all related to the Third World. Similarly, deaths associated with migration induced by extreme events are estimated at 50 million.

The other major impact area is death from diseases transmitted by insect vectors affected by climatic factors, such as mosquitoes. Today, 2400 million people are at risk for malaria (McMichael *et al.*, IPCC, 1996a, Chapter 18), and an additional 400 million are expected to be at risk due to greenhouse-warming expansion of the habitat for malaria-carrying mosquitoes (Martens *et al.*, 1994). The expansion will involve subtropical and even some temperate

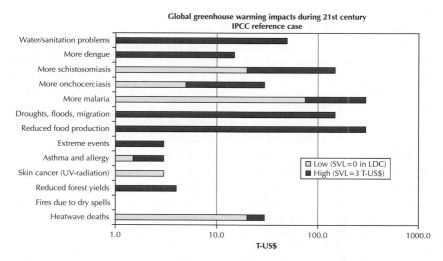

FIGURE 7.17   Impacts of global warming. From Table 7.2, logarithmic scale; SVL = statistical value of life, LDC = Third World countries (Kuemmel *et al.*, 1997).

regions, but in tropical regions the incidence of malaria is also predicted to increase. Assuming 100 million additional deaths from malaria during the 21st century, of which 75% are in the tropics, one arrives at the figures given in Table 7.2. Large uncertainties are associated with the possibilities of mitigating actions in the subtropical and temperate regions.

Considering the large impact of diseases like malaria, it is natural to compare damage costs with possible avoidance costs. In the past, relatively little money has been spent on control of vector-borne diseases, and it is recognized that, just as mosquitoes have developed resistance to pesticides, the malaria parasite (*Plasmodium* spp.) has become resistant to both the conventional medication chloroquine and newer drugs like mefloquine. Many scientists have doubts whether new genetically engineered agents can be introduced effectively into the mosquito population, and prospects for new, effective drugs are also unclear, although they are currently the subject of serious research. The early neglect by the pharmaceutical industry may reflect the low purchasing power of the populations most affected by malaria, e.g., rural African people (Butler *et al.*, 1997). Recent findings of maternal transmission of malaria antibodies raise hope for development of malaria vaccines (e.g., Fried *et al.*, 1998).

For other tropical diseases, such as onchocerciasis, vector populations are expected both to increase by 25% at current sites (Mills, 1995) and to spread to new sites (Walsh *et al.*, 1981), leading to 10 million additional cases. Schistosomiasis may spread into the subtropical regions, whereas dengue

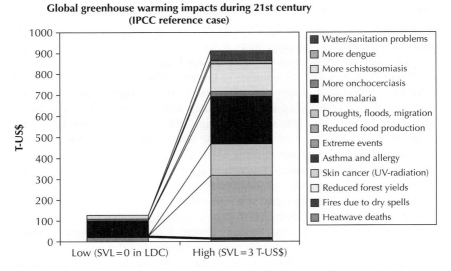

FIGURE 7.18 Impacts of global warming (same as Fig. 7.17, but on a linear scale).

and yellow fever are expected to remain in the tropics. Table 7.2 reflects these expectations through its distribution of impacts between developed and developing regions, and it also gives an estimate of deaths occurring due to sanitation problems connected with saltwater intrusion into drinking water (due to sea-level rise) and larger migrating populations, mainly in the developing countries (assuming that immigration into industrialized countries continues to be controlled). The economic consequences of other identified impacts, such as loss of species, effects on tourism, etc., have not been estimated.

The overall impacts are of the order of magnitude 10^{15} US$, when using the SVL of industrialized countries, and one order of magnitude lower if Third World impacts are valued at or near zero. This spells out the greenhouse-impact dilemma that 90% of the damage is in the Third World, a proportion much higher than their share in causing the problem. The IPCC Working Group II identification of a number of impacts specifically occurring in low-latitude regions explains why the impact estimates are so much higher than those of early work based upon, and extrapolated from, industrialized regions (Cline, 1992; Frankhauser, 1990; Nordhaus, 1990). Other factors contributing to this result include the high SVL and the assumption of less ability to mitigate impacts in the Third World (by switching food crops, building dikes to avoid floods, and so on). If the impacts of the present study, evaluated with the uniformly high SVL, are distributed equally over the 100-year period, the annual cost of greenhouse warming is found to be roughly 40% of the current global GNP. In Figs. 7.17 and 7.18, the total estimated greenhouse-warming impacts, and the part occurring outside the Third World (SVL = 0 in less developed countries, LDC), are shown on both a logarithmic and a linear scale.

Estimating Greenhouse-warming Externalities for Fuel Combustion

The translation of the 10^{15} US$ impact from greenhouse gas emissions into externalities for specific energy activities may be done as follows. It is assumed that the impacts identified correspond to the IPCC business-as-usual scenario IS92a, and that the situation in the year 2060, when CO_2 has doubled relative to pre-industrial levels, is a representative average of the entire 21st century. The average CO_2 concentration and forcing components are shown in Figs. 7.14 and 7.16. It is seen from Fig. 7.16 that the CO_2 forcing in the year 2060 is 4.2 W m^{-2}, or 95.4% of the total anthropogenic forcing. One may therefore make the simplifying assumption that 95.4% of the impacts determined above are associated with CO_2 emissions alone. The non-CO_2 emissions are dominated by emissions arising from activities other than burning of fossil fuels (notably land-use changes, such as felling of tropical forests), and it does not appear right to give an externality credit to the sulfate aerosols, although they may partially cancel some of the warming trends caused by CO_2. This is a question open for debate, but the view here is that human interference with climate in either direction should be given a negative externality tag. The possible cost of sulfate aerosol formation is not estimated, because in the energy-use scenarios considered adequate, this source of atmospheric contamination disappears within a few decades, owing to fuel shifts or to stronger SO_2 emission control. Similarly, the impacts of disappearing tropical forests are disregarded, in the hope that this trend is halted but also because it is not an externality of the energy system. Also, it should be mentioned that some CO_2 emissions do not arise from the energy sector. The IPCC IS92a scenario emission prediction is shown in Fig. 7.19, along with averaged historical data.

Integrating the CO_2 emissions from 1990 to 2060, one obtains a total of 814×10^{12} kg C. If the integration in made from 1765 to 2060, one gets 1151×10^{12} kg C. There are now two ways of assigning the externality cost. One is the straightforward division of 95% of 1000×10^{12} US\$ by the total emissions from pre-industrial times to the point of CO_2 doubling, giving 0.83 US\$ per kg of C or 0.23 US\$ per kg of CO_2. This means distributing the impacts felt through the entire 21st century over the emissions leading to doubling of CO_2 in the atmosphere, which, apart from the uncertainty in assigning the impacts to a specific time period, represents impacts per unit of the causative offense. However, in the climate debate, it is often argued that one cannot do anything about historical emissions but only regulate future emissions, and one should distribute the impacts only over future emissions, i.e., from 1990 to 2060. This is called the *grandfathering principle*. It gives a specific externality of 1.17 US\$ per kg of C or 0.32 US\$ per kg of CO_2.

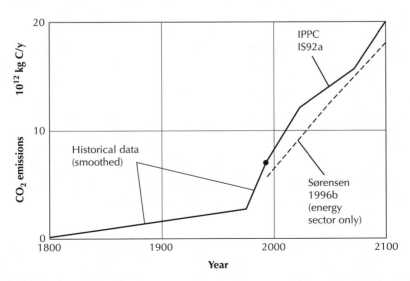

FIGURE 7.19 Historical and future global CO_2 emissions for the scenarios used in estimating greenhouse-warming externalities (Kuemmel *et al.*, 1997; based on IPCC, 1996b; Sørensen, 1997a).

To further illustrate the width of choice, we mention the method used in several earlier publications (Sørensen, 1997a,c). Here only the energy part of CO_2 emissions were considered, and it was assumed that doubling of CO_2 would occur in 2040 (cf. dashed line in Fig. 7.19). This leads to total energy-related CO_2 emissions for the period 1990–2060 amounting to 433×10^{12} kg of C. Assuming that the part of the impact externality caused by the energy sector is 75% (or 750×10^{12} US$), one obtains, using the grandfathering principle, a specific externality of 1.72 US$ per kg of C or 0.47 US$ per kg of CO_2. The range of estimates can then be summarized as in Table 7.3.

In the following examples, Estimate I from Table 7.3 is used as the central estimate, and the Estimates II and III are used as defining an uncertainty range, without claiming that values outside this range could not be defended (e.g., owing to different views taken with respect to monetizing).

The externalities for specific types of fossil-fuel combustion can now be evaluated. Among different types of fossil power production, present coal-fired power stations typically emit 0.27 kg of C kWh_{elec}^{-1}, gas-fired ones emit 0.16 kg of C kWh_{elec}^{-1}, and oil-fired plants emit 0.21 kg of C kWh_{elec}^{-1}, The transportation sector mainly uses oil products, and, for gasoline-driven automobiles, typical emissions would be 660 g of C $liter^{-1}$ or about 49 g of C per vehicle-km at an average of 13.5 km per liter of gasoline (approximately the current value).

Greenhouse Warming Impacts	Estimate I (IPCC, IS92a) Grandfathering	Estimate II (IPCC, IS92a) No Grandfathering	Estimate III (Sørensen, 1997a,c) Grandfathering
Cause (see text for further explanation)	All CO_2 emitted 1990–2060: 814×10^{12} kg of C	All CO_2 emitted 1765–2060: 1151×10^{12} kg of C	Energy-related CO_2 emitted 1990–2060: 433×10^{12} kg of C
Effect (throughout 21st century)	95% of total impact: 954×10^{12} US$	95% of total impact: 954×10^{12} US$	75% of total impact: 750×10^{12} US$
Specific externality	1.17 US$/kg of C 0.32 US$/kg of CO_2 0.26 ECU/kg of CO_2	0.83 US$/kg of C 0.23 US$/kg of CO_2 0.18 ECU/kg of CO_2	1.72 US$/kg of C 0.47 US$/kg of CO_2 0.38 ECU/kg of CO_2

TABLE 7.3   Summary of greenhouse-warming impact estimates (Kuemmel *et al.*, 1997).

For biofuels like biogas or methanol, one should not count the CO_2 emissions, owing to the short time interval between the carbon assimilation by plants and the subsequent release (assuming that even woody biomass for methanol production comes from relatively short-rotation crops), but emissions of other greenhouse gases like N_2O or CH_4 from burning of biomass should be included in the calculation. In those cases where heat and electricity are co-generated, one may assign the externality to the fuel input in order to avoid confusion over the distribution between multiple outputs.

LCA of Power-production Chains

This section gives examples of LCA calculations for various power-production chains, using conventional fuels (for reference) or renewable energy.

Coal-fired Power Station

Most present production of power takes place at fuel-based power stations, implying that the chains defining the life cycles start with fuel extraction and go through conversion steps until the final disposal of residues and release of pollutants to the air. The practice is currently changing as a result of more efficient use of the fuels' combined production of power and heat, sometimes in combined-cycle plants where energy is extracted from the fuel in more than one step. Also changing is the level of effluent control (currently capturing most SO_2 and NO_x from flue gases, in the future possibly also CO_2 for subsequent controlled storage). A traditional fuel chain for coal-based electricity production is shown in Fig. 7.9.

Recent studies have attempted to evaluate the impacts of each step in the fossil-fuel power chains, using a bottom-up approach based on particular installations at specific places. The results for coal are summarized in Table 7.4. A common feature is the dominating influence on total life-cycle externalities coming from global warming. The estimates made for greenhouse gas emissions above are used, but one should keep in mind that they pertain to the specific scenario of doubling greenhouse gas concentrations in the atmosphere by the mid-21st century. Selecting the correct externality to include in scenarios involving a reduction of fossil-fuel usage is a methodological problem. Many studies use the same values for different scenarios, which would lead to an underestimation of the benefits of substituting other forms of energy for those emitting greenhouse gases. The coal-fired power plant considered in Table 7.4 is located at Lauffen, Germany. The quantified impacts are mainly from a study performed under the EC Joule Programme (European Commission, 1995c), except for the greenhouse-warming impacts, which are estimated on the basis of Tables 7.2 and 7.3. The EC study looks at two power plants: one in Germany and a plant proposed for West Burton in England, with 99.7% particle removal, fairly decent SO_2 scrubbing, but poor NO_x emission control. The German plant has better NO_x cleaning than the British plant, but is situated in a densely populated area, and the calculation includes impacts from pollution across national borders, in contrast to the EC's treatment of the British plant, which deals only with pollution impacts inside the United Kingdom.

One of the central assumptions in the EC study is to monetize deaths using a statistical value of life (SVL) amounting to 2.6 M euro or 3.25 million US$ (see discussion in section 7.3.1). Other monetizing assumptions used in the study are shown in Table 7.5. The impacts of the coal fuel cycle are dominated by the global

	Type of Impact	Uncertainty*	Monetized Value (US cents/kWh)	Uncertainty and Ranges*
Environmental Impacts	Emissions (g/kWh)			
1. Plant construction/decommissioning	NA		NA	
2. Plant operation				
CO_2	880	L		
SO_2 (may form aerosols)	0.8	M		
NO_x (may form aerosols)	0.8	M		
particulates	0.2	M		
CH_4	3	M		
N_2O	0.5	H		26–53
Greenhouse warming (cf. Table 7.3)	From CO_2, CH_4, …		35.8	H, r, n
Degradation of building materials	From acid rain		0.03	
Reduced crop yields	From acid rain, ozone		0	
Forest and ecosystem impacts	**Cases:**	H	0	
Ozone impacts	0.1 per TWh		0.03	
European-wide Impacts:		H		H, r, n
Mortality from primary particles (PM_{10})	0.8 per TWh		0.25	H, r, n
from secondary aerosols	3.0 per TWh		1.0	
from chronic effects	14 per TWh		NQ	
Morbidity from dust and aerosols,				
major acute	17.4 per TWh		0	M, r, n
minor acute	187 000 workdays lost per TWh		0.25	M, r, n
chronic cases	760 per TWh		0.01	M, r, m
Noise (from power plant)			0.01	M, l, n
Occupational Health and Injury				
1. Mining diseases & death due to radon exp.	0.8 per TWh		0	M, l, m
Mining accidents, death	0.2 per TWh		0.06	L, l, n
major injury	6.8 per TWh		0.15	L, l, n
minor injury	70.5 per TWh		0.01	H, l, n
2. Transport, death	0.03 per TWh		0.01	L, l, n
major injury	0.31 per TWh		0	M, l, n
minor injury	9.8 per TWh		0	H, l, n
3. Construction/decommissioning	0 per TWh		0	M, l, n
4. Operation (injury)	0.08 per TWh		0	L, l, n
Economic Impacts				
Direct economy			3.1–5.6	
Resource use	Low but finite		NQ	
Labour requirements			NQ	
Import fraction (German plant)	Local coal assumed		NQ	
Benefits from power			6.5–19	
Other Impacts				
Supply security	Many import options		NQ	
Robustness (against technical error, planning errors, assessment changes)	Fairly low for large plants		NQ	
Global issues	Competition		NQ	
Decentralization and consumer choice	Not possible		NQ	
Institution building	Modest		NQ	

TABLE 7.4 Impacts from site-specific German coal fuel chain (Kuemmel *et al.*, 1997; occupational and environmental impacts, except for greenhouse warming, are based on European Commission, 1995c).

Notes: NA = not analyzed, NQ = not quantified. Values are aggregated and rounded (to zero if below 0.01 US c kWh^{-1}).
*L, M, H: low, medium, and high uncertainty. l, r, g: local, regional, and global impact. n, m, d: near, medium, and distant time frame. Monetary impacts originally given in euro are translated using 1 euro = 1.25 US$.

Health Effect	Euro	US$
Mortality (SVL)	2 600 000	3 250 000
Respiratory hospital admission	6600	8250
Emergency room visits	186	233
Bronchitis	138	173
One day with activity restricted	62	78
Asthma attack	31	39
One day with symptoms	6	8

TABLE 7.5 Valuation assumption of the European Commission study (European Commission, 1995a–f).

warming effects, where an SVL of 3 M US$ was used (slightly lower than the EC value at current exchange rates). However, the impacts associated with other emissions may be a factor of 10 lower than for the average coal-fired plant in some countries, as inferred from comparing the emission standards assumed here with typical emissions from current installations in the United Kingdom.

Several of the impacts mentioned in Table 7.4 are not quantified, so an assessment must deal with the qualitative impacts as well. In those cases where it has been feasible, uncertainties are indicated (L: low, within about a factor of 2; M: medium, within an order of magnitude; and H: high, more than an order of magnitude), and the impacts are labeled according to whether they are local, regional, or global (l, r, or g), as well as whether they appear in the near term (n: under 1 year), medium term (m: 1–100 years), or distant term (d: more than 100 years in the future).

A similar calculation for a combined heat and power plant on the island of Funen, Denmark, shows smaller impacts than those of the German counterpart, despite higher CO_2 emissions. The reason is lower population density and higher wind speeds (carrying emission outside the region studied).

Nuclear Fuel Chain

For nuclear power plants, the assessment in principle proceeds as in the fossil case, but one expects more important impacts to come from plant construction, fuel treatment, and final storage of nuclear waste. Because of the very long time horizons associated with radiation-induced health problems and deposition of nuclear waste, the impact assessment period by far exceeds the plant operation period, and questions of discounting and of changes in technological skills over periods of several centuries play a decisive role. The monetary issues are discussed in section 7.1. Another issue particularly (but not exclusively) important for nuclear power is accident-related impacts, the treatment of which would use risk analysis.

The observed number of large nuclear accidents involving core meltdown in commercial reactors is currently two (Three Mile Island and Chernobyl). The implied "probability" in the nuclear case is illustrated in Table 7.6.

Counting two accidents over the accumulated power production to the end of 1994 (when this estimate was made) one gets a 10^{-4} per TWh frequency, or a 5×10^{-5} per TWh frequency for an accident with severe external consequences. At the time of the Chernobyl accident, the estimate would have been 3–4 times higher, owing to the much lower accumulated power production of reactors worldwide by 1986. Today it would again be lower, due to more accumulated production and despite the little new nuclear capacity added. For comparison, the built-in probability for an accident with Chernobyl-type consequences for a new, state-of-the-art light-water nuclear reactor is calculated by fault-tree analysis to be about 1.25×10^{-6} per TWh (ST2-accident; cf. European Commission, 1995e; Dreicer, 1996). The factor 40 difference between the two numbers comes partly from the difference between a state-of-the-art reactor and the average stock and partly from the difference between the probability of anticipated accidents and the actual frequency estimate that includes unanticipated events.

Historical Evidence and Interpretation

Accumulated experience before Three Mile Island accident in 1979	3000 TWh
Accumulated experience before Chernobyl accident in 1986	5800 TWh
Accumulated experience to end of 1994	20 000 TWh
Implied order of magnitude for frequency of core-melt accident	1×10^{-4} TWh^{-1}
A. Implied order of magnitude for accident with Chernobyl-type releases	5×10^{-5} TWh^{-1}
Chernobyl dose commitment (UNSCEAR, 1993)	560 000 person-sievert

Valuation	million US$	million euro
Induced cancers (SVL = 2.6 M euro, no discounting)	250 000	200 000
Birth defects	25 000	20 000
Emergency teams, clean-up teams, security teams	63	50
Early radiation deaths (SVL = 2.6 M euro)	125	100
Evacuation and relocation	125	100
Food bans and restrictions	125	100
Unplanned power purchases	1000	1000
Capacity loss and reduced supply security	12 000	10 000
Cost of encapsulation and clean-up (at plant and elsewhere)	212 000	170 000
Increased decommissioning costs	125 000	100 000
Impact on nuclear industry (reputation, reduced new orders)	125 000	100 000
Monitoring, experts' and regulators' time	12	10
Concerns in general public (psychosomatic impacts)	125	100
B. Total estimate of Chernobyl accident costs	750 000 M US$	600 000 M euro
Average large-accident cost of generating nuclear power (**A** times **B**)	3.8 US cents/kWh	3.0 M euro/kWh

TABLE 7.6 Frequency of, and damage by, large nuclear accidents (Sørensen, 1995c).
Exchange rate assumptions: 1 M euro = 1.25 US cents = 1.25 ¥

The outcome of an LCA chain analysis (but excluding side-chains) of the French state-of-the-art nuclear power plant performed within the EC study (European Commission, 1995e) is shown in Table 7.7, except that the impacts of major accidents are taken from the estimate of Table 7.6. The emphasis is on impacts from release of radioisotopes, and again monetizing involves the assumptions stated in the notes to Table 7.4. The largest normal-operation impacts are from the reprocessing step. The magnitude is based on estimates made by the operator of the plant at La Hague, COGEMA. However, their appraisal has recently been challenged as constituting a gross underestimate, compared with values derived from actual samples obtained at sea near the plant (Butler, 1997). As regards the deposition of high-level waste, no quantitative estimate was made of incidents that may occur over the required long deposition periods. Also, the impacts of proliferation of nuclear materials and know-how have not been quantified.

The use of historical data (like those for Chernobyl) for the nuclear-accident analysis may be criticized for not taking into account technological progress that may have occurred later (cf. the discussion above on average and state-of-the-art technology). Recent disclosure of a possible earthquake occurring at a critical time of the Chernobyl accident sequence (DR, 1997) casts doubt on the causes of the accident, but the reactor design peculiarities certainly contributed to the temporal distribution and high temperatures of radioactivity releases (Sørensen, 1987a). However, a risk study does depend on the type of use to be made of the assessment, which may not be to characterize emerging technology, but rather to analyze an average existing one. For the few countries where nuclear power plants are currently built, there are questions of the standards of operational safety (e.g., absence of provisions for informing the population about the need to stay indoors with controlled closure and opening of windows, about evacuation, about food bans, etc.). Preparations for such events, with associated drills, have not been carried out in all countries currently using large amounts of nuclear power.

The impacts in Table 7.7 are for a nuclear power station located at Tricastin in France, with fuel manufacture and reprocessing also taking place at specific French sites. The impacts include those from use of fossil fuels (e.g., in the construction of nuclear power plants), because the evaluation performed is based on a marginal addition of one nuclear power station to the presently existing system. The impacts from side-chains yielding inputs to the main fuel cycle are considered small compared to the largest impacts included, just as in the fossil-fuel case, although this may not apply as well in the nuclear case because of its higher proportion of construction costs in the total.

Environmental Emissions	Dose Commitment (Person-sievert/TWh)	Uncertainty*	Monetized Value (US cents/kWh)	Uncertainty and Ranges*
Plant construction/decommissioning	NA		NA	
CO_2, SO_2, NO_x, particles	NA		NA	
Noise, smell, visual impact	NA		NA	
Radioactivity (by distance from event):				
1. Fuel extraction and refinement,				
local	0.1	L	0.01	M, l, n
regional	0.1	L	0	M, r
global	0	L	0	M, g
2. Normal power-plant operation,				
local	0.4	M	0.01	M, l, m
regional	0	M	0	M, r
global	1.9	M	0.04	M, g, d
3. Power-plant accidents (cf. Table 7.6),				
local	5	H	0.25	H, l, m
regional	10	H	0.4	H, r, m
global	15	H	0.6	H, g, d
4. Reprocessing and waste handling,				
local	0.2	H	0	H, l, d
regional	0	H	0	H, r, d
global	10.2	H	0.24	H, g, d
Social Impacts				
Occupational injuries	0		NQ	
Occupational radioactivity				
1. Fuel extraction and refinement	0.1	L	0	M
2. Construction and decommissioning	Over 0.02	M	0	M
3. Transport	0	L	0	L
4. Normal power-plant operation	0	M	0	M
5. Power-plant accidents	0	M	0	H, l, n
6. Reprocessing and waste handling	0	H	0	H
Accident handling (evacuation, food ban, clean-up, back-up power, cf. Table 7.6)			1.9	H, r, m
Indirect accident impacts (expert time, mistrust, popular concern, cf. Table 7.6)			0.63	H, g, m
Economic Impacts				
Direct costs			3.75–7.50	L
Resource use	Not sustainable without breeders		NQ	
Labour requirements	Low		NQ	
Import fraction (for France)	Low		NQ	
Benefits from power (consumer price)			5.6–16.9	L
Other Impacts				
Supply security	Medium		NQ	
Robustness (technical, planning)	Important		NQ	
Global issues (proliferation and weapons)	Important		NQ	
Decentralization and local choice	Not possible		NQ	
Institution building (safety and control)	Fairly high		NQ	

TABLE 7.7 Impacts of nuclear fuel cycle (European Commission, 1995e; for accident evaluation, Sørensen, 1995c).

Notes: NA = not analyzed, NQ = not quantified. Values are aggregated and rounded (to zero if below 0.01 US cent kWh^{-1}).
*L, M, H: low, medium, and high uncertainty. l, r, g: local, regional, and global impact. n, m, d: near, medium, and distant time frame. Monetary impacts originally given in euros are translated using 1 euro = 1.25 US$.

Renewable Energy Chains
Wind Power

Life-cycle analysis has been done for wind turbines of unit size around 500 kW installed in Denmark and elsewhere. These units typically feature three-bladed fiberglass rotors mounted on steel-tube or concrete towers, transmitting power via a gearbox to an induction generator, which is attached to the standard utility grid lines via an electronic control box. Several of the impacts depend on the natural, social, and human settings. This should be considered when transferring data between settings.

The highest penetration of wind power is in the Danish electricity system, where already in the late 1990s, it was about 12%. The current (2010) penetration is above 20%, and the question of energy storage may have to be addressed and corresponding impacts included in the LCA. However, for grid systems like the European one, characterized by strong international power transmission links, e.g., from Denmark to the Norwegian hydro system, based on annual storage cycles, the cost of dealing with the variability of wind energy production could be very low (Sørensen, 1981a; Meibom et al., 1997).

The overall cost of producing wind energy at the best locations in Denmark is currently about 0.25 DKr kWh^{-1} (4 US cents kWh^{-1}, evaluated by Sørensen, 1995a; Aastrand et al., 1996), with operation and maintenance (O&M) constituting an average of 0.07 DKr kWh^{-1} (1 US cent kWh^{-1}) out of this, averaged over an assumed 20-year lifetime. Because capital cost dominates, there is much less uncertainty in cost after the turbine is installed than there would be for a fuel-based system. The import fraction of the capital cost is 28% and that of the running cost, 15% (Sørensen, 1986). The employment factor (full-time equivalent per million DKr spent) is about 3. Whether labor is considered a positive or negative impact within an economy depends on whether there is unemployment; for the individual, it depends on how tightly social assets are bound to work in a given society and on that individual's attitude to work (work as a satisfying activity versus work as just the means of funding life activities). In any case, creating jobs within a society is often viewed as preferable to creating jobs abroad.

Land use associated with wind turbines may be assessed in the following way. In order to avoid reduced wind access, wind turbines in a wind park have to be placed several rotor diameters apart. Danish wind turbines are placed both in park configurations and individually. The land between them and right up to the towers may be used for agriculture, so the largest cost, in terms of land use, is often the access roads needed for service purposes. Typical values are 10 m^2 per kW of rated power (Sørensen, 1986). Land resources may be returned after decommissioning.

The visual impact of wind turbines also depends the particular location occupied by individual turbines or arrays of turbines. Aesthetically, the slender towers and white blades have been found to receive a positive reception among viewers. A factor in the assessment is also the long history of wind power in Denmark, which causes wind turbines to be more easily accepted as parts of the visual landscape, along with the fact that farmhouses and churches often have towers as conspicuous as those of wind turbines. For wind parks, architects are usually employed to ensure a visually acceptable integration into a particular landscape (Sørensen, 1981b). Further expansion of Danish wind power supply will increasingly involve offshore siting, owing to the dense population of the country, and several suitable locations have been identified.

The mechanical noise from Danish wind turbines erected in 1993 is 97 dB(A) at the point of emission.[*] The noise mostly originates from the gearbox and can be reduced by sound insulation. Aerodynamic noise from the blades depends on wind speed and is similar to that of other structures or vegetation (Sørensen, 1981b). The noise from a single wind turbine becomes inaudible just a few rotor diameters from the site, and even for wind farms the increment over background noise is less than 2 dB(A) at a distance of 1.5 km (European Commission, 1995f). Average background noise is typically 35–37 dB(A) in quiet rural locations, and legislation in Denmark requires that this level is not noticeably increased by human activities.

Telecommunication interference by wind turbines has been studied and has been found to be similar to that of static structures (e.g., buildings), with the exception of frequency modulations propagated in specific shapes of mountain locations (Sørensen, 1986).

Extraction of power from wind has a slight influence on the microclimate below and behind the turbines, but otherwise there is no pollution associated with the operation of wind turbines, assuming lubricants are contained. The chief potential source of pollution is in the manufacture and maintenance operations, which in Table 7.8 have been assumed to employ the current mix of fuels for the contribution from energy inputs (to both manufacture and transportation). The impacts will decrease in proportion to the possible future transition to renewable energy sources. Available accident statistics show there is a negligible risk that members of the general public will be hit by expelled rotor blades in cases of rotor failure (Sørensen, 1981b).

The work environment at manufacturers of wind-turbine components is similar to that of other equipment manufacturers, while the work environment for tower building, assembly, and maintenance is similar to that of work in the building industry, with open-air activities and scaffold work at a height. With proper safety precautions, the work generally offers a varied and challenging environment. The accident rates assumed in Table 7.8 are taken from industrial statistical data for the relevant sectors. Historically, one significant component has been accidental electrocution during improper maintenance attempts, usually by the owner, rather than by professional teams.

---

[*]dB(A) is an average logarithmic sound scale weighted to reflect the human perception of noise.

Health problems are primarily associated with the industrial part of manufacture, including in particular the use of epoxy resins in blade manufacture. Modern production technology should confine this step of production to closed spaces with fully automated facilities, so that employees do only control-room work and the risk of exposure to harmful chemicals would basically be via accidents. Numerical estimates of health and accident risks are uncertain because of the aggregate nature of available statistical data and because of the different production techniques used by different wind-turbine manufacturers.

The social benefit of wind turbines comes from using the electric power produced. As discussed above, this benefit might decrease if, at higher penetration of wind energy in the power-supply system, the variations of wind-produced power led to occasional dumping of excess power. As explained, this is unlikely to become an issue for Denmark, even for large wind energy penetrations.

Wind energy can affect the infrastructure of the electricity supply system, in a positive manner if the sites of wind turbines are dispersed and demands on the power transmission network are reduced owing to shorter average distances to consumers but, on the other hand, demands may be increased as a result of connecting distant wind farms, e.g., at offshore locations.

Security of supply is generally high, with the qualifications related to the variability of wind conditions made in the preceding subsection. Failures occur in all parts of the system, but not according to any particular pattern. Around 10% of Danish wind turbines experience a component failure in any given month. About 4% of the failures result in replacement of a component (e.g., a blade, the generator, the control unit), according to the newsletter *Windpower Monthly/Windstats Quarterly*. The failures are represented in the overall O&M costs, which include an insurance premium. A wind-power system has little sensitivity to individual turbine failures because of the modular nature of the system. Because of the short time lag between deciding to build a turbine and its operation, wind technology is much less sensitive to planning errors (e.g., wrong forecast of future loads) and changes in criteria used for selecting technology than are systems with several years' lag between decision and operation.

Wind power is consistent with keeping options open for the future. It might be abandoned without repercussions (in contrast to nuclear installations and waste repositories, which have to be maintained long after decommissioning). Wind power is consistent with the goal of creating an energy system with as little environmental impact as possible. It is a national solution, and it avoids global issues, such as the overuse of resources by rich countries, while at the same time it ensures national independence of energy supply, in proportion to its penetration.

Wind power allows for broad participation in decision-making in the energy field, and as one of several possible decentralized solutions, it permits different local communities to choose different solutions. Few institutional constructions are needed if ownership is dispersed, but one possible set-up would be for present utility companies to own the wind turbines, in which case the regulatory implications are no different from current ones. With dispersed ownership,

	Impact Type	Uncertainty*	Monetized Value (US cents/kWh)	Uncertainty and Ranges*
Environmental Impacts	Emissions (g/kWh)			
Releases from fossil energy currently used:				
1. Turbine manufacture (6.6 GJ/kW rated)				
CO_2 (leading to greenhouse effect)	12.1	L	0.39	0.28–0.57
SO_2 (leading to acid rain and aerosols)	0.05	L	0.01	H, r, n
NO_x (possibly aerosols and health impacts)	0.04	L	0	H, r, n
particulates (lung diseases)	0.002	L	0.01	H, r, n
2. Operation (2.2 GJ/kW over 20-year lifetime)				
CO_2 (leading to greenhouse effect)	3.8	L	0.12	0.09–0.18
SO_2 (leading to acid rain and aerosols)	0.01	L	0	
NO_x (possibly aerosols and health impacts)	0.02	L	0	
particulates (lung diseases)	0	L	0	
	Other:			
Noise increase from gearbox at about 1 km distance from wind-blade interaction	<1 dB(A) <3 dB(A)		0.01 total	H, l, n
Land use	10 m²/Kw		NQ	
Visual intrusion	+/− like buildings, church towers, etc.		NQ	
Social Impacts				
Occupational injuries (manufacturing and materials):				
1. Turbine manufacture, death	0.03/TWh	L	0	L, l, n
major injury	0.9/TWh	L	0.01	L, l, n
minor injury	5/TWh	M	0	M, l, n
2. Operation (same categories combined)			0	M, l, n
Economic Impacts				
Direct costs			5–11	
Resource use (energy payback time given)	1.1 y	L	NQ	
Labour requirements (manufacture)	9 person y/MW	L	NQ	
Import fraction (for Denmark)	0.28	L	NQ	
Benefits from power sold (penetration < 30%)			5–15	
Other Impacts				
Supply security (variability in wind is high, entry based on plant availability)	High		NQ	
Robustness (up-front investment binds, entry based on technical reliability)	High		NQ	
Global issues (non-exploiting)	Compatible		NQ	
Decentralization & choice (less with large size)	Good		NQ	
Institution building (grid required)	Modest		NQ	

TABLE 7.8 Impacts from current Danish wind energy systems (Sørensen, 1996e; with use of data from Sørensen, 1981b, 1993b; Meyer *et al.*, 1994; European Commission, 1995f).

Notes: NA = not analyzed, NQ = not quantified. Values are aggregated and rounded (to zero if below 0.01 US cent kWh⁻¹).
*L, M, H: low, medium, and high uncertainty. l, r, g: local, regional, and global impact. n, m, d: near, medium, and distant time frame.
Monetary impacts originally given in euros are translated using 1 euro = 1.25 US$.

essential questions are free access to the power distribution grid and negotiating prices for buying from and selling to the national grid. In Denmark, access is at present ensured by legislation, and price negotiations are conducted between the utility companies and the wind turbine owner's organization. Before 2002, they were overseen by the Department of Energy, which was supposed to intervene if fair pricing was not obtained between the unevenly sized institutions. Some of the social and systemic impacts discussed above cannot be quantified; they are flagged in Table 7.8.

Photovoltaic Power

Even more than for wind power, the LCA for the PV power system is dominated by the manufacturing process. The manufacture of crystalline silicon panels involves the main steps depicted in Fig. 7.20, whereas, for amorphous silicon panels, the wafer formation, cutting, and surface doping steps are replaced by direct vapor deposition of doped gases onto a substrate. Different types of solar panels differ with respect to use of substrates, glass cover material, and/or films. The cells may have surface textures and elaborate grooving patterns for conductors, and modules may incorporate reflectors and, in some cases, may bypass diodes and even inverters (which otherwise would be system components). Finally, mounting of panels in arrays may involve dedicated support structures or may be building integrated.

The system side may require additional transformer equipment as well as battery storage or back-up devices in the case of stand-alone systems. Decommissioning and dismantling of the solar equipment is expected to follow recycling and re-use patterns emerging for the building industry in general, probably as a front-runner industry.

Assessing PV impacts is aided by experience with similar production-process steps found throughout the microelectronics industry. The basic raw material for silicon cells is silicon dioxide (sand, quartzite). It is reduced to metallurgical-grade silicon in arc furnaces. Both mining and reduction may produce dust (and hence risk of silicoses). The arc furnaces also produce carbon monoxide and a range of silicon-containing compounds in dust that might be inhaled during cleaning operations (Boeniger and Briggs, 1980). At its current low penetration, the PV industry has used scrap material obtained inexpensively from the microelectronics industry but, in the future, dedicated solar-grade material (much less expensive to produce than microelectronics grade) will be used.

Amorphous cells require production of silane (SiH_4). For crystalline cells, usually trichlorosilane is produced in a fluidized bed and then is purified to semiconductor-grade multicrystalline silicon. Subsequent processes are doping and growth of monocrystalline ingots. These are then ground to cylindrical shape and sliced into wafers, which are then cleaned. Multicrystalline cells may be produced by slicing ingots made of cast multicrystal silicon in a process similar to that for crystalline ingots, or they may be formed by vapor deposition, as is done for amorphous cells, albeit at considerably higher deposition temperatures. The material used for monocrystalline cells is currently expensive, including input energy, but future production is expected to use thin-film technology for monocrystalline cells.

FIGURE 7.20   Main steps in LCA chain for monocrystalline silicon photovoltaic cells (Sørensen, 1995c).

Chlorosilane production involves hydrochloric acid, and the chlorosilanes themselves are corrosive, a skin and lung irritant, and toxic. Workers are required to use protective clothing and face masks with filters. Further risks are posed by hydrogen–air mixtures, which could ignite and explode. One such case has been reported (Moskowitz *et al.*, 1994). For amorphous silicon, special precautions are needed for handling silane gas, as it ignites spontaneously. One solution is never to store large quantities of silane gas and to use special containers designed to avoid leakage even with strong pressure increases. Also, fabrication sites are equipped with automatically releasing fire-extinguishing devices.

Vacuum growth of crystalline material may involve dispersal of oily aerosols that have to be controlled by wet scrubbers and electrostatic filters (CECSMUD, 1982). Doping of p-type material may involve boron trichloride, which reacts with water vapor to form acids easily absorbed through the skin, or diborane, which is a strong irritant and flammable as well. The n-type doping at the top layer of a crystalline cell uses phosphorus diffusion of $POCl_3$ or P_2O_5 in sealed environments, whereas the n-type doping of amorphous cells may involve phosphine (PH_3), a highly toxic substance widely used in the semiconductor industry (Watt, 1993).

Grinding and cleaning of wafers produce silicon-containing slurry with remains of the detergents used. An alternative is ribbon-growth, which avoids these problems (CECSMUD, 1982). Amorphous cell manufacture also involves a number of cleaning agents. Etching of surface textures may employ a variety of techniques, selected on the basis of concern for recycling of chemicals and reduction of the use of toxic substances (Watt, 1993). Workers have to wear protective clothing, and high levels of ventilation are required. Drying uses liquid nitrogen and may be fairly energy intensive.

Screen printing of electric circuits involves possible work environment problems familiar to those in the microelectronics industry (caused by metal particles and organic solvents). Laser grooving involves the laser safety precautions for radiation and fires, and the application of coatings like titanium oxide or silicon dioxide is considered relatively harmless. Cell testing and light soaking of amorphous cells (in order to avoid restructuring degradation) should be done in special rooms because of the risk of exposure to ultraviolet radiation. Personnel replacing bulbs should wear safety masks and gloves if pressurized krypton lamps are used. Polymer coatings, such as ethylene vinyl acetate (EVA) or polyvinyl fluoride (Tedlar), may have some health impacts during their manufacture. If soldering is used in module assembly, fumes should be controlled. The tendency is for increasing use of robots in the manufacturing process, which leads to reductions in health impacts for remaining workers. Cells based on cadmium telluride and gallium arsenide involve different types of potential impacts (Moskowitz et al., 1995), which could lead to very high life-cycle impacts (Alsema and Engelenburg, 1992). However, these cells are based on expensive materials, the natural abundance of which is much smaller than that of silicon, and therefore total recycling is required. A large fraction of the calculated impacts is involved with releases of toxic materials in fires involving solar panels. The photovoltaic industry is aware of the problems of current pilot production lines and aims at controlling or replacing the chemicals identified as troublesome (Patterson, 1997).

There are few impacts during the operation of photovoltaic installations. Land use may be an issue for central plants, but not for building-integrated systems.

	Impact Type	Uncertainty*	Monetized Value (US cents/kWh)	Uncertainty and Ranges*
Environmental Impacts	Emissions (g/kWh)			
Releases from fossil energy used in the steps of the PV conversion cycle:				
CO_2 (m-Si ¤)	75, 30	L		H, g, m
(a-Si ¤)	44, **11**	L		H, r, n
SO_2 and NO_x				
(m-Si ¤)	0.3, 0.1	L		H, r, n
(a-Si ¤)	0.2, **0.04**	L		H, r, n
Greenhouse effect from fossil-fuel emissions (m-Si)			2.4, 1.0	0.7–3.5
(a-Si, for both see ¤)			1.4, **0.4**	0.3–2.1
(if non-fossil energy is used in PV production)			or **0**	0
Mortality and morbidity from fossil-fuel air pollution described above				
(m-Si, ¤)			0.05, 0	H, r, n
(a-Si, ¤)			0.03, **0**	H, r, n
Land use	**0**		**0**	L, l, n
Visual intrusion			NQ	
Social Impacts				
Occupational injuries:				
1. From fossil fuel use				
(m-Si ¤)			0.01, 0	L, l, n
(a-Si ¤)			0, **0**	L, l, n
2. From panel manufacture			NA	
3. From construction and decommissioning (differential from using other building materials)			**0**	L, l, n
4. From operation			**0**	L, l, n
Economic Impacts				
Direct costs (early 2000s)			38–76	
(future ¤)			**3.8–11.3**	H
Energy payback time (¤, a-Si)	3 y, 0.5 y		NQ	
Labor requirements (¤)	40, 4 man-y/MW		NQ	
Benefits from power sold (penetration < 20%)			**5–15**	H
Other Impacts				
Supply security (plant availability)	High		NQ	
Robustness (technical reliability)	High		NQ	
Global issues (non-exploiting)	Compatible		NQ	
Decentralization and choice	Good		NQ	
Institution building (grid required)	Modest		NQ	

TABLE 7.9 Estimated impacts from present and future rooftop photovoltaic energy systems based on multicrystalline (m-Si) or amorphous (a-Si) silicon cells, placed at average European locations (Sørensen and Watt, 1993; Yamada *et al.*, 1995; Sørensen, 1994b, 1995c, 1997a). The boldface numbered alternatives pertain to what are considered the most likely technologies for the future.

Notes: NA = not analyzed, NQ = not quantified. Values are aggregated and rounded (to zero if below 0.01 US cent kWh^{-1}).
¤ First entry for early 2000s, second for mature industry having reached cost goals.
*L, M, H: low, medium, and high uncertainty. l, r, g: local, regional, and global impact. n, m, d: near, medium, and distant time frame. Monetary impacts originally given in euros are translated using 1 euro = 1.25 US$.

Albedo changes caused by the presence of the panels are not significantly different from those of alternative roof surfaces, in the case of roof-integrated cells, but could have climatic impacts in case of large, centralized solar plants located in desert areas, for example. Reflections from panels located in cities could be annoying, and consideration of visual impacts will generally require careful architectural integration of panels. In some areas, cleaning of solar panel surfaces for dust may be required, and electronic control equipment, such as inverters, may cause radiofrequency disturbances if they are not properly shielded. As mentioned for non-silicon cells, the behavior of panels during fires is an important consideration. Recycling of solar cell equipment has also been mentioned as a requirement (Sørensen, 1993b).

For a silicon-based photovoltaic system integrated into a building, the LCA impact evaluation presented in Table 7.9 shows modest negative impacts, most of which occur during the manufacturing phase, and substantial positive impacts on the local and global society. The impacts during manufacturing largely result from the use of fossil fuels for mining, manufacture, and transport, according to the marginal approach taken in the references used. A comprehensive analysis of a renewable energy scenario would instead use the new energy system to determine indirect energy inputs, with substantially altered results as a consequence.

The source used for energy payback times and carbon dioxide emissions is Yamada et al. (1995). Other estimates give current energy payback times for a-Si and c-Si (monocrystalline silicon) systems as 2–3 years and future ones as less than 1 year (Alsema, 1997; Frankl et al., 1997), while current greenhouse gas emissions have been estimated at 100–200 g of CO_2 equivalent per kWh of power produced, declining to some 40 g of CO_2-eq. kWh^{-1} after the year 2010 (Dones and Frischknecht, 1997).

The photovoltaic case is special, because its direct cost at present is far higher than that of the alternatives. If the cost goes down in the future, many of the impacts will also go down, because they are associated with material use or processes that will have to be eliminated or optimized in order to reach the cost goals. The lower cost estimate for 2010 is based on the stacked-cell concept of Wenham et al. (1995a). Yamada et al. (1995) quote about 6.7 US cents kWh^{-1} for amorphous cell systems integrated into roofs. The cost per kWh produced obviously also depends on the location of the building.

The different spans of economic benefits from the power sold, exhibited in Tables 7.4–7.9, are meant to reflect the differences in load-following capability of the three different types of plants. For a variable resource, such as solar energy, there will be additional costs if the penetration becomes large compared with the size of the grid system (say, over 30%), because, in that case, additional equipment would likely have to be introduced to deal with the fluctuating power production.

Biogas Plants

As described in Chapter 4, biogas can be produced from several biomass resources, such as manure, plant residues, or household waste. Table 7.10

shows a typical composition of waste in Denmark. The biogas plant at Ribe in Denmark is a typical example of current biogas technology that accepts such waste. It annually converts about 110 million kg of manure and 30 million kg of other organic waste into 100 TJ of biogas. This is at the lower end of conversion efficiencies for the ten large biogas plants operated in Denmark, ranging in energy terms from 30% to 60%, depending on feedstock (Danish Energy Agency, 1996; Nielsen and Holm-Nielsen, 1996). A future average conversion efficiency of 50% can be expected.

Biogas consists of methane (CH_4) plus varying amounts of carbon dioxide (CO_2). Methane in particular has a high greenhouse-warming potential (GWP), so that leakages during production and transport, e.g., via pipelines or in containers, have to be avoided or kept at a low value. Feedstock for the Ribe biogas plant is mainly slurry from farms that is transported by road to the plant, but there is a small solid component.

Substance	Weight (percent)	Range
Carbon	25	15–35
Oxygen	18	12–24
Hydrogen	3	2–5
Nitrogen	0.6	0.2–1.0
Sulfur	0.003	0.002–0.6
Chlorine	0.7	0.5–1.0
Water	20	15–35
Ash	25	15–40
Lower heating value, municipal solid waste	8.8 GJ t^{-1}	8.4–9.2 GJ t^{-1}
Lower heating value, industrial waste	13.7 GJ t^{-1}	8.7–19.0 GJ t^{-1}

TABLE 7.10 Elementary analysis and energy content of Danish waste. Source: Formidlingsrådet (1989).

Table 7.11 gives some LCA-relevant technical data for the Ribe plant. One issue of concern is the presence of small amounts of hydrogen sulfide (H_2S), which could contribute to acid-rain formation. At one Danish biogas plant (at Fangel), a cleaning process aimed at reducing the H_2S content to between 700 and 1500 ppm has been tested (Danish Energy Agency, 1995). This is equivalent to a weight percentage of 0.09 to 0.2, the upper value being equal to the one assumed in Table 7.11. More stringent sulfur-emission limits are likely to be implemented if biogas production reaches an important penetration in future energy systems.

Other important emissions are emissions of methane liberated from the slurry tanks before or during slurry collection and emissions while the slurry is being transported to the biogas plant. It is estimated that collecting the slurry and using it in a biogas plant will have lower methane emission levels than the emission levels from current practice, where the manure is stored for several months before being spread onto fields. Nielsen and Holm-Nielsen (1996) calculate the reduction for the Ribe plant to be 160 t of CH_4 per year, minus 40 t due to leakage from storage at the biogas plant. The greenhouse-warming potential of these amounts of methane should be credited to the biogas energy system, because the agricultural sector life-cycle impacts are not part of the investigation and thus any change must be included as an indirect cost or benefit to the energy sector. The total methane emissions from Danish livestock in 1990 were about 160 million kg (from 11.7 million animals, mainly cattle and pigs).

Parameter	Value	Source	Remarks
Technical Data			
Specific investment	2750 US$/kW	Danish Energy Agency, 1995	45.3 M DKr total
O&M	9.6% y^{-1}	Danish Energy Agency, 1995	4.6 M DKr annually
Net capacity	2.7 MW	Nielsen & Holm-Nielsen, 1996	10 000 m^3 per day
Annual load	8700 h	Nielsen & Holm-Nielsen, 1996	
Lifetime	20 y	Nielsen & Holm-Nielsen, 1996	
Lifetime generation	469.8 GWh		
Overall net efficiency	≈35%	Danish Energy Agency, 1995	30%–60%, dep. on feedstock
Input and Composition			
Biomass	410 t d^{-1}	Nielsen & Holm-Nielsen, 1996	60% cow manure, 20% pig slurry, rest is other waste
Biomass Transport			
Average total	32 km	Nielsen & Holm-Nielsen, 1996	
Average animal slurry	22 km	Nielsen & Holm-Nielsen, 1996	
Biogas Composition			
CH_4	64.8%	Nielsen & Holm-Nielsen, 1996	
CO_2	35%	Nielsen & Holm-Nielsen, 1996	
Rest (H_2, N_2, H_2S) taken as H_2S	0.2%	Fritsche & Rausch, 1993	Generic database
Combustion value (MJ m^{-3})	23.4	Nielsen & Holm-Nielsen, 1996	
Material Demand			Energy used for manufacture
Steel	5 t	Fritsche & Rausch, 1993	Generic data, 22.2 GJ/t
Concrete	10 t	Fritsche & Rausch, 1993	Generic data, 4.6 GJ/t
Transport of Materials			
Steel by truck	150 km	European Commission, 1995c	For construction
Steel by railway	50 km	European Commission, 1995c	For construction
Concrete by truck	50 km	European Commission, 1995c	For construction
Process Demands	**per MJ biogas**		
Process heat	0.12	Nielsen & Holm-Nielsen, 1996	
Electricity	0.01	Fritsche & Rausch, 1993	Generic data
Emissions	**per MJ biogas**		
CH_4 from storage at plant	0.4 g	Nielsen & Holm-Nielsen, 1996	
CH_4 avoided at farm tanks	1.6 g	Nielsen & Holm-Nielsen, 1996	Impacts to be subtracted
Miscellaneous			
Area demand	1 ha		Estimate

TABLE 7.11 Ribe Biogas plant technical data (Kuemmel *et al.*, 1997).

	Impact Type	Uncertainty*	Monetized Value (US cents/kWh)	Uncertainty and Ranges*
Environmental Impacts	Emissions (g/kWh)			
From fossil energy currently used in plant construction and operation:				
CO_2 equiv. (leading to greenhouse effect)				
Plant and truck construction:	23	L	0.7	0.5–1.1
Transportation of feedstock/residues	90	L	2.9	2.0–4.2
Methane leaks (incurred minus avoided)	−285**	M	−9.1	−6 to −13
SO_2 (leading to acid rain and aerosols)	0.25	L	0.07	H, r, n
NO_x (possibly aerosols and health impacts)	0.36	L	0.14	H, r, n
particulates (lung diseases)	0.03	L	0.01	H, r, n
Land use			NQ	
Social Impacts				
Occupational health damage (manuf. & operation):	Cases per TWh			
death	1.7	L	0.55	L, l, n
major injury	2.2	L	0.02	L, l, n
minor injury	0.7	M	0.01	M, l, n
reduced span of life	6.0	M	1.95	M, l, n
Economic Impacts				
Direct costs			5–20	
Resource use (energy payback time given)	2.1 y	L	NQ	
Labor requirements (manufacture)	17 person y/MW	L	NQ	
Import fraction (for Denmark)	0.1	L	NQ	
Benefits from energy sold			4–11	
Other Impacts				
Supply security (variability in wind is high, entry based on plant availability)	High		NQ	
Robustness (up-front investment binds, entry based on technical reliability)	High		NQ	
Global issues (non-exploiting)	Compatible		NQ	
Decentralization & choice (less with large size)	Good		NQ	
Institution building (collection management)	Modest		NQ	

TABLE 7.12   Impacts from large biogas plant at Ribe, Denmark, per kWh of biogas (Kuemmel *et al.*, 1997).

Notes: NA = not analyzed, NQ = not quantified. Values are aggregated and rounded (to zero if below 0.01 US cent kWh⁻¹).
*L, M, H: low, medium, and high uncertainty. l, r, g: local, regional, and global impact. n, m, d: near, medium, and distant time frame. Monetary impacts originally given in euros are translated using 1 euro = 1.25 US$.
**The negative impact is due to a reduction of impacts outside the energy sector that would not otherwise be counted (see text).

This amount is the contribution from enteric fermentation. A similar amount, 125 million kg, is emitted from manure (Fenhann and Kilde, 1994).

These emissions have so far been considered in the greenhouse balance of the agricultural sector. They may also be divided between the agricultural and the energy sectors, in proportion to the revenues from the two types of operation. However, they should, of course, not appear as an additional impact. This may serve as a warning against doing LCA for one sector and neglecting impacts on other sectors. A meaningful LCA either must include all sectors of society affected or must treat what happens outside the system's boundaries as indirect impacts for the compartments included in the LCA.

Emissions from manure spread on the soil are low, and for the biogas plant residues returned to the fields, nearly zero.

Some Danish biogas plants are currently accepting household refuse and waste from food industries. General use of industrial biomass waste is permitted only if there are no significant residues of heavy metals in the slurry.

The biogas chain impacts shown in Table 7.12 are estimated on the basis of emissions and energy production given above. The methane emissions from bovine metabolism are not included, as they are considered to be the same as would occur without the diversion of manure to the biogas plants. The avoided methane emissions from farm storage of manure are seen to be the dominating contribution to the LCA costs. It varies from installation to installation and has to be assessed for the actual conditions.

LCA for biofuel chains may be found in Kaltschmitt *et al.* (1996), Sørensen (2005), with use of renewable energy-based hydrogen for vehicles, in Njakou *et al.* (2006), and, with general LCA of road traffic, in Sørensen (2004a; 2004b; 2006).

LCA of Energy Systems

To make an LCA analysis of an entire energy system, one proceeds by using LCA's of each component and then adds the impacts and their valuations. Consistent treatment of imports and exports must be dealt with (see section 7.3.1), as must any synergies, positive or negative, between the individual ingredients of the energy system.

Although analysis at the system level involves aggregation (see section 7.3.1) and for this reason may have lower accuracy than the LCA analysis of a particular piece of equipment or installation, it should be remembered that such an approach is a necessary and welcome input to the decision-making process related to energy policy. More details and examples of LCA's for definite scenarios for the future (such as those described in Chapter 6) may be found in Kuemmel *et al.* (1997) and a forthcoming update (Sørensen, 2011b).

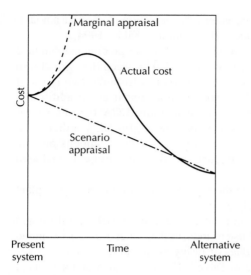

FIGURE 7.21　Schematic illustration of the difference between marginal appraisal methods and the scenario method, in describing a transition between two local cost minima separated by a cost barrier. The cost may or may not comprise LCA externalities. (Based upon Kuemmel *et al.*, 1997.)

The reason for basing life-cycle assessments of future energy systems on scenarios is explained in section 6.1.1. Figure 7.21 illustrates the advantage over a marginal approach in a case where the total system cost initially goes up and has to cross a barrier before reaching the lower minimum, once the old system and its LCA externalities are gone. The introduction of renewable energy systems often exhibits this kind of behavior.

Chapter 8

Integrated Approaches

In preceding chapters, this book gives the background for renewable energy sources and describes their occurrence, devices for using them, and the role of renewable energy in current and future energy systems, depending on social values and the economic rules used to judge the viability of competing solutions. This concluding chapter discusses the status of renewable energy development in the context of current political focus, and generally gives the author's personal evaluation of further development requirements, technically as well as institutionally.

8.1 GREENHOUSE WARMING MITIGATION

Climate impacts caused by human enterprise in general and human energy systems in particular are touched upon in several of the preceding chapters. The excess greenhouse warming caused by anthropogenic emissions related to fossil-fuel use is starting to get political attention, after 50 years of scientific warnings (cf. section 2.3). It was hoped that political intervention would be forthcoming before the damage became excessive, as in the case of ozone depletion by CFC-related gases, but this has not happened (presumably because energy provision has a much more influential role in society than spray-cans did, spanning from powerful financial interests to corporate enterprises of the energy industry). Depending on one's political outlook, one may attribute the delay to politicians' being marionettes in the hands of influential business interests or to politicians' not daring to make significant changes in a sector of the economy that has decisive importance for the well-being of nations and their inhabitants.

8.1.1 Proposed Fossil-fuel Phase-out Route

Recognizing that some countries refuse to set targets for limiting greenhouse gas emissions, and many other countries embrace this resistance by stating that they will not make reductions unless everyone follows suit, international climate negotiations are currently at an impasse. I have proposed a different approach that may be more appealing: to introduce a scheme based on automatically calculated reductions on a per capita basis. This would not require negotiated targets for each country or region and would not hinder the several populous countries that are embarking, or hope to embark, on a rapid route to industrialization, without compromising on reaching the stabilization goal before the end of the 21st century (Sørensen, 2008d). (The advantages of a per capita approach have been aired in connection with previous climate summits, but no political action has been pursued.) The approach is to calculate emission pathways for given levels of temperature stabilization by use of a simple model that is consistent with the circulation models and data sets surveyed by the Intergovernmental Panel on Climate Change. The allowable emissions would be detailed on a regional level and on individual country level.

The alternative to the per capita approach is individually decided, country-by-country CO_2-emission targets, such as those recently suggested by the EU, which are bound to create problems, as concerns agreement, extendibility to the rest of the world, and probability of actually being carried through. Individual targets are based on factors like current economic situation and growth, and, to some extent, current use of renewable or nuclear energy. When noncompliance becomes evident, offenders will most likely offer excuses like "our economic growth was higher (or lower) than assumed in setting the targets" or "our targets were set too high in comparison with those of Romania."

In contrast, the approach I propose is both simple and easy to understand, and therefore has a chance of working in the real world. It consists of introducing a common ceiling on per capita greenhouse gas emissions, along with optimal pathways allowing each country to reach the ceiling before a specified date. First, I select a model that allows calculation of the implications of various policies, with sufficient accuracy but without the long computing time of the general circulation models described in section 2.3.

The climate models reviewed by IPCC try to establish a connection between stabilization of the global temperature average (and other climate impacts) at a given level, the corresponding equilibrium concentration of greenhouse gases in the atmosphere, and allowable CO_2 emissions and other climate-altering human activities that policies have to ensure in order to reach stabilization at the desired level. The general circulation models used in climate science proceed from given emissions to calculated atmospheric concentrations, from which temperature changes are determined. Thus, one would have to look at many calculations with specific emission assumptions in order to find those for which the average temperature reaches say 1.5 °C above the year-2000 average temperature level (which translates to some 2.1 °C above the pre-industrial level). They would require a stable 450 ppm level of CO_2 in the atmosphere ($\approx$500 ppm if other greenhouse gases are included), with the uncertainty induced mainly by model accuracy but also by the different residence time of different greenhouse gases and the changes in cooling associated with particulate emissions from industry or from volcanoes. It follows that anthropogenic emissions of greenhouse gases would have to be radically reduced (IPCC, 2007; 2007b). The reverse calculation from a given temperature stabilization to the implied allowable emissions profiles (emissions as function of time) is not unique but requires additional assumptions, such as mitigation costs and technology readiness, in order to allow the determination of an optimal path to stabilization.

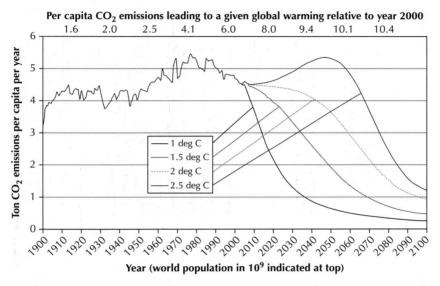

FIGURE 8.1 Per capita CO_2 emissions leading to global warming between 1 °C and 2.5 °C relative to year 2000. Population development is indicated at top (Sørensen, 2008d).

Figure 8.1 shows global average stabilization routes calculated in terms of CO_2 emission profiles of various policies for reaching a stable situation before year 2100, with average surface temperature increases in the range of 1.0 °C to 2.5 °C. These temperature rises are relative to year 2000, and the model assumes similar reduction in other greenhouse gases. The emissions in Fig. 8.1 are shown per capita, i.e., global emissions divided by global population for a given year. The historic data are those used by the IPCC, and the 21st-century values are calculated using the simple model called the Java Climate Model (Matthews, 2008). The model very well reproduces the results of earlier IPCC assessments and the results given in Fig. 8.1 are consistent with the simulations presented in the most recent IPCC report (IPCC, 2007b, Chapter 3). Detailed comparison is not possible because the IPCC report only gives ranges of results for a selection of literature emission scenarios, whereas the present calculation determines the emissions leading to a particular level of warming dynamically by iterating the approach to stabilization.

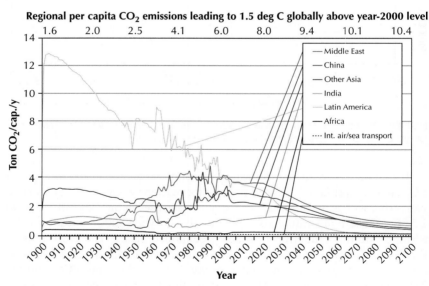

FIGURE 8.2   Regional per capita CO_2 emissions leading to stabilization of global warming 1.5 °C above the level in 2000. *Upper panel*: current high-emission regions. *Lower panel*: present low-emission regions (Sørensen, 2008d).

The calculation uses a population development reaching 10^{10} by the end of the 21st century, as in the IPCC "B2" scenario (IPCC, 2007b). The calculation behind Fig. 8.1 further assumes that per capita emissions converge toward a common value and reach it at, or shortly after, year 2100. The emission pathway to be followed by different regions of the world is depicted in Fig. 8.2, for the 1.5 °C warming case of Fig. 8.1. The same pathway, presented on an individual country level year-by-year, can be downloaded from the Elsevier journal website in the Reference list in connection with Sørensen (2008d). The input data for the calculation take into account the varying ability of different economies to mitigate greenhouse gas emissions, and mitigation therefore starts a little earlier in the most technologically advanced regions.

Stabilization by 2100 at 1.5 °C above 2000 Level	Accumulated Emission Allowance 2000–2100 (Gt CO_2)	Years to Zero CO_2 Allowance at Constant 2007 Emissions
United States	255	42
Canada, Australia, New Zealand	44	30
Japan	65	49
Western Europe	179	50
Eastern Europe B[*]	34	49
Russia, Ukraine, Belarus	78	39
Middle East	161	79
China	250	64
Other Asia	155	76
India	183	144
Latin America	86	41
Africa	213	115
International ship and air transportation	80	95
World	1783	61

TABLE 8.1 Total allowable emissions for the entire 21st century and the number of years that the allowed CO_2 emission of each region could continue at constant 2007 level, provided that it is reduced to zero after that time span (Sørensen, 2008d).

[*]For countries included in "Eastern Europe B" see Matthews (2008).

The stabilization corresponding to the 1.5 °C case of Fig. 8.1 can thus be achieved by imposing the increasingly strict (as a function of time) per capita emission limits, as shown in Fig. 8.2 for each region of the world or for each country. The limits start at the current level of emissions and converge to the common goal value by the end of the 21st century. Legislation can be introduced in each country to impose the diminishing ceiling either individually or collectively. The time-integrated emissions on a regional basis are given in Table 8.1, but it would be wise to monitor compliance by requiring the sliding targets to be met by the end of each decade. If noncompliance is detected, a penalty (such as a trade embargo) could be implemented, provided that a penalty system has been agreed upon. The concurrent reduction of non-CO_2 greenhouse gas emissions is shown in Fig. 8.3, for the 1.5 °C case.

Per capita CO_2 emissions leading to 1.5 deg C global warming relative to year 2000

FIGURE 8.3 Per capita total emissions, expressed as CO_2-equivalents, for the model calculation leading to global warming stabilizing at 1.5 °C above the year-2000 average temperature. For comparison, the CO_2 emissions alone, used in Figs. 8.1 and 8.2, are also shown (Sørensen, 2008d).

Clearly, not only emissions from national activities (raw materials extraction, production, transport, and consumption) should be considered, but also emissions from international air and ship traffic. To include these emissions in the national emission budget of each country, which also includes the corresponding transport activity in its activity measure (such as GNP), is a straightforward extension of the current method of accounting in the international climate negotiations. It would substantially increase emission counts for countries with large shipping industries, such as Denmark, but would similarly show up as economic activity in GNP accounting. There are at least two other ways of allocating emissions: 1) to ascribe emissions to the financial owners of each activity, or 2) to ascribe the emissions to those using or benefiting from the product or service causing the emissions (along its life cycle). Providing data for these alternative accounting principles is more difficult than for the method currently in use, but not impossible, given the wealth of statistical data available (for option 1, emissions from enterprises would be distributed on shareholders according to country affiliation, while for 2, the emissions from production and transport would be distributed on the final consumers according to country or residence, as derived from trade statistics). Method 2 would increase emission allocations to countries outsourcing energy-intensive production to other parts of the world, and would decrease emissions for developing countries carrying out such energy-intensive production but exporting most of the products. The calculations presented here all use conventional accounting by country of performed activity.

Possible Objections to the Per Capita Approach

Looking at the historical part of the per capita emissions in Fig. 8.1, one notes that per capita emissions rose remarkably little during the 20th century, obviously due to the explosive growth of the world's population and massive poverty in many parts of the world, with the associated smallness of per capita emissions. Population increase is a main reason for the lack of success in creating economic welfare for all the inhabitants of the world, and economic disparity has increased rather than decreased. Could rich countries use population growth to obtain a green light for continued high greenhouse gas emissions? The answer is that this does not seem to be the case even when the regional per capita emissions are used to regulate future emissions, as assumed in Fig. 8.2. Australia and countries in North America are already so high above their allocated emissions that they will have to use their technological skills to accomplish a rapid but realistic decrease in emissions, while less-developed countries initially have smaller per capita reductions prescribed for them. For these countries, there is also reason to use mitigating technology, because the current inefficient use of land and energy already leads to emissions above the target set for just a few years ahead.

Developing countries may complain that the scheme does not allow them to increase their emissions to American levels before doing something about the problem, but that is precisely the intent: they will move directly toward the stable situation without detours that may form bad habits clouding their future efforts. The proposed ceilings are proper and realistic because a country seeking high economic growth must invest in new equipment and processes anyway, and thus might as well choose the correct solution from the beginning. Furthermore, it is a mistake to think that the proper solutions are more expensive, because the expenses will have to be paid anyway and may be higher the longer one waits. It has also been suggested that long-industrialized countries should pay for the sins of the past (grandfathering, see section 7.3.3), but precise knowledge of the causes of global warming has been available for only about 40 years and the suggested scheme demands a very rapid mitigation from the rich nations anyway.

Would a scheme like the proposed one encourage nations to increase their population in order to get higher emission allowances? It is unlikely, because the greenhouse mitigation costs are small compared to the efforts needed to create economic welfare for more people. There is also no reason to compensate regions with lower than average population density, because it is usually a consequence of land areas' being hard to populate (examples are Arctic or desert regions). In any case, avoiding overpopulation should be a primary target in any region of the world. Population stabilization must be a strong international priority, with efforts at least as serious as in the case of greenhouse gas emissions. The fact that, until now, food production growth has on average kept up with population increases (although not necessarily the distribution of food to all) does not mean that it will continue to be possible or environmentally sustainable (Friedlingstein, 2008). The emerging food crisis must be solved separately from greenhouse policy, and the solution includes increased emphasis on education that can lead to increased industrial wealth as well as better understanding of family planning. The per capita emission target does not make this more difficult.

The tradable emission permits envisaged by some current greenhouse abatement schemes would not be needed for the scheme suggested here. Tradable permits aim at having investments made in the order of increasing abatement cost and would typically move investments from the countries with large emissions to less developed nations, where the equipment installed may not be used properly and thus not achieve the proposed emission reductions. The time-scale on which greenhouse gas emissions have to be dramatically reduced is so short, that fiddling with the succession of investments is unimportant. Fossil-fuel resources must be phased out faster than dictated by depletion, which is particularly relevant for coal used without decarbonization. The high-emission countries have the skills and should reduce their own emissions as quickly as possible, which is exactly what the proposed scheme can accomplish. Tradable permits constitute an attempt to force the greenhouse problem into a framework of 19th-century liberal economic theory, assuming that the market will solve any and all problems. We know that this is not true, and the global warming damage is a perfect example of the need to base political action on more than antiquated economic dogmas.

8.2 GREENHOUSE WARMING ADAPTATION

Several economists have addressed whether it is cheaper to let the impacts of global warming happen and then try to reduce the consequences (by adapting to the warmer world) than to avoid the impacts in the first place (by mitigation, such as replacing fossil fuels with renewable energy). This proposed "market trade-off" approach is flawed by the fact that changing the climate has very many types of impacts and it is far from certain that we have identified even the most important ones. Tol (2005) may be right that development aid could reduce the mortality rates for infectious diseases in Africa more than a few degrees' warming will increase their death toll, but this disregards the uncertainty of what life conditions will prevail in a warmer climate with changed weather patterns, altered water balances, and thus the possibility of radically different conditions for agriculture. Similarly, damage caused by increases in ocean-water levels can be reduced by building higher dikes around low-lying areas (IPCC, 2007c), but the issues of dike failure and more frequent occurrence of extreme storm and tide combinations still have to be faced. Avoiding more greenhouse warming by introduction of efficient energy conversion systems and renewable energy is better than dealing with the consequences, because the latter may hold surprises that we have not yet been able to identify.

Arguments for the marginal cost advantage of adaptation as compared to abstract reference to mitigation appear even more suspect when one considers the energy efficiency measures not implemented. Several studies have shown that energy demands could be reduced by about a factor of four by efficiency measures fully available and costing less than the energy they displace, rendering exactly the same energy service to the user (cf. Chapter 6; Weizsäcker et al. 1997; Sørensen 1982b; 1992a; 2008b). Reducing energy demand in this way without any cost increase has the same effect as a worldwide negotiated agreement to reduce greenhouse gas emissions by a factor of four. The cost of the first doubling of energy efficiency is far below what would have had to be paid for the energy saved, and only when demand is lowered 4- to 5-fold does the cost become comparable with that of providing the saved energy. In other words, governments should implement the regulation that will bring about improved energy efficiency. This can be done without any international agreement and has only positive effects on the national economy (reducing energy costs and at the same time supporting industry involved in the technical solutions for bringing about efficiency improvements), and the debate on mitigation versus adaptation is pushed some 50 years into the future.

8.3 ECOLOGICAL SUSTAINABILITY

Ecological considerations are gaining an increased role in the manufacture of consumer products. Decades ago, ecological food and personal hygiene products (*organic products* as they were called) were sold in specialty shops that charged a 100% or more premium for these exotic goods, which, moreover, were often produced by amateurs unable to ensure the products' wholesomeness. Today, at least in Europe, ecological products go through a certification process that includes tests for the presence of pesticides and other unhealthy chemical compounds, and they are sold in any supermarket, at prices typically 10%–30% above those of their non-ecological counterparts. In several countries, ecologically grown food products have a market share of 20%–30%, and ecological health products (soap, skin creams, etc.), textiles, and furniture show a growing market penetration.

In the case of energy, utilities sell "green electricity" and gas stations offer biodiesel, but these are still surpassed in volume of energy by private installations of solar heat or power, shares in wind farms, biogas installations, and so on. Generally, this is a good start and signals an interest, but the enthusiasts willing to make such investments are still a minority, and for the majority, who would rather not think about energy, legislative measures that force nonpolluting energy sources into the general energy supply appear to be the best way to achieve penetrations approaching 100%. As mentioned in section 8.1, there will be opposition to such political intervention from vested interests wanting to maintain fossil-fuel and nuclear technologies as long as resource availability allows. In reality, there is no reason to make use of the last drop of oil when better solutions are available. It is appropriate here to make a few observations regarding whether the different renewable energy solutions are ready to take up a dominating place on the energy scene.

Status of Renewable Energy Technologies

For hydropower, development is expected to be largely restricted to efforts to deal with its environmental and social impacts. This could lead to emphasis on smaller, cascading schemes, instead of the very large dam installations of the past. Still, the basic technology must be considered as fully developed, and because the cost of environmentally sound hydro schemes is rarely lower than that of fossil-fuel plants, the global expansion rate of hydropower has slowed to nearly zero. There are still possible new sites available, and if social acceptance can be regained after the disasters of the past (flooding the homes of hundreds of thousands people and destroying treasures of cultural heritage—most recently in Turkey and China), then environmentally integrated hydro could add a further substantial amount to the global renewable energy supply.

In the case of wind turbines, further development along current trends, particularly in materials, will allow somewhat larger horizontal-axis turbines to be built. Although other wind conversion techniques could become viable, the success of the horizontal axis is likely to keep it in a leading position. Adjustments in blade profile and regulation technology are likely to follow the widespread interest in offshore wind. Currently, essentially the same turbines are sold for deployment on land and at sea. However, the wind conditions at sea are so different that development of different designs appears warranted (cf. the difference between the two patterns in Figs. 6.25 and 6.26 or in Figs. 6.67 and 6.68, as regards the number of hours where the cut-off ceiling in power is reached). But these concerns still involve only small adjustments to a technology already close to maturity: the possibility of working with different blade designs for different locations and control of pitch (e.g., by being able to turn the entire blade along its axis and using smart control software). The latter has been tried in commercial turbines and gives a few percent improvement in average efficiency. Thus, the main challenge in wind turbines is improvement of materials, particularly for blades, but also for cost reduction of foundations (which is particularly relevant for offshore turbines, where foundation costs constitute a fairly large fraction of the total). Such advances in blade construction will allow turbines with rated power somewhat larger than the 2−5 MW typical at present. Furthermore, as with any mature technology, continued small improvements in cost per produced kWh with time can be expected. In summary, current wind power technology is ready for massive market expansion in all the regions of the world identified as suitable (see Chapters 3, 4, and 6).

Of course, entirely new methods could be imagined for harvesting the still higher wind power levels found at open sea, i.e., winds much further offshore than those exploited by current machines standing on the sea floor. These could be combined wave- and wind-capturing devices, floating but kept within a restricted area by an advanced navigation system. The wind turbines could be shrouded or ducted constructions, while the wave energy part is likely to use dual air turbines. No developments in this direction have been successful so far. Wave energy devices for near-shore employment have not successfully struck a balance between the necessary strength and economic power production, and no large mid-ocean wave converters have reached interesting sizes for energy production (only niche devices, such as Masuda's air pump for buoy light generators (Fig. 4.41), have shown acceptable performance).

A number of biomass conversion methods are close enough to direct economic viability for the inclusion of indirect economics in cost comparisons to make them acceptable. This is true of biogas plants, particularly on a communal scale and as part of general waste-management schemes. Various schemes for hydrogen or liquid biofuel production are promising viability in the short-term future, but their market penetration will at least initially depend on the valuation used for the externalities of current fuel-based energy systems. It is difficult to tell which of the biofuels and associated technologies (gasification routes, enzymatic routes, etc.) will first reach acceptable costs, but probably there will be room for more than one solution.

The entire bioenergy field may be in for radical restructuring if current trends toward valuing high food quality continue and spread to regions not currently concerned. This may alter agricultural practices, both for plant and for animal production, and lead to new concepts of integrated food and energy production based upon ecological principles, as distinct from the concept of dedicated energy crops.

Geothermal resources used sustainably can be a stable source of low-temperature heat, and costs already appear acceptable, with some consideration of externalities for the alternatives. Many existing district heating systems could benefit from this new source of heat. The market for geothermal solutions depends on the viability of district heating. If the full potential for making existing and new buildings highly energy efficient is realized, there will be very few locations in the world where new district heating schemes are economically attractive (probably only city centers with dense, high-rise buildings).

For solar energy, thermal applications are economically viable in lower-latitude regions, as judged in an assessment including externalities for conventional solutions. But this also depends on whether the building is energy efficient in the first place. A large heat requirement in winter is hardly compatible with solar coverage, but if the passive and active efficiency features of the building are optimal, the heat load may be dominated by hot water use, which is more amenable to solar supply. This is particularly so for niche markets, such as vacation areas (e.g., in the Mediterranean region or similar places), where there may be lots of building space being inhabited only during the summer season.

For solar electricity, the situation is different, as the current costs of photovoltaic power systems are still too high even with generous consideration of indirect costs. However, technical development is rapid in this area, and novel solutions under current development for conventional solar cells may bring about economically attractive solutions within the next one or two decades. In addition, a multitude of novel techniques for converting solar energy could play a decisive role in the future. While organic dyes, polymer cells, and other low-efficiency solutions do not seem practical (because neither structural support nor building integration is compatible with such low efficiencies, either due to cost or to mounting areas required), some of the thin-film technologies not using monocrystalline semiconductors appear to have a chance in the long run.

Above all, as mentioned several times in this book, both direct and, even more, a life-cycle approach to economic valuation will make efficient use of energy more attractive in nearly all cases, and the implied lowering of energy demand per capita will generally improve the conditions for all renewable energy types by making it possible to supply all required energy from renewables with minimal strain on resources.

Energy Storage and Auxiliaries

The intermittency of several renewable energy flows makes it imperative to include features in systems relying on renewable energy inputs that can ensure supply–demand matching in time and place. For less than 100% renewable energy systems, the nonrenewable units can be used for back-up, but for purely renewable energy systems, storage or suitable exchange of energy is necessary. To some extent, storable biofuels can serve this purpose, but it is highly likely that additional components will be required.

This brings into focus bulk storage development, as well as means for transmission to and from such devices, and between load locations and supply locations, especially when the closest ones may be unavailable. Possible candidates for a flexible conversion between electric power (i.e., the output energy form of variable renewable sources like wind turbines or photovoltaic converters) and storable energy are (preferably reversible) fuel cells combined with hydrogen storage. In certain areas, pumped hydro (or managed operation of combined hydro and intermittent renewable energy systems) can be used. Other storage forms, such as flywheels and batteries, evidently are not suited for bulk or long-term storage. This leaves storage of hydrogen in pressure containers or, much more economically, in underground caverns and aquifers as a central candidate for a general solution to the supply–demand mismatch problem.

Considerable hopes have been placed on utility uses of advanced batteries, and after 40 years of development this could finally be about to happen. During the last two decades, advanced batteries have finally emerged, first on a small scale suited for consumer appliances (portable computers, cameras, mobile telephones, light sources, etc.), but increasingly moving in the direction (price, performance) of bulk applications, such as for hybrid vehicles.

A range of hydrogen storage techniques are emerging as suitable for medium-term energy storage, in combination with fuel-cell conversion. These include underground storage of hydrogen (in aquifers or flushed-out salt domes). Small-scale compressed hydrogen storage in containers for use in the transportation sector may in the future be replaced by storage in media, such as metal hydrides, capable of easily incorporating and releasing hydrogen. Use of hydrogen as an energy carrier is linked to the development of dependable and affordable fuel-cell energy converters.

Fuel-cell technology has not yet fulfilled its promise (see, for example, Daimler-Chrysler-Ballard, 1998), but the substantial price reductions needed could still be forthcoming. Fuel-cell technology has many similarities to battery technology, so one might expect similar disappointments and similar successes. Pure fuel-cell vehicles are less likely to make it than intelligent hybrids, starting with diesel-engine/fuel-cell hybrids and moving toward plug-in battery/fuel-cell hybrid vehicles. The requirement for substantial infrastructure changes in order to use hydrogen both in vehicles and for stationary uses is a further impediment to the development of economically viable supply systems (cf. Sørensen, 2005).

The Global Political Dimension

The conditions for energy transitions away from fossil fuels are different in different parts of the world. In some countries, there is broad political consensus on not redistributing wealth through taxation. This implies a need to keep the price of basic goods, including energy, at such a low level that a large fraction of the population is not pushed below the poverty line. In other parts of the world, there is a tradition for significant income redistribution through taxes, and consumers have for many years been accustomed to energy prices far above the direct cost, made possible because the redistribution policy allows such prices to prevail without marginalizing large groups of society. In recent years, general taxation has increasingly been replaced by "fair taxation," meaning that the governments levy taxes on energy according to estimated *de facto* externality costs of pollution, climate impact, supply insecurity, etc. Countries committed to keeping energy prices low are more likely to consider extreme measures, such as going to war, in order to secure access to cheap energy sources from politically unstable parts of the world. Similar remarks can be made in reference to the efforts to encourage more efficient use of energy, which are often seen as politically unglamorous, although the extra cost of efficiency improvements, as mentioned, is generally much lower than the cost of introducing new energy sources.

From this perspective, there is little reason to be optimistic based just on the spectacular results obtained by modest investments in renewable energy technology. Technical feasibility, decent prices, and general public acceptance seem insufficient to persuade policy makers and the industrial lobby in many parts of the world. What additional arguments should be brought forward? Air pollution, greenhouse gas emissions, fossil resource exhaustion, nuclear accidents, and radioactive waste have all failed to affect actual political decisions, although few would disagree that all are problems "in the long run." The impetus behind this behavior is probably political governance's being increasingly motivated by short-term goals, probably at least in part due to the influence of the media, media that are no longer independent and unbiased in their analysis and criticism but are owned by, and serving, special interests. So it is tempting to conclude that, while several renewable energy solutions are already ready for the market, the rules of the marketplace have to be changed to include long-term human interests if it is to become a level playing field. Again, this calls for political action that is unlikely to be taken unless the rules of the political game are modified to include long-term human interests.

PART III: MINI PROJECTS, DISCUSSION ISSUES, AND EXERCISES

III.1

Discuss the possibility that solar electricity could play the roles considered for wind electricity in section 6.5.2.

What is the diurnal and seasonal match between variations in solar energy generation and electricity load? For example, construct power duration curves or power-minus-load curves and discuss the need for short- and long-term storage of energy, in comparison with the corresponding information for wind energy systems in geographic regions of good wind access.

Are there climatic regimes that, from a system structure point of view (as distinct from a purely economic point of view), would make solar electricity the preferable solution?

Does it make any difference to the answers to the above questions whether the solar system in mind is based on solar cells or on concentrating collectors followed by thermal conversion?

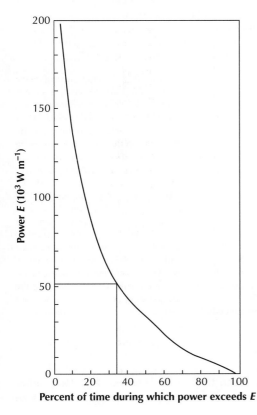

FIGURE III.1 Annual power duration curve for omnidirectional oscillating-vane wave-power converter of diameter 16 m, using the efficiency curve given in Fig. 4.45. The wave data are for Station India in the northeast Atlantic (cf. Fig. 3.59), having an average power level of 91 kW m^{-1}. The thin straight lines indicate the availability of the average power production (average efficiency 56%). (Based on Mollison *et al.*, 1976.)

III.2

Referring to Fig. III.1, discuss the potential of wave power in the North Atlantic Ocean for electricity production. Figure III.2 is the annual power duration curve for one particular wave device (a Salter duck, Fig. 4.44) with the efficiency curve given in Fig. 4.45. The diameter of the device is 16 m, and it is assumed to be omnidirectional, in the absence of directional data (Mollison *et al.*, 1976). The corresponding power duration curve for the waves themselves is given in Fig. 3.59.

Compare the availability of the average power level with that of the wind-power systems considered in section 6.5.2. Why does the wave power go right to the 100% time fraction in Fig. III.1, in contrast to the single wind turbine power duration curves (e.g., Fig. 6.69)?

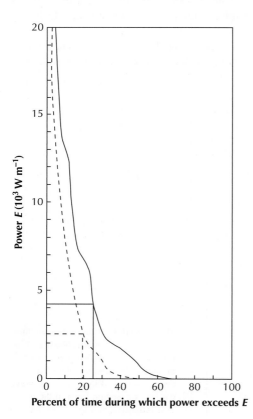

FIGURE III.2   Annual power duration curves for omnidirectional (*solid, heavy line*) or one-directional (west, *dashed line*) oscillating-vane wave power converter of diameter 6 m, using the efficiency curve given in Fig. 4.45. The wave data are for Station Vyl in the North Sea (for 1971, courtesy of Nielsen, 1977), having an average power level of 5.35 kW m^{-1}. The omnidirectional device has an average efficiency of 79%. If the device were assumed to accept two opposite directions (east and west), the power duration curve would lie about midway between the two curves drawn. Diminishing the diameter of the device does not increase the maximum availability.

III.3

Try to answer the same questions as in III.2 on the basis of Fig. III.2 rather than Fig. III.1. Figure III.2 is based on data from the North Sea, and the size of the Salter device (6 m) is chosen to give the maximum annual power production.

Consider the statement that "waves accumulate wind energy over time and over fetch distance, hence wave power is more stable and dependable than wind power."

III.4

Wind energy converters might be placed on platforms at sea. What advantages could be expected from such an arrangement? The structures associated with wave-power devices might conceivably be able to support wind energy converters on top. Would there be advantages in such a combination?

III.5

What size solar collector would be able to provide power for a small automobile (requiring, say, 25 kW peak power)?

Such a collector is not for integration in the roof of the car, but consider a car with a suitable storage facility (e.g., flywheel) that could be charged by electricity from solar cells covering a garage rooftop, for example.

The (annual) average power requirement of the car is much lower than the peak power. Assume an average net power requirement of 300 W (roughly corresponding to 1300 W of fuel for the more efficient types of automobiles produced at present), and estimate the solar cell area that would be required in order to produce power at about this level for most of the year (depending on latitude).

Flywheel storage, which has also been considered for "peak shaving" in present generations of cars, has the advantage of being capable of delivering very high peak power levels for starting and acceleration (if this feature of petrol-driven car engines is worth preserving).

III.6

Try to build a renewable energy supply–demand scenario for your country or region, based on the maps and data given in this chapter and possibly supplemented by additional local data (considering that the geographical resolution of about 50 × 50 km may not be sufficient for the study of a particular region).

Based on this work, determine whether your region is likely to be an energy-importing or energy-exporting area.

III.7

Compare direct cost evaluations for a definite renewable energy system, such as wind turbines, in which the annual interest rate has been taken alternatively as 15%, 10%, 5%, and 0%.

III.8

Discuss differences and similarities between economic models based on dynamic simulation models (e.g., the simplified one presented by Forrester, 1971), models based on quasi-static assumptions (such as the one outlined in section 7.1.2) or on the textbook input–output models (e.g., Cockrane *et al.*, 1974), and finally, models based on scenarios and life-cycle assessment, as described in section 7.3.

III.9

Consider a wind energy system with storage. Try to find a relation between the installed wind energy capacity (expressed as a fraction I_{wind} of the average load) and the size of storage most appropriate (this in particular implies deciding on the capacity credit factor A to aim for, as a function of I_{wind} and coupled with expectations of future escalation of fuel prices). Once this is at hand, construct accumulated present-value curves as functions of installed wind capacity (that is, I_{wind}).

III.10

Compare centralized and decentralized generation of electricity by means of solar cells. For the centralized system, one may assume support structures made of concrete or of steel frames. The projected cost for support structures and land acquisition is responsible for roughly half the total cost. Alternatively, consider decentralized units mounted on individual roofs of existing buildings. The cost of support structures and land would be saved. On the other hand, mounting and controls may be more expensive.

Try to collect costs for the relevant items that may be different for the two systems, and discuss the indirect economic factors relevant to the issue.

Extend the economic discussion to include the requirements for energy storage, based on data for variations in expected solar energy production at a location of your choice.

If wind converter production data for the same region are available, discuss the advantages that may be obtained by having the solar cells and wind energy converters feed into the same energy storage system.

III.11

Discuss the economy of central energy conversion followed by transmission, relative to decentralized conversion at the load areas, separately for heat and for electricity.

As an example, construct a small model including possible differences in costs of conversion equipment, depending on unit size, as well as possible differences in conversion efficiency. Add to this assumptions regarding the cost of, and losses associated with, energy transmission of heat or of electricity.

Identify factors of indirect economy that may have particular relevance for the discussion of centralized and decentralized conversion (e.g., supply security in various cases of system failure).

III.12

Try to carry through a life-cycle analysis of some energy system of interest to your location (province, country). First make an inventory of the possible types of social and environmental impacts and the pieces of equipment responsible for these impacts, then try to quantify each in physical terms, and, finally, discuss the possible monetization of impacts. Did the results become more accurate by monetizing, or less accurate?

References

Aarkrog, A. (1971). Prediction models for Strontium-90 and Caesium-137 levels in the human food chain. *Health Physics, 20,* 297–311.

Aastrand, C., Mose, O., & Sørensen, B. (1996). Wind power: Valuation and finance. In *1996 European Union Wind Energy Conference* (pp. 138–139). Bedford: Stephens and Association.

Abbasi, T., & Abbasi, S. (2010). Biomass energy and the environmental impacts associated with its production and utilization. *Renewable and Sustainable Energy Reviews, 14,* 919–937.

Achard, P., Lecomte, D., & Mayer, D. (1981). Characterization and modelling of test units using salt hydrates. In *Proceedings of the International Conference on Energy Storage* (Vol. 2, pp. 403–410). Cranfield, UK: BHRA Fluid Engineering.

Ackermann, T. (2002). Transmission systems for offshore wind farms. *Renewable Energy World, 5*(4), 49–61.

Adolfson, W., Mahan, J., Schmid, E., & Weinstein, K. (1979). In *Proceedings of the 14th Intersociety Energy Conversion Engineering Conference* (pp. 452–454). Washington, DC: American Chemical Society.

Alakangas, E., Hillring, B., & Nikolaisen, L. (2002). Trade of solid biofuels and fuel prices in Europe. In *12th European Biomass Conference*erence (pp. 62–65). ETA Firenze & WIP Munich.

Alcamo, J., Shaw, R., & Hordijk, L. (Eds.). (1990). *The RAINS model of acidification: Science and strategies in Europe.* Dordrecht: Kluwer Academic Publishers.

Alexandratos, N. (1995). *World agriculture: Towards 2010. An FAO study.* Chichester: Wiley.

Alfven, H. (1965). Antimatter and the development of the metagalaxy. *Reviews of Modern Physics, 37,* 652–665.

Alich, J., & Inman, R. (1976). *Energy* (Vol. 1, pp. 53–61). London: Pergamon Press.

Allen, C. W. (1973). *Astrophysical quantities.* London: The Athlone Press.

Almquist, E. (1974). An analysis of global air pollution. *Ambio, 3,* 161–167.

Alsema, E. (1997). *Appendix to environmental aspects of PV power systems* (Utrecht Universiteit VNS Report no. 97072). (irreg. pag.)

Alsema, E., & Engelenburg, B. (1992). Environmental risks of CdTe and CIS solar cell modules. In *Proceedings of the 11th European Commission Photovoltaics Solar Energy Conference, Montreux* (pp. 995–998). Newark, NJ: Harwood Academic Publishers.

Alyea, F., Cunnold, D., & Prinn, R. (1975). Stratospheric ozone destruction by aircraft-induced nitrogen oxides. *Science, 188,* 117–121.

Amann, M., & Dhoondia, J. (1994). *Regional air pollution information and simulation (RAINS–Asia), user's manual* (p. 34). Laxenburg: World Bank and IIASA.

Andersen, B., Eidorff, S., Lund, H., Pedersen, E., Rosenørn, S., & Valbjørn, O. (1974). *Meteorological data for design of building and installation: A reference year* (Report No. SBI–89). Danish Govt. Build. Res. Inst.

Andersen, J. (1997). Private communication of data from NORDPOOL ASA.

André, H. (1976). Operating experience with bulb units at the Rance tidal power plant and other French hydro-electric sites. *IEEE Transactions on Power Apparatus and Systems, PAS-95*(4), 1038–1044.

Andreen, H., & Schedin, S. (Eds.). (1980). *Den nya energin*. Göteborg: Centrum for Tvärvetenskap/ Forlaget Tvartryk.

Angrist, S. (1976). *Direct energy conversion* (3rd ed.). Boston: Allyn and Bacon.

Anonymous. (1980, September 11). News. *New Scientist, 87*, 782.

Anonymous. (2002). Electricity store. *ABB Review,* (4), 62–65.

APACE. (1982). *Technical information bulletin PA/111/1*. Hawkesbury, NSW, Australia: Apace Research Ltd.

Appropriate Technology Development Organization. (1976). *Gobar gas*. Islamabad (undated): Govt. of Pakistan.

Aqua-Media. (1997). Hydropower & dams. Map prepared for *The International Journal on Hydropower & Dams*. Sutton, UK: Aqua-Media Int. Ltd.

Armstrong, R., & Bruce, P. (1996). Synthesis of layered $LiMnO_2$ as electrode for rechargeable lithium batteries. *Nature, 381*, 499–500.

Arnason, G. (1971). Estuary modeling. In W. Matthews, F. Smith, & E. Goldberg (Eds.), *Man's impact on the terrestrial and oceanic ecosystems* (pp. 429–447). Cambridge, MA: MIT Press.

Asmussen, J. (1975). Wind system research project for the city of Hart, Michigan. In F. Eldridge (Ed.), *Proceedings 2. Workshop on Wind Energy Conversion Systems* (pp. 112–120), USA: MITRE Corp.

Athey, R. (1976). *Solar Energy, 18*, 143–147.

ATS. (2003). *Spheral solar technology*. http://www.spheralsolar.com

ATV. (1975). *Wind power*. Prepared by C. Estrup, J. Fischer, F. Hvelplund, M. Jensen, M. Johansson, S. Kofoed, *et al.* for the Danish Academy of Technical Sciences, Lundtofte.

ATV. (1976). *Wind power 2*. Prepared by F. Hvelplund, H. Jels, M. Johansson, S. Kofoed, N. Meyer, B. Pedersen, *et al.* for the Danish Academy of Technical Sciences, Lundtofte.

Baader, W., Dohne, E., & Brenndorfer. (1978). *Biogas in theoric und praxis*. Darmstadt: Landwirtschaftsverlag.

Badsberg, U., Jørgensen, A., Gasmar, H., Led, J., Hammerstad, J., Jespersen, L., *et al.* (1996). Solution structure of reduced plastocyanin from the blue-green alga *Anabaena variabilis*. *Biochemistry, 35*, 7021–7031.

Bahadori, M. (1977, July). *Solar energy utilization for developing countries*. International Solar Building Technology Conference, London.

Bahatyrova, S., Frese, R., Siebert, C., Olsen, J., Werf, K. van der, Grondelle, R. van, *et al.* (2004). The native architecture of a photosynthetic membrane. *Nature, 430*, 1058–1062.

Bahcall, J. (1969, July). Neutrinos from the Sun. *Scientific American, 221*(1), 28–37.

Bahcall, J., & Davis, R. (1976). Solar neutrinos: A scientific puzzle. *Science, 191*, 264–267.

Baker, A. (1987). Tidal power. *UK IEE Proceedings, 134*, 392–398.

Bakken, K. (1981). System Tepidus, high capacity thermochemical storage/heat pump. In *Proceedings International Conference on Energy Storage* (Vol. I, pp. 23–28). Cranfield, UK: BHRA Fluid Engineering.

Ballhausen, C., & Gray, H. (1965). *Molecular orbital theory*. New York: Benjamin.

Balling, N. (1976). Geothermal models of the crust and uppermost mantle of the Fennoscandian shield in south Norway and the Danish embayment. *Journal of Geophysics, 42*, 237–256.

Banas, J., & Sullivan, W. (1975). Sandia Laboratories Energy Report, SAND75–0530.

Banas, J., & Sullivan, W. (Eds.). (1976). *Sandia vertical-axis wind turbine program* (Quarterly Report Oct.–Dec. 1975, SAND76–0036).

Bandel, W. (1981). A review of the possibilities of using alternative fuels in commercial vehicle engines. In *International Conference on Energy Use Management, Berlin, 1981*, Session H-3, Daimler-Benz AG, Stuttgart.

Banner, M., & Melville, W. (1976). On the separation of air flow over water waves. *Journal of Fluid Mechanics, 77*, 825–842.

Barbier, E. (1999). Geothermal energy: A world overview. *Renewable Energy World, 2*, 148–155.

Barnes, C. (1971). In M. Baranger & E. Vogt (Eds.), *Advances in nuclear physics* (Vol. 4, pp. 133–204). New York: Plenum Press.

Barnett, T., & Kenyon, K. (1975). Recent advances in the study of wind waves. *Reports on Progress in Physics, 38*, 667–729.

Barnett, T., & Wilkerson, J. (1967). On the generation of ocean wind waves as inferred from airborne measurements of fetch-limited spectra. *Journal of Marine Research, 25*, 292–328.

Barnola, J., Raynauld, D., Korotkevich, Y., & Lorius, C. (1987). Vostok ice core provides 160,000-year record of atmospheric CO_2. *Nature, 329*, 408–414.

Bartels, J. (1957). In S. Flüügge (Ed.), *Handbuch der Physik (Geophysik II)* (Vol. 48, pp. 734–774). Berlin: Springer Verlag.

Basore, P. (1991). *PC1D v.3 manual and user guide* (Report 0516/rev. UC-274). Albuquerque: Sandia National Laboratory.

Baumgartner, T. (1993). Product life-cycle analysis. Current practice and methodological implications. In *Life-cycle analysis of energy systems* (pp. 172–181). Workshop Proceedings. OECD Publications, Paris.

Bay of Fundy Tidal Power Review Board. (1977). *Reassessment of fundy tidal power.* New Brunswick.

Beale, W. (1976). As quoted in Hughes, W. (Ed.). (1976). *Energy for rural development.* Washington, DC: United States National Academy of Sciences.

Bechinger, C., Ferrere, S., Zaban, A., Sprague, J., & Gregg, B. (1996). Photoelectrochromic windows and displays. *Nature, 383*, 608–610.

Beck, E. (1975). Ocean thermal gradient hydraulic power plant. *Science, 189*, 293–294.

Becke, A. (1993). Density-functional thermochemistry. III: The role of exact exchange. *Journal Chemical Physics, 98*, 5648.

Beckman, W., Schoffer, P., Hartman, W., & Lof, G. (1966). Design considerations for a 50-watt photovoltaic power system using concentrated solar energy. *Solar Energy, 10*, 132–136.

Beesch, S. (1952). *Industrial & Engineering Chemistry, 44*, 1677–1682.

Bell, B. (1998, September). *Looking beyond the internal combustion engine: The promises of methanol fuel cell vehicles.* Paper presented at Fuel Cell Technology Conference, IQPC Ltd., London.

Benemann, J., & Weare, N. (1974). Hydrogen evolution by nitrogen-fixing Anabaene cylindrica cultures. *Science, 184*, 174–175.

Berezin, I., & Varfolomeev, S. (1976). Conversion of solar energy by photosynthetic production of molecular hydrogen. *Geliotekhnika, 12*, 60–73.

Berger, W. (1982). Climate steps in ocean history—lessons from the Pleistocene. In W. Berger & J. Crowell (Eds.), *Climate in Earth history* (pp. 43–54). Washington, DC: National Academy Press.

Berglund, M., & Börjesson, P. (2002). Energy efficiency in different types of biogas systems. In *12th European Biomass Conference* (pp. 219–222). ETA Firenze & WIP Munich.

Berkner, L., & Marshall, L. (1970). In R. Fairbridge (Ed.), *The encyclopedia of geochemistry and environmental sciences* (Vol. 4A, pp. 845–861). New York: Van Nostrand.

Bertness, K., Kurtz, S., Friedman, D., Kibbler, A., Kramer, C., & Olsson, J. (1994). 29.5% efficient GaInP/GaAs tandem solar cells. *Applied Physics Letters, 65*, 989–991.

Besant, R., Dumont, R., & Schoenau, G. (1979, May). *Solar Age*, 18–24.

Betz, A. (1959). *Stromunglehre*. G. Brown Verlag, Karlsruhe, BRD.

Beyer, H., *et al.* (1994). *Journal Wind Engineering & Industrial Aerodynamics, 51*, 111.

Bilton, T., Flowers, E., McCormick, R., & Kurfis, K. (1974), pp. 61–67 in Turner (1974).

Biomass Energy Institute. (1978). *Biogas production from animal manure*. Winnipeg, Manitoba.

Birck, C., & Gormsen, C. (1999). Recent developments in offshore foundation design. In Petersen, Jensen, Rave, Helm, & Ehmann (Eds.), *Proceedings of the European Wind Energy Conference, Nice* (pp. 365–368). London: James & James.

Biswas, D. (1977). *Solar Energy, 19*, 99–100.

Bjørnholm, S. (1976). *Energy in Denmark 1990, 2005* (Report No. 7). Copenhagen: The International Federation of Institutes for Advanced Study.

Blackwell, B., & Reis, G. (1974). *Sandia Laboratories*. Report SLA–74–0154. Albuquerque, USA.

Blaxter, K. (1974, February 14). *New Scientist*, 400–403.

Blegaa, S., Hvelplund, F., Jensen, J., Josephsen, L., Linderoth, H., Meyer, N., *et al.* (1976). *Skitse til Alternativ Energiplan for Danmark, OOA/OVE*, Copenhagen (English summary in *Energy Policy*, June 1977, pp. 87–94).

Bligh, T. (1976, October/November). *Building Systems Design*, 1–22.

Blomquist, C., Tam, S., & Frigo, A. (1979). In *Proceedings of the 14th Intersociety Energy Conversion Engineering Conference* (pp. 405–413). Washington, DC: American Chemical Society.

Bockris, J. (1975). *Energy: The solar-hydrogen alternative*. Brookvale, Australia: Australia and New Zealand Book Co.

Bockris, J., & Reddy, A. (1973). *Modern electrochemistry*. New York: Plenum Press.

Bockris, J., & Shrinivasan, S. (1969). *Fuel cells: Their electrochemistry*. New York: McGraw-Hill.

Boeck, W. (1976). *Science, 193*, 195–198.

Boeniger, M., & Briggs, T. (1980). Potential health hazards in the manufacture of photovoltaic solar cells. In Rom & Archer (Eds.), *Health implications of new energy technologies* (Chap. 43). Ann Arbor, MI: Ann Arbor Science Publ.

Bohr, A., & Mottelson, B. (1969). *Nuclear structure* (Vol. 1). New York: Benjamin.

Boldt, J. (1978). *Solar powered water pump for the rural third world* (internal report). Danish Technical University, Lab. for Energetics.

Bolin, B. (1970, September). *Scientific American, 223*, 124–135.

Bolin, B. (1974). *Ambio, 3*, 180–188.

Bolle, H.-J. (1972). In G. Fiocco (Ed.), *Topics in atmospheric structure and composition* (Lecture notes). Frascati, Italy: The International School of Applied Physics.

Bonefeld, J., Jensen, J., & Anon. (2002). Horns rev 160 MW offshore wind. *Renewable Energy World, 5*(3), 77–87.

Bosanac, M., Sørensen, B., Katic, I., Sørensen, H., & Badran, J. (2001). *Photovoltaic/thermal solar collectors and their potential in Denmark* (Report EFP 1713/00-0014 to the Danish Energy Agency, Copenhagen).

Boussinesq, J. (1877). *Mémoires présentées par div. savants a l'Acad. des Sciences de Paris, 23*, 46 (As quoted by Hinze, 1975).

Brehm, N., & Mayinger, F. (1989). *A contribution to the phenomenon of the transition from deflagration to detonation* (VDI–Forschungsheft No. 653/1989, pp. 1–36). http://www.thermo–a.mw.tu–muenchen.de/lehrstuhl/foschung /eder_gerlach.html

Brettel, K., & Leibl, W. (2001). Electron transfer in photosystem I. *Biochimica et Biophysica Acta, 1507*, 100–114.

Broda, E. (1975). *Naturwissenschaftliche Rundschau (Stuttgart), 28*, 365–372.

Brooks, L., & Niiler, P. (1977). *Journal of Marine Research, 35*, 163–175.

Brown, J., & Cronin, J. (1974). *Proceedings of the 9th Intersociety Energy Conversion Engineering Conference* (USA), Paper No. 749139.

Brown, C., & Warne, D. (1975). *Wind power report* (Report P.2062/G). Ontario Res. Found.

Brown, S. (1998, March 30). The automakers' big-time bet on fuel cells. *Fortune Magazine*, p. 12. http://www.pathfinder.com/fortune/1998/980330

Brüel, P., Schiøler, H., & Jensen, J. (1976). *Foreløbig redegørelse for solenergi til boligopvarmning* (Project report). Copenhagen.

Bryan, K. (1969). *Monthly Weather Review, 97*, 806–827.

Bryan, K., & Cox, M. (1967). *Tellus, 19*, 54–80.

Bryson, R. (1971). *Climate modification by air pollution*, As quoted by Wilson and Matthews (1971).

BTM. (2001, March). *International wind energy development*. BTM Consult Aps. http://www.btm.dk

BTM. (2002, March). *International wind energy development*. BTM Consult Aps.

BTM. (2009). *International wind energy development*. BTM Consult Aps. http://www.btm.dk

Buch, H. (1954). *Hemispheric wind conditions during the year 1950* (Final Report, part 2 of contract AF-19(122)-153). Cambridge, MA: Deptartment of Meteorology, MIT.

Buchmann, I. (1998). *Understanding your batteries in a portable world*. Canada: Cadex Inc. http://www.cadex.com/cfm

Buchner, H. (1980). Thermal energy storage using metal hydrides. In J. Silverman (Ed.), *Energy storage*. Oxford: Pergamon Press.

Budyko, M. (1974). *Climate and life*. New York and London: Academic Press.

Burbridge, E., Burbridge, G., Fowler, W., & Hoyle, F. (1957). *Reviews of Modern Physics, 29*, 547–650.

Butler, D. (1997). Cogema's arrogance adds to la Hague's problems. *Nature, 387*, 839.

Butler, D., Maurice, J., & O'Brien, C. (1997). Time to put malaria control on the global agenda. *Nature, 386*, 535–541.

Cacciola, G., Giodano, N., & Restuccia, G. (1981). The catalytic reversible (de)hydrogenation of cyclohexane as a means for energy storage and chemical heat pump. In *Proceedings of the International Conference on Energy Storage* (pp. 73–89). Cranfield, UK: BHRA Fluid Engineering.

Caldwell, M. (1975). In J. Cooper (Ed.), *Photosynthesis and productivity in different environments* (pp. 41–73). Cambridge, UK: Cambridge University Press.

Callen, H. (1960). *Thermodynamics*. New York: Wiley.

Calvin, M. (1974). *Science, 184*, 375–381.

Calvin, M. (1977). Chemistry, population and resources. In *Proceedings of the 26th Meeting of the International Union of Pure Applied Chemistry,* Tokyo. Also Lawrence Berkeley Laboratory Report LBL–6433.

Camara-Artigas, A., Williams, J., & Allen, J. (2001). Structure of cytochrome c_2 from *Rhodospirillum centenum*. *Acta Crystallographica, Biology and Crystallization, D57*, 1498–1505.

Canny, M. (1977). *Annual Review of Fluid Mechanics, 9*, 275–296.

Carslaw. H., & Jaeger, J. (1959). *Conduction of heat in solids*. Oxford: Clarendon Press.

Casida, M., Casida, K., Jamorski, C., & Salahub, D. (1998). *Journal Chemical Physics, 108*, 4439.

Cavanagh, J., Clarke, J., & Price, R. (1993). Ocean energy systems. In T. Johansson, H. Kelly, A. Reddy, & R. Williams (Eds.), *Renewable energy sources for fuels and electricity* (pp. 513–547). Washington, DC: Island Press.

CECSMUD. (1982). *Sacremento municipal utility district 100 MW photovoltaic plant* (Draft Environmental Impact Report). Sacremento: Californian Energy Commission, State Clearing House # 81111253.

Ceder, G., Chiang, Y.-M., Sadoway, D., Aydinol, M., Jang, Y.-I., & Huang, B. (1998). Identification of cathode materials for lithium batteries guided by first-principles calculations. *Nature, 392,* 694–696.

Chapman, P. (1974, December 19). *New Scientist,* 866–869.

Chapman, S. (1943). *Reports on Progress in Physics, 9,* 92–100.

Chapman, D., & Pollack, H. (1975). *Earth and Planetary Science Letters, 28,* 23–32.

Charney, J., & Stone, P. (1975). *Science, 187,* 434–435.

Chartier, P., & Meriaux, S. (1980). *Recherche, 11,* 766–776.

Chelton, D., Schlax, M., Freilich, M., & Milliff, R. (2004). Satellite measurements reveal persistent small-scale features in ocean winds. *Science, 303,* 978–983.

Li, C. (1951). *Kho Hsueh Thung Paro (Science Correspondent) 2*(3), 266.

Cheremisinoff, N., Cheremisinoff, P., & Ellerbusch, F. (1980). *Biomass: Applications, technology, and production.* New York: Marcel Dekker.

China Reconstructs. (1975). December issue, pp. 24–27 (anonymous).

Chitnis, P. (2001). Photosystem I: Function and physiology. *Annual Review of Plant Physiology and Plant Molecular Biology, 52,* 593–626.

Chiu, H.-Y. (1964). *Annals of Physics, 26,* 364–410.

Christensen, P. (1981). *Kemiske Varmelagre* (Heat Storage Project Report No. 10). Danish Dept. of Energy, Copenhagen.

Christensen, L., & Gudmundsen, H. (1993). *Bilismens fremtid* (Danish Technology Council Report 93/3). Copenhagen.

CIESIN. (1997). *Gridded population of the world* (NCGIA Technical Report TR-95-6). David Simonett Center for Space Studies at University of Santa Barbara, California. Consortium for International Earth Science Information Network. http://www.ciesin.org/datasets/gpw/globdem.doc.html

Claesson, S. (1974). Jordbrukstekniska Inst., Sweden, Medd. No. 357.

Clarke, F. (1981). Wave energy technology. In R. Mayer & J. Olsson (Eds.), *Long-term energy sources* (pp. 1269–1303). Boston: Pittman Publ.

Claude, G. (1930). *Mechanical Engineering, 52,* 1039–1044.

Clayton, D. (1968). *Principles of stellar evolution and nucleosynthesis.* New York: McGraw-Hill.

Clayton, R. (1965). *Molecular physics in photosynthesis.* New York: Blaisdell.

Cline, W. (1992). *Global warming: The economic stakes.* Washington, DC: Institute for International Economics, 399.

Clot, A. (1977, March). *La Recherche, 8,* 213–222.

Cloud, P., & Gibor, A. (1970, September). *Scientific American,* 111–123.

Cockrane, J., Gubins, S., & Kiker, B. (1974). *Macro-economics.* London: Scott, Foresman and Co.

Cokelet, E. (1977). *Nature, 267,* 769–774.

Collet, P., Frederiksen, E., Hoffmann, T., & Madsen, G. (1976). Boligers luftskifte, Danish Technological Institute, Copenhagen.

Consoli, F., *et al.* (Eds.). (1993). *Guidelines for life-cycle assessment: A code of practice.* Society of Environmental Toxicology and Chemistry (SETAC).

Cooper, P. (1969). *Solar Energy, 12,* 3–8.

Coste, W. (1976). *Evaluation of wind power with energy storage* (Northeastern Utilities (M. Lotker) internal report (Hartford CT)).

Coty, U., & Dubey, M. (1976). *The high potential of wind as an energy source.* Paper presented at the 2nd Annual Energy Symposium, Los Angeles Council of Engineers and Scientists.

Coulson, K. (1975). *Solar and terrestrial radiation.* New York and London: Academic Press.

Cox, J., & Giuli, R. (1968). *Principles of stellar structure* (Vol. 1). New York: Gordon and Breach.

Crafoord, C. (1975). *An estimate of the interaction of a limited array of windmills* (Report DM–16). Department of Meteorology, University of Stockholm.

Craig, R. (1965). *The upper atmosphere.* New York and London: Academic Press.

CRES. (2002). *Wave energy utilization in Europe* (Project report for the European Commission). Pikermi: Centre for Renewable Energy Sources.

Crowley, T. (1998). Significance of tectonic boundary conditions for paleoclimate simulations. In T. Crowley & K. Burke (Eds.), *Tectonic boundary conditions for climate reconstructions* (pp. 3–17). New York: Oxford University Press.

Cultu, N. (1989a). Energy storage systems in operation. In B. Kilkis & S. Kakac (Eds.), *Energy storage systems* (pp. 551–574). Dordrecht: Kluwer.

Cultu, N. (1989b). Superconducting magnetic energy storage. In B. Kilkis & S. Kakac (Eds.), *Energy storage systems* (pp. 551–574). Dordrecht: Kluwer.

Cutchis, P. (1974). *Science, 184,* 13–19.

Daimler-Chrysler-Ballard. (1998). *Fuel-cell development programme.* http://www.daimler–benz. com/research/specials/necar/necar_e.htm

Damen, K., Faaij, A., Walter, A., & Souza, M. (2002). Future prospects for biofuel production in Brazil. In *12th European Biomass Conference,* (Vol. 2, pp. 1166–1169). ETA Firenze & WIP Munich.

Danish Association of Boiler Owners (Kedelforeningen). (1977). Danmarks Energibalance 1976, Copenhagen.

Danish Department of Energy. (1979). *Sæsonlagring af varme i store vandbassiner* (Heat Storage Project Report No. 2). Copenhagen.

Danish Department of Environment and Energy. (1996). Danmarks Energifremtider, Copenhagen.

Danish Department of Public Works. (1987). *Trafikundersøgelse 1986.* Copenhagen.

Danish Electricity Supply Undertakings. (1971). *Elværks-statistik 1970/71.* Copenhagen.

Danish Energy Agency. (1992). *Update on centralized biogas plants* (p. 31). Copenhagen: Author.

Danish Energy Agency. (1993). *District Heating in Denmark* (p. 58). Copenhagen.

Danish Energy Agency. (1995). *Progress report on the economy of centralized biogas plants* (p. 34). Copenhagen.

Danish Energy Agency. (1996). *Biomass for energy—Danish solutions* (p. 34). Copenhagen: Author.

Danish Energy Agency. (1999). *Windpower in Denmark.* http://www.ens.dk

Danish Energy Agency. (2002). *Energy Statistics 2001.* Copenhagen: Author. http://www.ens.dk

Danish Meteorological Institute. (1973). *Oceanographic observations from Danish light-vessels and coastal stations, 1972.* Denmark: Charlottenlund.

Danish Road Directorate. (1981). *Trafikulykker 1980.* Copenhagen.

Danish Statistical Office. (1993). *Miljø* (Report 93/1). Copenhagen.

Danish Statistical Survey. (1968). *Statistical inquiries, no. 22* (Agricultural Statistics 1900–1965, Vol. 1). Copenhagen.

Danish Statistical Survey. (1972). *Statistical information no. 9* (Agricultural Statistics 1971). Copenhagen.

Danish Technology Council. (1993). *Trafikkens Pris*. Copenhagen.

Danish Transport Council. (1993). *Externaliteter i transportsektoren* (Report 93-01). Copenhagen.

Danish Windpower Industry Association. (2003). http://www.windpower.dk

Danish wind turbine Price Lists and Overviews. (1981–2002). Alternately published by Danish Energy Agency, Renewable Energy Information Secretariat at Danish Institute of Technology, and Energi- og Miljødata Inc. http://www.emd.dk

Dansgaard, W. (1977). *Naturens Verden* (pp. 17–36). Copenhagen: Rhodos Publ. Co.

Dansgaard, W., Johnsen, S., Reeh, N., Gundestrup, N., Clausen, H., & Hammer, C. (1975). *Nature, 255*, 24–28.

Darmstadter, J., Teitelbaum, P., & Polach, J. (1971). *Energy in the world economy*. Baltimore: Johns Hopkins University Press.

d'Arsonval, J. (1881, September). *Revue Scientifique, 17*, 370–372.

DATV. (1995). *Life cycle screening of food products*. Lundtofte: Danish Academy of Technical Sciences.

Davidson, B., *et al.* (1980). *IEEE Proceedings, 127*, 345–385.

Davis, N. (1974, May 9). *New Scientist, 319*–320.

DEA Wave Program. (2002). Final report from Danish Wave Power Committee (in Danish). Virum: Rambøll Inc. http://www.waveenergy.dk

Deb, S. (1998). Recent developments in high efficiency photovoltaic cells. *Renewable Energy, 15*, 467–472.

Defant, A. (1961). *Physical oceanography* (2 Vols). Oxford: Pergamon Press.

Delwicke, C. (1970, September). *Scientific American, 136*–147.

Demler, E., & Zhang, S.-C. (1998). Quantitative test of a microscopic mechanism of high-temperature superconductivity. *Nature, 396*, 733–735.

Dennison, B., & Mansfield, V. (1976). *Nature, 261*, 32–34.

De Renzo, D. (1978). *European technology for obtaining energy from solid waste*. Park Ridge, NJ: Noyes Data Corp.

Desrosiers, R. (1981). In T. Reed (Ed.), *Biomass gasification* (pp. 119–153). Park Ridge, NJ: Noyes Data Corp.

Dietz, A. (1954). *Diathermarous materials and properties of surfaces* (As quoted by Duffie and Beckman, 1974).

Digby, S. (1954). In C. Singer, E. Holmyard, & A. Hall (Eds.), *A history of technology* (Vol. I, p. 735). Oxford: Clarendon Press.

Dones, R., & Frischknecht, R. (1997). Life cycle assessment of photovoltaic systems: Results of Swiss studies on energy chains. In Nieuwlaar & Alsema (Eds.), (1997) (p. 8).

Dones, R., Hirschberg, S., & Knoepfel, I. (1994, October 4–7). *Greenhouse gas emission inventory based on full energy chain analysis*. Beijing: IAEA Workshop.

Douce, R., & Joyard, J. (1977, June). *La Recherche, 527*–537.

DONG. (2003). *Gas stores*. http://www.dong.dk/dk/publikationer/lagerbrochure/

Douglas, D. (1974). *Transactions of IEEE, 37*–41.

DR. (1997, June 18). Testimony of K. Chekerov, Kurchatov Institute, Moscow, in the documentary Den skjulte faktor, Danish TV1.

Draper, L., & Squire, E. (1967). *Transactions of the Institution of Naval Architects, 109*, 85.

Dreicer, M. (1996, December). *Evaluation of the impacts from the production of electricity: A method for the nuclear fuel cycle*. Thesis, l'Ecole des Mines de Paris.

Drewes, P. (2003). Spheral solar—a completely different PV technology. In *Proceedings of the PV in Europe Conference 2002*, Roma.

Drift, A. van der. (2002). An overview of innovative biomass gasification concepts. In *12th European Biomass Conference* (pp. 381–384). ETA Firenze & WIP Munich.

DTI. (2000). *1996 Danish reference year* (obtained from Danish Technological Institute, Tåstrup).

Duffie, J., & Beckman, W. (1974). *Solar energy thermal processes*. New York: Wiley.

Duffie, J., & Beckman, W. (1991). *Solar energy thermal processes* (2nd ed.). New York: Wiley.

DuRy, C. (1969). *Völker des Alten Orient*. Baden-Baden: Holle Verlag.

Dutta, J., Wu, Z., Emeraud, T., Turlot, E., Cornil, E., Schmidt, J., *et al.* (1992). Stability and reliability of amorphous silicon pin/pin encapsulated modules. In L. Guimaraes, W. Palz, C. Reyff, H. Kiess, & P. Helm (Eds.), *11th EC PV Solar Energy Conference, Montreux* (pp. 545–548). Chur: Harwood Academic Publ.

Dwayne-Miller, R., *et al.* (2001). High density optical data storage and information retrieval using polymer nanostructures. In *Proceedings of the 10 th International Conference on Unconventional Photoactive Systems* (p. I-27), Diablerets.

EC. (1994). M. Ruiz-Altisent (Ed.), *Biofuels*. DG XII Report EUR 15647 EN, European Commission.

EC-ATLAS. (2003). European commission DG energy: ATLAS programme. http://europa.eu.int/energy_transport/atlas/htmlu/lbpot2.html

Eckart, C. (1958). *American Journal of Science, 256*, 225–240.

Eckert, E. (1960). Tidskrift for Landøkonomi (Copenhagen), No. 8.

Ehricke, K. (1971, November). *Bulletin of the Atomic Scientists, 18–26*.

Einstein, A. (1905). *Annalen der Physik, 17*, 549–560.

Eisenstadt, M., & Cox, K. (1975). *Solar Energy, 17*, 59–65.

Ekman, V. (1902). *Nyt Magasin for Naturvidenskab, 40*, 1 (as discussed by Sverdrup, 1957).

El-Hinnawi, E., & El-Gohary, F. (1981). In E. El-Hinnawi & A. Biswas (Eds.), *Renewable sources of energy and the environment* (pp. 183–219). Dublin: Tycooly International Press.

Ellehauge, K. (1981). *Solvarmeanlæg til varmt brugsvand*. Copenhagen: Danish Department of Energy Solar Heat Program, Report No. 16.

Elliott, D., & Elliott, R. (1976). *The control of technology*. London: Wykeham Publ. Co.

Ellis, H., & Pueschel, R. (1971). *Science, 172*, 845–846.

Eltra/Elkraft. (2001). *Time series of Danish wind power production 2000*. http://www.eltra.dk and http://www.elkraft-system.dk

Eltra. (2003). *Søkabel, Danish power utility*. http://www.eltra.dk

Emiliani, C. (1978). The cause of the ice age. *Earth and Planetary Science Letters, 37*, 349–352.

Engelenburg, van., & Nieuwlaar, E. (1993). *Methodology for the Life-cycle Assessment of Energy Technologies*. Report 93011. Dept. Science, Technology and Society, Utrecht University.

England, M., Godfrey, J., Hirst, A., & Tomczak, M. (1993). The mechanism of Antarctic intermediate water renewal in a world ocean model. *Journal of Physical Oceanography, 23*, 1553–1560.

Epstein, E., & Norlyn, J. (1977). *Science, 197*, 249–251.

Erdman, N., *et al.* (2002). The structure and chemistry of the TiO_2-rich surface of SrTiO3(001). *Nature, 419*, 55–58.

European Commission. (1985). *Test reference years TRY, DGXII: Science, research & development*, Report EUR 9765, p. 48. Brussels.

European Commission. (1995a). *ExternE: Externalities of Energy* (Vol. 1: *Summary*) prepared by ETSU for DGXII: Science, Research & Development, Study EUR 16520 EN, Luxembourg, p. 179.

European Commission. (1995b). *ExternE: Externalities of Energy* (Vol. 2). *Methodology*, prepared by ETSU and Metronomica for DGXII: Science, Research & Development, Study EUR 16521 EN, Luxembourg, p. 571.

European Commission. (1995c). *ExternE: Externalities of Energy* (Vol. 3). *Coal and lignite*, prepared by ETSU and IER for DGXII: Science, Research & Development, Study EUR 16522 EN, Luxembourg, p. 571.

European Commission. (1995d). *ExternE: Externalities of Energy* (Vol. 4). *Oil and gas,* prepared by ETSU and IER for DGXII: Science, Research & Development, Study EUR 16523 EN, Luxembourg, p. 426.

European Commission. (1995e). *ExternE: Externalities of Energy* (Vol. 5). *Nuclear*, prepared by CEPN for DGXII: Science, Research & Development, Study EUR 16524 EN, Luxembourg, p. 337.

European Commission. (1995f). *ExternE: Externalities of Energy* (Vol. 6): *Wind and hydro*, prepared by EEE and ENCO for DGXII: Science, Research & Development, Study EUR 16525 EN, Luxembourg, p. 249.

European Commission. (1997). Energy in Europe, 1997—Annual Energy Review, DGXII: Science, Research & Development, Special Issue, p. 179.

European Commission. (2001). Concerted action on offshore wind energy in Europe. Final Report from EC project NNE5-1999-562. Delft: Delft University Wind Energy Research Institute.

Eurostat. (1975). *Energy Statistics, 1970–1974*. Luxembourg: European Community Publ. Office.

Evans, D. (1976). *Journal of Fluid Mechanics, 77*, 1–25.

Faaij, A., & Hamelinck, C. (2002). Long term perspectives for production of fuels from biomass; integrated assessment and R&D priorities. In *12th European Biomass Conference* (Vol. 2, pp. 1110–1113). ETA Firenze & WIP Munich.

Fabritz, G. (1954). Wasserkraftmaschinen. In *Hiitte Maschinenbau* (Vol. IIA, pp. 865–961). Berlin: Wilhelm Ernst and Sohn.

Fairhall, A. (1973). *Nature, 245*, 20–23.

Falnes, J., & Løvseth, J. (1991, October). Ocean wave energy. *Energy Policy*, 768–775.

FAO. (2003). *UN food and agricultural organization statistics*. http://apps.fao.org

FAO-Asia. (2003). *Biomass energy technology: Wood energy database*. Regional Wood Energy Development Programme in Asia. http://www.rwedp.org/technobc.html

Federal Energy Administration. (1974). *Project independence, solar energy report*. Washington, DC: U.S. National Science Foundation.

Feld, B. (1976, June). *The Bulletin of the Atomic Scientists*, 10–13.

Fenhann, J., & Kilde, N. (1994). *Inventory of emissions to the air from Danish sources* (p. 111). Roskilde: Systems Analysis Dept., Risø National Laboratory.

Fernandes, R. (1974). In *Proceedings of the 9th Intersociety Energy Conversion Engineering Conference*, US, Paper No. 749032.

Ferreira, K., Iverson, T., Maghlaoui, K., Barber, J., & Iwata, S. (2004). Architecture of the photosynthetic oxygen-evolving center. *Science, 303*, 1831–1835.

Fittipaldi, F. (1981). Phase change heat storage. In G. Beghi (Ed.), *Energy storage and transportation* (pp. 169–182). Dordrecht, Holland: D. Reidel.

Flack, H., & Morikofer, W. (1964). In *Proceedings of the UN Conference on New and Renewable Energy Sources*, Rome 1961, Paper S/74. New York: UN Printing Office.

Fleagle, R., & Businger, J. (1963). *An introduction to atmospheric physics*. New York and London: Academic Press.

Foong, S. (2006). An accurate analytical solution of a zero-dimensional greenhouse model. *European Journal of Physics, 27*, 933.

Formidlingsrådet. (1989). *Forbehandling af biobrændsel. Decentrale kraftvarmeværker*. Report prepared by Krüger Engineers Inc., Industry and Trade Board, Danish Department of Commerce, Copenhagen, 93 pp. + maps.

Forrester, J. (1971). *World dynamics*. Cambridge, MA: Wright-Allen Press.

Fowler, W., & Hoyle, F. (1964). *The Astrophysical Journal Supplement series, 9,* No. 91.

Frankhauser, S. (1990). *Global warming damage costs—some monetary estimates*. CSERGE GEC Working paper 92/29. Norwich: University of East Anglia.

Frankl, P., Masini, A., Gamberale, M., & Toccaceli, D. (1997). Simplified LCA of PV systems in buildings, Appendix to *Environmental Aspects of PV Power Systems*, Utrecht Universiteit VNS Report no. 97072. (irreg. pag.)

Freeman, M., Haveman, R., & Kneese, A. (1973). *The economics of environmental policy*. New York: Wiley.

Freilich, M. (probably 1997). SeaWinds Algorithm Theoretical Basis Document (undated). Available as atbd-sws-01.pdf from the Jet Propulsion Laboratory website. Retrieved 2006, from http://podaac.jpl.nasa.gov/quicksat/qscat-doc/html.

Frenkiel, S. (1964). In *Proceedings of the UN Conference on New and Renewable Energy Sources*, Rome 1961, Paper W/33. New York: UN Printing Office.

Fried, M., Nosten, F., Brockman, A., Brabin, B., & Duffy, P. (1998). Maternal antibodies block malaria. *Nature, 395,* 851–852.

Friedli, H., Lotscher, H., Oeschger, H., Siegenthaler, U., & Stauffer, B. (1986). Ice core record of the $^{13}C/^{12}C$ ratio of atmospheric carbon dioxide in the past two centuries. *Nature, 324,* 237.

Friedlingstein, P. (2008). A steep road to climate stabilisation. *Nature, 451,* 297–298.

Frisch, M., *et al.* (1998). *Gaussian 98 software*, Revision A.9. Pittsburgh PA: Gaussian, Inc.

Fritsche, H. (1977). In W. Spear (Ed.), *Proceedings of the 7th International Conference on Amorphous and Liquid Semi-conductors* (pp. 3–15). Edinburgh: University of Edinburgh Press.

Fritsche, U., & Rausch, L. (1993). *TEMIS: Total emission model for integrated systems, User's guide for version 2.0*. Darmstadt: Oeko–Institute.

Fromme, P., Jordan, P., & Krauss, N. (2001). Structure of photosystem I. *Biochimica Et Biophysica Acta, 1507,* 5–31.

Fromme, P., Melkozernov, A., Jordan, P., & Krauss, N. (2003). Structure and function of photosystem I: Interaction with its soluble electron carriers and external antenna systems. *FEBS Letters, 555,* 40–44.

Frost, W. (1975). In M. Changery (Ed.), *Initial wind energy data assessment study* (pp. 80–106). Report No. NSF–RA–N–75–020. USA: National Science Foundation.

Frost, W., Maus, J., & Fichtl, G. (1974). *Boundary Layer Meteorology, 1,* 123–456.

Fujiwara, L., Nakashima, Y., & Goto, T. (1981). *Energy Conversion and Management, 21,* 157–162.

Furbo, S. (1982). *Communication No. 116*. Lyngby, Denmark: Thermal Insulation Laboratory, Technical University.

Gale, R. (1987). The environmental movement comes to town: A case study of an urban hazardous waste controversy. In B. Johnson, & V. Covello (Eds.), *The social and cultural construction of risk* (pp. 233–250). Dordrecht: Reidel.

Gates, D. (1966). *Science, 151,* 523–529.

Gates, D. (1968). *Annual Review of Plant Physiology, 19,* 211–238.

Gates, W., Henderson–Sellers, A., Boer, G., Folland, C., Kitoh, A., McAvaney, B., *et al.* (1996). Climate models—evaluation. *Climate change 1995: The science of climate change* (Chap. 5). IPCC Working Group I Report. Cambridge, MA: Cambridge University Press.

Geiger, R. (1961). *Das Klima der bodennahen Luftschicht*. Braunschweig: Vieweg and Sohn.

Gerlach, L. (1987). Protest movements and the construction of risk. In B. Johnson, & V. Covello (Eds.), *The social and cultural construction of risk* (pp. 103–145). Dordrecht: Reidel.

Giebel, G. (1999). Effects of distributing wind energy generation over Europe. In *European wind energy conference, Nice 1999* (pp. 417–420). London: James & James.

Gírio, F., Fonseca, C., Carvalheiro, F., Duarte, L., Marques, S., & Bogel-Lukasik, R. (2010). *Hemicelluloses for fuel ethanol: A review. Bioresource Technology*, doi:10.1016/j.biortech. 2010.01.088.

Glaser, P. (1977). *Journal of Energy, 1*, 75–84.

Glauert, H. (1935). In W. Durand (Ed.), *Aerodynamic theory* (Vol. 4, div. L, pp. 169–360). Berlin: J. Springer.

Goedheer, J., & Hammans, J. (1975). *Nature, 256*, 333–334.

Goguel, J. (1976). *Geothermics*. New York: McGraw-Hill.

Gøbel, B., Bentzen, J., Hindsgaul, C., Henriksen, U., Ahrenfeldt, J., Houbak, N., *et al.* (2002). High performance gasification with the two-stage gasifier. In *12th European Biomass Conference* (pp. 289–395). ETA Firenze & WIP Munich.

Gold, T., & Soter, S. (1980). *Scientific American*, 130–137.

Golibersuch, D., Bundy, F., Kosky, P., & Vakil, H. (1976). Thermal energy storage for utility applications. In J. Berkowitz, & H. Silverman (Eds.), *Proceedings of the Symposium on Energy Storage*. Princeton, NJ: The Electrochemical Society, Inc.

Govindjee (Ed.). (1975). *Bioenergetics of photosynthesis*. New York and London: Academic Press.

Gramms, L., Polkowski, L., & Witzel, S. (1971). *Transactions of the American Society of Agricultural Engineers, 1*, 7.

Granqvist, C., *et al.* (1998). *Solar Energy, 63*, 199–216.

Grasse, W. (1981). *Sunworld, 5*, 68–72.

Gray, T., & Gashus, O. (1972). *Tidal power*. New York: Plenum Press.

Green, M. (1992). Crystalline and polycrystalline silicon solar cells. In Johansson *et al.* (Eds.), *Renewable energy sources for fuels and electricity* (pp. 337–360). Washington, DC: Island Press.

Green, M. (1994). *Silicon solar cells: Advanced principles and practice*. Sydney: PV Special Research Centre, University of New South Wales.

Green, M. (2002). Third-generation photovoltaics: Recent theoretical progress. In McNelis, Palz, Ossenbrink, & Helm (Eds.), *17th European Photovoltaic Solar Energy Conference, Munich 2001* (Vol. I, pp. 14–17).

Green, M., Zhao, J., & Wang, A. (1998). 23% module and other silicon solar cell advances. In J. Schmid *et al.* (Eds.), *Proceedings 2nd World Conference on PV Energy Conversion, Vienna* (pp. 1187–1192). Luxembourg: JRC European Commission EUR 18656 EN.

Gregg, M. (1973, February). *Scientific American*, 65–77.

Gregory, D., & Pangborn, J. (1976). *Annual Review of Energy, 1*, 279–310.

Grimson, J. (1971). *Advanced fluid dynamics and heat transfer*. London: McGraw-Hill.

Gringarten, A., Wintherspoon, P., & Ohnishi, Y. (1975). *Journal of Geophysical Research, 80*, 1120–1124.

Grove, W. (1839). On voltaic series and the combination of gases by platinum. *Philosophical Magazine, 14*, 127–130.

Grubb, M. (1996). Purpose and function of IPCC. *Nature* 379, 108; response to news items in *Nature* 378, 322 (1995) and *Nature* 378, 119 (1995).

Gutenberg, G. (1959). In J. Mieghem (Ed.), *Physics of the Earth's interior, International Geophysics Series* (Vol. 1). New York and London: Academic Press.

Hadley, G. (1735). *Philosophical Transactions, 29*, 58–62.

Hagen, D., Erdman, A., & Frohrib, D. (1979). In *Proceedings of the 14th Intersociety Engineering Conference* (pp. 368–373). Washington, DC: American Chemical Society.

Hagenmuller, P. (1977). *La Recherche, 81*, 756–768.

Hagerman, G., & Heller, T. (1988). Wave energy: A survey of twelve near-term technologies. In *Proceedings of the International Renewable Energy Conference* (pp. 98–110). *Honolulu*, HI.

Hahn, D., & Manabe, S. (1975). *Journal of Atmospheric Science, 32*, 1515–1541.

Hall, C., Swet, C., & Temanson, L. (1979). In *Sun II, Proceedings of the Solar Energy Society Conference, New Delhi, 1978* (pp. 356–359). London: Pergamon Press.

Hamakawa, Y. (1998). A technological evolution from bulk crystalline age to multilayer thin film age in solar photovoltaics. *Renewable Energy, 15*, 22–31.

Hamakawa, Y., Tawada, Y., & Okamoto, H. (1981). *International Journal of Solar Energy, 1*, 125.

Hambraeus, G. (Ed.). (1975). *Energilagring* (Report No. IVA–72). Stockholm: Swedish Academy of Engineering Sciences .

Hammond, A. (1976). *Science, 191*, 1159–1160.

Hammond, A., Metz, W., & Maugh, T., II. (1973). *Energy and the future*. Washington: American Association for the Advancement of Science.

Handley, D., & Heggs, P. (1968). *Transactions of Economics and Engineering Reviews, 5*, 7.

Hanneman, R., Vakil, H., & Wentorf , R., Jr. (1974). Closed loop chemical systems for energy transmission, conversion and storage. In *Proceedings of the 9th Intersociety Energy Conversion Engineering Conference*. New York: American Society of Mechanical Engineers.

Hansen, H. (1985). *Poul la Cour* (biography) (p. 516). Askov Højskoles Forlag,

Hansen, J. (1975). *Journal of Atmospheric Science, 26*, 478–487.

Hara, K., Sayama, K., Ohga, Y., Shinpo, A., Suga, S., & Arakawa, H. (2001). A coumarin-derivative dye-sensitised nanocrystalline TiO_2 solar cell having a high solar-energy conversion efficiency up to 5.6%. *Chemical Communications, 6*, 569–570.

Harth, R., Range, J., & Boltendahl, U. (1981). EVA–ADAM System: A method of energy transportation by reversible chemical reactions. In G. Beghi (Ed.), *Energy storage and transportation* (pp. 358–374). Dordrecht, Boston, London: Reidel.

Hartline, F. (1979). *Science, 206*, 205–206.

Hasselmann, K. (1962). *Journal of Fluid Mechanics, 12*, 481–500.

Hasselmann, K. (1967). *Proceedings of the Royal Society (London), A299*, 77–100.

Hasselmann, K., Barnett, T., Bouws, E., Carlson, H., Cartwrigt, D., Enke, K., *et al.* (1973). *Deutschen Hydrographischen Zeitschrift*, Erganzungsheft, Reihe A(8°), Nr. 12 (Hamburg).

Hasumi, H., & Emori, S. (2004). K-1 Coupled GCM (MIROC) Description. University of Tokyo Internal Report.

Hauch, A., *et al.* (2001). New photoelectrochromic device. http://www.fmf.uni-freiburg.de/~biomed/FSZ/anneke2.htm

Haug, G., Sigman, D., Tiedemann, R., Pedersen, T., & Sarntheim, M. (1999). Onset of permanent stratification in the subarctic Pacific Ocean. *Nature, 401*, 779–782.

Havelock, T. (1919). *Proceedings of the Royal Society (London), A95*, 38–51.

Hay, H., & Yellot, J. (1972). *Mechanics and Engineering, 92*(1), 19–23.

Hayashi, C. (1966). *Annual Review of Astronomy and Astrophysics, 4*, 171–192.

Hayes, D. (1977). *Worldwatch Paper No. 11*. Washington, DC: Worldwatch Institute.

Hays, J., Imbrie, J., & Shackleton, N. (1976). *Science, 183*, 1121–1131.

Hein, R. (1974). *Science, 185*, 211–222.

Heppenheimer, T., & Hannah, E. (1975). *R & D Requirements for initial space colonization*, Contribution No. 2664 from Div. Geol. and Planetary Science. Pasadena, CA: California Institute of Technology.

Herbig, G. (1967, August). *Scientific American*, 30–36.

Hermes, J., & Lew, V. (1982). Paper CF9/V III/9. In *Proceedings of the UNITAR Conference on Small Energy Resources, Los Angeles 1981*. New York: United Nations Institute for Training and Research.

Heronemus, W. (1975). Proposal to United States National Science Foundation. Report NSF/RANN/S E/GI–34979/TR/75/4.

Herrick, C. (1982). *Solar Energy, 28*, 99–104.

Hespanhol, I. (1979, November–December). *Energia, 5*, Quoted from In E. El-Hinnawi & A. Biswas (Eds.), *Renewable Sources of Energy and the Environment*. Dublin: Tycooly International Press.

Hesstvedt, E. (1973). *Water, Air, and Soil Pollution, 2*, 49–60.

Hewson, E. (1975). *Bulletin of the American Meteorological Society, 56*, 660–675.

Hide, R. (1958). *Philosophical Transactions of the Royal Society (London), A250*, 441–478.

Hildebrandt, A., & Vant–Hull, L. (1977). *Science, 197*, 1139–1146.

Hind, G., & Olson, J. (1968). *Annual Review of Plant Physiology, 19*, 249–282.

Hinze, J. (1975). *Turbulence*. New York: McGraw–Hill.

Hippel, F. von, Fels, M., & Krugmann, H. (1975, April). *FAS Professional Bulletin* (Washington, DC) *3*, 5–6.

Hirsch, R., Gallagher, J., Lessard, R., & Wesselhoft, R. (1982). *Science, 215*, 121–127.

Hjalte, K., Lidgren, K., & Stahl, I. (1977). *Environmental policy and welfare economics*. Cambridge, UK: Cambridge University Press.

Hobbs, P., Harrison, H., & Robinson, E. (1974). *Science, 183*, 909–915.

Hoffmann, P. (1998a). ZEVCO unveils fuel cell taxi. *Hydrogen and Fuel Cell Letter*, feature article, August. http://www.mhv.net/~hfcletter/letter.

Hoffmann, P. (1998b). Fuel processors. Record attendance highlights fuel cell seminar in November at Palm Springs. *Hydrogen and Fuel Cell Letter*, feature article, December. http://www.mhv.net/~hfcletter/letter.

Hofmeister, A. (1999). Mantle values of thermal conductivity and the geotherm from phonon lifetimes. *Science, 283*, 1699–1706.

Hoganson, C., & Babcock, G. (1997). A metalloradical mechanism for the generation of oxygen from water in photosynthesis. *Science, 277*, 1953–1956.

Hohmeyer, O. (1988). *Social costs of energy consumption* (p. 126). Berlin: Springer-Verlag.

Holdren, J. (1974, October). *Bulletin of Atomic Science*, 14–23.

Holloway, J., & Manabe, S. (1971). *Monthly Weather Review, 99*, 335–370.

Holten, T. van. (1976). In *International Symposium on Wind Energy Systems, Cambridge 1976*, Paper E3. Cranfield: British Hydromechanical Research Association.

Honsberg, C. (2002). A new generalized detailed balance formulation to calculate solar cell efficiency limits. In McNelis, Palz, Ossenbrink, Helm (Eds.), *17th European Photovoltaic Solar Energy Conference, Munich 2001* (Vol. I, pp. 3–8).

Hooker, C., & Hulst, R. van. (1980). Institutionalizing a high quality conserver society. *Alternatives, 9*, 25–36.

Hordijk, L. (1991). Use of the RAINS models in acid rain negotiations in Europe. *Environmental Science Technology, 25*, 596–603.

Horn, M., *et al*. (1970). *Journal of the American Ceramic Society, 53*, 124.

Hoven, I. van der. (1957). *Journal of Meteorology, 14*, 160–164.

Hoyle, F. (1975). *Astronomy and cosmology*. San Francisco: W. Freeman and Co.

Hughes, W. (Ed.). (1976). *Energy for rural development*. Washington, DC: United States National Academy of Sciences.

Hulme, M., Zorita, E., Stocker, T., Price, J., & Christy, J. (2010). IPCC: Cherish it, tweak it or scrap it? *Nature, 463*, 730–732.

Hütte (a Berlin-based association, Ed.). (1954). *Des Ingenieurs Taschenbuch* (Vols. I and II). Berlin: Wilhelm Ernst.

Hütter, U. (1976). In *Proceedings of the Workshop on Advanced Wind Energy Systems*, Stockholm, 1974. Stockholm: Swedish Development Board/Vattenfall.

Hütter, U. (1977). *Annual Review of Fluid Mechanics, 9*, 399–419.

Hvelplund, F. (1975). *Indexing, resource utilization and capital crisis*. Internal Report. Aarhus, Denmark: School of Economics and Business Administration.

Hvelplund, F., Illum, K., Jensen, J., Meyer, N., Nørgård, J., & Sørensen, B. (1983). *Energy for the future—alternative energy plan* (in Danish) (p. 96). Copenhagen: Borgen Publisher.

Iben, I. (1970). *Scientific American,* July, 26–39.

Iben, I. (1972). In H.-Y. Chiu & A. Muriel (Eds.), *Stellar evolution* (pp. 1–106). Cambridge, MA: MIT Press.

Ichikawa, Y. (1993). Fabrication technology for large-area a-Si solar cells. In *Technical Digest of 7th International PV Science and Engineering Conference Nagoya* (pp. 79–95). Nagoya Institute of Technology.

IEA. (1999). Photovoltaic power systems in selected IEA member countries. 3rd survey report of the Power Systems Programme Task 1. Paris: International Energy Agency.

IEA. (2002). Key world energy statistics. Paris: International Energy Agency. http://www.iea.org

IEA-PVPS. (2001). Trends in photovoltaic applications, Report T1-10:2001, International Energy Agency Photovoltaic Power Systems Programme.

IEA-PVPS. (2002). Trends in photovoltaic applications, Report T1-11:2002, International Energy Agency Photovoltaic Power Systems Programme.

IEA-PVPS. (2008). Trends in photovoltaic applications, Report T1-17:2002, International Energy Agency Photovoltaic Power Systems Programme.

Inaba, A., Shimatani, T., Tabata, S., Kawamura, S., Shibuya, S., Iwase, Y., *et al.* (1993). *Journal of Chemical and Engineering Japan, 19*, 809–817.

Ion, D. (1975). *Availability of world energy resources*. London: Graham and Trotman.

IPCC. (1992). In J. Houghton, B. Callander, & S. Varney (Eds.), *Climate change 1992: The supplementary report to the IPCC scientific assessment* (p. 198). Cambridge, MA: Cambridge University Press.

IPCC. (1994). In S. Greco, R. Moss, D. Viner, & R. Jenne (Eds.), *Climate scenarios and socioeconomic projections for IPCC WGII assessment consortium for International Earth Science Info. Network*, Washington, DC (limited distribution working document).

IPCC. (1995). Climate change 1994. In J. T. Houghton, L. G. Meira Filho, J. Bruce, Hoesung Lee, B. A. Callander, E. Haites, *et al.* (Eds.), *Radiative forcing of climate change and an evaluation of the IPCC IS92 emission scenarios* (p. 339). Cambridge, MA: Cambridge University Press.

IPCC. (1996a). Climate change 1995: Impacts, adaptation and mitigation of climate change: Scientific–technical analysis. In R. T. Watson, M. C. Zinyowera, R. H. Moss, D. J. Dokken (Eds.), Contribution of Working Group II to the *Second assessment report of the intergovernmental panel on climate change* (p. 878). Cambridge, MA: Cambridge University Press.

IPCC. (1996b). Climate change 1995: The science of climate change. In J. T. Houghton, L. G. Meira Filho, B. A. Callander, N. Harris, A. Kattenberg, & K. Maskell (Eds.), Contribution of Working Group I to the *Second assessment report of the intergovernmental panel on climate change* (p. 572). Cambridge, MA: Cambridge University Press.

IPCC. (1996c). Climate change 1995. Economic and social dimensions of climate change. In J. Bruce, H. Lee, & E. Haites (Eds.), Contribution of Working Group III to the *Second assessment report of the intergovernmental panel on climate change* (p. 448). Cambridge, MA: Cambridge University Press.

IPCC. (2007a). Climate change 2007. *The Physical Science Basis.* In S. Solomon, *et al.* (Eds.), Contribution of Working Group I to the *Fourth assessment report of the intergovernmental panel on climate change*. Cambridge, MA: Cambridge University Press. http://www.ipcc.ch/publications_and_data

IPCC. (2007b). Climate change 2007. *Mitigation.* In B. Metz, *et al.* (Eds.), Contribution of Working Group III to the *Fourth assessment report of the intergovernmental panel on climate change*. Cambridge, MA: Cambridge University Press. http://www.ipcc.ch/publications_and_data

IPCC. (2007c). Climate change 2007. *Impacts, adaptation and vulnerability.* In M. Parry, *et al.* (Eds.), Contribution of Working Group II to the *Fourth assessment report of the intergovernmental panel on climate change*. Cambridge, MA: Cambridge University Press. http://www.ipcc.ch/publications_and_data

IPCC. (2010). Scenario data for the atmospheric environment. See Environmental data > Atmospheric data at http://www.ipcc-data.org

IPCC. Data Distribution Centre (1998–2010). Global climate model archive maintained by Deutsches Klimarechenzentrum at http://www.dkrz.de/ipcc/ddc/html/ dkrzmain.html, and at primary site http://www.ipcc-data.org/ar4.

Isaacs, J., Castel, D., & Wick, G. (1976). *Ocean Engineering, 3,* 175–187.

ISO. (1997). *Life Cycle Assessment—Principles and Framework,* Standard 14040, International Standards Organisation.

ISO. (1999). *Life Cycle Assessment—Life Cycle Impact Assessment,* Standard 14040, International Standards Organisation.

Iwata, S., Lee, J., Okada, K., Lee, J., Iwata, M., Rasmussen, B., *et al.* (1998). Complete structure of the 11-subunit bovine mitochondrial cytochrome bc_1 complex. *Science, 281,* 64–71.

Jacobsen, T. (1973). As quoted in *World Almanac 1974* (p. 427). New York: Newspaper Enterprise Association.

Jansen, E., *et al.* (2007). Paleoclimate. In S. Solomon, *et al.* (Eds.), *Climate change 2007: The physical science basis* (Chap. 6). Cambridge, MA: Cambridge University Press. http://www.ipcc.ch/publications_and_data

Jamshidi, M., & Mohseni, M. (1976). In G. VanSteenkiste (Ed.), *System simulation in water resources* (pp. 393–408). Amsterdam: North-Holland Publ. Co.

Jensen, J. (1981). Improving the overall energy efficiency in cities and communities by the introduction of integrated heat, power and transport systems. In J. P. Millhom & E. H. Willis (Eds.), *Proceedings of the IEA Inl. New Energy Conservation Technologies Conference* (p. 2981). Berlin: Springer-Verlag.

Jensen, J., & Sørensen, B. (1984). *Fundamentals of energy storage* (p. 345). New York: Wiley.

Jensen, M. (1962). In *Vindkraftudvalgets Betaœnkning* (pp. 54–66) and personal communication. Copenhagen: Danish Electricity Supply Undertakings.

Jensen, N. (1994). PARK computer program, described in Beyer, H., *et al. Journal Wind Engineering & Industrial Aerodynamics 51,* 111.

Jet Propulsion Laboratory. (2007). SeaWinds on QuickSCAT Products. Retrieved 2007, from http:// podaac.jpl.nasa.gov/cgi-bin/dcatalog/.

Johansson, T. (2002). Energy for sustainable development—a policy agenda for biomass. In *12th European Biomass Conference* (Vol. 1, pp. 3–6). ETA Firenze & WIP, Munich.

Johnson, B., & Covello, V. (eds.) (1987). *The social and cultural construction of risk.* Dordrecht: Reidel.

Johnson, C., Holland, M., & Amonier, S. (1994). Assessment of incinerating or recycling newspaper. In *SETAC Workshop on Integrating Impact Assessment into LCA, Brussels.*

Johnson, J., & Hinman, C. (1980). *Science, 208,* 460–463.

Johnston, H. (1977). *Journal of Geophysical Research.*

Jordan, P., Fromme, P., Witt, H., Klukas, O., Saenger, W., & Krauss, N. (2001). Three-dimensional structure of cyanobacterial photosystem I at 2.5 Å resolution. *Nature, 411,* 909–917.

Jørgensen, L., Mikkelsen, S., & Kristensen, P. (1980). *Solvarmeanlæg i Greve,* Danish Department of Energy Solar Heat Program, Report No. 6. Copenhagen (with follow up: Report No. 15, 1981).

Joukowski, N. (1906). *Bull. de l'Inst. Aeronaut. Koutchino,* Fasc. I, St. Petersburg.

Junge, C. (1963). *Atmospheric chemistry and radioactivity.* New York and London: Academic Press.

Juul, J. (1961). In *Proceedings of the UN Conference on New and Renewable Energy Sources of Energy,* Rome, Vol. 7, paper W/21 (Published 1964 as paper E/Conf. 35 by UN Printing Office, New York).

Kahn, J. (1996). Fuel cell breakthrough doubles performance, reduces cost. Berkeley Lab. Research News, 29. May. http://www.lbl.gov/science–articles/archive/ fuel–cells.html.

Kalinin, G., & Bykow, V. (1969). *Impact of Science on Society 19,* (2) [As quoted by Lindh, G. (1972). *Ambio 1,* 185–201].

Kalkstein, L. (1993). Health and climate change: Direct impacts in cities. *The Lancet, 342,* 1397–1399.

Kalkstein, L., & Smoyer, K. (1993). The impact of climate change on human health: Some international implications. *Experientia, 49,* 969–979.

Kalnay, E., Kanamitsu, M., Kistler, R., Collins, W., Deaven, D., Gandin, L., *et al.* (1996, March). The NCEP/NCAR 40-year reanalysis project. *Bulletin of the American Meteorological Society.*

Kaltschmitt, M., Reinhardt, G., & Stelzer, T. (1996). LCA of biofuels under different environmental aspects. In P. Chartier, G. Ferrero, U. Henius, S. Hultberg, J. Sachau, & M. Wiinblad (Eds.), *Biomass for energy and the environment* (Vol. 1, pp. 369–386). Oxford: Pergamon/Elsevier.

Kamat, P., & Vinodgopal, K. (1998). Environmental photochemistry with semiconductor nanoparticles. In V. Ramamurthy & K. Schanze (Eds.), *Organic and inorganic photochemistry* (Chap. 7).

Kamiya, N., & Shen, J-R. (2003). Crystal structure of oxygen-evolving photosystem II from *Thermosynechococcus vulcanus* at 3.7-Å resolution. *Proceedings of the National academy of Sciences (USA), 100,* 98–103.

Kasahara, A., & Washington, W. (1967). *Monthly Weather Review, 95,* 389–402.

Kaspar, F., & Cubasch, U., 2007. Simulation of the Eemian inter-glacial and the subsequent glacial inception with a coupled ocean-atmosphere general circulation model. In F. Sirocko, M. Claussen, M. Goñi, T. Litt, (Eds.), *The climate of past interglacials.* Developments in quaternary sciences (Vol. 7, pp. 499–515). Elsevier.

Kasten, F. (1977). *Solar Energy, 19,* 589–593.

Kato, K., Yamada, K., Inaba, A., Shimatani, T., Tabata, S., Kawamura, S., *et al.* (1993). *Journal of Chemical Engineering Japan, 20,* 261–267.

Kaye, G., & Laby, T. (1959). *Tables of physical and chemical constants* (12th ed.). London: Longmans.

Keenan, J. (1977). *Energy Conversion, 16,* 95–103.

Kertz, W. (1957). In S. Flügge (Ed.), *Handbuch der Physik* (Vol. 48, pp. 928–981) (Geophysik II). Berlin: Springer-Verlag.

Keynes, J. (1936). *The general theory of employment, interest and money.* New York: McMillan.

Khaselev, O., & Turner, J. (1998). *Science, 280,* 425–427.

Kiilerich, A. (1965). *Oceanografi.* Copenhagen: Gjellerup.

King Hubbert, M. (1969). In *Resources and man* (Chap. 8), (United States National Academy of Sciences and National Science Foundation). San Francisco: W. Freeman.

King Hubbert, M. (1971, September). *Scientific American,* 60–87.

Kinsman, B. (1965). *Wind waves.* Englewood Cliffs, NJ: Prentice-Hall.

Kira, T. (1975). In J. Cooper (Ed.), *Photosynthesis and productivity in different environments* (pp. 5–40). Cambridge, UK: Cambridge University Press.

Kitaigorodskii, S. (1970). *The physics of air-sea interaction,* Israel Program for Scientific Translation, Jerusalem 1973.

Klein, S. (1975). *Solar Energy, 17,* 79–80.

Knöepfel, I. (1993). The Swiss national research project Total Pollution of Energy Systems. In *Life-cycle analysis of energy systems,* Workshop proceedings (pp. 238–247). Paris: OECD Publications.

Kohn, W., & Sham, L. (1965). Self-consistent equations including exchange and correlation effects. *Physical Review, 140,* A1133.

Kok, B., Forbush, B., & McGloin, M. (1970). *Photochemistry and Photobiology 11,* 457–475.

Kok, B., Radmer, R., & Fowler, C. (1975). In M. Avron (Ed.), *Proceedings of the 3rd International Congr. on Photosynthesis, Rehovoth* (Vol. I, pp. 485–496).

Kondratyev, K., & Podolskaya, E. (1953). *Izvestiya USSR Acad. Sci. Nauk,* Ser. *Geophys.,* No. 4.

Kondratyev, K., & Fedorova, M. (1976). In *Proceedings of the UNESCO/WMO Solar Energy Symposium. Genève 1976.* paper ENG.S/Doc. 2.

Koppen, C. van, Fischer, L., & Dijkamns, A. (1979). In *Sun II, Proceedings of the International Solar Energy Society Conference, New Delhi* (pp. 294–299). London: Pergamon Press.

Korn, J. (Ed.). (1972). *Hydrostatic transmission systems.* London: International Text Book Co.

Kraemer, F. (1981). *En model for energiproduktion og økomomi for centrale anlæg til produktion af biogas.* Report from Physics Laboratory 3. Lyngby, Denmark: Danish Technical University.

Krasovec, O., *et al.* (2001). Nanocrystalline WO_3 layers for photoelectrochromic and energy storage dye sensitised solar cells. In *Proceedings of the 4th International Symposium on New Materials,* Montréal (pp. 423–425).

Kreider, J. (1979). *Medium and high temperature solar processes.* New York: Academic Press.

Kuemmel, B. (2000). A global clean fossil scenario. *International Journal of Global Energy Issues, 13,* 181–195.

Kuemmel, B., Nielsen, S. K., & Sørensen, B. (1997). *Life-cycle analysis of energy systems* (p. 216). Copenhagen: Roskilde University Press.

Kurisu, G., Zhang, H., Smith, J., & Cramer, W. (2003). Structure of the cytochrome b_6f complex of oxygenic photosynthesis: Tuning the cavity. *Science, 302,* 1009–1014.

Kussmaul, K., & Deimel, P. (1995). Materialverhalten in H_2-Hochdrucksystemen. *VDI Berichte, 1201,* 87–101.

Kutta, W. (1902, July). Auftriebkrafte in stromende Flüssigkeiten, *Ill. aeronaut. Mitteilungen,.*

La Cour, P. (1900). *The experimental mill* (Vol. 1). Copenhagen: Nordiske Forlag; Vols. 2–4 (1901–3).

Ladisch, M., & Dyck, K. (1979). *Science, 205*, 898–900.

Ladisch, M., Flickinger, M., & Tsao, G. (1979). *Energy, 4*, 263–275.

Landau, L., & Lifshitz, F. (1960). *Mechanics*. Oxford: Pergamon Press.

Landberg, L., Joensen, A., Giebel, G., Watson, S., Madsen, H., Nielsen, T., *et al.* (1999). Implementing short-term prediction at utilities. Final report of EC JOULE project JOR3–CT95–0008; a preliminary report dated 1996 was used by Meibom, P., Svendsen, T., Sørensen, B. (1997; 1999).

Laquer, H. (1976). In W. Bostick, V. Nardi, & O. Zucker (Eds.), *Energy storage, compression and switching* (pp. 279–305). New York: Plenum Press.

Laskin, A., Gaspar, D., Wang, W., Hunt, S., Cowin, J., Colsom, S., *et al.* (2003). Reactions at interfaces as a source of sulphate formation sea-salt particles. *Science, 301*, 340–344.

Lau, K. (1987). Electricity forecasting in Denmark: Conflict between ministries and utilities. In T. Baumgartner & A. Midttun (Eds.), *The politics of energy forecasting* (Chap. 8). Oxford: Oxford University Press.

Leakey, M. (1975). As quoted in *World Almanac* (1977), p. 432.

Lees, L. (1970) In C. Wilson & W. Matthews (Eds.), As quoted in *Man's impact on the global environment* (p. 63). Cambridge, MA: MIT Press.

Lehmann, J. (1981). Air storage gas turbine power plants: A major distribution for energy storage. In *Proceedings of the International Conference on Energy Storage* (pp. 327–336). Cranfield, UK: BHRA Fluid Engineering.

Leichman, J., & Scobie, G. (1975). *The development of wave power—a techno-economic study* (Pt. 2). East Kilbride, Scotland: UK Department of Industry, and Nat. Eng. Lab.

Levitus, S., & Boyer, T. (1994). *World Ocean Atlas 1994*, Vol. 4: *Temperature*, NOAA Atlas NESDIS 4. Washington, DC: United States Department of Commerce.

Lewis, J. (1974, March). *Scientific American*, 51–65.

Libowitz, G. (1974). In *Proceedings of the 9th Intersociety Energy Conversion Conference*, US, Paper No. 749025.

Linden, S. van der. (2003). The commercial world of energy storage: A review of operating facilities. Presentation for 1st *Annual Conference of Energy Storage Council*, Houston TX.

Lindh, G. (1972). *Ambio 1*, 185–201.

Lisiecki, L., & Raymo, M. (2005). A Pliocene-Pleistocene stack of 57 globally distributed benthic d18O records. *Paleoceanography, 20*, PA1003, doi:10.1029/2004PA001071. Data available from World Data Center for Paleoclimatology. http://www.ncdc.noaa.gov/paleo

Lissaman, P. (1976). In *Proceedings of the International Symposium on Wind Energy Systems, Cambridge*. Cranfield, UK: Paper C2: BHRA Fluid Engineering.

Little, Arthur D. (Inc.) (1975, March). Economic Comparison of Base-load Generation Alternatives for New England Electric. S. M. Stoller Corp.,.

Ljungström, O. (1975). *Det svenska vindenergiprogram*, Swedish Development Board (STU), Report No. VES–1975: 33.

Locquet, J.–P., Perret, J., Fompeyrine, J., Mächler, E., Seo, J., & Tendeloo, G. (1998). Doubling the critical temperature of $La_{1.9}Sr_{0.1}CuO_4$ using epitaxial strain. *Nature, 394*, 453–456.

Loferski, J. (1956). *Journal of Applied Physics, 27*, 777–784.

Longair, M., & Sunyaev, R. (1969). *Astrophysics Letters, 4*, 65–70.

Loomis, R., & Gerakis, P. (1975). In J. Cooper (Ed.), *Photosynthesis and productivity in different environments* (pp. 145–172). Cambridge, UK: Cambridge University Press.

Lorenz, E. (1967). *The nature and theory of the general circulation of the atmosphere*. World Meteorological Organization, WMO publ. No. 218TP115.

Lorenz, E. (1968). *Meteorol. Monographs, 8*(30), 1–3.

Losciale, M. (2002). Technical experiences and conclusions from introduction of biogas as a vehicle fuel in Sweden. In *12th European Biomass Conference* (Vol. 2, pp. 1124–1127). ETA Firenze & WIP Munich.

Lostao, A., Daoudi, F., Irún, M., Ramón, Á., Fernández-Cabrera, C., Romero, A., *et al.* (2003). How FMN binds to *Anabaena* apoflavodoxin. *Journal of Biological Chemistry, 278,* 24053–24061.

Lotker, M. (1974). In *Proceedings of the 9th Intersociety Energy Conversion Engineering Conference*, US, Paper No. 749033.

Lovins, A. (1977). *Soft energy paths.* New York: Ballinger.

Lund, H., *et al.* (1985). *Test reference year TRY.* Report EUR 9765 to DG12, European Commission from Cenergia, Ballerup.

Luzzi, A. (1999). IEA Annual Report from Agreement on Production and Utilization of Hydrogen (pp. 35–42).

Lvovitch, M. (1977). *Ambio 6,* 13–21.

Ma, W., Saida, T., Lim, C., Aoyama, S., Okamoto, H., & Hamakawa, Y. (1995). The utilization of microcrystalline Si and SiC for the efficiency improvement in a-Si solar cells. In *1994 First World Conference on Photovoltaic Energy Conversion, Kona* (pp. 417–420). Washington, DC: IEEE.

Maag, W. (1993). *Sonnenenergie und Wärmetechnik,* No. 5/93, 16–19.

MacGregor, I., & Basu, A. (1974). *Science, 185,* 1007–1011.

Machta, L. (1971). As quoted in Wilson, C., Matthews, W. *Inadvertent Climate Modification* (p. 235). Report of the Study of Man's Impact on Climate (SMIC). MIT Press, Cambridge, MA.

Machta, L. (1976). In *Proceedings of the UNESCO/WMO Solar Energy Symposium, Genève 1976,* paper ENGELS./Doc. 1.

Makhijani, A. (1977, August). *Economic and Political Weekly* (Bombay), Special issue, pp. 1451–1464.

Manabe, S. (1969). *Monthly Weather Review, 97,* 739–774; 775–805.

Manabe, S. (1971). In W. Matthews, W. Kellogs, & G. Robinson (Eds.), *Man's Impact on Climate* (pp. 249–264). Cambridge, MA: MIT Press.

Manabe, S., & Holloway, J. (1975). *Journal of Geophysical Research, 80,* 1617–1649.

Manabe, S., & Stouffer, R. (1988). Two stable equilibria of a coupled ocean–atmosphere model. *Journal of Climate, 16,* 185–192.

Manabe, S., & Strickler, R. (1964). *Journal of Atmospheric Science, 21,* 361–385.

Manabe, S., & Wetherald, R. (1967). *Journal of Atmospheric Science, 24,* 241–259.

Manabe, S., & Wetherald, R. (1975). *Journal of Atmospheric Science, 32,* 3–15.

Manabe, S., Hahn, D., & Holloway, J. (1974). *Journal of Atmospheric Science, 31,* 43–83.

Manassen, J., Cahen, D., Hodes, G., & Sofer, A. (1976). *Nature, 263,* 97–100.

Marchetti, C. (1973). *Chemical Economy and Engineering Review, 5,* 7.

Mardon, C. (1982). High-rate thermophilic digestion of cellulosic wastes. Paper presented at the 5th Australian Biotechnology Conference, Sydney, August 1982.

Margen, P. (1980). *Sunworld, 4,* 128–134.

Marks, S. (1983). *Solar Energy, 30,* 45–49.

Maron, S., & Prutton, C. (1959). *Principles of physical chemistry.* New York: Macmillan.

Marsh, G. (2002). RE storage—the missing link. Elsevier Advanced Technology. http://www.re-focus.net/mar2002_4.html

Martens, W., Rotmans, J., & Niessen, L. (1994). Climate change and malaria risk. GLOBO Report 3–461502003. Global Dynamics & Sustainable Development Programme, RIVM, Bilthoven.

Marx, K. (1859). *A critique of political economy*. London.

Marzano, F., Roberti, L., & Mugnai, A. (2000). Impact of rainfall incoherent backscattering upon radar echoes above 10 GHz. *Physics and Chemistry of the Earth (B)*, *25*, 943–948.

Maslow, A. (1943). A theory of human motivation. *Psychological Review*, *50*, 370–396.

Masterson, K., & Seraphin, B. (1975). Inter-laboratory comparison of the optical characteristics of selective surfaces for photo-thermal conversion of solar energy, National Science Foundation (USA) Report NSF/RANN–GI–36731X.

Masuda, Y. (1971). Paper presented at International Colloquium on Exploitation of the Oceans, Bordeaux, France.

Mathew, X., *et al.* (2001). In *Proceedings of the 4th International Symposium on New Materials, Montréal* (pp. 420–421).

Matthews, B. (2008). The Java Climate Model, v5. Software. Retrieved 2008, from http://www.climate.be

Maugh, T. II. (1979). *Science*, *206*, 436.

McAdams, W. (1954). *Heat transmission*. New York: McGraw-Hill.

McBride, J. (1981). Chemical heat pump cycles for energy storage and conversion. In *Proceedings of the International Conference on Energy Storage* (Vol. 2, pp. 29–46). Cranfield, UK: BHRA Fluid Engineering.

McCormick, M. (1973). *Ocean engineering wave mechanics*. New York: Wiley.

McCormick, M. (1976). *Ocean Engineering*, *3*, 133–144.

McCoy, E. (1967). *Transactions of the American Society of Agricultural Engineers* (6), 784.

McGowan, J. (1976). *Solar Energy*, *18*, 81–92.

McGuire, A., Melillo, J., Kicklighter, D., Pan, Y., Xiao, X., Helfrich, J., *et al.* (1997). Equilibrium responses of global net primary production and carbon storage to doubled atmospheric carbon dioxide: Sensitivity to changes in vegetation nitrogen concentrations. *Global Biogeochemical Cycles*, *11*, 173–189.

McKay, R., & Sprankle, R. (1974). In *Proceedings of the Conference on Research for the Development of Geothermal Energy Resources* (pp. 301–307). Pasadena, CA: Jet Propulsion Laboratory, California Institute of Technology.

Meibom, P., Svendsen, T., & Sørensen, B. (1997). Import/eksport-politik som redskab til optimeret udnyttelse af el produceret på VE–anlæg. IMFUFA Texts No. 343, p. 84.

Meibom, P., Svendsen, T., & Sørensen, B. (1999). Trading wind in a hydro-dominated power pool system. *International Journal of Sustainable Development*, *2*, 458–483.

Meinel, A., & Meinel, M. (1972). *Physics Today*, *25*, No. 2, 44–50.

Meinel, A., & Meinel, M. (1976). *Applied solar energy*. Reading, MA: Addison-Wesley.

Mekel, O., & Huppes, G. (1990). Environmental effects of different packaging systems for fresh milk. Centrum voor Milieukunde, Rijksuniversitet Leiden, CML–meeddelingen # 70.

Melillo, J., & Helfrich, J. (1998). NPP database created under NASA and EPRI grants, kindly put at the author's disposal.

Melillo, J., McGuire, A., Kicklighter, D., Moore, B. III, Vorosmarty, C., & Schloss, A. (1993). Global climate change and terrestrial net primary production. *Nature*, *363*, 234–240.

Merzbacker, E. (1970). *Quantum mechanics* (2nd ed.). New York: Wiley International.

Mesarovic, M., & Pestel, E. (1974). *Mankind at the turning point*. New York: Dutton and Co.

Meyer, H., Morthorst, P., Schleisner, L., Meyer, N., Nielsen, P., & Nielsen, V. (1994). Costs of environmental externalities of energy production (in Danish), Report R–770. Risø National Laboratory.

Mie, G. (1908). *Ann. der Phys., 25*, 377.

Milankovich, M. (1941). *K. Serb. Akad. Beogr.*, Spec. Publ. No. 132 (translation: Israel Program for Scientific Translations, Jerusalem, 1969).

Miles, J. (1957). *Journal of Fluid Mechanics, 3*, 185–204.

Miller, L., & Douglas, B. (2004). Mass and volume contributions to twentieth-century global sea level rise. *Nature, 428*, 406–409.

Milliff, R., Large, W., Morzel, J., Danabasoglu, G., & Chin, T. (1999). Ocean general circulation model sensitivity to forcing from scatterometer winds. *Journal of Geophysical Research, 104*, 11337–11358.

Milliff, R., Morzel, J., Chelton, D., & Freilich, M. (2004). Wind stress curl and wind stress divergence biases from rain effects on QSCAT surface wind retrievals. *Journal of Atmospheric and Oceanic, 21*, 1216–1231.

Millner, A. (1979, November). *Technology Review*, 32–40.

Mills, D. (1995). A climate water budget approach to blackfly population dynamics. *Publications in Climatology, 48*, 1–84.

Minami, E., Kawamoto, H., & Saka, S. (2002). Reactivity of lignin in supercritical methanol studied with some lignin model compounds. In *Proceedings of the 12th European Biomass Conference* (pp. 785–788). ETA Firenze & WIP, Munich.

Mishan, E. (1969). *Growth: The Price We Pay*. London: Staples Press.

Mitchell, J., John, T., Gregory, J., & Trett, S. (1995). Climate response to increasing levels of greenhouse gases and sulphate aerosols. *Nature, 376*, 501–504.

Mitchell, J., & Johns, T. (1997). On modification of global warming by sulphate aerosols. *Journal of Climate, 10*, 245–267. Model output (12 months by 240 years) has been since 1998 available at IPCC Data Distribution. http://www.dkrz.de/ipcc/ddc/html/gfdlrun2.html

Mitsui, A., & Kumazawa, S. (1977). In A. Mitsui, S. Miyachi, A. San Pietro & S. Tamura (Eds.), *Biological solar energy conversion* (pp. 23–51). New York and London: Academic Press.

Mitsuyasu, H. (1968). *Rep. Res. Inst. Appl. Mech.*, Kyushu University, *16*, 459–482.

Mock, J., Tester, J., & Wright, P. (1997). Geothermal energy from the Earth: Its potential impact as an environmentally sustainable resource. *Annual Review of Energy and the Environment 22*, 305–356.

Möller, F. (1957). In S. Flügge (Ed.), *Handbuch der Physik* (Vol. 48, pp. 155–253) (Geophysik III). Berlin: Springer-Verlag.

Molina-Heredia, F., Wastl, J., Navarro, J., Bendall, D., Hervás, M., Howe, C., et al. (2003). A new function for an old cytochrome? *Nature, 424*, 33–34.

Mollison, D., Buneman, O., & Salter, S. (1976). *Nature, 263*, 223–226.

Molly, J. (1977). In *Proceedings of the International Symposium on Wind Energy Systems, Cambridge 1976*. Cranfield, UK: British Hydromechanical Research Association, and *Wind Engineering 1*, 57–66 (1977).

Monin, A., & Obukhov, A. (1954). *Trudy Geofizicheskova Instituta, Akademiya Nauk, SSSR*, No. 24 (151).

Monin, A., & Yaglom, A. (1965). *Statistical Fluid Mechanics*, part I. Moscow: Izdatel'skvo Nauka.

Moore, W. (1972). *Physical chemistry* (5th ed.). London: Longman Group.

Morales, R., Chron, M., Hudry-Clegeon, G., Pélillot, Y., Nørager, S., Medina, M., et al. (1999). Refined X-ray structures of the oxidized, at 1.3 Å, and reduced, at 1.17 Å, [2Fe-2S] ferredoxin from the cyanobacterium *Anabaene* PCC7119 show redox-linked conformation changes. *Biochemistry*, *38*, 16764–15773.

Morishima, M. (1976). *The economic theory of modern society*. Cambridge, UK: Cambridge University Press.

Morse, R. (1977). *Ambio 6*, 209–215.

Moser, J. (1887). Notitz über Verstärkung photoelektrischer Ströme durch optische Sensibilisierung. *Monatshefte für Chemie und verwandte Teile anderer Wissenschaften*, *8*, 373.

Moskowitz, L. (1964). *Journal of Geophysical Research*, *69*, 5161–5179.

Moskowitz, P., Buchanan, W., & Shafarman, W. (1994). Lessons learned from a hydrogen explosion at a photovoltaic research facility. Brookhaven National Laboratory (unpublished report).

Moskowitz, P., Steinberger, H., & Thumm, W. (1995). Health and environmental hazards of CdTe photovoltaic module production, use and decommissioning. In *1994 IEEE First World Conference on Photovoltaic Energy Conversion, Kona* (pp. 115–118). Washington, DC: IEEE.

Müller, C., Falcou, A., Reckefuss, N., Rojahn, M., Wiederhirn, V., Rudati, P., *et al.* (2003). Multicolour organic light-emitting displays by solution processing. *Nature*, *421*, 829–833.

Müller, S., Kells, K., Litsios, J., Kumbein, U., Schenk, A., & Fichtner, W. (1992). *SIMUL v.1 Manual*. Zürich: Integrated Systems Lab., Technical Highschool.

Munk, W., & MacDonald, G. (1960). As quoted by King Hubbert, M. (1969). In *Resources and Man* (United States National Academy of Sciences and National Science Foundation), (Chap. 8). San Francisco: W. Freeman.

Munk, W. (1980). Affairs at the sea. *Annual Review of Earth and Planetary Sciences 8*, 1–16.

Murphy, D., & Christian, P. (1979). *Science, 205*, 651.

Murphy, L., Borch, K., McFarland, K., Bohlin, C., & Westh, P. (2010). A calorimetric assay for enzymatic saccharification of biomass. *Enzyme and Microbial Technology, 46*, 141–146.

Musgrove, P. (1976). *Nature, 262*, 206–207.

Nagaura, T. (1990). Paper for 3rd *International* Battery Seminar, Dearfield Beach, FL.

Naik, S., Goud, V., Rout, P., & Dalai, A. (2010). Production of first and second generation biofuels: A comprehensive review. *Renewable and Sustainable Energy Reviews, 14*, 578–579.

Nakicenovic, N., Grübler, A., Ishitani, H., Johansson, T., Marland, G., Moreira, J., *et al.* (1996). Energy primer. pp. 75–92 in IPCC (1996b).

NASA. (1971). Report No. R–351 and SP–8005, May.

NASA. (1997). Surface solar energy data set v1.0, NASA Langley Research Center EOSDIS Distributed Active Archive Center. http://eosweb.larc.nasa.gov/

NASA. (1998). Ozone data from ADEOS satellite. http://jwocky.gsfc.nasa.gov

NASA. (2004). Sea surface height anomalies from TOPEX/POSEIDON spacecraft measurements, date series. http://sealevel.jpl.nasa.gov/science

Nathan, G., Rajasooria, G., Tan, K., & Tan, T. (1976). In *Proceedings of the International Symposium on Wind Energy Systems, Cambridge 1976*, Paper A5, Cranfield, UK: British Hydromechanical Research Association.

Naturlig Energi. (1998). *Wind power production statistics* (pp. 14–17), June issue (in Danish).

Nazeeruddin, M., *et al.* (1993). Conversion of light to electricity by cis-X2Bis(2,2'-bipyridyl-4,4'-dicarboxylate) ruthenium-(II) charge transfer sensitizers (X = Cl-, Br-, I-, CN- and SCN-) on nanocrystalline TiO_2 electrodes. *Journal of the American Chemical Society, 115*, 6382–6390.

Nazeeruddin, M., *et al.* (2001). Engineering of efficient panchromatic sensitizers for nanocrystalline TiO_2-based solar cells. *Journal of the American Chemical Society, 123*, 1613–1620.

NCAR. (1997). The NCAR Community Climate Model CCM3 with NCAR/CSM Sea Ice Model. University Corporation for Atmospheric Research, National Center for Atmospheric Research, and Climate and Global Dynamics Division. http://www.cgd.ucar.edu:80/ccr/bettge/ice

NCAR. (2006). QSCAT/NCEP Blended Ocean Winds ds744.4. National Center for Atmospheric Research, Data Support Section. The not blended QSCAT dataset is ds744.2. Data download site. Retrieved 2006, from http://dss.ucar.edu/datasets/ds744.4/.

NCEP–NCAR. (1998). The NOAA NCEP-NCAR Climate Data Assimilation System I, described in Kalnay *et al.* (1996), data available from University of Columbia. http://ingrid.ldgo.columbia.edu

Neftel, A., Moor, E., Oeschger, H., & Stauffer, B. (1985). Evidence from polar ice cores for the increase in atmospheric CO_2 in the past two centuries. *Nature, 315*, 45–47.

Neiburger, M., Edinger, J., & Bonner, W. (1973). *Understanding our atmospheric environment.* San Francisco: W. Freeman.

Nelson, J., Haque, A., Klug, D., & Durran, J. (2001). Trap-limited recombination in dye-sensitised nanocrystalline metal oxide electrodes. *Physical Review B, 63*, 205321-9.

Neumann, G. (1949). *Deutscher Hydrogr. Zeitschrift, 2*, 187–199.

Neumann, G. (1956). *Deutscher Hydrogr. Zeitschrift, 9*, 66–78.

Neumann, G., & Pierson, W. (1966). *Principles of physical oceanography.* Englewood Cliffs, NJ: Prentice-Hall.

Newell, R., Vincent, D., Dopplick, T., Ferruzza, D., & Kidson, J. (1969). In G. Corby (Ed.), *The global circulation of the atmosphere* (pp. 42–90). London: Royal Meteorological Society.

Newman, M., & Rood, R. (1977). Internal report.

Niel, C. van. (1941). *Advances in Enzymology, 1*, 263.

Nielsen, P. S., & Holm-Nielsen, J. B. (1996). CO_2 balance in production of energy based on biogas. Poster presented at the European conference: Environmental Impact of Biomass for Energy, November 1996, Nordwijkerhout.

Nielsen, S., & Sørensen, B. (1996). Long-term planning for energy efficiency and renewable energy. Paper presented at Renewable Energy Conference, Cairo April 1996; revised as: Interregional power transmission: A component in planning for renewable energy technologies. *International Journal of Global Energy Issues 13* (1–3) (2000), 170–180.

Nielsen, S., & Sørensen, B. (1998). A fair market scenario for the European energy system. In LTI–research group (Ed.), *Long-term integration of renewable energy sources into the European energy system* (pp. 127–186). Heidelberg: Physica–Verlag.

Nielsen, S. (1977). Records from Nautical Dept. of the Danish Met. Inst., personal communication.

Nieuwlaar, E., & Alsema, E. (1997). *Environmental aspects of PV power systems.* Utrecht Universiteit VNS Report no. 97072. (irreg. pag.)

Nihoul, P. (1977). *Applied Mathematical Modelling, 1*, 3–8.

Niiler, P. (1977). *La Recherche, 79*, 517–526.

Njakou, S., Blumberga, D., & Sørensen, B. (2006). Gaseous fuels from biomass. Paper presented at World Renewable Energy Congress IX, Firenze, WREC, Brighton UK.

NOAA. (1998). Global Vegetation Index, weekly, monthly, seasonal and annual data from NOAA-NESDIS-NCDC-SDSD collaboration, available from UNEP (1998).

NOAA. (2010). World data center for paleoclimatology. http://www.ncdc.noaa.gov /paleo

Nordel (Nordic Electricity Supply Undertakings). (1975). Drifttekniska specifikationer for varmekraft, Oslo.

Nordhaus, W. (1990). *To slow or not to slow: The economics of the Greenhouse Effect.* New Haven CT: Yale University Press.

Nordhaus, W. (1994). *Managing the global commons: The economics of climate change* (p. 213). Cambridge, MA: MIT Press.

Norman, R. (1974). *Science, 186*, 350–352.

Norris, R., & Röhl, U. (1999). Carbon cycling and chronology of climate warming during the Palaeocene/Eocene transition. *Nature, 401*, 775–778.

NSES. (2001). Novator Solar Energy Simulator v1.2, software from Novator Advanced Technology Consulting, Gilleleje.

Obasi, G. (1963). *Journal of Atmospheric Science, 20*, 516–528; Atmospheric momentum and energy calculations for the Southern hemisphere during the IGY, Report No. 6, contract AF19(604)–6108, Cambridge, MA: Department of Meteorol., MIT.

Odum, E. (1972). *Ecology*. New York: Holt, Rinehart and Winston.

OECD. (2002). OECD Agricultural Outlook 2002–2007. Paris: OECD.

OECD/IEA. (2002a). Energy balances of OECD countries, 1999–2000. Paris: International Energy Agency.

OECD/IEA. (2002b). Energy balances of Non-OECD countries, 1999–2000. Paris: International Energy Agency.

Oetjen, R., Bell, E., Young, J., & Eisner, L. (1960). *Journal of the Optical Society of America, 50* (12), 1308.

Offshore Windenergy Europe. (2003). News item dated December 2002, Technical University Delft, posted on the website. http://www.offshorewindenergy.org

Ogilvie, T. (1963). *Journal of Fluid Mechanics, 16*, 451–472.

Öhgren, K., Vehmaanperä, J., Siika-Aro, M., Galbe, M., Viikari, L., & Zacchi, G. (2007). High temperature enzymatic prehydrolysis prior to simultaneous sacchrification and fermentation of steam pretreated corn stover for ethanol production. *Enzyme and Microbial Technology, 40*, 834–839.

Olsen, H. (1975). *Jordbrugsteknisk Institut Medd. No. 24*. Copenhagen: Royal Veterinary and Agricultural University.

O'Neill, G. (1975). *Science, 190*, 943–947.

Onsager, L. (1931). *Physical Review, 37*, 405–426.

Oort, A. (1964). *Monthly Weather Review, 92*, 483–493.

O'Regan, B., & Grätzel, M. (1991). *Nature, 353*, 737.

Osterle, J. (1964). *Applied Scientific Research Section A, 12*, 425–434.

Oswald, W. (1973). Progress in water technology. *Water Quality Management, and. Pollution Control, 3*, 153.

Ottinger, R. (Ed.). (1991). *Environmental costs of electricity*. New York: Oceana Publications.

Ovshinsky, S. (1978, 30 November). *New Scientist*, 674–677.

Pacific Solar Inc. (1997). Annual company review, Sydney, p. 20.

Parker, E. (1964, April). *Scientific American*, 66–76.

Parry, M., & Rosenzweig, C. (1993). Food supply and the risk of hunger. *The Lancet, 342*, 1345–1347.

Pasachoff, J. (1973, October). *Scientific American*, 68–79.

Patil, P. (1998). The US DoE fuel cell program. Investing in clean transportation. Paper presented at Fuel Cell Technology Conference, London, September, IQPC Ltd, London.

Patten, B. (Ed.). (1971). *Systems analysis and simulation in ecology* (Vol. I). New York and London: Academic Press.

Patten, B. (Ed.). (1972). *Systems analysis and simulation in ecology* (Vol. II). New York and London: Academic Press.

Patterson, M. (1997). The management of wastes associated with thin film PV manufacturing. Paper presented at IEA Workshop on Environmental Aspects of PV Power Systems, Utrecht, 25–27 June.

Pattie, R. (1954). *Nature, 174*, 660.

Peebles, P. (1971). *Physical cosmology.* Princeton, NJ: Princeton University Press.

Peixoto, J., & Crisi, A. (1965). Deptartment of Meteorology, Massachusetts Institute of Technology, Sci. Report No. 6, contract AF19(628)–2408, As quoted by Lorenz (1967).

Pelser, J. (1975). In F. Eldridge (Ed.), *Proceedings of the 2nd Workshop on Wind Energy Conversion Systems* (pp. 188–195). Report NSF–RA–N–75–050. McLean VA: The Mitre Corp.

Peltier, W. (1994). Ice age paleotopography. *Science, 265,* 195–201.

Peltier, W. (2004). Global glacial isostasy and the surface of the ice-age Earth: The ICE-5G (VM2) model and GRACE. *Annual Review of Earth Planetary Science, 32,* 111–149.

Penman, H. (1970, September). *Scientific American,* 98–109.

Penndorf, R. (1959). Air Force Cambridge Res. Center, AFCRC–TN–59–608. Report No. 1.

Perego, C., & Bianchi, D. (2010). Biomass upgrading through acid-base catalysis. *Chemical Engineering Journal,* doi:10.1016/j.cej.2010.01.036.

Perrin, G. (1981). *Verkehr und Technik,* (9).

Petersen, E. (1974). Risø Report No. 285. Copenhagen: Danish Atomic Energy Commission; and personal communication.

Petersen, H. (2001). Evaluation of wind turbine performance 2000, HP Consult Report from project ENS-51171/00-0016 of the Danish Energy Agency.

Petersen, H., & Lundsager, P. (1998). Comparison of wind turbines based on power curve analysis. Report for the Danish Energy Agency, HP Consult and Darup Associates Ltd.

Petersen, J. (1972). Statens Byggeforskningsinstitut (Danish Building Res. Inst.), Notat No. 20.

Petersen, P. (1991). *Life cycle analysis of local CHP-plants—energy and environmental assessment from cradle to grave.* Danish Ministry of Environment.

Pfister, G., & Scher, H. (1977). In W. Spear (Ed.), *Proceedings of the 7th International Conference on Amorphous and Liquid Semiconductors* (pp. 197–208). Edinburgh: University of Edinburgh Press.

Phillips, O. (1957). *Journal of Fluid Mechanics, 2,* 417–445.

Phillips, O. (1966). *The dynamics of the upper ocean.* Cambridge, UK: Cambridge University Press.

Pierson, W. (1955). In H. Landsberg (Ed.), *Advances in geophysics* (Vol. 2, pp. 93–178). New York and London: Academic Press.

Pigford, T. (1974). In R. Wilson & W. Jones (Eds.), *Energy, ecology and the environment* (pp. 343–349). New York: Academic Press.

Pines, D. (1994). Understanding high-temperature superconductivity, a progress report. *Physica B, 199–200,* 300–309.

Pinker, R., & Laszlo, I. (1992). Modelling surface solar irradiance for satellite applications on a global scale. *Journal of Applied Meteorology, 31,* 194–211.

Piper, A. (1975). *Nature 258,* December 4, news column.

Pippard, J. (1966). *The elements of classical thermodynamics.* Cambridge, UK: Cambridge University Press.

Pollack, H., & Chapman, D. (1977). *Earth and Planetary Science Letters, 34,* 174–184.

Pommer, K. *et al.* (1991). Miljømæssig vurdering af mælkeemballage. Miljøprojekt 168. Danish Environmental Agency.

Pond, S. (1971). *Eos (Transactions of the American Geophysics Society) 52,* 389–394.

Popel, F. (1970). *Landtechnische Forschung*, Heft 5, BRD.

Post, R., & Post, S. (1973, December). *Scientific American*, 17–23.

Post, R., & Ribe, F. (1974). *Science, 186*, 397–407.

Potter, G., Ellsaesser, H., MacCracken, M., & Luther, M. (1975). *Nature, 258*, 697–698.

Pöpperling, R., Schwenk, W., & Venkateswarlu, J. (1982). Abschätzung der Korrosionsgefärdung von Behältern und Rohrleitungen aus Stahl für Speicherung von Wasserstoff und wasserstofhältigen Gasen unter hohen Drücken. VDI Zeitschriften Reihe 5, No. 62.

Prandtl, L. (1932). *Beitrage Physik Freien Atmos. 19*(3) (Bjerknes Festschrift).

Prengle, H., & Sun, C.-H. (1976). *Solar Energy, 18*, 561–567.

Priestley, C. (1951). *Australian Journal of Science Research, A4*, 315–328.

Prigogine, I. (1968). *Introduction to thermodynamics of irreversible processes.* New York: Wiley-Interscience.

Pringle, H. (1998). The slow birth of agriculture. *Science, 282*, 1446–1450.

Proton Energy Systems. (2003). Unigen. http://www.protonenergy.com

Putnam, P. (1953). *Energy in the future.* New York: Van Nostrand.

Rabenhorst, D. (1976). In J. Savino (Ed.), *Wind energy conversion systems.* United States National Science Foundation, Report NSF/RA/W–73–006.

Rabl, A., & Nielsen, C. (1975). *Solar Energy, 17*, 1–12.

Ragauskas, A., *et al.* (2006). The path forward for biofuels and biomaterials. *Science, 311*, 484–489.

Raich, J., Rastetter, E., Melillo, J., Kicklighter, D., Steudler, P., Peterson, B., *et al.* (1991). Potential net primary productivity in South America: Application of a global model. *Ecological Applications, 1*, 399–429.

Rajeswaran, G., *et al.* (2000). In *SID 00 Digest*, No. 40 (4 pp)

Ralph, E. (1966). *Solar Energy, 10*, 67–71.

Rambøll Inc. (1999). Wave energy conditions in Danish North Sea. Report (in Danish) from Danish Energy Agency Project 51191/97-0014. Virum: Rambøll Inc.

Raschke, E., Haar, T., Bandeen, W., & Pasternak, M. (1973). *Journal of Atmospheric Science, 30*, 341–364.

Rasool, S., & Schneider, S. (1971). *Science, 173*, 138.

Ravn, H. (2001). Communication of preliminary time series from Danish Utility Public Service Obligation project.

Ratcliffe, J. (Ed.). (1960). *Physics of the upper atmosphere.* New York and London: Academic Press.

Reck *et al.* (1974). As quoted by Manabe and Holloway (1975).

Reed, T. (Ed.). (1981). *Biomass gasification.* Park Ridge, NJ: Noyes Data Corp.

Reeves, H. (1965). In L. Aller & D. McLaughlin (Eds.), *Stellar Structure* (Vol. 8, pp. 113–193). Chicago: University of Chicago Press.

Reeves, H. (1975). *La Recherche, 60*, 808–817.

Reeves R., Lom, E., & Meredith, R. (1982). Stable hydrated ethanol distillate blends in diesels. Hawkesbury, NSW, Australia: Apace Res. Ltd.

Reid, J. (1971). In W. Matthews, F. Smith, & E. Goldberg (Eds.), *Man's impact on terrestrial and oceanic ecosystems.* Cambridge, MA: MIT Press.

Reilly, J., & Wiswall, R. (1974). *Inorganic chemistry, 13*, 218.

Ren, Y., Zhang, Z., Gao, E., Fang, S., & Cai, S. (2001). A dye-sensitized nanoporous TiO_2 photochemical cell with novel gel network polymer electrolyte. *Journal of Applied Electrochemistry 31*, 445–447.

Ricaud, A. (1999). Economic evaluation of hybrid solar systems in private houses and commercial buildings. In *Photovoltaic/Thermal Solar Systems, IEA Solar Heating & Cooling/Photovoltaic Power Systems Programmes, Amersfoort* (p. 13). Utrecht: Ecofys.

Rijnberg, E., Kroon, J., Wienke, J., Hinsch, A., Roosmalen, J., Sinke, W., *et al.* (1998). Long-term stability of nanocrystalline dye-sensitized solar cells In J. Schmid, H. Ossenbrink, P. Helm, H. Ehman, & E. Dunlop (Eds.), *2nd World Conference on PV Solar Energy Conversion, Vienna* (pp. 47–52). Luxembourg: European Commission.

Risien, C., & Chelton, D. (2006). A satellite-derived climatology of global ocean winds. *Remote Sensing of Environment, 105*, 221–236.

Robinson, J. (Ed.). (1980). *Fuels from biomass.* Park Ridge, NJ: Noyes Data Corp.

Robinson, J. (1982). Apples and horned toads: On the framework-determined nature of the energy debate. *Policy Sciences, 15*, 23–45.

Robinson, J. (1988). Unlearning and backcasting: Rethinking some of the questions we ask about the future. *Technologial Forecasting Social Change, 33*, 325–338.

Robinson, N. (Ed.). (1966). *Solar Radiation.* Amsterdam: Elsevier.

Roll, H. (1957). In S. Flügge (Ed.), *Handbuch der Physik* (Vol. 48, pp. 671–733). Berlin: Springer-Verlag.

Rosa, M. de la, Navarro, J., Díaz-Quintana, A., Cerda, B. de la, Molina-Heredia, F., Balme, A., *et al.* (2002). An evolutionary analysis of the reaction mechanisms of photosystem I reduction by cytochrome c_6 and plastocyanin. *Bioelectrochemistry, 55*, 41–45.

Roseen, R. (1978). Central solar heat station in Studsvik, AB Atomenergi Report ET–78/77, Studsvik.

Rosenzweig, C., Parry, M., Fischer, G., & Frohberg, K. (1993). Climate change and world food supply. Report 3, Environmental Change Unit. Oxford: Oxford University.

Ross, M., & Williams, R. (1975). Assessing the potential for fuel conservation. Inst. Public Policy Alternatives, State University of New York.

Ross, M., & Williams, R. (1976, November). *The Bulletin of the Atomic Scientists,* 30–38.

Rossby, C.-G. (1932). *MIT Meteorological Papers* (Vol. 1, (4)).

Rowley, J. (1977, January). *Physics Today,* 36–45.

Royal Dutch Meteorological Institute (2006). HYDRA Project. Retrieved 2006, from http://www.knmi.nl/samenw/hydra

Rubey, W. (1951). *Geological Society of America Bulletin, 62*, 111.

Rubins, E., & Bear, F. (1942). *Soil Science, 54*, 411.

Russell, F., & Chew, S. (1981). In *Proceedings of the International Conference on Energy Storage* (pp. 373–384). Cranfield, UK: BHRA Fluid Engineering.

Saeki, T. (1975). In J. Cooper (Ed.), *Photosynthesis and productivity in different environments* (pp. 297–322). Cambridge, UK: Cambridge University Press.

Sagan, C., & Mullen, G. (1972). *Science, 177*, 52–56.

Sah, C., Noyce, R., & Shockley, W. (1957, September). *Proceedings of the IRE* (Vol. 45, pp. 1228–1243).

Sakai, H. (1993). Status of amorphous silicon solar cell technology in Japan. In *Technical Digest of 7th International PV Science and Engineering Conference Nagoya* (pp. 169–172). Nagoya Institute of Technology.

Salter, S. (1974). *Nature, 249*, 720–724.

Sandwell, D., Smith, W., Smith, S., & Small, C. (1998). Measured and estimated seafloor topography. Available together with land topography. http://ingrid.ldgo.columbia.edu/sources/

Sariciftci, N., Smilowitz, L., Heeger, A., & Wudl, F. (1992). *Science, 258,* 1474.

Savonius, S. (1931). *Mechanics and Engineering, 53,* 333–338.

Schlieben, E. (1975). In *Proceedings of the 1975 Flywheel Technology Symposium* (pp. 40–52). *Berkeley, CA.* Report ERDA 76.

Schmidt, A., Christiansen, K., & Pommer, K. (1994). Livscyklusmodel til vurdering af nye materialer—Metoder, vurdering og fremgangsmåde, dk–TEKNIK Report, Søborg, Denmark, p. 202.

Schneider, S., & Dennett, R. (1975). *Ambio, 4,* 65–74.

Schramm, D. (1974, January). *Scientific American,* 69–77.

Schulten, R., Decken, C. van der, Kugeler, K., & Barnert, H. (1974). Chemical latent heat for transport of nuclear energy over long distances. In *Proceedings of the British Nuclear Energy Society International Conference, The High Temperature Reactor and Process Applications.* London: BNES.

Schumpeter, J. (1961). *Capitalism, socialism and democracy.* London: Allen and Unwin.

Schwarzschild, M. (1958). *Structure and evolution of stars.* Princeton, NJ, Princeton University Press.

Scrosati, B. (1995). Challenge of portable power. *Nature, 373,* 557–558.

Seeger, P., Fowler, W., & Clayton, D. (1965). *The Astrophysical Journal Supplement Series, 11*(97).

Seitel, S. (1975). *Solar Energy, 17,* 291–295.

Sekera, Z. (1957). In S. Flügge (Ed.), *Handbuch der Physik* (Vol. 48 (Geophysik II), pp. 288–328). Berlin: Springer-Verlag.

Seliger, H., & McElroy, W. (1965). *Light: Physical and biological action.* New York and London: Academic Press.

Sellers, P., Dickinson, R., Randall, D., Betts, A., Hall, F., Berry, J., *et al.* (1997). Modelling the exchanges of energy, water, and carbon between continents and the atmosphere. *Science, 275,* 502–509.

Sellers, W. (1965). *Physical climatology.* Chicago: University of Chicago Press.

Service, R. (1998). Superstrong nanotubes show they are smart, too. *Science, 281,* 940–942.

Sforza, P. (1976). In *Proceedings of the International Symposium on Wind Energy Systems, Cambridge, 1976,* paper E1. Cranfield: British Hydromech. Res. Assoc.

Shaheen, S., *et al.* (2001). 2.5% Efficient organic plastic solar cells. *Applied Physics Letters, 78,* 841–843.

Shakleton, N., & Opdyke, N. (1973). *Quaternary Research* (New York) *3,* 39.

Shelton, J. (1975). *Solar Energy, 17,* 137–143.

Shigley, J. (1972). *Mechanical engineering design* (2nd ed.). New York: McGraw-Hill.

Shirland, F. (1966). *Advanced Energy Conversion, 6,* 201–221.

Shklover, V., Ovchinnikov, Y., Braginsky, L., Zakeeruddin, S., & Grätzel, M. (1998). Structure of organic/inorganic interface in assembled materials comprising molecular components. Crystal structure of the sensitizer bis[4,4'-carboxy-2,2'-bipyridine)(thiocyana-to)]ruthenium(II). *Chemistry of Materials, 10,* 2533–2541.

Shockley, W. (1950). *Electrons and holes in semiconductors.* New York: Van Nostrand.

Shubik, M. (Ed.). (1991). *Risk, organizations and society.* Boston, MA: Kluwer Academic Publishers.

Siegel, R., & Howell, J. (1972). *Thermal radiation heat transfer.* New York: McGraw-Hill.

Siegenthaler, U., & Oeschger, H. (1987). Biospheric CO_2 emissions during the past 200 years reconstructed by deconvolution of ice core data. *Tellus, 39B,* 140–154.

Silverstein, S. (1976). *Science, 193,* 229–231.

Sjoblom, C.-A. (1981). Heat storage in phase transitions of solid electrolytes. In *Proceedings of the 16th Intersociety Energy Conversion Engineering Conference* (Vol. I, paper 8 19441). New York: The American Society of Mechanical Engineers.

Skyllas-Kazacos, M. (2003). Novel vanadium chloride/polyhalide redox flow battery. *Journal of Power Sources, 124*, 299–302.

Slama-Schwok, A., Ottolenghi, M., & Avnir, D. (1992). Long-lived photoinduced charge separation in a redox system trapped in a sol–gel glass. *Nature, 355*, 240–242.

Slavîk, B. (1975). In J. Cooper (Ed.), *Photosynthesis and productivity in different environments* (pp. 511–536). Cambridge, UK: Cambridge University Press.

Slesser, M. (1975). *Nature, 254*, 170–172.

Slotta, L. (1976). In R. Cohen & M. McCormick (Eds.), *Workshop on wave and salinity gradient energy conversion*, Paper H, US ERDA Report No. C00–2946–1.

SMAB. (1978). Metanol sam drivmedel, Annual Report, Svensk Metanol-utveckling AB, Stockholm.

Smagorinsky, J., Manabe, S., & Holloway, J. (1965). *Monthly Weather Review, 93*, 727–768.

Smil, V. (1977, February). *The Bulletin of the Atomic Scientists*, 25–31.

Smith, D. (1974). *Earth and Planetary Science Letters, 23*, 43–52.

Snyder, R., & Cox, C. (1966). *Journal of Marine Research, 24*, 141–178.

Sørensen, B. (1975a). Computer simulation of ^{131}I transfer from fallout to man. *Water, Air, and Soil Pollution, 4*, 65–87.

Sørensen, B. (1975b). Energy and resources. *Science, 189*, 255–260; and in P. Abelson & A. Hammond (Eds.), *Energy: Use, conservation and supply* (Vol. II, pp. 23–28). Washington, DC: American Association for the Advancement of Science (1978).

Sørensen, B. (1975c). Vindkraft (Appendix, pp. 41–67). Lyngby, Denmark: Danish Academy of Technical Sciences. (Summary report *Wind power* available in English.)

Sørensen, B. (1976a). A simple model of economic growth or decline under the influence of resource depletion. *Applied Mathematical Modelling, 1*, 24–28.

Sørensen, B. (1976b). Solar heat systems for use at high latitudes. In *Proceedings of the UNESCO-WMO Solar Energy Symposium, Geneve, 1976*, WMO Paper No. 477, 1977.

Sørensen, B. (1976c). Dependability of wind energy generators with short-term energy storage. *Science, 194*, 935–937.

Sørensen, B. (1976d). Wind energy. *The Bulletin of the Atomic Scientists, 32*, September, 38–45; and In R. Williams (Ed.). *Toward a solar civilization*. Cambridge, MA: MIT Press (1978).

Sørensen, B. (1977). Direct and indirect economics of wind energy systems relative to fuel based systems. In *Proceedings of the International Symposium on Wind Energy Systems, Cambridge, 1976*, Paper D2. Cranfield: British Hydromechanical Research Association; and *Wind Engineering 1*, 15–22.

Sørensen, B. (1978a). On the fluctuating power generation of large wind energy converters with and without storage facilities. *Solar Energy, 20*, 321–331.

Sørensen, B. (1978b). The regulation of an electricity supply system including wind energy generators. In *Proceedings of the 2nd International Symposium on Wind Energy Systems, Amsterdam 1978*, Paper G1. Cranfield: British Hydromechanical Research Association.

Sørensen, B. (1978c). Wind—large scale utilisation. In *Proceedings of the Conference on Wind, Wave and Water, Dublin 1978*, . Dublin: Industrial Development Authority of Ireland, IIRS and Solar Energy Society.

Sørensen, B. (1979). *Renewable energy* (p. 683). London: Academic Press.

Sørensen, B. (1981a, March). A combined wind and hydro power system. *Energy Policy*, 51–55.

Sørensen, B. (1981b). Wind energy. In E. El-Hinnawi & A. Biswas (Eds.), *Renewable sources of energy and the environment* (pp. 97–116). Dublin: Tycooli International Press.

Sørensen, B. (1982). Comparative risk assessment of total energy systems. In *Health impacts of different sources of energy* (pp. 455–471). IAEA Publ. SM–254/105, Vienna.

Sørensen, B. (1982b). Energy choices: Optimal path between efficiency and cost. In S. Yuan (Ed.), *Energy, resources and environment* (pp. 279–286). New York: Pergamon Press.

Sørensen, B. (1983). Physics in Society, Lecture notes. IMFUFA Texts 129, Roskilde University, p. 56. Reprinted as Part I of *Physics revealed*, cf. Sørensen (2001a).

Sørensen, B. (1986). A study of wind-diesel/gas combination systems. Energy Authority of New South Wales, EA86/17. Sydney (p. 82).

Sørensen, B. (1987a). Chernobyl accident: Assessing the data. *Nuclear Safety*, 28, 443–447.

Sørensen, B. (1987b). *Superstrenge—en teori om alt og intet* (p. 184). Copenhagen, Denmark: Munksgaard Publisher.

Sørensen, B. (1988). Optimisation of wind-diesel systems. In *Proceedings of the Asian and Pacific Area Wind Energy Conference, Shanghai* (pp. 94–98).

Sørensen, B. (1991). Energy conservation and efficiency measures in other countries. *Greenhouse studies series* No. 8 (p. 118). Canberra: Australian Department of the Arts, Sport, the Environment, Tourism and Territories.

Sørensen, B. (1992a). The impact of energy efficiency on renewable energy options. In Sayigh (Ed.), *Renewable energy technology and the environment* (pp. 2654–2664). Oxford, England: Pergamon Press.

Sørensen, B. (1992b). Methods and models for estimating the global circulation of selected emissions from energy conversion. *Proceedings of the Technical Committee meeting*, Vienna 11–15. May, International Atomic Energy Agency, Wien (also Texts from IMFUFA, No. 226, Roskilde University, p. 47).

Sørensen, B. (1992c). The future of renewable energy. *Ecodesicion*, 4, 54–56.

Sørensen, B. (1993a). Technology change: The actor triangle. *Philosophy and Social Action*, 19, 7–12.

Sørensen, B. (1993b). What is life-cycle analysis? In *Life-cycle analysis of energy systems* (pp. 21–53). Workshop Proceedings. Paris: OECD Publications.

Sørensen, B. (1994a). Strategy for a rich, fulfilling and sustainable society. In Lundgren, Nilsson, & Schlyter (Eds.), *Expanding environmental perspectives* (pp. 83–101). Lund: Lund University Press.

Sørensen, B. (1994b). Life-cycle analysis of renewable energy systems. *Renewable Energy*, 5(Pt. II), 1270–1277.

Sørensen, B. (1994c). Model optimization of photovoltaic cells. *Solar Energy Materials and Solar Cells*, 34, 133–140.

Sørensen, B. (1995a). History of, and recent progress in, wind-energy utilization. *Annual Review of Energy & Environment*, 20, 387–424.

Sørensen, B. (1995b). Life-cycle analysis. In R. Paehlke (Ed.), *Conservation and environmentalism—an encyclopedia* (pp. 413–415). New York: Garland Publishers.

Sørensen, B. (1995c). Life-cycle approach to assessing environmental and social costs of photovoltaics. *Applied Energy*, 52(Suppl.), 357–374.

Sørensen, B. (1996a). Non-linear modelling of integrated energy supply and demand matching systems. Text 327 from IMFUFA (Roskilde University), p. 12, and *International Journal of Global Energy Issues* 12, 131–137 (1999).

Sørensen, B. (1996b). Life-cycle approach to assessing environmental and social externality costs. In *Comparing energy technologies* (Chap. 5, pp. 297–331). Paris: International Energy Agency, IEA/OECD.

Sørensen, B. (1996c). The use of life-cycle analysis to address energy cycle externality problems. In *Comparison of Energy Sources in Terms of Their Full Energy–Chain (FENCH) Emission Factors of Greenhouse Gases* (pp. 115–131). IAEA Tecdoc–892, Vienna, Austria.

Sørensen, B. (1996d). Does wind energy utilization have regional or global climate impacts? In (A. Zervos, H. Ehmann, & P. Helm (Eds.), *1996 European Union Wind Energy Conference, Göteborg* (pp. 191–194). Felmersham: H. Stephens & Associates.

Sørensen, B. (1996e). The role of life-cycle analysis in risk assessment. *International Journal of Environment and Pollution, 6*, 729–746.

Sørensen, B. (1996f). Issues of data collection and use for quantifying the impacts of energy installations and systems. In *Electricity, health and the environment: Comparative assessment in support of decision making* (pp. 123–137). IAEA Report SM–338/21, Vienna.

Sørensen, B. (1997a). Impacts of energy use. In Diesendorf & Hamilton (Eds.), *Human ecology, human economy* (pp. 243–266). New South Wales: Allen and Unwin.

Sørensen, B. (1997b). Renewable energy and environmental policy. In R. Venkata (Ed.), *Rural and renewable energy: Perspectives from developing countries* (pp. 49–66). Delhi: TERI, Tata Energy Research/Rajkamal Press.

Sørensen, B. (1997c). Externality estimation of greenhouse warming impacts. *Energy Conversion and Management, 38*, S643–S648.

Sørensen, B. (1999a). Wave energy. In *Encyclopedia of desalination and water resources*, UNESCO project. Oxford: EOLSS Publ., by subscription at website. http://www.desware.net

Sørensen, B. (1999b), Long-term scenarios for global energy demand and supply: Four global greenhouse mitigation scenarios. Final Report from a project performed for the Danish Energy Agency, IMFUFA Texts 359 (pp. 1–166). Roskilde University.

Sørensen, B. (2000a). Role of hydrogen and fuel cells in renewable energy systems. In *Renewable energy: The energy for the 21st century, Proceedings of the World Renewable Energy Conference VI, Reading* (Vol. 3, pp. 1469–1474). Amsterdam: Pergamon.

Sørensen, B. (2000b). PV power and heat production: An added value. In H. Scheer, *et al.* (Eds.), *16th European Photovoltaic Solar Energy Conference* (Vol. 2, pp. 1848–1851). London: James & James.

Sørensen, B. (2001a). Physics revealed. The methods and subject matter of physics. IMFUFA Texts 129bis, 392 and 394, Roskilde University, p. 284, Downloadable from, http://mmf.ruc.dk/ ~boson under Fysik E.

Sørensen, B. (2001b). The need for global modelling, Editorial. *International Journal of Global Energy Issues, 13*, 1–3.

Sørensen, B. (2001c). see NSES (2001).

Sørensen, B. (2002a). Handling fluctuating renewable energy production by hydrogen scenarios. In *14th World Hydrogen Energy Conference*, Montreal, CD published by CogniScience Publ. for l'Association Canadienne de l'Hydrogène, Revised CD issued 2003, p. 9.

Sørensen, B. (2002b). Modelling of hybrid PV-thermal systems, in *Proceedings of the 17th European Photovoltaic Solar Energy Conference, Munich 2001* (Vol. 3, pp. 2531–2538). Florence: WIP and ETA.

Sørensen, B. (2002c). Biomass for energy: How much is sustainable? In *12th European Biomass Conference*, Amsterdam 2002 (Vol. 2, pp. 1394–1397). WIP-Munich and ETA, Florence.

Sørensen, B. (2002d). Replacing subsidies by fair life-cycle costing. European Parliament Workshop on Energy Subsidies in the EU; http://energy.ruc.dk/downloads.htm

Sørensen, B. (2003a). Understanding photoelectrochemical solar cells. In *PV in Europe*, Rome 2002 (pp. 3–8). WIP-Munich and ETA, Florence.

Sørensen, B. (2003b). Progress in nanostructured photoelectrochemical solar cells. In *Third World Conference on Photovoltaic Energy Conversion, Osaka 2003*, paper 1\1O-B13-04, CDROM.

Sørensen, B. (2003c). Scenarios for future use of hydrogen and fuel cells. In *Towards a Greener World—Hydrogen and Fuel Cells Conference*, Vancouver (p. 12). Vancouver: CD published by Fuel Cells Canada.

Sørensen, B. (2003d). Time-simulations of renewable energy plus hydrogen systems. In M. Marini & G. Spazzafumo (Eds.), *Proceedings of the Hypothesis V Hydrogen Conference*, Porto Conte, Italy (pp. 35–42), Servizi Grafici Ed., Padova (also on CDROM).

Sørensen, B. (2004). Surface reactions in photoelectrochemical cells. In *Proceedings of the 19th European PV Solar Energy Conference, Paris*. Paper 1AV.2.13, CDROM. WIP, Munich and ETA, Florence.

Sørensen, B. (2004a). *Renewable energy* (3rd ed.). Burlington, MA: Elsevier-Academic Press.

Sørensen, B. (2004b). Total life-cycle assessment of PEM fuel cell car. In *Proceedings of the 15th World Hydrogen Conference, Yokohama*. Paper 09K-09 on CDROM, Hydrogen Energy Systems Society of Japan

Sørensen, B. (2005). *Hydrogen and fuel cells: Emerging technologies and applications*. Burlington, MA: Elsevier-Academic Press, Sustainable World Series.

Sørensen, B. (2006). Comparison between hydrogen fuel cell vehicles and bio-diesel vehicles. Paper # 111 in *Proceedings of the 16th World Hydrogen Energy Conference, Lyon*. Paris: CDROM, French, European and International Association for Hydrogen Energy.

Sørensen, B. (2007). Geological hydrogen storage. In *Proceedings of the World Hydrogen Technologies Convention, Montecatini*, CDROM, International Hydrogen Energy Association, IT Forum, Roma.

Sørensen, B. (2008a). A new method for estimating off-shore wind potentials. *International Journal of Green Energy, 5*, 139–147.

Sørensen, B. (2008b). A sustainable energy future: Construction of demand and renewable energy supply scenarios. *International Journal of Energy Research, 32*, 436–470.

Sørensen, B. (2008c). A renewable energy and hydrogen scenario for northern Europe. *International Journal of Energy Research, 32*, 471–500.

Sørensen, B. (2008d). Pathways to climate stabilisation. *Energy Policy, 36*, 3505–3509. Supplementary material at the website. http://www.elsevier.com/locate/enpol under doi:10.1016/j.enpol.2008.05.028.

Sørensen, B. (2008e). Mapping potential renewable energy resources in the Mediterranean region. Paper presented at 3rd International Symposium Environment, Athens, May 22–25.

Sørensen, B. (2009). Energy use by Eem Neanderthals. *Journal of Archaeological Science, 36*, 2201–2205.

Sørensen, B. (2010a). Demography and the extinction of the European Neanderthals. Submitted for publication.

Sørensen, B. (Ed.), (2010b). *Renewable energy reference book set* (Four volumes of reprints, each volume with introduction by BS). Earthscan, being processed.

Sørensen, B. (2010c). On the road simulation of battery, hydrogen and hybrid cars. In G. Pistoia, (Ed.), Chapter in *Battery, hybrid and fuel cell vehicles, performance, market and environmental issues*. Elsevier.

Sørensen, B. (2011a). *A history of energy. The case of Denmark from Stone Age to present*. Monograph in preparation.

Sørensen, B. (2011b). *Life-cycle analysis of energy systems: From methodology to applications*. Cambridge, MA: Royal Society of Chemistry Publishing (in preparation).

Sørensen, B., & Meibom, P. (1998). A global renewable energy scenario. IMFUFA Texts 354, (p. 112). Roskilde University.

Sørensen, B., & Meibom, P. (2000). A global renewable energy scenario. *International Journal of Global Energy Issues, 13*(1–3), 196–276, based on Sørensen and Meibom (1998).

Sørensen, B., Petersen, A., Juhl, C., Ravn, H., Søndergren, C., Simonsen, P., *et al.* (2001). Project report to Danish Energy Agency (in Danish): Scenarier for samlet udnyttelse af brint som energibærer i Danmarks fremtidige energisystem, *IMFUFA Texts* No. 390, p. 226, Roskilde University; Report download site: http://mmf.ruc.dk/energy

Sørensen, B., Petersen, A., Juhl, C., Ravn, H., Søndergren, C., Simonsen, P., *et al.* (2004). Hydrogen as an energy carrier: Scenarios for future use of hydrogen in the Danish energy system. *International Journal of Hydrogen Energy, 29*, 23–32; based upon Sørensen et al. (2001).

Sørensen, B., Meibom, P., Nielsen, H., Karlsson, K., Pedersen, A., Lindboe, H., *et al.* (2008). Comparative assessment of hydrogen storage and international electricity trade for a Danish energy system with wind power and hydrogen/fuel cell technologies. EECG Research Paper 1/08. Roskilde University Digital Archive. http://rudar.ruc.dk/handle/1800/2431.

Sørensen, B., & Watt, M. (1993). Life-cycle analysis in the energy field. In *Proceedings of the 5th International Energy Conference "Energex '93,"* (Vol. 6, pp. 66–80). Seoul: Korea Institute of Energy Research.

Sørensen, B., Nielsen, L., Petersen, S., Illum, K., & Morthorst, P. (1994). Future renewable energy system—light green or dark green? Danish Technology Council, Report 1994/3 (in Danish). http://www.tekno.dk/ udgiv/943/943all.htm

Sorensen, K., & MacLennan, C. (1974). *IEEE Transactions,* 62–68, US Institute of Electrical and Electronics Engineers.

Spear, W., & Le Comber, P. (1975). *Solid State Communications, 17,* 1193–1196.

Spitzer, H. (1954). In *Hütte, Maschinenbau A* (pp. 556–566). Berlin: Wilhelm Ernst and Sohn.

Sproul, A., Shi, Z., Zhao, J., Wang, A., Tang, Y., Yun, F., *et al.* (1995). Characterization and analysis of multilayer solar cells. In *1994 First World Conference on Photovoltaic Energy Conversion, Kona* (Vol. 2, pp. 1410–1412). IEEE.

Squires, A. (1974). *Ambio 3,* 2–14.

Stafford, D., Hawkes, D., & Horton, R. (1981). *Methane production from waste organic matter.* Boca Raton, FL: CRC Press.

Stambolis, C. (1976). In *Proceedings of the UNESCO/WMO Solar Energy Symposium, Geneva 1976,* Paper ENG.S/Doc. 3, WMO Paper No. 477 (1977).

Stansell, J. (1978). *New Scientist* 16. February, 420–421.

Starr, C. (1969). Social benefit versus technological risk. *Science, 165,* 1232–1238.

Steinberg-Yfrach, G., Rigaud, J.–L., Durantini, E., Moore, A., Gust, D., & Moore, T. (1998). Light–driven production of ATP catalysed by F_0F_1-ATP synthase in an artificial photosynthetic membrane. *Nature, 392,* 479–482.

Steinhart, J., & Steinhart, C. (1974). *Science, 184,* 306–316.

Stewart, D., & McLeod, R. (1980). *New Zealand Journal of Agriculture,* Sept., 9–24.

Stewart, G., Hawker, J., Nix, H., Rawlins, W., & Williams, L. (1982). *The potential for production of hydrocarbon fuels from crops in Australia.* Melbourne: Commonwealth Scientific and Industrial Research Organization.

Stewart, R. (1969, September). *Scientific American,* 76–105.

STI. (2002). Sustainable Technology International, Queanbeyan, NSW Australia. http://www.sta.com.au

Stock, M., Carr, A., & Blakers, A. (1996). Texturing of polycrystalline silicon, *Solar Energy Materials and Solar Cells, 40,* 33.

Stokes, G. (1880). *Math. and Phys. Papers* (Vol. 1, p. 225). As quoted by Wehausen and Laitone (1960).

Stolwijk, J., & Canny, P. (1991). Determinants of public participation in management of technological risk. pp. 33–47 in Shubik (1991).

Stommel, H. (1965). *The gulf stream: A physical and dynamical description* (2nd ed.). Berkeley, CA: University of California Press.

Street, R., & Mott, N. (1975). *Physical Review Letters, 39,* 1293–1295.

Strickland, J. (1975). Sandia Laboratories Energy Report SAND75–0431, Albuquerque, New Mexico.

Stroebel, D., Choquet, Y., Popot, J., & Picot, D. (2003). An atypical haem in the cytochrome b_6f complex. *Nature, 426,* 413–418.

Svendsen, S. (1980). Effektivitetsprøvning af solfangere, Danish Technical University, Laboratory for Heat Insulation, Communication No. 107, Lyngby, Denmark.

Sverdrup, H. (1957). In S. Flügge (Ed.), *Handbuch der Physik* (Vol. 48 (Geophysik II), pp. 608–670). Berlin: Springer-Verlag.

Sverdrup, H., Johnson, M., & Fleming, F. (1942). *The oceans, their physics, chemistry and general biology.* Englewood Cliffs, NJ: Prentice–Hall.

Sverdrup, H., & Munk, W. (1947). Wind, Sea and Swell. United States Hydrogr. Office Publ. No. 601 (also No. 604, 1951).

Swanson, R., Smith, R., & Johnson, C. (1975). In F. Eldridge (Ed.), *Proceedings of the 2nd Workshop on Wind Energy Conversion Systems* (pp. 92–97). Report No. NSF–RA–N–75–050. McLean, VA: The Mitre Corp.

Swedish Association of Boiler Owners (Ångpanneföreningen) (1973). Verksamhedsberättelse. Stockholm.

Swedish Ministries of Foreign Affairs and of Agriculture. (1971). Air Pollution across National Boundaries, case study for the United Nations Conference on the Human Environment, Stockholm.

Swisher, R. (1995). Wind power a strong contender in US marketplace. In *The World Directory of Renewable Energy Suppliers and Services* (pp. 190–196). London: James & James.

Tabor, H. (1967). *Solar Energy, 7,* 189.

Tafdrup, S. (1993). Environmental impact of biogas production from Danish centralized plants. Paper presented at IEA Bioenergy Environmental Impact Seminar, Elsinore, 1993.

Taiganides, E. (1974). *Agricultural Engineering (American Society Agricultural Engineering), 55*(4).

Takahashi, K. (1998). Development of fuel cell electric vehicles. Paper presented at Fuel cell technology conference, London, September, IQPC Ltd., London.

Takamoto, T., Ikeda, E., Kurita, H., & Yamaguchi, M. (1997). Over 30% efficient InGaP/GaAs tandem solar cells with InGaP tunnel junction. In *14th European PV Solar Energy Conference* (pp. 970–975). Bedford: HS Stephens & Associates.

Talbert, S., Eibling, J., & Löf, G. (1970). *Manual on Solar Distillation of Saline Water,* United States Department of the Interior, Office of Saline Water, R&D Report No. 546; cf. also *Solar Energy, 13,* 263–276.

Tang, C. (2001). Organic light emitting diodes. In *Proceedings of the 10th International Conference On Unconventional Photoactive Systems* (pp. I-23). Diablerets.

Tarascon, J., & Armand, M. (2001). Issues and challenges facing rechargeable lithium batteries. *Nature, 414,* 359–367.

Taylor, G., *et al.* (1993). Full scale measurements in wind turbine arrays, In *EC Wind Energy Conference, Travemünde* (pp. 755–758). Felmersham: H. Stephens & Associates.

Telegadas, K. (1971). In *Fallout Program Qua. Summary Rep., Mar.–Jun.*, (pp. I–2 to I–88). Report HASL–243. Stanford: Health and Safety Lab.

Telkes, M. (1952). *Industrial and Engineering Chemistry, 44,* 1308.

Telkes, M. (1976). In C. Stein (Ed.), *Critical materials problems in energy production.* New York and London: Academic Press.

Templin, R. (1976). In O. Ljungstrom (Ed.), *Proceedings of the Workshop on Advanced Wind Energy Systems, Stockholm 1974.* Stockholm: Swedish Development Board/ Vattenfall.

Teplyakov, D., & Aparisi, R. (1976). *Geliotekhnika, 12,* No. 3, 35–48.

Thekaekara, M. (1976). *Solar Energy, 18,* 309–325.

Thorpe, T. (2001). Current status and development in wave energy. In *Proceedings of the Conference Marine Renewable Energies* (pp. 103–110). UK: Institute of Marine Engineers.

Tideman, J., & Hawker, J. (1981). *Search, 12,* 364–365.

Tol, R. (1995). The damage costs of climate change: Towards more comprehensive calculations. *Environmental & Resource Economics, 5,* 353–374.

Toland, R. (1975). In *Proceedings of the 1975 Flywheel Technology Symposium* (pp. 243–256). Berkeley, CA, Report ERDA 76–85.

Tomaschek, R. (1957). In S. Flügge (Ed.), *Handbuch der Physik* (Vol. 48 (Geophysik II), pp. 775– 845). Berlin: Springer-Verlag.

Tonne, F., & Normann, W. (1960). *Zeitschrift für Meteorologie, 14,* 166.

Toyota. (1996). High-performance hydrogen-absorbing alloy, http://www.toyota.co.jp/e/november_96/ electric_island/press.html

Trebst, A. (1974). *Annals Review Plant Physiology, 25,* 423–447.

Trenberth, K., & Solomon, A. (1994). The global heat balance: Heat transfer in the atmosphere and ocean. *Climate Dynamics, 10,* 107–134.

Trenkowitz, G. (1969). Die Värmepumpe, Verein Deutscher Ingenieurs Berichte Nr. 136.

Trinidade, S. (1980). Energy crops—the case of Brazil, in *International Conference on Energy from Biomass*, Brighton, UK, 1980, Centro de Tecnologia Promon, Rio de Janeiro

Troen, I., & Petersen, E. (1989). *European wind Atlas.* Roskilde: Risø National Laboratory.

Trombe, F. (1973). Centre Nationale de Récherche Scientifique, Report No. B–1– 73–100.

Trombe, F., & Vinh, A. (1973). Thousand kW solar furnace, built by the National Center of Scientific Research in Odeillo (France). *Solar Energy, 15,* 57–61.

Tsang, C., Lippmann, M., & Wintherspoon, P. (1979). In *Sun II, Proceedings of the Solar Energy Society Conference, New Delhi 1978* (pp. 349–355). London: Pergamon Press.

Tsubomura, H., Matsumura, M., Nomura, Y., & Amamiya, T. (1976). Dye sensitised zinc oxide: aqueous electrolyte:platinum photocell. *Nature, 261,* 402–403.

Turkenburg, W., *et al.* (2000). Renewable energy technologies. *World energy assessment* (Chap. 7, pp. 219–272). New York: United Nations Development Programme.

Turner, C. (Ed.), (1974). Solar Energy Data Workshop, United States National Science Foundation and Nat. Oceanic and Atm. Adm., Silver Spring, Report No. NSF–RA–N–74–062.

Tuttle, J., Ward, J., Duda, A., Berens, A., Contreras, M., Ramanathan, K., *et al.* (1996). The performance of Cu(In,Ga)Se$_2$-based solar cells in conventional and concentrator applications. *Proceedings of the Material Research Society Symposium, 426*, 143.

Twomey, S., & Wojciechowski. (1969). *Journal of Atmospheric Science, 26*, 684–688.

Ulaby, F., Moore, R., & Fung, A. (1986). *Microwave remote sensing*, Vol. 3. Reading, PA: Addison-Wesley. Vol. 1 (1981), Vol. 2 (1982), Vol. 3 (1986).

Ulrich, R. (1973). In D. Schramm & W. Arnett (Eds.), *Explosive Nucleosynthesis* (pp. 139–167). Austin, TX: University of Texas Press.

Ulrich, R. (1975). *Science, 190*, 619–624.

UMIP. (1996). Miljøvurdering af produkter, Project on the development of environmentally sound industrial products (Wenzel *et al.*), Institut for produktudvikling, Report, Danish Technical University.

UNEP. (1998). GRID data website: http://www.unep.ch/noaagnv22.html

UNEP. (2009). *Towards sustainable production and use of resources: Assessing biofuels*. (S. Bringezu *et al.*), Paris, http://www.unep.org

United Kingdom Energy Technology Support Unit. (1976). Wave Energy on U.K. Coasts and North Sea. UKAEA Harwell.

United Kingdom Meteorological Office. (1997). Unified model (User guide and support documents), London.

United Nations. (1981). *World energy supplies*. United Nations, New York.

United Nations. (1996). Populations 1996, 2015, 2050. United Nations Population Division and UNDP. http://www.undp.org/popin/wdtrends/pop /fpop.htm

United Nations. (1997a). *1995 World energy supplies*. United Nations, New York.

United Nations. (1997b). UN urban and rural population estimates and projections as revised in 1994. United Nations Population Division and UNDP, Washington. Website: http://www.undp.org/popin/wdtrends/urban.html

United Nations. (2010). *World population prospects: The 2008 revision*. Department of Economic and Social Affairs, Population Division. http://esa.un.org/unpp/

United States Bureau of Mines. (1975). As quoted by Ross, M., Williams, R. Assessing the potential for fuel conservation, Inst. Public Policy Alternatives, State University of New York.

United States Department of Energy. (1979). Peat prospectus. Washington, DC: United States Department of Energy.

United States Environmental Protection Agency. (1980). US Environmental Protection Agency Report EPA–600/7–80–040 (D. deAngelis *et al.*), Washington, DC; also earlier reports EPA–600/2–76–056 and EPA–600/7–77–091.

United States Government. (1962). United States Standard Atmosphere. Guideline for meteorological observational methodology, Washington, DC.

University of Columbia. (1998). Selected climate data sets, available at http://ingrid.ldgo.columbia.edu

UNSCEAR. (1993). *Sources and Effects of Ionizing Radiation*. United Nations Scientific Committee on the Effects of Atomic Radiation, Report to the General Assembly, with scientific annexes, E.94.IX.2. United Nations, New York.

Vermeulen, L., & Thompson, M. (1992). Stable photoinduced separation in layered viologen compounds. *Nature, 358*, 656–658.

Vindeløv, S. (1994, December). Research activities in wave energy. *Sustainable Energy News* (7), 12–13.

Vohra, K. (1982). Rural and urban energy scenario of the developing countries and related health assessment. In *Proceedings of the International Symposium on Health Impacts of Different Sources of Energy, Nashville, 1981*, pp. 79–96. International Atomic Energy Agency, Vienna, Paper No. IAEA–SM–254/102.

Volfin, P. (1971, September). *La Récherche*, 741–755.

Vollmer, R. (1977). *Nature, 270*, 144–147.

Wagner, U., Geiger, B., & Schaefer, H. (1998), Energy life cycle analysis of hydrogen systems. *International Journal of Hydrogen Energy, 23*, 1–6.

Walsh, J., Davis, J., & Garms, R. (1981). Further studies on the reinvasion of the onchocerciasis control programme by *Simulus damnosum* s.l. *Tropical Medicine & Parasitology, 32*(4), 269–273.

Wan, E., Simmins, J., & Nguyen, T. (1981). In T. Reed (Ed.), *Biomass gasification* (pp. 351–385). Park Ridge, NJ: Noyes Data Corp.

Wang, J., Wang, D., Smith, K., & Hermes, J. (1982). In *The future of small energy resources* (pp. 465–472). New York: McGraw-Hill.

Ward, R. (1982). *Solar Energy, 29*, 83–86.

Warming, S. (1996). Valuation of environmental impacts from coal-based power production. ELSAM Power Utility, (p. 17), (unpublished note in Danish).

Wassink, E. (1975). In J. Cooper (Ed.), *Photosynthesis and productivity in different environments* (pp. 675–687). Cambridge, UK: Cambridge University Press.

Watt, M. (1993). *Environmental & health considerations in the production of cells and modules.* Centre for Photovoltaic Devices & Systems Report # 1993/02. Sydney: University of New South Wales.

WEA. (2000). Energy and the challenge of sustainability. *World energy assessment*, UNDP, UNDESA, WEC, New York.

Webb, W. (1976). In S. Flügge (Ed.), *Handbuch der Physik* (Vol. 49/5 (Geophysik III, K. Rawer, Ed.) pp. 117–176. Berlin: Springer-Verlag.

Weber, O. (1975). *Brown Boveri Mitt., 62*(7/8), 332–337.

Webster, A. (1974, August). *Scientific American*, 26–33.

WEC. (1991). District heating/combined heat and power. London: World Energy Council.

Wehausen, J., & Laitone, E. (1960). In S. Flügge (Ed.), *Handbuch der Physik* (Vol. 9 (Stromungs-mechanik III), pp. 446–778). Berlin: Springer-Verlag.

Weiner, S. (1977). The sodium–sulphur battery: Problems and promises. In J. Goodenough & M. Whittingham (Eds.), *Solid state chemistry of energy conversion and storage* (Chap. 12), Advances in chemistry series 163. Washington, DC: American Chemical Society.

Weinstein, J., & Leitz, F. (1976). *Science, 191*, 557–559.

Weissglas, P. (1977). Solcellskraftverk i Sverige, Report No. E77–5011. Stockholm: Institute of Mikrovågsteknik.

Weizsäcker, E. von, Lovins, A., & Lovins, L. (1997). *Factor 4: Doubling wealth—halving resource use.* Sydney: Allen and Unwin.

Wenham, S., Green, M., Edminston, S., Campbell, P., Koschier, L., Honsberg, C., et al. (1995a). Limits to efficiency of silicon multilayer thin film solar cells. In *1994 First World Conference on Photovoltaic Energy Conversion, Kona* (Vol. 2, pp. 1234–1241). IEEE.

Wenham, S., Robinson, S., Dai, X., Zhao, J., Wang, A., Tang, Y., et al. (1995b). Rear surface effects in high efficiency silicon solar cells. In *1994 First World Conference on Photovoltaic Energy Conversion, Kona* (Vol. 2, pp. 1278–1282). Washington, DC: IEEE.

Wentink, T. (1976). Study of Alaskan wind power and its possible applications. Report No. NSF/RANN/SE/AER–00239/FR–76/1.

Wentorf, R., & Hanneman, R. (1974). *Science, 185*, 311–319.

West, C. (1974). Fluidyne Heat Engine, Harwell Report AERE–R6775, U.K. Atomic Energy Agency.

Wetherald, R., & Manabe, S. (1972). *Monthly Weather Review, 100*, 42–59.

Wetherald, R., & Manabe, S. (1975). *Journal of Atmospheric Science, 32*, 2044–2059.

Whittingham, M. (1976). *Science, 192*, 1126–1127.

Wilbur, P., & Mancini, T. (1976). *Solar Energy, 18*, 569–576.

Williams, R. (1975). Working paper No. 21. New Jersey: Center for Environmental Studies, Princeton University.

Wilson, C., & Matthews, W. (1970). *Man's impact on the global climate*, Report of the Study of Critical Environmental Problems (SCEP). Cambridge, MA: MIT Press.

Wilson, C., & Matthews, W. (1971). *Inadvertent climate modification*, Report of the Study of Man's Impact on Climate (SMIC). Cambridge, MA: MIT Press.

Wilson, R., & Lissaman, P. (1974). Applied aerodynamics of wind power machines. Oregon State University, Report No. NSF-RA-N-74-113.

Windpower Monthly. (2003). Wind market status 2003. Editorial, March issue, http://www.wpm.co.nz

Winsberg, S. (1981). *Sunworld, 5*, 122–125.

Wise, D. (1981). *Solar Energy, 27*, 159–178.

Wittenberg, L., & Harris, M. (1979). In *Proceedings of the 14th Intersociety Energy Conversion Engineering Conference* (pp. 49–52). Washington, DC: American Chemical Society.

Wizelius, T. (1998, December). Potential for offshore transmission. *Windpower Monthly*, p. 25.

Wofsy, S., McElroy, M., & Sze, N. (1975). *Science, 187*, 535–537.

Wolf, M. (1963). *Proceedings of the IEEE. Institute of Electrical and Electronics Engineers, 51*, 674–693.

Wolf, M. (1971). *Energy Conversion, 11*, 63–73.

Wolfbauer, G. (1999). The electrochemistry of dye sensitised solar cells, their sensitisers and their redox shuttles, Ph. D. thesis, Victoria, Australia: Monash University.

World Almanac. (1977). (G. Delury, ed.), Newspaper Enterprise Ass., New York.

World Energy Conference. (1974). *Survey of energy resources*, privately published in London.

World Energy Conference. (1995). *Survey of energy resources* (17th ed.). London.

World Energy Conference. (2009). *Survey of energy resources. Interim update 2009*. London: World Energy Council. www.worldenergy.org

World Meteorological Organization. (1964). In *Proceedings of the UN Conference on Alternative Energy Sources, Rome 1961*, Paper W/11. New York: United Nations Printing Office.

WRI. (2008). *World resources 2008*. Washington, DC, available online at www.wri.org

Wrighton, M., Ellis, A., & Kaiser, S. (1977). Conversion of visible light to electrical energy: Stable cadmium selenide photoelectrodes in aqueous electrolytes. In J. B. Goodenough & M. S. Whittingher (Eds.), *Solid state chemistry of energy conversion and storage* (pp. 71–92). Advances in Chemistry Series 163. Washington, DC: American Chemical Society.

Wronski, C., & Carlson, D. (1977). In W. Spear (Ed.), *Proceedings of the 7th International Cont; on Amorphous and Liquid Semiconductors* (pp. 452–456). UK: University of Edinburgh.

Wulff, H. (1966). *The traditional crafts of Persia*. Cambridge, MA: MIT Press.

Wurster, R. (1997). PEM fuel cells in stationary and mobile applications. Paper for Biel Conference (http://www.hyweb.de/knowledge).

Wynne, B. (1984). The institutional context of science, models, and policy. *Policy Sciences, 17,* 277–320.

Wysocki, J., & Rappaport, P. (1960). *Journal of Applied Physics, 31,* 571–578.

Yamada, K., Komiyama, H., Kato, K., & Inaba, A. (1995). Evaluation of photovoltaic energy systems in terms of economic, energy and CO_2 emissions. University of Tokyo, Report.

Yamaguchi, M., Horiguchi, M., Nakanori, T., Shinohara, T., Nagayama, K., & Yasuda, J. (2000). Development of large-scale water electrolyzer using solid polymer electrolyte in WE-NET. In Mao and Veziroglu, (Eds.), *Hydrogen energy progress XIII* (Vol. 1, pp. 274–281). Beijing: IAHE.

Yamamoto, K., Yoshimi, M., Tawada, Y., Okamoto, Y., & Nakajima, A. (1999). Cost effective and high performance thin film Si solar cell towards the 21st century. In T. Saitoh (Ed.), *Technical digest of the International PVSEC-11, Sapporo* (pp. 225–228). Tokyo: Tokyo University of Agriculture and Technology.

Yartym, J., *et al.* (2001). In *Proceedings of the 4th International Symposium on New Materials,* Montréal (pp. 417–419).

Yasukawa, S., Tadokoro, Y., Sato, O., & Yamaguchi, M. (1996). Integration of indirect CO_2 emissions from the full energy chain. In *Comparison of energy sources in terms of their Full Energy-Chain (FENCH) emission factors of greenhouse gases* (pp. 139–150). IAEA Tecdoc–892, Vienna, Austria.

Yazawa, Y., Tamura, K., Watahiki, S., Kitatani, T., Minemura, J., & Warabisako, T. (1996). GaInP single-junction and GaInP/GaAs two-junction thin-film solar cell structures by epitaxial lift-off. In *Technical digest of 9th PV Science & Engineering Conference, Miyazaki,* p. 865. Tokyo Institute of Technology.

Yeh, A-I., Huang, Y-C., & Chen, S. (2010). Effect of particle size on the rate of enzymatic hydrolysis of cellulose. *Carbohydrate Polymers, 79,* 192–199.

Yen, J. (1976). In *Proceedings of the International Symposium on Wind Energy Systems, Cambridge 1976,* Paper E4. Cranfield: British Hydro-Mechanical Research Association

Yoneda, N., Ito, S., & Hagiwara, S. (1980). Study of energy storage for long term using chemical reactions, 3rd International Solar Forum, Hamburg, Germany, June 24–27.

Yu, G., Gao, J., Hummelen, J., Wudl, F., & Heeger, A. (1995). Polymer photovoltaic cells. *Science, 270,* 1789–1791.

Zachos, J., *et al.* (2001). Trends, rhythms, and aberrations in global climate 65 Ma to present. *Science, 292,* 686–693.

Zakeeruddin, S., *et al.* (1997). Molecular engineering of photosensitizers for nanocrystalline solar cells: Synthesis and characterization of Ru dyes based on phosphonated terpyridines. *Inorganic Chemistry, 36,* 5937–5944.

Zakhidov, A., Baughman, R., Iqbal, Z., Cui, C., Khayrullin, I., Dantas, S., *et al.* (1998). Carbon structures with three-dimensional periodicity at optical wavelengths. *Science, 282,* 897–901.

Zener, C. (1973, January). *Physics Today,* 48–53.

Zener, C., & Fetkovich, J. (1975). *Science, 189,* 294–295.

Zhang, Y., Suenaga, K., Colliex, C., & Iijima, S. (1998). Coaxial nanocables: Silicon carbide and silicon oxide sheathed with boron nitride and carbon. *Science, 281,* 973–975.

Zhao, J., Wang, A., Altermatt, P., Wenham, S., & Green, M. (1995). 24% efficient solar cells. In *1994 First World Conference on Photovoltaic Energy Conversion, Kona* (Vol. 2, pp. 1477–1480). Washington, DC: IEEE.

Zhao, J., Wang, A., & Green, M. (1998). 19.8% efficient multicrystalline silicon solar cells with "honeycomb" textured front surface. In J. Schmid, *et al.* (Eds.), *Proceedings of the 2nd*

World Conference on PV Energy Conversion, Vienna (pp. 1681–1684). Luxembourg: JRC European Commission EUR 18656 EN.

Zittel, W., & Wurster, R. (1996). Hydrogen in the energy sector. Ludwig-Bölkow-ST Report: http://www.hyweb.de/knowledge/w–i–energiew–eng

Zlatev, Z., & Thomsen, P. (1976). Numerical institute. Internal Reports No. 76–9/10. Lyngby, Denmark: Danish Technical University.

Zouni, A., Witt, H., Kern, J., Fromme, P., Krauss, N., Saenger, W., *et al.* (2001). Crystal structure of photosystem II from *Synechococcus elongatus* at 3.8Å resolution. *Nature, 409,* 739–743.

Zuidema, G., Born, J. van der, Alcoma, J., & Kreileman, G. (1994). Simulating changes in global land cover as affected by economic and climatic factors. *Water, Air and Soil Pollution, 76,* 163–198.

Subject Index

Page numbers in *italics* indicate figures and tables

A

Absorber, *see* Collector, Solar energy
    conversion
Absorptance, 453, *455*
    hemispherical, 455
Absorption
    of light in solar cell, 422–428
    of radiation in Earth-atmosphere system,
      58, *166*
    of ocean waves, 150
Absorption coefficient, 45–46, *85*
Absorption cooling cycle, 480–481, *481*
Absorption cross-section, 166, 424
Acceleration of gravity *g*, 233
Acceptor level, 417
Accumulated present value, *797*, 797–798
Activity coefficient, 286
Adenosine diphosphate/triphosphate
    (ADP/ATP), 298
Adiabatic temperature gradient, 49
Adsorption, 79, 841
Aerobic process, 298
Aerosols, *see also* Particles
    effect on climate, 124
    in atmosphere, 167
Agricultural use of energy, 163, 524, *671*
Air exchange in buildings, 628
Airfoil theory, 366, *366*, 370
Aggregation in life-cycle assessment, 831–835
Albedo
    change of Earth's, 157
    influence on rejected solar radiation, 216
    of Earth–atmosphere system, *59*, 165, *166*
    of land surfaces, 67, *68*
    of water surface, 67, *68*
Alkaline electrolyte batteries, 592–593
Algae, 318, 322, *328*
Alternating current transmission, 538
Amorphous solar cells, 436–439
Amylose, 530, *531*
Anaerobic process, 506, 522, 602
Angular momentum transport, 102–103, *104*
Annuity, 796
Annuity loan, 798, 811

Appropriate technology, *513*, 809–812
Aquifers, 175, 578
Aquifer storage, 551–555, 584–585
Area use, 6, *672*
Atmosphere
    energy conversion in, 188
    particulate matter in, 78
    processes in, 188
    properties of, 293
    temperature distribution in, *169*
Atmospheric circulation, 98–103, 106–110
    energy in, 104, 145, 188, 232
Atmospheric electricity, 293–294
Attenuation of solar radiation, 645
Availability of power, 266, 730, *730*, 896
Available potential energy, 145, 232
Avogadro's number N_A, 44
Axial interference factor, 364, 391

B

Back–payment on loan, *see* Cost evaluation,
    Cost profile
Balance of foreign payments, 807
Base load, 727
Basic needs, 621
Batteries, 484, 589–596
Benson-Bassham-Calvin cycle, 300
Bernoulli's theorem, 343
Betz limit, 392–393, 397
Bi-angular reflectance, 215
Bias, of semiconductor junction, 422
Big Bang theory, 55, 292
Bio-energy, *see* Biological energy conversion
    and storage
Biogas, 7, 506–517, *877–878*
Biological energy conversion and storage,
    294–329
    efficiency of conversion, 316–318, 496
    liquid fuel production, 522–530
    gaseous fuel production, 505–521
    green plant photosynthesis, 295–298
    heat production, 499–502
    limiting factors, 320–327
    productivity, 319–329

943

FIGURE 4.74 Calculated electron density difference between the first excited singlet and ground state of coumarin-derivative dye (*top*: positive values enhanced; *below*: negative values enhanced; Sørensen, 2001a, 2003a).

10-nm-thick space charge regions in surfaces of solids. The conclusion from the present calculation is that both dye and surface are modified, with the dye "arm" being bent to better attach to the surface, and the surface atoms to accommodate the dye particle (but even in the absence of the dye, the surface of the lattice structure would be different from the regular interior used as starting point for the optimization).

A further "handle" on the quantum chemical calculations is the comparison of measured spectra (IR, NMR, etc.) with those predicted by the calculations. Understanding of complex molecules has progressed rapidly as a result of the combination of spectral studies in the laboratory with quantum model calculations.

The photo-electrochemical dye and nanostructure technique has several applications beyond the formation of solar cells. Among them are smart windows (Bechinger *et al.*, 1996; Granqvist *et al.*, 1998; STI, 2002), energy storage (Hauch *et al.*, 2001; Krasovec *et al.*, 2001), environmental monitors (Kamat and Vinodgopal, 1998; Yartym *et al.*, 2001), hydrogen production (Khaselev and Turner, 1998; Luzzi, 1999; Mathew *et al.*, 2001), computer and TV screens also suitable for outdoor use (using the dye as light emitter rather than absorber; Rajeswaran *et al.*, 2000; Tang, 2001; Müller *et al.*, 2003), and three-dimensional data storage (Dwayne-Miller *et al.*, 2001).

FIGURE 4.75   Attachment of coumarin-derivative dye to and modification of anatase surface. *Left:* initial guess, *right:* optimized (the depth of TiO$_2$ surface layer distortion cannot be determined by modeling just the top layer; Sørensen, 2003a, 2004b).

4.4.3 Solar Thermal Conversion

Heat Generation

Conversion of solar energy to heat requires a light-absorbing material, a *collector*, which is able to distribute the absorbed radiant energy over internal degrees of freedom associated with kinetic energy of motion at the molecular level (e.g., lattice vibrations in the case of a solid). The Earth and its atmosphere are examples of such collectors, as discussed in sections 2.2 and 2.3. Absorption of solar energy will raise the temperature of the collector or transfer energy to a reservoir, if the collector is connected to one. The collector will also emit radiation, and it may lose heat energy by conduction and convection processes. The frequency spectrum of the emitted radiation will correspond to the Planck spectrum (2.3) for the collector temperature T_c, if the collector is in a state allowing the definition of a thermodynamic temperature.

Man-made collectors may try to achieve a large absorption by minimizing reflection and transmission and to achieve small losses, e.g., by operating the collector at temperatures not much above ambient air temperatures or, if higher load temperatures are required, by reducing the heat loss rates by suitable transparent covers and insulation.

One may distinguish between *passive* and *active* systems, according to whether energy is specifically added (pumps, etc.) in order to bring the collector heat gain to the load areas. A passive system need not be characterized by the absence of definite heat flow paths between collectors and load areas, but such flows should be natural, i.e., they should not depend on other energy inputs provided by man. There may be borderline cases in which the term *natural circulation* would be difficult to define.

FIGURE 4.76a,b,c Solar-wall passive heating and cooling system. Top cross-sections show air flow during (*a*) winter and (*b*) summer. Curves below show, for a selected period of a few days' operation, the air flow-rate and temperature [at the location X in (*a*)], ambient temperature outside the house, and solar radiation. The inside wall temperature remains above 20 °C during the period considered. The house is situated at latitude 43°N. (Based on Trombe, 1973; Stambolis, 1976.)

Examples of passive solar heat systems are ordinary windows in buildings, which transmit a large fraction of the solar radiation (if the angle of incidence is small, i.e., if the direction of the incident light does not make a small angle with the pane). The room behind the window may, to a large extent, act like a black-body, absorbing practically all of the radiation transmitted through the window and re-emitting only a small fraction to the outside (providing that the total window area of the room is not large).

Another kind of passive solar heat system uses the heat capacity of walls facing the sun during the daytime. The walls absorb radiation and accumulate

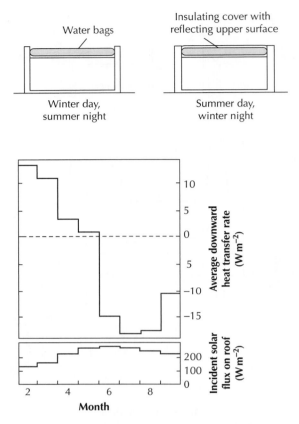

FIGURE 4.77 Solar-roof passive heating and cooling system. The top cross-sections show operation of the cover panel during the day and the night, during summer and winter. Below are heat-transfer rates through the roof (positive downward) and solar radiation, both averaged month by month during an 8-month period. The house is situated at latitude 35 °N. (Based on Hay and Yellot, 1972; Stambolis, 1976.)

it (the heat capacity of building materials increases roughly in proportion to mass), and at night they lose heat to their colder surroundings, including the inside area, which is thus heated. The wall's heat capacity also serves to cool the building during at least the first part of the daytime, if the wall temperature after the night's cooling off is lower than the next day's ambient temperature. More elaborate versions of solar-wall and solar-roof passive systems, directing the natural convection according to conditions (night/day, summer/winter), are shown in Figs. 4.76 and 4.77. These systems require a little "active" help in opening and closing vents, shutters, or covers. Because they depend on a daily cycle, these systems are most applicable in climatic regions where the daily solar input is substantial during the cold season as well.

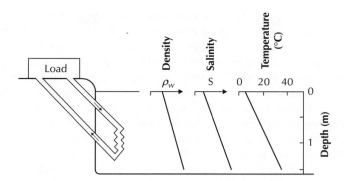

FIGURE 4.78 Schematic cross-section through a solar pond (cf. Tabor, 1967). The temperature profile may have a secondary peak within the top hundredths of a meter, due to absorption of ultraviolet radiation in this layer, and depending on the re-radiation conditions (effective sky temperature relative to temperature of pond surface layer).

Greenhouses are also passive solar-collecting systems, as are water heaters based on water bags (e.g., placed on roofs) or flat-plate collectors (see below) delivering heat to water, which is transferred by natural circulation to a storage tank lying higher than the collector. A saline pond can play the role of the collector, with a heat exchanger at the bottom of the pond transferring the collected energy to a working fluid, which can be circulated to the load area by natural or forced circulation. The solar pond itself (see Fig. 4.78) contains water with a high content of dissolved salts, causing the formation of a salinity and density gradient, preventing the mixing of surface and bottom water. The water absorbs some solar radiation, but, if the pond is shallow, most of the absorption takes place at the bottom, thereby creating a stable temperature gradient increasing toward the bottom, because heat is transferred upward only by slow processes.

Flat-plate Collectors

The term *flat-plate collector* is used for absorbers with a generally flat appearance, although the collecting surface need not be flat in detail (it might have V-shaped carvings or even a focusing substructure). The side of the collector facing the sun may have a cover system (e.g., one or more layers of glass), and a mass flow J_m^c of some fluid (e.g., water or air) passes the absorber and is supposed to carry the heat, which is transferred from the absorber plate to the mass flow and then to the load area (i.e., the place of actual usage) or to some temporary energy storage. The general layout is shown in Fig. 4.79.

The absorber is characterized by an absorptance, $\alpha_\lambda(\Omega)$, that may depend on wavelength and on the direction of incident light (cf. section 3.1.4). In many situations, it may be assumed in a first approximation that the absorptance is independent of direction and of wavelength (gray surface). For a black-painted surface, α may be around 0.95, but, if the surface is structureless, the assumption that α is independent of direction breaks down when the angle between the normal to the surface and the direction of incidence exceeds about 60°. Toward 90°, $\alpha_\lambda(\Omega)$ actually approaches zero (Duffie and Beckman, 1974). According to (3.16), the emittance ε equals the absorptance (both are properties of the surface in the gray-surface approximation), and high absorptance thus implies high emittance for all wavelengths, including those characterizing the thermal emission from the absorber of temperature T_c. In order to reduce this loss, use is made of surface treatments causing the emittance to assume two distinct values, a high one for the short wavelengths of the solar spectrum (implying a high solar absorptance α^{sw}) and a low one (α^{lw}) for the longer wavelengths characterizing the emission from typical absorber temperatures, assumed to have a spectrum approximately equal to the blackbody spectrum for that temperature. Such surfaces are called *selective surfaces*, and their wavelength-dependent absorptance/emittance may resemble that shown in Fig. 4.80, exhibiting regions that can be characterized by $\alpha^{sw} \approx 0.95$ and $\alpha^{lw} \approx 0.1$. Further discussion of selective surface technologies may be found in Meinel and Meinel (1976).

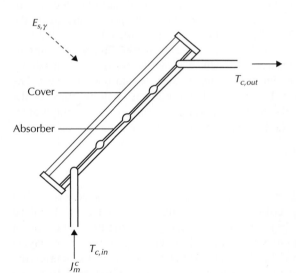

FIGURE 4.79 Example of a flat-plate solar collector. The performance may not be given simply by the net incident radiation flux $E_{s,\gamma}$ (s is tilt angle and γ is azimuth angle), because the transmission–absorption product of the cover system may be different for different components of the incident radiation.

As indicated in Fig. 4.79, the absorber may be covered by material that reduces heat losses and at the same time transmits most of the incoming radiation. The cover may be, for example, one or more layers of glass. The transmittance of glass depends on the type of glass and in particular on its minority constituents, such as Fe_2O_3. Figure 4.81 gives an example of the wavelength dependence of the transmittance, τ_λ, through ordinary window glass, defined as the fraction of incident light that is neither reflected [cf. (3.12)] nor absorbed,

$$\tau_\lambda = 1 - \rho_\lambda - \alpha_\lambda. \tag{4.111}$$

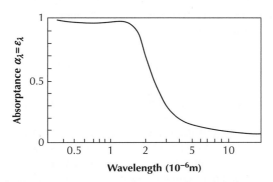

FIGURE 4.80 Spectral absorptance (or emittance) of a commercial "black-chrome" type of selective surface. (Based on Masterson and Seraphin, 1975.)

FIGURE 4.81 Spectral transmittance, $(1 - \rho_\lambda - \alpha_\lambda)$, of 2.8×10^{-3} m thick glass layer with a Fe_2O_3 content of 0.1. (Based on Dietz, 1954.)

For a collector of the type considered, multiple reflections may take place between the absorber and the cover system as well as within the cover system layers, if there is more than one. The reflections are mostly specular (angle of exit equal to angle of incidence) at glass surfaces, but mostly diffuse (hemispherically isotropic) at the absorber surface. Thus, the amount of radiation absorbed by the absorber plate, in general, cannot be calculated from knowledge of the total incoming flux but must be calculated for each incident direction from which direct, scattered, or reflected light is received, as well as for each wavelength. The total absorbed flux is then obtained by integration.

In many cases, a sufficiently good approximation can be obtained by considering only the two wavelength intervals denoted *short* and *long* wavelengths above (the dividing wavelength being about 3×10^{-6} m), in view of the near-constancy of the absorptance and transmittance of the relevant materials (cf. Figs. 4.80 and 4.81). The net short-wavelength energy gain by the collector may then be written

$$E_c^{sw} = A \int E_{c+}^{sw}(\Omega) \, P^{t.a.}(\Omega) \, d\Omega, \qquad (4.112)$$

in terms of the incoming flux from a given direction Ω, $E_{c+}^{sw}(\Omega)$, and the *transmission–absorption product*, $P^{t.a.}(\Omega)$, describing the fraction of incident short-wavelength radiation from the direction Ω that gets transmitted through the cover system and gets absorbed by the absorber plate, the area of which is denoted A. It has been assumed that A serves as a simple proportionality constant, implying that incident radiation as well as transmission–absorption properties are uniform over the entire collector area. For a cover system consisting of N layers, e.g., of glass, characterized by a refraction index relative to air, n (~1.5 for glass), and an extinction coefficient, x, such that the reduction in intensity of radiation from traversing a distance d through the material is $\exp(-xd)$, the total thickness of each cover layer being L, $P^{t.a.}(\Omega)$ may be approximately written (see, for example, Duffie and Beckman, 1974)

$$P^{t.a.}(\Omega) = \frac{1-\rho}{1+(2N-1)\rho} e^{-xNL\cos\theta} \frac{\alpha^{sw}}{1-(1-\alpha^{sw})\rho_d(N)}, \qquad (4.113)$$

where θ is the polar angle between the incident direction Ω and the normal to the collector surface, and ρ is the reflectance of one cover layer, given by the Fresnel formula in the case of unpolarized light,

$$\rho = \frac{1}{2} \left(\frac{\sin^2(\theta'-\theta)}{\sin^2(\theta'+\theta)} + \frac{\tan^2(\theta'-\theta)}{\tan^2(\theta'+\theta)} \right),$$

with the polar angle of refraction, θ', given by

$$\theta' = \text{Arc} \sin \left(\frac{\sin\theta}{n} \right).$$

The factorization of (4.113) into a factor describing the multiple reflection through the N-layer cover system in the absence of absorption, times a factor describing the attenuation of the intensity from passage through glass, and a final factor describing the absorption in the absorber plate, after multiple reflections back and forth between the plate and cover system, is a valid approximation for most practical applications, owing to the smallness of the extinction product xL. The final factor in (4.113) may be rewritten as

$$\frac{\alpha^{sw}}{1-(1-\alpha^{sw})\rho_d} = \alpha^{sw}\sum_{i=1}^{\infty}((1-\alpha^{sw})\rho_d)^i,$$

revealing that it is the sum of terms corresponding to i reflections back and forth between the plate and cover (considering only the lower cover surface), and assuming the non-absorbed fraction, $(1-\alpha^{sw})$, from each impact on the absorber plate to be diffusely reflected from the cover inside surface with a reflectance ρ_d different from that of specular reflection, ρ. In order to make up for processes in which light reflected back from the absorber gets transmitted to and reflected from cover layers other than the lowest one, and then eventually reaches back to become absorbed by the absorber plate, effective values of ρ_d depending on N may be used, as indicated in (4.113). Duffie and Beckman (1974) suggest that the diffuse reflectance ρ_d may be approximated by the specular reflectance for an incident polar angle of 60°, using values of 0.16, 0.24, 0.29, and 0.32 for $N = 1, 2, 3$, and 4.

For the direct part (3.1) of the incident radiation flux (3.15), only one direction Ω is permitted [given by (3.2)], and the integration in (4.113) should be left out. For specular reflection on to the collector, the same is true in simple cases (cf. Seitel, 1975), but for complicated geometry of the specularly reflecting objects in the surroundings, more than one direction may have to be taken into account. In general cases of reflection on to the collector, as well as for scattered radiation from the atmosphere [see (3.6)], the integration in (4.112) would have to be kept, but approximations, such as multiplying the total scattered *or* reflected flux at the outside collector surface, (3.6) or (3.8) and (3.11) or (3.14), by the transmission–absorption product for an "average" angle of incidence, such as $\theta = 60°$, are sometimes used, permitting the total short-wavelength gain by the collector plate to be written

$$E_c^{sw} = A(D_{s,\gamma}P^{t.a.}(\theta) + (d_{s,\gamma}+R_{s,\gamma})<P^{t.a.}>), \tag{4.114}$$

e.g., with $<P^{t.a.}> = P^{t.a.}(\theta = 60°)$.

The net long-wavelength energy gain by the collector is the difference between the long-wavelength radiation received from the surroundings and the long-wavelength radiation emitted by the collector system. Considering the temperature T_c of the absorber plate to be constant, the thermal radiation from the plate is of the form (3.17), and the net gain of an absorber without cover would be

$$E_c^{lw}(N = 0) = A\varepsilon^{lw}\sigma(T_e^4 - T_c^4), \tag{4.115}$$

where the temperature T_e of the environment, in general, is lower than the ambient air temperature, T_a, as discussed in section 3.1.4.

Finally, sensible heat may be exchanged between the collector and its surroundings through conduction and convection processes. The back and edge heat transfers are usually expressed in the same way as heat losses from buildings,

$$E_{back}^{sens} = -A\, U_{back}(T_c - T_b), \tag{4.116}$$

where U_{back} is a constant depending on the insulating properties of the materials separating the back of the absorber plate from the environment of temperature T_b, which may be equal to the ambient temperature, T_a, if the collector is mounted freely, or equal to the indoor temperature, T_L, if the collector forms an integral part of a building wall or roof. U_{back} may be assumed to include the "edge" heat losses from the sides of the collector, mainly of conduction type.

The net exchange of sensible energy at the front of the collector depends on the number of cover layers and on the wind speed, V, since the convection of heat is largely limited by the rate at which warm air at the top of the front cover is removed. In the absence of cover, the front (or top) exchange of heat may be written

$$E_{back}^{sens}(N = 0) = -A f_1 (T_c - T_a), \tag{4.117}$$

where f_1 is a polynomial in V with empirical coefficients, as suggested by Duffie and Beckman (1991),

$$f_1 = max\,(5.0;\ 3.95\, V[\text{ms}^{-1}]^{0.6})\,[\text{W K}^{-1}\text{m}^{-2}]. \tag{4.118}$$

With glass covers, expressions (4.115) and (4.117) must be modified, for example, by explicitly considering the heat transfer equations connecting each layer and solving for steady-state solutions with constant (but different) temperatures of each layer. Klein (1975) has parametrized these solutions in an explicit form, given by

$$E_c^{lw} = \frac{A\sigma(T_e^4 - T_c^4)}{(\varepsilon_p^{lw} + c_1 N(1 - \varepsilon_p^{lw}))^{-1} + N\,(2 + (f_2 - 1)/N) - N)/\varepsilon_g^{lw}}, \tag{4.119}$$

where ε_p^{lw} and ε_g^{lw} are the long-wavelength emittances of the absorber plate and of the cover glasses, $c_1 = 0.05$, and

$$f_2 = (1 + 0.089 f_1 - 0.1166 f_1\, \varepsilon_p^{lw})\,(1 + 0.07866 N), \tag{4.120}$$

with f_1 inserted from (4.118) in the units indicated (Duffie and Beckman, 1991). The relation replacing (4.117) is

$$E_{top}^{sens} = -A(T_c - T_a)\cdot \left\{ \left(\frac{N T_c}{f_3} f_4^{-c_2} + f_1^{-1}\right)^{-1} + \sigma(T_c + T_e)(T_c^2 + T_e^2)\right.$$

$$\left. \left((\varepsilon_p^{lw} + 0.00591 N f_1)^{-1} + \frac{2N + f_2 - 1 + 0.133 \varepsilon_p^{lw}}{\varepsilon_g^{lw}} - N\right)^{-1}\right\}. \tag{4.121}$$

Here $c_2 = 0.43(1 - 100/T_c)$ and

$$f_3 = 520 \, [\text{K}] \, (1 - 0.000051s^2), \tag{4.122}$$

with s being the tilt angle of the collector or 70°, whichever is smaller. Furthermore,

$$f_4 = (T_c - T_a)/(N + f_2), \tag{4.123}$$

unless this expression becomes smaller than 1, in which case f_4 should be equal to 1. All temperatures in (4.119) and (4.121) should be inserted in K. The parametrizations are clearly phenomenological, and substantial changes between 1974 and 1991 editions of the Duffie and Beckman book, notably affecting winter results, emphasize the *ad hoc* nature of the procedure.

The total net energy gain of the absorber plate of the collector is the sum of the contributions specified above,

$$E_c^{gain} = E_c^{sw} + E_c^{lw} + E_{top}^{sens} + E_{back}^{sens}. \tag{4.124}$$

Stalled and Operating Collector

If the mass flow $J_m^c = \mathrm{d}m/\mathrm{d}t$ through the solar collector is zero, the entire energy gain (4.124) will be used to raise the temperature of the collector. This situation, a *stalled collector*, is of considerable interest for determining the maximum temperature that the materials used in constructing the collector must be able to withstand, for example, in a situation of pump failure or complete loss of circulating fluid. Denoting the heat capacity of the "dry" collector C', taken

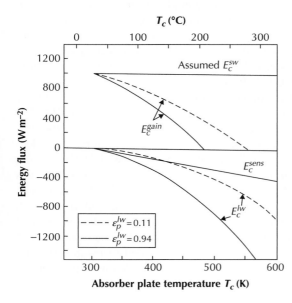

FIGURE 4.82 Energy flux components for a stalled collector, as a function of plate temperature T_c for a selective and a nonselective absorber surface (*dashed* and *solid lines*). There are two cover glasses of emittance $\varepsilon_g^{lw} = 0.88$, there is no wind (the curves are sensitive to wind speed changes), and all environmental temperatures have been taken as 300 K. The plate absorptance is $\alpha^{sw} = 0.94$, and the collector is placed horizontally. The curve E_c^{sens} represents the sum of fluxes reaching the top and back of the collector (both positive toward the collector).

per unit area of collector, the equation determining the time development of the stalled-collector plate temperature becomes

$$AC' \, dT_c/dt = E_c^{sens}(T_c). \tag{4.125}$$

As T_c rises, the negative terms in (4.124) increase, and if the incident energy can be regarded as constant, an equilibrium situation will result, in which the heat losses exactly balance the gains and the temperature thus remains constant,

$$E_c^{gain}(T_{c,max}) = 0. \tag{4.126}$$

FIGURE 4.83 Plate temperature for a stalled, dry collector as a function of time. The incident flux is constant $E_c^{sw} = 1000$ W m^{-2}, and the heat capacity C' in (4.93) has been taken as 10^4 J m^{-2} K^{-1}, corresponding to a fairly lightweight construction. Other parameters are as in Fig. 4.82. About 7 min are required to reach operating temperatures (some 50 K above ambient). The time scale is proportional to C', so results for other values of C' are easily inferred.

The determination of $T_{c,max}$ and the magnitude of the individual terms in (4.124), as a function of T_c, are illustrated in Fig. 4.82 for a high incoming-flux E_c^{sw}, environmental temperatures $T_e = T_a = T_L = 300$ K, and no wind, based on the explicit expressions (4.116)–(4.124). Two collectors are considered, one with a selective surface ($\varepsilon_p^{lw} = 0.11$) and one with nonselective black surface ($\varepsilon_p^{lw} = \alpha^{sw} = 0.94$). Both have two layers of cover glass. The corresponding equilibrium temperatures are approximately $T_{c,max} = 550$ K and $T_{c,max} = 480$ K. If wind were present, the convective heat loss (4.121) would be much higher and the maximum temperature correspondingly lower.

The time required to reach a temperature T_c close to $T_{c,max}$ can now be calculated from (4.105) by integration, as illustrated in Fig. 4.83. While the maximum temperature is independent of the heat capacity of the collector, the time scales linearly with C'. The value $C' = 10^4$ J m^{-2} K^{-1} used in Fig. 4.83 corresponds to a fairly light collector (e.g., absorber plate of extruded aluminum), in which case the asymptotic temperature region is reached after about 1 h of constant high incoming radiation. The shortness of this period makes the assumption of a constant E_c^{sw} acceptable. The amount of time needed to reach operating temperatures of about 50 °C is roughly 7 min. It would double if the ambient temperature were at the freezing point of water and would further increase if the radiation were lower than the 1000 W m^{-2} assumed in the example used in Figs. 4.82 and 4.83. The short response time of the collector is also relevant in situations of variable radiation (e.g., caused by frequent cloud passage).

In the case of an *operating collector*, J_m^c is no longer zero and a fluid carries energy away from the absorber (cf. Fig. 4.79). The temperature of the outgoing fluid, $T_{c,out}$, will in this case be higher than that of the incoming fluid, $T_{c,in}$, and it is no longer consistent to assume that the plate temperature T_c is uniform over the plate. The change in plate temperature must be at least as large as the change in fluid temperature, between the points where the fluid is receiving its first and last heat transfer contribution from the absorber plate. Because of the nonlinear relation between plate temperature and heat loss, it is not correct to evaluate the losses at the mean plate temperature, T_c. Still, this is often done in approximate calculations, choosing

$$\overline{T}_c = \frac{1}{2}(T_{c,in} + T_{c,out}),\tag{4.127}$$

and assuming further that the average fluid temperature is also the average plate temperature. Incomplete heat transfer between absorber plate and fluid channels, as well as the nonlinearity of loss functions, may be simulated by adding a constant term to (4.127), or by using a factor larger than 1/2. The transfer of heat to the fluid may be effectuated by passing the fluid along the absorber plate (in most collectors, air is used as the working fluid) or by the fluid's passage through pipes in highly heat-conducting contact with the absorber plate (most often when the working fluid is water, oil, etc.). In the latter case, the non-uniform heat extraction constitutes a further cause of non-uniform plate temperature (the temperature gradients toward the pipe locations being, of course, the reason for obtaining a heat flow from any plate position to the relatively small area of contact with the fluid-carrying pipes).

If a fluid of fixed inlet temperature $T_{c,in}$ is allowed to pass only once through the collector (e.g., for hot water production), and if the simplifying assumptions mentioned above are made, then the equation replacing (4.105) for an operating collector will be

$$\overline{E}_c^{gain}(T_c) = AC'd\overline{T}_c/dt + J_m^c C_p^c(T_{c,out} - T_{c,in}),\tag{4.128}$$

or, in a steady-state situation, just

$$E_c^{gain}(\overline{T}_c) = J_m^c C_p^c(T_{c,out} - T_{c,in}).$$

C_p^c is the heat capacity of the fluid flowing through the collector. This equation should determine $T_{c,out}$, but the left-hand side also depends on $T_{c,out}$, through (4.127) or some generalization of this relation between the flow inlet and outlet temperatures and the effective average plate temperature. Therefore, it is conveniently solved by iteration,

$$T_{c,out}^{(i+1)} = T_{c,in} + E_c^{gain}(\frac{1}{2}(T_{c,in} + T_{c,out}^{(i)}))/(J_m^c C_p^c).\tag{4.129}$$

A better approximation for the transfer of heat from the absorber plate to the fluid channels may be obtained along the lines described below for the general problem of heat exchange in solar heat-supply systems, such as in cases involving heat storage that is not part of the collector flow circuit.

Flat-plate Collector with Heat Storage

The general layout of a flat-plate solar collector with heat storage is shown in Fig. 4.84. There are two heat exchangers on the collector side and two on the load side. If the actual number is less, for example, if the storage is in terms of a fluid that can be circulated directly through the collector, then the corresponding heat transfer coefficient [h in (4.37)] should be taken as infinitely large. While the relationship between temperatures at every heat exchanger may be taken from (4.37), the net energy transfer to storage is associated with a change in the storage temperature (the average value of which is $\overline{T}_s$), which may be obtained from the equation accounting for the relevant energy fluxes, assuming the storage temperature to be uniform,

$$SC^s \, dT_s/dt = J_m^c C_p^c (T_{c,out} - T_{c,in}) - J_m^L C_p^L (T_{L,in} - T_{L,out}) - h_s(\overline{T}_s - T_0).$$
(4.130)

Here S is the storage volume (or mass), C^s is the heat capacity on a volume (or mass) basis of the material used for storage (assuming for simplicity storage in the form of sensible heat), superscripts c and L distinguish the flows in the collector and the load circuits, and finally, T_0 is the temperature of the surroundings of the storage volume (e.g., soil in underground storage), to which the storage is losing energy through its walls. Expression (4.130) must be supplemented by (4.128) and a similar equation for the load area, e.g., of the form

$$LC^L \, dT_L/dt = J_m^L C_p^L (T_{L,in} - T_{L,out}) - h_L(\overline{T}_L - T_0),$$
(4.131)

assuming that the load area is a building of volume (or mass) L and average temperature $\overline{T}_L$, which may lose energy to surroundings of ambient air temperature T_a. If the load is in the form of water heating, it may be described by a term of the same form as the first one on the right-hand side of (4.131), but with $T_{L,out}$ being the water inlet temperature and $T_{L,in}$ being the water temperature of the water supplied by the system for usage.

Together with the four heat exchanger equations (4.37) with h_c, h_{cs}, h_{sL}, and h_L (see Fig. 4.21), (4.128), (4.130), and (4.131) constitute seven equations for determining the seven temperatures appearing in Fig. 4.84. If the load circuit is not activated, the four equations for $T_{c,in}$, $T_{c,out}$, $\overline{T}_s$, and $\overline{T}_c$ remain. Instead of the heat exchanger equation (4.37) for the collector, which for $h_c \to \infty$ gives $T_{c,out} = \overline{T}_c$ owing to the assumption that the reservoir of temperature $\overline{T}_c$ is large compared with the amounts of heat exchanged, relations such as (4.127) may be used, or a more elaborate calculation, taking into account the geometry of the collector and fluid pipes or channels, may be performed.

FIGURE 4.84 Solar heating system with storage. h denotes a heat exchanger, P denotes a circulation pump, and the Ts are temperatures.

The efficiency of the collector itself may be defined as

$$\eta_c = J_m^c C_p^c (T_{c,out} - T_{c,in})/E_{s,\gamma}^{sw}, \tag{4.132}$$

where the numerator may be obtained from (4.128) and the denominator from (3.15). Only if the heat capacity of the collector is neglected, or if the collector is operating at a constant temperature, can the collector heat output (to the fluid) be taken as the collector gain $E_c^{gain}(T_c)$ from (4.124) directly.

If an energy store is used in connection with the collector, the relevant efficiency is that of delivering the heat to the storage, which is formally the same as (4.132). In practice, however, the introduction of energy storage may completely alter the efficiency by changing $T_{c,in}$, which again causes a change in $T_{c,out}$. When the storage temperature, $\overline{T}_s$, is low, the average collector plate temperature $\overline{T}_c$ will be fairly low and the loss terms correspondingly small. Since $T_{c,in}$ will always be higher than T_s (but not always higher than $\overline{T}_s$, since often the fluid circuit is made to leave the storage at a place of low temperature, if T_s is not uniform), the accumulation of stored energy and associated increases in T_s and $T_{c,in}$ will cause the collector loss terms to increase, implying a diminishing efficiency. Examples of this are given in Chapter 6, in the discussion of simulations of complete solar heating systems.

Another important factor in determining the efficiency of energy collection is the rate of fluid circulation through the collector circuit, J_m^c. Figure 4.85 gives an example of the performance of the same system with two different values of J_m^c. With low fluid velocity, the exit temperature from the collector, $T_{c,out}$, is about 45 °C above the inlet temperature, but the energy transferred to storage is much less than in the case of a higher fluid velocity, causing a smaller difference between $T_{c,out}$ and $T_{c,in}$, but larger net collector gain, owing to the smaller loss terms. The optimum fluid flow rate for a given heat exchange capability h_{cs} (assuming the heat transfer at the collector to be nearly perfect) may be found by calculating the collection efficiency as a function of J_m^c or, as a more generally applicable criterion, by calculating the solar heat system's actual coverage of the actual energy demand. If a given solar heat system is considered, the percentage of the total heat load over an extended period, such as a year, may be evaluated as function of J_m^c, as done in Fig. 4.86.

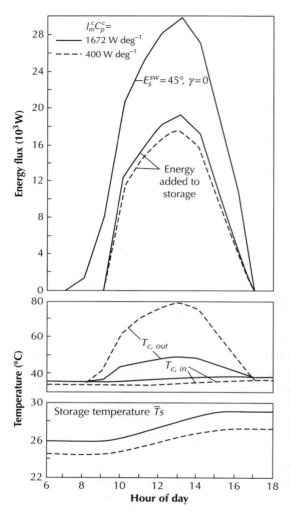

FIGURE 4.85 Single-day performance of a solar heating system, based on a simulation model using the Danish Reference Year (latitude 56 °N). The day is 8 February, the collector area $A = 40$ m^2, and the storage volume $S = 25$ m^3 (water equivalent). The collector has one glass cover layer and a selective surface ($\varepsilon_c^{lw} = 0.11$), and the heat transfer coefficient $h_{cs} = 4000$ W K^{-1}. Two different flow rates are considered for the collector circuit, $J_m^c C_p^c = 1672$ or 400 W K^{-1} corresponding to 0.4 or 0.096 kg of water per second.

If 100% coverage is aimed at, the parameter to be optimized may be taken as the minimum size (e.g., collector area A) that will allow the full load to be covered at any time. The example shown in Fig. 4.86 relates to the heating and hot water requirements of a one-family dwelling at latitude 56 °N. For each value of the heat exchange coefficient h_{cs} between the collector circuit and the storage, the optimum flow rate is determined. With increasing h_{cs}, the region of "acceptable" values of J_m^c becomes larger, but the increased J_m^c indicated for large h_{cs} does not significantly improve the performance of the system, and it will soon become uneconomical due to the energy spent in pumping the fluid through the circuit (the energy required for the pump is negligible for low fluid speeds, but the pipe resistance increases with fluid speed, and the resistance through the heat exchanger increases when h_{cs} does). The results of Fig. 4.86 may then be interpreted as suggesting a modest increase in fluid velocity, from about 0.1 to about 0.2 kg (or liter) s^{-1}, when the heat exchange coefficient is increased from 400 to 4000 W K^{-1} (for the particular system considered).

The optimization problem outlined in this example is typical of the approach to the concept of an "efficiency" for the total solar heat system, including storage and load sections. Because the time distribution of the load is, in general, barely correlated with that of solar energy collection (in fact, it is more "anti-correlated" with it), an instantaneous efficiency, defined as the ratio between the amount of energy delivered to the load area and the amount of incident solar energy, is meaningless. A meaningful efficiency may be defined as the corresponding ratio of average values, taken over a sufficient length of time to accommodate the gross periodicity of solar radiation and of load, i.e., normally a year:

$$\overline{\eta}_{system} = \overline{c}/\overline{E}_{s,\gamma}^{sw}, \qquad (4.133)$$

where $\overline{c}$ is the average load covered by the solar energy system and $\overline{E}_{s,\gamma}^{sw}$ is the average incident short-wavelength radiation on the surface of the collector (its cover, if there is one). Conventionally, the long-wavelength or other heat contributions to the incident energy flux are not included in the efficiency calculations (4.133) or (4.132).

Equivalent water flow in collector circuit (kg s⁻¹)

FIGURE 4.86 Annual average coverage of heating and hot water load for a one-family dwelling (average load 1.91 kW) under conditions set by the Danish Reference Year (56 °N) (cf. section 6.3.1) as a function of water flow in collector circuit and heat exchanger transfer coefficient. The collector is as described in the caption to Fig. 4.85 and placed on a south-facing roof with tilt angle $s = 45°$.

4.4.4 Concentrators and Solar-thermal Electricity Generators

Various advanced versions of the flat-plate collector could increase the amount of absorbed energy and decrease the amount of heat lost. The incident average flux may be increased by replacing the fixed installation (of tilt and azimuth angles s and γ) with a movable one, which can be made to track the direction toward the Sun. Fully tracking (i.e., in both height and azimuth angle) devices may be mounted in a manner similar to the one used for (star)tracking astronomical instruments, and incompletely tracking systems (e.g., only following the azimuth angle of the Sun) might combine a simpler design with some improvement in energy gain. An idea of the gain made possible by tracking the Sun can be obtained by comparing the power duration curves in Fig. 3.22. The scattered and (diffusely) rejected fluxes do not change much on average by going from a fixed to a tracking collector, but the direct part (3.1) is replaced by the normal incidence flux S_N. For a flat-plate collector, the maximum average gain from a fully tracking system (less than a factor 2) would rarely justify the extra cost, which, at least at present, exceeds the cost of doubling the collector area.

Other possible improvements to the flat-plate collector system include coating the cover layers to reduce reflection losses on the outside and increase reflection (back on to the absorber) on the inside surfaces. The mainly convective heat loss E_{top}^{sens} may be reduced by installation of shields that diminish the wind velocity outside the top cover layer. The strong dependence of convection losses on wind speed is clear from (4.121). The convective losses may be more radically diminished by evacuating the space between the absorber plate and the cover system. In vacuum, heat transport can take place only by radiation. Evacuation of the substantial space between absorber and cover typical of collector designs of the type shown in Fig. 4.79 is not practical, and systems using evacuation-type insulation would be built differently, e.g., with units of evacuated cylinders or spheres containing the absorbers and possibly the concentrating equipment.

Focusing, or (more generally) concentrating, devices constitute a separate class of solar collectors. They are necessary if very high temperatures are required, but may, in principle, also be considered for heating purposes involving modest temperature rises over the ambient. For a flat-plate collector without any of the special loss-reducing features mentioned above, the net gain is a rapidly decreasing function of the plate temperature, as seen in Fig. 4.82. Since the losses are proportional to the absorber area, there is a special interest in reducing this area relative to the collector area A determining the incident radiation energy.

One extreme is a point-focusing device, such as the parabolic reflectors in Fig. 4.87 or the lenses in Fig. 4.89. If the absorber is placed in the focus on the symmetry axis and its dimension corresponds to that of the image of the Sun when the Sun is in the direction of the symmetry axis (optical axis), then direct radiation will be accepted, but only as long as the optical axis is made to track the Sun. Imperfections and, in the case of lenses, the wavelength dependence of the refraction index may further enlarge the image dimensions.

FIGURE 4.87 Parabolic- and Fresnel-type reflectors, for use in heliostats and tower absorbers, respectively.

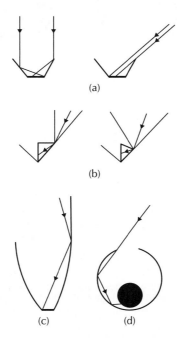

(a)

(b)

(c)   (d)

FIGURE 4.88a,b,c,d Examples of concentrating collectors.

The absorber dimension would have to be chosen so that it covered the bulk of the image under most conditions.

The total short-wavelength radiation received by the absorber, E_c^{sw} may be written in the same form as for the flat-plate collector, (4.112) or (4.114), but with a transmission–absorption product, $P^{t.a.}(\Omega)$, suitable for the focusing geometry rather than (4.113). If reflections and re-absorptions at the absorber surface are neglected, the transmission–absorption product for a focusing device may be written as the product of the absorptance α^{sw} of the absorber surface and a transmission function $t^{sw}(\Omega)$, which expresses the fraction of the incident short-wavelength radiation from the direction Ω that reaches the absorber surface (directly or after one or more reflections or refractions within the device),

$$P^{t.a.}(\Omega) = \alpha^{sw} t^{sw}(\Omega). \tag{4.134}$$

For the parabolic reflector (Fig. 4.87) and the lens (Fig. 4.89), the idealized transmission function is zero, unless the absorber is placed exactly in the focal point corresponding to incoming radiation from the direction Ω. Thus, either the collector has to fully track the Sun or the absorber has to be moved to the focal point corresponding to a given direction to the Sun. For scattered radiation, which by definition does not come from the direction of the Sun (the same being normally the case for reflected radiation), $t^{sw}(\Omega)$ is zero, and such radiation will not be available for these devices. As a compromise, the acceptance angle may be increased at the expense of a high concentration ratio. Some scattered and reflected radiation will be accepted, and complete tracking may not be necessary.

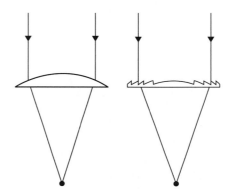

FIGURE 4.89   Simple Fresnel-type lenses.

Figure 4.88 gives a number of examples of such devices. The inclined "booster" mirrors along the rim of a flat-plate collector (Fig. 4.88a) result in a modest increase in concentration for incident angles not too far from the normal to the absorber plate, but a decrease (due to shadows) for larger incident angles. The V- or cone-shaped collector with the absorber in the shape of a wall or an upright line (cylinder) in the middle (Fig. 4.88b) has an acceptance angle that depends on the distance from the convergence point of the cone. At this point, the acceptance angle equals the opening angle, but, at points further out along the cone sides, the acceptance angle is smaller (depending on the length of the absorber line, or wall height) relative to that of the cone. A fairly high concentration can be achieved with a cusp construction (Fig. 4.88c), comprising two parabolic sections, each of which has its focus at the point of intersection between the opposite parabolic section and the absorber. Figure 4.88d shows a "trapping" device consisting of a cylinder (or sphere) with a window, and an absorber displaced from the center. The dimensions of the absorber are such that a large fraction of the radiation passing through the window will be absorbed after (multiple) reflections on the reflecting inside walls of the cylinder. Also, for the long-wavelength radiation emitted by the absorber, the reflecting walls will lead a substantial fraction back to the absorber, so that the loss-to-gain ratio is improved, although there is no actual concentration (if the dimension of the absorber is as large as that of the window).

Energy Collection from Focusing Systems

A survey of a number of optically focusing systems for solar energy collection may be found in Meinel and Meinel (1976).

In the absence of cover systems, the long-wavelength emission from a focusing system may be written in the form (4.83),

$$-E_c^{lw} = A_a \varepsilon^{lw} \sigma (T_c^4 - T_e^4), \tag{4.135}$$

where A_a is the absorber area.

Similarly, the sensible heat exchange with the surroundings of ambient temperature T_a may be given in analogy to (4.116) and (4.117),

$$E_c^{sens} = -A_a U (T_c - T_a), \tag{4.136}$$

where U may depend on wind speed in analogy to (4.118). The collector gain is now given by $E_c^{gain} = E_c^{sw} + E_c^{lw} + E_c^{sens}$, where the first term is proportional to A, while the two last terms are proportional to A_a. Assuming now a fluid flow J_m^c through the absorber, the expression for determination of the amount of energy extracted becomes [cf. (4.128)]

$$J_m^c C_p^c (T_{c,out} - T_{c,in}) = E_c^{gain}(\overline{T}_c) - A_a C' \mathrm{d}\overline{T}_c/\mathrm{d}t$$

$$= E_c^{sw} + E_c^{lw}(\overline{T}_c) + E_c^{sens}(\overline{T}_c) - A_a C' \, \mathrm{d}\overline{T}_c/\mathrm{d}t), \tag{4.137}$$

where $A_a C'$ is the heat capacity of the absorber and the relation between $\overline{T}_c$ and $T_{c,out}$ may be of the form (4.129) or (4.37). Inserting (4.114), with use of (4.134), (4.135), and (4.136), into (4.137), it is apparent that the main additional parameters specifying the focusing system are the area concentration ratio,

$$X = A/A_a, \tag{4.138}$$

and the energy flux concentration ratio,

$$C^{flux} = A t^{SW}(\Omega)/A_i, \tag{4.139}$$

where t^{SW} is given in (4.134) and A_i is the area of the actual image (which may be different from A_a). Knowledge of A_i is important for determining the proper size of the absorber, but it does not appear in the temperature equation (4.137), except indirectly through the calculation of the transmission function t^{SW}.

For a stalled collector, $J_m^c = 0$ and $\mathrm{d}\overline{T}_c/\mathrm{d}t = 0$ give the maximum absorber temperature $T_{c,max}$. If this temperature is high, and the wind speed is low, the convective heat loss (4.138) is negligible in comparison with (4.135), yielding

$$\sigma (T_{c,max}^4 - T_c^4) = X D \alpha^{sw} \varepsilon^{lw},$$

where D is the incident short-wavelength flux in the direction of the Sun [cf. (4.114)], assuming the device accepts only the flux from this direction and assuming perfect transmission to the absorber ($t^{SW} = 1$). With the same conditions, the best performance of an operating collector is obtained from (4.139), assuming a steady-state situation. The left-hand side of (4.139) is the amount of energy extracted from the collector per unit of time, E^{extr}:

$$E^{extr} = A\alpha^{SW}D - A_a\varepsilon^{lw}\sigma(\overline{T}_c^4 - T_c^4), \qquad (4.140)$$

and the efficiency corresponding to the idealized assumptions

$$\eta_c^{ideal} = \frac{E^{extr}}{AD} = \alpha^{SW}\left(1 - \frac{\varepsilon^{lw}\sigma}{X\alpha^{SW}D}(\overline{T}_c^4 - T_e^4)\right). \qquad (4.141)$$

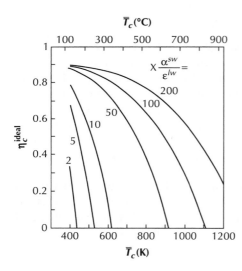

FIGURE 4.90 Ideal collector efficiency for concentrating collectors, evaluated for a direct radiation incident flux of $D = 800$ W m^{-2}, an environmental temperature $T_e = 300$ K, and $\alpha^{SW} = 0.9$.

This relation is illustrated in Fig. 4.90, assuming $D = 800$ W m^{-2}, $T_e = 300$ K, and $\alpha^{SW} = 0.9$, independent of absorber temperature. It is clear that the use of a selective absorber surface (large ratio $\alpha^{SW}/\varepsilon^{lw}$) allows a higher efficiency to be obtained at the same temperature and that the efficiency drops rapidly with decreasing light intensity D. If the total short-wavelength flux on the collector is substantially different from the direct component, the actual efficiency relative to the incident flux $E_{s,\gamma}^{SW}$ of (3.15),

$$\eta_c = E^{extr}/E_{s,\gamma}^{SW},$$

may be far less than the "ideal" efficiency (4.141).

Solar-thermal Electricity Generation

Instead of using photovoltaic or photo-electrochemical devices, one can convert solar radiation into electric energy in two steps: first, by converting radiation to heat, as described in section 4.4.3, or more likely by use of concentrating devices, as described above; and second, converting heat into electricity, using one of the methods described in sections 4.1.2 and 4.2.2. (Two-step conversion using chemical energy rather than heat as the intermediate energy form is also possible, for instance, by means of the photogalvanic conversion scheme briefly mentioned in section 4.5.2.) Examples of the layout of systems capable of performing solar heat to electric power conversion using thermodynamic cycles are given below, with discussion of efficiencies believed attainable.

Photo-thermoelectric Converter Design

A two-step conversion device may be of the general form shown in Fig. 4.91. The solar collector may be the flat-plate type or it may perform a certain measure of concentration, requiring partial or full tracking of the Sun. The second step is performed by a thermodynamic engine cycle, for example, a Rankine cycle with expansion through a turbine, as indicated in the figure. Owing to the vaporization of the engine-cycle working fluid in the heat exchanger linking the two conversion steps, the heat exchange performance does not follow the simple description leading to (4.37). The temperature behavior in the heat exchanger is more like the type shown in Fig. 4.92. The fluid in the collector circuit experiences a uniform temperature decrease as a function of the distance traveled through the heat exchanger, from x_1 to x_2. The working fluid entering the heat exchanger at x_2 is heated to the boiling point, after which further heat exchange is used to evaporate the working fluid (and perhaps superheat the gas), so that the temperature curve becomes nearly flat after a certain point.

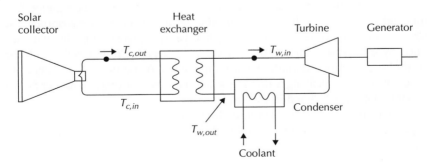

FIGURE 4.91 Photo-thermoelectric generator based on concentrating solar collectors.

The thermodynamic engine cycle (the right-hand side of Fig. 4.91) can be described by the method outlined in sections 4.1.1 and 4.1.2, with the efficiency of heat to electricity conversion given by (4.19) [limited by the ideal Carnot process efficiency (4.4)],

$$\eta_w = E^{electric}/J^w_{Q,in} \leq (T_{w,in} - T_{w,out})/T_{w,in},$$

where $E^{electric} = -J_q F_q$ is the electric power output [cf. (4.14)] and

$$J^w_{Q,in} = J^c_m C^c_p (T_{c,out} - T_{c,in})$$

is the rate of heat transfer from the collector circuit to the working-fluid circuit in the heat exchanger. The right-hand side is given by (4.137) for a concentrating solar collector and by (4.128) and (4.124) for a flat-plate collector. The overall conversion efficiency is the product of the efficiency η_c of the collector system and η_w,

$$\eta = \eta_c \eta_w.$$

Determination of the four temperatures $T_{c,in}$, $T_{c,out}$, $T_{w,in}$, and $T_{w,out}$ (cf. Fig. 4.91) requires an equation for collector performance, for heat transfer to the collector fluid, for heat transfer in the heat exchanger, and for processes involving the working fluid of the thermodynamic cycle. If collector performance is assumed to depend only on an average collector temperature, $\overline{T}_c$ [e.g., given by (4.127)], and if $T_{w,in}$ is considered to be roughly equal to $\overline{T}_c$ (Fig. 4.92 indicates that this may be a fair first-order approximation), then the upper limit for η_w is approximately

$$\eta_w \lesssim (\overline{T}_c - T_{w,out})/\overline{T}_c.$$

Taking $T_{w,out}$ as 300 K (determined by the coolant flow through the condenser in Fig. 4.91), and replacing η_c by the idealized value (4.141), the overall efficiency may be estimated as shown in Fig. 4.93, as a function of $\overline{T}_c$. The assumptions for η_c^{ideal} are as in Fig. 4.90: no convective losses (this may be important for flat-plate collectors or collectors with a modest concentration factor) and an incident radiation flux of 800 W m^{-2} reaching the absorber (for strongly focusing collectors this also has to be direct radiation). This is multiplied by η_w^{ideal} (the Carnot limit) to yield η_{total}^{ideal} depicted in Fig. 4.93, with the further assumptions regarding the temperature averaging and temperature drops in the heat exchanger that allow the introduction of $\overline{T}_c$ as the principal variable both in η_w^{ideal} and in η_c^{ideal}.

In realistic cases, the radiation reaching the absorber of a focusing collector is perhaps half of the total incident flux $E^{SW}_{s,\gamma}$ and η_w is maybe 60% of the Carnot value, i.e.,

$$\eta_{total} \approx 0.3 \, \eta_{total}^{ideal}.$$

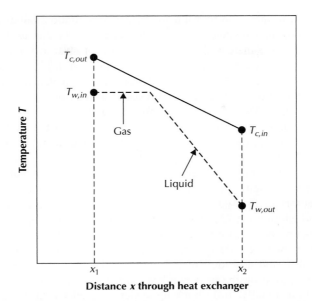

FIGURE 4.92 Temperature behavior in heat exchanger for two-phase working fluid of temperature T_w and fluid in the collector circuit of temperature T_c. (Based on Athey, 1976.)

FIGURE 4.93 Ideal overall efficiency of photo-thermoelectric converter, based on the assumptions underlying Figs. 4.90 and 4.92 and with the Rankine cycle efficiency replaced by an ideal Carnot cycle efficiency.

This estimate may also be valid for small concentration values (or, rather, small values of the parameter $X\alpha^{sw}/\varepsilon^{lw}$ characterizing the curves in Figs. 4.93 and 4.90), since the increased fraction of $E_{s,\gamma}^{sw}$ being absorbed is compensated for by high convective heat losses. Thus, the values of η_{total}^{ideal} between 6% and 48%, obtained in Fig. 4.93 for suitable choices of $\overline{T}_c$ (which can be adjusted by altering the fluid flow rate J_m^c), may correspond to realistic efficiencies of 2%–15%, for values of $X\alpha^{sw}/\varepsilon^{lw}$ increasing from 1 to 200.

The shape of the curves in Fig. 4.93 is brought about by the increase in η_w^{ideal} as a function of $\overline{T}_c$, counteracted by the accelerated decrease in η_w^{ideal} as a function of $\overline{T}_c$, which is seen in Fig. 4.90.

Photo-thermoelectric conversion systems based on flat-plate collectors have been discussed, for example, by Athey (1976); systems based on solar collectors of the type shown in Fig. 4.88d have been considered by Meinel and Meinel (1972). They estimate that a high temperature on the absorber can be obtained by use of selective surfaces, evacuated tubes (e.g., made of glass with a reflecting inner surface except for the window shown in Fig. 4.88d), and a crude Fresnel lens (see Fig. 4.89) concentrating the incoming radiation on the tube window. Molten sodium is suggested as the collector fluid. Finally, fully tracking systems based on the concept shown on the right-hand side of Fig. 4.87 ("power towers") have been proposed by Teplyakov and Aparisi (1976) and by Hildebrandt and Vant-Hull (1977). This concept is used in several installations, the best known of which is the Odeillo plant in southern France (Trombe and Vinh, 1973), which originally aimed at achieving temperatures for metalwork that exceeded those possible at fossil-fuel furnaces.

Optical Subsystem and Concentrators

As indicated by the devices shown in Figs. 4.57 and 4.60, optical manipulation of incoming solar radiation is used in non-concentrating solar cells. These measures serve the purpose of reducing the reflection on the surface to below a few percent, over the wavelength interval of interest for direct and scattered solar radiation, as seen from Fig. 4.94. Their collection efficiency is far better than that shown on Fig. 4.54, exceeding 90% for wavelengths between 0.4 and 1.05 μm, and exceeding 95% in the interval 0.65–1.03 μm.

For large-factor concentration of light onto a photovoltaic cell much smaller than the aperture of the concentrator, the principles mentioned for thermal systems (discussion of Fig. 4.90) apply unchanged. Most of these concentrators are designed to focus the light onto a very small area, and thus tracking the Sun (in two directions) is essential, with the implications that scattered radiation largely cannot be used and that the expense and maintenance of nonstationary components have to be accepted.

One may think that abandoning very high concentration would allow the construction of concentrators capable of accepting light from most directions (including scattered light). However, this is not easy. Devices that accept all

FIGURE 4.94 Front reflection for cell of the type shown in Fig. 4.57 (single-layer anti-reflection coating) and the same with double-layer coating (Zhao *et al.*, 1995).

Fate of unit incoming radiation flux in concentrator

Absorber: Flat, horizontal.
Cover: Inverted pyramids,
Refraction index $n=1.6$

Angle of inverted pyramids:

Concentrator shape:
Catenary $A=1$, $B=0.4$.
Ratio between aperture and absorber surface (concentration factor) $X=7$.

The total radiation on the concentrator aperture is assumed to vary as cosine to the incident angle.

FIGURE 4.95 Ratio of ray intensity on the absorber surface with and without concentrator (times cosine of incident angle, in order to reflect the different intensity of rays from different directions) for the $X = 7$ concentrator with a refracting, inverted pyramid cover, as illustrated in Fig. 4.96.

angles have a concentration factor of unity (no concentration), and even if the acceptance angular interval is diminished to, say, 0°–60°, which would be suitable because light from directions with a larger angle is reduced by the incident cosine factor, only a very small concentration can be obtained (examples like the design by Trombe are discussed in Meinel and Meinel, 1976).

The difficulty may be illustrated by a simple two-dimensional ray-tracing model of an (arbitrarily shaped) absorber, i.e., the PV cell, sitting inside an arbitrarily shaped concentrator (trough) with reflecting inner sides, and possibly with a cover glass at the top that could accommodate some texturing or lens structure.

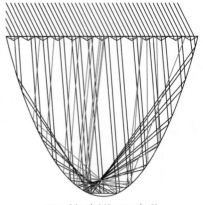

ncover=1.6 ngel=1 13 pyrm.angle=30
catenary A=1, B=0.4 conc.X=7
inc.angle=0 conc.obtained=0.14

ncover=1.6 ngel=1 13 pyrm.angle=30
catenary A=1, B=0.4 conc.X=7
inc.angle=25 conc.obtained=1.93

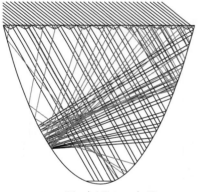

ncover=1.6 ngel=1 13 pyrm.angle=30
catenary A=1, B=0.4 conc.X=7
inc.angle=50 conc.obtained=0.59

FIGURE 4.96 Two-dimensional ray-tracing through a concentrator with glass cover textured with 30° inverted pyramids, for incident angles 0°, 25°, and 50°. Only for incident angles around 25° does the absorber at the bottom receive more light than would have hit it if there had been no concentrator at all.

The one component that unfortunately is not available is a semitransparent cover that fully transmits solar radiation from one side and fully reflects the same wavelengths of radiation hitting the other side. With such a cover, nearly 100% of the radiation coming from any direction could reach the absorber, the only exception being rays being cyclically reflected into paths never reaching the absorber.

Figure 4.96 shows some ray-tracing experiments for a catenary-shaped concentrator holding a flat absorber at the bottom, with a concentration factor $X = 7$ (ratio of absorber and aperture areas, here lines). The cover has an inverted pyramid shape. This is routinely used for flat-plate devices (see Fig. 4.57), with the purpose of extending the path of the light rays inside the semiconductor material so that absorption is more likely. For the concentrator, the purpose is only to make it

(a)

(b)

FIGURE 4.97a,b,c Intensity of rays on the absorber surface relative to that experienced in the absence of the concentrator (times cosine of incident angle, in order to reflect the different intensity of rays from different directions) for different depths of the $X = 5$ concentrator without cover, but with a refracting inside material having pyramid texture on top, as illustrated in Fig. 4.99.

FIGURE 4.97c   See preceding page for details.

more likely that some of the rays reach the absorber, by changing their direction from one that would not have led to impact onto the absorber. However, at the same time, some rays that otherwise would have reached the absorber may now become diverted. Thus, not all incident angles benefit from the concentrator, and it has to be decided whether the desired acceptance interval is broad (say, 0°–60°) or can be narrowed, e.g., to 0°–30°. As illustrated, the concentration of rays toward the center absorber is paid for by up to half of the incident rays' being reflected upward after suffering total reflection on the lower inside of the cover glass.

The result of varying some of the parameters is shown in Fig. 4.95. An inverted pyramid angle of 0° corresponds to a flat-plate glass cover, and the figure shows that in this situation the concentration is limited to incident angles below 20°, going from a factor of four to unity. With increasing pyramid angle, the maximum concentration is moved toward higher incident angles, but its absolute value declines. The average concentration factor over the ranges of incident angles occurring in practice for a fixed collector is not even above unity, meaning that a PV cell without the concentrating device would have performed better.

Another idea is to direct the cover pyramids upward, but to avoid "stretching out" the rays again at the lower glass boundary by having the entire trough filled with the refractive substance (which might be a gel or plastic material, textured at the upper surface). This is illustrated in Figs. 4.97a–c for an X = 5 concentrator. For a deep catenary-shaped collector, all rays up to 10° may be led to the absorber, but in this case (Figs. 4.97a and 4.98), incident angles above 25° are not accepted. To get reasonable acceptance, pyramid angles around 30° would again have to be chosen, and Figs. 4.97a–c show that the concentration penalty in trying to extend the acceptance interval to angles above 35° (by decreasing the depth of the concentrator, see Fig. 4.99) is severe. The best choice is still the deep

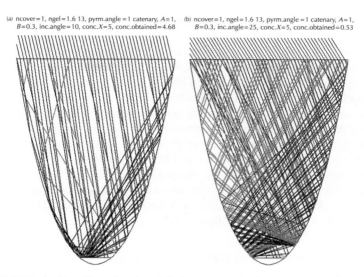

(a) ncover=1, ngel=1.6 13, pyrm.angle =1 catenary, A=1, B=0.3, inc.angle=10, conc.X=5, conc.obtained=4.68

(b) ncover=1, ngel=1.6 13, pyrm.angle =1 catenary, A=1, B=0.3, inc.angle=25, conc.X=5, conc.obtained=0.53

FIGURE 4.98a,b Two-dimensional model of concentrator without cover but filled with a refracting material with slight pyramid texture at top. Incident angles are 10° and 25°.

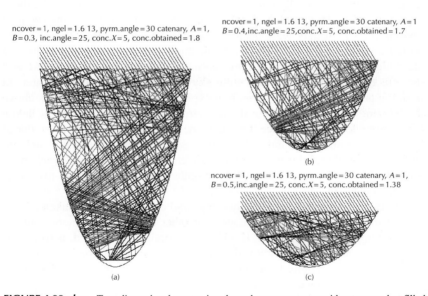

ncover=1, ngel=1.6 13, pyrm.angle=30 catenary, A=1, B=0.3, inc.angle=25, conc.X=5, conc.obtained=1.8

ncover=1, ngel=1.6 13, pyrm.angle=30 catenary, A=1 B=0.4,inc.angle=25,conc.X=5, conc.obtained=1.7

(b)

ncover=1, ngel=1.6 13, pyrm.angle=30 catenary, A=1, B=0.5,inc.angle=25, conc.X=5, conc.obtained=1.38

(a)

(c)

FIGURE 4.99a,b,c Two-dimensional ray-tracing through a concentrator without cover, but filled with a refracting material, having its top pyramid textured, for an incident angle of 25° and three catenary depths.

catenary shape, with maximum concentration between 2 and 3 and average concentration over the interesting interval of incident angles (say, to 60°) no more than about 1.5. This is better than the inverted pyramid structure in Fig. 4.96, but it is still doubtful whether the expense of the concentrator is warranted. The collector efficiency, as measured relative to total aperture area, is the ratio of absorber intensity and X, i.e., considerably less than the cell efficiency.

Generation 24 h a day at constant radiation intensity could be achieved by placing the photovoltaic device perpendicular to the direction of the Sun at the top of the Earth's atmosphere. Use of photovoltaic energy systems is, as mentioned, common on space vehicles and also on satellites in orbit around the Earth. It has been suggested that, if a large photovoltaic installation were placed in a geosynchronous orbit, the power generated might be transmitted to a fixed location on the surface of the Earth, after having been converted to microwave frequency (see, for example, Glaser, 1977).

4.4.5 Solar Cooling and Other Applications

A variety of other applications of solar energy thermal conversion have been considered. Comfortable living conditions and food preservation require cooling in many places. As in the case of heating, the desired temperature difference from the ambient is usually only about 10 °C–30 °C. Passive systems like those depicted in Figs. 4.76 and 4.77 may supply adequate cooling in the climatic regions for which they were designed. Radiative cooling is very dependent on the clearness of the night sky (cf. section 3.1.4). In desert regions with a clear night sky, the difference between the winter ambient temperature and the effective temperature of the night sky has been used to freeze water. In Iran, the ambient winter temperature is typically a few degrees Celsius above the freezing point of water, but the temperature T_e of the night sky is below 0 °C. In previous centuries, ice (for use in the palaces of the rich) was produced from shallow ponds surrounded by walls to screen out daytime winter sunlight, and the ice produced during the night in such natural icemakers ("yakhchal," see Bahadori, 1977) could be stored for months in very deep (say, 15 m) underground containers.

Many places characterized by hot days and cool nights have taken advantage of the heat capacity of thick walls or other structures to smooth out the diurnal temperature variations. If additional cooling is required, active solar cooling systems may be employed, and "cold storage" may be introduced to cover the cooling requirement independent of variations in solar radiation, in analogy to "hot storage" of solar heating systems.

A solar cooling system may consist of flat-plate collectors delivering the absorbed energy to a "hot storage," which is then used to drive an absorption cooling cycle (Fig. 4.100), drawing heat from the "cool storage," to which the load areas are connected (see, for example, Wilbur and Mancini, 1976). In principle, only one kind of storage is necessary, but, with both hot and cold storage, the system can simultaneously cover both heating needs (e.g., hot water) and cooling needs (e.g., air conditioning).

The absorption cooling cycle (Fig. 4.100) is achieved by means of an absorbent–refrigerant mix, such as LiBr–H_2O or H_2O–NH_3. The lithium–bromide–water mix is more suitable for flat-plate solar collector systems, having a higher efficiency than the water–ammonia mix for the temperatures characteristic of flat-plate collectors. LiBr is hygroscopic, i.e., it can mix with water in any ratio. Solar heat is used in a "generator" to vaporize some water from the mix. The resulting vapor is led to a condenser unit, using a coolant flow, and is then expanded to regain a gaseous phase, whereby it draws heat from the area to be cooled, and subsequently is returned to an "absorber" unit. Here it becomes absorbed in the LiBr–H_2O mix with the help of a coolant flow. The coolant inlet temperature would usually be the ambient one, and the same coolant passes through the absorber and the condenser. The coolant exit temperature would then be above ambient, and the coolant cycle could be made closed by exchanging the excess heat with the surroundings (e.g., in a "cooling tower"). The refrigerant-rich mix in the absorber is pumped back to the generator and is made up for by recycling refrigerant-poor mix from the generator to the absorber (e.g., entering through a spray system). In order not to waste solar heat or coolant flow, these two streams of absorbent–refrigerant mix are made to exchange heat in a heat exchanger. The usefulness of a given absorbent–refrigerant pair is determined by the temperature dependence of vaporization and absorption processes.

FIGURE 4.100 Solar absorption cooling system using a mix of absorbent (e.g., LiBr) and refrigerant (e.g., H_2O). P denotes pump. The coolant may be recycled after having rejected heat to the environment in a cooling tower.

In dry climates, a very simple method of cooling is to spray water into an airstream (evaporative cooling). If the humidity of the air should remain unchanged, the air has first to be dried (spending energy) and then cooled by evaporating water into it, until the original humidity is again reached.

In principle, cooling by means of solar energy may also be achieved by first converting the solar radiation to electricity by one of the methods described in sections 4.4.1, 4.4.2, or 4.4.4 and then using the electricity to drive a thermodynamic cooling cycle, such as the Rankine cycle in Fig. 4.3.

The same applies if the desired energy form is mechanical work, as in the case of pumping water from a lower to a higher reservoir (e.g., for irrigation of agricultural land). In practice, however, the thermodynamic cycles discussed in section 4.1.2 are used directly to make a pump produce mechanical work on the basis of the heat obtained from a solar collector. Figure 4.101 shows two types of devices based on the Stirling cycle, using air or another gas as a working fluid (cf. Fig. 4.3). In the upper part of the figure, two ball valves ensure an oscillatory, pumping water motion, maintained by the tendency of the temperature gradient in the working fluid (air) to move air from the cold to the hot region. The lower part of the figure shows a "free piston Stirling engine," in which the oscillatory behavior is brought about by the presence of two pistons of different mass, the heavy one being delayed by half an oscillatory period with respect to the lighter one. The actual water pumping is done by the "membrane movements" of a diaphragm, but if the power level is sufficient, the diaphragm may be replaced by

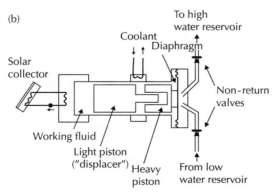

FIGURE 4.101a,b Solar water pumps based on Stirling cycles. In (a), air and water should not mix. The ball valves produce cyclic operation. In (b), the light and heavy pistons should be out of phase by a half cycle. (Based on West, 1974; Beale, 1976.)

a piston pump. For Stirling cycle operation, the working fluid is taken as a gas. This is an efficient cycle at higher temperatures (e.g., with use of a focusing solar collector), but if the heat provided by the (flat-plate) solar collector is less than about 100 °C, a larger efficiency may be obtained by using a two-phase working fluid (corresponding to one of the Rankine cycles shown in Fig. 4.3). If focusing solar collectors are used, the diaphragm may be replaced by a more efficient piston pump.

Two Rankine-type solar pumps, based on a fluid-to-gas and gas-to-fluid cycle, are outlined in Fig. 4.102. The one shown in the upper part of the figure is based on a cyclic process. The addition of heat evaporates water in the container on the left, causing water to be pumped through the upper one-way valve. When the vapor reaches the bottom of the U-tube, all of it moves to the container on the right and condenses. New water is drawn from the bottom reservoir, and the pumping cycle starts all over again. This pump is intended for wells of shallow depth, below 10 m (Boldt, 1978).

FIGURE 4.102a,b Solar water pumps based on Rankine cycles. In (*a*), the formation of vapor, its transfer through the U-tube, and its condensation produce a cyclic pressure–suction variation, which draws water from the lower reservoir and delivers water to the high reservoir. (Based on Boldt, 1978.) In (*b*), a turbine-type pump is shown, the working fluid of which may be an organic compound (cf. Meinel and Meinel, 1976).

Figure 4.102*b* shows a pump operated by a conventional Rankine engine, expanding the working fluid to the gas phase and passing it through a turbine, and then condensing the gas using the pumped water as coolant, before returning the working fluid to the heat exchanger receiving solar absorbed heat. For all the pumps based on a single thermodynamic cycle, the maximum efficiency (which cannot be reached in a finite time) is given by (4.4) and the actual efficiency is given by an expression of the form (4.19).

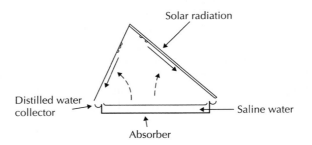

FIGURE 4.103   Solar still.

Distillation of saline water (e.g., seawater) or impure well water may be obtained by a solar still, a simple example of which is shown in Fig. 4.103. The side of the cover system facing the Sun is transparent, whereas the part sloping in the opposite direction is highly reflective and thus remains at ambient temperature. It therefore provides a cold surface for condensation of water evaporated from the saline water surface (kept at elevated temperature due to the solar collector part of the system). In more advanced systems, some of the heat released by condensation (about 2.3×10^6 J kg^{-1}, which is also the energy required for vaporization) is recovered and used to heat more saline water (Talbert *et al.*, 1970).

4.5 ELECTROCHEMICAL ENERGY CONVERSION

Electrochemical energy conversion is the direct conversion of chemical energy, i.e., free energy of the form (4.8), into electrical power or vice versa. A device that converts chemical energy into electric energy is called a *fuel cell* (if the free energy–containing substance is stored within the device rather than flowing into the device, the term *primary battery* is sometimes used). A device that accomplishes the inverse conversion (e.g., electrolysis of water into hydrogen and oxygen) may be called a *driven cell*. The energy input for a driven cell need not be electricity, but could be solar radiation, for example, in which case the process would be photochemical rather than electrochemical. If the same device can be used for conversion in both directions, or if the free energy–containing substance is regenerated outside the cell (energy addition required) and recycled through the cell, it may be called a *regenerative* or *reversible fuel cell* and, if the free energy–containing substance is stored inside the device, a *regenerative* or *secondary battery*.

The basic ingredients of an electrochemical device are two electrodes (sometimes called *anode* and *cathode*) and an intermediate electrolyte layer capable of transferring positive ions from the negative to the positive electrode (or negative ions in the opposite direction), while a corresponding flow of electrons in an external circuit from the negative to the positive electrode provides the desired power. Use has been made of solid electrodes and fluid electrolytes (solutions), as well as fluid electrodes (e.g., in high-temperature batteries) and solid electrolytes (such as ion-conducting semiconductors). A more detailed treatise on fuel cells may be found in Sørensen (2005).

4.5.1 Fuel Cells

The difference in electric potential, $\Delta\phi_{ext}$, between the electrodes (cf. the schematic illustration in Fig. 4.104) corresponds to an energy difference $e\Delta\phi_{ext}$ for each electron. The total number of electrons that could traverse the external circuit may be expressed as the product of the number of moles of electrons, n_e, and Avogadro's constant N_A, so the maximum amount of energy emerging as electrical work is

$$\Delta W^{(elec)} = n_e\, N_A\, e\, \Delta\phi_{ext}, \tag{4.142}$$

where $\mathscr{F} = N_A e = 96\,400$ C mol^{-1} (Faraday's constant) is sometimes introduced. This energy must correspond to a loss (conversion) of free energy,

$$-\Delta G = \Delta W^{(elec)} = n_e\, \mathscr{F}\, \Delta\phi_{ext}, \tag{4.143}$$

FIGURE 4.104 Schematic picture of a hydrogen–oxygen fuel cell. In this case, the electrodes are porous, so that the fuel gases may diffuse through them.

which constitutes the total loss of free energy from the "fuel" for an ideal fuel cell. This expression may also be derived from (4.8), using (4.2) and $\Delta Q = T \Delta S$, because the ideal process is reversible, and $\Delta W = -P \Delta V + \Delta W^{(elec)}$.

Figure 4.104 shows an example of a fuel cell, based on the free energy change $\Delta G = -7.9 \times 10^{-19}$ J for the reaction

$$2H_2 + O_2 \rightarrow 2H_2O$$

[cf. (3.50)]. Hydrogen gas is led to the negative electrode, which may consist of a porous material, allowing H^+ ions to diffuse into the electrolyte, while the electrons enter the electrode material and may flow through the external circuit. If a catalyst (e.g., a platinum film on the electrode surface) is present, the reaction at the negative electrode

$$2H_2 \rightarrow 4H^+ + 4e^- \tag{4.144}$$

may proceed at a much enhanced rate (see, for example, Bockris and Shrinivasan, 1969). Gaseous oxygen (or oxygen-containing air) is similarly led to the positive electrode, where a more complex reaction takes place, the net result of which is

$$O_2 + 4H^+ + 4e^- \rightarrow 2H_2O. \tag{4.145}$$

This reaction may be built up by simpler reactions with only two components, such as oxygen first picking up electrons or first associating with a hydrogen ion. A catalyst can also stimulate the reaction at the positive electrode. Instead of porous electrodes, which allow direct contact between the input gases and the electrolyte, membranes can be used (cf. also Bockris and Shrinivasan, 1969), like those found in biological material, i.e., membranes that allow H^+ to diffuse through but not H_2, etc.

The drop in free energy (4.143) is usually considered to be mainly associated with reaction (4.145), expressing G in terms of a chemical potential (3.36), e.g., of the H^+ ions dissolved in the electrolyte. Writing the chemical potential μ as Faraday's constant times a potential ϕ, the free energy for n moles of hydrogen ions is

$$G = n\mu = n\mathscr{F}\phi = nN_A e\phi. \tag{4.146}$$

When the hydrogen ions "disappear" at the positive electrode according to reaction (4.145), this chemical free energy is converted into electrical energy (4.142) or (4.143), and, since the numbers of electrons and hydrogen ions in (4.145) are equal, $n = n_e$, the chemical potential μ is given by

$$\mu = \mathscr{F}\phi = \mathscr{F}\Delta\phi_{ext}. \tag{4.147}$$

Here ϕ is the quantity usually referred to as the *electromotive force* (e.m.f.) of the cell, or *standard reversible potential* of the cell, if taken at standard atmospheric pressure and temperature. From the value of ΔG quoted above, corresponding to -2.37×10^5 J per mole of H_2O formed, the cell's e.m.f. becomes

$$\phi = -\Delta G/n\mathscr{F} = 1.23 \text{ V}, \tag{4.148}$$

with $n = 2$ since there are two H^+ ions for each molecule of H_2O formed. The chemical potential (4.147) may be parametrized in the form (3.37), and the cell e.m.f. may thus be expressed in terms of the properties of the reactants and the electrolyte [including the empirical activity coefficients appearing in (3.37) as a result of generalizing the expression obtained from the definition of the free energy, (4.6), assuming P, V, and T to be related by the ideal gas law, $PV = \mathscr{R}T$, valid for one mole of an ideal gas (cf., for example, Angrist, 1976)].

The efficiency of a fuel cell is the ratio between the electrical power output (4.142) and the total energy lost from the fuel. However, it is possible to exchange heat with the surroundings, and the energy lost from the fuel may thus be different from ΔG. For an ideal (reversible) process, the heat added to the system is

$$\Delta Q = T \Delta S = \Delta H - \Delta G,$$

and the efficiency of the ideal process thus becomes

$$\eta^{ideal} = -\Delta G/(-\Delta G - \Delta Q) = \Delta G/\Delta H. \tag{4.149}$$

For the hydrogen–oxygen fuel cell considered above, the enthalpy change during the two processes (4.144) and (4.145) is $\Delta H = -9.5 \times 10^{-19}$ J, or -2.86×10^5 J per mole of H_2O formed, and the ideal efficiency is

$$\eta^{ideal} = 0.83.$$

There are reactions with positive entropy change, such as $2C + O_2 \rightarrow 2CO$, which may be used to cool the surroundings and at the same time create electric power with an ideal efficiency above one (1.24 for CO formation).

In actual fuel cells, a number of factors tend to diminish the power output. They may be expressed in terms of "expenditure" of cell potential fractions on processes not contributing to the external potential,

$$\Delta\phi_{ext} = \phi - \phi_1 - \phi_2 - \phi_3 - \cdots,$$

where each of the terms $-\phi_i$ corresponds to a specific loss mechanism. Examples of loss mechanisms are blocking of pores in the porous electrodes [e.g., by piling up of the water formed at the positive electrode in process (4.145)], internal resistance of the cell (heat loss), and the buildup of potential barriers at or near the electrolyte–electrode interfaces. Most of these mechanisms limit the reaction rates and thus tend to place a limit on the current of ions that may flow through the cell. There will be a limiting current, I_L, beyond which it will not be possible to draw any more ions through the electrolyte, because of either the finite diffusion constant in the electrolyte, if the ion transport is dominated by diffusion, or the finite effective surface of the electrodes at which the ions are formed. Figure 4.105 illustrates the change in $\Delta\phi_{ext}$ as a function of current,

$$I = \Delta\phi_{ext} R_{ext},$$

expressed as the difference between potential functions at each of the electrodes, $\Delta\phi_{ext} = \phi_c - \phi_a$. This representation gives some indication of whether the loss mechanisms are connected with the positive or negative electrode processes, and, in this case, the largest fraction of the losses is connected with the more complex positive electrode reactions. For other types of fuel cells, it may be negative ions that travel through the electrolyte, with corresponding changes in characteristics.

It follows from diagrams of the type in Fig. 4.105 that there will be an optimum current, usually lower than I_L, for which the power output will be maximal,

$$\max(E) = I^{opt} \Delta\phi_{ext}^{opt}.$$

Dividing by the rate at which fuel energy ΔH is added to the system in a steady-state situation maintaining the current I^{opt}, one obtains the actual efficiency of conversion maximum,

$$\max(\eta) = I^{opt} \Delta\phi_{ext}^{opt}/(dH/dt). \tag{4.150}$$

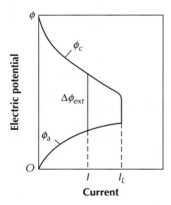

FIGURE 4.105 Fuel-cell negative electrode potential ϕ_a and positive electrode potential ϕ_c as a function of current. The main cause of the diminishing potential difference $\Delta\phi_{ext}$ for increasing current is at first incomplete electrocatalysis at the electrodes; also, for larger currents, ohmic losses in the electrolyte solution; and, finally, a lack of ion transport. (Based on Bockris and Shrinivasan, 1969.)

The potential losses within the cell may be described in terms of an internal resistance, the existence of which implies non-zero energy dissipation, if energy is to be extracted in a finite time. This is the same fundamental type of loss as that encountered in solar cells (4.106) and in the general thermodynamic theory described in section 4.1.1.

Reverse Operation of Fuel Cells

A driven-cell conversion based on the dissociation of water may be accomplished by electrolysis, using components similar to those of the hydrogen–oxygen fuel cell to perform the inverse reactions. Thus, the efficiency of an ideal conversion may reach 1.20, according to (4.150), implying that the electrolysis process draws heat from the surroundings.

If a fuel cell is combined with an electrolysis unit, a regenerative system is obtained, and if hydrogen and oxygen can be stored, an energy storage system, a *battery*, results. The electric energy required for the electrolysis need not come from the fuel cell, but may be the result of an intermittent energy conversion process (e.g., wind turbine, solar photovoltaic cell, etc.).

Direct application of radiant energy to the electrodes of an electrochemical cell has been suggested, aiming at the achievement of the driven-cell process (e.g., dissociation of water) without having to supply electric energy. The electrodes could be made of suitable p- and n-type semiconductors, and the presence of photo-induced electron–hole excitations might modify the electrode potentials ϕ_a and ϕ_c in such a way that the driven-cell reactions become thermodynamically possible. Owing to the electrochemical losses discussed above, additional energy would have to be provided by solar radiation. Even if the radiation-induced electrode processes have a low efficiency, the overall efficiency may still be higher for photovoltaic conversion of solar energy into electricity followed by conventional electrolysis (Manassen *et al.*, 1976). In recent years, several such photo-electrochemical devices have been constructed. Some of them are described in section 4.4.2. Direct hydrogen production by photo-electrochemical devices is discussed in Sørensen (2005).

Fuel-cell Design

The idea of converting fuel into electricity by an electrode–electrolyte system originated in the 19th century (Grove, 1839). The basic principle behind a hydrogen–oxygen fuel cell is described above. The first practical applications were in powering space vehicles, starting during the 1970s. During the late 1990s, fuel-cell road vehicles were predicted to be ready within a decade, but it did not happen, due to problems with durability and cost. Hybrid vehicles allow a smaller (and thus less costly) fuel cell rating, but if combined with traction batteries the lowering of cost may turn out as difficult as for pure fuel cell vehicles. However, hybrids between either battery electric vehicles or fuel cell vehicles and conventional internal combustion engine vehicles may smooth the transition.

First, hybrids between conventional internal combustion engines and batteries reduce pollution in cities (they use batteries that are charged by the engine at convenient times or that are charged by off-road plug-in power supplies), while using the fuel-based engine for long highway trips will relieve demand on battery size. Second, if and when (presumably) both battery and fuel-cell technologies have come down in cost, the hybrid car could replace the internal combustion engine with a fuel cell plus electric motor, thus allowing total reliance on primary energy from hydrogen or grid electricity, which may both be renewable (see Sørensen, 2005; 2006; 2010c).

Fuel technology is described in detail in Sørensen (2005). Here, only some general remarks are made about the development from early fuel-cell concepts to the present-day situation.

Originally developed for stationary power applications, *phosphoric acid cells* use porous carbon electrodes with a platinum catalyst and phosphoric acid as electrolyte and feed hydrogen to the negative electrode, with electrode reactions given by (4.144) and (4.145). Their operating temperature is 175 °C– 200 °C, and water is continuously removed.

Alkaline cells use KOH as electrolyte and have electrode reactions of the form

$$H_2 + 2OH^- \rightarrow 2H_2O + 2e^-,$$

$$1/2 O_2 + H_2O + 2e^- \rightarrow 2OH^-.$$

Their operating temperature is 70 °C–100 °C, but specific catalysts require maintenance of fairly narrow temperature regimes. Also, the hydrogen fuel must be very pure and, notably, not contain any CO_2. Alkaline fuel cells have been used extensively on spacecraft and recently in an isolated attempt for road vehicles (Hoffmann, 1998a). Their relative complexity and use of corrosive compounds requiring special care in handling make it unlikely that the cost can be reduced to levels acceptable for general-purpose use.

The third fuel cell in commercial use is the *proton exchange membrane (PEM) cell*. It has been developed over a fairly short period of time and is considered to hold the greatest promise for economic application in the transportation sector. It contains a solid polymer membrane sandwiched between two gas diffusion layers and electrodes. The membrane material may be polyperfluorosulfonic acid. A platinum or Pt–Ru alloy catalyst is used to break hydrogen molecules into atoms at the negative electrode, and the hydrogen atoms are then capable of penetrating the membrane and reaching the positive electrode, where they combine with oxygen to form water, again with the help of a platinum catalyst. The electrode reactions are again (4.144) and (4.145), and the operating temperature is 50 °C–100 °C (Wurster, 1997).

Figure 4.106 shows a typical layout of an individual cell. Several of them are then stacked on top of each other. This modularity implies that PEM cells can be used for applications requiring little power (1 kW or less). PEM cell stacks dominate the current wealth of demonstration projects in road transportation, portable power, and special applications. The efficiency of conversion for small systems is between 30% and 50%, but a 50 kW system has recently shown an efficiency in the laboratory of near 60%. As indicated in Fig. 4.107, an advantage of particular importance for automotive applications is the high efficiency at partial loads, which alone gives a factor of 2 improvement over current internal combustion engines.

FIGURE 4.106 Layout of a PEM cell layer, several of which may be stacked.

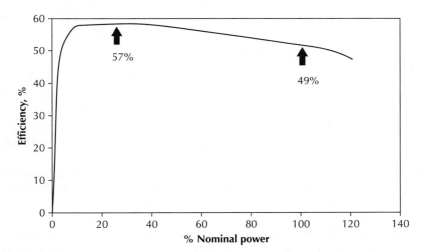

FIGURE 4.107 Expected partial load efficiencies for a 50 kW PEM cell, projected from measurements involving 10–20 cell test stacks (Patil, 1998).

For use in automobiles, compressed or liquefied hydrogen is limited by low energy density and safety precautions for containers, but most of all by the requirement for a new infrastructure for fueling. The hydrogen storage problem, which until recently limited fuel-cell projects to large vehicles like buses, may be solved by use of metal hydride or carbon nanofiber stores (see Chapter 5). In order to avoid having to make large changes to gasoline and diesel filling stations, methanol has been proposed as the fuel to be distributed to the vehicle fuel tank. The energy density of methanol is 4.4 kWh liter^{-1}, which is half that of gasoline. This is quite acceptable owing to methanol's higher efficiency of conversion. Hydrogen is then produced onboard, by a methanol reformer, before being fed to the fuel cell to produce the electric power for an electric motor. The set-up is illustrated in Fig. 4.108. Prototype vehicles with this set-up have been tested (cf. Takahashi, 1998; Brown, 1998), but the reformer technology turned out to have many problems and most current vehicle fuel-cell concepts plan to use hydrogen directly, thereby also avoiding methanol emissions and the health issues connected with the methanol infrastructure.

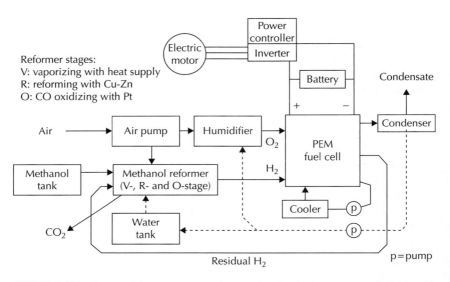

FIGURE 4.108 Layout of the power system for a methanol-to-hydrogen powered vehicle with fuel-cell conversion and an electric motor. The power controller allows a shift from direct drive to battery charging.

Methanol, CH_3OH, may, however, still find use in fuel-cell development, in a direct methanol PEM fuel cell, without the extra step of reforming to H_2. PEM fuel cells that accept a mixture of methanol and water as feedstock have been developed (Bell, 1998) but still have lower energy conversion efficiency than hydrogen-fueled PEM cells. Because methanol can be produced from biomass, hydrogen may be eliminated from the energy system. On the other hand, handling of surplus production of wind or photovoltaic power might still conveniently involve hydrogen as an intermediate energy carrier, since it may have direct uses and thus may improve the system efficiency by avoiding the losses incurred in methanol production. The electric power to hydrogen conversion efficiency is about 65% in high-pressure alkaline electrolysis plants (Wagner *et al.*, 1998; Sørensen, 2005), and the efficiency of the further hydrogen (plus CO or CO_2 and a catalyst) to methanol conversion is around 70%. Also, the efficiency of producing methanol from biomass is about 45%, whereas higher efficiencies are obtained if the feedstock is methane (natural gas) (Jensen and Sørensen, 1984; Nielsen and Sørensen, 1998).

For stationary applications, fuel cells of higher efficiency may be achieved by processes operating at higher temperatures. One line of research has been *molten carbonate fuel cells*, with electrode reactions

$$H_2 + CO_3^{2-} \rightarrow CO_2 + H_2O + 2e^-,$$
$$1/2O_2 + CO_2 + 2e^- \rightarrow CO_3^{2-}.$$

The electrolyte is a molten alkaline carbonate mixture retained within a porous aluminate matrix. The carbonate ions formed at the positive electrode travel through the matrix and combine with hydrogen at the negative electrode at an operating temperature of about 650 °C. The process was originally aimed at hydrogen supplied from coal gasification or natural gas conversion. Test of a new 250 kW system is taking place at a power utility company in Bielefeld, Germany (Hoffmann, 1998b). The expected conversion efficiency is about 55%, and some of the additional high-temperature heat may be utilized. Earlier experiments with molten carbon fuel cells encountered severe problems with material corrosion.

Considerable efforts are being dedicated to *solid electrolyte cells*. Solid oxide fuel cells (SOFC) use zirconia as the electrolyte layer to conduct oxygen ions formed at the positive electrode. Electrode reactions are

$$H_2 + O^{2-} \rightarrow H_2O + 2e^-,$$
$$1/2O_2 + 2e^- \rightarrow O^{2-}.$$

The reaction temperature is 700 °C–1000 °C. The lower temperatures are desirable, due to lower corrosion problems, and may be achieved by using a thin electrolyte layer (about 10 μm) of yttrium-stabilized zirconia sprayed onto the negative electrode as a ceramic powder (Kahn, 1996). A number of prototype plants (in the 100-kW size range) are in operation. Current conversion efficiency is about 55%, but it could reach 70%–80% in the future (Hoffmann, 1998b).

Particularly for vehicle applications of hydrogen-based technologies, efforts are needed to ensure a high level of safety in collisions and other accidents. Because of the difference between the physical properties of hydrogen fuel and the hydrocarbon fuels currently in use (higher diffusion speed, flammability, and explosivity over wider ranges of mixtures with air), there is a need for new safety-related studies, particularly where hydrogen is stored onboard in containers at pressures typically of 20–50 MPa. A few such studies have already been done (Brehm and Mayinger, 1989; see also Sørensen, 2005).

4.5.2 Other Electrochemical Energy Conversion

This chapter describes a number of general principles that have been, or may become, of use in designing actual conversion devices, primarily those aimed at utilizing the renewable flows of energy. Clearly, not all existing or proposed devices have been covered.

The present energy system is dominated by fuels. Solar, wind, and hydro energy bring us away from fuels, but fuels may still play an important role, e.g., in the transport sector. A number of technologies are available to convert fuels among themselves and into electricity and heat or motive power. Many of the technologies can also be used for fuels derived from renewable energy sources. (Fuels derived from biomass are dealt with separately in section 4.6.) For renewable energy sources, such as wind and solar power, one way of dealing with fluctuating production would be to convert surplus electricity to a storable fuel. This could be hydrogen obtained by electrolysis (section 5.3.3 and Sørensen, 2005) or by reversible fuel cells (section 5.3.4). The further conversion of hydrogen to the energy forms in demand can be accomplished by methods already available, for example, for natural gas, such as boilers, engines, gas turbines, and fuel cells.

So far, this chapter describes at least one example of conversion into a useful form of energy for each of the renewable resources described in Chapter 3. This section contains an example of converting more speculative sources of energy, such as the salinity differences identified in section 3.4.3, as renewable energy resources of possible interest.

Conversion of Ocean Salinity Gradients

As discussed in section 3.4.3 (Fig. 3.69), a salinity difference like the one existing between fresh (river) and saline (ocean) water, may be used to drive an osmotic pump, and the elevated water may in turn be used to generate electricity by an ordinary turbine (cf. section 4.3.4).

An alternative method, aiming directly at electricity production, takes advantage of the fact that the salts of saline water are ionized to a certain degree and thus may be used to derive an electrochemical cell of the type discussed in section 4.5.1 (Pattie, 1954; Weinstein and Leitz, 1976).

The electrochemical cell shown in Fig. 4.109 may be called a *dialytic battery*, since it is formally the same as the dialysis apparatus producing saline water with electric power input. The membrane allows one kind of ion to pass and thereby reach the corresponding electrode (in contrast to the osmotic pump in Fig. 3.69, where water could penetrate the membrane but ions could not). The free energy difference between the state with free Na^+ and Cl^- ions (assuming complete ionization) and the state with no such ions is precisely the difference between saline and fresh water, calculated in section 3.4.3 and approximately given by (3.38). Assuming further that each Na^+ ion reaching electrode A neutralizes one electron, and that correspondingly each Cl^- ion at electrode B gives rise to one electron (which may travel in the external circuit), then the electromotive force, i.e., the electric potential ϕ of the cell, is given in analogy to (4.146), with the number of positive ions per mole equal to n_{Na^+}, and ϕ is related to the change in free energy (3.38), just as (4.146) is related to (4.143),

FIGURE 4.109 Schematic picture of dialytic battery.

$$n_e \mathscr{F} \phi = n_{Na^+} \mathscr{F} \phi = \delta G \approx 2\mathscr{R} T n_{Na^+},$$

or

$$\max{(\Delta\phi_{ext})} = \phi \approx 2\mathscr{R}T/\mathscr{F}. \qquad (4.151)$$

Inserting the numerical values given in section 3.7.2, and $T = 300$ K, the cell electromotive force becomes $\phi \approx 0.05$ V. The actual value of the external potential, $\Delta\phi_{ext}$, may be reduced as a result of various losses, as described in section 4.5.1 (e.g., Fig. 4.105). ϕ is also altered if the "fresh" water has a finite amount of ions or if ions other than Na^+ and Cl^- are present.

If the load resistance is R_{ext}, the power delivered by the cell is given by the current $I = \Delta\phi_{ext} R_{ext}^{-1}$,

$$E = I\Delta\phi_{ext},$$

and, as usual, the maximum power corresponds to an external potential difference $\Delta\phi_{ext}$, which is smaller than the open-circuit value (4.151). Internal losses may be represented by an internal resistance R_{int}, defined by

$$\Delta\phi_{ext} = \phi - IR_{int} = \phi - \Delta\phi_{ext}(R_{int}/R_{ext}).$$

Thus, the power output may also be written

$$E = \phi^2 R_{ext}/(R_{int} + R_{ext})^2.$$

R_{int} depends on electrode properties, as well as on n_{Na^+} and n_{Cl^-} and their variations across the cell.

Several anode-membrane plus cathode-membrane units may be stacked beside each other in order to obtain a sufficiently large current (which is necessary because the total internal resistance has a tendency to increase strongly if the current is below a certain value). Figure 4.110 shows the results of a small-scale experiment (Weinstein and Leitz, 1976) for a stack of about 30 membrane-electrode pairs and an external load, R_{ext}, close to the one giving maximum power output. The power is shown, however, as a function of the salinity of the fresh water, and it is seen that the largest power output is obtained for a fresh-water salinity that is not zero but 3%–4% of the seawater salinity. The reason is that, although the electromotive force (4.151) diminishes with increasing fresh-water salinity, the decrease is at first more than compensated for by the improved conductivity of the solution (decrease in R_{int}) when ions are also initially present in the fresh-water compartment.

FIGURE 4.110 Measured performance of a dialytic battery, containing about 30 membrane pairs in a stack. The external load was $R_{ext} = 10\ \Omega$ (close to power optimization), and electrode losses had been compensated for by insertion of a balancing power source. (Based on Weinstein and Leitz, 1976.)

Needless to say, small-scale experiments of the kind described above are a long way from a demonstration of viability of the salinity gradient conversion devices on a large scale, and it is by no means clear whether the dialysis battery concept will be more or less viable than power turbines based on osmotic pumps. The former seems likely to demand a larger membrane area than the latter, but no complete evaluation of design criteria has been made in either case.

4.6 BIOENERGY CONVERSION PROCESSES

A number of conversion processes aim at converting one fuel into another that is considered more versatile. Examples are charcoal production from wood (efficiency of conversion about 50%), liquefaction, and gasification (Squires, 1974). Figure 4.111 gives an overview of the nonfood uses of biomass, to be discussed in more detail in the following.

Fresh biomass is a large potential source of renewable energy that, in addition to use for combustion, may be converted into a number of liquid and gaseous biofuels. The conversion may be achieved by thermochemical or biochemical methods, as described below.

However, biomass is not just stored energy; it is also a store of nutrients and a potential raw material for a number of industries. Therefore, bioenergy is a topic that cannot be separated from food production, timber industries (serving construction purposes, paper and pulp industries, etc.), and organic feedstock-dependent industries (chemical and biochemical industries, notably). Furthermore, biomass is derived from plant growth and animal husbandry, linking the energy aspect to agriculture, livestock, silviculture, aquaculture, and quite generally to the management of the global ecological system. Thus, procuring and utilizing organic fuels constitute a profound interference with the natural biosphere, and understanding the range of impacts as well as elimination of unacceptable ones should be an integral part of any scheme for diversion of organic fuels to human society.

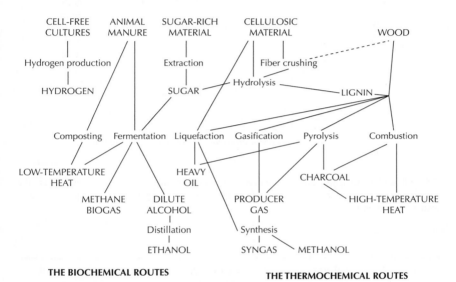

THE BIOCHEMICAL ROUTES

THE THERMOCHEMICAL ROUTES

FIGURE 4.111 Overview of nonfood energy uses of biomass.

4.6.1 Combustion and Composting of Biomass

Traditional uses of organic fuels as energy carriers comprise the use of fresh biomass, i.e., storage and combustion of wood fuels, straw, and other plant residues, and, in recent centuries, notably storage and combustion of fossilized biomass, i.e., fuels derived from coal, oil, and natural gas. While the energy density of living biomass is in the range of 10–30 MJ per kg of dry weight, several million years of fossilization processes typically increase the energy density by a factor of 2 (see Fig. 4.112). The highest energy density is found for oil, where the average density of crude oil is 42 MJ kg^{-1}.

The known reserves of fossil fuels that may be economically extracted today, and estimates of further resources exploitable at higher costs, are indicated in Table 4.2. The finiteness of such resources is, of course, together with the environmental impact issues, the reason for turning to renewable energy sources, including renewable usage of biomass resources.

FIGURE 4.112   Coal to peat classification and selected properties. *Dashed lines* indicate the fractional content of oxygen and of fixed carbon. The remaining fraction consists of volatile matter. (Based on U.S. DoE, 1979.)

Source	Reserves (EJ)	Add'l Resources (>50% Prob.)	Occurrence (Speculative)	1990 Use (EJ)	Accum. Use 1860–1990
Coal				91	5 203
hard coal	15 000	100 000	900 000		
brown coal/					
lignite	4 000	30 000	90 000		
peat	0	4 000	4 000		
Oil				128	3 343
conventional	5 000	3 000	13 000		
unconventional	7 000	9 000	20 000		
Natural gas				71	1 703
conventional	5 000	5 000	15 000		
unconventional	7 000	20 000	25 000		
in hydrates	0	0	100 000		

TABLE 4.2   Fossil reserves, additional resources, and consumption in EJ (UN, 1981; Jensen and Sørensen, 1984; Nakicenovic *et al.*, 1996).

Producing Heat by Burning Biomass

Heat may be obtained from biological materials by burning, eventually with the purpose of further conversion. Efficient burning usually requires the reduction of water content, for example, by sun-drying. The heat produced by burning cow dung is about 1.5×10^7 J per kg of dry matter, but initially only about 10% is dry matter, so the vaporization of 9 kg of water implies an energy requirement of 2.2×10^7 J, i.e., the burning process is a net energy producer only if substantial sun-drying is possible. Firewood and other biomass sources constitute stores of energy, since they can be dried during the summer and used during winter periods when the heating requirement may be large and the possibility of sun-drying may not exist. The heat produced by burning 1 kg of dry wood or sawmill scrap is about 1.25×10^7 J (1 kg in these cases corresponds to a volume of approximately 1.5×10^{-3} m^3), and the heat from burning 1 kg of straw (assumed water content 15%) is about 1.5×10^7 J (Eckert, 1960). Boilers used for firing with wood or straw have efficiencies that are often considerably lower than those of oil or gas burners. Typical efficiencies are in the range 0.5–0.6 for the best boilers. The rest of the enthalpy is lost, mostly in the form of vapor and heat leaving the chimney (or other smoke exit), and therefore is not available at the load area.

Combustion is the oxidation of carbon-containing material in the presence of sufficient oxygen to complete the process

$$C + O_2 \rightarrow CO_2.$$

Wood and other biomass is burned for cooking, for space heating, and for a number of specialized purposes, such as provision of process steam and electricity generation. In rural areas of many Third World countries, a device consisting of three stones for outdoor combustion of twigs is still the most common. In the absence of wind, up to about 5% of the heat energy may reach the contents of the pot resting on top of the stones. In some countries, indoor cooking on simple chulas is common. A chula is a combustion chamber with a place for one or more pots or pans on top, resting in such a way that the combustion gases will pass along the outer sides of the cooking pot and leave the room through any opening. Indoor air quality is extremely poor when chulas are in use, and village women in India using chulas for several hours each day are reported to suffer from severe cases of eye irritation and respiratory diseases (Vohra, 1982).

Earlier, most cooking in Europe and its colonies was done on stoves made of cast iron. These stoves, usually called European stoves, had controlled air intake and both primary and secondary air inlets, chimneys for regulation of gas passage, and several cooking places with ring systems allowing the pots to fit tightly in holes, with a part of the pot indented into the hot gas stream. The efficiency was up to about 25%, counted as energy delivered to the pots divided by wood energy spent, but such efficiency would only be reached if all holes were in use and if the different temperatures prevailing at different boiler holes could be made useful, including after-heat. In many cases, the average efficiency would hardly have exceeded 10%, but, in many of the areas in question, the heat lost to the room could also be considered useful, in which case close to 50% efficiency (useful energy divided by input) could be reached. Today, copies of the European stove are being introduced in several regions of the Third World, with the stoves constructed from local materials, such as clay and sand–clay mixtures, instead of cast iron.

Wood-burning stoves and furnaces for space heating have conversion efficiencies from below or about 10% (open furnace with vertical chimney) up to 50% (oven with two controlled air inlets and a labyrinth-shaped effluent gas route leading to a tall chimney). Industrial burners and stokers (for burning wood scrap) typically reach efficiencies of about 60%. Higher efficiencies require a very uniform fuel without variations in water content or density.

Most biological material is not uniform, and pretreatment can often improve both the transportation and storage processes and the combustion (or other final use). Irregular biomass (e.g., twigs) can be chopped or cut to provide unit sizes fitting the containers and burning spaces provided. Furthermore, compressing and pelletizing the fuel can make it considerably more versatile. For some time, straw compressors and pelletizers have been available, so that bulky straw bundles can be transformed into a fuel with volume densities approaching that of coal. Other biomass material can conceivably be pelletized advantageously, including wood scrap, mixed biomass residues, and even aquatic plant material (Anonymous, 1980). Portable pelletizers are available (e.g., in Denmark) that allow straw to be compressed in the growing fields, so that longer transport becomes economically feasible and so that even long-term storage (seasonal) of straw residues becomes attractive.

A commonly practiced conversion step is from wood to charcoal. Charcoal is easier to store and to transport. Furthermore, charcoal combustion—for example, for cooking—produces less visible smoke than direct wood burning and is typically so much more efficient than wood burning that, despite wood-to-charcoal conversion losses, less primary energy is used to cook a given meal with charcoal than with wood.

Particularly in rich countries, a considerable source of biomass energy is urban refuse, which contains residues from food preparation and discarded consumer goods from households, as well as organic scrap material from commerce and industry. Large-scale incineration of urban refuse has become an important source of heat, particularly in Western Europe, where it is used mainly for district heating (de Renzo, 1978).

For steam generation purposes, combustion is performed in the presence of an abundant water source (*waterwall incineration*). In order to improve pollutant control, fluidized bed combustion techniques may be utilized (Cheremisinoff *et al.*, 1980). The bed consists of fine-grain material, for example, sand, mixed with material to be burned (sawdust is particularly suitable, but any other biomass, including wood, can be accepted if finely chopped). The gaseous effluents from combustion, plus air, fluidize the bed as they pass through it under agitation. The water content of the material in the bed may be high (in which case steam production occurs). Combustion temperatures are lower than for flame burning, and this partly makes ash removal easier and partly reduces tar formation and salt vaporization. As a result, reactor life is extended and air pollution can better be controlled.

In general, the environmental impacts of biomass utilization through combustion may be substantial and comparable to, although not entirely of the same nature as, the impacts from coal and oil combustion (see Table 4.3). In addition, ashes will have to be disposed of. For boiler combustion, the sulfur dioxide emissions are typically much smaller than for oil and coal combustion, which would give 15–60 kg t^{-1} in the absence of flue gas cleaning. If ash is re-injected into the wood burner, higher sulfur values appear, but these values are still below the fossil-fuel emissions in the absence of sulfur removal efforts.

Substance Emitted	Emissions (kg/10^3kg)
Particulates	12.5–15.0
Organic compounds[a]	1.0
Sulphur dioxide	0–1.5[b]
Nitrogen oxides	5.0
Carbon monoxide	1.0

TABLE 4.3 Uncontrolled emissions from biomass combustion in boilers (kg per ton of fuel, based on woody biomass; U.S. EPA, 1980).
[a]*Hydrocarbons including methane and traces of benzo(a)pyrine.*
[b]*Upper limit is found for bark combustion. Ten times higher values are reported in cases where combustion ashes are re-injected.*

Particulates are not normally regulated in home boilers, but for power plants and industrial boilers, electrostatic filters are employed, with particulate removal up to over 99%. Compared to coal burning without particle removal, wood emissions are 5–10 times lower. When a wood boiler is started, there is an initial period of very visible smoke emission, consisting of water vapor and high levels of both particulate and gaseous emissions. After the boiler reaches operating temperatures, wood burns virtually without visible smoke. When stack temperatures are below 60 °C, during start-up and during incorrect burning, severe soot problems arise.

Nitrogen oxide emissions are typically 2–3 times lower for biomass burning than for coal burning (per kilogram of fuel), but often similar emissions occur per unit of energy delivered.

Particular concern should be directed at organic-compound emissions from biomass burning. In particular, benzo(a)pyrene emissions from biomass are found to be up to 50 times higher than for fossil-fuel combustion, and the measured concentrations of benzo(a)pyrene in village houses in Northern India ($1.3–9.3 \times 10^{-9}$ kg m^{-3}), where primitive wood-burning chulas are used for several (6–8) hours every day, exceed German standards of 10^{-11} kg m^{-3} by 2–3 orders of magnitude (Vohra, 1982). However, boilers with higher combustion temperatures largely avoid this problem, as indicated in Table 4.3.

The lowest emissions are achieved if batches of biomass are burned at optimal conditions, rather than regulating the boiler up and down according to heating load. Therefore, wood heating systems consisting of a boiler and a heat storage unit (gravel, water) with several hours of load capacity will lead to the smallest environmental problems (Hermes and Lew, 1982). This example shows that there can be situations where energy storage would be introduced entirely for environmental reasons.

The occupational hazards that arise during tree felling and handling should be mentioned. The accident rate among forest workers is high in many parts of the world, and safer working conditions for forest workers are imperative if wood is to be used sensibly for fuel applications.

Finally, although CO_2 accumulates in the atmosphere as a consequence of fossil fuel combustion, CO_2 emissions during biomass combustion are balanced in magnitude by the CO_2 assimilation by plants, so that the atmospheric CO_2 content is not affected, at least by the use of biomass crops in fast rotation. However, the lag time, e.g., for trees, may be decades or centuries, and, in such cases, temporary CO_2 imbalance may contribute to climatic alterations.

Composting

Primary organic materials form the basis for a number of energy conversion processes other than burning. Since they produce liquid or gaseous fuels, plus associated heat, they are dealt with in the following sections on fuel production. However, conversion aiming directly at heat production has also been utilized, with noncombustion processes based on manure from livestock animals and in some cases on primary biomass residues.

Two forms of composting are in use, one based on fluid manure (less than 10% dry matter), and the other based on solid manure (50%–80% dry matter). In both cases the chemical process is bacterial decomposition under aerobic conditions, i.e., the compost heap or container has to be ventilated in order to provide a continuous supply of oxygen for the bacterial processes. The bacteria required for the process (of which lactic acid producers constitute the largest fraction; cf. McCoy, 1967) are normally all present in manure, unless antibiotic residues that kill bacteria are retained after some veterinary treatment. The processes involved in composting are very complex, but it is believed that decomposition of carbohydrates [the inverse of reaction (3.45)] is responsible for most of the heat production (Popel, 1970). A fraction of the carbon from the organic material is found in new-bred microorganisms.

A device for treating fluid manure may consist of a container with a blower injecting fresh air into the fluid in such a way that it becomes well distributed over the fluid volume. An exit air channel carries the heated airflow to, say, a heat exchanger. Figure 4.113 shows an example of the temperature of the liquid manure, along with the temperature outside the container walls, as a function of time. The amount of energy required for the air blower is typically around 50% of the heat energy output, and is in the form of high-quality mechanical energy (e.g., from an electrically driven rotor). Thus, the simple efficiency may be around 50%, but the second law efficiency (4.20) may be quite low.

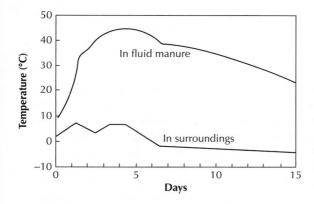

FIGURE 4.113 Temperature development in composting device based on liquid manure from pigs and poultry and with a blower providing constant air supply. Temperature of surroundings is also indicated. (Based on Popel, 1970.)

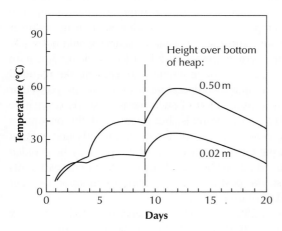

FIGURE 4.114 Temperature development at different locations within a solid manure composting heap (total height 0.8 m). After nine days, air is added from below. (Based on Olsen, 1975.)

Heat production from solid manure follows a similar pattern. The temperature in the middle of a manure heap (dunghill) may be higher than that of liquid manure, owing to the low heat capacity of the solid manure (see Fig. 4.114). Air may be supplied by blowers placed at the bottom of the heap, and, in order to maintain air passage space inside the heap and remove moisture, occasional re-stacking of the heap is required. A certain degree of variation in air supply can be allowed, so that the blower may be driven by a renewable energy converter, for example, a windmill, without storage or back-up power. With frequent re-stacking, air supply by blowing is not required, but the required amount of mechanical energy input per unit of heat extraction is probably as high as for liquid manure. In addition, the heat extraction process is more difficult, demanding, for example, that heat exchangers be built into the heap itself (water pipes are one possibility). If an insufficient amount of air is provided, the composting process will stop before the maximum heat has been obtained.

Owing to the bacteriological decomposition of most of the large organic molecules present, the final composting product has considerable value as fertilizer.

Metabolic Heat

Metabolic heat from the life processes of animals can also be used by man, in addition to the heating of human habitats by man's own metabolic heat. A livestock shed or barn produces enough heat, under most conditions of winter occupancy, to cover the heating requirements of adjacent farm buildings, in addition to providing adequate temperature levels for the animals. One reason for this is that livestock barns must have a high rate of ventilation in order to remove dust (e.g., from the animal's skin, fur, hair, or feathers) and water vapor. Therefore, utilization of barn heat may not require extra feeding of the animals, but may simply consist of heat recovery from air that for other reasons has to be removed. Utilization for heating a nearby residence building often requires a heat pump (see section 4.6.1), because the temperature of the ventilation air is usually lower than that required at the load area, and a simple heat exchanger would not work.

In temperate climates, the average temperature in a livestock shed or barn may be about 15 °C during winter. If young animals are present, the required temperature is higher. With an outside temperature of 0 °C and no particular insulation of walls, the net heat production of such barns or sheds is positive when the occupants are fully grown animals, but negative if the occupants are young individuals and very much so if newborn animals are present (Olsen, 1975). In chicken or pig farms, the need for heat input may be avoided by having populations of mixed age or heat exchange between compartments for young and adult animals. The best candidates for heat extraction to other applications might then be dairy farms.

A dairy cow transfers about 25% of the chemical energy in fodder to milk and a similar amount to manure (Claesson, 1974). If the weight of the cow remains constant, the rest is converted to heat and is rejected as heat radiation, convection, or latent heat of water vaporization. The distribution of the heat production in sensible and latent forms of heat is indicated in Fig. 4.115. It is strongly dependent on the air temperature in the barn. At 15 °C, about two-thirds of the heat production is in the form of sensible heat. Heat transfer to a heat pump circuit may take place from the ventilation air exit. Water condensation on the heat exchanger surface involved may help to prevent dust particles from accumulating on the heat exchanger surface.

FIGURE 4.115 Average heat production and form of heat for a "standard" cow (the heat production of typical cows of red or black-spotted breeds is about 20% higher, while that of typical Jersey cows is about 30% lower). (Based on Petersen, 1972.)

4.6.2 Biological Conversion into Gaseous Fuels

Fuels are biological material (including fossilized forms), hydrocarbons, or just hydrogen, which can be stored and used at convenient times. *Usage* traditionally means burning in air or pure oxygen, but other means of conversion exist, for example, fuel cells (see section 4.5).

Before discussing the conversion of fresh biomass, the gasification of coal is briefly discussed because of its importance for possible continued use of fossil biomass (coal being the largest such source) and also because of its similarity to processes relevant for other solid biomass.

Inefficient conversion of gasified coal to oil has historically been used by isolated coal-rich but oil-deficient nations (Germany during World War II, South Africa). Coal is gasified to carbon monoxide and hydrogen, which is then, by the Fischer-Tropsch process (passage through a reactor, e.g., a fluidized bed, with a nickel, cobalt, or iron catalyst), partially converted into hydrocarbons. Sulfur compounds have to be removed because they would impede the function of the catalyst. The reactions involved are of the form

$$(2n+1)H_2 + nCO \rightarrow C_nH_{2n+2} + nH_2O,$$

$$(n+1)H_2 + 2nCO \rightarrow C_nH_{2n+2} + nCO_2$$

and conversion efficiencies range from 21% to 55% (Robinson, 1980). Further separation of the hydrocarbons generated may then be performed; for instance, gasoline corresponding to the range $4 \leq n \leq 10$ in the above formulae.

Alternative coal liquefaction processes involve direct hydrogenation of coal under suitable pressures and temperatures. Pilot plants have been operating in the United States, producing up to 600 t a day (slightly different processes are named *solvent refining, H-coal process,* and *donor solvent process*; cf. Hammond, 1976). From an energy storage point of view, either coal or the converted product may be stockpiled.

For use in the transportation sector, production of liquid hydrocarbons, such as ethanol, methanol, and diesel-like fluids, is considered. The liquid hydrocarbons can be produced from fossil or renewable biomass sources as described below, often with thermochemical gasification as the first conversion step.

Biogas

Conversion of (fresh) biological material into simple hydrocarbons or hydrogen can be achieved by a number of anaerobic fermentation processes, i.e., processes performed by suitable microorganisms and taking place in the absence of oxygen. The anaerobic "digestion" processes work on most fresh biological material, wood excepted, provided that proper conditions are maintained (temperature, population of microorganisms, stirring, etc.). Thus, biological material in forms inconvenient for storage and use may be converted into liquid or gaseous fuels that can be utilized in a multitude of ways, like oil and natural gas products.

The "raw materials" that may be used are slurry or manure (e.g., from dairy farms or "industrial farming" involving large feedlots), city sewage and refuse, farming crop residues (e.g., straw or parts of cereal or fodder plants not normally harvested), or direct "fuel crops," such as ocean-grown algae or seaweeds, water hyacinths (in tropical climates), or fast-growing bushes or trees. The deep harvesting necessary to collect crop residues may not be generally advisable, owing to the role of these residues in preventing soil erosion.

Among the fermentation processes, one set is particularly suited for producing gas from biomass in a wet process (cf. Fig. 4.111). It is called *anaerobic digestion*. It traditionally used animal manure as biomass feedstock, but other biomass sources can be used within limits that are briefly discussed in the following. The set of biochemical reactions making up the digestion process (a term indicating the close analogy to energy extraction from biomass by food digestion) is schematically illustrated in Fig. 4.116.

Anaerobic digestion has three discernible stages. In the first, complex biomass material is decomposed by a heterogeneous set of microorganisms, not necessarily confined to anaerobic environments. These decompositions comprise hydrolysis of cellulosic material to simple glucose, using enzymes provided by the microorganisms as catalysts. Similarly, proteins are decomposed to amino acids and lipids to long-chain fatty acids. The significant result of the first stage is that most of the biomass is now water-soluble and in a simpler chemical form, suited for the next stage.

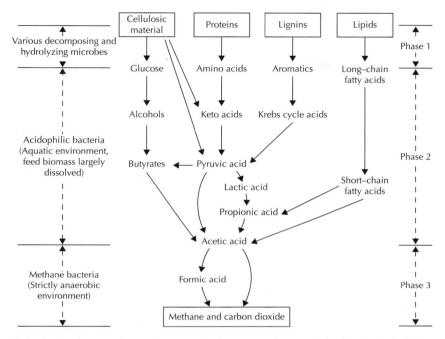

FIGURE 4.116   Simplified description of biochemical pathways in anaerobic digestion of biomass. (Based on Stafford *et al.*, 1981.) Used with permission from D. Stafford, D. Hawkes, and R. Horton, *Methane Production from Waste Organic Matter*. Copyright 1981 by The Chemical Rubber Co., CRC Press, Inc., Boca Raton, FL.

The second stage involves dehydrogenation (removing hydrogen atoms from the biomass material), such as changing glucose into acetic acid, carboxylation (removing carboxyl groups) of amino acids, and breakdown of the long-chain fatty acids into short-chain acids, again obtaining acetic acid as the final product. These reactions are fermentation reactions accomplished by a range of acidophilic (acid-forming) bacteria. Their optimum performance requires an environmental pH of 6–7 (slightly acid), but because the acids already formed will lower the pH of the solution, it is sometimes necessary to adjust the pH (e.g., by adding lime).

Finally, the third stage is the production of biogas (a mixture of methane and carbon dioxide) from acetic acid by a second set of fermentation reactions performed by methanogenic bacteria. These bacteria require a strictly anaerobic (oxygen-free) environment. Often, all processes are made to take place in a single container, but separation of the processes into stages allows greater efficiencies to be reached. The third stage takes on the order of weeks, while the preceding stages take on the order of hours or days, depending on the nature of the feedstock.

Starting from cellulose, the overall process may be summarized as

$$(C_6H_{10}O_5)_n + nH_2O \rightarrow 3nCO_2 + 3nCH_4 + 19n \, J \, mol^{-1}. \tag{4.152}$$

The first-stage reactions add up to

$$(C_6H_{10}O_5)_n + nH_2O \rightarrow nC_6H_{12}O_6. \tag{4.153}$$

The net result of the second-stage reactions is

$$nC_6H_{12}O_6 \rightarrow 3nCH_3COOH, \tag{4.154}$$

with intermediate steps, such as

$$C_6H_{12}O_6 \rightarrow 2C_2H_5OH + 2CO_2 \tag{4.155}$$

followed by dehydrogenation:

$$2C_2H_5OH + CO_2 \rightarrow 2CH_3COOH + CH_4. \tag{4.156}$$

The third-stage reactions then combine to

$$3nCH_3COOH \rightarrow 3nCO_2 + 3nCH_4. \tag{4.157}$$

In order for the digestion to proceed, a number of conditions must be fulfilled. Bacterial action is inhibited by the presence of metal salts, penicillin, soluble sulfides, or ammonia in high concentrations. Some source of nitrogen is essential for the growth of the microorganisms. If there is too little nitrogen relative to the amount of carbon-containing material to be transformed, then bacterial growth is insufficient and biogas is production low. With too much nitrogen (a carbon–nitrogen ratio below 15), "ammonia poisoning" of the bacterial cultures may occur. When the carbon–nitrogen ratio exceeds about 30, gas production starts diminishing, but in some systems carbon–nitrogen values as high as 70 have prevailed without problems (Stafford *et al.*, 1981). Table 4.4 gives carbon–nitrogen values for a number of biomass feedstocks. It is seen that mixing feedstocks can often be advantageous. For instance, straw and sawdust would have to be mixed with some low C:N material, such as livestock urine or clover/lucerne (typical secondary crops that may be grown in temperate climates after the main harvest).

If digestion time is not a problem, almost any cellulosic material can be converted to biogas, even pure straw. Initially, one may have to wait for several months, until the optimum bacterial composition has been reached, but then continued production can take place, and, despite low reaction rates, an energy recovery similar to that of manure can be achieved with properly prolonged reaction times (Mardon, 1982).

Average manure production for fully grown cows and pigs (in Europe, Australia, and the Americas) is 40 and 2.3 kg wet weight d^{-1}, respectively, corresponding to 62 and 6.2 MJ d^{-1}, respectively. The equivalent biogas production may reach 1.2 and 0.18 m^3 d^{-1}. This amounts to 26 and 3.8 MJ d^{-1}, or 42% and 61% conversion efficiency, respectively (Taiganides, 1974). A discussion of the overall efficiency, including transportation of biomass to the production plant, is given in Berglund and Börjesson (2002), who found maximum acceptable transport distances of 100–150 km.

The residue from the anaerobic digestion process has a higher value as a fertilizer than the feedstock. Insoluble organics in the original material are, to a large extent, made soluble, and nitrogen is fixed in the microorganisms.

Pathogen populations in the sludge are reduced. Stafford *et al.* (1981) found a 65%–90% removal of *Salmonella* during anaerobic fermentation, and there is a significant reduction in streptococci, coliforms, and viruses, as well as an almost total elimination of disease-transmitting parasites like *Ascaris*, hookworm, *Entamoeba*, and *Schistosoma*.

Material	Ratio
Sewage sludge	13:1
Cow dung	25:1
Cow urine	0.8:1
Pig droppings	20:1
Pig urine	6:1
Chicken manure	25:1
Kitchen refuse	6–10:1
Sawdust	200–500:1
Straw	60–200:1
Bagasse	150:1
Seaweed	80:1
Alfalfa hay	18:1
Grass clippings	12:1
Potato tops	25:1
Silage liquor	11:1
Slaughterhouse wastes	3–4:1
Clover	2.7:1
Lucerne	2:1

TABLE 4.4 Carbon–nitrogen ratios for various materials. (Based on Baader *et al.*, 1978; Rubins and Bear, 1942.)

For this reason, anaerobic fermentation has been used fairly extensively as a cleaning operation in city sewage treatment plants, either directly on the sludge or after algae are grown on the sludge to increase fermentation potential. Most newer sewage treatment plants make use of the biogas produced to fuel other parts of the treatment process, but, with proper management, sewage plants may well be net energy producers (Oswald, 1973).

The other long-time experience with biogas and associated fertilizer production is in rural areas of a number of Asian countries, notably China and India. The raw materials here are mostly cow dung, pig slurry, and human waste that is referred to as night soil, plus in some cases grass and straw. The biogas is used for cooking, and the fertilizer residue is returned to the fields. The sanitary aspect of pathogen reduction lends strong support to the economic viability of these schemes.

The rural systems are usually based on simple one-compartment digesters with human labor for filling and emptying of material. Operation is either in batches or with continuous new feed and removal of some 3%–7% of the reactor content every day. Semi-industrialized plants have also been built during the last decade, for example, in connection with large pig farms, where mechanized and highly automated collection of manure has been feasible. In some cases, these installations have utilized separate acid and methanation tanks.

Figure 4.117 shows an example of a town-scale digester plant, where the biogas is used in a combined electric power and district heat-generating plant (Kraemer, 1981). Expected energy flows are indicated. Storage of biogas for

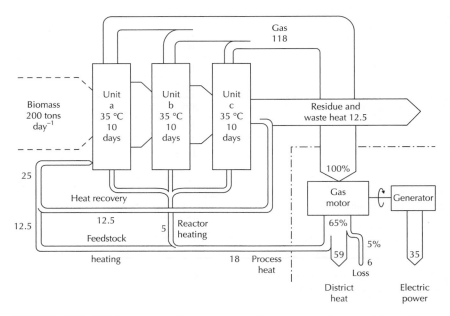

FIGURE 4.117 Calculated energy flows for a town biogas reactor plant, consisting of three successive units with 10 days' residence time in each. A biogas motor drives an electric generator, and the associated heat is, in part, recycled to the digestion process, while the rest is fed into the town's district heating lines. Flows (all numbers without indicated unit) are in GJ d^{-1}. (Based on energy planning for Nysted commune, Denmark; Kraemer, 1981.)

rural cooking systems is accomplished by variable-volume domes that collect the gas as it is produced (e.g., an inverted, water-locked can placed over the digester core). Biogas contains approximately 23 MJ m^{-3} and is therefore a medium-quality gas. CO_2 removal is necessary in order to achieve pipeline quality. An inexpensive way of achieving over 90% CO_2 removal is by water spraying. This method of producing compressed methane gas from biogas allows for automotive applications, such as a farmer producing his tractor fuel on site. In New Zealand, such uses have been developed since 1980 (see Fig. 4.118; Stewart and McLeod, 1980). Several demonstration experiences have recently been undertaken in Sweden (Losciale, 2002).

FIGURE 4.118 Schematic view of New Zealand scheme for methane production and vehicle use. (From Stewart, D., and McLeod, R., New Zealand Journal of Agriculture, Sept. 1980, 9-24. Used with permission.)

Storage of a certain amount of methane at ambient pressure requires over a thousand times more volume than the equivalent storage of oil. However, actual methane storage at industrial facilities uses pressures of about 140 times ambient (Biomass Energy Institute, 1978), so the volume penalty relative to oil storage would then be a factor of 9. Storage of methane in zeolitic material for later use in vehicles has been considered.

If residues are recycled, little environmental impact can be expected from anaerobic digestion. The net impact on agriculture may be positive, owing to nutrients being made more accessible and due to parasite depression. Undesirable contaminants, such as heavy metals, are returned to the soil in approximately the same concentrations as existed before collection, unless urban pollution has access to the feedstock. The very fact that digestion depends on biological organisms may mean that poor digester performance could serve as a warning signal, directing early attention to pollution of cropland or urban sewage systems. In any case, pollutant-containing waste, for example, from industry, should never be mixed with the energy-valuable biological material in urban refuse and sewage. Methane-forming bacteria are more sensitive to changes in their environment, such as temperature and acidity, than are acid-forming bacteria.

The digestion process itself does not emit pollutants if it operates correctly, but gas cleaning, such as H_2S removal, may lead to emissions. The methane gas itself shares many of the safety hazards of other gaseous fuels, being asphyxiating and explosive at certain concentrations in air (roughly 7%–14% by volume). For rural cooking applications, the risks may be compared with those of the fuels being replaced by biogas, notably wood burned in simple stoves. In these cases, as follows from the discussion in section 4.6.1, the environment is dramatically improved by the introduction of biogas digesters.

An example of early biogas plants for use on a village scale in China, India, and Pakistan is shown in Fig. 4.119. All the reactions take place in one compartment, which does not necessarily lead to optimal conversion efficiency. The time required for the acid-forming step is less than 24 h, whereas the duration of the second step should be 10–30 days. In many climatic regions, the heating of the fermentation tank required for application may be done by solar collectors, which could form the top cover of the tank. Alternatively, the top cover may be an inflatable dome serving as a store of gas, which can smooth out a certain degree of load variation. Some installations obtain the highest efficiency by batch operation, i.e., by leaving one batch of biological material in the tank for the entire fermentation period. The one shown in Fig. 4.119 allows continuous operation, i.e., a fraction of the slurry is removed every day and replaced by fresh biological material.

Examples of predicted biogas production rates, for simple plants of the type shown in Fig. 4.119 and based on fluid manure from dairy cows, pigs, or poultry, are shown in Table 4.5 and Fig. 4.120. Table 4.5 gives typical biogas production rates, per day and per animal, while Fig. 4.120 gives the conversion efficiencies measured, as functions of fermentation time (tank residence time), in a controlled experiment. The conversion efficiency is the ratio of the energy in the biogas (approximately 23 MJ m^{-3} of gas) and the energy in the manure that would be released as heat by complete burning of the dry matter (some typical absolute values are given in Table 4.5). The highest efficiency is obtained with pigs' slurry, but the high bacteriological activity in this case occasionally has the side-effect of favoring bacteria other than those helping to produce biogas, e.g., ammonia-producing bacteria, the activity of which may destroy the possibility of further biogas production (Olsen, 1975).

As mentioned, the manure residue from biogas plants has a high value as fertilizer because the decomposition of organic material followed by hydrocarbon conversion leaves plant nutrients (e.g., nitrogen that was originally bound in proteins) in a form suitable for uptake. Pathogenic bacteria and parasites are not removed to as high a degree as by composting, owing to the lower process temperature.

FIGURE 4.119 Single-chamber biogas plant. (Based on Chinese installations described by Appropriate Technology Development Organization, 1976.)

Source	Manure Per Day		Biogas Per Day	
	kg Wet Weight	MJ	m³	MJ
Cows	40	62	1.2	26
Pigs	2.3	6.2	0.18	3.8
Hens	0.19	0.9	0.011	0.26

TABLE 4.5 Manure and potential biogas production for a typical animal per day. (Based on Taiganides, 1974.)

Some city sewage plants produce biogas (by anaerobic fermentation of the sewage) as one step in the cleaning procedure, using the biogas as fuel for driving mechanical cleaning devices, etc. Thus, in many cases, it is possible to

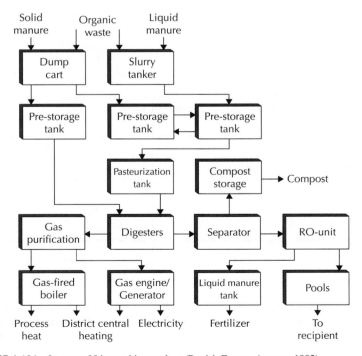

FIGURE 4.120 Measured conversion efficiencies (ratio between energy in biogas gas produced and energy in the manure) for simple biogas plants (like that in Fig. 4.119), with 10–13 kg of fresh manure added per day per m^3 of tank volume. (Based on Gramms *et al.*, 1971.)

avoid the need for other energy inputs and, in some cases, the plant becomes a net energy producer (Danish Energy Agency, 1992). Figure 4.121 shows the system layout for a biogas plant processing 300 t of biomass per day and accepting multiple types of feedstock; the plant is capable of delivering power, process and district heat, and fertilizer. Figure 4.122 gives the measured performance data for nine large prototype biogas plants in Denmark.

FIGURE 4.121   Layout of Lintrup biogas plant (Danish Energy Agency, 1992).

Energy Balance

The 1992 average production from large Danish biogas plants was 35.1 m^3 per m^3 of biomass (at standard pressure, methane content in the biogas being on average 64%), or 806 MJ m^{-3} (Tafdrup, 1993). In-plant energy use amounted to 90 MJ m^{-3}, distributed in 28 MJ of electricity and 50 MJ of heat, all produced by the biogas plant itself. Fuel used in transporting manure to the plant totalled 35 MJ, and the fertilizer value of the returned residue was estimated at 30 MJ. Thus, the net outside energy requirement was 5 MJ for a production of 716 MJ, or 0.7%, corresponding to an energy payback time of 3 days. If the in-plant biogas use is added, energy consumption in the process is 13%. The energy for construction of the plant, which has not been estimated, should also be added. However, the best plants roughly break even economically, indicating that the overall energy balance is acceptable.

Greenhouse Gas Emissions

Using, as in the energy balance section above, the average of large Danish plants as an example, the avoided CO_2 emission from having combined power and heat production use biomass instead of coal as a fuel is 68 kg per m^3 of biomass converted. Added should be emissions from transportation of biomass, estimated at 3 kg, and the avoided emissions from conventional production of fertilizer, which is replaced by biogas residue, estimated at 3 kg. Reduced methane emissions, relative to the case of spreading manure

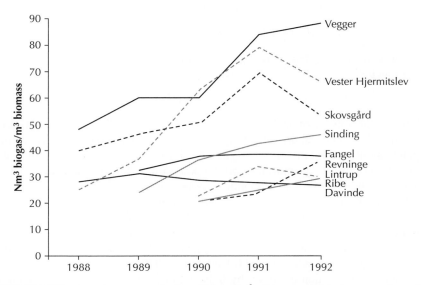

FIGURE 4.122 Annual average production efficiency (m^3 of biogas at standard pressure, denoted Nm^3, per m^3 of biomass feedstock) for nine community-size biogas plants in Denmark (Danish Energy Agency, 1992).

directly on fields, is of the order of 61 kg CO_2 equivalent (Tafdrup, 1993). The nitrous oxide balance is possibly improved by reducing denitrification in the soil, but high uncertainty makes actual estimates difficult at the present. The overall CO_2 reduction obtained by summing up the estimates given above is then 129 kg for each m^3 of biomass converted to biogas.

Other Environmental Effects

Compared to the current mix of coal, oil, and natural gas plants, biogas plants have two to three times lower SO_2 emission but a correspondingly higher NO_x emission. Higher ammonia content in the digested residue calls for greater care in using fertilizer from biogas plants, in order to avoid loss of ammonia. This is also true regarding avoiding loss of nutrients from fertilizer to waterways. Compared to unrefined manure, biogas-residue fertilizer has a marked gain in quality, including a better-defined composition, which contributes to using the correct dosage and avoiding losses to the environment. The dissemination of biogas plants removes the need for landfills, which is an environmental improvement. Odor is moved from the fields (where manure and slurry would otherwise be spread) to the biogas plant, where it can be controlled by suitable measures, such as filters (Tafdrup, 1993).

Hydrogen-producing Cultures

Biochemical routes to fuel production include a number of schemes not presently developed to a technical or economic level of commercial interest. Hydrogen is a fuel that may be produced directly by biological systems. Hydrogen enters in photosynthesis in green plants, where the net result of the process is

$$2H_2O + \text{solar radiation} \rightarrow 4e^- + 4H^+ + O_2.$$

However, the electrons and protons do not combine directly to form hydrogen,

$$4e^- + 4H^+ \rightarrow 2H_2,$$

but instead transfer their energy to a more complex molecule (NADPH$_2$; cf. Chapter 3), which can drive the CO_2 assimilation process. By this mechanism, plants avoid recombination of oxygen and hydrogen from the two processes mentioned above. Membrane systems keep the would-be reactants apart, and thus the energy-rich compound may be transported to other areas of the plant, where it takes part in plant growth and respiration.

Much thought has been given to modifications of plant material (e.g., by genetic engineering) in such a way that free hydrogen is produced on one side of a membrane and free oxygen on the other side (Berezin and Varfolomeev, 1976; Calvin, 1974; Hall et al., 1979).

While dissociation of water (by light) into hydrogen and oxygen (photolysis; cf. Chapter 5) does not require a biological system, it is possible that utilization of the process on a realistic scale can be more easily achieved if the critical components of the system, notably the membrane and the electron transport system, are of biological origin. Still, much more research is required before any thought can be given to practical application of direct biological hydrogen production cultures.

Thermochemical Gasification of Biomass

Conversion of fossil biomass, such as coal, into a gas is considered a way of reducing the negative environmental effects of coal utilization. However, in some cases, the effects have only been moved, not eliminated. Consider, for example, a coal-fired power plant with 99% particle removal from flue gases. If it were to be replaced by a synthetic gas-fired power plant with gas produced from coal, then sulfur could be removed at the gasification plant using dolomite-based scrubbers. This would practically eliminate the sulfur oxide emissions, but, on the other hand, dust emissions from the dolomite processing would represent particle emissions twice as large as those avoided at the power plant by using gas instead of coal (Pigford, 1974). Of course, the dust is emitted at a different location.

This example, as well as the health effects of both surface and underground coal mining (and the effects are not identical), has sparked interest in methods of gasifying coal *in situ*. Two or more holes are drilled. Oxygen (or air) is injected through one, and a mixture of gases, including hydrogen and carbon oxides, emerges at the other hole. The establishment of proper communication between holes, and suitable underground contact surfaces, has proved difficult, and recovery rates are modest.

The processes involved include

$$2C + O_2 \rightarrow 2CO, \tag{4.158}$$

$$CO + H_2O \leftrightarrow CO_2 + H_2. \tag{4.159}$$

The stoichiometric relation between CO and H_2 can then be adjusted using the shift reaction (4.159), which may proceed in both directions, depending on steam temperature and catalysts. This opens the way for methane synthesis through the reaction

$$CO + 3H_2 \rightarrow CH_4 + H_2O. \tag{4.160}$$

At present, the emphasis is on improving gasifiers using coal already extracted. Traditional methods include the Lurgi fixed-bed gasifier (providing gas under pressure from noncaking coal at a conversion efficiency as low as 55%) and the Koppers–Totzek gasifier (oxygen input, the produced gas unpressurized, also of low efficiency).

Improved process schemes include the hy-gas process, requiring a hydrogen input; the bi-gas concept of brute force gasification at extremely high temperatures; and the slagging Lurgi process, capable of handling powdered coal (Hammond, 1976).

FIGURE 4.123   Schematic diagram of catalytic gasification process (SNG is synthetic natural gas). (From Hirsch *et al.*, 1982. Reprinted from *Science*, **215**, 121-127, 8, January 1982, with permission. Copyright 1982 American Association for the Advancement of Science.)

Promising, but still at an early stage of development, is catalytic gasification (e.g., using a potassium catalyst), where all processes take place at one relatively low temperature, so that they can be combined in a single reactor (Fig. 4.123). The primary reaction here is

$$C + H_2O \rightarrow H_2 + CO \tag{4.161}$$

(H_2O being in the form of steam above 550 °C), to be followed by (4.156) and (4.160). The catalyst allows all processes to take place at 700 °C. Without a catalyst, the gasification would have to take place at 925 °C and the shift reaction and methanation at 425 °C, that is, in a separate reactor where excess hydrogen or carbon monoxide would be lost (Hirsch *et al.*, 1982).

In a coal gasification scheme, storage would be (of coal) before conversion. Peat can be gasified in much the same way as coal, as can wood (with optional subsequent methanol conversion, as described in section 4.8.3).

Fresh Biomass Gasification

Gasification of biomass, and particularly wood and other lignin-containing cellulosic material, has a long history. The processes may be viewed as "combustion-like" conversion, but with less oxygen available than needed for burning. The ratio of oxygen available and the amount of oxygen that would allow complete burning is called the *equivalence ratio*. For equivalence ratios below 0.1, the process is called *pyrolysis*, and only a modest fraction of the biomass energy is found in the gaseous product—the rest being in char and oily residues. If the equivalence ratio is between 0.2 and 0.4, the process is called a proper *gasification*. This is the region of maximum energy transfer to the gas (Desrosiers, 1981).

The chemical processes involved in biomass gasification are similar to reactions (4.158)–(4.161) for coal gasification. Table 4.6 lists a number of reactions involving polysaccharidic material, including pyrolysis and gasification. In addition to the chemical reaction formulae, the table gives enthalpy changes for idealized reactions (i.e., neglecting the heat required to bring the reactants to the appropriate reaction temperature).

Figure 4.124 gives the energy of the final products, gas and char, as a function of the equivalence ratio, still based on an idealized thermodynamic calculation. The specific heat of the material is 3 kJ g^{-1} of wood at the peak of energy in the gas, increasing to 21 kJ g^{-1} of wood for combustion at equivalence ratio equal to unity. Much of this sensible heat can be recovered from the gas, so that process heat inputs for gasification can be kept low.

Chemical Reaction	Energy Consumed (kJ g^{-1})[a]	Products / Process
$C_6H_{10}O_5 \rightarrow 6C + 5H_2 + 2.5\ O_2$	5.94[b]	Elements, dissociation
$C_6H_{10}O_5 \rightarrow 6C + 5H_2O(g)$	−2.86	Charcoal, charring
$C_6H_{10}O_5 \rightarrow 0.8\ C_6H_8O +$ 1.8 $H_2O(g) + 1.2\ CO_2$	−2.07[c]	Oily residues, pyrolysis
$C_6H_{10}O_5 \rightarrow 2C_2H_4 + 2CO_2 + H_2O(g)$	0.16	Ethylene, fast pyrolysis
$C_6H_{10}O_5 + \frac{1}{2}O_2 \rightarrow 6CO + 5H_2$	1.85	Synthesis gas, gasification
$C_6H_{10}O_5 + 6H_2 \rightarrow 6"CH_2" + 5H_2O(g)$	−4.86[d]	Hydrocarbons, –generation
$C_6H_{10}O_5 + 6O_2 \rightarrow 6CO_2 + 5H_2O(g)$	−17.48	Heat, combustion

TABLE 4.6 Energy change for idealized cellulose thermal conversion reactions. (Source: T. Reed (1981), in *Biomass Gasification* (T. Reed, ed.), reproduced with permission. Copyright 1981, Noyes Data Corporation, Park Ridge, NJ.)

[a]*Specific reaction heat.*
[b]*The negative of the conventional heat of formation calculated for cellulose from the heat of combustion of starch.*
[c]*Calculated from the data for the idealized pyrolysis oil C_6H_8O ($\Delta H_c = −745.9$ kcal mol^{-1}, $\Delta H_f = 149.6$ kcal g^{-1}, where $H_c =$ heat of combustion and $H_f =$ heat of fusion).*
[d]*Calculated for an idealized hydrocarbon with ΔH_c as above. H_2 is consumed.*

Figure 4.125 gives the equilibrium composition calculated as a function of the equivalence ratio. *Equilibrium composition* is defined as the composition of reaction products occurring after the reaction rates and reaction temperature have stabilized adiabatically. The actual processes are not necessarily adiabatic; in particular, the low-temperature pyrolysis reactions are not. Still, the

theoretical evaluations assuming equilibrium conditions serve as a useful guideline for evaluating the performance of actual gasifiers.

The idealized energy change calculation of Table 4.6 assumes a cellulosic
composition like the one in (4.153). For wood, the average ratios of carbon,
hydrogen, and oxygen are 1:1.4:0.6 (Reed, 1981).

FIGURE 4.124 Calculated energy content in gas and char produced by equilibrium processes between air (or oxygen) and biomass, as a function of equivalence ratio. (From Reed, 1981. Reprinted from *Biomass Gasification* (T. Reed, ed.), with permission. Copyright 1981, Noyes Data Corporation, Park Ridge, NJ.)

FIGURE 4.125 Calculated gas composition resulting from equilibrium processes between air and biomass, as a function of equivalence ratio. (From Reed, 1981. Reprinted from *Biomass Gasification* (T. Reed, ed.), with permission. Copyright 1981, Noyes Data Corporation, Park Ridge, NJ.)

Figure 4.126 shows three examples of wood-gasifiers: the updraft, the downdraft, and the fluidized-bed types. The drawback of the updraft type is a high rate of oil, tar, and corrosive chemical formation in the pyrolysis zone. This problem is solved by the downdraft version, where oils and other matter pass through a hot charcoal bed in the lower zone of the reactor and become simpler gases or char. The fluidized-bed reactor may prove superior for large-scale operations, because its passage time is smaller. Its drawback is that ash and tars are carried along with the gas and have to be removed later in cyclones and scrubbers. Several variations on these gasifier types have been suggested (Drift, 2002; Gøbel et al., 2002).

The gas produced by gasification of biomass is a "medium-quality" gas; meaning a gas with burning value in the range 10–18 MJ m^{-3}. This gas may be used directly in Otto or diesel engines, it may be used to drive heat pump compressors, or alternatively, it may be upgraded to pipeline-quality gas (about 30 MJ m^{-3}) or converted to methanol, as discussed in section 4.6.3.

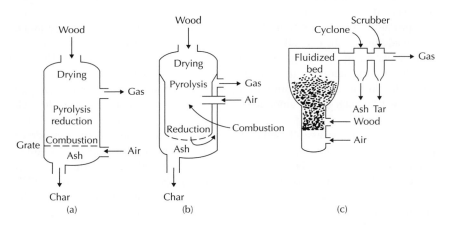

FIGURE 4.126a,b,c Gasifier types: (a) updraft, (b) downdraft, and (c) fluidized bed.

Environmental impacts of biomass gasifiers derive from biomass production, collection (e.g., forestry work), and transport to the gasification site, from the gasification and related processes, and finally from the use made of the gas. The gasification residues—ash, char, liquid wastewater, and tar—have to be disposed of. Char may be recycled to the gasifier, while ash and tars could conceivably be utilized in road building or the construction industry. The alternative, landfill disposal, is a recognized impact. Investigations of emissions from combustion of producer gas find lower emissions of nitrous oxides and hydrocarbons than in combustion of natural gas. In one case, carbon monoxide emissions were found to be higher than for natural gas burning, but it is believed that this can be rectified as more experience in adjusting air-to-fuel ratios is gained (Wang et al., 1982).

4.6.3 Biological Conversion into Liquid Biofuels

Anaerobic fermentation processes may also be used to produce liquid fuels from biological raw materials. An example is ethanol production (4.152) from glucose, known as standard yeast fermentation in the beer, wine, and liquor industries. It has to take place in steps, so that the ethanol is removed (by distillation or dehydrator application) whenever its concentration approaches a value (around 12%) that would impede reproduction of the yeast culture.

To reduce the primary biological material (e.g., molasses, cellulose pulp, or citrus fruit wastes) to glucose, hydrolysis (4.157) may be used. Some decomposition takes place in any acid solution, but in order to obtain complete hydrolysis, specific enzymes must usually be provided, either directly or by adding microorganisms capable of forming such enzymes. The yeast themselves contain enzymes capable of decomposing polysaccharides into glucose. The theoretical maximum efficiency of glucose-to-ethanol conversion (described in more detail below) is 97%. According to Calvin (1977), in 1974 the Brazilian alcohol industry obtained 14% of the energy in raw sugar input in the form of ethanol, which was produced by fermentation of just the molasses residues from sugar refining, i.e., in addition to the crystallized sugar produced. Currently, the figure is 25% of input sugar energy (see Fig. 4.128) for an optimized plant design (EC, 1994).

Mechanical energy inputs, e.g., for stirring, could be covered by burning the fermentation wastes in a steam power plant. In the European example illustrated in Fig. 4.128, such inputs amount to about a third of the energy inputs through the sugar itself.

Fermentation processes based on molasses or other sugar-containing materials produce acetone–butanol, acetone–ethanol, or butanol–isopropanol mixtures when the proper bacteria are added. In addition, carbon dioxide and small amounts of hydrogen are formed (see, for example, Beesch, 1952; Keenan, 1977).

Conversion of fossilized biomass into liquid fuels is briefly mentioned at the beginning of section 4.6.2, in conjunction with Fig. 4.111, which shows an overview of the conversion routes open for biofuel generation. Among the non-food energy uses of biomass, there are several options leading to liquid fuels that may serve as a substitute for oil products. The survey of conversion processes given in Fig. 4.111 indicates liquid end-products as the result of either biochemical conversion using fermentation bacteria or a thermochemical conversion process involving gasification and, for example, further methanol synthesis. The processes that convert biomass into liquid fuels that are easy to store are discussed below, but first the possibility of direct production of liquid fuels by photosynthesis is presented.

Direct Photosynthetic Production of Hydrocarbons

Oil from the seeds of many plants, such as rape, olive, groundnut, corn, palm, soybean, and sunflower, is used as food or in the preparation of food. Many of these oils will burn readily in diesel engines and can be used directly or mixed with diesel oil of fossil origin, as they are indeed in several pilot projects around the world. However, for most of these plants, oil does not constitute a major fraction of the total harvest yield, and use of these crops to provide oil for fuel use could interfere with food production. A possible exception is palm oil, because intercropping of palm trees with other food crops may prove advantageous, by retaining moisture and reducing wind erosion.

Therefore, interest is concentrated on plants that yield hydrocarbons and that, at the same time, are capable of growing on land unsuited for food crops. Calvin (1977) first identified the Euphorbia plant genus as a possibility. The rubber tree, *Hevea brasiliensis*, is in this family, and its rubber is a hydrocarbon–water emulsion, the hydrocarbon of which (polyisoprenes) has a large molecular weight, about a million, making it an elastomer. However, other plants of the genus *Euphorbia* yield latex of much lower molecular weight which could be refined in much the same way as crude oil. In the tropical rainforests of Brazil, Calvin found a tree, *Copaifera langsdorffii*, which annually yields some 30 liters of virtually pure diesel fuel (Maugh, 1979). Still, interest centers on species that are likely to grow in arid areas like the deserts of the southern United States, Mexico, Australia, and so on.

Possibilities include *Euphorbia lathyris* (gopher plant*)*, *Simmondsia chinensis* (jojoba), *Cucurbita foetidissima* (buffalo gourd), and *Parthenium argentatum* (guayule). The gopher plant has about 50% sterols (on a dry weight basis) in its latex, 3% polyisoprene (rubber), and a number of terpenes. The sterols are suited as feedstocks for replacing petroleum in chemical applications. Yields of first-generation plantation experiments in California are 15–25 barrels of crude oil equivalent or some 144 GJ ha^{-1} (i.e., per 10^4 m^2). In the case of *Hevea*, genetic and agronomic optimization has increased yields by a factor of 2000 relative to those of wild plants, so high hydrocarbon production rates should be achievable after proper development (Calvin, 1977; Johnson and Hinman, 1980; Tideman and Hawker, 1981). Other researchers are less optimistic (Stewart *et al.*, 1982; Ward, 1982).

Alcohol Fermentation

The ability of yeast and bacteria like *Zymomonas mobilis* to ferment sugar-containing material to form alcohol is well known from beer, wine, and liquor manufacture. If the initial material is cane sugar, the fermentation reaction may be summarized as

$$C_6H_{12}O_6 \rightarrow 2C_2H_5OH + 2CO_2. \qquad (4.162)$$

The energy content of ethanol is 30 MJ kg^{-1}, and its octane rating is 89–100. With alternative fermentation bacteria, the sugar may be converted into butanol, $C_2H_5(CH_2)_2OH$. In Brazil, the cost of ethanol has finally been lowered to equal that of gasoline (Johansson, 2002).

In most sugar-containing plant material, the glucose molecules exist in polymerized form, such as starch or cellulose, of the general structure $(C_6H_{10}O_5)_n$.

Starch or hemicellulose is degraded to glucose by hydrolysis (cf. Fig. 4.111), while lignocellulose resists degradation owing to its lignin content. Lignin glues the cellulosic material together to keep its structure rigid, whether it be crystalline or amorphous. Wood has a high lignin content (about 25%), and straw also has considerable amounts of lignin (13%), while potato and beet starch contain very little lignin.

Some of the lignin seals may be broken by pretreatment, ranging from mechanical crushing to the introduction of swelling agents causing rupture (Ladisch et al., 1979).

The hydrolysis process is given by (4.153). In earlier times, hydrolysis was always achieved by adding an acid to the cellulosic material. During both world wars, Germany produced ethanol from cellulosic material by acid hydrolysis, but at very high cost. Acid recycling is incomplete; with low acid concentration, the lignocellulose is not degraded, and, with high acid concentration, the sugar already formed from hemicellulose is destroyed.

Consequently, alternative methods of hydrolysis have been developed, based on enzymatic intervention. Bacteria (e.g., of the genus *Trichoderma*) and fungi (such as *Sporotrichum pulverulentum*) have enzymes that have proved capable of converting cellulosic material, at near ambient temperatures, to some 80% glucose and a remainder of cellodextrins (which could eventually be fermented, but in a separate step with fermentation microorganisms other than those responsible for the glucose fermentation) (Ladisch et al., 1979).

The residue left behind after the fermentation process (4.162) can be washed and dried to give a solid product suitable as fertilizer or as animal feed. The composition depends on the original material, in particular with respect to lignin content (small for residues of molasses, beets, etc., high for straws and woody material, but with fiber structure broken as a result of the processes described above). If the lignin content is high, direct combustion of the residue is feasible, and it is often used to furnish process heat to the final distillation.

The outcome of the fermentation process is a water–ethanol mixture. When the alcohol fraction exceeds about 10%, the fermentation process slows down and finally halts. Therefore, an essential step in obtaining fuel alcohol is to separate the ethanol from the water. Usually, this is done by distillation, a step that may make the overall energy balance of ethanol production negative. The sum of agricultural energy inputs (fertilizer, vehicles, machinery) and all process inputs (cutting, crushing, pretreatment, enzyme recycling, heating for different process steps from hydrolysis to distillation), as well as energy for transport, is, in existing operations, such as those of the Brazilian alcohol program (Trinidade, 1980), around 1.5 times the energy outputs (alcohol and fertilizer if it is utilized). However, if the inputs are domestic fuels, for example, combustion of residues from agriculture, and if the alcohol produced is used to displace imported oil products, the balance might still be quite acceptable from a national economic point of view.

If, further, the lignin-containing materials of the process are recovered and used for process heat generation (e.g., for distillation), then such energy should be counted not only as input but also as output, making the total input and output energy roughly balance. Furthermore, more sophisticated process design, with cascading heat usage and parallel distillation columns operating with a time displacement such that heat can be reused from column to column (Hartline, 1979), could reduce the overall energy inputs to 55%–65% of the outputs.

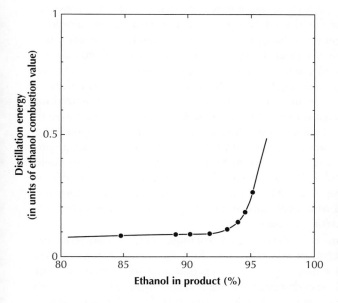

FIGURE 4.127   Distillation energy for ethanol–water mixture, as a function of ethanol content (assumed initial ethanol fraction 12%). (Based on Ladisch and Dyck, 1979.)

Energy balances would be radically improved if distillation could be replaced by a less energy-intensive separation method. Several such methods for separating water and ethanol have been demonstrated on a laboratory scale, including drying with desiccants, such as calcium hydroxide, cellulose, or starch (Ladisch and Dyck, 1979); gas chromatography using rayon to retard water, while organic vapors pass through; solvent extraction using dibutyl phthalate, a water-immiscible solvent of alcohols; and passage through semi-permeable membranes or selective zeolite absorbers (Hartline, 1979) and phase separation (APACE, 1982). The use of dry cellulose or starch appears particularly attractive, because over 99% pure alcohol can be obtained with less than 10% energy input, relative to the combustion energy of the alcohol. Furthermore, the cellulosic material may be cost-free, if it can be taken from the input stream to the fermentation unit and returned to it after having absorbed water (the fermentation reaction being "wet" anyway). The energy input of this scheme is for an initial distillation, bringing the ethanol fraction of the aqueous mixture from the initial 5%–12% up to about 75%, at which point the desiccation process is started. As can be seen from Fig. 4.127, the distillation energy is modest up to an alcohol content of 90% and then starts to rise rapidly. The drying process thus substitutes for the most energy-expensive part of the distillation process.

Ethanol fuel can be stored and used in the transportation sector much the same way as gasoline is used. It can be mixed with gasoline or can fully replace gasoline in spark ignition engines with high compression ratios (around 11). The knock resistance and high octane number of ethanol make this possible, and with preheating of the alcohol (using combustion heat that is recycled), the conversion efficiency can be improved. Several countries presently use alcohol–gasoline blends with up to 10% ethanol, and these blends do not require any engine modification. Altering gasoline Otto engines may be inconvenient in a transition period, but if alcohol distribution networks are implemented and existing gas stations modified, then car engines could be optimized for alcohol fuels without regard to present requirements. A possible alternative to spark ignition engines is compression ignition engines, where auto-ignition of the fuel under high compression (a ratio of 25) replaces spark or glow plug ignition. With additives or chemical transformation into acetal, alcohol fuels could be used this way (Bandel, 1981). Ethanol does not blend with diesel oil, so mixtures require the use of special emulsifiers (Reeves et al., 1982). However, diesel oil can be mixed with other biofuels without problems, e.g., the plant oils (rapeseed oil, etc.) presently in use in Germany.

A number of concerns about the environmental impact of the ethanol fermentation energy conversion chain must be considered. First, the biomass being used may have direct uses as food or may be grown in competition with food production. The reason is, of course, that the easiest ethanol fermentation is obtained by starting with a raw material with the highest possible elementary sugar content, that is, starting with sugar cane or cereal grain. Since sugar cane is likely to occupy prime agricultural land, and cereal production must increase with increasing world population, neither of these biomass resources should be used as fermentation inputs. However, residues from cereal production and from necessary sugar production (in many regions of the world, sugar consumption is too high from a health and nutrition point of view) could be used for ethanol fermentation, together with urban refuse, extra crops on otherwise committed land, perhaps aquatic crops, and forest renewable resources. The remarks made in Chapter 3 about proper soil management, recycling nutrients, and covering topsoil to prevent erosion also apply to the enhanced tillage utilization that characterizes combined food and ethanol production.

The hydrolysis process involves several potential environmental effects. If acids are used, corrosion and accidents may occur, and substantial amounts of water would be required to clean the residues for re-use. Most of the acid would be recovered, but some would follow the sewage stream. Enzymatic hydrolysis seems less risky. Most of the enzymes would be recycled, but some might escape with wastewater or residues. Efforts should be made to ensure that they are made inactive before release. This is particularly important when, as envisaged, the fermentation residues are to be brought back to the fields or used as animal feed. A positive impact is the reduction of pathogenic organisms in residues after fermentation.

Transport of biomass could involve dust emissions, and transport of ethanol might lead to spills (in insignificant amounts, as far as energy is concerned, but with possible local environmental effects), but, overall, the impact from transport would be very small.

Finally, the combustion of ethanol in engines or elsewhere leads to pollutant emissions. Compared with gasoline combustion, ethanol combustion in modified ignition engines has lower emissions of carbon monoxide and hydrocarbons, but increased emissions of nitrous oxides, aromatics, and aldehydes (Hespanhol, 1979). When special ethanol engines and exhaust controls are used, critical emissions may be controlled. In any case, the lead pollution from gasoline engines still ongoing in some countries would be eliminated.

FIGURE 4.128 Energy flows in the production of ethanol from sugar beets (EC, 1994). Energy inputs to biomass growth, harvesting, and transport are not indicated.

The energy balance of current ethanol production from biomass is not very favorable. A European study has estimated the energy flows for a number of feedstocks (EC, 1994). The highest yield, about 100 GJ ha^{-1}, is for sugar beets, shown in Fig. 4.128, but the process energy inputs and allotted life-cycle inputs into technical equipment are as large as the energy of the ethanol produced. A part of the process energy may be supplied by biogas co-produced with the ethanol, but the overall energy efficiency remains low.

In a life-cycle analysis of ethanol production (cf. Chapter 7), the fact that such production is currently based upon energy crops rather than on residues (sugar cane or beets, rather than straw and husk) means that all energy inputs and environmental effects from the entire agricultural feedstock production should be included, along with the effects pertaining to ethanol plants and downstream impact. Clearly, in this mode it is very difficult to balance energy outputs and inputs and to reach acceptable impact levels. Interest should therefore be limited to bioenergy processes involving only residues from an otherwise sensible production (food or wood).

Methanol from Biomass

There are various ways of producing methanol from biomass sources, as indicated in Fig. 4.111. Starting from wood or isolated lignin, the most direct routes are liquefaction and gasification. Pyrolysis gives only a fraction of energy in the form of a producer gas.

High-pressure hydrogenation transforms biomass into a mixture of liquid hydrocarbons suitable for further refining or synthesis of methanol (Chartier and Meriaux, 1980), but all methanol production schemes so far have used a synthesis gas, which may be derived from wood gasification or coal gasification. The low-quality "producer gas" resulting directly from wood gasification (used, as mentioned, in cars throughout Europe during World War II) is a mixture of carbon monoxide, hydrogen gas, carbon dioxide, and nitrogen gas (see section 4.6.2). If air is used for gasification, the energy conversion efficiency is about 50%, and if pure oxygen is used instead, some 60% efficiency is possible and the gas produced has a lower nitrogen content (Robinson, 1980). Gasification or pyrolysis could conceivably be performed with heat from (concentrating) solar collectors, for example, in a fluidized-bed gasifier maintained at 500 °C.

The producer gas is cleaned, CO_2 and N_2 as well as impurities are removed (the nitrogen by cryogenic separation), and methanol is generated at elevated pressure by the reaction

$$2H_2 + CO \to CH_3OH. \tag{4.163}$$

The carbon monoxide and hydrogen gas (possibly with additional CO_2) is called the *synthesis gas*, and it is usually necessary to use a catalyst in order to maintain the proper stoichiometric ratio between the reactants of (4.163) (Cheremisinoff *et al.*, 1980). A schematic process diagram is shown in Fig. 4.129.

An alternative is biogas production from the biomass (section 4.6.2) followed by the methane to methanol reaction,

$$2CH_4 + O_2 \to 2CH_3OH, \tag{4.164}$$

also used in current methanol production from natural gas (Wise, 1981). Change of the H_2/CO stoichiometric ratio for (4.163) is obtained by the *shift reaction* (4.159) discussed in section 4.6.2. Steam is added or removed in the presence of a catalyst (iron oxide, chromium oxide).

The conversion efficiency of the synthesis-gas-to-methanol step is about 85%, implying an overall wood to methanol energy efficiency of 40%–45%. Improved catalytic gasification techniques raise the overall conversion efficiency to about 55% (Faaij and Hamelinck, 2002). The efficiency currently achieved is about 50%, but not all life-cycle estimates of energy inputs have been included or performed (EC, 1994).

The octane number of methanol is similar to that of ethanol, but the heat of combustion is less, amounting to 18 MJ kg^{-1}. However, the engine efficiency of methanol is higher than that of gasoline, by at least 20% for current automobile car engines, so an "effective energy content" of 22.5 MJ kg^{-1} is sometimes quoted (EC, 1994). Methanol can be mixed with gasoline in standard engines, or used in specially designed Otto or diesel engines, such as a spark ignition engine run on vaporized methanol, with the vaporization energy being recovered from the coolant flow (Perrin, 1981). The uses of methanol are similar

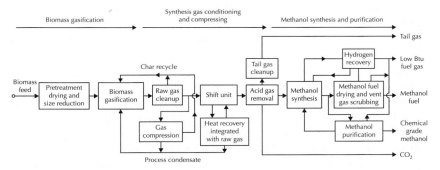

FIGURE 4.129 Schematic flow diagram for conversion of biomass to methanol. (From Wan, Simmins, and Nguyen, 1981. Reprinted from *Biomass Gasification* (T. Reed, ed.), with permission. Copyright 1981, Noyes Data Corporation, Park Ridge, NJ.)

to those of ethanol, but there are several differences in the environmental impact from production to use (e.g., toxicity of fumes at filling stations).

The future cost of methanol fuel is expected to reach US$ 8/GJ (Faaij and Hamelinck, 2002). The gasification can be done in closed environments, where all emissions are collected, as well as ash and slurry. Cleaning processes in the methanol formation steps will recover most catalysts in re-usable form, but other impurities would have to be disposed of, along with the gasification products. Precise schemes for waste disposal have not been formulated, but it seems unlikely that all nutrients could be recycled to agri- or silviculture, as in the case for ethanol fermentation (SMAB, 1978). However, production of ammonia by a process similar to the one yielding methanol is an alternative use of the synthesis gas. Production of methanol from *Eucalyptus* rather than from woody biomass has been studied in Brazil (Damen *et al.*, 2002). More fundamental research, aimed at a better understanding of how methanol production relies on degradation of lignin, is ongoing (Minami *et al.*, 2002).

4.6.4 Enzymatic Decomposition of Cellulosic Material

Sustainable use of biomass for fuels requires CO_2 neutrality and recycling of nutrients, as well as recognition of the use of biomass for other purposes, such as food production, construction materials, and other uses (e.g., fiber for materials). The different uses of biomass in society must be balanced, and meeting the food requirements of the world's large human population naturally has first priority.

The difference between edible and non-edible biomass is largely that the edible parts (cereals, fruits, etc.) contain amylose (starch) or fructose and the non-edible parts have a strong component of cellulose (which can be used as fiber for clothes and other materials or in construction materials). As illustrated in Fig. 4.130, amylose is made from repeating glucose segments, while in cellulose, each second glucose-molecule segment is turned 180°. This "small" difference makes cellulose fibers very strong, while starch pieces are curling, separating, fragile, and, first of all, digestible.

Agricultural and forestry residues, usually straw or woodchips plus some wet components, are mostly composed of cellulose (long β-(1,4)-glucan polymers), hemicellulose (short, branched polymers of sugars), and lignin (polymers derived from coumaryl, coniferyl and sinapyl alcohol). Transformation of these components to transportation fuels uses a variety of methods that exhibit different efficiencies for the three types of components. It has therefore been suggested that genetic modification of the primary plants could result in more amenable starting points for biofuel synthesis (Ragauskas *et al.*, 2006).

In contrast to starch sugar crops, such as cereal grains, currently used by the biofuel industry (chiefly in Brazil and the United States), which can be transformed into ethanol by fermentation, and the oil-containing plant parts (e.g., rapeseed) that can be transformed into biodiesel fuel by transesterification (in Germany), lignin and cellulosic polymers require more elaborate processes. The starch- and oil-based fuels are called *first-generation biofuels*, and they are in competition with food production. Those that are not are called *second-generation biofuels* (Naik *et al.*, 2010).

The second-generation biofuel processes currently receiving most attention entail: first, mechanical or chemical pretreatment to break down the fiber structure of the biomass feedstock, allowing hydrolytic transformation of cellulose or lignin into simple sugars; and, subsequently, a fermentation process to arrive at the biofuel. Hydrolysis is basically the splitting of water into H^+ ions and OH^- ions, but, in a biomass context, it describes the transformation of polysaccharide chains like those shown in Fig. 4.20 into simple soluble sugars. Hydrolysis can be monitored in real time by measuring heat flows associated with it (Murphy *et al.*, 2010).

Amylose

Cellulose

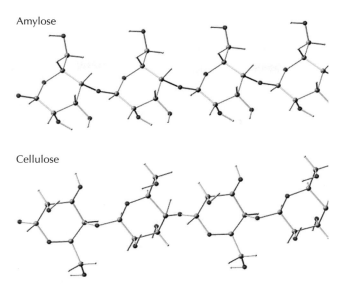

FIGURE 4.130   Differing structures of amylose (starch) and cellulose chains (4 segments generated by simple Hartree-Fock quantum molecular optimization). Atoms are oxygen (darkest), carbon, and hydrogen (smallest).

There are several alternative routes from biomass to biofuels, such as high-temperature processes employing pyrolysis of the biomass into bio-oil, liquefaction into heavy oil, or the Fischer-Tropsch syngas route (with the shift reactions mentioned above after gasification in order to produce hydrogen or, further, hydrocarbons). Biological fermentation into methane also may be used to arrive at ethanol as an end-product, using suitable enzymes, and chemical reactions other than those involved in hydrolysis can be used (Perego and Bianchi, 2010). At present, all of these alternatives have conversion efficiencies below 50%.

Pretreatment may include mechanical crushing, acid treatment, and various catalytic processes. Pretreatment processes are required to allow operation at modest temperatures (200 °C or lower) and to reach decent efficiencies (Öhgren et al., 2007; Yeh et al., 2010; Perego and Bianchi, 2010), but care should be taken in recovering the acids and catalysts used, because otherwise large-scale production is not environmentally acceptable (Abbasi and Abbasi, 2010).

Different saccharification enzymes have been tried, most of them cellulases, which are complex mixtures of cellulose-degrading agents. However, the enzymatic processes still require that pretreatment has taken place to open the fiber structure and separate lignin from the celluloses (alkali processes, weak or strong acid processes, and steam explosion have been tried; see Perego and Bianchi, 2010). Pretreatment is the step that contributes most to making second-generation biofuels more expensive to produce than first-generation biofuels.

Chapter 5

Energy Transmission and Storage

5.1 ENERGY TRANSMISSION

Transport of energy may, of course, be in the form of transportation of fuels to the site of conversion. With regard to renewable energy resources, transport is useful for biomass-derived energy, either by direct movement of the biological materials themselves or by movement after their conversion into biofuels (cf. section 4.6), which may be more convenient to move. For most other types of renewable energy, the resource itself cannot be "moved" (possible exceptions may exist, such as diverting a river flow to the place of hydropower utilization). Instead, an initial conversion process may be performed, and the emerging energy form may be transmitted to the load areas, where it may be used directly or subjected to a second conversion process before delivery to the actual users.

Like fuels (which represent chemical energy), heat and mechanical and possibly electrical energy may be stored, and the storage "containers" may be transported. Alternatively, energy may be transmitted through a suitable transmission system, which may be a pipeline (for heat, fuels, and certain types of mechanical energy, such as pressure or kinetic energy of a gas or a fluid), electric lines (for electricity), or a radiant transmission system (for heat or electricity).

Energy transmission is used not only to deliver energy from the sites of generation (such as where the renewable resources are) to the dominant sites of energy use, but also to deal with temporal mismatch between (renewable) energy generation and variations in demand. Therefore, energy transmission and energy storage may supplement each other. Some demands may be displaceable, while others are time-sensitive. The latter often display a systematic variation over the hours of the day and over the seasons. This may be taken advantage of by long-distance transmission of energy across time zones (east–west).

Renewable Energy, Fourth Edition
© 2011, 2004, 2000 Elsevier Ltd. All rights reserved.

5.1.1 Heat Transmission

District Heating Lines

Most heating systems involve the transport of sensible heat in a fluid or gas (such as water or air) through pipes or channels. Examples are the solar heating system illustrated in Fig. 4.84, the heat pump illustrated in Fig. 4.6, and the geothermal heating plants illustrated in Fig. 4.9. Solar heating systems and heat pumps may be used in a decentralized manner, with an individual system providing heat for a single building, but they may also be used on a community scale, with one installation providing heat for a building block, a factory complex, a whole village, or a city of a certain size. Many combined heat and power (CHP) systems involve heat transmission through distances of 10–50 km (decentralized and centralized CHP plants). In some regions, pure heating plants are attached to a district heating grid.

Assuming that a hot fluid or gas is produced at the central conversion plant, transmission may be accomplished by pumping the fluid or gas through a pipeline to the load points. The pipeline may be placed underground and the tubing insulated in order to reduce conduction and convection of heat away from the pipe. If the temperature of the surrounding medium can be regarded as approximately constant, equal to T_{ref}, the variation in fluid (or gas) temperature $T^{fluid}(x)$ along the transmission line (with the path-length co-ordinate denoted x) can be evaluated by the same expression (4.36) used in the simple description of a heat exchanger in section 4.2.3. Temperature variations across the inside area of the pipe perpendicular to the streamwise direction are not considered.

The rate of heat loss from the pipe section between the distances x and $(x + dx)$ is then of the form

$$dE/dx = J_m \, C_p^{fluid} \, dT^{fluid}(x)/dx = h'(T_{ref} - T^{fluid}(x)), \qquad (5.1)$$

where h' for a cylindrical insulated pipe of inner and outer radii r_1 and r_2 is related to the heat transfer coefficient λ^{pipe} [conductivity plus convection terms, cf. (3.34)] by

$$h' = 2\pi \lambda^{pipe}/\log(r_2/r_1).$$

Upon integration from the beginning of the transmission pipe, $x = x_1$, (5.1) gives, in analogy to (4.36),

$$T^{fluid}(x) = T_{ref} + (T^{fluid}(x_1) - T_{ref}) \exp(-h'(x-x_1)/J_m \, C_p^{fluid}), \qquad (5.2)$$

where J_m is the mass flow rate and C_p^{fluid} is the fluid (or gas) heat capacity. The total rate of heat loss due to the temperature gradient between the fluid in the pipe and the surroundings is obtained by integrating (5.1) from x_1 to the end of the line, x_2,

$$\Delta E = \int_{x_1}^{x_2} \frac{dE}{dx} dx = -J_m C_p^{fluid} (T^{fluid}(x_1) - T_{ref}) \left(1 - \exp\left(-\frac{h'(x_2 - x_1)}{J_m C_p^{fluid}} \right) \right).$$

The relative loss of heat supplied to the pipeline, along the entire transmission line, is then

$$\frac{\Delta E}{E} = -\left(1 - \exp\left(-\frac{h'(x_2 - x_1)}{J_m C_p^{fluid}} \right) \right), \tag{5.3}$$

provided that the heat entering the pipe, E, is measured relative to a reservoir of temperature T_{ref}.

The total transmission loss will be larger than (5.3) because of friction in the pipe. The distribution of flow velocity, v, over the pipe's inside cross-section depends on the roughness of the inner walls (v being zero at the wall), and the velocity profile generally depends on the magnitude of the velocity (say at the centerline). Extensive literature exists on the flow in pipes (see, for example, Grimson, 1971). The net result is that additional energy must be provided in order to make up for the pipe losses. For an incompressible fluid moving in a horizontal pipe of constant dimensions, flow continuity demands that the velocity field be constant along the pipeline, and, in this case, the friction can be described by a uniformly decreasing pressure in the streamwise direction. A pump must provide the energy flux necessary to compensate for the pressure loss,

$$\Delta E^{pump} = -Av \, \Delta P,$$

where A is the area of the inner pipe cross-section and ΔP is the total pressure drop along the pipeline.

Transmission lines presently in use for city district heating have average heat losses of 10%–15%, depending on insulation thickness (see, for example, WEC, 1991). The pump energy added is in the form of high-quality mechanical work, but with proper dimensioning of the tubes it may be kept to a small fraction of the heat energy transmitted. This energy is converted into heat by frictional dissipation, and some of the heat may actually be credited to the heat transmission. Maximum heat transmission distances currently considered economical are around 30 km, with the possible exception of some geothermal installations. Figure 5.1 shows an integrated design that allows much easier installation than earlier pipes with separate insulation. In countries with a high penetration of combined heat and power production, such as Denmark, a heat storage facility holding some 10 h of heat load is sometimes added to each plant, in order that the (fuel-based) plant does not have to generate more electric power than needed, say, at night, when the heating load is large (Danish Energy Agency, 1993).

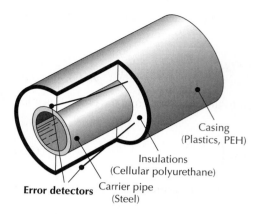

Casing (Plastics, PEH)

Insulations (Cellular polyurethane)

Error detectors   Carrier pipe (Steel)

FIGURE 5.1   Pre-insulated district heating pipe (Danish Energy Agency, 1993).

Heat Pipes

A special heat transmission problem arises where very large amounts of heat have to be delivered to, or removed from, a small region, for example, in connection with highly concentrating solar energy collectors. Such heat transfer may be accomplished by a heat pipe, a device described in section 4.2.1 in relation to thermionic generators.

5.1.2 Power Transmission

Normal Conducting Lines

At present, electric current is transmitted in major utility grids, as well as distributed locally to each load site by means of conducting wires. Electricity use is dominated by alternating current (AC), as far as utility networks are concerned, and most transmission over distances up to a few hundred kilometers is by AC. For transmission over longer distances (e.g., by ocean cables), conversion to direct current (DC) before transmission and back to AC after transmission is common. Cables are either buried in the ground (with appropriate electric insulation) or overhead lines are suspended in the air between masts, without electrical insulation around the wires. Insulating connections are provided at the tower fastening points, but otherwise the low electric conductivity of air is counted on. This implies that the losses will comprise conduction losses depending on the instantaneous state of the air (the "weather situation"), in addition to the ohmic losses connected with the resistance R of the wire itself, $E^{heat} = RI^2$, with I being the current. The leak current between the elevated wire and the ground depends on the potential difference as well as on the integrated resistivity (cf. section 3.4.5), such that the larger the voltage, the further the wires must be placed from the ground.

Averaged over different meteorological conditions, the losses in a standard AC overhead transmission line (138–400 kV, at an elevation of 15–40 m) are currently a little under 1% per 100 km of transmission (Hammond *et al.*, 1973), but the overall transmission losses of utility networks, including the finely branched distribution networks in the load areas, may for many older, existing grids amount to 12%–15% of the power production for a grid extending over a land area of about 10^4 km^2 (Blegaa *et al.*, 1976). Losses are down to 5%–6% for the best systems installed at present and are expected to decrease further, to 2%–3%, in the future, when the currently best technologies penetrate further (Kuemmel *et al.*, 1997). This loss is calculated relative to the total production of electricity at the power plants attached to the common grid, and thus includes certain in-plant and transformer losses. The numbers also represent annual averages for a power utility system occasionally exchanging power with other utility systems through interconnecting transmission lines, which may involve transmission distances much longer than the linear extent of the load area being serviced by the single utility system in question.

The trend is to replace overhead lines with underground cables, primarily for visual and environmental reasons. This has already happened for the distribution lines in Europe and is increasingly also being required for transmission lines. In Japan and the United States, overhead lines are still common.

Underground transmission and distribution lines range from simple coaxial cables to more sophisticated constructions insulated by compressed gas. Several transoceanic cables (up to 1000 km) have been installed in Scandinavia in order to bring the potentially large surpluses of hydropower production to the European continent. The losses through these high-voltage (up to 1000 kV) DC lines are under 0.4% per 100 km, to which should be added the 1%–2% transmission loss occurring at the thyristor converters on shore that transform AC into DC and vice versa (Chapter 19 in IPCC, 1996a). The cost of these low-loss lines is currently approaching that of conventional AC underwater cables (about 2 euro kW^{-1} km^{-1}; Meibom *et al.*, 1997, 1999; Wizelius, 1998).

One factor influencing the performance of underground transmission lines is the slowness of heat transport in most soils. In order to maintain the temperature within the limits required by the materials used, active cooling of the cable could be introduced, particularly if large amounts of power have to be transmitted. For example, the cable may be cooled to 77 K (liquid nitrogen temperature) by means of refrigerators spaced at intervals of about 10 km (cf. Hammond *et al.*, 1973). This allows increased amounts of power to be transmitted in a given cable, but the overall losses are hardly reduced, since the reduced resistance in the conductors is probably outweighed by the energy spent on cooling. According to (4.22), the cooling efficiency is limited by a Carnot value of around 0.35, i.e., more than three units of work have to be supplied in order to remove one unit of heat at 77 K.

Offshore Issues

The power from an offshore wind farm is transmitted to an onshore distribution hub by means of one or more undersea cables, the latter providing redundancy that, in the case of large farms, adds security against cable disruption or similar failures. Current offshore wind farms use AC cables of up to 150 kV (Eltra, 2003). New installations use cables carrying all three leads plus control wiring. In the interest of loss minimization for larger installations, it is expected that future systems may accept the higher cost of DC–AC conversion (on shore, the need for AC–DC conversion at sea depends on the generator type used), similar to the technology presently in use for many undersea cable connections between national grids (e.g., between Denmark and Norway or Sweden). Recent development of voltage source-based high-voltage DC control systems to replace the earlier thyristor-based technology promises better means of regulation of the interface of the DC link to the onshore AC system (Ackermann, 2002).

Superconducting Lines

For DC transmission, ohmic losses may be completely eliminated by use of superconducting lines. A number of elements, alloys, and compounds become superconducting when cooled to a sufficiently low temperature. Physically, the onset of superconductivity is associated with the sudden appearance of an energy gap between the *ground state*, i.e., the overall state of the electrons, and any excited electron state (similar to the situation illustrated in Fig. 4.48, but for the entire system rather than for individual electrons). A current, i.e., a collective displacement (flow) of electrons, will not be able to excite the system away from the ground state unless the interaction is strong enough to overcome the energy gap. This implies that no mechanism is available for the transfer of energy from the current to other degrees of freedom, and thus the current will not lose any energy, which is equivalent to stating that the resistance is zero. In order for the electron system to remain in the ground state, the thermal energy spread must be smaller than the energy needed to cross the energy gap. This is the reason why superconductivity occurs only below a certain temperature, which may be quite low (e.g., 9 K for niobium, 18 K for niobium–tin, Nb_3Sn). However, there are other mechanisms that in more complex compounds can prevent instability, thereby explaining the findings in recent years of materials that exhibit superconductivity at temperatures approaching ambient (Pines, 1994; Demler and Zhang, 1998).

For AC transmission, a superconducting line will not be loss-free, owing to excitations caused by the time variations of the electromagnetic field (cf. Hein, 1974), but the losses will be much smaller than for normal lines. It is estimated that the amount of power that can be transmitted through a single cable is in the gigawatt range. This figure is based on suggested designs, including the required refrigeration and thermal insulation components within overall dimensions of about 0.5 m (cable diameter). The power required for cooling, i.e., to compensate for heat flow into the cable, must be considered in calculating the total power losses in transmission.

For transmission over longer distances, it may, in any case, be an advantage to use direct current, despite the losses in the AC–DC and DC–AC conversions (a few percent, as discussed above). Intercontinental transmission using superconducting lines has been discussed by Nielsen and Sørensen (1996) and by Sørensen and Meibom (1998) (cf. scenario in section 6.7). The impetus for this is of course the location of some very promising renewable energy sites far from the areas of load. Examples are solar installations in the Sahara or other desert areas, or wind power installations at isolated rocky coastlines of northern Norway or in Siberian highlands.

Finally, radiant transmission of electrical energy may be mentioned. The technique for transmitting radiation and re-collecting the energy (or some part of it) is well developed for wavelengths near or above visible light. Examples are laser beams (stimulated atomic emission) and microwave beams (produced by accelerating charges in suitable antennas), ranging from the infrared to the wavelengths used for radio and other data transmission (e.g., between satellites and ground-based receivers). Large-scale transmission of energy in the form of microwave radiation has been proposed in connection with satellite solar power generation, but it is not currently considered practical. Short-distance transmission of signals, e.g., between computers and peripheral units, involves only minute transfers of energy and is already in widespread use.

5.1.3 Fuel Transmission

Fuels like natural gas, biogas, hydrogen, and other energy-carrier gases may be transmitted through pipelines, at the expense of only a modest amount of pumping energy (at least for horizontal transfer). Pipeline oil transmission is also in abundant use. Intercontinental transport may alternatively use containers onboard ships for solid fuels, oil, compressed gases or liquefied gases. Similar containers are used for shorter-distance transport by rail or road. Higher energy densities may be attained by the energy storage devices discussed in section 5.2.2, such as metal hydrides or carbon nanotubes, e.g., for hydrogen transport. Lightweight container materials are, of course, preferable in order to reduce the cost of moving fuels by vessel, whether on land or at sea.

Current natural gas networks consist of plastic distribution lines operated at pressures of 0.103 to about 0.4 MPa and steel transmission lines operated at pressures of 5–8 MPa. With upgraded valves, some modern natural gas pipelines could be used for the transmission of hydrogen (Sørensen et al., 2001). Certain steel types may become brittle with time, as a result of hydrogen penetration into the material, and cracks may develop. It is believed that H_2S impurities in the hydrogen stream increase the problem, but investigations of the steel types currently used for new pipelines indicate little probability of damage by hydrogen (Pöpperling *et al.*, 1982; Kussmaul and Deimel, 1995).

Mechanical devices have been used to transfer mechanical energy over short distances, but mechanical connections with moving parts are not practical for distances of transfer that may be considered relevant for transmission lines. However, mechanical energy in forms like hydraulic pulses can be transmitted over longer distances in feasible ways, as, for example, mentioned in connection with wave energy conversion devices placed in open oceans (section 4.3.6).

5.2 HEAT STORAGE

Storage of energy allows adjustment to variations in energy demand, i.e., it provides a way of meeting a load with a time-dependence different from that of generation. For fuel-type energy, storage can help burn the fuel more efficiently, by avoiding situations where demand variations would otherwise require regulation of combustion rates beyond what is technically feasible or economic. For fluctuating renewable energy sources, storage helps make the energy systems as dependable as fuel-based systems.

The characteristics of an ideal energy storage system include availability for rapid access and versatility of the form in which energy from the store is delivered. Conversion of one type of stored energy (initial fuel) into another form of stored energy may be advantageous for utilization. For example, production of electricity in large fossil fuel or nuclear power plants may involve long start-up times that incur additional costs when used for load leveling, while pumped water storage allows delivery upon demand in less than a minute. The economic feasibility of energy storage depends on the relative fixed and variable costs of the different converter types and on the cost and availability of different fuels. Another example is a fuel-driven automobile, which operates at or near peak power only during short intervals of acceleration. If short-term storage is provided that can accumulate energy produced off-peak by the automobile engine and deliver the power for starting and acceleration, then the capacity rating of the primary engine may be greatly reduced.

In connection with renewable energy resources, most of which are intermittent and of a fluctuating power level, supplementing conversion by energy storage is essential if the actual demand is to be met at all times. The only alternative would seem to be fuel-based back-up conversion equipment, but this, of course, is just a kind of energy storage that in the long run may require fuels provided by conversion processes based on renewable primary energy sources.

5.2.1 Heat Capacity Storage

Heat capacity, or "sensible heat" storage, is accomplished by changing the temperature of a material without changing its phase or chemical composition. The amount of energy stored by heating a piece of material of mass m from temperature T_0 to temperature T_1 at constant pressure is

$$E = m \int_{T_0}^{T_1} c_P \, dT, \tag{5.4}$$

where c_P is the specific heat capacity at constant pressure.

Energy storage at low temperatures is needed in renewable systems like solar absorbers delivering space heating, hot water, and eventually heat for cooking (up to 100 °C). The actual heat storage devices may be of modest size, aiming at delivering heat during the night after a sunny day, or they may be somewhat larger, capable of meeting demand during a number of consecutive overcast days. Finally, the storage system may provide seasonal storage of heat, as required at high latitudes where seasonal variations in solar radiation are large, and, furthermore, heat loads are to some extent inversely correlated with the length of the day.

Another aspect of low-temperature heat storage (as well as of some other energy forms) is the amount of decentralization. Many solar absorption systems are conveniently placed on existing rooftops; that is, they are highly decentralized. A sensible heat store, however, typically loses heat from its container, insulated or not, in proportion to the surface area. The relative heat loss is smaller, the larger the store dimensions, and thus more centralized storage facilities, for example, of communal size, may prove superior to individual installations. This depends on an economic balance between the size advantage and the cost of additional heat transmission lines for connecting individual buildings to a central storage facility. One should also consider other advantages, such as the supply security offered by the common storage facility (which would be able to deliver heat, for instance, to a building with malfunctioning solar collectors).

Water Storage

Heat energy intended for later use at temperatures below 100 °C may be conveniently stored as hot water, owing to the high heat capacity of water (4180 J kg^{-1} K^{-1} or 4.18×10^6 J m^{-3} K^{-1} at standard temperature and pressure), combined with the fairly low thermal conductivity of water (0.56 J m^{-1} s^{-1} K^{-1} at 0 °C, rising to 0.68 J m^{-1} s^{-1} K^{-1} at 100 °C).

Solar collector
To/From
Hot water
Cold water
Insulation
Heat exchanger
Safety valve
Pressure meter
Pump
One–way valve
Expansion device

FIGURE 5.2   Water container for heat storage with possibility of temperature stratification (e.g., for use in connection with solar collectors) (Ellehauge, 1981).

Most space heating and hot water systems for individual buildings include a water storage tank, usually in the form of an insulated steel container with a capacity corresponding to less than a day's hot water usage and often only a small fraction of a cold winter day's space heating load. For a one-family dwelling, a 0.1 m^3 tank is typical in Europe and the United States.

A steel hot water container may look like the one sketched in Fig. 5.2. It is cylindrical, with a height greater than the diameter, in order to make good temperature stratification possible, an important feature if the container is part of a solar heating system. A temperature difference of up to 50 °C between the top and bottom water can be maintained, with substantial improvements (over 15%) in the performance of the solar collector heating system, because the conversion efficiency of the collector declines (see Fig. 5.3) with the temperature difference between the water coming into the collector and the ambient outdoor temperature (van Koppen *et al.*, 1979). Thus, the water from the cold lower part of the storage tank would be used as input to the solar collector circuit, and the heated water leaving the solar absorber would be delivered to a higher-temperature region of the storage tank, normally the top layer. The take-out from the storage tank to load (directly or through a heat exchanger) is also from the top of the tank, because the solar system will, in this case, be able to cover load over a longer period of the year (and possibly for the entire year). There is typically a minimum temperature required for the fluid carrying heat to the load areas, and during winter, the solar collector system may not always be able to raise the entire storage tank volume to a temperature above this threshold. Thus, temperature stratification in storage containers is often a helpful feature. The minimum load-input temperatures are around 45 °C–50 °C for space heating through water-filled radiators and convectors, but only 25 °C–30 °C for water-filled floor heating systems and air-based heating and ventilation duct systems.

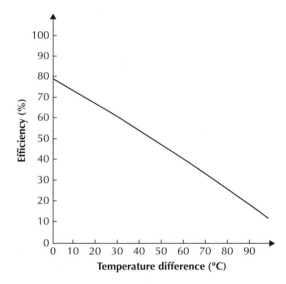

FIGURE 5.3   Efficiency curve for a flat-plate solar collector as a function of the difference between the average temperature of the fluid that removes heat from the collector and the ambient outside air temperature. The efficiency is a percentage of the absorbed solar radiation that is transferred to the working fluid. The curve is based on measurements for a selective-surface collector, tilted 45° and receiving 800 W m^{-2} of solar radiation at incident angles below 30°. Wind speed along the front of collector is 5 m s^{-1} (Svendsen, 1980).

For hot water provision for a single family, using solar collector systems with a few square meters of collectors (1 m^2 in sunny climates, 3–5 m^2 at high latitudes), a storage tank of around 0.3 m^3 is sufficient for diurnal storage, while a larger tank is needed if consecutive days of zero solar heat absorption can be expected. For complete hot water and space heating solar systems, storage requirements can be quite substantial if load is to be met at all times and the solar radiation has a pronounced seasonal variation.

Most solar collector systems aiming at provision of both hot water and space heating for a single-family dwelling have a fairly small volume of storage and rely on auxiliary heat sources. This is the result of an economic trade-off due to the rapid reduction in solar collector gains with increasing coverage, that is, the energy supplied by the last square meter of collector added to the system becomes smaller and smaller as the total collector area increases. Of course, the gain is higher with increased storage for a fixed collector area over some range of system sizes, but the gain is very modest (see section 6.5.1).

For this reason, many solar space-heating systems only have diurnal storage (for example a hot water storage). In order to avoid boiling, when the solar radiation level is high for a given day, the circulation from storage tank to collector is disconnected whenever the storage temperature is above some specified value (e.g., 80 °C), and the collector becomes stagnant. This is usually no problem for simple collectors with one layer of glazing and black paint absorber, but for multilayered covers or selective-surface coating of the absorber, the stagnation temperatures are often too high and materials would be damaged before these temperatures were reached. Instead, the storage size may be increased to such a value that continuous circulation of water from storage through collectors can be maintained during sunny days without violating maximum storage temperature requirements at any time during the year. If the solar collector is so efficient that it still has a net gain above 100 °C (such as the one shown in Fig. 5.3), this heat gain must be balanced by heat losses from piping and from the storage itself, or the store must be large enough for the accumulated temperature rise during the most sunny periods of the year to be acceptable. In high-latitude climatic regions, this can be achieved with a few square meters of water storage, for collector areas up to about 50 m^2.

Larger amounts of storage may be useful if the winter is generally sunny, but a few consecutive days of poor solar radiation do occur from time to time. For example, the experimental house in Regina, Saskatchewan (Besant et al., 1979) is superinsulated and is designed to derive 33% of its heating load from passive gains of south-facing windows, 55% from activities in the house (body heat, electric appliances), and the remaining 12% from high-efficiency (evacuated tube) solar collectors. The collector area is 18 m^2 and there is a 13 m^3 water storage tank. On a January day with outdoor temperatures between −25 °C (afternoon) and −30 °C (early morning), about half the building's heat loss is replaced by indirect gains, and the other half (about 160 MJ d^{-1}) must be drawn from the store, in order to maintain an indoor temperature of 21 °C (although the temperature is allowed to drop to 17 °C between midnight and 0700 h). On sunny January days, the amount of energy drawn from storage is reduced to about 70 MJ d^{-1}. Since overcast periods of much over a week occur very rarely, the storage size is such that 100% coverage can be expected, from indirect and direct solar sources, in most years.

The situation is very different in, for instance, Denmark, where the heating season has only 2700 Celsius degree days (due to Gulf Stream warming), compared with 6000 degree days for Regina, and there is very little solar gain (through windows or to a solar collector) during the months of November through February. A modest storage volume, even with a large solar collector area, is therefore unable to maintain full coverage during the winter period, as indicated by the variations in storage temperatures in a concrete case, shown in Fig. 5.4 (the water store is situated in the attic, losing heat to ambient air temperatures; this explains why the storage temperatures approach freezing in January).

FIGURE 5.4 Measured average daily temperature in a 5.5 m³ water storage tank (having 0.2 m of rock wool insulation) fed by 50 m² of solar flat-plate collector (two layers of glass cover, nonselective absorber, tilt angle 38°). The solar system covers 12% of the building's heat load. (Based on Jørgensen et al., 1980.)

In order to achieve near 100% coverage under conditions like those in Denmark, very large water tanks would be required, featuring facilities to maintain stable temperature stratification and containing so much insulation (over 1 m) that truly seasonal mismatch between heat production and use can be handled. This is still extremely difficult to achieve for a single house, but for a communal system with a number of reasonably close-set buildings and with a central storage facility (and maybe also central collectors), 100% coverage should be possible (see section 6.5.1).

Community-size Storage Facilities

With increasing storage container size, the heat loss through the surface—for a given thickness of insulation—will decrease per unit of heat stored. There are two cases, depending on whether the medium surrounding the container (air, soil, etc.) is rapidly mixing or not. Consider first the case of a storage container surrounded by air.

The container may be cylindrical, like the one illustrated in Fig. 5.2. The rate of heat loss is assumed to be proportional to the surface area and to the temperature difference between inside and outside, with the proportionality constant being denoted U. It is sometimes approximated in the form

$$U = 1/(x/\lambda + \mu),$$

where x is the thickness of insulation, λ is the thermal conductivity of the insulating material (about 0.04 W m^{-1} per °C for mineral wool), and μ (around 0.1 m^2 W^{-1} per °C) is a parameter describing the heat transfer in the boundary layer air [may be seen as a crude approximation to (4.121)].

The total heat loss rate is

$$P^{loss} = 2\pi R (R + L) U (T_s - T_a), \tag{5.5}$$

where R and L are radius and height of the cylinder, T_s is the average temperature of the water in the store, and T_a is the outside ambient air temperature. The fraction of the stored heat energy lost per unit time is

$$\frac{P^{loss}}{E^{sens}} = \frac{2U(1 + R/L)}{R \, c_P^{water} \rho^{water}}, \tag{5.6}$$

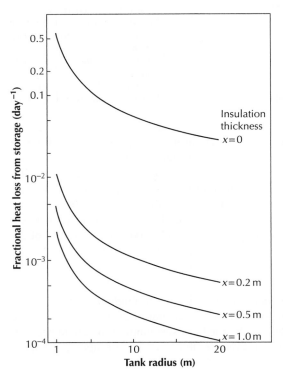

FIGURE 5.5   Fractional heat loss from a cylindrical storage tank with different degrees of insulation, assuming well-mixed conditions inside as well as outside the tank.

that is, the loss is independent of the temperatures and inversely proportional to a linear system dimension (c_P is heat capacity and ρ is density).

Figure 5.5 shows, for $L = 3R$, the ratio of heat loss rate P^{loss} and stored energy E^{sens}, according to (5.6), as a function of R and for different values of insulation thickness. This ratio is independent of the temperature difference, owing to the linear assumption for the heat loss rate. If the storage times required imply maximal fractional losses of 10%–20%, uninsulated tanks in the size range 5 m $\leq R \leq$ 20 m are adequate for a few days of storage. If storage times around a month are required, some insulation must be added to tanks in the size range under consideration. If seasonal storage is needed, the smallest tank sizes will not work, even with a meter of mineral or glass wool wrapping. Community-size tanks, however, may serve for seasonal storage, with a moderate amount of insulation (0.2 m in the specific example).

A hot water tank with $R = 11.5$ m and $L = 32$ m has been used since 1978 by a utility company in Odense, Denmark, in connection with combined production of electricity and heat for district heating (Jensen, 1981). The hot water store is capable of providing all necessary heating during winter electricity peak hours, during which the utility company wants maximal electricity production. With a hot water store two or three times larger, the co-generating power units could be allowed to follow the electricity demand, which is small relative to the heat demand during winter nights.

A hot water store of similar magnitude, around 13 000 m³, may serve a solar-heated community system for 50–100 one-family houses, connected to the common storage facility by district heating lines. The solar collectors may still be on individual rooftops, or they may be placed centrally, for example, in connection with the store. In the first case, more piping and labor are required for installation, but in the second case, land area normally has to be dedicated to the collectors. Performance is also different for the two systems, as long as the coverage by solar energy is substantially less than 100% and the auxiliary heat source feeds into the district heating lines. The reason is that when the storage temperature is below the minimum required, the central solar collector will perform at high efficiency (Fig. 5.3), whereas individual solar collectors will receive input temperatures already raised by the ancillary heat source and thus will not perform as well. Alternatively, auxiliary heat could be added by individual installations on the load side of the system, but, unless the auxiliary heat is electrically generated, generating auxiliary heat is inconvenient if the houses do not already have a fuel-based heating system.

Most cost estimates argue against placing storage containers in air. If the container is buried underground (possibly with its top facing the atmosphere), the heat escaping the container will not be rapidly mixed into the surrounding soil or rock. Instead, the region closest to the container will reach a higher temperature, and a temperature gradient through the soil or rock will slowly build up. An exception is soil with ground water infiltration, because the moving water helps to mix the heat from the container into the surroundings. However, if a site can be found with no ground water (or at least no ground water in motion), then the heat loss from the store will be greatly reduced, and the surrounding soil or rock can be said to function as an extension of the storage volume.

As an example, consider a spherical water tank embedded in homogeneous soil. The tank radius is denoted R, the water temperature is T_s, and the soil temperature far away from the water tank is T_0. If the transport of heat can be described by a diffusion equation, then the temperature distribution as function of distance from the center of the storage container may be written (Shelton, 1975)

$$T(r) = T_0 + (T_s - T_0)R/r, \tag{5.7}$$

where the distance r from the center must be larger than the tank radius R in order for the expression to be valid. The corresponding heat loss is

$$P^{sens} = \int_{sphere} \lambda\, \partial T(r)/\partial r\, \mathrm{d}A = -\lambda(T_s - T_0)4\pi R, \tag{5.8}$$

where λ is the heat conductivity of the soil and (5.8) gives the heat flux out of any sphere around the store of radius $r \geq R$. The flux is independent of r. The loss relative to the heat stored in the tank itself is

$$P^{loss}/E^{sens} = -3\lambda/(R^2\, c_P^{water}\, \rho^{water}). \tag{5.9}$$

The relative loss from the earth-buried store declines more rapidly with increasing storage size than the loss from a water store in air or other well-mixed surroundings [cf. (5.6)]. The fractional loss goes as R^{-2} rather than as R^{-1}.

Practical considerations in building an underground or partly underground water store suggest an upside-down obelisk shape and a depth around 10 m for a 20 000 m^2 storage volume. The obelisk is characterized by tilting sides, with a slope as steep as feasible for the soil type encountered. The top of the obelisk (the largest area end) would be at ground level or slightly above it, and the sides and bottom would be lined with plastic foil not penetrable by water. Insulation between lining and ground can be made with mineral wool foundation elements or similar materials. As a top cover, a sailcloth held in a bubble shape by slight overpressure is believed to be the least expensive solution. Top insulation of the water in the store can be floating foam material. If the sailcloth bubble is impermeable to light, algae growth in the water can be avoided (Danish Department of Energy, 1979).

Two community-size seasonal hot water stores placed underground are operating in Sweden. They are both shaped as cut cones. One is in Studsvik. Its volume is 610 m^3, and 120 m^2 of concentrating solar collectors float on the top insulation, which can be turned to face the sun. Heat is provided for an office building with a floor area of 500 m^2. The other system is in the Lambohov district of Linköping. It serves 55 semidetached one-family houses having a total of 2600 m² flat-plate solar collectors on their roofs. The storage is 10 000 m³ and situated in solid rock (excavated by blasting). Both installations have operated since 1979, and they furnish a large part of the heat loads of the respective buildings. Figure 5.6 gives the temperature changes by season for the Studsvik project (Andreen and Schedin, 1980; Margen, 1980; Roseen, 1978).

Another possibility is to use existing ponds or lake sections for hot water storage. Top insulation would normally be required, and, in the case of lakes used only in part, an insulating curtain should separate the hot and cold water.

Keeping the time derivative in (3.35), a numerical integration will add information about the time required for reaching the steady-state situation. According to calculations like Shelton's (1975), this may be about 1 year for typical large heat stores.

The assumption of a constant temperature throughout the storage tank may not be valid, particularly not for large tanks, owing to natural stratification that leaves the colder water at the bottom and the warmer water at the top. Artificial mixing by mechanical stirring is, of course, possible, but for many applications temperature stratification is preferable. It could be used to advantage, for example, with storage tanks in a solar heating system (see Fig. 4.84), by feeding the collector with the colder water from the lower part of the storage and thus improving the efficiency (4.132) of the solar collector.

FIGURE 5.6 Temperature profiles in the Studsvik hot water store, monthly during (*right*) charging period March (M) to August (A) and during (*left*) discharging period from August (A) to March (M). (Based on Roseen, 1978.)

An alternative would be to divide the tank into physically separated subunits (cf., for example, Duffie and Beckman, 1974), but this requires a more elaborate control system to introduce and remove heat from the different units in an optimized way, and, if the storage has to be insulated, the subunits should at least be placed with common boundaries (tolerating some heat transfer among the units) in order to keep the total insulation requirement down.

Multi-unit storage systems may be most appropriate for uninsulated storage in the ground, owing to the participation of the soil between the storage units. Figure 5.7 shows a possible arrangement (Brüel *et al.*, 1976) based on cylindrical heat exchangers placed in one or more circles around a central heat exchanger cylinder in the soil. If no active aquifers traverse the soil region, a steady-state temperature distribution with a high central temperature can be built up. For example, with a flat-plate solar collector, the fluid going to the collector circuit could come from a low-temperature source and the fluid returning from the solar collector could be delivered to successive heat exchangers of diminishing temperatures in order to maximize the collector efficiency as well as the amount of heat transferred to storage. Dynamic simulations of storage systems of this type, using a solution of (3.33) with definite boundary conditions and source terms, have been performed by Zlatev and Thomsen (1976).

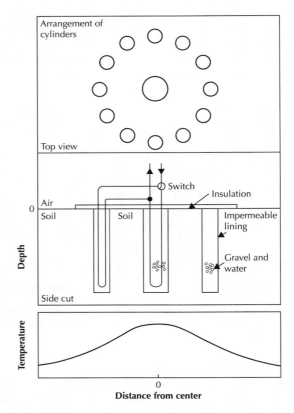

FIGURE 5.7   Uninsulated heat storage in soil based on an arrangement of cylindrical heat exchangers (placed in holes drilled by water-flushing drills, for example). (Based on Brüel *et al.*, 1976.)

Other materials, such as gravel, rock, or soil, have been considered for use in heating systems at temperatures similar to those relevant for water. Despite volume penalties of a factor of 2–3 (depending on void volume), these materials may be more convenient than water for some applications. The main problem is to establish a suitable surface for the transfer of heat to and from the storage. For this reason, gravel and rock stores have been mostly used with air as a transfer fluid, because large volumes of air can be blown through the porous material.

For applications in other temperature ranges, other materials may be preferred. Iron (e.g., scrap material) is suitable for temperatures extending to several hundred degrees Celsius. In addition, for rock-based materials, the temperatures would not be limited to the boiling point of water.

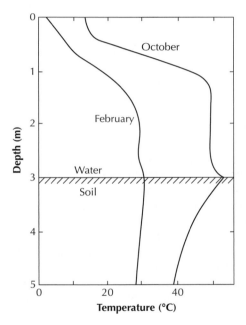

FIGURE 5.8   Temperature profiles for solar salt-gradient pond in Miamisburg, Ohio, for the months of October and February (maximum and minimum temperature in convective layer). The continuation of the profiles in the soil beneath the pond is also included. (From Wittenberg and Harris, 1979. Reprinted with permission from *Proceedings of 14th Intersociety Energy Conversion Engineering Conference.* Copyright 1979 American Chemical Society.)

Solar Ponds and Aquifer Storage

A solar pond is a natural or artificial hot water storage system much like the ones described above, but with the top water surface exposed to solar radiation and operating like a solar collector. In order to achieve both collection and storage in the same medium, layers from top to bottom have to be inversely stratified, that is, stratified with the hottest zone at the bottom and the coldest one at the top. This implies that thermal lift must be opposed, either by physical means, such as placing horizontal plastic barriers to separate the layers, or by creating a density gradient in the pond, which provides gravitational forces to overcome the buoyancy forces. This can be done by adding certain salts to the pond, taking advantage of the higher density of the more salty water (Rabl and Nielsen, 1975).

An example of a solar pond of obelisk shape is the 5200 m^3 pond in Miamisburg, Ohio. Its depth is 3 m, and the upper half is a salt-gradient layer of NaCl, varying from 0% at the top to 18.5% at a depth of 1.5 m. The gradient opposes upward heat transport and thus functions as a top insulation without impeding the penetration of solar radiation. The bottom layer has a fixed salt concentration (18.5%) and contains heat exchangers for withdrawing energy. In this layer, convection may take place without problems. Most absorption of solar radiation takes place at the pond's bottom surface (which is why the pond should be shallow), and the heat is subsequently released to the convective layer. At the top, however, some absorption of infrared solar radiation may destroy the gradient of temperature.

Figure 5.8 shows temperature gradients for the Miamisburg pond during its initial loading period (no load connected). After the start of operation in late August, the first temperature maximum occurred in October, and the subsequent minimum occurred in February. The two situations are shown in Fig. 5.8. The temperature in the ground just below the pond was also measured. In October, top-layer disturbance can be seen, but in February it is absent, due to ice covering the top of the pond.

Numerical treatment of seasonal storage in large top-insulated or solar ponds may be done by time simulation, or by a simple approximation, in which solar radiation and pond temperature are taken as sine functions of time, with only the amplitude and phase as parameters to be determined. This is a fairly good approximation because of the slow response of a large seasonal store, which tends to be insensitive to rapid fluctuations in radiation or air temperature. However, when heat is extracted from the store, it must be ensured that disturbance of the pond's temperature gradient will not occur, say, on a particularly cold winter day, where the heat extraction is large. Still, heat extraction can, in many cases, also be modeled by sine functions, and if the gradient structure of the pond remains stable, such a calculation gives largely realistic results.

In Israel, solar ponds are operated for electricity generation by use of Rankine cycle engines with organic working fluids that respond to the small temperature difference available. A correspondingly low thermodynamic efficiency must be accepted (Winsberg, 1981).

Truly underground storage of heat may also take place in geological formations capable of accepting and storing water, such as rock caverns and aquifers. In the aquifer case, it is important that water transport be modest, that is, that hot water injected at a given location stay approximately there and exchange heat with the surroundings only by conduction and diffusion processes. In such cases, it is estimated that high cycle efficiencies (85% at a temperature of the hot water some 200 °C above the undisturbed aquifer temperature—the water being under high pressure) can be attained after breaking the system in, that is, after having established stable temperature gradients in the surroundings of the main storage region (Tsang *et al.*, 1979).

Medium- and High-temperature Storage

In industrial processes, temperature regimes are often defined as medium in the interval from 100 °C to 500 °C and as high above 500 °C. These definitions may also be used in relation to thermal storage of energy, but it also may be useful to single out the lower-medium temperature range from 100 °C to about 300 °C, as the discussion below indicates.

Materials suitable for heat storage should have a large heat capacity, they must be stable in the temperature interval of interest, and it should be convenient to add heat to, or withdraw heat from, them.

The last of these requirements can be fulfilled in different ways. Either the material itself should possess good heat conductivity, as metals do, for example, or it should be easy to establish heat transfer surfaces between the material and some other suitable medium. If the transfer medium is a liquid or a gas, it could be passed along the transfer surface at a velocity sufficient for

Material	Temperature Interval (°C)	Mass Spec. Heat (kJ kg^{-1} °C^{-1})	Volume Spec. Heat (MJ m^{-3} °C^{-1})	Heat Conductivity (W m^{-1} °C^{-1})
Solids				
Sodium chloride	< 800	0.92	2.0	9[a,b]
Iron (cast)	< 1500	0.46	3.6	70[b]–34[c]
Rock (granite)	< 1700	0.79	2.2	2.7[b]
Bricks		0.84	1.4	0.6
Earth (dry)		0.79	1.0	1.0
Liquids				
Water	0–100	4.2	4.2	0.6
Oil ("thermal")	−50 to 330	2.4	1.9	0.1
Sodium	98 to 880	1.3	1.3	85[b]–60[c]
Diethylene glycol	−10 to 240	2.8	2.9	

TABLE 5.1 Heat capacities of various materials (Kaye and Laby, 1959; Kreider, 1979; Meinel and Meinel, 1976). All quantities have some temperature dependence. Standard atmospheric pressure has been assumed, that is, all heat capacities are c_P.

[a]*Less for granulates with air-filled voids.*
[b]*At 1000 °C.*
[c]*At 700 °C.*

the desired heat transfer, even if the conductivities of the transfer fluid and the receiving or delivering material are small. If the storage material is arranged in a finite geometry, such that insufficient transfer is obtained by a single pass, then the transfer fluid may be passed along the surface several times. This is particularly relevant for transfer media like air, which has very low heat conductivity. When air is used as a transfer fluid, it is important that the effective transfer surface be large, and this may be achieved for granular storage materials like pebble or rock beds, where the nodule size and packing arrangement can be such that air can be forced through and reach most of the internal surfaces with as small an expenditure of compression energy as possible.

These considerations lie behind the approaches to sensible heat storage that are exemplified by the range of potential storage materials listed in Table 5.1. Some are solid metals, where transfer has to be by conduction through the material. Others are solids that may exist in granular form, for heat transfer by air or another gas blown through the packed material. They exhibit more modest heat conductivity. The third group comprises liquids, which may serve as both heat storage materials and transfer fluids. The dominating path of heat transfer may be conduction, advection (moving the entire fluid), or convection (turbulent transport). For highly conducting materials, such as liquid sodium, little transfer surface is required, but for the other materials listed, substantial heat exchanger surfaces may be necessary.

Solid metals, such as cast iron, have been used for high-temperature storage in industry. Heat delivery and extraction may be accomplished by passing a fluid through channels drilled into the metal. For the medium- to high-temperature interval, the properties of liquid sodium (cf. Table 5.1) make it a widely used material for heat storage and transport, despite serious safety problems (sodium reacts explosively with water). It is used in nuclear breeder reactors and in concentrating solar collector systems, for storage at temperatures between 275 °C and 530 °C in connection with generation of steam for industrial processes or electricity generation. The physics of heat transfer to and from metal blocks and of fluid behavior in pipes is a standard subject covered in several textbooks (see, for example, Grimson, 1971).

FIGURE 5.9   Rock bed sensible heat store. Air of density ρ_a and velocity v_a may be blown through the bed cross-section A, traveling the length L of the bed.

Fixed beds of rock or granulate can be used for energy storage at both low and high temperatures. Normally, air is blown through the bed to transfer heat to and from the store. The pressure drop ΔP across a rock bed of length L, such as the one illustrated in Fig. 5.9, where air is blown through the entire cross-sectional area A, may be estimated as (Handley and Heggs, 1968)

$$\Delta P \approx \rho_a v_a^2 \, L \, d_s^{-1} \, m_s^2 \, (368 + 1.24 \, Re/m_s)/(Re(1 - m_s)^3), \tag{5.10}$$

where ρ_a and v_a are density and velocity of the air passing through the bed in a steady-state situation, d_s is the equivalent spherical diameter of the rock particles, and m, their mixing ratio, is one minus the air fraction in the volume $L \times A$. Re is the Reynolds number describing the ratio between "inertial" and "viscous" forces on the air passing between the rock particles. Re may be estimated as $\rho_a v_a d_a/\mu$, where μ is the dynamic viscosity of air. If the rock particles are not spherical, the equivalent diameter may be taken as

$$d_s = (6ALm_s/n)^{1/3},$$

where n is the number of particles in the entire bed volume. The estimate (5.10) assumes the bed to be uniform and the situation stationary. The total surface area of the particles in the bed is given by

$$A_s = 6m_s \, AL/d_s = n\pi d_s^2.$$

Optimal storage requires that the temperature gradient between the particle surfaces and their interior be small and that the pressure drop (5.10) also be small, leading to optimal particle diameters of a few centimeters and a void fraction of about 0.5 (implying that m_s is also about 0.5).

Organic materials, such as diethylene glycol or special oil products (Table 5.1), are suitable for heat storage between 200 °C and 300 °C and have been used in concentrating solar collector test facilities (Grasse, 1981). Above 300 °C, the oil decomposes.

Despite low-volume heat capacities, gaseous heat storage materials could also be considered. For example, steam (water vapor) is often stored under pressure, in cases where it is the form of heat energy to be used later (in industrial processes, power plants, etc.).

5.2.2 Latent Heat and Chemical Transformation Storage

The energy associated with a phase change for a material can be used to store energy. The phase change may be melting or evaporating, or it may be associated with a structural change, e.g., in lattice form, content of crystal-bound water, etc. When heat is added to or removed from a material, a number of changes may take place successively or in some cases simultaneously, involving phase changes as well as energy storage in thermal motion of molecules, i.e., both latent and sensible heat. The total energy change, which can serve as energy storage, is given by the change in enthalpy.

Solid–solid phase transitions are observed in one-component, binary, and ternary systems, as well as in single elements. An example of the latter is solid sulfur, which occurs in two different crystalline forms, a low-temperature orthorhombic form and a high-temperature monoclinic form (cf. Moore, 1972). However, the elementary sulfur system has been studied merely out of academic interest, in contrast to the practical interest in the one-component systems listed in Table 5.2. Of these systems, which have been studied for practical energy storage, Li_2SO_4 exhibits both the highest transition temperature T_t and the highest latent heat for the solid–solid phase change ΔH_{ss}. Pure Li_2SO_4 undergoes a transition from a monoclinic to a face-centered cubic structure with a latent heat of 214 KJ kg^{-1} at 578 °C. This is much higher than the heat of melting (-67 KJ kg^{-1} at 860 °C). Another one-component material listed in Table 5.2 is Na_2SO_4, which has two transitions at 201 °C and 247 °C, with the total latent heat of both transitions being ~ 80 KJ kg^{-1}.

Material	Transition Temperature T_t (°C)	Latent Heat ΔH_{ss} (kJ kg^{-1})
V_2O_2	72	50
FeS	138	50
KHF_2	196	135
Na_2SO_4	210, 247	80
Li_2SO_4	578	214

TABLE 5.2   Solid–solid transition enthalpies ΔH_{ss} (Fittipaldi, 1981).

Mixtures of Li_2SO_4 with Na_2SO_4 and K_2SO_4 with ZnSO have also been studied, as well as some ternary mixtures containing these and other sulfates (in a Swedish investigation, Sjoblom, 1981). Two binary systems ($Li_2SO_4–Na_2SO_4$, 50 mol% each, $T_t = 518$ °C; and 60% $Li_2SO_4–40\%$ $ZnSO_4$, $T_t = 459$ °C) have high values of latent heat, ~ 190 KJ kg^{-1}, but they exhibit a strong tendency for deformation during thermal cycling. A number of ternary salt mixtures based on the most successful binary compositions have been studied experimentally, but knowledge of phase diagrams, structures, and re-crystallization processes that lead to deformation in these systems is still lacking.

Salt Hydrates

The possibility of energy storage by use of incongruently melting salt hydrates has been intensely investigated, starting with the work of Telkes (1952, 1976). Molten salt consists of a saturated solution as well as some undissolved anhydrous salt because of its insufficient solubility at the melting temperature, considering the amount of released crystal water available. Sedimentation will develop, and a solid crust may form at the interface between layers. To control this, stirring is applied, for example, by keeping the material in rotating cylinders (Herrick, 1982), and additives are administered in order to control agglomeration and crystal size (Marks, 1983).

An alternative is to add extra water to prevent phase separation. This has led to a number of stable heat-of-fusion storage systems (Biswas, 1977; Furbo, 1982). Some melting points and latent heats of salt hydrates are listed in Table 5.3. Here we use as an example Glauber salt ($Na_2SO_4 \cdot 10H_2O$), the storage capacity of which is illustrated in Fig. 5.10, both for the pure hydrate and for a 33% water mixture used in actual experiments. Long-term verification of this and other systems in connection with solar collector installations have been carried out by the European Economic Community. For hot water systems, the advantage over sensible heat water stores is minimal, but this may change when space heating is included, because of the seasonal storage need (Furbo, 1982).

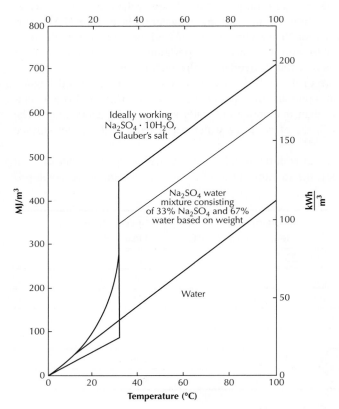

FIGURE 5.10 Heat storage capacity as a function of temperature for ideally melting Glauber salt, for Glauber salt plus extra water, and for pure water (Furbo, 1982).

Salt hydrates release water when heated and release heat when they are formed. The temperatures at which the reaction occurs vary for different compounds, ranging from 30 °C to 80 °C, which makes possible a range of storage systems for a variety of water-based heating systems, such as solar, central, and district heating. Table 5.3 shows the temperatures T_m of incongruent melting and the associated quasi-latent heat Q (or ΔH) for some hydrates that have been studied extensively in heat storage system operation. The practical use of salt hydrates faces physicochemical and thermal problems, such as supercooling, incongruent melting, and heat transfer difficulties imposed by locally low heat conductivities (cf., for example, Achard et al., 1981).

Generally, storage in a chemical system with two separated components, of which one draws low-temperature heat from the environment and the other absorbs or delivers heat at fairly low (30 °C–100 °C) or medium (100 °C–200 °C) temperature, is referred to as a *chemical heat pump* (McBride, 1981). Simply stated, the chemical heat pump principle consists in keeping a substance in one of two containers, although it would prefer to be in the other one. In a classical example, the substances are water vapor and sulfuric acid. Because the pressure over sulfuric acid is much lower than over liquid water (see Fig. 5.11), water vapor will move from the water surface to the H_2SO_4 surface and become absorbed there, with a heat gain deriving in part from the mixing process and in part from the heat of evaporation. The temperature of the mixture is then above what is needed at the water source. Heat is stored when the temperature of the sulfuric acid/water container is made still higher, so that the equilibrium pressure of vapor above the acid surface at this temperature becomes higher than that above the water surface at its temperature. The pressure gradient will therefore move water vapor back to the water surface for condensation.

A similar system, but with the water being attached as crystal water in a salt (i.e., salt hydration), is the sodium sulfide (Na_2S)/water system being developed

Hydrate	Incongruent Melting Point, T_m (°C)	Specific Latent Heat ΔH (MJ m^{-3})
$CaCl_2 \cdot 6H_2O$	29	281
$Na_2SO_4 \cdot 10H_2O$	32	342
$Na_2CO_3 \cdot 10H_2O$	33	360
$Na_2HPO_4 \cdot 12H_2O$	35	205
$Na_2HPO_4 \cdot 7H_2O$	48	302
$Na_2S_2O_3 \cdot 5H_2O$	48	346
$Ba(OH)_2 \cdot 8H_2O$	78	655

TABLE 5.3   Characteristics of salt hydrates.

in Sweden (*System Tepidus*; Bakken, 1981). Figure 5.12 shows this chemical heat pump, which is charged by the reaction

$$Na_2S + 5H_2O \, (\text{vapor}) \rightarrow Na_2S \cdot 5H_2O + 312 \, kJ \, mol^{-1}. \tag{5.11}$$

The heat for the evaporation is taken from a reservoir of about 5 °C, that is, a pipe extending through the soil at a depth of a few meters (as in commercial electric heat pumps with the evaporator buried in the lawn), corresponding to roughly 10 °C in the water container (B in Fig. 5.12) owing to heat-exchanger losses. The water vapor flows to the Na_2S container (A in Fig. 5.12) through a connecting pipe that has been evacuated of all gases other than water vapor and where the water vapor pressure is of the order of 1% of atmospheric pressure. During charging, the temperature in the sodium sulfide rises to 65 °C–70 °C, owing to the heat formed in the process (5.11). When the temperatures in containers A and B and the equilibrium pressures of the water vapor are such that they correspond to each other by a horizontal line in the pressure–temperature diagram shown in Fig. 5.13, flow stops and container A has been charged.

To release the energy, a load area of temperature lower than that of container A is connected to it, and heat is transferred through a heat exchanger. Lowering

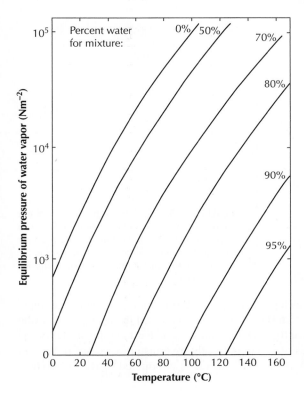

FIGURE 5.11 Equilibrium pressure of water vapor over sulfuric acid/water mixtures as a function of temperature and percentage of water in the mixture (Christensen, 1981).

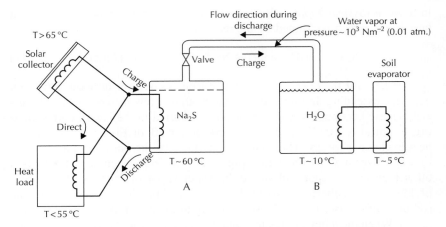

FIGURE 5.12   Schematic picture of a chemical heat pump operating between a sodium sulfide and a water container and based on the formation of the salt hydrate $Na_2S \cdot 5H_2O$. Typical temperatures are indicated. There is a low-temperature heat source connected to the water container and a high-temperature heat source (a solar collector) connected to the salt container, along with the heat demand (load). A switch allows either load or solar collector to be connected.

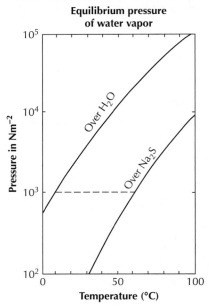

FIGURE 5.13   Equilibrium pressure of water vapor over water and over sodium sulfide as a function of temperature. For a given pressure (*dashed line*), the equilibrium temperatures in the water and the salt containers will differ by roughly 55 °C.

the temperature in A causes a pressure gradient to form in the connecting pipe, and new energy is drawn from B to A. In order to prevent the heat reservoir (the soil) from cooling significantly, new heat must be added to compensate for the heat withdrawn. This takes place continuously by transfer processes in the soil (solar radiation absorbed at the soil surface is conducted to the subsurface location of the evaporator pipes). However, in the long term, a lower temperature would develop in the soil environment if no active makeup heat were supplied. This is done by leading surplus heat from a solar collector to the sodium sulfide container when the solar heat is not directly required in the load area. When the temperature of container A is raised in this way, the pressure gradient above the salt will drive water vapor to container B, thereby removing some of the crystallization water from the salt.

The two actual installations of this type of chemical heat pump are a one-family dwelling with a storage capacity of 7000 kWh (started in 1979) and an industrial building (Swedish Telecommunications Administration) with 30 000 kWh of storage, started in 1980. Future applications may comprise transportable heat stores, since container A may be detached (after closing the valve indicated in Fig. 5.12) and carried to another site. Once the container is detached, the sensible heat is lost, as the container cools from its 60 °C to ambient temperatures, but this amounts to only 3%–4% of the energy stored in the $Na_2S \cdot 5H_2O$ (Bakken, 1981).

It should be mentioned that a similar loss occurs during use of the storage unit in connection with a solar heating system. This is because of the intermittent call upon the store. Every time it is needed, its temperature increases to 60 °C (i.e., every time the valve is opened), using energy to supply the sensible heat, and every time the need starts to decrease, there is a heat loss associated either with making up for the heat transfer to the surroundings (depending on the insulation of the container) to keep the container temperature at 60 °C or, if the valve has been closed, with the heat required to reheat the container to its working temperature. These losses could be minimized by using a modular system, where only one module at a time is kept at operating temperature, ready to supply heat if the solar collector is not providing enough. The other modules would then be at ambient temperatures except when they are called upon to recharge or to replace the unit at standby.

The prototype systems are not cost effective, but the estimated system cost in regular production is 4–5 euro per kWh of heat supplied to a 15 m^3 storage system for a detached house, with half the cost incurred by the solar collector system and the other half incurred by the store. For transport, container sizes equivalent to 4500 kWh of 60 °C heat are envisaged. However, the charge-rate capacity of roughly 1 W kg^{-1} may be insufficient for most applications.

Although the application of the chemical heat pump considered above is for heating, the concept is equally useful for cooling. Here the load would simply be connected to the cold container. Several projects are underway to study various chemical reactions of the gas/liquid or gas/solid type based on pressure differences, either for cooling alone or for both heating and cooling. The materials are chosen on the basis of their temperature requirements and long-term stability while allowing many storage cycles. For example, $NaI–NH_3$ systems have been considered for air conditioning purposes (Fujiwara et al., 1981). A number of ammoniated salts that react on heat exchange could also be considered.

Chemical Reactions

The use of high-temperature chemical heat reactions in thermal storage systems is fairly new and to some extent related to attempts to utilize high-temperature waste heat and to improve the performance of steam power plants (cf. Golibersuch et al., 1976). The chemical reactions that are used to store the heat allow, in addition, upgrading of heat from a lower temperature level to a higher temperature level, a property that is not associated with phase transition or heat capacity methods.

Conventional combustion of fuel is a chemical reaction in which the fuel is combined with an oxidant to form reaction products and surplus heat. This type of chemical reaction is normally irreversible, and there is no easy way that the reverse reaction can be used to store thermal energy. The process of burning fuel is a chemical reaction whereby energy in one form (chemical energy) is transformed into another form (heat) accompanied by an increase in entropy. In order to use a chemical reaction for storage of heat, it would, for example, in the case of hydrocarbon, require a reverse process whereby the fuel (hydrocarbon) could be obtained by adding heat to the reaction products carbon dioxide and water. So, use of chemical heat reactions for thermal energy storage requires suitable reversible reactions.

The change in bond energy for a reversible chemical reaction may be used to store heat, but although a great variety of reversible reactions are known, only a few have so far been identified as being technically and economically acceptable candidates. The technical constraints include temperature, pressure, energy densities, power densities, and thermal efficiency. In general, a chemical heat reaction is a process whereby a chemical compound is dissociated by heat absorption, and later, when the reaction products are recombined, the absorbed heat is again released. Reversible chemical heat reactions can be divided into two groups: thermal dissociation reactions and catalytic reactions. The thermal dissociation reaction may be described as

$$AB \ (+\Delta H \ \text{at} \ T_1 \ \text{and} \ p_1) \leftrightarrow A + B \ (-\Delta H \ \text{at} \ T_2 \ \text{and} \ p_2), \qquad (5.12)$$

indicating that dissociation takes place by addition of heat ΔH to AB at temperature T_1 and pressure p_1, whereas heat is released $(-\Delta H)$ in the reverse reaction at temperature T_2 and pressure p_2. The reciprocal reaction (from right to left) occurs spontaneously if the equilibrium is disturbed, that is, if $T_2 < T_1$ and $p_2 > p_1$. To avoid uncontrolled reverse reaction, the reaction products must therefore be separated and stored in different containers. This separation of the reaction products is not necessary in catalytic reaction systems, where both reactions (left to right and right to left) require a catalyst in order to obtain acceptable high reaction velocities. If the catalyst is removed, neither of the reactions will take place even when considerable changes in temperature and pressure occur. This fact leads to an important advantage, namely, that the intrinsic storage time is, in practice, very large and, in principle, infinite. Another advantage of closed-loop heat storage systems employing chemical reactions is that the compounds involved are not consumed, and because of the high energy densities (in the order of magnitude: 1 MWh m^{-3}, compared to that of the sensible heat of water at $\Delta T = 50$ K, which is 0.06 MWh m^{-3}), a variety of chemical compounds are economically acceptable.

The interest in high-temperature chemical reactions is derived from the work of German investigators on the methane reaction

$$Q + CH_4 + H_2O \leftrightarrow CO + 3H_2 \tag{5.13}$$

(Q being heat added), which was studied in relation to long-distance transmission of high-temperature heat from nuclear gas reactors (cf. Schulten et al., 1974). The transmission system called EVA-ADAM, an abbreviation of the German *Einzelrohrversuchsanlage und Anlage zur dreistufigen adiabatischen Methanisierung*, is being further developed at the nuclear research center at Jülich, West Germany. It consists of steam reforming at the nuclear reactor site, transport over long distances of the reformed gas (CO + 3H$_2$), and methanation at the consumer site, where heat for electricity and district heating is provided (cf., for example, Harth et al., 1981).

Closed-loop System	Enthalpy[a] ΔH^0 (kJ mol^{-1})	Temperature Range (K)
$CH_4 + H_2O \leftrightarrow CO + 3H_2$	206 (250)[b]	700–1200
$CH_4 + CO_2 \leftrightarrow 2CO + 2H_2$	247	700–1200
$CH_4 + 2H_2O \leftrightarrow CO_2 + 4H_2$	165	500–700
$C_6H_{12} \leftrightarrow C_6H_6 + 3H_2$	207	500–750
$C_7H_{14} \leftrightarrow C_7H_8 + 3H_2$	213	450–700
$C_{10}H_{18} \leftrightarrow C_{10}H_8 + 5H_2$	314	450–700

TABLE 5.4 High-temperature closed-loop chemical C-H-O reactions. (Hanneman et al., 1974; Harth et al., 1981).
[a]Standard enthalpy for complete reaction.
[b]Including heat of evaporation of water.

The reaction in (5.13) is a suitable candidate for energy storage that can be accomplished as follows: heat is absorbed in the endothermic reformer, where the previously stored low-enthalpy reactants (methane and water) are converted into high-enthalpy products (carbon monoxide and hydrogen). After heat exchange with the incoming reactants, the products are then stored in a separate vessel at ambient temperature conditions, and although the reverse reaction is thermodynamically favored, it will not occur at these low temperatures and in the absence of a catalyst. When heat is needed, the products are recovered from storage and the reverse, exothermic reaction (methanation) is run (cf. Golibersuch et al., 1976). Enthalpies and temperature ranges for some high-temperature closed-loop C-H-O systems, including reaction (5.13), are given in Table 5.4. The performance of the cyclohexane to benzene and hydrogen system (listed fourth in Table 5.4) has been studied in detail by Italian workers, and an assessment of a storage plant design has been made (cf. Cacciola et al., 1981). The complete storage plant consists of hydrogenation and dehydrogenation multistage adiabatic reactors, storage tanks, separators, heat exchangers, and multistage compressors. Thermodynamic requirements are ensured by independent closed-loop systems circulating nitrogen in the dehydrogenation unit and hydrogen in the hydrogenation unit.

A number of ammoniated salts are known to dissociate and release ammonia at different temperatures, including some in the high-temperature range (see, for example, Yoneda et al., 1980). The advantages of solid–gas reactions in general are high heats of reaction and short reaction times. This implies, in principle, high energy and power densities. However, poor heat and mass transfer characteristics in many practical systems, together with problems of sagging and swelling of the solid materials, lead to reduced densities of the total storage system.

Metal hydride systems are primarily considered as stores of hydrogen, as is discussed later. However, they have also been contemplated for heat storage, and in any case the heat-related process is integral in getting the hydrogen into and out of the hydride. The formation of a hydride MeH_x (metal plus hydrogen) is usually a spontaneous exothermic reaction:

$$Me + 1/2\,x\,H_2 \rightarrow MeH_x + Q, \qquad (5.14)$$

which can be reversed easily by applying the amount of heat Q,

$$MeH_x + Q \rightarrow Me + 1/2\,x\,H_2. \qquad (5.15)$$

Thus, a closed-loop system, where hydrogen is not consumed but is pumped between separate hydride units, may be used as a heat store. High-temperature hydrides, such as MgH_2, Mg_2NiH_2, and TiH_2, have, owing to their high formation enthalpies (e.g., for MgH_2, $\Delta H \geq 80$ kJ per mol of H_2; for TiH_2, $\Delta H > 160$ KJ per mol of H_2), heat densities of up to 3 MJ kg^{-1} or 6 GJ m^{-3} in a temperature range extending from 100 °C to 600 °C (cf. Buchner, 1980).

5.3 HIGH-QUALITY ENERGY STORAGE

A number of storage systems may be particularly suitable for the storage of "high-quality" energy, such as mechanical energy or electric energy. If the energy to be stored is derived from a primary conversion delivering electricity, for example, then one needs an energy storage system that will allow the regeneration of electricity with a high cycle efficiency, i.e., with a large fraction of the electricity input recovered in the form of electricity. Thermal stores, such as the options mentioned in the previous section, may not achieve this, even at $T = 800$ °C-1500 °C (metals, etc.), because of the Carnot limit to the efficiency of electricity regeneration, as well as storage losses through insulation, etc. Thermal stores fed via a heat pump drawing from a suitable reference reservoir may reach tolerable cycle efficiencies more easily.

Table 5.5 gives an indication of the actual or estimated energy densities and cycle efficiencies of various storage systems. The theoretical maximum of energy density is, in some cases, considerably higher than the values quoted.

	Energy Density		
Storage Form	kJ kg^{-1}	MJ m^{-3}	Cycle Efficiency
Conventional fuels			
Crude oil	42 000	37 000	
Coal	32 000	42 000	
Dry wood	12 500[a]	10 000	
Synthetic fuels			
Hydrogen, gas	120 000	10	0.4–0.6
Hydrogen, liquid	120 000	8 700	
Hydrogen, metal hydride	2000–9000	5000–15 000	
Methanol	21 000	17 000	
Ethanol	28 000	22 000	
Thermal – low quality			
Water, 100 °C → 40 °C	250	250	
Rocks, 100 °C → 40 °C	40–50	100–140	
Iron, 100 °C → 40 °C	~30	~230	
Thermal – high quality			
Rocks, e.g., 400 °C → 200 °C	~160	~430	
Iron, e.g., 400 °C → 200 °C	~100	~800	
Inorganic salts, heat of fusion > 300 °C	> 300	> 300	
Mechanical			
Pumped hydro, 100 m head	1	1	0.65–0.8
Compressed air		~15	0.4–0.5
Flywheels, steel	30–120	240–950	
Flywheels, advanced	> 200	> 100	~0.95
Electrochemical			
Lead–acid	40–140	100–900	0.7–0.8
Nickel–cadmium	~350	~350	varying
Lithium ion (other advanced batteries)	700 (> 400)	1400 (> 300)	0.7 (> 0.8)
Superconducting		~100	~0.85

TABLE 5.5 Energy density by weight and volume for various storage forms, based on measured data or expectations for practical applications. For the storage forms aimed at storing and regenerating high-quality energy (electricity), cycle efficiencies are also indicated. Hydrogen gas density is quoted at ambient pressure and temperature. For compressed air energy storage, both electricity and heat inputs are included on equal terms in estimating the cycle efficiency. (Compiled with use of Jensen and Sørensen, 1984.)

[a]*Oven-dry wood may reach values up to 20 000 kJ kg^{-1}.*

For comparison, energy densities of thermal stores and a number of fuels are also given. Some of the fuels (methanol, wood, and hydrogen) may be produced by conversion based on renewable energy sources (without having to wait for fossilization processes to occur). The cycle efficiency is defined with the assumption that the initial energy form is electricity or another high-quality energy form, and the value quoted for hydrogen is based on electrolysis of water as the first step in the storage cycle. Methanol may also be reversibly transformed into hydrogen and carbon oxide to play the role of a closed storage cycle (Prengle and Sun, 1976). The most striking feature is the low-volume energy density of nearly all the reversible storage concepts considered, relative to that of solid or liquid fossil fuels.

The magnitudes of stores that might be associated with renewable energy use (such as indicated in Fig. 2.91) are discussed in more detail in section 6.3, with a summary in the tables of section 6.7, in connection with future energy scenario construction. For comparison, reservoirs of fossil fuels may be found in Table 4.2.

5.3.1 Pumped Hydro Storage

The total exploitable hydro potential is of the order of 10^{12} W on average over the year (section 3.3.2), and only the fraction of this associated with reservoirs can be considered relevant for energy storage. The river flows that have to be tapped as they come may be interesting as energy sources, but not as energy storage options.

Hydro reservoirs feeding into turbine power plants may be utilized for storage of electric energy generated by non-hydropower plants (e.g., wind or photovoltaic energy converters), provided that all the power plants are connected by a common grid, and provided that transmission capacity is sufficient to accommodate the extra burden of load-leveling storage operation of the system. The storage function in a system of this kind is primarily obtained by displacement of load. This means that the hydropower units are serving as backup for the non-hydro generators by providing power when non-hydropower production falls short of load. The small start-up time for hydro turbines (½−3 min) makes this mode of operation convenient. When there is surplus power generation from the non-hydro units, then the hydro generation is decreased, and non-hydro-produced power is transmitted to the load areas otherwise served by hydropower (Sørensen, 1981a; Meibom et al., 1999). In this way, there is no need to pump water up into the hydro reservoirs, as long as the non-hydropower generation stays below the combined load of hydro and non-hydro load areas. To fulfill this condition, the relative sizes of the different types of generating units must be chosen carefully.

When the surplus energy to be stored exceeds the amounts that can be handled in the displacement mode described above, then upward pumping of water into the hydro reservoirs may be considered. It is accomplished by use of two-way turbines, so that the energy can be stored and recovered by the same installation. Alternatively, pumped storage may utilize natural or artificially constructed reservoirs not associated with any exploitable hydropower.

Figure 5.14 shows an example of the layout of a pumped storage facility. Installations where reservoirs are not part of a hydro flow system are typically intended for short-term storage. They may be used for load-leveling purposes, providing a few hours of peak load electric power per day, based on nighttime

FIGURE 5.14 Layout of pumped hydro storage system at Cruachan in Scotland.

FIGURE 5.15 Schematic layout of underground pumped hydro storage system. (From Blomquist et al., 1979. Reprinted with permission from the *Proceedings of the 14th Intersociety Energy Conversion Engineering Conference*. Copyright 1979 American Chemical Society.)

pumping. In terms of average load covered, the storage capacities of these installations are below 24 h. On the other hand, some of the natural reservoirs associated with hydro schemes have storage capacities corresponding to one or more years of average load (e.g., the Norwegian hydro system; cf. Sørensen, 1981a; Meibom *et al.*, 1999). Pumping schemes for such reservoirs could serve for long-term storage of energy.

If no natural elevated reservoirs are present, pumped storage schemes may be based on underground lower reservoirs and surface-level upper reservoirs. The upper reservoirs may be lakes or oceans. The lower ones should be excavated or should make use of natural cavities in the underground. If excavation is necessary, a network of horizontal mine shafts, such as the one illustrated in Fig. 5.15, may be employed in order to maintain structural stability against collapse (Blomquist *et al.*, 1979; Hambraeus, 1975).

The choice of equipment is determined primarily by the size of head, that is, the vertical drop available between the upper and the lower reservoir. Figure 4.37 shows in a schematic form the three most common types of hydro turbines. The Kaplan turbine (called a Nagler turbine if the position of the rotor blades cannot be varied) is most suited for low heads, down to a few meters. Its rotor has the shape of a propeller, and the efficiency of converting gravitational energy into shaft power is high (over 0.9) for the design water velocity, but lower for other water speeds. The efficiency drop away from the design water velocity is rapid for the Nagler turbine, less so for the Kaplan version. These turbines are inefficient for upward pumping, although they can be made to accept water flow from either side (André, 1976). A displacement pump may be used in a "tandem arrangement" (i.e., separate turbine and pump). The electric generator is easily made reversible, so that it may serve either as generator or as motor.

For larger heads, the Francis and Pelton turbines may be used. Francis turbines have a set of fixed guiding blades leading the water onto the rotating blades (the *runner*) at optimum incident angle. It can be used with water heads up to about 600 m in the simple version illustrated in Fig. 4.37*b*, but multistage versions have been considered, guiding the water through a number of runners (five for an actual French installation; cf. Blomquist *et al.*, 1979). In this way, heads above 1000 m can be accepted, and the arrangement may be completely reversible, with very modest losses. For either pumping or generating, the turbine efficiency at design water flow may be over 0.95, but for decreasing flow the efficiency drops. Typical overall efficiencies of the storage cycle (pumping water up by use of surplus electric power, regenerating electric power based on downward flow through turbines) are around 0.8 for existing one-stage Francis turbine installations. Shifting from pumping to generating takes about 1 min (Hambraeus, 1975). The total cycle efficiency of the multistage Francis turbines for heads of 1000–1500 m is about 0.7 (Blomquist *et al.*, 1979).

If the head is larger than the limit for single-stage, reversible Francis turbines, an alternative to the multistage Francis turbines is offered by tandem units consisting of separate impulse turbines and pumps. The pump units for pumping upward over height differences exceeding 1000 m are usually multistage pumps (six stages for an actual installation in Italy), with efficiency over 0.9 being achieved. The impulse turbine part is of Pelton type (see Fig. 4.37a), consisting of a bucket-wheel being driven by the impulse of one or more water jets created by passing the water through nozzles. The power for this process is the pressure force created by the column of water, from the turbine placed at the lower reservoir level to the upper reservoir level. The pressure energy can be converted partially or fully into linear kinetic energy according to the requirements of the different turbine types,

$$mg \, \Delta z = W_{initial}^{pot} = m' 1/2u^2 + (m - m')P\rho^{-1} = W^{kin} + H.$$

Here the initial potential energy associated with the head Δz is transformed into a kinetic energy part associated with partial mass m' moving with velocity u and a pressure energy part with the enthalpy H given by the pressure P over the density of water ρ, times the remaining mass $m - m'$. The conversion efficiency of Pelton turbines is about 0.9 over a wide range of power levels, and the tandem arrangement of separate turbine and pump (but with generator/motor, turbine, and pump all mounted on a common shaft) allows quick shifts between generating and pumping or vice versa.

The losses in conversion are associated in part with "leakage", that is, with water that passes around the turbine without contributing to power, and in part with energy dissipation in the form of heat, for example, due to friction (cf. Angrist, 1976). Further losses are associated with evaporation of water, especially from solar-exposed upper reservoirs.

Excavation for underground storage limits the application to short-term storage (up to about 24 h of average load) because the cost scales approximately linearly with storage capacity. For large natural reservoirs, seasonal energy storage can be considered, since the cost has a large component determined by the maximum load requirement and therefore becomes fairly independent of storage capacity beyond a certain point, as long as the reservoir is available.

5.3.2 Flywheels

Mechanical energy may be stored in the form of rotational motion under conditions of low frictional losses. A flywheel is such a rotating structure, capable of receiving and delivering power through its shaft of rotation. Friction is diminished by use of high-performance bearings, and the entire rotating structure may be enclosed in a space with partial vacuum or filled with an inert gas.

The amount of energy stored in a body of mass distribution $\rho\,(x)$ rotating about a fixed axis with angular velocity ω is

$$W = 1/2\,I\,\omega^2, \tag{5.16}$$

with the moment of inertia I given by

$$I = \int \rho\,(x)\,r^2\,dx.$$

It appears from these expressions that high angular velocity and a majority of the mass situated at a large distance r from the axis of rotation would lead to high amounts of energy stored. The relevant question to ask, however, is how to obtain the highest energy density, given material of a certain strength.

The strength of materials is quantified as the tensile strength, defined as the highest stress not leading to a permanent deformation or breaking of the material. If the material is inhomogeneous, the breaking stresses, and hence the tensile strengths, are different in different directions. For fibrous materials, there are characteristic tensile strengths in the direction of the fibers and perpendicular to the direction of the fibers, the former in some cases being orders of magnitude larger than the latter.

The components in x-, y-, and z-directions of the force per unit volume, f, are related to the stress tensor τ_{ij} by

$$f_i = \sum_j \partial\, t_{ij}/\partial x_j, \tag{5.17}$$

and the tensile strength σ_I in a direction specified by i is

$$\sigma_I = \max\left(\sum_j \tau_{ij}\,n_j\right), \tag{5.18}$$

where n is a unit vector normal to the "cut" in the material (see Fig. 5.16) and the maximum sustainable stress in the direction i is to be found by varying the direction of n, that is, varying the angle of slicing. In other words, the angle of the cut is varied until the stress in the direction i is maximum, and the highest value of this maximum stress not leading to irreversible damage defines the

FIGURE 5.16   Definition of internal stress force (Jensen and Sørensen, 1984).

tensile strength. If the material is isotropic, the tensile strength is independent of direction and may be denoted σ.

Consider now a flywheel like the one illustrated in Fig. 5.17, rotating with angular velocity ω about a fixed axis. The mass distribution is symmetric around the axis of rotation, that is, invariant with respect to rotations through any angle θ about the rotational axis. It is further assumed that the material is homogeneous, so that the mass distribution is fully determined by the mass density ρ and the variation of disk width $b(r)$ as a function of radial distance r from the axis, for a shape symmetric about the midway plane normal to the axis of rotation.

The internal stress forces (5.17) plus any external forces f_{ext} determine the acceleration of a small volume of the flywheel situated at the position x:

$$\rho \frac{d^2 x_i}{dt^2} = f_{ext,i} + \sum_j \frac{\partial \tau_{ij}}{\tau x_j}, \tag{5.19}$$

which is the Newtonian equation of motion. Integrated over some volume V, the force becomes

$$F_i = \int_V f_{ext,i} \, d\mathbf{x} + \int_V \sum_j \frac{\partial \tau_{ij}}{\partial x_j} \, d\mathbf{x} = F_{ext,i} + \int_V \sum_j \tau_{ij} \, n_j \, da, \tag{5.20}$$

where in the last line the volume integral over V has been transformed into a surface integral over the surface A enclosing the volume V, with $\mathbf{n}$ being a unit vector normal to the surface.

The Constant-stress Disk

For uniform rotation with constant angular velocity ω, the acceleration on the left-hand side of (5.19) is radial and given by $r\omega^2$ at the distance r from the axis. Disregarding gravitational forces, the centrifugal force alone must be balanced by the internal stresses, and one may proceed to find the conditions under which all parts of the material experience the same stress τ. If τ equals the tensile strength σ or a specified fraction of it (in order to have a safety margin, as one always would in practice), then the material is utilized optimally, and the energy stored is the maximum that can be achieved using the given material properties.

Taking the volume V as that enclosed between the radius r and $r + dr$ and between the center angle $\theta = -x/2$ and $\theta = x/2$, with the full widths being $b(r)$ and $b(r + dr)$, the balance between the centrifugal force and the internal stresses is obtained from (5.19) and (5.20),

$$2\rho \, r w^2 \, b(r) \, r \, dr = 2\tau((r + dr) \, b(r + dr) - r \, b(r) - b(r) \, dr). \tag{5.21}$$

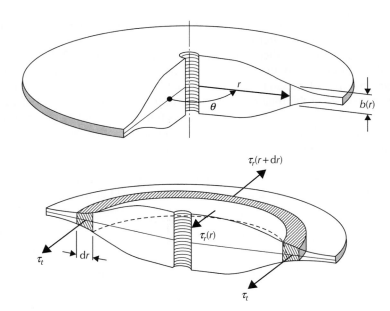

FIGURE 5.17   Co-ordinates and other quantities used in the description of flywheels. The lower part of the figure illustrates the half-torus shape, confined between radii r and $r + dr$, used in the evaluation of stresses in the direction perpendicular to the cut (Jensen and Sørensen, 1984).

The factors 2 come from the angular integrals over $\cos \theta$. The first two terms on the right-hand side of (5.21) are derived from the radial stresses, while the last term represents the tangential stresses on the cuts halving the torus shape considered (cf. Fig. 5.16). To first order in dr, (5.21) may be rewritten as

$$\rho\, r^2 \omega^2 b(r) = \tau r\, db(r)/dr \tag{5.22}$$

from which the disc thickness leading to constant stress is found as

$$b(r) = b_0 \exp\left(-\tfrac{1}{2}\rho\, r^2 \omega^2/\tau\right). \tag{5.23}$$

The optimal shape is seen to be an infinitely extending disc of exponentially declining thickness.

Other Flywheel Shapes

The approach used above may be generalized. Instead of assuming constant stress, the shape of the flywheel is assumed [i.e., $b(r)$ is known, the material still being homogeneous and the shape symmetrical around the axis of rotation as well as upon reflection in the midway plane perpendicular to the axis].

Then the stresses will have to be calculated as a function of rotational speed ω. Owing to the assumptions made, there are only two radially varying stress functions to consider, the radial stress $\tau_r(r)$ and the tangential stress $\tau_t(r)$, both depending only on the distance r from the axis of rotation. Stress components parallel to the axis of rotation are considered absent. Considering again a half-torus shape (see Fig. 5.17), the forces perpendicular to the cut plane may be written in a generalization of (5.21):

$$2\rho r^2 \omega^2 b(r) dr = 2(\tau_r(r+dr)b(r+dr)(r+dr) - r\tau_r(r)b(r)) - \tau_t(r)b(r)\,dr.$$
(5.24)

To first order in dr, this gives, after rearrangement,

$$\tau_t(r) = \rho\, r^2 \omega^2 + \tau_r(r) + d\tau_t(r)/dr + r\tau_r(r)\, db(r)/(b(r)\,dr).$$
(5.25)

This equation relates radial and tangential stresses. In order to determine the stresses, a second relation must be established. This is the relation between stresses and strains, corresponding to Hooke's law in the theory of elasticity. Introduce deformation parameters ε_t and ε_r for tangential and radial stretching by the relations

$$2\pi\Delta r = 2\pi\varepsilon_t,$$
(5.26)

$$d(\Delta r)/dr = \varepsilon_r = \varepsilon_t + r\,d\varepsilon_t/dr,$$
(5.27)

where (5.26) gives the tangential elongation of the half-torus confined between r and $r + dr$ (see Fig. 5.17), and (5.27) gives the implied radial stretching. Then the stress–strain relations may be written

$$\varepsilon_t(r) = Y^{-1}(\tau_t(r) - \mu\,\tau_r(r)),$$
(5.28)

$$\varepsilon_r(r) = Y^{-1}(\tau_r(r) - \mu\,\tau_t(r)),$$
(5.29)

where the strength of compression Y is called Young's module and μ is Poisson's ratio, being the ratio between Y and the corresponding quantity Z measuring the strength of expansion in directions perpendicular to the compression (Shigley, 1972). Eliminating the deformations from (5.27)–(5.29), a new relation between the stresses is obtained:

$$(1 + \mu)(\tau_r(r) - \tau_t(r)) = r\,d\tau_t(r)/dr - r\mu\,d\tau_r(r)/dr.$$
(5.30)

Inserting (5.25) into (5.30), a second-order differential equation for the determination of $\tau_r(r)$ results. The solution depends on materials properties through ρ, Y, and μ and on the state of rotation through ω. Once the radial stress is determined, the tangential one can be evaluated from (5.25).

As an example, consider a plane disc of radius r_{max}, with a center hole of radius r_{min}. In this case, the derivatives of $b(r)$ vanish, and the solution to (5.30) and (5.25) is

$$\tau_r(r) = (3+\mu)\rho\,\omega^2(r_{min}^2 + r_{max}^2 - r_{min}^2\,r_{max}^2/r^2 - r^2)/8$$

$$\tau_t(r) = (3+\mu)\rho\,\omega^2(r_{min}^2 + r_{max}^2 + r_{min}^2\,r_{max}^2/r^2 - (1+3\mu)r^2/(3+\mu))/8.$$

(5.31)

The radial stress rises from zero at the inner rim, reaches a maximum at $r = (r_{min}r_{max})^{1/2}$, and then declines to zero again at the outer rim. The tangential stress is maximal at the inner rim and declines outward. Its maximum value exceeds the maximum value of the radial stress for most relevant values of the parameters (μ is typically around 0.3).

Comparing (5.25) with (5.16) and the expression for I, it is seen that the energy density W in (5.16) can be obtained by isolating the term proportional to ω^2 in (5.25), multiplying it by $1/2r$, and integrating over r. The integral of the remaining terms is over a stress component times a shape-dependent expression, and it is customary to use an expression of the form

$$W/M = \sigma\,K_m/\rho,$$

(5.32)

where $M = \int\rho b(r)r\,d\theta\,dr$ is the total flywheel mass and σ is the maximum stress [cf. (5.18)]. K_m is called the shape factor. It depends only on geometry, if all stresses are equal, as in the constant-stress disc, but as the example of a flat disc has indicated [see (5.31)], the material properties and the geometry cannot generally be factorized. Still, the maximum stress occurring in the flywheel may be separated, as in (5.32), in order to leave a dimensionless quantity K_m to describe details of the flywheel construction (also, the factor ρ has to be there to balance ρ in the mass M, in order to make K_m dimensionless). Expression (5.32) may now be read in the following way: given a maximum acceptable stress σ, there is a maximum energy storage density given by (5.32). It does not depend on ω, and it is largest for light materials and for large design stresses σ. The design stress is typically chosen as a given fraction (*safety factor*) of the tensile strength. If the tensile strength itself is used in (5.32), the physical upper limit for energy storage is obtained, and using (5.16), the expression gives the maximum value of ω for which the flywheel will not fail by deforming permanently or by disintegrating.

Flywheel Performance

Some examples of flywheel shapes and the corresponding calculated shape factors K_m are given in Table 5.6. The infinitely extending disc of constant stress has a theoretical shape factor of unity, but for a practical version with

Shape	K_m
Constant-stress disc	1
Flat, solid disc (μ = 0.3)	0.606
Flat disc with center hole	~0.3
Thin rim	0.5
Radial rod	1/3
Circular brush	1/3

TABLE 5.6 Flywheel shape factors.

finite truncation, K_m of about 0.8 can be expected. A flat, solid disc has a shape factor of 0.6, but if a hole is pierced in the middle, the value reduces to about 0.3. An infinitely thin rim has a shape factor of 0.5 and a finite one, of about 0.4, and a radial rod or a circular brush (cf. Fig. 5.18) has K_m equal to one-third.

According to (5.32), the other factors determining the maximum energy density are the maximum stress and the inverse density, in the case of a homogeneous material. Table 5.7 gives tensile strengths and/or design stresses with a safety factor included and gives densities for some materials contemplated for flywheel design.

For automotive purposes, the materials with the highest σ/ρ values may be contemplated, although they are also generally the most expensive. For stationary applications, weight and volume are less decisive, and low material cost becomes a key factor. This is the reason for considering cellulosic materials (Hagen *et al.*, 1979). One example is plywood discs, where the disc is assembled from layers of unidirectional plies, each with different orientation. Using (5.32) with the unidirectional strengths, the shape factor should be reduced by a factor of almost 3. Another example in this category is paper-roll flywheels, that is, hollow, cylindrically wound shapes, for which the shape factor is $K_m = (1 + (r_{min}/r_{max})^2)/4$ (Hagen *et al.*, 1979). The specific energy density would be about 15 kJ kg^{-1} for the plywood construction and 27 kJ kg^{-1} for "super-paper" hollow torus shapes.

Unidirectional materials may be used in configurations like the flywheel illustrated in Fig. 5.18b, where tangential (or "hoop") stresses are absent. Volume efficiency is low (Rabenhorst, 1976). Generally, flywheels made from filament have an advantage in terms of high safety, because disintegration into a large number of individual threads makes a failure easily contained. Solid flywheels may fail by expelling large fragments, and for safety such flywheels are not proper in vehicles, but may be placed underground for stationary uses.

Material	Density (kg m^{-3})	Tensile Strength (10^6 N m^{-2})	Design Stress (10^6 N m^{-2})
Birch plywood	700	125	30
"Super-paper"	1100	335	
Aluminum alloy	2700	500	
Mild steel	7800		300
Maraging steel	8000	2700	900
Titanium alloy	4500		650
Carbon fiber (40% epoxy)	1550	1500	750
E-glass fiber (40% epoxy)	1900	1100	250
S-glass fiber (40% epoxy)	1900	1750	350
Kevlar fiber (40% epoxy)	1400	1800	1000

TABLE 5.7   Properties of materials considered for flywheels (Davidson *et al.*, 1980; Hagen *et al.*, 1979).

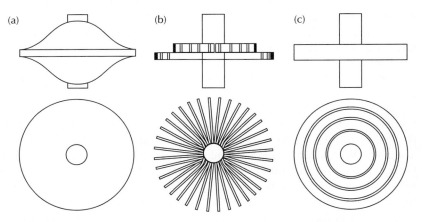

FIGURE 5.18a,b,c   Different flywheel concepts. The upper row gives side views, and the lower row gives top views.

Approximately constant-stress shapes (cf. Fig. 5.18*a*) are not as volume efficient as flat discs. Therefore, composite flywheels of the kind shown in Fig. 5.18*c* have been contemplated (Post and Post, 1973). Concentric flat rings (e.g., made of Kevlar) are separated by elastomers that can eliminate breaking stresses when the rotation creates differential expansion of adjacent rings. Each ring must be made of a different material in order to keep the variations in stress within a small interval. The stress distribution inside each ring can be derived from the expressions in (5.31), assuming that the elastomers fully take care of any interaction between rings. Alternatively, the elastomers can be treated as additional rings, and the proper boundary conditions can be applied (see, for example, Toland, 1975).

Flywheels of the types described above may attain energy densities of up to 200 kJ kg^{-1}. The problem is to protect this energy against frictional losses. Rotational speeds would typically be 3–5 revolutions per second. The commonly chosen solution is to operate the flywheel in near vacuum and to avoid any kind of mechanical bearings. Magnetic suspension has recently become feasible for units of up to about 200 t, using permanent magnets made from rare-earth cobalt compounds and electromagnetic stabilizers (Millner, 1979). In order to achieve power input and output, a motor generator is inserted between the magnetic bearing suspension and the flywheel rotor. If the motor is placed inside the vacuum, a brushless type is preferable.

For stationary applications, the weight limitations may be circumvented. The flywheel could consist of a horizontally rotating rim wheel of large dimensions and weight, supported by rollers along the rim or by magnetic suspension (Russell and Chew, 1981; Schlieben, 1975). Energy densities of about 3000 kJ kg^{-1} could, in principle, be achieved by using fused silica composites (cf. fibers of Table 5.7). The installations would then be placed underground in order to allow reduced safety margins. Unit sizes could be up to 10^5 kg.

5.3.3 Compressed Gas Storage

Gases tend to be much more compressible than solids or fluids, and investigations of energy storage applications of elastic energy on a larger scale have therefore concentrated on the use of gaseous storage media.

Storage on a smaller scale may make use of steel containers, such as the ones common for the compressed air used in mobile construction work. In this case, the volume is fixed and the amount of energy stored in the volume is determined by the temperature and the pressure. If air is treated as an ideal gas, the (thermodynamic) pressure P and temperature T are related by the equation of state

$$PV = \nu \mathscr{R} T, \tag{5.33}$$

where V is the volume occupied by the air, ν is the number of moles in the volume, and $\mathscr{R} = 8.315$ J K^{-1} mol^{-1}. The pressure P corresponds to the stress in the direction of compression for an elastic cube, except that the sign is reversed (in general, the stress equals $-P$ plus viscosity-dependent terms). The container may be thought of as a cylinder with a piston enclosing a given number of moles of gas, say, air. The compressed air is formed by compressing the enclosed air from standard pressure at the temperature of the surroundings, that is, increasing the force f_x applied to the piston, while the volume decreases from V_0 to V. The amount of energy stored is

$$W = A \int_{x_0}^{x} f_x \, dx = -\int_{V_0}^{V} P \, dV, \tag{5.34}$$

where A is the cylinder cross-sectional area, x and x_0 are the piston positions corresponding to V and V_0, and P is the pressure of the enclosed air.

For large-scale storage applications, underground cavities have been considered. The three possibilities investigated until now are salt domes, cavities in solid rock formations, and aquifers.

Cavities in salt deposits may be formed by flushing water through the salt. The process has, in practical cases, been extended over a few years, in which case the energy spent (and cost) has been very low (Weber, 1975). Salt domes are salt deposits extruding upward toward the surface, therefore allowing cavities to be formed at modest depths.

Rock cavities may be either natural or excavated, and the walls are properly sealed to ensure air-tightness. If excavated, they are much more expensive to make than salt caverns.

Aquifers are layers of high permeability, permitting underground water flows along the layer. In order to confine the water stream to the aquifer, there have to be encapsulating layers of little or no permeability above and below the water-carrying layer. Aquifers usually do not stay at a fixed depth, and thus, they can have slightly elevated regions where a certain amount of air can become trapped without impeding the flow of water. This possibility for air storage (under the elevated pressure corresponding to the depth involved) is illustrated in Fig. 5.19c.

Figure 5.19 illustrates the forms of underground air storage mentioned: salt, rock, and aquifer storage. In all cases, site selection and preparation are a fairly delicate process. Although the general geology of the area considered is known,

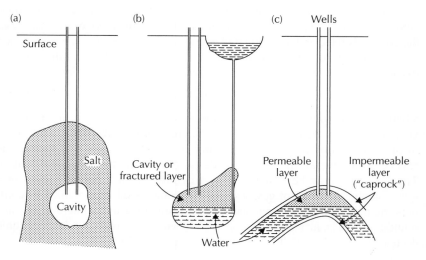

FIGURE 5.19a,b,c Types of underground compressed air storage: (a) storage in salt cavity, (b) rock storage with compensating surface reservoir, and (c) aquifer storage.

the detailed properties of the cavity will not become fully known until the installation is complete. The ability of the salt cavern to keep an elevated pressure may not live up to expectations based on sample analysis and pressure tests at partial excavation. The stability of a natural rock cave, or of a fractured zone created by explosion or hydraulic methods, is also uncertain until actual full-scale pressure tests have been conducted. For aquifers, decisive measurements of permeability can be made at only a finite number of places, so surprises are possible due to rapid permeability change over small distances of displacement (cf. Adolfson *et al.*, 1979).

The stability of a given cavern is influenced by two design features that operation of the compressed air storage system will entail: notably temperature and pressure variations. It is possible to keep the cavern wall temperature nearly constant, either by cooling the compressed air before letting it down into the cavern or by performing the compression so slowly that the temperature rises only to the level prevailing on the cavern walls. The latter possibility (isothermal compression) is impractical for most applications, because excess power must be converted at the rate at which it is produced. Most systems, therefore, include one or more cooling steps. With respect to the problem of pressure variations, when different amounts of energy are stored, the solution may be to store the compressed air at constant pressure but variable volume. In this case, either the storage volume itself should be variable, as it is with aquifer storage (when variable amounts of water are displaced), or the underground cavern should be connected to an open reservoir (Fig. 5.19*b*), so that a variable water column may respond to the variable amounts of air stored at the constant equilibrium pressure prevailing at the depth of the cavern. This kind of compressed energy storage system may alternatively be viewed as a pumped hydro storage system, with extraction taking place through air-driven turbines rather than through water-driven turbines.

Adiabatic Storage

Consider now the operation of a variable-pressure system. The compression of ambient air takes place approximately as an adiabatic process, that is, without heat exchange with the surroundings. If γ denotes the ratio between the partial derivatives of pressure with respect to volume at constant entropy and at constant temperature,

$$(\partial P/\partial V)_s = \gamma(\partial P/\partial V)_T, \tag{5.35}$$

and the ideal gas law (5.33) gives $(\partial P/\partial V)_T = -P/V$, so that for constant γ,

$$PV^\gamma = P_0 V_0^\gamma. \tag{5.36}$$

The constant on the right-hand side is here expressed in terms of the pressure P_0 and volume V_0 at a given time. For air at ambient pressure and temperature, $\gamma = 1.40$. The value decreases with increasing temperature and increases with increasing pressure, so (5.36) is not entirely valid for air. However, in the temperature and pressure intervals relevant for practical application of compressed air storage, the value of γ varies less than $\pm 10\%$ from its average value.

Inserting (5.36) into (5.34), we get the amount of energy stored,

$$W = -\int_{V_0}^{V} P_0 \left(\frac{V_0}{V}\right)^{\gamma} dV = \frac{P_0 V_0}{\gamma - 1}\left(\left(\frac{V_0}{V}\right)^{\gamma-1} - 1\right), \tag{5.37}$$

or, alternatively,

$$W = \frac{P_0 V_0}{\gamma - 1}\left(\left(\frac{P}{P_0}\right)^{(\gamma-1)/\gamma} - 1\right). \tag{5.38}$$

More precisely, this is the work required for the adiabatic compression of the initial volume of air. This process heats the air from its initial temperature T_0 to a temperature T, which can be found by rewriting (5.33) in the form

$$T/T_0 = PV/(P_0 V_0)$$

and combining it with the adiabatic condition (5.36),

$$T_0 = T_0 (P/P_0)^{(\gamma-1)/\gamma}. \tag{5.39}$$

Since desirable pressure ratios in practical applications may be up to about $P/P_0 = 70$, maximum temperatures exceeding 1000 K can be expected. Such temperature changes would be unacceptable for most types of cavities considered, and the air is therefore cooled before transmission to the cavity. Surrounding temperatures for underground storage are typically about 300 K for salt domes and somewhat higher for storage in deeper geological formations. With this temperature denoted as T_s, the heat removed if the air is cooled to T_s at constant pressure amounts to

$$H = c_P(T - T_s), \tag{5.40}$$

where c_P is the heat capacity at constant pressure. Ideally, the heat removed would be kept in a well-insulated thermal energy store, so that it can be used to reheat the air when it is taken up from the cavity to perform work by expansion in a turbine, with the associated pressure drop back to ambient pressure P_0. Viewed as a thermodynamic process in a temperature–entropy (T, S)-diagram, the storage and retrieval processes in the ideal case are as shown in Fig. 5.20. The process leads back to its point of departure, indicating that the storage cycle is loss-free under the idealized conditions assumed so far.

In practice, the compressor has a loss (of maybe 5%–10%), meaning that not all the energy input (electricity, mechanical energy) is used to perform compression work on the air. Some is lost as friction heat, and so on. Furthermore, not

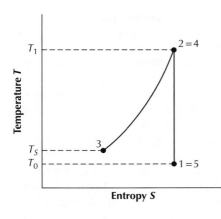

FIGURE 5.20 Idealized operation of adiabatic compressed air storage system. The charging steps are: 1-2 adiabatic compression, and 2-3 isobaric cooling to cavern temperature. The unloading steps are: 3-4 isobaric heating, and 4-5 adiabatic expansion through turbine. The diagram follows a given amount of air, whereas an eventual thermal energy store is external to the "system" considered. T_0 is surface ambient temperature, T_1 is temperature after compression, and T_s is the cavern temperature.

all the heat removed by the cooling process can be delivered to reheat the air. Heat exchangers have finite temperature gradients, and there may be losses from the thermal energy store during the time interval between cooling and reheating. Finally, the exhaust air from actual turbines has temperatures and pressures above ambient. Typical loss fractions in the turbine may be around 20% of the energy input, at the pressure considered above (70 times ambient) (Davidson *et al.*, 1980). If less than 10% thermal losses could be achieved, the overall storage cycle efficiency would be about 65%.

The real process may proceed as shown in Fig. 5.21, as function of temperature and entropy changes. The compressor loss in the initial process 1-2 modifies the vertical line to include an entropy increase. Furthermore, compression

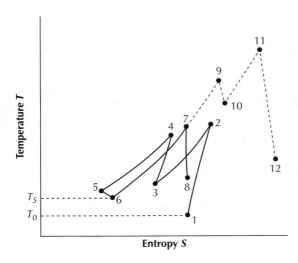

FIGURE 5.21 Operation of compressed air storage systems with finite losses. The solid path corresponds to a scheme with two compression and two cooling stages, temperature adjustment of the stored air, reheating, and a single turbine stage, with reject air sent into the atmosphere (open cycle 1-8). Alternatively, the path from step 7 to step 12 exhibits two heating and expansion steps, corresponding to the operating plant at Huntorf, Germany. See text for further details.

has been divided into two steps (1-2 and 3-4) in order to reduce the maximum temperatures. Correspondingly, there are two cooling steps (2-3 and 4-5), followed by a slight final cooling performed by the cavity surroundings (5-6). The work-retrieval process involves, in this case, a single step 6-7 of reheating by use of heat stored from the cooling processes (in some cases, more than one reheating step is employed). Finally, 7-8 is the turbine stage, which leaves the cycle open by not having the air reach the initial temperature (and pressure) before it leaves the turbine and mixes into the ambient atmosphere. Also, this expansion step shows deviations from adiabaticity, seen in Fig. 5.21 as an entropy increase.

There are currently only a few utility-integrated installations. The earliest full-scale compressed storage facility has operated since 1978 at Huntorf, Germany. It is rated at 290 MW and has about 3×10^5 m^3 storage volume (Lehmann, 1981). It does not have any heat recuperation, but it has two fuel-based turbine stages, implying that the final expansion takes place from a temperature higher than any of those involved in the compression stages (and also at higher pressure). This is indicated in Fig. 5.21 as 7-9-10-11-12, where steps 7-9 and 10-11 represent additional heating based on fuel, while steps 9-10 and 11-12 indicate expansion through turbines. If heat recuperation is added, as it is in the 110 MW plant operated by the Alabama Electric Corp. (USA) since 1991, point 7 moves upward toward point 9, and point 8 moves in the direction of 12, altogether representing an increased turbine output (Linden, 2003).

The efficiency calculation is changed in the case of additional fuel combustion. The additional heat input may be described by (5.40) with appropriate temperatures substituted, and the primary enthalpy input H_0 is obtained by dividing H by the fuel-to-heat conversion efficiency. In the case of a finite compressor efficiency η_c, the input work W_{in} to the compressor changes from (5.38) to

$$W_{in} = \frac{P_0 V_0}{\gamma - 1} \left(\left(\frac{P}{P_0} \right)^{(\gamma-1)/(\gamma \eta_c)} - 1 \right). \tag{5.41}$$

The work delivered by the turbine receiving air of pressure P_1 and volume V_1, and exhausting it at P_2 and V_2, with a finite turbine efficiency η_t, is

$$W_{out} = \frac{P_1 V_1}{\gamma - 1} \left(1 - \left(\frac{P_2}{P_1} \right)^{\eta_t (\gamma-1)/\gamma} \right), \tag{5.42}$$

which, except for the appearance of η_t, is just (5.38) rewritten for the appropriate pressures and volume.

When there is only a single compressor and a single turbine stage, the overall cycle efficiency is given by

$$\eta = W_{out}/(W_{in} + H_0). \tag{5.43}$$

FIGURE 5.22   Layout of the Huntorf compressed air storage facility. Compressors are denoted C, turbines are T, and burners are B. The subscripts H and L stand for high and low pressure, repectively.

FIGURE 5.23   Layout of an advanced compressed air storage facility, with heat recuperation and no fuel input. Compressors are denoted C and turbines are T. The subscripts H and L stand for high and low pressure, respectively.

For the Huntorf compressed air storage installation, η is 0.41. Of course, if the work input to the compressor is derived from fuel (directly or through electricity), W_{in} may be replaced by the fuel input W_0 and a fuel efficiency defined as

$$\eta_{fuel} = W_{out}/(W_0 + H_0). \qquad (5.44)$$

If W_{in}/W_0 is taken as 0.36, η_{fuel} for the example becomes 0.25, which is 71% of the conversion efficiency for electric power production without going through the store. The German installation is used for providing peak power on weekdays, and is charged during nights and weekends.

Figure 5.22 shows the physical layout of the German plant, and Fig. 5.23 shows the layout of a more advanced installation with no fuel input, corresponding to the two paths illustrated in Fig. 5.21.

Aquifer Storage

The aquifer storage system shown in Fig. 5.19c would have an approximately constant working pressure, corresponding to the average hydraulic pressure at the depth of the air-filled part of the aquifer. According to (5.34), the stored energy in this case simply equals the pressure P times the volume of air displacing water in the aquifer. This volume equals the physical volume V times the porosity p, that is, the fractional void volume accessible to intruding air (there may be additional voids that the incoming air cannot reach), so the energy stored may be written

$$W = pVP. \qquad (5.45)$$

Typical values are $p = 0.2$ and P around 6×10^6 N m^{-2} at depths of about 600 m, with useful volumes of 10^9 to 10^{10} m^3 for each site. Several such sites have been investigated for the possible storage of natural gas.

An important feature of an energy storage aquifer is the time required for charging and emptying. This time is determined by the permeability of the aquifer. The permeability is basically the proportionality factor between the flow velocity of a fluid or gas through the sediment and the pressure gradient causing the flow. The linear relationship assumed may be written

$$\nu = -K(\eta\rho)^{-1}\partial P/\partial S, \qquad (5.46)$$

where ν is the flow velocity, η is the viscosity of the fluid or gas, ρ is its density, P is the pressure, and s is the path-length in the downward direction. K is the permeability, being defined by (5.46). In metric (SI) units, the unit of permeability is m^2. The unit of viscosity is m^2 s^{-1}. Another commonly used unit of permeability is the *darcy*. One darcy equals 1.013×10^{12} m^2. If filling and emptying of the aquifer storage are to take place in a matter of hours, rather than days, permeability has to exceed 10^{11} m^2. Sediments, such as sandstone, have permeabilities ranging from 10^{10} to 3×10^{12} m^2, often with considerable variation over short distances.

In actual installations, losses occur. The cap-rock bordering the aquifer region may not have negligible permeability, implying a possible leakage loss. Friction in the pipes leading to and from the aquifer may cause a loss of pressure, as may losses in the compressor and turbine. Typically, losses of about 15% are expected in addition to those of the power machinery. Large aquifer stores for natural gas are in operation: the store at Stenlille, Denmark has 10^9 m^3 total volume, from which 3.5×10^8 m^3 of gas can be extracted at the employed pressure of 17 MPa (DONG, 2003). The same gas utility company built an excavated salt dome gas storage at Lille Thorup that has a slightly smaller volume but allows 4.2×10^8 m^3 of gas to be extracted at 23 MPa.

Hydrogen Storage

Hydrogen can be stored like other gases, compressed in suitable containers capable of managing the high diffusivity of hydrogen, as well as sustaining the pressures required to bring the energy density up to useful levels. However, the low energy density of gaseous hydrogen has brought alternative storage forms into the focus of investigation, such as liquid hydrogen and hydrogen trapped inside metal hydride structures or inside carbon-based or other types of nanotubes (cf. Sørensen, 2004a).

Hydrogen is an energy carrier, not a primary energy form. The storage cycle therefore involves both the production of hydrogen from primary energy sources and the retrieval of the energy form demanded by a second conversion process.

Hydrogen Production

Conventional hydrogen production is by catalytic steam-reforming of methane (natural gas) or gasoline with water vapor. The process, which typically takes place at 850 °C and 2.5×10^6 Pa, is

$$C_nH_m + n\,H_2O \rightarrow n\,CO + (n + m/2)\,H_2, \qquad (5.47)$$

followed by the catalytic shift reaction

$$CO + H_2O \rightarrow CO_2 + H_2. \qquad (5.48)$$

Finally, CO_2 is removed by absorption or membrane separation. The heat produced by (5.48) often cannot be directly used for (5.47). For heavy hydrocarbons, including coal dust, a partial oxidation process is currently in use (Zittel and Wurster, 1996). An emerging technology is high-temperature plasma-arc gasification, and a pilot plant based on this technology operates on natural gas at 1600 °C in Norway (at Kvaerner Engineering). The advantage of this process is the pureness of the resulting products (in energy terms, 48% hydrogen, 40% carbon, and 10% water vapor) and therefore low environmental impact. Since all the three main products are useful energy carriers, the conversion efficiency may be said to be 98% minus the energy needed for the process.

However, conversion of natural gas to carbon is not normally desirable, and the steam can be used only locally, so 48% efficiency is more meaningful.

Production of hydrogen from biomass may be achieved by biological fermentation or by high-temperature gasification similar to that used for coal. These processes are described in more detail in section 4.6.2. Production of hydrogen from (wind- or solar-produced) electricity may be achieved by conventional electrolysis (demonstrated by Faraday in 1820 and widely used since about 1890) or by reversible fuel cells, with current efficiencies of about 70% and over 90%, respectively.

Electrolysis conventionally uses an aqueous alkaline electrolyte, with the anode and cathode areas separated by a microporous diaphragm (replacing earlier asbestos diaphragms), so that

$$H_2O \rightarrow H_2 + 1/2\,O_2, \tag{5.49}$$

$$\Delta H = \Delta G + T\,\Delta S. \tag{5.50}$$

At 25 °C, the change in free energy, ΔG, is 236 kJ mol^{-1}, and electrolysis would require a minimum amount of electric energy of 236 kJ mol^{-1}, while the difference between enthalpy and free energy changes, $\Delta H - \Delta G$, in theory could be heat from the surroundings. The energy content of hydrogen (equal to ΔH) is 242 kJ mol^{-1} (lower heating value), so $T\,\Delta S$ could exceed 100%. However, if heat is added at 25 °C, the process is exceedingly slow. Temperatures used in actual installations are so high that the heat balance is positive and cooling has to be applied. This is largely a consequence of electrode overvoltage, mainly stemming from polarization effects. The cell potential V for water electrolysis may be expressed by

$$V = V_r + V_a + V_c + Rj, \tag{5.51}$$

where V_r is the reversible cell potential. The overvoltage has been divided into the anodic and cathodic parts V_a and V_c. The current is j, and R is the internal resistance of the cell. The last three terms in (5.51) represent the electrical losses, and the voltage efficiency η_V of an electrolyzer operating at a current j is given by

$$\eta_V = V_r/V, \tag{5.52}$$

while the thermal efficiency is

$$\eta_t = \Delta H / \Delta G = |\,\Delta H/(n\,\mathscr{F}V)\,|, \tag{5.53}$$

with the Faraday constant being $\mathscr{F} = 96\,493$ coulombs mol^{-1}, and n being the number of moles transferred in the course of the overall electrochemical reaction to which ΔG relates.

Efforts are being made to increase the efficiency above the current 50% to 80% (for small to large electrolyzers) by increasing the operating temperature and optimizing electrode materials and cell design; in that case, the additional costs should be less than for the emerging solid-state electrolyzers, which are essentially fuel cells operated in reverse, i.e., using electric power to produce hydrogen and oxygen from water in an arrangement and with reaction schemes formally the same as those of the fuel cells described in section 4.5. If the same fuel cell allows operation in both directions, it is called a *reversible fuel cell*.

A third route contemplated for hydrogen production from water is thermal decomposition of water. Because direct thermal decomposition of water molecules requires temperatures exceeding 3000 K, which is not possible with presently available materials, attempts have been made to achieve decomposition below 800 °C by an indirect route using cyclic chemical processes. Such thermochemical or water-splitting cycles were originally designed to reduce the required temperature to the low values attained in nuclear reactors, but could, of course, be used with other technologies generating heat at around 400 °C. An example of the processes studied (Marchetti, 1973) is the three-stage reaction

$$6FeCl_2 + 8H_2O \rightarrow 2Fe_3O_4 + 12HCl + 2H_2 \ (850\,°C)$$

$$2Fe_3O_4 + 3Cl_2 + 12HCl \rightarrow 6FeCl_3 + 6H_2O + O_2 \ (200\,°C)$$

$$6FeCl_3 \rightarrow 6FeCl_2 + 3Cl_2 \ (420\,°C).$$

The first reaction still requires a high temperature, implying a need for energy to be supplied, in addition to the problem of the corrosive substances involved. Researchers are still a long way from creating a practical technology.

The process of water photodissociation is described in section 3.5.1. There have been several attempts to imitate the natural photosynthetic process, using semiconductor materials and membranes to separate the hydrogen and oxygen formed by application of light (Calvin, 1974; Wrighton *et al.*, 1977). So far, no viable reaction scheme has been found, either for artificial photodissociation or for hybrid processes using heat and chemical reactions in side processes (Hagenmuller, 1977).

Processing of the hydrogen produced involves removal of dust and sulfur, as well as other impurities, depending on the source material (e.g., CO_2 if biogas is the source).

Hydrogen Storage Forms

The storage forms relevant for hydrogen are connected with its physical properties, such as the energy density shown in Table 5.5 for hydrogen in various forms. Combustion and safety-related properties are shown in Table 5.8 and are compared with those of methane, propane, and gasoline. Hydrogen's high diffusivity has implications for container structure, and hydrogen has a large range of flammability/explosivity to consider in all applications.

Property	Unit	Hydrogen	Methane	Propane	Gasoline
Minimum energy for ignition	10^{-3} J	0.02	0.29		0.24
Flame temperature	°C	2045	1875		2200
Auto-ignition temperature in air	°C	585	540	510	230–500
Maximum flame velocity	m s^{-1}	3.46	0.43	0.47	
Range of flammability in air	vol.%	4–75	5–15	2.5–9.3	1.0–7.6
Range of explosivity in air	vol.%	13–65	6.3–13.5		1.1–3.3
Diffusion coefficient in air	10^{-4} m^2 s^{-1}	0.61	0.16		0.05

TABLE 5.8   Safety-related properties of hydrogen and other fuels (with use of Zittel and Wurster, 1996).

Compressed Storage in Gaseous Form   The low volume density of hydrogen at ambient pressure (Table 5.5) makes compression necessary for energy storage applications. Commercial hydrogen containers presently use pressures of $20\text{–}30 \times 10^6$ Pa, with corresponding energy densities of $1900\text{–}2700 \times 10^6$ J m^{-3}, which is still less than 10% of that of oil. Research is in progress for increasing pressures to about 70 MPa, using high-strength composite materials such as Kevlar fibers. Inside liners of carbon fibers (earlier, glass/aluminum was used) are required to reduce permeability. Compression energy requirements influence storage-cycle efficiencies and involve long transfer times. The work required for isothermal compression from pressure p_1 to p_2 is of the form

$$W = A\,T\,\log(p_2/p_1),$$

where A is the hydrogen gas constant 4124 J K^{-1} kg^{-1} times an empirical, pressure-dependent correction (Zittel and Wurster, 1996). To achieve compression, a motor rated at Bm must be used, where m is the maximum power throughput and B, depending on engine efficiency, is around 2.

Liquid Hydrogen Stores   Because the liquefaction temperature of hydrogen is 20 K ($-253\,°C$), the infrastructure and energy requirements for liquefaction are substantial (containers and transfer pipes must be superinsulated). On the other hand, transfer times are low (currently 3 min to charge a passenger car). The energy density is still 4–5 times lower than for conventional fuels (see Table 5.5). The liquefaction process requires very clean hydrogen, as well as several cycles of compression, liquid nitrogen cooling, and expansion.

Metal Hydride Storage   Hydrogen diffused into appropriate metal alloys can achieve storage at volume densities over two times that of liquid hydrogen. However, the mass storage densities are still less than 10% of those of conventional fuels (Table 5.5), making this concept doubtful for mobile applications, despite the positive aspects of nearly loss-free storage at ambient pressures (0–6 MPa) and transfer accomplished by adding or withdrawing modest amounts of heat (associated with high safety in operation), according to

$$Me + \tfrac{1}{2} xH_2 \leftrightarrow MeH_x, \qquad (5.54)$$

where the hydride may be body-centered cubic lattice structures with about 6×10^{28} atoms per m^3 (such as $LaNi_5H_6$, $FeTiH_2$). The currently highest density achieved is for metal alloys absorbing two hydrogen atoms per metal atom (Toyota, 1996). The lattice absorption cycle also cleans the gas, because impurities in the hydrogen gas are too large to enter the lattice.

Methanol Storage One current prototype hydrogen-fueled vehicle uses methanol as storage, even if the desired form is hydrogen (because the car uses a hydrogen fuel cell to generate electricity for its electric motor; Daimler-Chrysler-Ballard, 1998). The reason is the simplicity of the methanol storage and filling infrastructure. In the long run, transformation of hydrogen to methanol and back seems too inefficient, and it is likely that the methanol concept will be combined with methanol fuel cells (cf. section 4.7), while hydrogen-fueled vehicles must find simpler storage alternatives.

Graphite Nanofiber Stores Current development of nanofibers has suggested wide engineering possibilities for both electric and structural adaptation, including the storage of foreign atoms inside "balls" or "tubes" of large carbon structures (Zhang *et al.*, 1998). Indications are that hydrogen may be stored in nanotubes in quantities exceeding the capacity of metal hydrides, and at a lower weight penalty, but no designs exist yet (Service, 1998).

Regeneration of Power from Hydrogen

Retrieval of energy from stored hydrogen may be by conventional low-efficiency combustion in Otto engines or gas turbines, or it may be through fuel cells at a considerably higher efficiency, as described in section 4.7 and in Sørensen (2004a).

5.3.4 Battery Storage

Batteries may be described as fuel cells where the fuels are stored inside the cell rather than outside it. Historically, batteries were the first controlled source of electricity, with important designs being developed in the early 19th century by Galvani, Volta, and Daniell, before Grove's discovery of the fuel cell and Planté's construction of the lead–acid battery. Today, available batteries use a wide range of electrode materials and electrolytes, but, despite considerable development efforts aimed at the electric utility sector, in practice battery storage is still restricted to small-scale use (consumer electronics, motor cars, etc.).

Efforts are being made to find systems with better performance than the long-employed lead–acid batteries, which are restricted by a low energy density (see Table 5.5) and a limited life. Alkaline batteries, such as nickel–cadmium cells, proposed around 1900 but first commercialized during the 1970s, are the second largest market for use in consumer electronics and, recently, in electric vehicles. Despite their high cost, lithium-ion batteries have had a rapid impact since their introduction in 1991 (see below). They allow charge topping (i.e., charging before complete discharge) and have a high energy density, suitable for small-scale portable electronic equipment.

Rechargeable batteries are called accumulators or *secondary batteries*, whereas use-once-only piles are termed *primary batteries*. Table 5.9 gives some important characteristics of various battery types. The research goals set in 1977 for high-power batteries have been reached in commercially available products, but the goals for high-energy-density cells have not quite been reached. One reason for the continued high market share of lead–acid batteries is the perfection of this technology that took place over the last decades.

Type	Electrolyte	Energy Efficiency (%)	Energy Density (Wh kg^{-1})	Power Densities Peak (W kg^{-1})	Sustained (W kg^{-1})	Cycle Life (Cycles)	Operating Temperatures (°C)
Commercial							
Lead–acid	H$_2$SO$_4$	75	20–35	120	25	200–2000	−20 to 60
Nickel–cadmium	KOH	60	40–60	300	140	500–2000	−40 to 60
Nickel–metal-hydride	KOH	50	60–80	440	220	<3000	10 to 50
Lithium-ion	LiPF$_6$	70	100–200	720	360	500–2000	−20 to 60
Under development							
Sodium–sulfur	β-Al$_2$O$_3$	70	120	240	120	2000	300 to 400
Lithium–sulfide	AlN	75	130	200	140	200	430 to 500
Zinc–chlorine	ZnCl$_2$	65	120	100			0
Lithium–polymer	Li–β-Alu	70	200			>1200	−20 to 60
1977 goal cells							
High energy		65	265		55–100	2500	
High power		70	60	280	140	1000	

TABLE 5.9 Characteristics of selected batteries (Jensen and Sørensen, 1984; Cultu, 1989a; Scrosati, 1995; Buchmann, 1998) and comparison with 1977 development goals (Weiner, 1977).

An electrochemical storage battery has properties determined by cell voltage, current, and time constants. The two electrodes delivering or receiving power to or from the outside are called e_n and e_p (*negative* and *positive electrode*). (The conventional terms *anode* and *cathode* are confusing in the case of rechargeable batteries.) Within the battery, ions are transported between the negative and positive electrodes through an *electrolyte*. This, as well as the electrodes, may be solid, liquid, or, in theory, gaseous. The electromotive force E_0 is the difference between the electrode potentials for an open external circuit,

$$E_0 = E_{ep} - E_{en}, \tag{5.55}$$

where it is customary to measure all potentials relative to some reference state. The description of open-cell behavior uses standard steady-state chemical reaction kinetics. However, when current flows to or from the cell, equilibrium thermodynamics is no longer applicable, and the cell voltage V_c is often parametrized in the form

$$V_c = E_0 - \eta IR, \tag{5.56}$$

where I is the current at a given time, R is the internal resistance of the cell, and η is a "polarization factor" receiving contributions from the possibly very complex processes taking place in the transition layer separating each electrode from the electrolyte. Figure 5.24 illustrates the different potential levels across the cell for open and closed external circuits (cf., for example, Bockris and Reddy, 1973).

The Lead–acid Battery

In the electrolyte (aqueous solution of sulfuric acid) of a lead–acid battery, three reactions are at work,

$$H_2O \leftrightarrow H^+ + OH^-,$$

$$H_2SO_4 \leftrightarrow 2H^+ + SO_4{}^{2-}, \tag{5.57}$$

$$H_2SO_4 \leftrightarrow H^+ + HSO_4{}^-,$$

and at the (lead and lead oxide) electrodes, the reactions are

negative electrode: $Pb + SO_4^{2-} \leftrightarrow PbSO_4 + 2e^-$

positive electrode: $PbO_2 + SO_4^{2-} + 4H^+ + 2e^- \leftrightarrow PbSO_4 + 2H_2O.$ (5.58)

The electrolyte reactions involve ionization of water and either single- or double-ionization of sulfuric acid. At both electrodes, lead sulfate is formed, from lead oxide at the positive electrode and from lead itself at the negative electrode. Developments have included sealed casing, thin-tube electrode structure, and electrolyte circulation. As a result, the internal resistance has been reduced, and the battery has become practically maintenance-free throughout its life. The energy density of the lead–acid battery increases with temperature and decreases with discharge rate (by about 25% when going from 10 to 1 h discharge, and by about 75% when going from 1 h to 5 min discharge; cf. Jensen and Sørensen, 1984). The figures given in Table 5.9 correspond to an average discharge rate and an 80% depth of discharge.

While flat-plate electrode grid designs are still in use for automobile starter batteries, tubular-plate designs have a highly increased cycle life and are used for electric vehicles and increasingly for other purposes. The projected life of tubular-plate batteries is about 30 years, according to enhanced test cycling. Charging procedures for lead–acid batteries influence battery life.

Alkaline Electrolyte Batteries

Among the alkaline electrolyte batteries, nickel–cadmium batteries, which have been used since about 1910, were based upon investigations during the 1890s by Jungner. Their advantage is a long lifetime (up to about 2000 cycles) and, with careful use, a nearly constant performance, independent of discharge rate and age (Jensen and Sørensen, 1984). However, they do not allow drip charging and easily drop to low capacity if proper procedures are not followed. During the period 1970–1990, they experienced an expanding penetration in applications for powering consumer products, such as video cameras, cellular phones, portable players, and portable computers, but they have now lost most of these markets to the more expensive lithium-ion batteries.

FIGURE 5.24 Potential distribution through an electrochemical cell: *solid line*, open external circuit; *dashed line*, load connected (Jensen and Sørensen, 1984).

Iron–nickel oxide batteries, which were used extensively in the early part of the 20th century in electric motorcars, have inferior cell efficiency and peaking capability, owing to their low cell voltage and high internal resistance, which also increases the tendency for self-discharge. Alternatives, such as nickel–zinc batteries, are hampered by low cycle life.

The overall reaction for NiCd batteries may be summarized as

$$2\text{NiOOH} + 2\text{H}_2\text{O} + \text{Cd} \leftrightarrow 2\text{Ni(OH)}_2 + \text{Cd(OH)}_2. \tag{5.59}$$

The range of cycle lives indicated in Table 5.9 reflects the sensitivity of NiCd batteries to proper maintenance, including frequent deep discharge. For some applications, it is not practical to have to run the battery down to zero output before recharging.

An alternative considered is nickel–metal hydride batteries, which exhibit a slightly higher energy density but so far have a shorter cycle life.

High-temperature Batteries

Research on high-temperature batteries for electric utility use (cf. Table 5.9) has been ongoing for several decades, without decisive breakthroughs. Their advantage would be fast, reversible chemistry, allowing for high current density without excess heat generation. Drawbacks include serious corrosion problems, which have persisted and have curtailed the otherwise promising development of, for example, the sodium–sulfur battery. This battery has molten electrodes and a solid electrolyte; it is usually of tubular shape and is made from ceramic beta-alumina materials. Similar containment problems have affected zinc–chlorine and lithium–sulfur batteries.

Lithium-ion Batteries

Lithium metal electrode batteries attracted attention several decades ago (Murphy and Christian, 1979), owing to their potentially very high energy density. However, explosion risks delayed commercial applications until the lithium-ion concept was developed by Sony in Japan (Nagaura, 1990). Its electrode materials are LiCoO_2 and Li_xC_6 (carbon or graphite), respectively, with an electrolyte of lithium hexafluorophosphate dissolved in a mixture of ethylene carbonate and dimethylcarbonate (Scrosati, 1995). The battery is built from alternating layers of electrode materials, between which the lithium ions oscillate cyclically (Fig. 5.25). The cell potential is high, 3.6 or 7.2 V. Lithium-ion batteries do not accept overcharge, and a safety device is usually integrated in the design to provide automatic venting in case of emergency. Owing to its high power density (by weight or volume) and easy charging (topping up is possible without loss of performance), the lithium-ion battery has rapidly penetrated to the high-end portable equipment sector. Its safety record is excellent, justifying abandoning the lithium metal electrode, despite some loss of power. The remaining environmental concern is mainly the use of cobalt, which is very toxic and requires an extremely high degree of recycling to be acceptable.

Ongoing research on lithium-ion batteries aims at bringing the price down and avoiding the use of cobalt, while maintaining the high level of safety and possibly improving performance. The preferred concept uses an all-solid-state structure with lithium–beta–alumina forming a layered electrolyte and

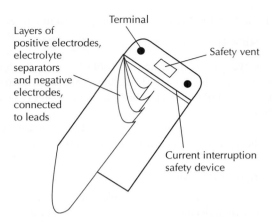

FIGURE 5.25   Layout of commercial lithium-ion rechargeable battery.

LiMn$_2$O$_4$ or LiMnO$_2$ as the positive electrode material (Armstrong and Bruce, 1996). The reaction at the positive electrode belongs to the family of intercalation reactions (Ceder *et al.*, 1998),

$$\text{Li}_{x1}\text{MO}_2 + (X_2 - X_1)\text{Li} \rightarrow \text{Li}_{x2}\text{MO}_2.$$

If the cost of lithium–polymer batteries can be reduced to about 20% of the cost of lithium-ion batteries, they could be used not just in small-scale portable equipment but also in electric vehicles. Currently, the ionic conductivity of solid polymer materials is too low to allow ambient temperature operation, but operating temperatures of around 80 °C should be acceptable for nonportable applications (Tarascon and Armand, 2001).

Mandatory recycling of batteries has already been introduced in several countries and is clearly a must for some of the recently emerging battery designs.

If batteries ever get to the point of being suitable for use in the grid-based electricity supply sector, it is likely to be for short-term and medium-term storage. Batteries have already been extensively incorporated into off-grid systems, such as rural power systems (whether based upon renewable energy or diesel sets) and stand-alone systems, such as security back-up facilities. Utility battery systems could serve as load-leveling devices and emergency units, as well as back-up units for variable primary supplies from variable renewable sources.

The small-scale battery options currently extremely popular in the consumer sector have one very promising possibility, namely, decoupling production and load in the electricity supply sector, which traditionally has been strongly focused upon a rigid expectation that any demand will be instantaneously met. If many household appliances become battery operated, the stringent requirements placed upon the electricity system to meet peak loads may become less burdensome. A considerable off-peak demand for charging rechargeable batteries could provide precisely the load control desired (as seen from the utility end), whether it be nearly a constant load pattern or a load following the variations in renewable energy supplies. The same goes for batteries serving the transportation sector (electric vehicles). Freedom from having to always be near an electric plug to use power equipment is seen as a substantial consumer benefit, which many people are willing to pay a premium price for.

Reversible Fuel Cells (Flow Batteries)

As mentioned above, fuel cells in reverse operation may replace alkaline electrolysis for hydrogen production, and if fuel cells are operated in both directions, they constitute an energy storage facility. Obvious candidates for further development are current PEM fuel cells, with hydrogen or methanol as the main storage medium. Presently, prototypes of hydrogen/oxygen or hydrogen/air reversible PEM fuel cells suffer from the inability of currently used membranes to provide high efficiency both ways. Whereas 50% efficiency is acceptable for electricity production, only the same 50% efficiency is achieved for hydrogen production, which is lower than the efficiency of conventional electrolysis (Proton Energy Systems, 2003). Fuel cells operated in reverse mode have reached nearly 100% hydrogen production efficiency using a few very large membranes ($0.25\ \mathrm{m}^2$), but such cells are not effective for power production (Yamaguchi *et al.*, 2000).

Early development of reversible fuel cells, sometimes called flow batteries and then referred to as redox flow batteries, used the several ionization stages of vanadium as a basis for stored chemical energy. The reactions at the two electrodes would typically involve

$$positive\ terminal\colon\ V^{4+} \to V^{5+}$$

and

$$negative\ terminal\colon\ V^{3+} \to V^{2+}.$$

The electrode material would be carbon fiber, and an ion-exchange membrane placed between the electrodes would ensure the charge separation. Specific energy storage would be about $100\ \mathrm{kJ\ kg^{-1}}$. Recently, Skyllas-Kazacos (2003) proposed replacing the positive electrode process with a halide-ion reaction, such as

$$Br^- + 2Cl^- \leftarrow BrCl^- + 2e^-$$

in order to increase the storage density. Other developments propose replacing expensive vanadium with cheaper sodium compounds, using reactions like

$$3NaBr \to NaBr_3 + 2Na^+ + 2e^-$$

$$2e^- + 2Na^+ + Na_2S_4 \to 2Na_2S_2$$

and the reverse reactions. A 120 MWh store with a maximum discharge rate of 15 MW, based upon these reactions, has just been completed at Little Barford in the United Kingdom. [A similar plant is under construction by the Tennessee Valley Authority in the United States (Anonymous, 2002).] Physically, the fuel-cell hall and cylindrical reactant stores take up about equal amounts of space. The cost is estimated at 190 euro per kWh stored or 1500 euro per kW rating (Marsh, 2002).

In the more-distant future, reversible-fuel-cell stores are likely to be based on hydrogen, because it offers the advantage of being useful as a fuel throughout the energy sector. A scenario exploring this possibility in detail is presented elsewhere (Sørensen, 2004a; and in condensed form in Sørensen *et al.*, 2003; Sørensen, 2002a, 2003c, 2003d), but the same idea is used in the energy scenarios presented in section 6.7.

5.3.5 Other Storage Forms

Direct Storage of Light

Energy storage by photochemical means is essential for green plants, but attempts to copy their technique have proved difficult, because of the nature of the basic ingredient: the membrane formed by suitably engineered double layers, which prevents recombination of the storable reactants formed by the solar-induced photochemical process. Artificial membranes with similar properties are difficult to produce at scales beyond that of the laboratory. Also, there are significant losses in the processes, which are acceptable in natural systems but which would negatively affect the economy of man-made systems.

The interesting chemical reactions for energy storage applications accepting direct solar radiation as input use a storing reaction and an energy retrieval reaction of the form

$$A + h\nu \rightarrow B,$$

$$B \rightarrow A + \text{useful energy}.$$

An example is absorption of a solar radiation quantum for the purpose of fixing atmospheric carbon dioxide to a metal complex, such as a complex ruthenium compound. If water is added and the mixture is heated, methanol can be produced,

$$[M]CO_2 + 2H_2O + 713 \text{ KJ mol}^{-1} \rightarrow [M] + CH_3OH + 1.5O_2.$$

The metal complex is denoted [M] in this reaction equation. Solar radiation is used to recycle the metal compound by reforming the CO_2-containing complex. The reaction products of a photochemical reaction are likely to back-react if they are not prevented from doing so. This is because they are necessarily formed at close spatial separation distances, and the reverse reaction is energetically favored because it is similar to the reactions by which the stored energy may be regained.

One solution to this problem is to copy the processes in green plants by having the reactants with a preference for recombination form on opposite sides of a membrane. The membranes could be formed using surface-active molecules. The artificial systems consist of a carbohydrate chain containing 5–20 atoms and, on one end, a molecule that associates easily with water ("hydrophilic group"). A double layer of such cells, with the hydrophilic groups facing in opposite directions, makes a membrane. If it is closed, for example, forming a shell (Fig. 5.26), it is called a *micelle*.

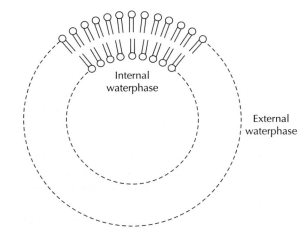

Internal waterphase

External waterphase

FIGURE 5.26 Double-shell type of micelle. The water-accepting compounds are illustrated by small circles, and the hydrocarbon chain molecules are illustrated by short lines.

Consider now the photochemical reaction bringing A to an excited state A*, followed by ionization,

$$A + h\nu \rightarrow A^*$$

$$A^* + B \rightarrow A^+ + B^-.$$

Under normal circumstances, the two ions would have a high probability of recombining, and the storage process would not be very efficient. But if A^+ can be made to form in a negatively charged micelle, the expelled electron would react with B to form B^- outside the micelle, and B^- will not be able to react with A^+. The hope is to be able to separate macroscopic quantities of the reactants, which would be equivalent to storing meaningful amounts of energy for later use (see, for example, Calvin, 1974), but efficiencies of artificial photosynthetic processes have remained low (Steinberg-Yfrach *et al.*, 1998). Recently, matrix structures have been identified that delay the back-reaction by hours or more, but as yet the retrieval of the stored energy is made correspondingly more difficult. The materials are layered viologen compounds [such as

N,N'-dimethyl-4,4'-bipyridinium chloride, methyl viologen (Slama-Schwok *et al.*, 1992), or zirconium phosphate–viologen compounds with Cl and Br inserts (Vermeulen and Thompson, 1992)].

Research on photochemical storage of energy is at a very early stage and certainly not close to commercialization. If the research is successful, a new set of storage options will be available. However, it should be stressed that storage cycle efficiencies will not be very high. For photo-induced processes, the same limitations exist as for photovoltaic cells: for example, only part of the solar spectrum is useful and losses are associated with transitions through definite atomic and molecular energy states. Further losses are involved in transforming the stored chemicals to the energy form demanded.

Superconducting Storage

A magnetic field represents energy stored. When a magnet is charged by a superconducting coil (e.g., a solenoid), heat losses in the coil may be practically zero, and the system constitutes an electromagnetic energy store with rapid access time for discharging. Maintenance of the coil materials (type II superconductors, such as NbTi) near the absolute zero temperature requires cryogenic refrigeration by, for example, liquid helium. Owing to the cost of structural support as well as protection against high magnetic flux densities around the plant, underground installation is envisaged. A storage level of 1 GJ is found in the scientific installation from the mid-1970s at the European physics laboratory CERN at Geneva (the installation aims at preserving magnetic energy across intervals between the accelerator experiments performed at the laboratory). A 100 GJ superconducting store has been constructed in the United States by the Department of Defense, who want to exploit the fact that superconducting stores can accumulate energy at modest rates but release it during very short times, as is required in certain antimissile defense concepts (Cultu, 1989b). This is still a prototype development. Economic viability is believed to require storage ratings of 10–100 TJ, and, ideally, high-temperature superconductors can be used to reduce the cooling requirements (however, limited success has been achieved in raising the critical temperature; cf. Locquet *et al.*, 1998).

Only type II superconductors are useful for energy storage purposes, since the superconducting property at a given temperature T (below the critical temperature that determines the transition between the normal and superconducting phase) disappears if the magnetic field exceeds a critical value, $B_c(T)$ (*magnetic field* here is taken to mean *magnetic flux density*, a quantity that may be measured in units of tesla = V s m^{-2}). Type II superconductors are characterized by a high $B_c(T)$, in contrast to type I superconductors.

A magnetic field represents stored energy, with the energy density w related to the magnetic flux density (sometimes called *magnetic induction*) B by

$$w = B^2/(2\mu_0),$$

where μ_0 (= 1.26×10^{-6} henry m^{-1}) is the vacuum permeability. When the superconducting coil is connected to a power source, the induced magnetic field represents a practically loss-free transfer of energy to the magnetic storage. With a suitable switch, the power source may be disconnected, and energy may later be retrieved by connecting the coil to a load circuit. Cycle efficiencies will not be 100%, owing to the energy required for refrigerating the storage. Since the magnetic field must be kept below the critical value $B_c(T)$, increased capacity involves building larger magnets.

Conventional circuit components, such as capacitors, also represent an energy store, as do ordinary magnets. Such devices may be very useful to smooth out short-term variations in electrical energy flows from source to user, but they are unlikely to find practical uses as bulk storage devices.

PART II: MINI PROJECTS, DISCUSSION ISSUES, AND EXERCISES

II.1

Find out what kinds of hourly and seasonal variations one can expect for the power production from a photovoltaic array mounted on an inclined rooftop in the area where you live.

You will need some solar radiation data for your local region. Satellite sources usually contain data only for horizontal surfaces, so it is better to find locally measured data for inclined surfaces suitable for solar panels. If such data are difficult to find, perhaps you can find separate solar radiation data for the direct and scattered parts (or for the direct and total radiation, from which you can derive the scattered radiation, at least if you assume it to be identical for any direction within a hemisphere). Such datasets are contained in *reference-year data*, which architects and consulting engineers use for design of buildings. Reference-year data exist for one or more urban locations in most countries of the world and allow you to calculate the total solar radiation falling on a particular inclined surface at a particular time of the day and the year. (Each particular time corresponds to a particular direction to the Sun, so that manipulating direct radiation becomes a simple geometric transformation. If you need help to calculate this relation, look in Chapter 3).

You will also have to make assumptions on the function of the solar photovoltaic system. Initially, assume that it has a fixed efficiency for all sunlight. If you want to go further, you could include the wavelength-dependence of the panel efficiency, but, in that case, you need to know the frequency composition of the incoming radiation, which is more likely to be available from satellite data that may be found on the Internet.

II.2

Discuss the production of heat of the temperature T_L from an amount of heat available, of temperature T above T_L, and access to a large reservoir of temperature T_{ref} below T_L.

Quantities of heat with temperatures T and T_{ref} may be simply mixed in order to obtain heat of the desired temperature.

Alternatively, a thermodynamic cycle may be used between the temperatures T and T_{ref} to produce a certain amount of mechanical or electrical energy, which then is used to power the compressor of a heat pump, lifting heat from the T_{ref} reservoir to the desired temperature T_L.

As a variant, use a thermodynamic cycle between the temperatures T and T_L, producing less work but on the other hand providing reject heat at the desired temperature. Again, a heat pump is used to provide more heat of temperature T_L on the basis of the work output from the thermodynamic cycle.

What are the simple and the second law efficiencies in the individual cases?

II.3

Discuss design aspects of a propeller-type wind energy converter for which no possibility of changing the overall pitch angle θ_0 is to be provided.

Consider, for instance, a design similar to the one underlying Figs. 4.17–4.22, but which is required to be self-starting at a wind speed of 2 m s^{-1} and to have its maximum power coefficient at a wind speed of 4 m s^{-1}. Find the tip-speed ratio corresponding to this situation for a suitable overall pitch angle θ_0 and for given assumptions on the internal torque $Q_0(\Omega)$, e.g., that Q_0 is independent of the angular velocity Ω, but proportional to the third power of the blade radius R, with the proportionality factor being given by

$$Q_0 = 10^{-3} \, \pi\rho R^3 u_{cut\text{-}in}^2,$$

with $u_{cut\text{-}in} = 2$ m s^{-1}.

Compare the power produced per square meter swept at a wind speed of 4, 6, and 10 m s^{-1} to the power that could be derived from a solar cell array of the same area.

For an actual design of wind energy converters, the blade shape and twist would not be given in advance, and one would, for a device with fixed angular velocity, first decide on this angular velocity (e.g., from generator and gearbox considerations) and then choose a design wind speed and optimize the blade shape $c(r)$ and blade twist $\theta(r)$, in order to obtain a large power coefficient at the design speed and possibly to avoid a rapid decrease in the power coefficient away from the design point.

The blade shape $c(r)$, of course, depends on blade number B, but to lowest order, the performance is independent of changes that leave the product of (cr) for the individual blades and B unchanged. Name some other considerations that could be important in deciding on the blade number.

II.4

In Section 4.3.1, a maximum power coefficient C_p far above unity is suggested for certain cases of sail-driven propulsion on friction-free surfaces. Is this not a contradiction, like taking more power out than there is in the wind?

(Hint: C_p is defined relative to the wind power per unit area, which is of immediate relevance for rotors sweeping a fixed area. But what is the area seen by a sailing vessel moving with velocity U?).

II.5

Based, for example, on local data, such as those suggested in problem 1, what is the ratio between the daily amount of solar radiation that on a clear day in different seasons intercepts a fully tracking collector system, and the amount reaching a flat plate of fixed, south-facing orientation and a certain tilt angle s?

What is the relative importance of east–west tracking alone and solar height tracking alone?

II.6

Discuss shore-based wave energy converters, e.g., converters based on letting the wave trains ascend an inclined (and maybe narrowing) ramp, so that kinetic energy is converted into potential energy of elevation, which may be used to drive a turbine.

The simple solutions for gravity waves in a deep ocean, which were considered in the beginning of section 2.5.2, cannot be used directly for waves approaching shallow coastal waters. However, it may be assumed that the wave period remains unchanged, but that the amplitude a of the wave surface increases in such a way that the total power remains constant as the velocity potential becomes altered due to the boundary condition of no motion at the sea floor. This assumption implies the neglect of frictional dissipation at the sea floor.

There is a maximum ramp height that will permit a wave to reach the top and cross it into an elevated reservoir. If the turbine used requires a larger height to function well, a narrowing ramp construction may be used to give some fraction of the water mass the desired elevation.

II.7

Combine biomass productivity data with estimated efficiencies of fuel production by anaerobic fermentation processes (section 4.6.2) to give overall conversion efficiencies of bioconversion of solar energy into fuels. (A global biomass productivity map may be found in B. Sørensen: *Biomass for energy: how much is sustainable?* The map can be downloaded from the download-section at http://energy.ruc.dk; otherwise assume 1%–2% of incoming solar radiation converted into biomass.)

Compare this to other methods of converting solar radiation into mechanical or electrical work, and discuss their relative virtues other than conversion efficiency.

II.8

Estimate the magnitude of, and seasonal variation in, energy contained in stored food, separately for standing crops on the field, for livestock to be used for food, and for actually stored food (in grain stores, supermarkets, freezers, etc.). Compare food energy stored to the energy value of emergency stores of oil and natural gas, for example, in your own country.

Hint: Data will have to be found at many different places. Some are in Chapter 6. You could estimate energy stored in supermarkets by your own spot checks of the declarations on typical food products, which should include their energy content.

II.9

Estimate the potential world production of equivalent crude oil from *Euphorbia* plants, assuming only arid land is utilized and annual harvest yields are 40 MJ m^{-2}. Areas of different types of land, including arid land, may be found on the Internet.

II.10

Consider a continuous-operation biogas digester for a one-family farm. The digester feedstock is pig slurry (collected daily) and straw (stored). The biogas is used for cooking, for hot water, and for space heating. Estimate the load curves for different seasons and calculate the volume of uncompressed gas storage required if load is to be met at any time during the year.

II.11

Consider pure methanol- and ethanol-driven cars, and, for comparison, a gasoline-driven car, weighing 800 kg and going on average 18 km per liter of gasoline and 500 km on a full tank. Calculate the mass penalties for methanol and ethanol fuel tanks, if the same operation range is required. Assume that the fuel usage varies linearly with total mass of the car.

II.12

Consider a thermal storage of temperature T (say, 200 °C) aimed at receiving electric energy through a heat pump (with access to a reservoir of temperature T_{ref}) and delivering electrical energy by means of a thermodynamic engine cycle (available coolant also of temperature T_{ref}). What is the theoretical maximum cycle efficiency, and what can be expected in practice?

II.13

Discuss energy and power levels for the following "natural hydro energy" concept. Behind a seashore mountain chain there are a number of elevated water reservoirs (e.g., 200 m above sea level). The sea is characterized by a high evaporation rate (from solar energy), and the prevailing winds carry the evaporated water toward the coast, where a high percentage of the water returns to the land as rain on the mountainous region's reservoirs (or imagine that precipitation is stimulated by seeding the clouds with some kind of condensation nuclei). Sample evaporation and precipitation rates may be found in meteorological tables or may be extracted from Chapter 2.

The water is returned from the reservoirs to the sea through hydropower plants. In a refined model, storage losses from the reservoirs due to evaporation should be included.

II.14

On the basis of time sequences of data for an automobile (e.g., your own), try to construct a load-duration curve for the entire period. Based on this, estimate a suitable level of rated power, if the engine should only cover average load, and specify the amount of storage needed for operating a hybrid car with a fuel-based engine rated at the average power needed, a battery store and an electric motor capable of delivering the maximum power required. The electric motor can be considered, to a good approximation, loss-free.

Hint: A load-duration curve shows the fraction of time during which power demand exceeds a given value as a function of that power value.

II.15

Construct the load-duration curve for space heating of a dwelling at your geographical location. Assume that this load is to be covered by an energy converter providing constant power all year and that loss-free heat storage is available. Determine the magnitude of the constant-converter output that, through use of storage, will suffice to cover the load at all times. Further, determine the minimum storage size needed. Compare the constant-converter output to the range of power provided by currently used heating systems. Use estimated losses for actual storage systems that may be relevant to assess what the required storage capacity would be in case the storage involved realistic losses.

II.16

A steam turbine power plant with a steam temperature of 700 K and a condenser temperature of 350 K has an efficiency of 0.36 (electric output energy divided by input steam energy), and the turbine efficiency is 20% less than the Carnot efficiency (ε_{max}). Calculate the efficiency of the electric generator.

II.17

An electric vehicle of mass 1000 kg (excluding the battery) is designed for a 50 km range in a city driving cycle. The vehicle is equipped with a new 500 kg battery, and the average energy consumption is measured to be 0.3 kWh km^{-1} (from the battery). At a speed of 36 km h^{-1}, the required acceleration is 2.0 m s^{-2}, and the power required to overcome the frictional losses is one-third of the total power required. Do the same calculation for an acceleration $a = 1.5$ m s^{-2}.

- Calculate the required average energy density of the new battery.
- Calculate the required power density of the new battery.
- What type of battery could fulfill these requirements (alone or in combination as a hybrid system like the one considered in problem II.14)?

II.18

Estimate the total hydro resource available for storage applications in a country or region. Characterize the natural reservoirs in terms of water volume and feasible turbine head, and identify the geological formations that could be considered for underground reservoir construction. In this way, arrive at rough storage capacities for short- and long-term storage separately. Estimation may be repeated for different regions and the results compared.

II.19

Use (5.39) to express the compressed storage energy (5.38) in terms of the temperature difference $T - T_0$. rather than in terms of the pressure P.

II.20

Compare the energy needed to compress a certain amount of air by an adiabatic process with that required for an isothermal and an isobaric process.

II.21

Calculate the shape factor for a thin-rim flywheel, as well as mass- and volume-specific energy density, assuming the material to be steel. Do the same calculation numerically for some of the flywheel types shown in Fig. 5.18 (using data from Tables 5.6 and 5.7) and compare the properties. Discuss priorities for various types of application.

Part III

Renewable Energy Impacts

Planning for Sustainability and Climate Change Aversion

Chapter 6

Energy System Planning

6.1 METHODOLOGY OF ENERGY PLANNING

In Chapter 4, energy conversion devices are largely viewed as separate entities. In this chapter, they are regarded as parts of larger systems, which may comprise several converter units, storage facilities, and transmission networks. The determinant of each system is, of course, the purpose it is meant to serve, i.e., the end-point of the energy conversion processes. Therefore, a systems study must include analysis of both the end-uses (*loads* or *demands*) and the chains of conversion and transport steps connecting the primary energy source extraction with the loads, as well as spatial and temporal variations of supply and demand.

It is important to agree on the definition of *end-use energy*. There is the energy delivered to the end-user, and after subtraction of the losses in the final conversion taking place at the end-user, there is a net amount of energy being made useful. However, the real demand is always a product or a service, not the energy in itself. This implies that true energy demand is often difficult to pinpoint, because novel ways of satisfying the non-energy final need could radically change the end-use energy "required" or "demanded." A typical definition used in practice is the lowest amount of energy associated with delivering the product or service demanded, as measured after the final conversion at the end-user and selected among all conversion schemes that can be realized with present knowledge. For some technologies, there will be a thermodynamic limit to energy conversion efficiency, but whether it is fundamental or may be circumvented by going to a different technology depends upon the nature of the final product or service. For many services, the theoretical minimum energy input is zero. Future advances in technology or new ideas for satisfying a particular need may change the minimum amount of end-use energy, and one should decide on which of the following to use in the definition:

- only technology available in the marketplace today, or
- only technology proven today and with a potential for market penetration, or
- any technology that can be imagined at present.

The outcome will be different, depending on the choice of technology level, and, in comparisons of system solutions, it is important to use the same definition of demand throughout. My preference is to use *best technology available today* in defining end-use, but any definition will do if it is used consistently.

6.1.1 Use of the Scenario Concept

This chapter gives examples of energy system simulations. Some of the examples simulate the function of existing systems, while others deal with hypothetical systems that may be created in the future. They will be analyzed using the scenario method, which is a way of checking the consistency of proposed energy systems as part of a view of the general future development of a society.

Scenario studies are meant to assist the decision-making process by describing a given energy system in its social context, in a way suitable for assessing systems not yet implemented. Simple forecasts of demand and supply based on economic modeling cannot directly achieve this, as economic theory deals only with the past and occasionally the present structure of society (cf. Chapter 7). To deal with the future, economic modelers may invoke the established quantitative relations between parts of the economic system and assume that they stay valid in the future. This produces a "business-as-usual" forecast. Because the relations between the ingredients of the economy, e.g., expressed through an input–output matrix, vary with time, one can improve the business-as-usual forecast by taking into account trends already present in past development. However, even trend forecasts cannot be expected to retain their validity for very long periods, and it is not just the period of forecasting time that matters, but also changes in the rules governing society. The rules may change due to abrupt changes in technology used (in contrast to the predictable, smooth improvements of technological capability or average rate of occurrence of novel technologies), or they may be changed by deliberate policy choices. It is sometimes argued that econometric methods could include such nonlinear behavior by replacing the input–output coefficients with more complex functions. However, prediction of these functions cannot be based on studies of past or existing societies, because the whole point in human choice is that options are available that are different from past trends. The nonlinear, nonpredictable relations that may prevail in the future, given certain policy interventions at appropriate times, must therefore be postulated on normative grounds. This is what the scenario method does, and any attempt to mend economic theory also implies invoking a scenario construction and analysis, so, in any case, this is what has to be done (Sørensen, 2001b).

It is important to stress that scenarios are not predictions. They should be presented as options that *may* come true *if* a prescribed number of actions are carried out. In democratic societies, this kind of change happens only if it is preceded by value changes that affect a sufficiently large portion of the society. Generally, in a democratic society, the more radically a scenario differs from present conditions, the larger the support it requires among the population. For non-democratic societies, where change may be implemented by decree, negative implications will likely entail.

The actual development may comprise a combination of some reference scenarios selected for analysis, with each reference scenario being a clear and perhaps extreme example of pursuit of a concrete line of political preference. It is important that the scenarios selected for political consideration be based on values and preferences that are important in the society in question. The value basis should be made explicit in the scenario construction. Although all analysis of long-term policy alternatives is effectively scenario analysis, particular studies may differ in their comprehensiveness of the treatment of future society. A simple analysis may make normative scenario assumptions only for the sector of society of direct interest in the study (e.g., the energy sector), assuming the rest to be governed by trend rules similar to those of the past. A more comprehensive scenario analysis will make a gross scenario for the development of society as a whole, as a reference framework for a deeper investigation of the sectors of particular interest. One may say that the simple scenario is one that uses trend extrapolation for all sectors of the economy except the one focused upon, whereas the more radical scenario will make normative, nonlinear assumptions regarding the development of society as a whole. The full, normative construction of future societies will come into play for scenarios describing an ecologically sustainable global society, whereas scenarios aiming only at avoiding or coping with one particular problem, such as the climate change induced by greenhouse warming, are often of the simpler kind. Both types are exemplified in the scenario work described below.

6.1.2 Treatment of the Time Variable

In order to predict the performance of systems consisting of one or several energy converters, stores, and transmission devices, a mathematical model of the energy flow may be constructed. Such a model is composed of a number of energy conversion and transport equations, including source and sink terms corresponding to the renewable energy input and the output to load areas, both of which vary with time. The conversion processes depend on the nature of the individual devices, and the description of such devices (cf. Chapter 4) aims at providing the necessary formulae for a sufficiently complete description of the processes involved. In a number of cases (e.g., among those considered in Chapter 4), only a steady-state situation is studied, and the energy outputs are calculated for a given level of energy input. In a time-dependent situation, this type of calculation is insufficient, and a dynamic description must be introduced in order to evaluate the response time and energy flow delay across the converter (see, for example, section 4.3.6). Similar remarks apply to the description of storage systems, and, finally, the transmission network introduces a further time dependence and a certain delay in the energy flow reaching the load areas. The transmission network is often in the form of pipelines carrying a flow of some fluid (e.g., natural gas, hydrogen, or hot water) or an electric conductor carrying a flow of electric current. Additional transport of energy may take place in containers (e.g., oil products or methanol carried as ship, rail, or road cargo).

In order to arrive at manageable problems, it is, in most cases, necessary to simplify the time dependence for some parts of the system. First, short-term fluctuations in the source energy flow may, in some circumstances, be left out. This is certainly possible if the conversion device is itself insensitive to fluctuations of sufficiently high frequency. It could be so in a wind energy converter due to the inertia of the rotating mass, or, in a solar heat collector, due to the time constant for temperature changes in the absorber plate (and also in the circulating fluid). It may also be a valid approximation if short-term variations in the energy flux from the source can be regarded as random and if the collection system consists of a large number of separate units placed in such a way that no coherence in the fluctuating inputs can be expected.

Second, the performance of the conversion devices may often be adequately described in terms of a quasi-steady-state approximation. This consists of calculating an instantaneous energy output from the converter on the basis of an instantaneous energy input as if the input flux were permanent, i.e., doing a steady-state calculation for each moment of time. This excludes an assessment of the possible time delay between the input flux and the output flux. If a solid mechanical connection transfers the energy through the converter (e.g., the rotor–shaft–gearbox–electric generator connections in a horizontal-axis wind energy converter), the neglect of time delays is a meaningful approximation. It may also be applicable for many cases of non-rigid transfer (e.g., by a fluid) if short-term correlations between the source flux and the load variations are not essential (which they seldom are in connection with renewable energy sources). For the same reason, time delays in transmission can often be neglected. The flow received at the load points may be delayed by seconds or even minutes, relative to the source flow, without affecting any of the relevant performance criteria of the system.

On the other hand, delays introduced by the presence of energy storage facilities in the system are essential features that cannot, and should not, be neglected. Thus, storage devices will have to be characterized by a time-dependent level of stored energy, and the input and output fluxes will, in general, not be identical. The amount of energy $W(S_i)$ accumulated in the storage S_i can be determined from a differential equation of the form

$$\frac{dW(S_i)}{dt} = \sum_j E_{ji}^+ - \sum_k E_{ik}^- - E_i^{loss}, \qquad (6.1)$$

or from the corresponding integral equation. The individual terms in the two expressions involving summation on the right-hand side of (6.1) represent energy fluxes from the converters to and from the storage devices. The loss term E_i^{loss} may depend on the ingoing and outgoing fluxes and on the absolute amount of energy stored in the storage in question, $W(S_i)$.

In practice, the simulation is performed by calculating all relevant quantities for discrete values of the time variable and determining the storage energy contents by replacing the time integral of (6.1) by a summation over the discrete time points considered. This procedure fits well with the quasi-steady-state approximation, which at each integration step allows the calculation of the converter outputs (some of which are serving as storage inputs E_{ji}^+) for given renewable energy inputs, and similarly allows the calculation of conversion processes in connection with the storage facilities, and the energy fluxes E_{ij}^- to be extracted from the storage devices in order to satisfy the demands at the load areas. If the time required for conversion and transmission is neglected, a closed calculation can be made for each time integration step. Interdependence of storage inputs and outputs, and of the primary conversion on the system variables in general (e.g., the dependence of collector performance on storage temperature for a flat-plate solar collector), may lead to quite complex calculations at each time step, such as the solution of nonlinear equations by iteration procedures (section 4.4.3).

If finite transmission times cannot be neglected, they may be included, to a first approximation, by introducing simple, constant delays, such that the evaluations at the mth time step depend on the values of certain system variables at earlier time steps, $m - d$, where d is the delay in units of time steps. The time steps need not be of equal length, but may be successively optimized to obtain the desired accuracy with a minimum number of time steps by standard mathematical methods (see, for example, Patten, 1971, 1972).

The aim of modeling may be to optimize either performance or system layout. In the first case, the system components are assumed to be fixed, and the optimization aims at finding the best control strategy, i.e., determining how best to use the system at hand ("dispatch optimization"). In a multi-input–multi-output conversion system, this involves choosing which of several converters to use to satisfy each load and adjusting inputs to converters in those cases where this is possible (e.g., biofuels and reservoir-based hydro as opposed to wind and solar radiation). For system optimization, the structure of the conversion system may also be changed, with recognition of time delays in implementing changes, and the performance over an extended period would be the subject of optimization. For simple systems (without multiple inputs or outputs from devices), linear programming can furnish a guaranteed optimum dispatch of existing units, but, in the general case, it is not possible to prove the existence of an optimum. Still, there are systematic ways to approach the optimization problem, e.g., by using the method of steepest descent for finding the lowest minimum of a complex function, combined with some scheme for avoiding shallow, secondary minima of the function to minimize (Sørensen, 1996a).

6.2 DEMAND SCENARIO CONSTRUCTION

The demand for products or services using energy, taken at the end-user after any final conversion done there, is basically determined by the structure of society, the technology available to it, and the activities going on within it— and, behind these, the aims, desires, and habits of the people constituting the society in question.

Enumerating the energy demands coming out of such appraisals can be done methodologically, as in section 6.2.3, but first it may be illustrative to paint a broad picture of possible developments of societies, as they pertain to energy end-use demand and to the intermediary energy demands of energy conversion processes made necessary by the structure of the energy system (Sørensen, 2008b). Both of these are sketched below for a limited range of development scenarios, without claiming completeness and clearly basing the construction on normative positions. The end-use scenarios are called *precursor scenarios*, to distinguish them from the actual scenarios created later that specify the energy use quantitatively. The corresponding scenarios for intermediary conversion system scenarios are usually characterized by their emphasis on efficiency. Of course, the overall structure of the energy system can be fixed only after both end-use demands and the supply options to be exploited have been decided, and the intersociety (whether regions within a country or international) setting (trade, grid connections, etc.) is known.

6.2.1 End-use Precursor Scenarios

Runaway Precursor Scenario

In the runaway scenario, the energy demand grows at least as quickly as the overall economic activity (measured, for example, by the gross national product). This has historically been the case during periods of exceptionally low energy prices, notably in the years around 1960. Conditions for this scenario, in addition to low energy prices, include: in the transportation sector increased passenger-kilometers (facilitated by more roads, cheap air connections, and decentralization of the locations of homes, workplaces, and leisure facilities) and increased ton-kilometers of freight haul, both locally and globally (facilitated by decentralization of component production and inexpensive worldwide shipment of parts and products); in the building sector, more square meters of living space and more square meters per unit of economic activity; and, in the electricity sector, more appliances and other equipment. Building-style developments could create a perceived need for air conditioning and space cooling. For industry, there could be increased emphasis on energy-intensive production in countries with low labor costs, and the opposite in countries with high labor costs. However, service-sector activities and their energy use could increase substantially, with greatly enlarged retail shopping areas and use of much more light and other energy-demanding displays for business promotion. For leisure activities, traditional nature walks or swimming could be replaced by motocross, speedboat use, and other energy-demanding activities.

High-energy-growth Precursor Scenario

The high-energy-growth scenario is similar to the runaway scenario, but with a slower, but still significant, increase in energy demand. In the transportation sector, this could be due to a certain saturation tendency in transport activities, caused by the higher value placed on time lost in traveling on more congested roads and in more congested air space. For industry, continued decrease in energy-intensive production may lead to demand growing more slowly than the economic activity. In buildings, heat use may increase less than floor area, due to zoning practices, etc. Generally, activity level and energy demand may undergo a certain amount of decoupling, reflecting the fact that the primary demands of a society are goods and services, and that these can be provided in different ways with different energy implications. An effect of this type damps the energy demand in the high-energy-growth scenario, as compared to the runaway scenario, but due more to technological advances and altered industry mix than to a dedicated policy aimed at reducing energy demand.

Stability Precursor Scenario

The stability scenario assumes that end-use energy demand stays constant, despite rearrangements in specific areas. Specifically, in the building sector, energy demand is assumed to saturate (that is, the number of square meters per person occupying the building, whether for work or living, will not continue to increase, but reach a natural limit with enough space for the activities taking place, without excessive areas to clean and otherwise maintain). In the industrial sector, increasingly knowledge-based activity will reduce the need for energy-intensive equipment, replacing it primarily by microprocessor-based equipment suited for light and flexible production. Industrial energy use will decline, although industries in the service and private sectors will continue to add new electronic equipment and computerized gadgets. In other sectors, dedicated electricity demand will increase substantially, but, in absolute terms, it will be more or less compensated for by the reductions in the industrial sector. For transportation, saturation is assumed both in number of vehicles and number of passenger- or ton-kilometers demanded, for the reasons outlined above in the section on the high energy-growth scenario. Explanations for this could include the replacement of conference and other business travel by video conferencing, so that an increase in leisure trips may still be possible. Presumably, there have to be transportation-related strategies implemented for this to be realistic, including abandoning tax rebates for commercially used vehicles and for business travel, and possibly also efforts in city planning to avoid the current trend toward increased travel distances for everyday shopping and service delivery. The stability scenario was used as the only energy demand scenario in some earlier studies on the possibilities for hydrogen use in the Danish energy system (see Sørensen *et al.*, 2001, 2004; Sørensen, 2005).

Low-energy-demand Precursor Scenario

In the low-energy-demand scenario, full consideration is paid to the restructuring of industry in countries making a transition from a goods orientation to service provision. Today, many enterprises in Europe or the United States already only develop new technology (and sometimes test it on a limited domestic market): once the technology is ready for extended markets, production is transferred to low-wage companies, currently in Southeast Asia. This change in profit-earning activities has implications for the working conditions of employees. Much of the information-related work can be performed from home offices, using computer equipment and electronic communications technology, and thereby greatly reducing the demand for physical transportation.

In the retail food and goods sector, most transactions between commerce and customer will be made electronically, as is already the case in a number of subsectors today. An essential addition to this type of trade is the market for everyday products, where, before, customers made limited use of electronic media to purchase grocery and food products, probably because of their perceived need to handle the goods (for example, to examine fruit to see if it is ripe) before purchasing them. Clearly, better electronic trade arrangements, with video inspection of actual products, could change customers' practices. If everyday goods are traded electronically, their distribution will also be changed toward an optimal dispatch that requires considerably less transport energy than current shopping practices do. All in all, a substantial reduction in energy demand will result from these changes, should they come true. In the remaining manufacturing countries, these changes are expected to take place equally fast, if not faster, due to workers' less-flexible time schedules for shopping. Thus, economic development is further decoupled from energy use and may continue to exhibit substantial growth.

Catastrophe Precursor Scenario

In the catastrophe scenario, reduced energy demand is due to failure to achieve desirable economic growth. Reasons could be the over-allocation of funds to the financial sector, causing major periodic economic recessions. In some European and North American countries, additional reasons could include the current declining public and political interest in education, particularly in those areas most relevant to a future knowledge society. In this scenario, there would be a decreased number of people trained in skills necessary for participating in international industrial and service developments, and the opportunity to import people with these intellectual skills would have been lost by the imposition of immigration policy unfavorable to precisely the regions of the world producing a surplus of people with technical and related creative high-level education.

Although there are pessimists who view this scenario as the default if there is no immediate change in the political climate in the regions with such problems, the stance here is that the traditional openness of many Western societies will work to overcome the influence of certain negative elements. This is particularly true for small countries like Denmark, if it can maintain a distance from globalization pressures from the European Union, the U.S.-dominated World Trade Organization, etc. Even if Denmark should choose to continue to concentrate on less education-demanding areas, such as coordination and planning jobs in the international arena (which require primarily language and overview skills), these could easily provide enough wealth to a small nation of open-minded individuals willing to serve as small wheels in larger international projects. The current economic decline would become a passing crisis, to be followed by a niche role for Denmark, which in energy terms would imply returning to one of the central scenarios described above. Only if Denmark became internationally isolated would this option fade away and the catastrophe scenario become a reality.

This discussion uses Denmark, a small country, as an example. In other countries, some aspects of these same problems are also evident, and if a country that lowers skills through poor information policies (e.g., in news media) and bad consumer choices (influenced by the curricula offered by schools and universities) is large, then a downturn may be more difficult to avoid.

6.2.2 Intermediary System Efficiency

Many countries, particularly in Europe, have a long tradition of placing emphasis on efficient conversion of energy. Following the 1973/4 energy crises, detached homes (where the occupants are also the owners making decisions on investments) in particular were retrofitted to such an extent that overall low-temperature heat use dropped by almost a third over a decade. Carbon-emission and pollution taxes on electricity and transportation fuels (taxes that are not always clearly labeled) have influenced the mix of appliances being installed and car models bought, with the lowest-energy-consuming equipment dominating the market in some countries. This trend is only partial, because there are still substantial sales of luxury cars and 4-wheel-drive special utility vehicles not serving any apparent purpose in countries with hardly any unpaved roads. Transmission losses are fairly low in Europe, where underground coaxial cables have been installed, mainly to avoid the vulnerability of overhead lines during storms.

Following are three precursor-scenario outlines illustrating the planning considerations made implicitly or explicitly by planners and legislators.

Laisser Faire Precursor Scenario

In the *laisser faire* scenario, conversion efficiencies are left to the component and system manufacturers, which would typically be international enterprises (such as vehicle and appliance manufacturers, power station and transmission contractors, and the building industry). The implication is that efficiency trends follow an international common denominator, which at least in the past has meant lower average efficiency than suggested by actual technical advances and sometimes even lower than the economic optimum at prevailing energy prices. Still, efficiency does increase with time, although often for reasons not related to energy (for example, computer energy use has been lowered dramatically in recent years, due to the need to avoid component damage by excess heat from high-performance processors). The gross inadequacies of the current system, deriving from tax-exemption of international travel and shipping by sea or air, and its impact on choice of transportation technology, are assumed to prevail.

Rational Investment Precursor Scenario

In the rational investment scenario, the selection of how many available efficiency measures will actually be implemented, through the technologies chosen at each stage in the time development of the energy system, is based on a lifetime economic assessment. This means that the efficiency level is not chosen according to a balancing of the cost of improving efficiency with the current cost of energy used by the equipment, but with the present value of all energy costs incurred during the lifetime of the equipment. This assessment requires an assumption of future average energy costs and it is possible, by choice of the cost profile, to build in a certain level of insurance against unexpectedly high energy prices. The important feature of the rational investment scenario is that it forces society to adopt a policy of economic optimization in the choice of energy-consuming equipment and processes. This policy is partially implemented at present, e.g., through energy provisions in building codes, through appliance labeling, and through vehicle taxation. In the last case, there is a distinction between the efficiency optimization for a vehicle with given size and performance, considered here, and the question of proper vehicle size and performance characteristics dealt with in the previous section on end-use energy. The energy taxation currently used in many countries for passenger cars does not make such a distinction, and the tax reduction for commercially used vehicles actually counters rational economic considerations; it should be characterized as taxpayers' subsidies for industry and commerce.

Maximum-efficiency Precursor Scenario

One version of the maximum-efficiency scenario could require that every introduction of new energy-consuming equipment should use the best currently available technical efficiency. This implies selection of the highest-efficiency solution available in the marketplace, or even technology ready for, but not yet introduced into, the commercial market. Higher-efficiency equipment under development and not fully proven would, however, not be implemented, except as part of demonstration programs. This "best-current-technology" approach was used in several previous scenario studies (Sørensen, 2005; Sørensen and Meibom, 2000; Sørensen *et al.*, 1999, 2001, 2004, and for some of the scenarios later in this chapter) with the rationale that current most-efficient technology can serve as a good proxy for average-efficiency technology 50 years in the future. Depending on assumptions regarding future energy prices, the rational investment precursor scenario could be less efficient, of similar efficiency, or of higher efficiency than the "best-current-equipment" approach.

Therefore, a more appropriate definition of the maximum-efficiency scenario would be based on projecting typical average efficiency improvements over the entire planning period, and then insisting that the best technology at each instant in time is used for all new equipment introduced at that moment in time. While projections of future efficiency of individual pieces of equipment may be uncertain and sometimes wrong, it seems reasonable to assume that average efficiencies over groups of related equipment can be extrapolated reliably.

6.2.3 Load Structure

Loads are end-use energy demands as experienced by the energy system. From the consumer's perspective, loads satisfy needs by delivering goods and services that require energy. Energy delivery is often constrained to certain time profiles and locations. This is the reason that energy systems often contain energy storage facilities, as well as import-export options, whether for transport by a grid or by vessels or vehicles. Some energy storage could be at the end-user and form part of the equipment used to satisfy the end-user's demand (e.g., solar thermal systems with hot water tanks). Utilization of energy stores can modify both time variations and total amounts of energy demanded.

On a regional or national level, loads are customarily divided onto sectors, such as energy demands of industry, agriculture, commerce, services and residences, and transportation. A sector may require one or more types of energy, e.g., heat, electricity, or a "portable source" capable of producing mechanical work (typically a fuel). The distribution of loads on individual sectors depends on the organization of the society and on climatic conditions. The latter influence the need for space heating or cooling, while the former includes settlement patterns (which influence the need for transportation), building practices, health care and social service levels, types of industry, etc. A systematic method of treating energy demands is described in the following, for use in the scenarios and system simulations in sections 6.5 to 6.7.

The development of energy demands is sometimes discussed in terms of marginal changes relative to current patterns. For changes over extended periods of time, this is not likely to capture the important issues. Another approach, called the *bottom-up model*, looks at human needs, desires, and goals and then builds up, first, the corresponding material demands requirement and then the energy required to fulfill these demands, under certain technology assumptions (Kuemmel *et al.*, 1997, appendix A). This approach is based on the view that certain human needs are basic needs, i.e., nonnegotiable, while others are secondary needs that depend on cultural factors and stages of development and knowledge and could turn out differently for different societies, subgroups, or individuals within a society. Basic needs include adequate food, shelter, security, and human relations. Then there is a continuous transition to more negotiable needs, which include material possessions, art, culture, human interactions, and leisure. [Maslow (1943) was among the first to discuss the hierarchy of human needs.] Energy demand is associated with satisfying several of these needs, with manufacture and construction of the equipment and products used in fulfillment of the needs, and with procuring the materials needed along the chain of activities and products.

In normative models like scenarios for the future, the natural approach to energy demand is to translate needs and goal satisfaction into energy requirements, consistent with environmental sustainability in case that is part of the normative assumptions. For market-driven scenarios, basic needs and human goals play an equally important role, but secondary goals are more likely to be influenced by commercial interest rather than by personal motives. It is interesting that the basic needs approach is routinely taken in discussion of the development of societies with low economic activity, but rarely in discussions of industrialized countries.

The methodology developed here is first to identify needs and demands (in short, human goals), and then to discuss the energy required to satisfy them in a series of steps tracing backward from the goal-satisfying activity or product to any required manufacture and then further back to materials. The outcomes are expressed on a per capita basis (involving averaging over differences within a population), but separates different geographical and social settings as required for local, regional, or global scenarios.

Analysis may begin by assuming 100% goal satisfaction, from which energy demands in societies that have not reached this level can subsequently be determined. Of course, a normative scenario may include the possibility that some members of the society do not enjoy full goal satisfaction. This is true for liberalism, which considers inequality a basic economic driver in society. Nevertheless, the end-use approach assumes that it is meaningful to specify the energy expenditure at the end-user without considering the system responsible for delivering the energy. This is only approximately true. In reality, there may be couplings between the supply system and the final energy use, and therefore in some cases end-use energy demand is dependent on the overall system choice. For example, a society rich in resources may undertake production of large quantities of resource-intensive products for export, while a society with fewer resources may instead focus on knowledge-based production; both societies are interested in balancing their economy to provide satisfaction of the goals of their populations, but possibly with quite different implications for energy demand.

End-use energy demands are distributed both on usage sectors and on energy qualities, and they may be categorized as follows:

1. Cooling and refrigeration 0 °C–50 °C below ambient temperature
2. Space heating and hot water 0 °C–50 °C above ambient temperature
3. Process heat below 100 °C
4. Process heat in the range 100 °C–500 °C
5. Process heat above 500 °C
6. Stationary mechanical energy
7. Electric energy (no simple substitution possible)
8. Energy for transportation (mobile mechanical energy)
9. Food energy

The goal categories used to describe the basic and derived needs can then be selected as follows:

A: Biologically acceptable surroundings
B: Food and water
C: Security
D: Health
E: Relations and leisure
F: Activities
    f1: Agriculture
    f2: Construction
    f3: Manufacturing industry
    f4: Raw materials and energy industry
    f5: Trade, service, and distribution
    f6: Education
    f7: Commuting

Here categories A–E refer to direct goal satisfaction, f1–f4 to primary derived requirements for fulfilling needs, and, finally, f5–f7 to indirect requirements for carrying out the various manipulations stipulated. Ranges of estimated energy requirements for satisfying all needs identified by present societies are summarized in Table 6.1 (Kuemmel *et al.*, 1997), with a more detailed, regional distribution given in Table 6.2 for a 2050 scenario-study by Sørensen and Meibom (1998). The same study includes an analysis of year-1994 energy end-uses, using the same principles as for the future scenario, and this analysis is given in Table 6.3. The table shows the low average efficiencies of the final conversion steps in 1994. The detailed assumptions behind the "full goal satisfaction" energy estimates are given below.

For use in the year-2050 global scenario described in section 6.7, the end-use energy components for each category are estimated on the basis of the actually assumed partial goal fulfillment by the year 2050 and are given in Table 6.4 on a regional basis. This analysis assumes population development based on the United Nations population studies (UN, 1996), using the alternative corresponding to high economic growth (in the absence of which population is estimated to grow more). The population development is in good agreement with the central choice used in the energy emission scenarios created for the IPCC process (IPCC, 2007).

Figures 6.1 and 6.2 show the present and the assumed year-2050 population density, including the effect of increasing urbanization, particularly in developing regions, leading to 74% of the world's population living in urban conglomerates by the year 2050 (UN, 1997b). Discussion of energy use categories A to F above is expanded in the following survey.

	1. Cooling & refrigeration	2. Space heating	3. Process heat under 100 °C	4. Process heat 100–500 °C	5. Process heat over 500 °C	6. Stationary mechanical energy	7. Electric appliances	8. Transportation work	9. Food energy	TOTAL
A. Biologically acceptable surroundings	0–200	0–650	0	0	0	0	0	0	0	0–650
B. Food and water	21	0	5	7	0	0	0	0	120	153
C. Security	(0)	(0)	0	0	0	(0)	0	1	0	1
D. Health	0	0	115	17	0	0	(0)	(0)	0	132
E. Relations, leisure	(0)	(0)	0	0	0	0	60–82	25–133	0	85–215
F. Activities:										
Construction	0	0	0	0	0	30–60	0	7–15	0	37–75
Trade, service and distribution	1–80	0–150	0–10	0	0	5	20	30–100	0	56–325 5–16
Agriculture	0	0	0	0–6	0	1–3	1	3–6	0	90–445
Manufacturing industry	1–40	0–150	10–100	20–70	12–30	20–40	20–40	7–15	0	
Raw Materials and energy industry	0	0	0–30	0–30	0–250	0–170	0–30	0–20	0	0–530 1–67
Education	0–20	0–65	0	0	0	0	1–2	0	0	0–30
Commuting	0	0	0	0	0	0	0	0–30	0	
TOTAL	23–361	0–1015	130–260	44–130	12–280	56–278	102–175	73–320	120	560–2639

TABLE 6.1 Global end-use energy demand based upon bottom-up analysis of needs and goal satisfaction in different parts of the world, using best available currently available technologies (average energy flow in W cap^{-1}). (From Kuemmel *et al.*, 1997.)

Biologically Acceptable Surroundings

Suitable breathing air and shelter against wind and hot or cold temperatures may require energy services both indirectly, in the manufacture of clothes and habitable structures, and directly, in the provision of an active heat supply or a cooling system. Insulation by clothing makes it possible for humans to stay in cold surroundings with a modest increase in food supply (which serves to heat the layer between the body and the clothing). The main heating and cooling demands occur in extended spaces (buildings and sheltered walkways, etc.) intended for human occupation without the inconvenience of clothing or temperatures that would impede activities, such as manual labor.

Rather arbitrarily, but within realistic limits, it is assumed that fulfillment of goals related to shelter requires an average space of 40 m^2 times a height of 2.3 m to be at the disposal of each individual in society and that this space should be maintained at a temperature of 18 °C–22 °C, independent of outside temperatures and other relevant conditions. As a "practical" standard of housing technology, for the heat loss P from this space I shall use the approximate form $P = C \times \Delta T$, where ΔT is the temperature difference between the desired indoor temperature and the outside temperature.

Regions &/ Energy Quality:	1. USA, Canada	2. Western Europe, Japan, Australia	3. Eastern Europe, Ex-Soviet, Middle East	4. Latin America, SE Asian "tigers"	5. China, India, Rest of Asia	6. Africa	Average & Total
Space heating*	205	212	164	19	96	10	
Other low-temp. heat	150	150	150	150	150	150	
Medium-temp. heat	50	50	50	50	50	50	
High-temp. heat	40	40	40	40	40	40	
Space cooling*	9	2	43	37	71	44	
Other refrigeration	35	35	35	35	35	35	
Stationary mechanical	150	150	150	150	150	150	
Electric applicances	150	150	150	150	150	150	
Transpor-tation	200	150	200	150	150	150	
Food energy	120	120	120	120	120	120	
Total 2050 end-use energy	**1109** **420**	**1059** **559**	**1102** **1146**	**901** **1243**	**1012** **4008**	**899** **1834**	**986 W/cap.** **9210 GW**
Population 2050, mill.	379	528	1040	1380	3960	2040	**9340**
Total area million km²	20.1	15.4	28.3	26.3	20.1	30.9	**141.1**

TABLE 6.2   Per capita energy use for "full goal satisfaction" in W cap^{-1} and total in GW for the assumed 2050 population stated. Rows marked * are based on temperature data for each cell of geographical area (0.5° longitude-latitude grid used). Manufacturing and raw materials industries are assumed to be distributed in proportion to population between regions. (Reprinted from Sørensen and Meibom, 1998, with permission.) A full list of the countries included in each region is given in Sørensen and Meibom (1998).

Region: 1994 End-use Energy	1. USA, Canada	2. Western Europe, Japan, Australia	3. Eastern Europe, Ex-Soviet, Middle East	4. Latin America, SE Asian "tigers"	5. China, India, Rest of Asia	6. Africa	Average & Total	
Space heating	186	207	61	2	19	1	**46**	W/cap
	52	116	41	1	48	1	**260**	GW
Other low-temp. heat	120	130	40	15	18	10	**36**	W/cap
	34	73	27	12	47	7	**199**	GW
Medium-temp. heat	40	50	30	10	10	5	**17**	W/cap
	11	28	20	8	26	3	**97**	GW
High-temp.heat	35	40	30	10	10	3	**16**	W/cap
	10	22	20	8	26	2	**88**	GW
Space cooling	9	1	13	2	3	0	**4**	W/cap
	2	1	9	2	8	0	**22**	GW
Other refrigeration	29	23	14	2	2	0	**7**	W/cap
	8	13	9	1	5	0	**37**	GW
Stationary mechanical	100	130	80	25	5	4	**34**	W/cap
	28	73	53	21	13	3	**191**	GW
Electric appliance	110	120	50	20	5	4	**29**	W/cap
	31	67	33	16	13	3	**164**	GW
Transportation	200	140	40	20	5	3	**34**	W/cap
	56	79	27	16	13	2	**193**	GW
Food energy	120	120	90	90	90	90	**95**	W/cap
	34	67	60	74	233	61	**530**	GW
Total end-use	**948**	**962**	**448**	**195**	**167**	**121**	**318**	**W/cap**
energy	**268**	**540**	**298**	**160**	**432**	**83**	**1781**	**GW**
Population 1994	282	561	666	820	2594	682	**5605**	million
Region area	20	15	28	26	20	31	**141**	million km²

TABLE 6.3 Estimated end-use energy in 1994. (Reprinted from Sørensen and Meibom, 1998, with permission.) Due to the nature of available statistical data, the categories are not identical to those used in the scenarios. Furthermore, some end-use energies are extrapolated from case studies. These procedures aim to provide a more realistic scenario starting point. However, as the scenario assumptions are based upon basic principles of goal satisfaction, the inaccuracy of current data and thus of scenario starting points does not influence scenario reliability, but only stated differences between now and the future.

Regions: / Energy Quality:	1. USA, Canada	2. Western Europe, Japan, Australia	3. Eastern Europe, Ex-Soviet, Middle East	4. Latin America, SE Asian "tigers"	5. China, India, Rest of Asia	6. Africa
Space heating	0.9 (0.9)	1.0 (0.96)	0.75 (0.25)	0.67 (0.08)	0.63 (0.16)	0.15 (0.10)
Other low-temp. heat	0.87 (0.8)	1.0 (0.87)	0.53 (0.27)	0.6 (0.10)	0.67 (0.12)	0.13 (0.07)
Medium-temp. heat	0.9 (0.8)	1.0 (1.0)	0.8 (0.6)	0.8 (0.2)	0.8 (0.2)	0.1 (0.1)
High-temp. heat	0.88 (0.88)	1.0 (1.0)	0.75 (0.75)	0.75 (0.25)	0.75 (0.25)	0.13 (0.08)
Refrigeration	1.0 (0.83)	1.0 (0.67)	0.4 (0.4)	0.5 (0.05)	0.33 (0.05)	0.05 (0.01)
Stationary mechanical	0.93 (0.67)	1.0 (0.87)	0.53 (0.53)	0.67 (0.17)	0.67 (0.03)	0.13 (0.03)
Electric applicances	1.0 (0.73)	1.0 (0.8)	0.53 (0.33)	0.67 (0.13)	0.67 (0.03)	0.1 (0.03)
Transportation	0.9 (1.0)	1.0 (0.93)	0.35 (0.20)	0.67 (0.13)	0.33 (0.03)	0.1 (0.02)
Food energy	1.0 (1.0)	1.0 (1.0)	1.0 (0.75)	1.0 (0.75)	1.0 (0.75)	0.83 (0.75)
Averages of the above	**0.93 (0.85)**	**1.0 (0.90)**	**0.62 (0.45)**	**0.68 (0.21)**	**0.64 (0.18)**	**0.20 (0.13)**

TABLE 6.4 The fraction of "full goal satisfaction" assumed in the year-2050 scenario, with estimated values for 1994 given in parentheses. (Reprinted from Sørensen and Meibom, 1998, with permission.) These estimates involve assumptions about conditions for development in different parts of the world: on the one hand, positive conditions, such as previous emphasis on education, which are a prerequisite for economic development, and, on the other hand, negative conditions, such as social instability, frequent wars, corrupt regimes, lack of tradition for democracy, for honoring human rights, and so on. It is recognized that these assumptions are considerably subjective. As an example, United Nations projections traditionally disregard non-economic factors and, for instance, assume a much higher rate of development for African countries.

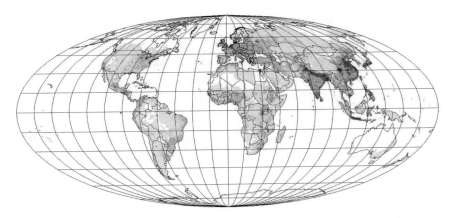

Population density
cap km^{-2}

3000	to 6000
1500	to 3000
800	to 1500
550	to 800
380	to 550
250	to 380
150	to 250
80	to 150
40	to 80
20	to 40
10	to 20
0.1 to	10
all others	

FIGURE 6.1 1994 population density (cap km^{-2}) (based on CIESIN, 1997). The geographical projection of Mollweide is used in order to obtain a faithful reproduction of equal areas (but at the expense of angular relations).

FIGURE 6.2 Population density assumed in 2050 scenario (cap km^{-2}) (consistent with UN, 1996, 1997; using an urbanization model described in Sørensen and Meibom, 1998). Note that the higher population densities in large cities are inconspicuous, while the lower densities in rural areas, relative to 1994, show up clearly.

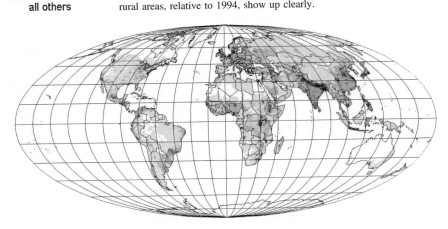

The constant C denotes a contribution from heat losses through the external surfaces of the space, plus a contribution from exchanging indoor air with outside air at a minimum rate of about once every 2 h. Half of the surfaces of the "person space" are considered external, with the other half being assumed to face another heated or cooled space. Best-current-technology solutions suggest that heat-loss and ventilation values of $C = 0.35$ (heat loss) + 0.25 (air exchange) = 0.6 W $°C^{-1}$ per m^2 of floor area can be attained. The precise value, of course, depends on building design, particularly window area. The air exchange assumed above is about 40% lower than it would be without use of heat exchangers in a fraction of the buildings. The 40-m^2 cap^{-1} assumed dwelling space is augmented below with 20 m^2 cap^{-1} for other activities (work, leisure). Light is treated under activities.

Now, energy needs[*] for heating and cooling, at a given location and averaged over the year, can be calculated with the use of climate tables giving the ambient temperature, e.g., hour by hour, for a typical year. If there are periods where temperatures are such that constant heating or cooling is required, the corresponding energy needs can be determined from the average temperatures alone. Heat capacity of the building will smooth out short-term variations, so that it is often a good approximation to determine heating and cooling demands from daily or even monthly average temperatures. An example of such estimations is shown in Figs. 6.3 and 6.4, for the annual heating and cooling requirements, separately, as a function of geographic location. The assumption is that space cooling is required for outdoor temperatures above 24 °C and heating is required for temperatures below 16 °C. Heat from indoor activities, combined with the thermal properties of suitable building techniques and materials, can provide indoor temperatures within the desired range of 18 °C–22 °C. Local space heating and cooling requirements are then obtained by folding the climate-related needs with the population density (taken from Fig. 6.2 for the year 2050 and used in the scenarios for the future described in section 6.7).

Here are a few examples: For Irkutsk, Siberia, the annual average temperature of -3 °C gives an average dwelling energy requirement for heating of 651 W (per capita and neglecting the possibility of heat gain by heat exchangers included in Figs. 6.3 and 6.4). For Darwin, Australia, no heating is needed. These two values are taken as approximate extremes for human habitats in the summary table. Very few people worldwide live in harsher climates, such as that of Verkhoyansk (also in Siberia: average temperature, -17 °C; heating

---

[*]Several studies use a thermodynamic concept called free energy (F). It is related to the conventional, conserved energy P by the relation $F = P \Delta T/T$, where T is the absolute temperature of the space to be heated or cooled and the temperature of the surroundings is $T_0 = T - \Delta T$ (all temperatures measured in Kelvin). The free energy concept incorporates the quality of energy, which is less than the conserved energy for heat and equal to it only for mechanical or electric energy. The rate of free-energy supply needed to provide the heat energy flux P in a space of temperature T embedded in surroundings of temperature T_0 is $F = P \Delta T/T = C (\Delta T)^2/T$. This expression is valid for both heating and cooling. In the latter case, both ΔT and P are negative.

requirement, 1085 W cap^{-1}). Other examples are $P = 225$ W cap^{-1} (New York City), $P = 298$ W cap^{-1} (Copenhagen), and P very nearly zero for Hong Kong. Cooling needs are zero for Irkutsk and Copenhagen, while for Darwin, with an annual average temperature of 29 °C, the cooling energy requirement is $-P = 209$ W cap^{-1}, assuming that temperatures above 22 °C are not acceptable. The range of cooling energy demands is assumed to be roughly given by these extremes, from $-P = 0$ to $-P = 200$ W cap^{-1}. For New York City, the annual average cooling requirement is 10 W cap^{-1} (typically concentrated within a few months), and for Hong Kong it is 78 W cap^{-1}.

Food and Water

The energy from food intake corresponding to full satisfaction of food needs is about 120 W cap^{-1}. Today, 28% of food intake is in the form of meat and other animal products (Alexandratos, 1995), but in the 2050 scenario presented below it is assumed that this fraction declines to 23%, partly owing to better balance in the diet of the industrialized regions and partly owing to an increased meat fraction in those regions presently having the lowest (e.g., East Asia, 9%). The distribution of food energy is shown in Figs. 6.5–6.6.

To store food adequately, the use of short- and long-term refrigeration is assumed. The average per capita food intake is of the order of 2×10^{-5} kg s^{-1}, of which 0.8×10^{-5} kg s^{-1} is assumed to have spent 5 days in a refrigerator at a temperature $\Delta T = 15$ °C below the surrounding room temperature, and 0.4×10^{-5} kg s^{-1} is assumed to have spent 2 months in a freezer at $\Delta T = 40$ °C below room temperature. The heat-loss rate through the insulated walls of the refrigerator or freezer is taken as 2×10^{-2} W °C^{-1} per kg of stored food. The energy requirement then becomes

$$P \sim 0.8 \times 10^{-5} \times 5 \times 24 \times 3600 \times 2 \times 10^{-2} \times 15 = 1.04 \text{ W/cap (refrigerator)}$$
$$P \sim 0.4 \times 10^{-5} \times 2 \times 720 \times 3600 \times 2 \times 10^{-2} \times 40 = 16.6 \text{ W/cap (freezer),}$$

plus the energy needed to bring the food down to the storage temperatures,

$$P \sim 0.72 + 2.12 = 2.84 \text{ W cap}^{-1}$$

(assuming a heat capacity of 6000 J kg^{-1} °C^{-1} above 0 °C and half that value below the freezing point, and a phase change energy of 350 kJ kg^{-1}). Energy is assumed to be delivered at the storage temperatures. Some energy could be regained from the melting of frozen food.

Cooking food requires further energy. Assuming that 40% of food intake is boiled at $\Delta T = 70$ °C above room temperature, and that 20% of food intake is fried at $\Delta T = 200$ °C above room temperature, the energy needed to bring the food up to the cooking temperatures is $P \sim 3.36 + 4.80 = 8.16$ W cap^{-1}, and the energy required for keeping the food cooking is $P \sim 1.45 + 2.08 = 3.53$ W cap^{-1}, assuming daily cooking times of 30 minutes for boiling and 15 minutes for frying (some food cultures certainly use more), and heat losses from the pot/pan/oven averaging 1 W $°C^{-1}$ for the quantities of food cooked per person per day.

Provision of water involves pumping and cleaning/purification. The pumping and treatment energy needs are negligible on a per capita basis, but both are included in the industry sector considered below.

Security

Heating and cooling of buildings used by courts, police, military, and other security-related institutions are included as part of the 40 m^2 area accorded each person. Remaining energy use for personal and national security would be for transportation and energy depreciation of materials and would hardly amount to more than 1 W cap^{-1}, except for very belligerent or crime-ridden nations or nations with badly disliked regimes.

Health

Hot water for personal hygiene is taken as 50 liters day^{-1} cap^{-1} at $T = 40\,°C$ above the waterworks' supply temperature, implying a rate of energy use averaging roughly $P = 97$ W cap^{-1}. Some of this heat could be recycled. Clothes washing and drying may amount to treating about 1 kg of clothes per day per capita. Washing is assumed to handle 5 kg of water/kg of clothes, at $T = 60\,°C$ (in practice, often more water is used, at different temperatures, some of which are closer to inlet temperature), or an average energy of $P = 15$ W cap^{-1}. For drying, it is assumed that 1 kg of water has to be evaporated (heat of evaporation about 2.3×10^6 J kg^{-1}) per day per capita, at an effective temperature elevation of 80 °C (the actual temperature is usually lower, but mechanical energy is then used to enhance evaporation by blowing air through rotating clothes containers). Local air humidity plays a considerable role in determining the precise figure. Condensing dryers recover part of the evaporation heat, say, 50%. Energy use for the case considered is then 17 W cap^{-1}. Hospitals and other buildings in the health sector use energy for space conditioning and equipment. These uses are included in household energy use (where they contribute 1%–2%).

Relations

Full goal satisfaction in the area of human relations involves a number of activities that are dependent on cultural traditions, habitats, and individual preferences. One possible example of a combination of energy services for this sector is used to quantify energy demands.

The need for lighting depends on climate and habitual temporal placement of light-requiring activities. Taking 40 W of present "state-of-the-art" commercial light sources (about 50 lumen per watt) per capita for 6 h/day, an average energy demand of 10 W cap^{-1} results. Still, radiant energy from light sources represents about 10 times less energy, and more efficient light sources are likely to become available in the future.

Radio, television, telecommunication, music, video, and computing, etc., are included in the scenario in section 6.7 and are assumed to take 65–130 W cap^{-1} (say 3 h a day on average for each appliance), or an energy flux of 8–16 W cap^{-1}. Add-on devices (scanners, printers, DVD burners, plus some not invented yet) increase demand, but increased efficiency of the new technology is assumed to balance the increased number of devices. Examples are flat-screen displays, which use 10–50 times less energy than the CRT screens that dominated at least television equipment up to the 1990s. Stand-by power for much of the equipment, as well as the power for computer screens and peripherals, was often not energy optimized by the end of the 1990s (cf. Sørensen, 1991). It is assumed that this will change in the future. Still, because the hoped-for balance between the increase in new equipment and their energy use is not proven, an additional energy expenditure of 30 W cap^{-1} was included in the section 6.7 scenario. The newer scenarios presented in section 6.5 have a much higher energy demand in this area (200 W cap^{-1}), considering that, although efficiency increases, so do the sizes of television and computer screens, etc. In addition, permanent Internet access and automatization of a number of maintenance jobs in buildings could be among the reasons for increased energy use, although it would be coupled to reductions elsewhere in society. Social and cultural activities taking place in public buildings are assumed to be included above (electricity use) or under space conditioning (heating and cooling).

Recreation and social visits can entail a need for transportation, by surface or by sea or air. A range of 25–133 W cap^{-1} is taken to be indicative of full goal satisfaction: the higher figure corresponds to traveling 11 000 km y^{-1} in a road-based vehicle occupied by two persons and using for this purpose 100 liters of gasoline equivalent per year per person. This amount of travel could be composed of 100 km weekly spent on short trips, plus two 500 km trips and one 5000 km trip a year. Depending on habitat and where friends and relatives live, the shorter trips could be reduced or made on bicycle or foot, and there would be variations between cultures and individuals. Road congestion and crowded air space increase the likelihood of flattening out the current increase in transportation activity (as would teleconferencing and videophones), and transport is increasingly seen as a nuisance, in contrast to the excitement associated with early motoring and air travel. Road and air guidance systems would reduce the energy spent in stop-and-go traffic and in aircraft circling airports on hold. Hence, the lower limit of energy use for recreation and social visits is between 5 and 6 times less than the upper limit.

Activities

Education (understood as current activities plus lifelong continuing education required in a changing world) is assumed to entail building energy needs corresponding to 10% of the residential value, i.e., an energy flux of 0–20 W cap^{-1} for cooling and 0–65 W cap^{-1} for heating.

Construction is evaluated on the basis of 1% of structures being replaced per year, but the rate would be higher in periods of population increase. Measuring structures in units of the one-person space as defined above under *Biologically acceptable surroundings*, it is assumed that there are 1.5 such structures per person (including residential, cultural, service, and work spaces). This leads to an estimated rate of energy spending for construction amounting to 30–60 W cap^{-1} of stationary mechanical energy plus 7–15 W cap^{-1} for transportation of materials to building sites. The energy hidden in materials is deferred to the industrial manufacture and raw materials industry.

Agriculture, which includes fishing, the lumber industry, and food processing, in some climates requires energy for food crop drying (0–6 W cap^{-1}), for water pumping in irrigation and other mechanical work (about 3 W cap^{-1}), for electric appliances (about 1 W cap^{-1}), and for transport (tractors and mobile farm machinery, about 6 W cap^{-1}).

The distribution and service (e.g., repair or retail) sector is assumed, depending on location, to use 0–80 W cap^{-1} of energy for refrigeration, 0–150 W cap^{-1} for heating of commerce- or business-related buildings, about 20 W cap^{-1} of electric energy for telecommunications and other electric appliances, and about 5 W cap^{-1} of stationary mechanical energy for repair and maintenance service. Transportation energy needs in the distribution and service sectors, as well as energy for commuting between home and workplaces outside home, depend strongly on the physical location of activities and on the amount of planning that has been done to optimize such travel, which is not in itself of any benefit. Estimated energy spending is 30–100 W cap^{-1}, depending on these factors. All the energy estimates here are based on actual energy use in present societies, supplemented with reduction factors pertaining to the replacement of existing equipment by technically more efficient types, according to the "best-available-and-practical technology" criterion, accompanied by an evaluation of the required energy quality for each application.

In the same way, energy use in manufacturing can be deduced from present data, once the volume of production is known. Assuming the possession of material goods to correspond to the present level in the United States or Scandinavia, and a replacement rate of 5% per year, leads to a rate of energy use of about 300 W cap^{-1}. Less materialistically minded societies would use less. Spelled out in terms of energy qualities, there would be 0–40 W cap^{-1} for cooling and 0–150 W cap^{-1} for heating and maintaining comfort in factory buildings, 7–15 W cap^{-1} for internal transportation, and 20–40 W cap^{-1} for electric appliances. Most of the electric energy would be used in the production processes, for computers and for lighting, along with another 20–40 W cap^{-1} used for stationary mechanical energy. The assumed lighting efficiency is 50 lumen/W. Finally, the process heat requirement would include

10–100 W cap^{-1} below 100 °C, 20–70 W cap^{-1} from 100 °C to 500 °C and 12–30 W cap^{-1} above 500 °C, all measured as average rates of energy supply over industries very different in regard to energy intensity. Some consideration is given to heat cascading and reuse at lower temperatures, in that the energy requirements at lower temperatures have been reduced by what corresponds to about 70% of the reject heat from the processes in the next higher temperature interval.

Most difficult to estimate are the future energy needs of the resource industry. This is for two reasons: one is that the resource industry includes the energy industry and thus will vary greatly depending on what the supply option or supply mix is. The second reason is the future need for primary materials: will it be based on new resource extraction, as is largely the case today, or will recycling increase to near 100% (for environmental and economic reasons connected with depletion of mineral resources)?

As a concrete example, let us assume that renewable energy sources are used, as in the scenario considered in section 6.4. Extraction of energy by the mining and oil and gas industries as we know them today will disappear, and activities related to procuring energy will take a quite different form, related to renewable energy conversion equipment, which in most cases is more comparable to present utility services (power plants, etc.) than to a resource industry. This means that energy equipment manufacturing will become the dominant energy-requiring activity.

For other materials, the ratios of process heat, stationary mechanical energy, and electricity use depend on whether mining or recycling is the dominant mode of furnishing new raw materials. In the ranges given, not all maxima are supposed to be realized simultaneously, and neither are all minima. The numbers are assumed to comprise both the energy and the material provision industries. The basis assumption is high recycling, but for the upper limits not quite 100%. There will therefore be a certain requirement for adding new materials for a growing world population. The assumed ranges are 0–30 W cap^{-1} for process heat below 100 °C as well as for 100 °C–500 °C, 0–250 W cap^{-1} above 500 °C, 0–170 W cap^{-1} of stationary mechanical energy, 0–30 W cap^{-1} of electric energy and 0–20 W cap^{-1} of transportation energy.

Summary of End-use Energy Requirements

Table 6.5 summarizes, by main energy qualities, the estimates of energy delivered to end-users in the scenario discussed for year 2050. The 2050 per capita energy services are more than twice those of today, and yet the average energy delivered to each end-user is only half that of today, owing to higher efficiency of each conversion step. In section 6.7, the primary energy required for the global 2050 scenario is determined, while the regional studies in section 6.6 use updated end-use estimates.

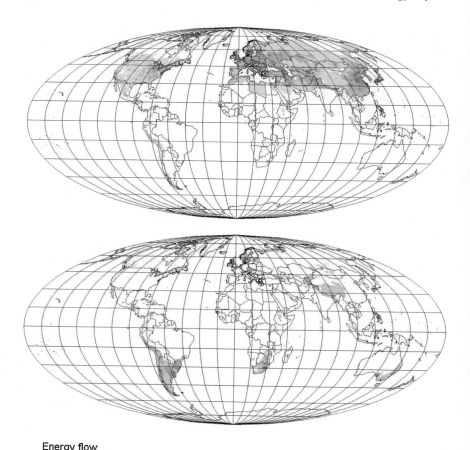

Energy flow
W m^{-2}

5	to 10
2	to 5
1	to 2
0.5	to 1
0.2	to 0.5
0.1	to 0.2
0.05	to 0.1
0.02	to 0.05
0.01	to 0.02
0.005	to 0.01
0.002	to 0.005
0.0001	to 0.002
all others	

FIGURE 6.3 Energy requirement for January (*above*) and July (*below*) space heating (W m^{-2}), assuming the year-2050 population model (UN, 1996, 1997) and full satisfaction of space-heating needs in an average space of 60 m^2 (floor area) by 2.3 m (ceiling height) per capita, including domestic as well as other indoor spaces in use (Sørensen and Meibom, 1998).

Energy flow
W m^{-2}

5	to 10
2	to 5
1	to 2
0.5	to 1
0.2	to 0.5
0.1	to 0.2
0.05	to 0.1
0.02	to 0.05
0.01	to 0.02
0.005	to 0.01
0.002	to 0.005
0.0001	to 0.002
all others	

FIGURE 6.4 Energy requirement for January (*above*) and July (*below*) space cooling (W m^{-2}), assuming the year 2050 population model (UN, 1996, 1997) and full satisfaction of space-cooling needs in an average space of 60 m^2 (floor area) by 2.3 m (ceiling height) per capita, including domestic as well as other indoor spaces in use (Sørensen and Meibom, 1998).

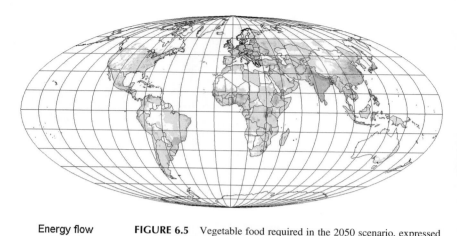

Energy flow
W m⁻²

5	to 10
2	to 5
1	to 2
0.5	to 1
0.2	to 0.5
0.1	to 0.2
0.05	to 0.1
0.02	to 0.05
0.01	to 0.02
0.005	to 0.01
0.002 to	0.005
0.0001 to	0.002
all others	

FIGURE 6.5   Vegetable food required in the 2050 scenario, expressed in W m^{-2} and including 30% losses in preparation at the end-user.

FIGURE 6.6   Animal-based food required in the 2050 scenario, expressed in W m^{-2} and including 20% losses in preparation at the end-user.

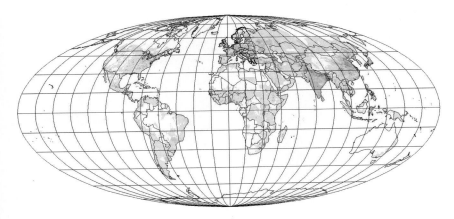

Energy flow
W m⁻²

5	to 10
2	to 5
1	to 2
0.5	to 1
0.2	to 0.5
0.1	to 0.2
0.05	to 0.1
0.02	to 0.05
0.01	to 0.02
0.005	to 0.01
0.002	to 0.005
0.0001 to	0.002
all others	

FIGURE 6.7 Average energy input to heat pumps used for space conditioning and low-temperature heat production, required in 2050 scenario and expressed in W m⁻².

FIGURE 6.8 Electric and mechanical energy plus medium- to high-temperature heat required in the 2050 scenario, expressed in W m⁻² (the heat-pump inputs of Fig. 6.7 are excluded).

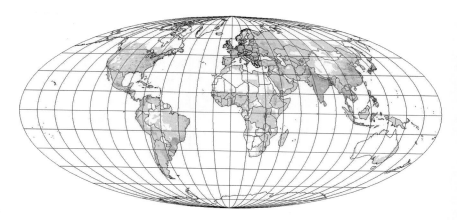

**Energy flow
W m⁻²**

■	5	to 10
■	2	to 5
■	1	to 2
	0.5	to 1
	0.2	to 0.5
	0.1	to 0.2
	0.05	to 0.1
	0.02	to 0.05
	0.01	to 0.02
	0.005	to 0.01
	0.002	to 0.005
	0.0001 to	0.002
	all others	

FIGURE 6.9 Energy for all forms of transportation required in the 2050 scenario and expressed in W m⁻² (including losses in vehicle energy conversion and storage equipment).

FIGURE 6.10 Total energy required in the 2050 scenario for delivery to the end-user, expressed in W m⁻² (food energy and environmental heat inputs to heat pumps included).

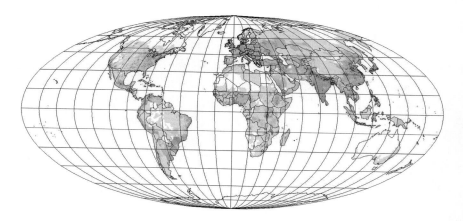

Regions: / Energy Quality:	1. USA, Canada	2. Western Europe, Japan, Australia	3. Eastern Europe, Ex-soviet, Middle East	4. Latin America, SE Asian "tigers"	5. China, Rest of Asia	6. Africa	Average & Total	
Food based on animals	30	30	30	25	25	20	**23**	%
	45	45	45	37	37	25	**36**	W/cap
	17	24	47	52	148	51	**339**	GW
Food based on grain & vegetables	70	70	70	75	75	80	**77**	%
	119	119	119	128	128	114	**123**	W/cap
	45	63	124	177	506	232	**1148**	GW
Gross transportation energy	359	299	140	201	99	30	**125**	W/cap
	136	158	146	277	392	61	**1170**	GW
Heat pump input for low-temp. heat and cooling	110	110	87	43	80	22	**65**	W/cap
	58	58	90	60	318	45	**610**	GW
Environmental heat	240	256	203	100	186	51	**151**	W/cap
	91	135	210	140	741	105	**1421**	GW
Direct electric & all other energy	420	424	245	288	283	47	**240**	W/cap
	153	224	255	398	1116	96	**2242**	GW
Total delivered energy	**1272**	**1252**	**838**	**800**	**814**	**290**	**742**	W/cap
	482	**661**	**871**	**1104**	**3225**	**591**	**6934**	GW
Population 2050	379	528	1040	1380	3960	2040	**9340**	millions

TABLE 6.5 Energy delivered to end-user in 2050 scenario, including environmental heat (Sørensen and Meibom, 1998). The energy delivered differs from end-use energy by losses taking place at the end-users' location.

Table 6.5 gives the gross energy input to transportation needs, accumulating the individual demands for personal transportation, for work-related transport of persons and goods, and for transport between home and work ("commuting"). The net energy required to overcome frictional resistance and the parts of potential energy (for uphill climbing) and acceleration that are not reclaimed is multiplied by a factor of 2 to arrive at the gross energy delivery. This factor reflects an assumed 2050 energy conversion efficiency for fuel-cell-driven (with electric motor) road vehicles of 50%, as contrasted with about 20% for present-day combustion engines. For the fraction of vehicles (presumably an urban fleet) using batteries and electric drives, the 50% efficiency is meant to reflect storage-cycle losses (creep current discharge during parking and battery-cycle efficiency). The geographical distribution of transportation energy demands is shown in Fig. 6.9.

The row "direct electric and all other energy" in Table 6.5 comprises medium- and high-temperature heat, refrigeration other than space cooling (which is included in the space-heating and space-cooling energy), stationary mechanical energy, and dedicated electric energy for appliances and motors outside the transportation sector. For these forms of energy, no end-use efficiency is estimated, as the final service efficiency depends on factors specific to each application (re-use and cascading in the case of industrial process heat, sound- and light-creating technologies, computing, and display and printing technologies, all of which are characterized by particular sets of efficiency considerations). The geographical distribution of these energy requirements in the 2050 scenario is shown in Fig. 6.8. The distribution of the electric energy input to heat pumps (COP = 3), which in the scenario covers space heating, cooling, and other low-temperature heat demands, is shown in Fig. 6.7, and Fig. 6.10 adds all the scenario energy demands, including environmental heat drawn by the heat-pump systems. The low COP assumed for heat pumps reflects the location of the corresponding loads at high latitude, where suitable low-temperature reservoirs are difficult to establish.

Estimations of future energy demands are obviously inaccurate, due to both technical and normative factors: on the one hand, new activities involving energy use may emerge, and, on the other hand, the efficiency of energy use may be further increased by introduction of novel technology. Yet it is reassuring that the gross estimate of energy demands associated with full goal satisfaction (for a choice of goals not in any way restrictive) is much lower than present energy use in industrialized countries. It demonstrates that bringing the entire world population, including little-developed and growing regions, up to a level of full goal satisfaction is not precluded for any technical reasons. The scenarios in section 6.4 are consistent with this, in assuming energy-efficiency gains of about a factor of 4.

Figure 6.11 illustrates the reference scenario assumptions in different regions by comparing end-use energy, with and without food energy, between 2050 and 1994 populations. Compared to many other scenarios (see, for example, WEA, 2000), the 2050 scenario used here places more emphasis on using the most efficient technology. The reason for this is an assumption of economic rationality that strangely contrasts with observed behavior: Current consumers often choose technology that entails a much higher cost for energy inputs during the operation life than the increased cost they would need to pay at purchase for the most energy-efficient brand of the technology in question. This economic irrationality pervades our societies in several other ways, including the emphasis in RD&D programs on supply solutions that are much more expensive than solutions directed at diminishing energy use. Thus, the "growth paradigm" seems to be so deeply rooted in current economic thinking that many people deem it "wrong" to satisfy their needs with less energy and "wrong" to invest in technology that may turn their energy consumption on a declining path. This problem is rarely discussed explicitly in economic theory, but is buried in terms like "free consumer choice," and is coupled with encouraging the "freedom" of advertisement, allowing substandard equipment to be marketed without mention of the diseconomies of its lifetime energy consumption.

One wonders when consumers will, instead, respond with the thought that, "If a product needs to be advertised, there must be something wrong with it"? Either the product does not fulfill any need, or it is of lower quality than other products filling the same need with lower energy consumption. New, expensive energy-production technology, such as solar or fuel cells, get much more media attention than freezers with a four-times-lower-than-average energy consumption. Given that world development at present is not governed by economic rationality, the supply–demand matching exercises presented in the following subsections, which are based on the scenarios defined according to the principles set forth above, will take up the discussion of robustness against underestimates of future energy demand.

Specification of the energy form required at the end-user first amounts to distributing the sector loads on electric, mechanical, heat, radiant energy, etc., but the heat energy demand must be further specified, as described above, in order to provide the type of information needed for planning and selecting the energy systems capable of serving the loads. As an example of the investigations needed, Fig. 6.12 shows the temperature distribution of heat demand for the Australian food-processing industry.

Patterns of Time Variations

In a complex system with decisions being taken by a large number of individual consumers, the time pattern of the load cannot be predicted precisely. Yet the average composition and distribution of the load are expected to change smoothly, with only modest ripples appearing relative to the average. However, there may be extreme situations, but with a low probability of occurrence, that should be taken into account in designing energy system solutions.

Fluctuations in electricity demand have routinely been surveyed by utility companies, many using simulation models to optimize use of their supply system. Figure 6.13 gives two examples of seasonal variations in electricity loads, one being for a high-latitude area, exhibiting a correlation with the number of dark hours in the day, and the other being for a metropolitan area in the southern United States, reflecting the use of electricity for space-cooling appliances.

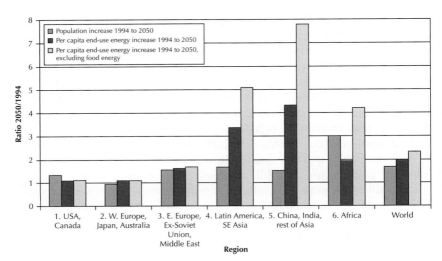

FIGURE 6.11   Ratio of the year-2050 scenario end-use energy to estimated 1994 value.

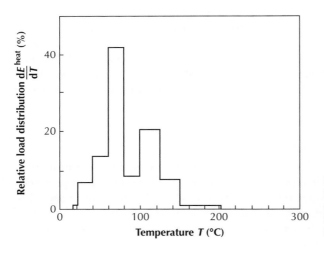

FIGURE 6.12   Temperature distribution of heat demand in the Australian food-processing industry. (Based on Morse, 1977.)

Superimposed on seasonal variations is a daily cycle, which often exhibits different patterns on weekdays versus weekends. The example given in Fig. 6.14 (also from the United States) shows a further irregularity for Mondays (a "peak working day"), probably reflecting a higher energy requirement for the "cold starts" of industrial processes relative to continual operation during the following days. Yet not all regions exhibit a pronounced "peak working day" load like the one in Fig. 6.14. Figure 6.15 shows the daily load variations at Danish utilities for a summer and a winter weekday. The ratio between the maximum and the minimum load is considerably higher than in the American

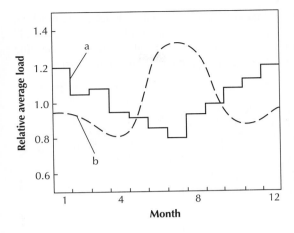

FIGURE 6.13 Seasonal variations in relative electricity load in areas with dominating components of lighting and air conditioning, respectively. *Solid line*: Denmark, latitude 56 °N (Danish Electricity Supply Undertakings, 1971); *dashed line*: metropolitan area in southern United States. (Based on Brown and Cronin, 1974.)

example in Fig. 6.14, presumably because a higher proportion of industries in the United States require continuous operation.

In connection with renewable energy sources, a particularly attractive load is one that need not be performed at a definite time but can be performed anywhere within a certain time span. Examples of such loads (well known from historical energy use) are flour milling, irrigation, and various household tasks, such as bathing and washing. The energy-demand patterns for many such tasks rest on habits that, in many cases, developed under the influence of definite types of energy supply and that may change as a result of a planned or unplanned transition to different energy sources. Distinction between casual habits and actual goals requiring energy conversion is not always easy, and the differentiation is a worthy subject for social debate.

Data on renewable energy flows (or reservoirs) that may serve as inputs to the energy conversion system are a major theme of Chapter 2 and particularly of Chapter 3.

As mentioned in section 6.1, the performance of most conversion systems can be calculated with sufficient accuracy by a "quasi-steady-state approximation," and most of these systems possess "buffering mechanisms" that make them insensitive to input fluctuations of short duration (such as high-frequency components in a spectral decomposition). For this reason, input data describing renewable energy fluxes may be chosen as time averages over suitable time intervals. What constitutes a "suitable" time interval depends on the response time (time constant) of the conversion system, and it is likely that the description of some systems will require input data of a very detailed nature (e.g., minute averages), whereas other systems may be adequately described using data averaged over months.

A situation often encountered is that the data available are not sufficient for a complete system simulation on the proper time scale (i.e., with properly sized time steps). Furthermore, interesting simulations usually concern the performance of the system some time in the future, whereas available data by necessity pertain to the past. Demand data require that a model be established, while, for resource data, it may be possible to use "typical" data. "Typical" means not just average data, but data containing typical variations in the time scales that are important for the contemplated energy conversion system.

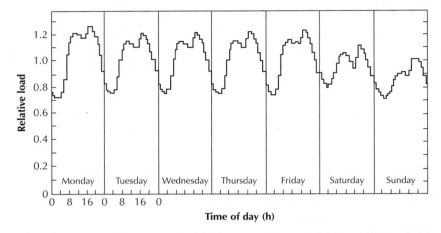

FIGURE 6.14   Relative electricity load (hourly averages) during a typical week for a utility system in the northeastern United States. (Based on Fernandes, 1974.)

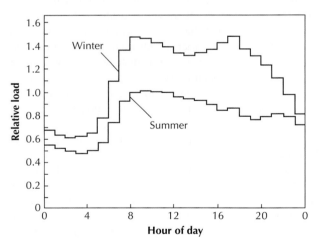

FIGURE 6.15   Relative electricity load (hourly averages) on a typical summer and a typical winter weekday for the Danish utility systems (Danish Electricity Supply Undertakings, 1971).

Examples of such data are hourly values of wind velocities selected from past data, which, incorporated into a simulation model, may serve to estimate system performance for future years. Clearly, this type of "reference year" should not contain features too different from long-term statistics.

For some renewable energy systems, e.g., hydroelectricity, the variations between years are often important system design parameters.

Synthetic databases (e.g., covering a year) have been constructed by combining sequences of actual data picked from several years so that averages of important variables take values close to their long-range averages, but such that "typical" fluctuations are preserved in the data. *Reference year* solar radiation and other meteorological data, which are made available for engineering purposes for selected locations (notably certain cities in Europe and North America, cf. section 3.1.3), are an example.

For computer simulations, standard procedures exist that will generate random numbers with a specified distribution (e.g., Gaussian) once the fluctuation (or "half-width") and average value of the distribution are specified. This approach may be described as a *stochastically generated database*. It has found widespread application, e.g., in prediction of wind turbine production characteristics (Sørensen, 1986; Troen and Petersen, 1989).

6.3 SUPPLY SCENARIO CONSTRUCTION

In the construction of scenarios, it is necessary to have data for the energy supply sources intended for inclusion, and it is often convenient to create a source database of potential production profiles to pick from when piecing together a scenario. An overview of common elements of such a resource-exploitation database is given below.

6.3.1 Photovoltaic Power Production

A solar radiation model is constructed on the basis of data for solar radiation incident on a horizontal plane. One set of data, from NASA, is based on analysis (Pinker and Laszlo, 1992) of satellite measurements of radiation, albedo (reflectance), cloud cover, and attenuation of radiation in the air (NASA, 1997). The other study considered here collects many types of weather and climate data (NCEP-NCAR, 1998) and uses balance equations and a global circulation model (horizontal resolution about 210 km and 28 vertical levels) to improve the data's consistency (Kalnay *et al.*, 1996). However, for solar radiation, this study does not use ground-based observations, so, again, the solar radiation data are based on top-of-the-atmosphere fluxes and various absorption, reflection, and attenuation processes plus the balance requirements at the Earth's surface. The two types of analysis give similar results for solar radiation. The balance equations (which take the difference between upward and downward short- and long-wavelength radiation and heat fluxes) do suggest too high an albedo over oceans, but the land data that we are using appear to be reliable. Figure 3.1 shows the radiation on a horizontal plane at the Earth's surface for selected months of 1997 (NCEP-NCAR, 1998). In order to calculate the solar radiation incident on inclined surfaces, such as the ones characterizing most solar installations, ideally one would have hourly data for direct and scattered radiation (or equivalently for direct and total global radiation).

Then, to estimate the value of scattered radiation for differently inclined surfaces, one would assume that the scattered radiation is uniform. For direct radiation, ray calculations with the appropriate angles between the direction to the sun and the normal to the collector surface have to be performed hourly. Such calculations have been performed and compared with measurements on inclined surfaces at several major solar installations and some weather stations (see Chapter 3). Such data are not globally available but can be approximately inferred from available radiation data, as is done in Figs. 3.14–3.15. These data will be used in the scenarios discussed in section 6.6.

The task is to estimate the radiation on a surface tilted either north or south by an angle approximately equal to the latitude (as this gives the optimum performance) on the basis of the horizontal surface solar radiation data available. Only a modest accuracy is required because of the characteristics of the actual solar installations used in the scenarios:

1. Building-integrated solar panels will be influenced by the structure of the building: some solar panels will be mounted on vertical facades, while others will be placed on roofs that are either flat or tilted, typically 30°, 45°, or 60°. In all cases, the panels' actual orientation may not be precisely south or north, although the resource estimates are here given only for buildings facing approximately correctly and not being exposed to strong shadowing effects from other structures. The penalty for incorrect orientation and tilt angle is usually modest and mostly influences seasonal distribution of power (cf. Chapter 3).

2. Centralized solar farms are generally oriented optimally, using total production maximization for panels not tracking the sun. However, the majority of locations suited for central photovoltaic (PV) installations are the desert areas of the Sahara, the Arabian Peninsula, the Gobi, and Australian inland locations. Because they are all fairly close to the Equator, there is modest seasonal variation in solar radiation, and horizontal-surface data are often quite representative.

The early scenarios presented in section 6.7 used the following simple approximation, originating from the analysis of the Danish (latitude 56 °N) data given in Fig. 3.17. Here the radiation on a latitude-inclined surface in January and July very nearly equals the horizontal radiation in October and April, whereas the horizontal-surface data for January and July are lower and higher than the inclined-surface measurements, respectively. The October and April horizontal data are thus used as a proxy for January and July inclined-surface radiation, and the April and October inclined-surface values are constructed as simple averages of the January and July adopted values. This procedure, which works well for the Danish latitude, is also less inaccurate for low latitudes, because of the relative independence of seasons mentioned above, and we simply use it for all locations.

Figure 2.24 shows the NCEP-NCAR (1998) data for a horizontal surface. An alternative calculation using the European general circulation model HADCM2-SUL (Mitchell and Johns, 1997) gave very similar monthly 1997 solar radiation values.

Energy flow
W m⁻²

5	to	10
2	to	5
1	to	2
0.5	to	1
0.2	to	0.5
0.1	to	0.2
0.05	to	0.1
0.02	to	0.05
0.01	to	0.02
0.005	to	0.01
0.002	to	0.005
0.0001 to		0.002
all others		

FIGURE 6.16 Potential power production from building-integrated solar cells in the 2050 scenario after subtraction of transmission and storage-cycle losses. The January estimates are shown above and the July estimates are shown below.

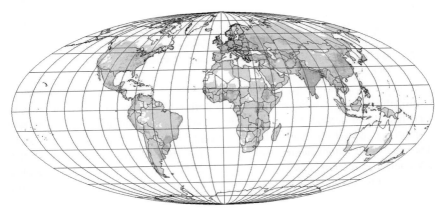

In order to derive the actual energy extracted from PV solar panels, a fixed conversion efficiency of 15% is assumed. This is a fairly conservative estimate for the year-2050 technology, considering that the current efficiency of the best monocrystalline cells is above 20% and that of amorphous cells is near 10%, with multicrystalline cells falling in between. The 2050 choice is likely to be thin-film technology, but not necessarily amorphous or group III–V based, as new techniques allow crystalline or multicrystalline silicon material to be deposited on substrates without the complicated process of ingot growth and cutting. The efficiency of such cells is lower than that of the best current crystalline cells, but increasing, and since 2002 multicrystalline cells have surpassed monocrystalline cells in worldwide shipments.

Energy flow
W m⁻²

1.8 to 2	
1.6 to 1.8	
1.4 to 1.6	
1.2 to 1.4	
1 to 1.2	
0.8 to 1	
0.6 to 0.8	
0.4 to 0.6	**FIGURE 6.17** Potential annual average power production for the 2050
0.2 to 0.4	scenario from centralized photovoltaic plants, placed on marginal land, after
0.01 to 0.2	subtraction of transmission and storage-cycle losses. Seasonal variations are
all others	small.

Finally, the availability of sites for mounting PV panels either in a decentralized fashion or centrally in major energy parks is estimated: the decentralized potential is based on availability of suitably inclined, shadow-free surfaces. The area of suitably oriented surfaces that may be used for building-integrated PV energy collection is assumed to be 1% of the urban horizontal land area plus 0.01% of cropland area. The latter reflects the density of farmhouses in relation to agricultural plot sizes and is based on European estimates, roughly assuming that 25% of rural buildings may install solar panels. Potential PV production is thus taken as 15% of radiation times the fraction of the above two area types and the stated percentage for each. A final factor of 0.75 is applied in order to account for transmission and storage-cycle losses, assuming about 5% transmission losses and that roughly half the energy goes through a storage facility of 60% round-trip efficiency (e.g., reversible fuel cells). The flow of energy reduced by all these factors represents the energy delivered to the end-use customers in the form of electricity. It is shown in Fig. 6.16 for two months, with regional sums being discussed in section 6.7.

For the centralized PV possibility, the potential is taken as that derived from 1% of all rangeland plus 5% of all marginal land (deserts and scrubland), again times 15% of the incoming radiation and times 0.75 to account for transmission and storage losses. As shown in Fig. 6.17, this is a huge amount of energy, and it shows that such centralized PV installations can theoretically cover many times the demand of our 2050 scenario. Setting aside 1% of rangeland incurs only an insignificant reduction in other uses of the land, even for a small number of large plants. The same is true for desert and marginal land, where only the 5% most suitable area needs to be used, which in reality might be 10% when the entire installation is included (including frames, access roads), of which the solar panels are only part. A centralized installation in the huge area of the Sahara Desert would suffice to supply more than the entire world's need for energy. Of course, it would require intercontinental transmission, which may be realistic by the year 2050, e.g., through superconducting trunk lines (Nielsen and Sørensen, 1996).

6.3.2 Wind-power Production

There have been substantial advances in data availability for estimating wind-power production at any location on the Earth. The scenarios in section 6.6 use scatterometer and new re-analysis data discussed below, available on a 6-hour basis for 0.5° longitude or latitude grids. When the earlier scenarios of section 6.7 were simulated, only monthly averaged wind data were available, and the following coarse approximation was used to arrive at useful data:

The basic data used in the 1998 calculation of potential wind-power production are from a 1997 re-analysis of meteorological station data and altitudinal data from balloons and aircraft, analyzed according to the method of Kalnay *et al.* (1996) and using a general circulation model to improve consistency (NCEP-NCAR, 1998). For the scenario, simple average pressure levels of 1000 mb and 925 mb were used to represent the wind speeds at typical turbine hub heights of around 70 m. However, as the data are monthly mean wind speeds $<v>$ constructed on the basis of zonal and meridional winds, we need a model of the relationship between $<v>^3$ and $<v^3>$ to go from wind speeds to power in the wind. Simple models imply a rough proportionality between these two quantities, as used, for example, in the U.S. wind atlas prepared by the Pacific NW Laboratory (Swisher, 1995) and also in the Weibull distribution approach of the European wind atlas (Troen and Petersen, 1989). Inspired by these sources, one may estimate the relation between the power in wind P_w and the wind speed v (m s^{-1}) from the relation

$$P_w = 0.625 < v^3 > \approx 1.3 <v>^3 \text{ W m}^{-2}.$$

The power in the wind obtained in this way is illustrated in Fig. 3.27 for the four seasons. The results in the figure may be compared with the outcome of using the new 6-hour data below, in which case no *ad hoc* approximation is needed to get from wind speeds to power in the wind, because the data are not averages over 6 hours, but instantaneous values measured every 6 hours (well, actually every 12 hours, see the reason below).

Going from the power in the wind to the power that may be produced by a wind turbine, the nonlinear response of wind turbines has to be taken into consideration. Turbines of modern design aim at a high annual production and typically start producing only at around 5 m s^{-1}; they reach a fixed maximum production at around 12 m s^{-1}. Thus, there would not be any production for a monthly average wind speed below 5 m s^{-1}, if it were made up of nearly constant values throughout the time range. However, actual time-series of wind speeds reflect the passage of weather fronts, typically oscillate with periods of about 2 weeks, and entail power production from the above type of wind turbine at practically all monthly average wind speeds, down to zero. The approach is then to parameterize the average power production from wind turbines as

$$W_p = \text{Minimum}(0.33\, P_w, 500) \text{ W m}^{-2},$$

representing contemporary wind turbine constructions through a power factor slightly above current levels. Owing to the coarseness of the GIS[*] grids used in 1998 (for wind-speed data for approximately 300 km cells at the Equator, diminishing by the cosine of the latitude toward the poles), the method does not at present compete with wind atlas methods available in certain parts of the world. The 2008 scatterometer plus re-analysis data are better, but still with a grid size of 25 km, which leads to an expected power production accuracy of around 100 km. This is, however, sufficient for planning purposes. The use of general circulation models in recalculating the data ensures that both surface roughness and general circulation determinants are included.

It should be noted that the data from low-height meteorological stations (which are often at sheltered locations at or under 10 m and are a substantial source of the data underlying the NCEP-NCAR data), relative to elevated data at typical wind turbine hub heights, may lead to a poor representation

---

[*]GIS = geographical information system.

of actual wind conditions in a given area. For this reason, the European wind atlas used only selected station data believed to be more relevant for wind turbine exploitation. However, comparison with the Hadley model data, which have only a weak dependence on station data because they are generated from a general circulation model using topological features of wind roughness (friction), indicates that the method is quite reliable. The reason for this is presumably that we use pressure-level data for both of the two lowest model pressure levels, rather than the 10 m data also available in the NCEP-NCAR database. Repeating the calculation for the 10 m data indeed gives a considerably stronger dependence on the dubious ground-level measurements.

The final step in estimating potential production from wind turbines in the scenario is to appraise the locations suited for erection of wind turbines. They are divided into three categories: decentralized wind-power production on farmland, and centralized production in wind parks, either onshore or offshore. For decentralized production, we assume that a (vertical) turbine-swept area amounting to 0.1% of cropland and rangeland areas may be used for wind production. Only the part of cropland designated "open cropland" (cf. section 6.3.3) is used. This excludes mixed forest and cropland, where trees are likely to impede wind flow. The assumed density of turbines corresponds roughly to the current density of wind turbines in many parts of Denmark, although the current turbines' average size is smaller than that of the turbines envisaged for the mid-21st century.

Currently, new turbines are typically of unit size above 2 MW and have a hub-height of at least 50–60 m. Owing to the combination of swept area increasing as the square of the linear dimension and wind speeds increasing with height, the 2–5 MW turbines envisaged in the 2050 scenario will not be conspicuously larger than those of the current generation.

In terms of ownership, the density of decentralized turbines in the scenario would correspond to 1 out of 5–10 farmers owning a wind turbine (with large variations due to farm size variations between regions). Current wind-power utilization in the country with highest per capita wind-power production, Denmark, still collects a substantial amount of power from farm-situated wind turbines. In many other countries, the emphasis has been on centralized production from turbine clusters. The amount of decentralized wind-power potentially available is shown in Figs. 6.18 and 6.19. Strong seasonality is evident, and repeating the evaluation using data for years other than 1997 indicates that variation between years is also large.

Centralized wind-power is here understood as power not produced in connection with individual farms. For land-based wind parks, only placement on marginal land is accepted in the global scenario. In contrast to photovoltaic power, there is a fundamental limit to the fraction of land that can be used for wind-power extraction, owing to shadowing and interference between wake flows that create turbulence and reduce the power output.

(a)

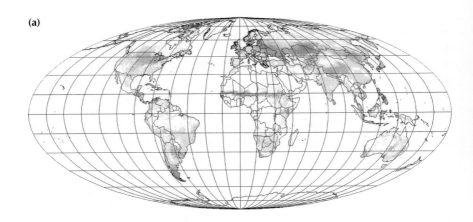

Energy flow
W m^{-2}

5	to 10
2	to 5
1	to 2
0.5	to 1
0.2	to 0.5
0.1	to 0.2
0.05	to 0.1
0.02	to 0.05
0.01	to 0.02
0.005	to 0.01
0.002	to 0.005
0.0001 to	0.002
all others	

FIGURE 6.18a,b Potential wind-power production from decentralized wind turbines placed near farms for January 1997 (*a*) and for April 1997 (*b*). Transmission and storage-cycle losses (estimated at 25%) have been subtracted.

(b)

(c)

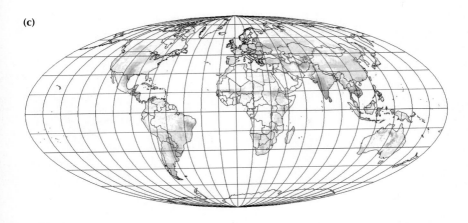

Energy flow
W m⁻²

5	to 10
2	to 5
1	to 2
0.5	to 1
0.2	to 0.5
0.1	to 0.2
0.05	to 0.1
0.02	to 0.05
0.01	to 0.02
0.005	to 0.01
0.002	to 0.005
0.0001 to	0.002
all others	

FIGURE 6.18c,d Potential wind-power production from decentralized wind turbines placed near farms for July 1997 (*c*) and for October 1997 (*d*). Transmission and storage-cycle losses (estimated at 25%) have been subtracted.

(d)

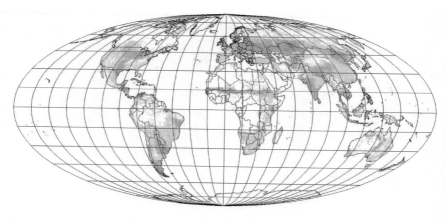

Energy flow
W m⁻²

■ 5   to 10
■ 2   to 5
■ 1   to 2
■ 0.5   to 1
■ 0.2   to 0.5
■ 0.1   to 0.2
■ 0.05   to 0.1
  0.02   to 0.05
  0.01   to 0.02
  0.005 to 0.01
  0.002 to 0.005
  0.0001 to 0.002
all others

FIGURE 6.19 Potential annual average power production from decentralized wind turbines placed near cropland or rangeland. An assumed 25% transmission and storage-cycle loss has been subtracted.

FIGURE 6.20 Potential annual average power production from centralized wind turbine clusters placed on marginal land. An assumed 25% transmission and storage-cycle loss has been subtracted.

Height (m)

5.000 to 9.000
3.000 to 5.000
2.000 to 3.000
1.000 to 2.000
 500 to 1.000
 200 to   500
  20 to   200
   0 to   20
 -20 to   0
-200 to   -20
-500 to -200
-1.000 to -500
-2.000 to -1.000
-3.000 to -2.000
-5.000 to -3.000
-9.000 to -5.000

FIGURE 6.21   Land and ocean topography. (Based on data from Sandwell *et al.*, 1998.)

Energy flow
W m^{-2}

5   to 10
2   to 5
1   to 2
0.5   to 1
0.2   to 0.5
0.1   to 0.2
0.05   to 0.1
0.02   to 0.05
0.01   to 0.02
0.005   to 0.01
0.002   to 0.005
0.0001 to 0.002
all others

FIGURE 6.22   Potential net annual average power production from offshore wind turbine parks (at water depths under 20 m).

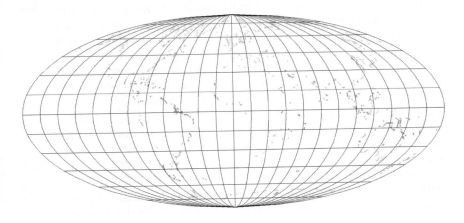

This limit would be a vertical swept area of about 1% of the horizontal land area (Sørensen, 1995a). The scenario assumes that only 10% of the marginal land areas are used for placement of wind turbines at such spacing, implying a maximum swept area constituting 0.1% of the land area. This is small enough not to interfere with other uses of the marginal land, such as the solar cell parks considered in section 6.3.1. Figure 6.20 gives the annual production that potentially may be derived from such wind parks.

Finally, the scenario considers offshore wind parks, meaning turbines placed in shallow waters at a distance from the shore that minimizes visual impact. This mode of wind utilization has become popular in recent years, as innovative low-cost solutions to the offshore foundation problem have been found (Sørensen, 1995a). In order to obtain an estimate for the potential, offshore water bodies with a depth under 20 m are first identified (Fig. 6.21). Of these, it is assumed that only 10% can be used for wind-power generation, owing to competing uses (fishing, military, ship routes, etc.). Again taking into account the 1% maximum utilization rate imposed by turbine interference, one arrives at a 0.1% maximum use of offshore locations for wind-power extraction. The average amount of power that can be derived from these locations is shown in Fig. 6.22. There is an additional number of inland water bodies that could also be used for wind-power production, but they are disregarded here due to their possible recreational value.

Industry Development

The assumptions about future hub height, power characteristics, and capacity factors made in the scenarios are based on the choices made by current manufacturers and trends appearing in the history of the wind industry, with emphasis on recent behavior.

Wind turbines have been in substantial use for more than 1000 years, with early text evidence in China and India and the earliest pictures being from Persia around 1300. The earliest were all vertical axis machines, spreading to the Mediterranean region of Europe, while horizontal axis machines were developed in England, Holland, Denmark, the Baltic region, and Russia (cf. overview of wind history in Sørensen, 1995a). In the 16th century, cities like Copenhagen had about 100 turbines with a hub height of greater than 10 m that were used for grinding flour just outside the city walls, and, in Holland, some wind turbines were used to pump water as part of the drainage system for lowlands. Around 1900, wind turbines played a significant role in powering new settlements in the North American plains and Australian outposts, and the first hydrogen- and electricity-producing wind turbines were tested in Denmark (la Cour, 1900; cf. Hansen, 1985). The DC electricity-producing wind turbines kept competing with coal-based power during the first decades of the 20th century, but they were gradually used only on island sites and eventually they lost there, too, as undersea cables became extended to most islands. By the end of the

1920s, Denmark had about 30 000 wind turbines, three times as many as today. Experiments with AC-generating wind turbines were made before and after World War II, with a comeback in several European countries following the 1957 closure of the Suez Canal (blocking transport of the Middle East oil then used in European power stations). The 200 kW Gedser turbine was technically successful (Juul, 1961), and it served as a model for the subsequent resurrection of wind turbine manufacture following the oil supply cuts in 1973/1974. Since 1974, development has experienced decreasing cost and rapidly increasing installed capacity, as stated in section 1.1.

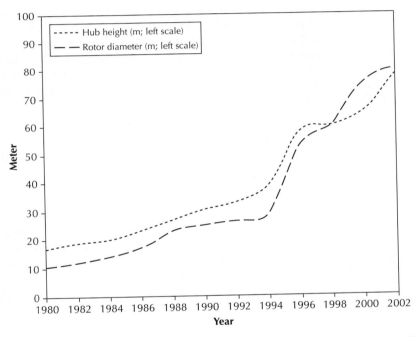

FIGURE 6.23 Development in average height and rotor diameter of Danish wind turbines. For recent years, turbines typically used in new, large wind parks are represented, rather than an average of all turbines sold, which would be less indicative of trends, due to shifting policies for replacement of old wind turbines on land. (Adapted from Danish wind turbine Price Lists and Overviews, 1981–2002.)

Figure 6.23 shows the development in turbine size (height and swept diameter) of commercial wind turbines since 1980, and Fig. 6.24 shows corresponding trends in rated capacity and specific power production at the locations of typical installation. By 2010, manufacturers offered turbines up to 4 MW and prototypes of 5 MW machines were undergoing tests. These large turbines have extraordinary demands for blade materials, causing the currently most used glass- or wood-fiber blades to be replaced by advanced fiber technologies.

Also, advanced computerized control and generator technologies have made possible the extra generation potential associated with use of variable-speed rotors, allowing variable wind speeds to generate AC of varying frequency, which is subsequently transformed to fixed frequency, typically by way of DC as intermediary power form.

During the late 1970s, efforts were directed at several wind technologies: vertical axis machines in the United States and Canada, megawatt-size one- and two-bladed horizontal axis machines in the United States and Germany, and rotors with tip-vanes in the Netherlands. Yet, the winning technology was the Danish three-bladed, horizontal axis machine, starting in 1975 at a small unit size of 25–50 kW. The reason for this was primarily that the initial design was a scaled-down version of the 1957 fixed-pitch 200 kW Gedser turbine with blade stall control, gearbox, and asynchronous generator, i.e., an already proven design, and that further development was for a long time taken in very small, safe enlargement steps, gradually reaching megawatt size, rather than facing the problems of a totally new design in one jump (cf. Figs. 6.23 and 6.24).

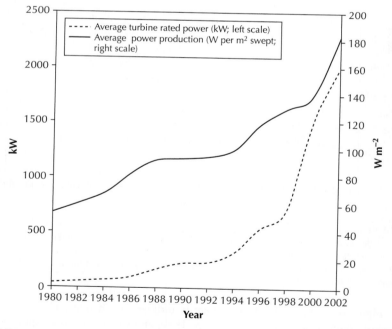

FIGURE 6.24 Development in average rated power and power production of Danish wind turbines. For recent years, turbines typically installed in new, large wind parks are represented, rather than an average based on all sales, which would be less indicative of trends, due to shifting overseas markets and Danish policies for replacement of old wind turbines on land. (Adapted from Danish Windpower Industry Association, 2003; Danish Price Lists and Overviews, 1981–2002.)

Scaling up means controlling the most critical components, such as the rotor blades, which must be able to survive the forces on long pieces of suitable material, experiencing wind loads with gust effects that can cause the blades to bend in different directions and exert large forces on the fastening point of the blades. Also pitch change of blade angles has been introduced and thereby new options for operation or shutdown in high winds. Current production of megawatt units spreads over a number of different design principles. Interesting questions are being asked about whether there is a limit to the length of blades. Considerations should include the blade operating life aimed for. Currently, it is designed to be equal to the lifetime of the entire turbine, in contrast to faster rotating helicopter rotor blades, which have to be replaced at short intervals.

Figure 6.23 shows that a major technology shift, increased blade length, took place around 1994. In Denmark, this reflects the transition from mainly individually placed wind turbines to utility-owned wind parks or wind farms. Individual owners, mainly farmers, had carried the development of wind-power technology to maturity with essentially no help from established utility companies. This may be the chief explanation for the early high penetration of wind energy in Denmark. Later development has been dominated by utility wind parks, onshore or offshore. Utility privatization has caused the penetration to stay at the 20% already reached a decade ago. Fig. 6.24 indicates that a significant increase in turbine rated power has appeared since 1998, but that the average production did not increase as much as the rated power. In this period, installations started to go offshore and the distribution of wind speeds over water was not sufficiently known when machines rated at 1–2 MW were designed. Earlier machines, placed at favorable land sites, were very nearly optimized to give the maximum energy production over the year.

Design optimization is not a trivial matter, as it involves considerations of average energy produced, number of hours operating, and maximum loads exerted upon the structure. Figures 6.25 and 6.26 show the calculated power output at onshore and offshore locations in the inner Danish region, indicating that the imposed power ceiling is reached more often offshore. This situation is expected to occur more frequently at offshore locations in the North Sea, such as that of the 160 MW Hornsrev wind farm completed in 2002. It suggests that either the choice of blade profiles, the sizing of the power generator, or the regulation strategies used in current megawatt-size machines could be improved to capture more of the potential power in the wind for turbines placed at the reserved Danish offshore sites, at least during periods of high winds. Deficiencies in aerodynamic behavior can be determined by repeating the calculations with different power curves, but it is already clear that, at most offshore sites, just a larger generator would increase the annual energy output.

The model calculation assumes a constant C_p at high wind speeds, up to a maximum wind speed of 25 m s^{-1}, where the turbine is brought to a stop.

FIGURE 6.25　Power output calculated for a wind turbine placed at Stigsnæs (near Næstved on inner-Baltic coast of Denmark), based on wind speeds measured during 1995 (Risø, 1999) and a turbine power curve rising from 0 to 400 W m^{-2} between wind speeds of 5 and 12.27 m s^{-1} and then staying at 400 W m^{-2} until 25 m s^{-1} (typical of the behavior of actual 2 MW Danish turbines; cf. Petersen, 2001).

FIGURE 6.26　Power output from wind turbine placed at Vindeby (offshore North of Lolland, in the inner-Baltic coast of Denmark slightly South of Stigsnæs), based on wind speeds measured during 1995 and the same turbine power curve as used in Fig. 6.25. Some missing data sequences (from a few hours to about a week long) have been replaced by measured sequences for the same hours of 1994.

It seems that manufacturers may have let their designs be influenced by complaints from the power companies that large surpluses were hard to accept, because the auction-based selling schemes used in the newly established liberalized power markets tend to substantially lower the value of such surpluses. On the other hand, the turbine owner can boast of a high power factor, i.e., the ratio

FIGURE 6.27 Efficiency as a function of wind speed for commercial 2 MW wind turbine (cf. Petersen, 2001; Bonefeld *et al.*, 2002).

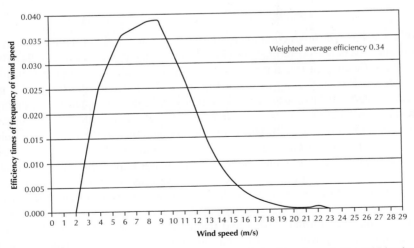

FIGURE 6.28 The efficiency curve of the commercial wind turbine shown in Fig. 6.27 has here been folded with the wind-speed frequency distribution at Hornsrev, site of a 160 MW Danish offshore wind farm. (Based on data from Bonefeld *et al.*, 2002.)

Installed cost	Onshore	Offshore	Unit
Turbines	0.77	0.77	2000-euro/W$_{rated}$
Foundations	0.06	0.35	2000-euro/W$_{rated}$
Grid connection[a]	0.15	0.27	2000-euro/W$_{rated}$
Engineering & administration	0.04	0.04	2000-euro/W$_{rated}$
Land rights, access roads, facilities	0.09	0.07	2000-euro/W$_{rated}$
Operation and maintenance (present value of 20 years' cost)	0.25	0.36	2000-euro/W$_{rated}$
Total capital cost including O&M	**1.36**	**1.86**	**2000-euro/W$_{rated}$**
Assumed power production	2.6[b]	3.8	kWh/y/W$_{rated}$
Cost of power produced	**0.035**	**0.033**	**2000-euro/kWh**

TABLE 6.6 Installation price and cost of energy produced (in year-2000 fixed prices), for wind parks on land or offshore. Estimates have been made on the basis of current price lists for 2 MW turbines, tender data and trend analysis, with use of data from Danish wind turbine Price Lists and Overviews (2002), Danish Energy Agency (1999), European Commission (2001). A depreciation time of 20 years and a real interest rate of 3% per year (roughly the developed world average taken over the entire 20th century) have been assumed. Currently, public display of wind turbine costs has been discontinued, so the table cannot be updated easily.

[a]The source data on onshore grid connection costs are relatively low because of fairly short distances to grids of sufficient strength for Danish wind parks used in the studies underlying the value given. Thus, the additional cost of offshore cables appears more dominant than would be the case in areas less well served with existing grids.
[b]On the very best Danish inland locations, power production may approach 3.5 kWh/y/W$_{rated}$.

of average power production to constant production at full turbine rated power. For the 2 MW Vestas turbines used at Hornsrev, the power factor is 0.54. The actual efficiency of this turbine (power produced over power in the wind) is shown in Fig. 6.27, and again in Fig. 6.28, folded with the frequency of wind speeds. It is seen that the efficiency is high at frequently occurring wind speeds below 9 m s^{-1}, and the weighted efficiency (the integral under the curve) is a decent 0.34. However, the discontinuity in the curve and rapid fall-off for wind speeds greater than 9 m s^{-1} indicate that improved design could greatly increase annual production by better exploiting wind at speeds 9–15 m s^{-1}, which might be done without serious rotor load problems for blades allowed to operate at up to 25 m s^{-1} winds. It is clear that, if the electricity system of a given region can be made to accept larger variations in power production, then turbine design and regulation may be changed in directions that immediately would increase energy production and significantly lower the cost of each kilowatt-hour of energy produced. At present, wind turbine manufacturers insist on offering the same design for use in all wind-speed regimes, except for a possible choice of tower height and, for manufacturers interested in arctic markets, some reinforcement for operation in an environment with high peak wind speed and icy conditions.

Current wind turbine unit size is around 2 MW, but prototypes of 4–5 MW have been developed. The economic breakdown of current costs is estimated in Table 6.6. There are, of course, variations of installation and grid connection costs depending on location, as there are variations in the cost of the electric energy produced, depending on the actual average power production at the location of the turbines and on financing conditions. Details of payments occurring at different times and of interest rates are discussed in Chapter 7.

New Method for Estimating Offshore Wind-power Production

Because meteorological measuring stations are mostly placed on land, estimation of potential production from offshore sites is more difficult and has been unreliable in predicting the precise advantage of the more consistent and higher wind speeds known to prevail over the oceans. The new method for offshore wind-power prediction presented here (Sørensen, 2008a) is based on sweeping microwave radar (scatterometer) measurements by satellite, which provide radar wind data from reflections by the sea surface or airborne water particles over the sea surface. The original purpose of these satellite missions was quite different: to explore sea–atmosphere interactions important for climate modeling (Chelton et al., 2004).

Current offshore wind-power extraction typically involves propeller-type turbines with hub heights up to about 100 m above sea level and placed on foundations at a water depth of up to 50 m. During the last decade, satellite-based microwave scatterometer measurements of the effects of wind stresses on ocean surfaces have reached continuous, global coverage and improved accuracy in deriving near-surface wind speeds and directions from the radar reflection signals received at the satellite. The Seawinds instrument on the QuickSCAT satellite has provided such data since mid-1999 (NCAR, 2006). The sweeping pencil beams of microwave radiation emitted from the satellite have a frequency of 13.4×10^9 s^{-1}, corresponding to a wavelength of 0.022 m. This frequency lies in a band known for the smallness of atmospheric attenuation, allowing a simplified subtraction of such effects en route from, and back to, the satellite (Freilich, 1997).

The footprint of each pencil beam from the spinning parabolic antenna of the Seawinds system has an approximately 30 km × 35 km elliptical area at the ocean surface. From a volume with this area and a certain height, the satellite receives a reflected radar signal. In fact, there are two transmitters and receivers operating at different azimuth angles, offering two looks at the same ocean area at short time intervals, thereby allowing an over-determination of wind velocities that is used to eliminate ambiguities in the directional solutions inherent in the algorithms used and to improve accuracy.

Conventionally, the reflections from the ocean surfaces are considered to arise from wind-generated capillary ripples on top of wave patterns, as well as from the wave surface and ocean surfaces in general. Capillary waves may be defined (see section 2.5.3) as those with wavelengths under some 0.02 m, i.e., with wavelength lower than, or similar to, that of the microwave beam used. However, additional sources of reflection may be whitecaps and water spray particles, which are often fairly abundant at heights up to around 20 m above the sea surface, but radiation at the beam frequency used would be fairly insensitive (Ulaby *et al.*, 1981) to water particles much smaller than 0.01 m. Of the sources mentioned, the capillary waves are most directly linked to instantaneous wind stress, whereas gravity waves represent accumulated wind influence over periods that may involve many hours or even days, as evidenced by waves persisting long after the wind has quieted down. Like all moving sources, ocean currents induce a Doppler shift, unrelated to wind, in the microwave radar reflections. Rain will, depending on its intensity, blur the radar signal en route to or from the satellite, and it is customary to eliminate, or use models to correct, wind measurements affected by strong rain (Marzano *et al.*, 2000; Milliff *et al.*, 2004).

Because the backscattered signal averages over all of the mentioned processes within a volume equal to the kilometer-sized footprint times a height of some meters, one cannot identify a single type of reflection process as the source, but one does note a high sensitivity to the combined expressions of near-surface winds, implying that the wind vector extracted by use of the radar equation may have interesting applications.

Going from the wind speeds near the ocean surface to the productivity of megawatt wind turbines involves two additional calculations. One is to determine the height profile of the wind speeds and the other is to fold the wind-speed data with the conversion characteristics of a typical wind turbine. The height profile is that of a flow over a surface of a certain level of roughness. It may be described by the Monin-Obukhov relations (cf. section 3.2.1), which are derived from classical fluid dynamics and contain two parameters: the friction velocity of the air (which has as its dimension length, the square of which equals the surface wind stress over the air density) and a second parameter describing the stability of the air. The air-stability parameter more precisely describes the transport of heat across a unit area in the atmosphere. If this transport is zero, the atmosphere is said to be stable and, in that case, the height dependence of wind speeds is a simple logarithmic scaling law.

To estimate power production from wind turbines, it is customary to scale wind speeds to the turbine hub height by using the scaling law for a stable atmosphere, because the error is considered low due to insensitivity at both low and high wind speeds. The output from a typical propeller-type horizontal-axis wind turbine is zero for wind speeds below a threshold (say 5 ms^{-1}), then rises to a maximum (at about 12 ms^{-1} wind speed), and stays there at higher wind speeds,

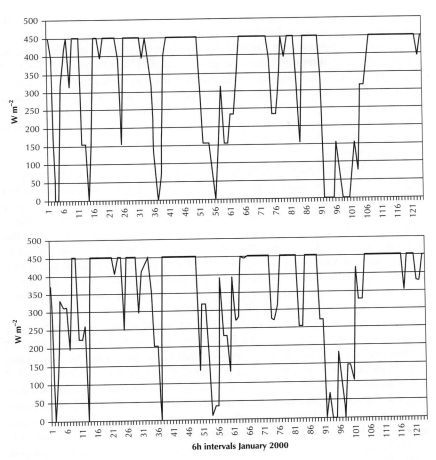

FIGURE 6.29 Measured time-series of wind-power output at the offshore Dutch location Station 252 K3 during January 2000 (*upper panel*, Royal Dutch Meteorological Institute, 2006) and corresponding 6-hour wind turbine power calculated from scatterometer blended data for (*lower panel*). The average power for the month is 336 W m^{-2} (measured) and 346 W m^{-2} (calculated).

until the turbine is shut down for safety reasons at high wind speeds (say above 25 ms^{-1}).

Monthly year-2000 6-hour satellite scatterometer data downloaded from NCAR (2006) were uniformly scaled (by 1.3 for the illustrations presented here) and converted to wind turbine outputs by a power curve linearly rising from 0 to 450 Wm^{-2} between 5 ms^{-1} and 12 ms^{-1}, staying at 450 Wm^{-1} until 25 ms^{-1} and then dropping to zero again. The roughness parameter used in the scaling law for the Dutch offshore site (see Fig. 6.29) is 0.0001 m. Monthly and annual average values were placed on a geographical grid of

0.5°-resolution on both axes and displayed using the Mollweide projection of the Earth in order that the areas with a given wind potential are faithfully represented (at the expense of a variable vertical-line distortion), and thus can be simply added or integrated. For the plots including wind direction (such as Fig. 6.31), a Mercator projection was used in order to achieve some similarity to conventional wind rose plots. Wind directions are not used in the potential power production calculations, assuming wind turbines with perfect yaw. This assumption is fulfilled in praxis, as any yawing adjustment to a new wind direction can be achieved in at most a few minutes by present-generation, large wind turbines.

Figure 6.29 compares a January 2000 time-series of wind turbine outputs calculated from the scatterometer satellite data by the methods described above with outputs calculated from measured wind speeds for the same sequence of 6-hour intervals (Royal Dutch Meteorological Institute, 2006). The location is about 100 km offshore from the Dutch coast, at one of the few offshore locations where data are available at a height suitable for wind-power exploitation (74 m above the sea). The wind-speed scaling factor that makes the satellite data reproduce the total power production over the period derived from the measured wind speeds is 1.3. The 6-hour variations are reproduced very well by the proposed methodology.

While no other publicly available data that would allow similar direct comparison were found, the 1995 data from measurements performed as a prelude to the installation of Danish offshore wind farms are shown in Fig. 6.26. They have been compared to the predictions calculated from the year-2000 satellite data (Sørensen, 2008b). There is, of course, no detailed similarity between passages of weather fronts during a given calendar period in different years, but the overall picture of variability is similar, and the wind-speed scaling factor of 1.3 again reproduces the accumulated power production with an accuracy that suggests that satellite-based predictions using this scaling factor may be used generally for the identification of suitable sites for offshore wind installations.

The spatial resolution of the primary extracted scatterometer winds (Freilich, 1997) is roughly 25 km. Due to the intermittent passage of the satellite over a given location, these data have been blended with calculated wind speeds from general circulation models adapted to measured wind data on land (Kalney et al., 1996; Milliff et al., 1999), in such a way that high-frequency variability from the scatterometer data is retained, while circulation model data are used for interpolation between satellite sweeps over a given location (the 1800-km-wide QuickSCAT sweeps take 101 minutes and a given location is measured at about 12-hour intervals). The resulting 6-hour data thus have high-quality input at each second time step, but only over water. For land sites, the radar reflections may come from a multitude of sources and the satellite data cannot easily be used to estimate wind speeds.

The blended data are presented on a grid of 0.5° × 0.5°, combining the scatterometer swath resolution approaching 25 km with the results from the general circulation model, which, if used alone, would have an output resolution of the order of 250 km (for an input integration-step grid with $0.5° = 56$ km latitude grid-sides and for longitudinal grid-sides multiplied by the cosine of the latitude). For locations on land, the circulation model data are clearly unsuited for specific placement of wind turbines, but may give a fair indication of the wind potential of an extended region (cf. preceding section). The ocean and near-shore (from the sea side) blended data, on the other hand, offer a clear improvement over any existing method of estimating wind-power potential at these locations. Because of the finite grid size used in presenting the blended data, some blended data points along a given shoreline (on both land and water sides) could represent circulation-model calculations of low resolution rather than the high-resolution scatterometer readings. The offshore location used in Fig. 6.29 is sufficiently far from land to exclude any such problems.

Having established a scaling factor close to 1.3 in going from the primary scatterometer data to a height of 74 m over the sea surface, one can use the Monin-Obukhov relations to determine the average effective height z_1 associated with the scatterometer data, a quantity that is otherwise not easy to assess, because the wind extraction method used in analyzing scatterometer data involves a search for the best solution under assumptions of simple stochastic distributions for the backscattered rays and without considering a particular height (Freilich, 1997). To estimate z_1, one may use the logarithmic scaling law describing stable atmospheric situations,

$$v_2/v_1 = \log((z_2 + z_0)/z_0)/\log((z_1 + z_0)/z_0),$$

where v_1 and v_2 are wind speeds at heights z_1 and z_2, log is the natural logarithm, and z_0 the roughness length for the ocean surface, (3.24), related to the friction velocity and, as stated, being about 0.0001 m for the Dutch site (Royal Dutch Meteorological Institute, 2006). Because the Dutch measurements include this estimate of the roughness length z_0 at the particular offshore location, the scaling law for a neutral atmosphere given above contains just one remaining unknown quantity, the reference height, which is then found to be 3 m (representing the height over the average sea surface that the scatterometer winds on average pertain to). This estimate is considered more accurate than one based upon speculations about the particular scattering objects involved, because it is based on long-term averages of a power calculation sensitive only to the most common mid-size wind speeds.

The 6-hour time-series constructed from the satellite data may be used to assess the variability of wind-power with time and location, and hence, for example, any need for inter-regional power trading or for establishing energy stores (Sørensen, 2008c). Short-term fluctuations (under 1 hour) are immediately smoothed by dispersing wind turbine locations over a few hundred kilometers, but smoothing over longer periods requires power transmission between regions of different weather (and hence wind) regimes, implying distances of some 500 km or more. In addition to determining full time-series, averages over time are taken, because they can be used to assess wind-power production potentials over periods deemed important, from weeks to years.

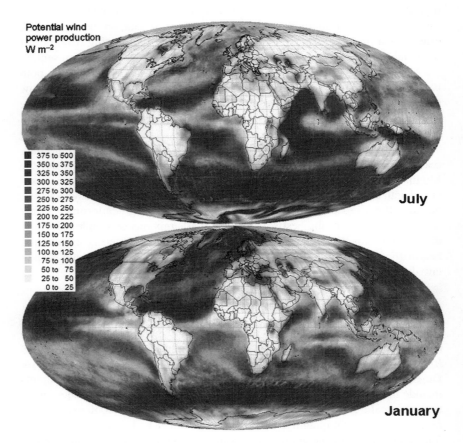

FIGURE 6.30   Global map of wind resources expressed as average potential power production during two months in year 2000, based on blended scatterometer data model with a scaling factor of 1.3 and a power conversion curve typical of current wind turbines. The unit is average watts per m^2 of swept turbine area (Sørensen, 2008e).

Decisions to build wind turbine parks are often taken on the basis of expected average annual production, and a global map of production potentials is shown in Fig. 6.30. As stated above, the accuracy over land is low and values shown should be taken as indicative only for large regions. One would expect scaling